51st American Association of Swine Veterinarians Annual Meeting 2020

A Vision for the Future

Atlanta, Georgia, USA
7-10 March 2020

ISBN: 978-1-7138-2281-3

Printed from e-media with permission by:

Curran Associates, Inc.
57 Morehouse Lane
Red Hook, NY 12571

CURRAN ASSOCIATES INC. proceedings.com

Some format issues inherent in the e-media version may also appear in this print version.

American Association of Swine Veterinarians
830 26th Street
Perry, IA 50220-2328
USA

Phone: 515-465-5255
Fax: 515-465-3832

aasv@aasv.org

Additional copies of this publication are available from:

Curran Associates, Inc.
57 Morehouse Lane
Red Hook, NY 12571 USA
Phone: 845-758-0400
Fax: 845-758-2633
Email: curran@proceedings.com
Web: www.proceedings.com

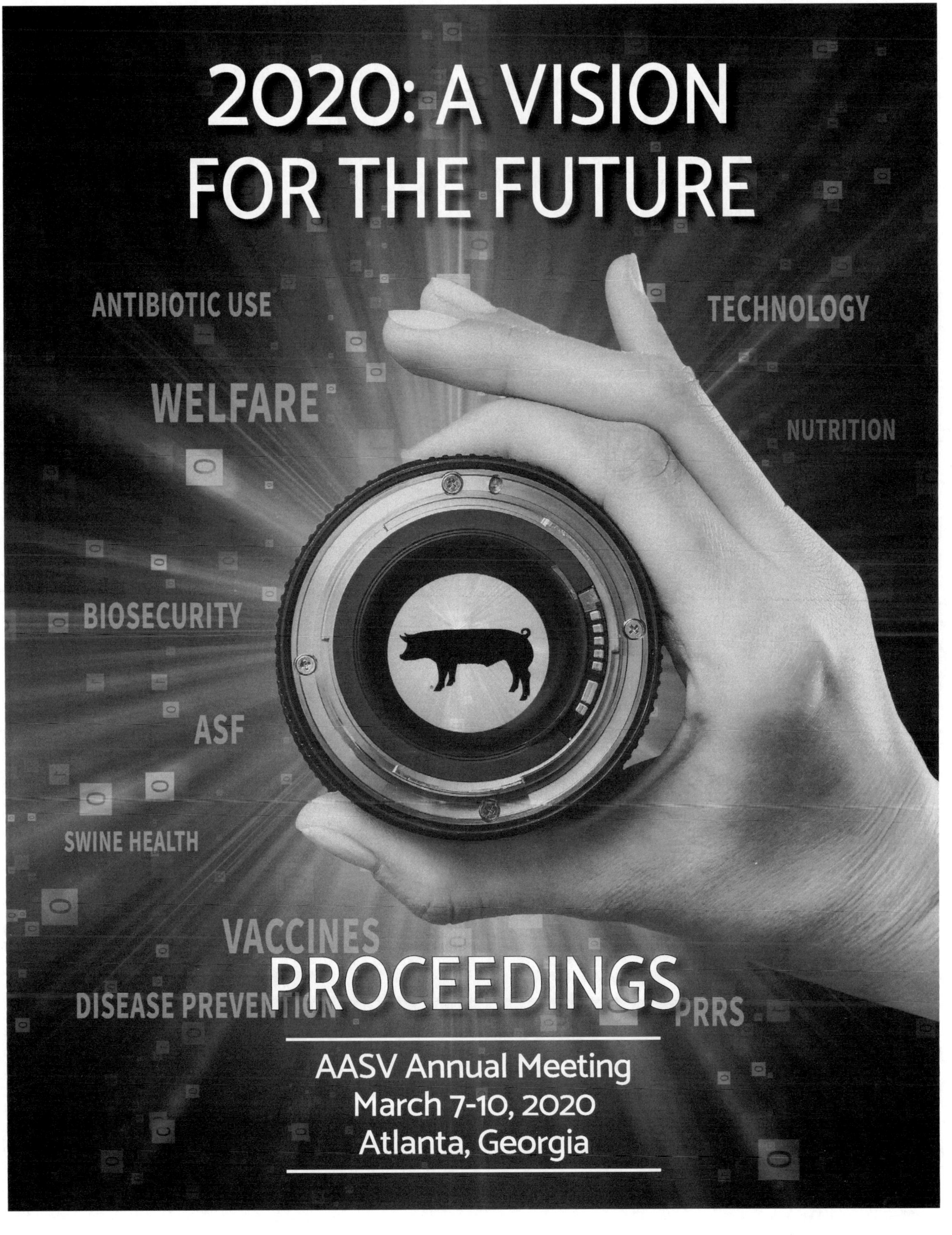
2020: A VISION
FOR THE FUTURE
ANTIBIOTIC USE
TECHNOLOGY
WELFARE
NUTRITION
BIOSECURITY
ASF
SWINE HEALTH
VACCINES
PROCEEDINGS
DISEASE PREVENTION
PRRS
AASV Annual Meeting
March 7-10, 2020
Atlanta, Georgia

Table of Contents

Howard Dunne Memorial Lecture

Alex Hogg Memorial Lecture

Sunday: Research Topics

Sunday Concurrent Session #1: Student Seminar

Sunday Concurrent Session #2: Industrial Partners

Sunday Concurrent Session #3: Industrial Partners

Sunday Concurrent Session #4: Industrial Partners

Poster Session: Veterinary Students

Poster Session: Research Topics

Poster Session: Industrial Partners

Monday General Session: 2020: A Vision for the Future

Monday Concurrent Session #1: Disease Prevention, Control, and Elimination

Monday Concurrent Session #2: Biosecurity: Keeping Bad Stuff Out

Monday Concurrent Session #3: Pharmaceutical Issues

Tuesday General Session: Swine Welfare; Foreign Animal Disease Prevention

Howard Dunne Memorial Lecture

Trust the people

Bret D. Marsh, DVM

In remembrance and honor of
Howard W. Dunne

Dr. Howard W. Dunne received his veterinary degree from Iowa State University in 1941. He was in engaged in private practice for one year and industry for four years before entering graduate school at Michigan State University. After receiving his PhD in 1951, he spent two years as deputy chief of veterinary microbiology at the Chemical Corps Biological Laboratories in Camp Detrick, MD. In 1953, he joined the faculty of Pennsylvania State University, where he headed the veterinary science research program and the large animal diagnostic laboratory. Dr. Dunne was internationally known for his work on swine diseases, particularly hog cholera and the swine reproductive viruses then referred to as the SMEDI syndrome. Dr. Dunne authored more than 90 scientific publications, and is perhaps best known for initiating and editing the textbook *Diseases of Swine*, which is still widely considered the "bible" of swine disease.

In 1971, the American Veterinary Medical Association presented Dr. Dunne with the XII International Veterinary Congress Prize to recognize his contributions to the international understanding of veterinary medicine. At the 1974 Congress of the International Pig Veterinary Society (IPVS), Dr. Dunne was elected president of the 1976 IPVS Congress, to be held in Ames, Iowa. His sudden death at the age of 61 in September of 1974 saddened the swine veterinary community, and the 1976 IPVS Congress was dedicated to his memory.

At the suggestion of Dr. Allen Leman, a financial gift made by the 1976 IPVS Congress organizers to the fledgling American Association of Swine Practitioners was combined with additional association funds to endow this prestigious annual lecture in remembrance and honor of Dr. Howard W. Dunne.

HOWARD DUNNE MEMORIAL LECTURERS

1978	Michael Muirhead	1989	Steven Henry	2000	David Reeves	2010	James F. Lowe
1979	W. J. (Bill) Smith	1990	Stanley Curtis	2001	Max Rodibaugh	2011	Locke A. Karriker
1980	N.C. Nielsen	1991	Roy Schultz	2002	Tim Blackwell	2012	R. Dean Boyd
1981	Allen Leman	1992	Joseph Connor	2003	Lonnie King	2013	Craig Rowles
1982	Michael Wilson	1993	Camille Moore	2004	Robert Desrosiers	2014	Daryl Olsen
1983	Edward Bohl	1994	Robert Morrison	2005	Scott Dee	2015	Gregory Stevenson
1984	Richard Ross	1995	Rodney Johnson	2006	Gregg BeVier	2016	John Harding
1985	Kent Kislingbury	1996	Paul Yeske	2007	Tom Burkgren	2017	Jeff Zimmerman
1986	Ralph Vinson	1997	Barbara Straw	2008	Timothy J. Loula	2018	William R. DuBois
1987	Gary Dial	1998	Mike Daniel	2009	John U. Thomson	2019	John T. Waddell
1988	Thomas Alexander	1999	Larry Rueff				

Trust the people

Bret D. Marsh, DVM
Indiana State Board of Animal Health, Indianapolis, Indiana

One of the highlights of my professional career is to stand before you today. I am grateful for the invitation to speak, and I am especially pleased to have received the invitation from a great colleague, Dr. Jeffrey Harker.

Why pigs?

Dr. Howard Walter Dunne was born in Omaha in 1913. I did not have the good fortune to meet Dr. Dunne, but I have, as is the case with all of you, admired his dedicated service to veterinary medicine and his efforts to advance our collective understanding of the diseases of swine. But I have often wondered, why pigs? I suppose Dr. Dunne could have edited a textbook on the diseases of dairy or beef cattle, but he didn't. He could have edited a textbook on the diseases of sheep and goats, but he didn't. So, why pigs? Was it a result of his early exposure to pigs on the family farm in Nebraska? Or was it because of his education at Iowa State University where he earned a degree in veterinary medicine? Or was it because of his pursuit of a PhD at Michigan State University? Or his service on the faculty at Penn State University?

To find the answer to the question why pigs? I reached out to Dr. Hank Harris. When I posed the question to Dr. Harris, he believes Dr. Dunne's role as editor of the first edition of *Diseases of Swine* in 1958 was likely a result of discussions at the annual meeting of the Conference of Research Workers in Animal Diseases (CRWAD) held in Chicago. CRWAD celebrated its 100th Anniversary in November 2019, and it continues to serve as a major networking center. Incidentally, Dr. Harris also reported that he and Dr. Bob Glock met with Dr. Dunne at a CRWAD meeting when they were asked by Dr. Dunne to write the chapter on Swine Dysentery for the Fifth Edition of *Diseases of Swine*.

But for whatever reasons Dr. Dunne chose to write about pigs and their unique diseases, we are all beneficiaries of his expertise, and it strikes me that each of us here today has our own answer to the question, why pigs? For me, my first experiences with pigs didn't take place in the farrowing house, nor on the finishing floor, nor in the fairgrounds showring. My experiences with pigs started in a more uncommon place, the Stockyards. For many years, my Dad was a hog buyer at the Indianapolis Stockyards. As a young boy, on one special day each year, generally near the end of the summer before classes started in the fall, I would receive a tap on my shoulder very early in the morning, and I knew what that meant. I would spring from my bed, because it was my turn to go to the Stockyards with Dad. I would walk the alleys of the Stockyards with him while he bought hogs from the various commission companies, often in full view of the producer who had brought his hogs to the Stockyards. After buying the hogs, they were then sorted to meet packer specifications, reweighed and penned for the truckers to load and haul them to packing plants, often located in the eastern United States (US). I would arrive home exhausted with hog dust on my clothes and on my breath and with $1 in my pocket, a token of appreciation from one of Dad's colleagues. Those trips to the Stockyards with my Dad remain one of my most cherished memories.

Little did I know then that I would one day serve as Director of Swine Health Programs, and later as Indiana State Veterinarian at the Indiana State Board of Animal Health, and I would draw upon those early experiences in the dusty alleys of the Stockyards.

For all the various reasons we have chosen pigs, I agree, and you likely do as well, with Winston Churchill who said, "I like pigs. Dogs look up to us. Cats look down on us. But pigs, they treat us as equals."

Trust the people

As Director of Swine Health Programs, I was responsible for the pseudorabies (PRV) eradication program. Each quarter for many years, the Indiana Swine Health Advisory Committee, a committee not unlike those in many pork producing states, met to discuss how to move the program forward. Through those early experiences, I learned the value of seeking input from the people to determine how to develop animal health policy that made sense and would be effective. The process may have been slow, tedious and sometimes frustrating, but the final product of those careful deliberations was one that endured. I was blessed to have many patient mentors during those early years, including many swine veterinarians, Drs. Rodibaugh, Harker, Gillespie, Rueff, Ackerman, Villani, Keffaber, Long, Baker, Hendrickson, Lemmon, Jackman, Bush and so many others. As I became increasingly engaged in the national PRV eradication effort, I met veterinarians from other states who were dedicated to the cause. Drs. Hill, Yeske, Connor, McKean, Sundberg, Pyburn, Webb, Wagstrom, Becton and many others. And the greatest lesson I learned was to trust the people. Trust the veterinarians who by virtue of their professional training and enduring integrity make front-line decisions on behalf of the producers they serve. Trust the producers who have taken all the risk and put their livelihoods on the line. Trust the market operators and packers because we often don't fully understand all the challenges they endure to continue to produce a safe and wholesome product for consumers worldwide. I knew very early that without the input of these key citizens, any program I may propose would inevitably fail.

I have witnessed dozens of acts of selfless service by veterinarians in all animal health programs. Whether the task was eradication of highly pathogenic avian influenza (HPAI), bovine tuberculosis, piroplasmosis or monkey pox, the contributions of veterinarians were always reliable, timely and professional. Throughout the PRV eradication program, veterinarians bled thousands of pigs to support the effort, and as a result of their toil, I have taken a little gentle ribbing from them. For example, while talking to a veterinarian on the phone one day he said, "Dr. Marsh, you're going to have to speak up, I've bled so many hogs for you, I can't hear anything anymore!" Another veterinarian sent me the hospital bill for his knee replacement surgery with a note that read, "Dr. Marsh, here's the bill for my new knees. Figured you should pay for them considering all the stooping I did bleeding pigs for you!"

I admire them all for what they have done for me and for this nation. I trust them. They have given so much and demanded so little while making this profession and our country better.

"One never notices what has been done, one can only see what remains to be done." Madame Marie Curie

The theme for this year's American Association of Swine Veterinarians (AASV) conference is "2020: A Vision for the Future," and we are therefore encouraged to cast our eyes to the horizon and beyond. Madame Marie Curie said, "One never notices what has been done, one can only see what remains to be done." I believe it is appropriate for us to pause for just a moment to consider the remarkable changes that have taken place over the last few years in support of this massive, dynamic animal agriculture industry. I have chosen to focus on a few of what I believe to be some of the most significant advancements.

1. The National Strategy for the Physical Protection of Critical Infrastructures and Key Assets was released by the Bush White House in February 2003. This document lists agriculture and food as a critical infrastructure for the US, and this is the first time agriculture and food were identified in such a manner. It, therefore, places the sector at a significantly higher level of visibility for planning and preparedness.
2. Homeland Security Presidential Directive (HSPD), #9, was signed by President Bush on January 30, 2004. This HSPD is focused entirely on the agriculture and food sector, and it calls upon the Secretary of the US Department of Agriculture (USDA) to establish an agriculture and food defense plan for the US.
3. The National Animal Health Laboratory Network (NAHLN*)*, is a network of state and university-associated veterinary diagnostic laboratories across the US that improve the country's surveillance capabilities for foreign and emerging diseases. Further, the NAHLN laboratories provide surge capacity when a foreign animal disease (FAD) is diagnosed in the US. Launched in May 2002 with twelve laboratories tasked with targeting diagnostic capability for 8 high priority FADs, the network has grown to over 60 state and university-associated laboratories.
4. The National Centers for Animal Health (NCAH), Ames, Iowa, combines the critical functions of the National Veterinary Services Laboratory (NVSL), National Animal Disease Center (NADC), and the Center for Veterinary Biologics (CVB). They have been rebuilt, retooled and renamed to become an internationally recognized national reference laboratory. The final phase of this multi-million-dollar project was completed in 2009.
5. The National Bio and Agro-Defense Facility (NBAF), Manhattan, Kansas, is a state-of-the-art biocontainment laboratory for the study of diseases that threaten both America's animal agricultural industry and public health. The US Department of Homeland Security is building the $1.25B facility to standards that fulfill the mission needs of the USDA which will own, manage and operate the NBAF once construction and commissioning activities are complete. The NBAF will be the first laboratory facility in the US to provide biosafety level 4 (BSL-4) laboratories capable of housing cattle and other large livestock. The current functions at the Plum Island Animal Disease Center and the Foreign Animal Disease Diagnostic Laboratory (FADDL) will transition to the NBAF, and the facility is expected to be fully operational in December 2022.
6. Although a North American foot-and-mouth disease (FMD) Vaccine Bank has existed for several decades, the 2018 Farm Bill specifically identified funding for a National Animal Vaccine and Veterinary Countermeasures Bank. The funding will be used to establish a US-only FMD vaccine bank.

These infrastructure investments, funded by our citizens through either national or state appropriations, are remarkable, and these investments, largely unknown to most US citizens, have changed the course of America's ability to effective address disease threats to the swine industry.

Government can work for the people

As you likely know, I decided to enter public service right out of veterinary college, and my classmates thought I was throwing my degree away! Now they call me and ask, "Bret, do you have anything for me down at the state office?"

I believed then, and still believe today, government can work for the people, and I was determined to make my small slice of the government the very best it could be. I believe the people deserve the best. My colleagues at the Indiana State Board of Animal Health, of whom I am enormously proud, never forget who they work for, and that's you!

Further, I believe, and have learned from first-hand experience, that an effective government, and in our case, its role in developing animal health policy, is only possible with an engaged citizenry.

I believe in the power of the deliberative process.

I believe in organizations, like AASV, that provide a forum for thoughtful debate to determine the best course for progress.

I believe in the value, the enormous value, of the networking opportunities AASV provides.

I believe in our people. Our citizens. Our veterinarians, whether they are in private or public practice, employed at a diagnostic laboratory, an academic institution, or a pharmaceutical company. We need each other to succeed.

I believe Winston Churchill when he said, "Democracy is the worst form of government, except for all the other forms that have been tried from time to time."

And I believe, that you believe in these same principles.

And, therefore, if these devastating viruses that continue to ravage the swine populations of countries worldwide, reach our shores, we will prevail.

General David Hackworth said, "If you find yourself in a fair fight, you didn't plan your mission properly." I'm not interested in a fair fight with these viruses, and I don't believe you are either.

Are we prepared to make transformational change before a foreign disease devastates the US pork industry?

Are we willing to make the tough choices in peacetime to preserve and protect this industry during times of war?

Are we determined to succeed in the face of seemingly insurmountable obstacles?

How much time and money will be spent before we do what we already know needs to be done?

We must immediately address critical gaps in our preparedness planning. We need to establish an indemnity policy, plan for the use of certified samplers, better understand permitted movements, research the best methods for depopulation and disposal, explore implementing and releasing a 72-hour standstill, appropriately resource our laboratories, enhance transport vehicle biosecurity and expand the role of accredited veterinarians in FAD investigations. While some of the gaps are being rapidly addressed, we must do all we can to finish the task.

One of these gaps is what I refer to as the pseudorabies (PRV) gap. Indiana was declared free of the PRV in 2002, eighteen years ago. As the PRV eradication program wound down, state animal health officials (SAHOs) began to lose contact with the pork industry, while it, understandably, continued to innovate and grow. Therefore, the current level of understanding of the pork industry by SAHOs is frankly poor, and bridges of communication and collaboration must be reestablished to effectively address an FAD threat.

The swine industry has experienced remarkable changes in the last decade. The trend of consolidation in the pork industry continues, as it has for decades, and the percentage of pork produced for the export market continues to climb. But just as we have seen consolidation in production and packing, we more recently have experienced consolidation in the delivery of veterinary medical services to the industry. Long-standing veterinary practices in Indiana have been sold to out-of-state interests or they have established broader partnerships with practices throughout the country. As another example of the consolidation of services, just four veterinary diagnostic laboratories account for nearly 98% of the swine diagnostics performed in the US. Therefore, it is incumbent upon me in my role as State Veterinarian to reach out to these new partners to make sure they understand who we are as an agency and how we operate. It is imperative we work together in advance of a catastrophic event, because the SAHOs will be responsible for establishing quarantines and control areas, issuing permits for movement and providing the primary communication to all stakeholders. Indiana has hosted several meetings with pork industry stakeholders to get to know each other, all over again. When at one point several years ago we encouraged all veterinarians in our state to get fax machines for more rapid delivery of information, today we provide text alerts, email messaging, webinars, podcasts, as well as face-to-face meetings, teleconferences and offers to speak at producer meetings. We are grateful for the opportunity to learn more about the industry, and we have received enormous support from the veterinary community.

There are many significant efforts underway to better prepare the pork industry, and they serve as excellent examples of synergistic collaborations between industry and government.

Animal Disease Traceability (ADT), which includes premises registration, premises identification number (PIN) tags and agreements for swine to move in interstate commerce within a production system, has greatly improved the nation's ability to rapidly respond to a disease threat. The pork industry has been a national leader in the adoption of premises registration. Premises must be validated to provide accurate data on the locations of swine, and although there has been significant progress in this area, more must be done. Further, premises identification must be incorporated into all movement documentation and laboratory accessions for more rapid traceability.

There has been significant progress in the advancement of diagnostic tools for African swine fever (ASF), classical swine fever (CSF) and foot-and-mouth disease (FMD), and yet there is much more to be done. Validation of oral fluids, as well as processing fluids, is underway. There is a great need to better understand the testing algorithms necessary for determining the status of a population of pigs, particularly related to approved movements. To expand the sampling capacity of the country, a plan for utilizing certified samplers is being closely evaluated. The protocol for certifying a sampler requires an accredited veterinarian to select and train those farm employees who can effectively and efficiently collect samples for diagnostic testing. Further, there is a realization that accredited veterinarians must be utilized to collect samples for FAD diagnostic purposes. The policy that all suspected FADs must be investigated by a trained Foreign Animal Disease Diagnostician (FADD) is being reevaluated to more closely align with field-level reality.

At the close of 2018, through the adoption of electronic methods to prepare certificates of veterinary inspection (CVIs), 93% of all livestock leaving Indiana were documented on electronic certificates. The adoption of this technology by veterinarians has been remarkable, and the improvement in the ability to rapidly trace animal movements has been greatly enhanced.

Funded by US pork producers, the mission of the Swine Health Information Center (SHIC) is to protect and enhance the health of the US swine herd through coordinated global disease monitoring, targeted research investments that minimize the impact of future disease threats, and analysis of swine health data. This unique center does not exist with any other commodity, and its ability to provide rapid answers to complex research topics is an essential service.

The Secure Pork Supply (SPS) plan is the foundation of all state preparedness planning for the pork industry. Biosecurity plans developed as a part of the SPS plan may be required by some states to be sent to the state office, while others will be satisfied with the plan being on file at the production sites.

Formally announced in January 2019, the National Swine Disease Council (NSDC) is comprised of key leaders from the pork industry whose top priority is rapid and effective response to disease events and promoting acceptance of recommended actions from the NSDC throughout the US pork industry.

A 14-state working group (WG) of the State Animal Health Officials (SAHOs) of the 14 most populous pork producing states continues to meet with the goal of standardizing, as much as possible, the response of states to an FAD incident.

In the final analysis, in my view, the US pork industry, with significant input from the veterinary community, must determine its own fate regarding FAD preparedness, and in recent years the industry has made enormous strides toward that end. For too long, state and federal animal health officials, myself included, told livestock and poultry producers the government had a plan for not only indemnity, but for depopulation and disposal for an FAD, and therefore left the impression with producers and veterinarians that, "We've got it." But the reality today is, the size and complexity of modern pork production requires producers, collaborating with veterinarians and SAHOs, to determine the best methods for depopulation and disposal on their premises, how to request a permit for movement from a control area and what an effective national FAD surveillance program must include. The pork industry must establish an effective national forum for the careful deliberation of these critical issues and thereby engage the entire industry in the development of national policy.

We do all of this for one purpose, and one purpose only. To preserve and protect the agricultural assets of this nation.

Professional passion

As AASV launches into the second half of its first century, I hope you take great pride in the role you have played throughout the decades. There are those of you in the room today who will participate in AASV's 100th Anniversary, just 49 short years from now. The world will have changed many times by then and yet the same principles and ideals will still apply. First and foremost, each member of every new class of veterinary college graduates must accept the challenge of handing this profession on to those who follow in better condition than they found it. Without this careful nurturing, the profession will not reach its potential. It requires those of us in the profession to preserve and protect its guiding principles, just as those who have gone before us have done.

The decisions that have guided this profession over the decades are the foundation upon which we stand, and they have been as pivotal to the continuing success of the profession as any advancement of science. But beyond all of the investments in infrastructure, beyond all of the science and research, and beyond all of the preparedness planning, there remains an intangible part to our profession, and it is the part, I believe, that has sustained us through the decades. Simply put, we love this profession. It can't be explained further. I have witnessed this passion demonstrated over and over through countless acts of selfless service. It is the single most impressive characteristic of the individuals I have met and served with throughout the years. We must keep the passion, because this passion is the very essence of the profession, and it can't be manufactured, bottled, bought or sold, because it lives within our hearts. We must continue to apply the same principle to our care of this profession that William McCay used when he described the land. He said, "The land is not something you inherit from your parents, it's something you borrow from your children." And so, we will gently pass on this great profession to those who will follow us so they too can carefully plan and shape its future.

Alex Hogg Memorial Lecture

Choosing a pathway forward in swine practice

William L. Hollis, DVM

In remembrance and honor of Alexander Hogg

Alexander Hogg
1920-2006

A native of Scotland, Alex (pronounced "Alec") Hogg immigrated to the United States with his family when he was six years old. After reaching adulthood, he served in the US Navy during World War II before entering veterinary school at Kansas State University. He received his DVM degree in 1950, and was a mixed animal practitioner in Iowa for twenty years. Dr. Hogg returned to college – this time, Iowa State University – to complete a Masters Degree in 1972. He spent the following years serving and educating swine veterinarians and producers in his roles as Nebraska Swine Extension veterinarian and professor at the University of Lincoln-Nebraska. Upon "retirement" from the university in 1990, Alex worked as a technical services consultant for MVP Laboratories. Ever an enthusiastic learner, he graduated from the University of Illinois Executive Veterinary Program in 1995.

An active and longstanding member of the AASV, Dr. Hogg joined what was then the American Association of Swine Practitioners shortly after it was formed. He served as secretary/treasurer of the association from 1974 to 1976, and was elected president in 1979.

The many awards and honors bestowed upon him over the years included several presented by the AASV: The Howard Dunne Memorial Award in 1985, the Meritorious Service Award in 1996, and in 2005, the first-ever AASV Foundation Heritage Award for a lifetime of achievement in veterinary medicine.

ALEX HOGG MEMORIAL LECTURERS

2011 Joseph F. Connor
2012 Lisa Tokach
2013 John T. Waddell
2014 Mark J. Engle
2015 C. Scanlon Daniels
2016 Peggy Anne Hawkins
2017 Matthew Turner
2018 Rodger Main
2019 Deborah Murray

Choosing a pathway forward in swine practice

William L. Hollis, DVM
Carthage Veterinary Service, Ltd., Carthage, Illinois

Thank you for asking

You will forget - My family will *not* forget - My business partners will *not* forget - *I will not forget*

I have three messages:

1. Why I see the world the way I do
2. Where we may go – production, health, service, products, education
3. What might help you along the way – people, books, experiences

Fred Rogers has said the distance between the listener and the speaker is "hallowed ground." There is more going on in the mind of the listener than can ever be transferred by the speaker or the slide show. Fred Rogers produced what some called the first Reality TV show (Mister Roger's Neighborhood). He was also a Presbyterian minister and won accolades for taking his small children's TV show of puppets and stories to a decades' long public television program. My including Mister Rogers is meant more for the realization that what I say is not nearly as important as what you think and what you do with the information gathered and shared here at the AASV meeting. The distance between the speaker and the listener is hallowed ground. SO – what are you thinking about?

Please allow me to share a few facts and figures in support of CHOOSING A PATHWAY FORWARD.

Knowing what is going to happen in the future is tough: There have been many quoted to have shared this discovery from 1970's computer models, "Forecasting is very difficult especially about the future," or 1950's baseball player Yogi Berra, "I never make predictions, especially about the future," or maybe we go as far back as a Danish Proverb found carved into a boat, "It is difficult to prophesy, particularly about the future."

FIRST let me start with a perspective on *practice* – we as veterinarians are ALL in practice. In fact, since the name change of our organization from Swine Practitioner to Swine Veterinarian...the career path of those working for swine farms directly and those working for private practice have both been categorized in the same group. WE ALL are working for the health, welfare, and care of pigs. Education supports the development of the caregiver. Research supports the development of products and services. We are ALL in practice.

Given that we are ALL in practice, it is imperative that we ALL be willing to challenge our current scientific understandings and assumptions, acting and reporting results to further improve the quality of medicine we practice. The term "practice" denotes an evolving process, a pathway through which we learn and improve. Don't be afraid to challenge current dogma and tribal knowledge; only in this way do we improve the quality of services we provide to both our client and their pigs. Clients like decisive action... your assessment of their current clinical state gives you tremendous feedback into what's working and what's not working. When you take calculated action based on all available evidence and report your results, you make invaluable contributions to our profession. Reporting your results can be time consuming and intimidating. Fortunately, there are numerous mechanisms to do so, review the various communication pathways and land on one that works best for your time availability and preferred communication style. Larger veterinary groups have tended to create this opportunity without formal structure. Individual veterinarians in production, technical service, or practice thrive on the discussions both formal and informal to share valuable experiences. Options for sharing results range from peer-reviewed journal articles, to hallway rumors during meetings, to calling friends while talking to the windshield, and everything in between. The mission of the AASV is to provide education. In order for there to be education, we must challenge old ways of thinking and be willing to report our outcomes, good, bad, or indifferent.

Why I see the world the way I do

We each see the opportunities and experiences before us through our own lens into the world. In order to understand my perspective on swine veterinary medicine, please allow me to share some of my experiences. During veterinary school at the University of Illinois in the early 1990's, there were many great opportunities to learn about pigs. Dr. Leroy Biehl was putting together the first EVP class. He was kind enough to offer me a student job (putting together notebooks and making coffee). He introduced me to many leaders in the swine veterinary community and allowed me to sit in on occasion when speakers presented to the group. Dr. Biehl introduced me to my first internship at Premium Standard Farms in Missouri working for Illinois veterinary school graduates Dr. Gregg BeVier and Dean Dau. Other internships at Murphy Family Farms with Dr. Howard Hill and student externships with Dr. Larry Rueff and Dr. Joe Connor shaped my interest in swine medicine. I am one of the few who are still at the same location I started 24 years ago. Dr. Connor and his business partners Dr. Kenneth Wright and Dr. Jack Coleman started to invest in swine production operations. Their investments along with several of our local clients started the pathway to building sow farms for producers who wanted someone else to deal with the labor issues, health and biosecurity, and genetics. Our first gilt multiplier was 1,200 sows and had 43 shareholders. We had three herds by the end of 1998 for a total of 4,200 sows. We also had clients asking us to make larger and larger commitments of time to their operations. We could see the focus of swine specific service was going require our single focus if we were to help this part of the business to grow.

My family background also guided my choices in practice (private business ownership – similar to my dad's ownership of the local funeral home and similar to mom's study of blindness and teaching blind and visually impaired). I wanted to learn veterinary medicine because I wanted to own a small business, and because I wanted to understand more about biology and animals. Involvement in 4-H and the FFA installed a desire to build teams and work on large projects. Swine medicine in the late 90's and early 2000's was all about building teams and large projects. Our production management business grew to over 100 shareholders with over 100,000 sows by 2010 and is on pace to double again by the end of 2020.

I also like old cars. One famous quote in particular from Henry Ford stands out to me for this presentation. Henry Ford is alleged to have said, "If I had asked my customers what they wanted they would have said faster horses." Harvard Business Review (HBR) has disputed the validity of the quote confirming there is no evidence Henry Ford actually said the now famous quote for innovation and marketing.[1] HBR also goes on to explain that Henry Ford certainly did think along those line. There are famous successes to this mode of thinking (efficiency of the assembly line and rapid expansion to capture market share). There are also colossal failures for the shareholders and the company as a result of ignoring the customer (in 1921 Ford had 67% market share of all automobile sales in the US but by the end of that decade they only claimed 15%). Some Ford family members also made this proudful mistake. Have you heard of the Edsel? From my perspective, I like old cars. I like fast cars. I also like strong, sustainable shareholders and positive customer relationships. Our industry could suffer if we refuse to engage with our customers.

My perspective – Swine veterinarians are very fortunate

1. Swine veterinarian salaries have remained the highest of our peers and stayed there for two decades.[2,3]
2. Swine veterinarians are entrepreneurs with ownership potential in veterinary practice, veterinary products, swine production or often all the above.
3. Swine veterinarians have recruited, developed, and invested in a strong base of students with interest in growing with the profession.

Salaries are incredibly difficult to compare as not all jobs are created the same. Fortunately, the AASV has regularly conducted a salary survey kept internal to our membership.[3] A high participation rate from our members has provided us with valuable data on wages and benefits to compare to our peers and other industries. AVMA Economics and Salary studies support that swine veterinarians are paid well and operate strong businesses.

Investment professionals often encourage multiple income streams. Veterinary medicine and a healthy entrepreneurial spirit offer the swine veterinarian multiple opportunities to grow wealth as they grow in service to their clients. Dr. KT Wright suggested in 2009 if the potential for losses keep you up at night then investments may not be for you. Investment in swine production or in swine products is not without risk. We are in an exciting time with a clear view of the needs and opportunities for pig farms. My perspective is there will continue to be product breakthroughs and production opportunities for those with available capital to invest and a desire to pursue. This is based on current total membership of 1348, which includes a few complimentary members, all international members, and all Affiliate (non-DVM) members (Figure 1, 2).

Swine veterinary students have become swine veterinarians with increasing number and have stayed in the professions with an increasing percentage.

We need to be proactive and well trained – intense consolidation will continue

Dr. John Waddell shared just last year that in the twenty years 1998 through 2018 the top 50 largest swine operations have only 15 remaining companies in 2018.[4] Said another way, over the last two decades there have been 35 large swine production companies either consolidated into other businesses or exited all together. Competition among our swine clients will lead to continued concentration of services. Building networks and collaboration has proven valuable to swine veterinarians as these businesses grow. Veterinary medicine has also seen intense competition. Small animal medicine has seen the entrance and collection of large veterinary clinic corporations pulled together through a combination of improved efficiency and investment strategy. Mars Inc., originally known as M&M candy and the Mars family fortune, is now the single largest veterinary practice owner in the world with over 2,000 clinics in the US and Canada and hundreds in Europe. AVMA Economics fellow Dr. Melissa Maddux puts it this way, consolidation is likely due to the strong returns in veterinary practice ownership.[5] The example of a $1MM practice gross annual revenue creating a $200,000 annual return to the owner would have traditionally driven a 5 or 6× multiple in practice value ($200K × 6 = $1.2MM value). Investment capital funds or large family estate funds are commonly trying to beat a 7% return. Given a lower target return they could pay up to a 14X multiple. Short message – private practices retain great value if they continue to generate great returns for their owners. Other examples also exist in the greater global veterinary community for providing a network of education and veterinary service. Dr. Chris Richards of Bendigo, Australia has fostered the growth of the largest veterinary service company in Australia. Apiam Animal Health, Ltd. has grown to include a network of veterinary clinics and services from all species.[6] My greater concern in swine veterinary medicine is the growing intrusion into veterinary decision making by accounting or production teams given greater authority than medicine. We should never expect to make all

Table 1: Swine veterinary numbers

Category	Number	Percent
100% swine practitioner	384	28.5%
> 50% swine practice	114	8.4%
50% swine practice	32	2.4%
< 50% swine practice	110	8.2%
Industry	296	22%
Regulatory – state	15	1.1%
Regulatory – federal	11	0.8%
Education	102	7.5%
Research	108	8.0%
Public health	5	0.4%
Retired	89	6.6%
Other	47	3.5%
Not specified (blank)	35	2.6%

Figure 1: AASV investment in students.

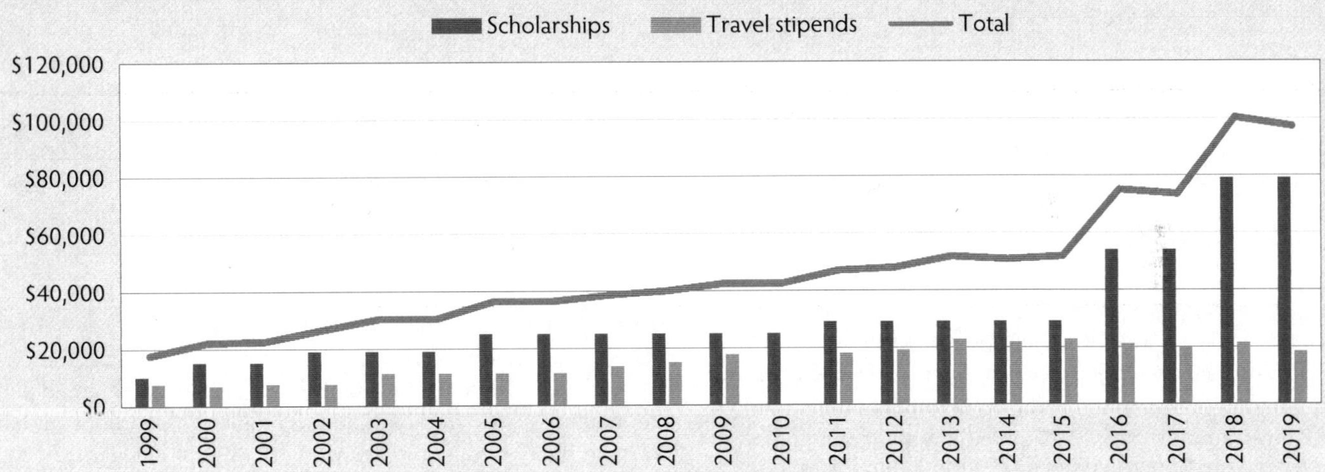

the decisions. If we do not bring strong value or concise direct actions, founded in good financial planning, we will not be in the room with the decision makers. Intense competition and industry consolidation will continue.

Where we may be going – choosing a pathway forward

Production – veterinarians will continue to have opportunities in pig production

Veterinarians in swine production companies today are often placed in dual production roles. Three veterinarians in our Carthage System have taken the opportunity to serve in a full-time production capacity reporting to the production team for the first year out of veterinary school. This may mean serving for periods of time as a breeding project coordinator, serving as a farrowing house trainer, or even as a sow farm manager for an open position during a manager transition. In every case, I can confidently say Dr. Will Fombelle, Dr. Neal Benjamin, and Dr. Claire LeFevre have each proven to gain from this program and bring excellence to the company. The strength of character and work ethic that comes with a commitment to complete the DVM is why Dr. Joe Connor first established this program. Currently the several associate veterinarians report to Dr. Dyneah Classen as the Director of Health of our Professional Swine Management service company. You can ask her about

Figure 2: AASV student members and percentage retained.

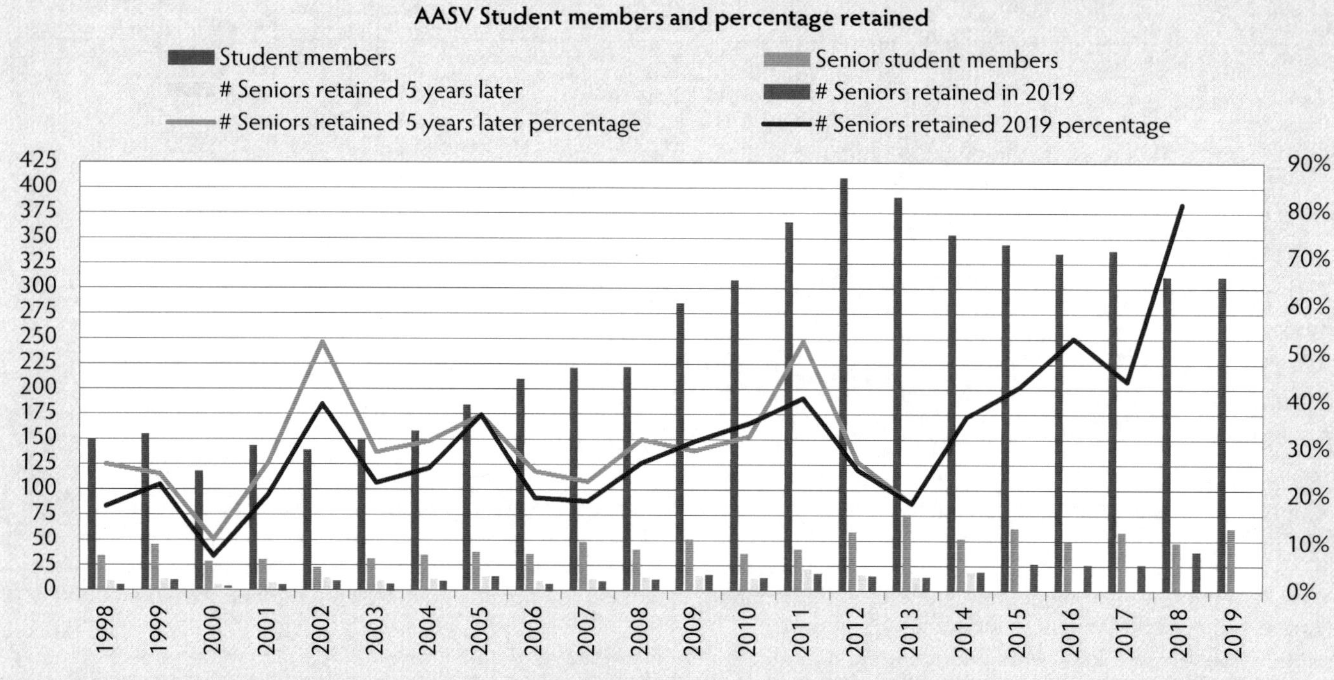

the success of the program. We all questioned if the production teams would really welcome the vets. We also had more than one occasion where production staff failed to teach, coach, or develop the vet. Gaps in training and communication were brought to the attention of the veterinarian mentors and solved with time and talent from the new vets. My point in this is that our new veterinarians are often challenged and questioned by production teams. Even simple vocabulary like "gestation snake, pig processing, and feedback" can cause confusion. New veterinarians are quick to figure this out and strong on commitment and work ethic. I encourage others to offer production opportunities and responsibilities in order to provide new swine veterinarians interested in swine production medicine the chance dive in and learn hands on.

Swine practices in support of production service have changed dramatically over the last two decades. In 1996 when I started, we were somewhat in total control of decision making with producers laying everything at our feet (HR, environmental, everything). Our first board members did want to tour the farm but more often to learn for their own staff management, ventilation, and construction for their own businesses. Over the years our shareholder groups and clients have greater access to information. Their own businesses have grown considerably. We must bring in talented decision makers to our farms, our offices, and our suppliers to elevate our service.

A few small examples of changes in service include pregnancy diagnostics and farm audits. Twenty years ago, the veterinarian would perform pregnancy diagnostics on farm and tighten the breeding forecasts. While this is still excellent service for breeding teams, the more effective farm service is training the farm staff and monitoring closely the results. Likewise, we have seen audits more recently move to support staff with training and people management skills honed to bring accountability and farm standardization. Our veterinarians are now training the system auditors and following up with classes for farm staff where shortcomings are identified.

Part of serving the swine industry and swine clients is having a thick skin. I have heard derogatory comments toward professional, well-trained veterinarians simply because of their desire to dig for results or understand a problem completely. One of the dumb things I have heard, "never known a vet who could balance his own checkbook." What I can confidently share is Dr. KT Wright knows how to balance a checkbook (refer to his article in AASV 2009 "The Wright Stuff" if you want an education in successful financial planning as a veterinarian). The producer was challenging the National Pork Board Animal Science committee with the idea that veterinarians can get sidetracked on problem solving without proper focus on finance. A lack of focus can lead us to attempt to retain services much better in the hands of trained and respected farm staff or other science professionals. We should become the trainer and support the development of teams for success on farm. In this manner, we are actually more cost effective and elevated to a higher level of decision making as a result (which in time leads to more opportunities). Educational opportunities exist today for veterinarians to become better trained in new areas. EVP is an excellent catalyst for the veterinarian looking to broaden their network.[7] I would challenge those in EVP to also ask "what will you do with this new understanding of the industry." Put the learning to work in your practice and for your clients. University master's programs and pathology programs also further the intensity and focus of vets interested in growing their roles in production or management with a greater understanding of nutrition, genetics, or disease. Dr. Rodger Main challenged me several years ago to ask, "What are you going to do with the education?" and "How will you put your education to work for you?"

My brother Rich Hollis has taught me the value of enthusiasm in training. You have to know his background to appreciate the change he recently made. Rich was comfortably travelling the US helping Rabo Agrifinance to place the right banker in the right opportunity. He was asked to travel more internationally when he met Bradley Wolters of The Maschhoff's. Rich decided to make a move into human resource development in the pig business. My explanation to Rich when he talked with me about considering this opportunity was, "that sounds like moving from structure and organization into the belly of hell, maybe into complete chaos." Not a reflection of The Maschhoff's in any way, but the reality of the turnover and the struggles we have with people development in the pig business. His first senior staff meeting sounded much like any production meeting with any swine client in the US – 90% production, analysis, and metrics followed by a brief update on turnover and percent fully staffed. Rich says we have a lot of work to do in our industry. As veterinarians and senior production members, we are obligated for the good of the pig and the success of our clients to continue to build better training and staff development. Rich would say culture change is like religion. We are only going to convert one person at a time. We can hold revivals and celebrate successes, but the individual on-farm contact (right at the center of the Veterinary-Client-Patient Relationship)[8] is how we will make a difference in animal care and quite likely in the lives of the caregiver to the pigs. Future production opportunities will go to excellent communicators willing to build people up while holding people accountable.

Health – technology is both a blessing and a curse for the veterinarian

The blessings from technology include improved diagnostic assays, electronic records and databases with rapid search and reporting functions. We also now have the ability to include photographs and videos as part of a consultation and farm assessment. Its curses include much of the same, improved diagnostic assays raise new disease questions with poorly understood organisms. Rapid data analysis yields more challenges than we often have resources to address. Photographs and videos raise concerns of public perception and aesthetics. We also must be mindful of the legalities and regulatory unknowns of telemedicine. Simply put, the world is getting more complicated every day, and while technologies bring us transparency we may not always know how to react to the transparency of new questions and challenges. One of the dumb things I have heard in the health of pigs was, "Circovirus, more like circus virus. We do not see it." That was me in 2005 before PCV2 moved into the industry and when there were no vaccines. My advice from that mistake, keep an open mind on disease transmission and disease presentation. When someone else is having significant health problems in their region, get to know the situation well. It may be headed your way.

It is obvious that our customers and the public in general expect the science of practicing medicine to improve at the speed of technology. The new diagnostic test is supposed to instantly lead to intervention strategies which solve the problems. The new production metrics platform should provide insight into where improvements are needed and those challenges should be instantly corrected when identified. While these expectations are unrealistic, we can't expect them to change. When you have a health problem it remains top of mind until you feel confident there is an effective solution in place. No matter whether these health problems occur within yourself, a family member, a pet or your livestock, the health problem is the most important thing in your life until you feel confident and an effective remediation plan is in place.

Recently our family shared experiences with a human medical professional and their 12-year-old dog. My wife was having a significant medical procedure. We were, of course, nervous. The medical professional wanted to ease the tension talking about her dog. The short version is specialty high-end emergency clinic with diagnostics and medications missed a significant tumor in the abdomen of the dog which was simply identified by a local veterinarian when the owner asked for a second opinion and took her dog to her hometown local veterinarian. Experience taught the local veterinarian how to palpate properly and how to predict and identify the growing mass.

My point, there will always be demand for locally available and experienced veterinary service. Those practicing veterinarians that are willing to pick up the phone and answer the call for their clients will continue to build an unyielding loyalty. Observation skills and good clinical judgement cannot be replaced.

While "fire engine" services will always be needed, we should constantly challenge ourselves to eliminate disease wherever possible and economically feasible. Think of the analogy of our "fire engine" services as the emergency room section of swine practice. We should invest aggressively in the reduction of endemic disease through preventative medicine approaches that keep our patients well and minimize the impact of emergency room visits. Our new diagnostic assays provide for a remarkably improved ability to study the epidemiology of old diseases such as *Strep suis*, HPS, *M hyorhinis* and many other pathogens which have been endemic in the global swine herd for decades.

We should leverage these learnings to attempt novel control and/or elimination approaches, with emphasis on reporting our results to quicken the pace of medicine evolution throughout our practices.[9]

We should be creative in attempting novel disease elimination approaches for diseases such as *M hyopneumoniae* and PRRS, diseases which have well-established elimination procedures but which, unfortunately, are not feasible nor cost effective for some producers. Lastly, we should be perseverant in our efforts to eradicate disease from targeted geographies, honing these techniques and sharing them with veterinarians throughout the world in an effort to take back so much of the "buffer" we have lost from African swine fever.

In my human health experience above, I was also impressed by the professional collaboration necessary in human medicine. We pale by comparison to a robust and continual collaboration for the good of our clients. We have made advances in Vet-to-Vet communications for pig flows, pig purchases, and genetic changes. We often bristle, or worse become outright protective, when our clients search for second opinions to help solve difficult problems.

As we consider the management of both old diseases and new, we should involve experts in our AASV membership as well as other technical disciplines. Collaboration will help further improve our understanding of the impact of nutrition, genetics and the environment on disease. Our expertise lies primarily in understanding pathogen life cycles and dynamics within the pig, and the immune responses typically associated with these host-pathogen interactions.[10] We all appreciate the impact that nutrition, genetics and environment can have on host-pathogen interactions, but we don't do enough to involve technical experts in these areas in disease evaluations. Particular in unusual cases in which the host-pathogen interactions and outcomes are unexpected and not consistent with our current scientific understanding, we should challenge this science in part by including other scientific disciplines in the evaluation. The future of pig production will be dominated by those that best understand the interactions between health, nutrition, genetics, and environment. Position yourself and your business to be on the cutting edge of these new learnings by routinely communicating with those outside your scientific area of expertise.

Creative veterinarians with a drive for healthy production will have additional opportunities to understand epidemiology. Rotavirus is an example of an old disease in dire need of new effort. Epidemiology of the disease within the population and among multiple related herds may be the key. Without collaboration and investigation among systems, we all continue to struggle learning to battle this old disease on our own. I have become fixed in my old methods of identification and control while riding the ups and downs of this performance-robbing piglet enteric virus. I am just now starting to look into sow diets, virus changes, and piglet exposure to chip away at alternatives to the activity of the virus. We have much to learn here. Piglet losses in the farrowing house are embarrassing at some high-health herds. Creative and persistent veterinarians will need to work together to tackle this one.

Public service – Packer needs drive future pork production investment

The fall of 2019 has again been a lesson on the value of a close working relationships with meat packer processors. Our clients with strong marketing agreements and packer relationships have enjoyed nearly double the average contract price going into these market lows. Producers and their veterinarians who fail to recognize the needs and wishes of packers are already being left behind. My perspective on choices available to the swine veterinarian interested in food safety and public service include building relationships with meat packers. We must engage regularly in animal welfare and antibiotic usage discussions as these are high on the wish list of meat packers and their marketing department. Swine veterinarians have the best training to advocate for the animal in these issues. These are large enough issues that casual conversations and meeting resolutions will not bring solutions. Veterinarians will need to engage in actual production system programs in order to create and provide for the demands of the industry. Programs which call for changes in animal housing or antibiotic use will continue to require investigation. Searching the AASV e-letter for job openings will tell you the opportunities exist as meat packing companies are searching for veterinarians.[11] Additionally, veterinarians in regulatory service are growing and replacing the "old guard." AASV would be smart to engage in these groups and share membership priorities. For the independents, like most of us, asking our clients for contract details and operational demands will further support the packer relationships. Producers will need financial analysis for these contract strategies. Producers also will need production and operational support for the analysis of antibiotic therapy. The AASV and our membership communications will help us learn together.

The AVMA as a whole has veterinarians in industry and public health participating in AVMA membership at a much higher rate than those choosing to join AASV. Our association would be well served to recruit membership from regulatory and public health groups. Journal articles and annual meeting sessions could be an opening for communication. My perspective is these veterinarians are engaged in problem solving for the industry and simply do not see swine practice or the AASV as an outlet to collaborate. We should fix that. There is a strong incentive for us to build on the "One Health" initiatives. The incentive is simply part of our charge to ensure proper animal care. Additionally, we are in the best position to understand the health and safety of farm staff to the potential risks of infectious disease. The United States Centers for Disease Control and Prevention (CDC) states that three out of four emerging infectious diseases are acquired from animals.[12] I am aware these communications may have started on the wrong foot. There was "one health" human health and the vets were simply directed to fall in line and do what they are told. Fortunately, there has been progress on the two-way communications needed to foster trust in public health and food safety.[12] Our friends, family, neighbors, and communities will continue to eat meat. We have the unique skills in clinical disease diagnostics and trusted relationships on-farm to further improve the safety and wellbeing of the animals raised for food. Swine veterinarians with the relationships and experiences of production companies, food companies, and animal health companies could be a strong voice to those lost in the wilderness of elimination of animal health products. Veterinarians in the meat packing business or in regulatory medicine may help us build the bridge in this communication.

Products – anything that helps reduce labor requirements or reduce antibiotic use

In 1998, Dr. Ralph Vinson was kind enough to mentor Dr. Jim Lowe and I as new partners in Carthage Veterinary Service. He sold his business to us and took us individually to each client. Dr. Vinson introduced the families and staff to us and helped transition the business to the next generation. Those car rides in his big white Cadillac were good lessons in both production management and business planning. One thing I remember Dr. Vinson sharing was his belief that he never should have given up product sales. Dr. Vinson was very honest with his business planning advice. He never wanted to be in a moral dilemma of having product sales color his judgement. He also hated the thought of a producer going broke and having all that accounts receivable lost in bankruptcy. Dr. Vinson shared there are ways to manage both of these concerns and still be a conduit for products in order to keep independent producers competitive. He also shared that the product sales keep veterinarians in touch with innovation.

Consolidation in purchasing patterns and large group negotiations will continue. Veterinarians working together for the benefit of the clients and the pigs will continue to grow opportunities for keeping the independent veterinary practice competitive. These collaboration and product purchase patterns will no doubt bring forward opportunities for new products as well. Enterprising veterinarians with a strong sense of persistence need to search for answers to some of our big questions. Here are a few of the ones I would like answered:

1. Why can't we get sow herds to less than 10% total piglet mortality before weaning?
2. Why do some herds only lose 5% of their sows annually and others suffer 15% annual mortality?
3. Can we determine feed ordering without climbing the bin or installing expensive equipment?
4. Is it possible to wash trailers robotically? Someone please build me a military grade "Roomba."

Another dumb thing I once heard, "Large systems will only need one veterinarian." In fact, our AASV was for a time discussing the real possibility we would only have 300 members based on the animal inventory and projected service requirements.[13] My current opinion is intelligent, driven swine veterinarians will never run out of work. It is possible they may no longer carry products or sell nearly the inventory or items carried on most vet clinic walls; however, it will be the enterprising veterinarian who develops a product, medication, or solution that presents to the AASV in the future. Product sales should not be given up in the private practice. Likewise, problem-solving veterinarians should also think and act accordingly if they identify important solutions for their clients or employers.

Education – teach them to fish, feed them for a lifetime

Animal agriculture continues to move toward a model where the execution of health management strategies occurs at the farm level by on-farm labor and not through physical labor applied through the veterinarian. Swine production and medicine are transitioning in parallel with dairy, poultry, beef, and small ruminants down this pathway, and as such, education will continue to become more and more important in our practices. The better job we do educating our animal caretakers, the better job they will do in executing our health management strategies, and therefore the better we will collectively improve the health of the pig.

Education is most valued when measured appropriately and some method of evaluation follows the training. It is not acceptable to simply check the box and report back. We had a client purchase a new herd where the farm staff said we used to get report after report saying the PRRS is still here and nothing was being done about it. While we are not the final decision maker, we are most often a trusted advocate for change. When we give information and guide decisions, our success will be measured by the client. We will be given the benefit of a return phone call if we can help implement a method to measure success before leaving the farm. Checking the boxes and leaving reports is no longer acceptable. Measuring performance and asking for results over time are necessary to monitor improvement.

The demand for education exists far beyond the boundaries of our United States. Countries throughout the world strive for food independence. No matter where you are, you value the ability to produce your own food domestically, providing for your own people without fear of being "cut off" from your food source via embargos and/or trade restrictions. Dr. Clayton Johnson shared experiences of heavily armored riot police surrounding the attendees at a recent Asian Swine Disease conference with over 1000 attendees.[14] This desire to be food independent means both developing and developed countries must modernize both their facilities and practices. Our veterinary education system in the United States is second to none throughout the world. Many production companies have dozens and even hundreds of veterinarians on staff, but their lack of training and education provides us an opportunity to improve their technical skills through both formal and informal training programs. Training requests come in all shapes and sizes. Just within our practice we provide anything from intensive, 13-week training programs to dozens of dedicated students working seven days a week and living in dormitories within steps from our training center to targeted meetings with specific individuals that may last for only a couple of hours.

While the demand for veterinary education is great, the demand for production management education to develop farm staff leaders is even greater. There are schools to create professionals needed in almost every position within a swine veterinary production system and/or practice. Schools not only create qualified and trained veterinarians, but also can be counted on as a recruiting pipeline to fill roles in accounting, human resources, IT and finance. While there are a few college programs that still do have swine management curriculum and produce graduates with some swine technical skills, producers of all sizes both domestically and internationally are increasingly focused on developing their own people into the next generation of production leadership. There is demand for a range of leadership roles, from departmental leads to farm managers to production managers and even production directors in large systems. Curriculum must be tailored to each production role, but for those willing to invest in developing educational programs for production management there is no shortage of demand.

Regardless of who is being educated and where this education is occurring, we still primarily rely on the classic educational model we're all comfortable with from our primary school experiences. A content expert creates materials which are reviewed in a classroom setting and/or delivered via written training materials. A limited time is allowed for question and answer between student and educator, and generally even less time is spent doing experiential learning with the student working inside a farm, hand-in-hand and under the supervision of the educator. This sort of educational approach works for those with a deep passion for science as is evidenced by the large volume of health professionals produced by universities each year. That being said, I think we lag behind in using technology to improve the educational process. Sure, chalkboards have been replaced with projector screens and hand-written lecture notes have been replaced with fancy PowerPoint slides; but that's using technology to replace parts of the educational process, not using technology to change the educational process itself. I think this is critically important for those who are kinesthetic learners, and from my experience the vast majority of people who work in our pig farms are kinesthetic learners (for those looking for barn speak – that is hands-on learning). Never have I taught a group of farm workers who requested the classic classroom approach to learning – the response I normally get is closer to "No death by PowerPoint!" They want to be directly involved in hands-on learning, and they want to physically experience doing the tasks associated with all we do in swine medicine and production. How can we include technology for the kinesthetic learner? I think it starts with having the technology present in our pig barns also present in our training areas.[15]

It's extremely difficult to describe the principles of ventilation, yet when you're standing in front of a controller and can see the changes being made to fans, inlets, cool cells and heaters as you practice manipulating its controls, most people tend to catch on fairly quickly. Investing in training facilities that allow hands-on practice using the same technology and equipment present in our farm environments will increase both the pace of learning and the recall ability of the learner, allowing you as the educator to train more effectively and on more material in the same amount of time. The next time you are asked to explain something difficult while on farm, please step back a second and ask them to gather the important people involved, to watch you, then to do it themselves, and also make sure they bring something to take good notes.

What might help you along the way

People – my clients have been some of my best teachers

Professional Swine Management (PSM) exists because pig farm owners did not want to organize large groups of staff people. Producers wanted PSM to put together the business model for them to receive large groups of weaned pigs, and they did not have to deal with sow herd organization. The lesson for me in that process was to learn from their business planning. Producers helped me understand the impact of a balance sheet on long-term business planning, and the influence of income statement for operational efficiency and a cash flow statement on meeting routine ongoing expenses. Our producer shareholders have held PSM accountable for watching the books. As a result, we have also struggled with them through difficult times in health, performance, and marketing. Some of my clients have also been some of my best teachers.

Members in the AASV would also be some of my greatest teachers and motivators. Internship opportunities have been available within the membership of our organization for decades. Before formal internships, I have heard stories of vet students sleeping in the back room of our clinic in order to hit the trail with Dr. Connor before sunrise. Practices, technical service companies, and production companies all provide meaningful experiences for students today. The student presentations are some of the best attended and hotly contested AASV talks each year. In order for you to choose a pathway forward as a swine veterinarian, I encourage you to watch these young people. Invite a couple of interns to your business and help them challenge your way of thinking. We are also blessed in Carthage with a diverse group of swine veterinarians I am proud to call my friends. Dr. Aaron Lower has been elected by our group to guide the vet team in working together. Dr. Doug Groth and I have raised our families together from the time they started kindergarten. Our kids are just now starting college and have grown up in the same 4H club. The Gallup polling organization has been proving for decades the value of building strong friendships at work. Whether you are fortunate enough to have them in the same building with you, or you build your network over the miles and states through the AASV, we all have many people with the shared goals of helping each other grow.[16] Choosing a pathway forward does not get any easier with age. Dr. Wright told me this year that starting a practice is easy compared to trying to figure out how to stop. While struggling through cancer recovery, he shared that his wife Betty would maintain the front desk and bring difficult cases by the house to talk with him. We can all hope to build such strong, lasting relationships with our clients and employers.

In order for our membership to remain viable and growing, we must welcome diverse pathways. We need AASV members willing to support the development of these paths. Too often we talk about production teams becoming barn blind. I believe our association can become blind to the needs of our membership if we focus only on the same pathways we have always trod. One of my classmates is a veterinarian in lab animal medicine. She challenged me that we have no idea what biosecurity is compared to the animal care and housing requirements in her laboratories.

After a brief conversation about air distribution and routine infectious disease monitoring, I believe she may be right. AASV could be well served by sending delegates to learn from other disciplines and report back to our own committee structure. The American Association of Lab Animal Sciences maintains over 264 online courses on their website. They have over 360,000 subscribers to their organizations learning library. We have more growing to do in AASV and certainly could learn from that exciting group.[17]

Advanced degrees may become highly valuable as we see our largest production companies and client needs evolve. I admire those who have re-invested in their own education in order to gain access to the right board room or in order to more completely serve in education or research. We are blessed with strong university faculty interested in swine medicine. We need to support these people with future leaders willing to enter the lab, the classroom, or the board room with a growing number of young talent. My challenge has always been how to count the cost of education. Fortunately, we now have growing endowments and foundations developed by the founders of our AASV association committed to support young leader development for the future. Our own AASV budget is tiny compared to the big fish of the AVMA or the National Pork Board. Even so, the AASV leaders targeted for future staff development signal an important step our founders have started. It is in all of our best interest to carry on the plans they have made.

Books that I find useful – in addition to *Diseases of Swine* and regular swine periodicals

Business owner book: *The Millionaire Next Door* by Thomas J. Stanley

Every town in America has successful businesses. Mr. Stanley wrote this book when I was just getting started in practice. The exciting thing for me to learn from this book was that every town in America has people build successful businesses, who save well, and who live comfortably late in life.

Career advice book: *What Color is your Parachute* by Richard Bolles

Even if you stay right where you are this guidebook gives you control. Mr. Bolles realized during periods of downturn in the economy there are many people searching without any idea of what they really like to do. He jokes in the introduction that executive level teams create "golden parachutes" for themselves. If you plan your career path today while things are going well, you might just have a better time of it, should life or your employer throw you a curveball.

Persistence book: *Get Well Soon* by Jennifer Wright

History's Worst Plagues and the Heroes Who Fought Them. This book is written as multiple short stories. Interesting to me is the author is not a medical professional. She is most interested in how and why the medical professionals involved remained persistent. Society celebrates athletes and artists. Actually, the author contends, many of us may not even exist if it were not for the persistence of some of these intelligent scientists.

Positive World View book: *Factfulness* by Hans Rosling

Ten Reasons were wrong about the world – Why things are better than you think. It was interesting for me to hear Dr. Steve Henry and I have both enjoyed this book. I consider Dr. Henry one of our deepest thinkers in the AASV (I am not). We both like this book, I believe, because the author brings strong facts and data to the reality that the world is getting safer, smarter, and more prosperous than media or even our own governments would have us believe. Dr. Rosling has now passed away. His family believes so strongly in his mission that they continue to share his personal works and presentations with the world. He shares many good reasons to hold a positive outlook for the poorest people on earth.

Loyalty book: Paul's Letter to Timothy in the *Bible*

My Christian faith started at home. My parents brought us up in the church and fostered our development with kind Sunday School teachers, youth fellowship programs, and good examples of how to treat people. The Apostle Paul didn't have any of that. It is very interesting that after 2000 years of human development, and we are still struggling with many of the same simple behaviors. What Paul did have was a good teacher in Jesus. Paul tries to guide Timothy to build the church and develop a sense of loyal care for its people.

Experiences – tough thing about experience, you take the test first and then get the lesson

Over the last two decades, pig producers have welcomed investment and partnership with their swine veterinarian. While the opportunities may change over time, the production industry may still have openings to perceived barriers to entry. Facilities require land, and land takes significant capital. Other opportunities to gain experience may come from taking on significant health strategies or elimination programs and holding responsibility for the outcome. As production systems grow in size there are multiple decisions to be made within the operation. Creation and practice of using a partial budget in decision making is a good start on bringing solutions to the problem. Even if you do not own pigs, accepting a stretch goal to experience market responsibility may help you appreciate the volatility that still remains in livestock production. We have producers today willing to share pig ownership or sell shares in their operation. Preparing for this presentation, I also went back to review a "what-if" scenario for comparison of investment in land, sow shares, or the stock of three publicly traded meat packers. I assumed for this short study that I had $50,000 to invest in 1996 (which at that time would not have been true). Still for our purposes today, let's use $50,000

Table 2: Stock investment $50,000 after 24 years

	May 96	Shares	Oct 19		Inflation $80,000	Growth	23 years	Annual
Tyson*	$16.00	3125	$93.00	$435,938		$355,938	$15,476	9.53%
Smithfield†	$7.00	7143	$34.00	$971,429		$891,429	$38,758	13.61%
Hormel‡	$26.00	1923	$41.00	$621,929		$541,929	$23,562	11.34%

* Tyson split 1:5 to 1 in 1997 then sat nearly flat for almost 15 years, 4688 with significant growth last 5 years
† Smithfield split 2:1 twice between 1996 and 2013, 28571 new total of SFD stock then done in 2013
‡ Hormel split 2:1 three times in 2000, 2011, 2016, 15169 final shares after three splits

Table 3: Farm land value $50,000 initial investment

	1996				2019				
Central IA	$1,500 per acre price		33.33	Acres	$7,500 per acre			$250,000	Value
	Cash rent	$100	$3,333	Rent		$225	Rent	$119,167	Income EST.
						$7,500	Income		
Central IL	$2,000 per acre price		25	Acres	$7,500 per acre			$187,500	Value
	Cash rent	$80	$2,000	Rent		$250	Rent	$90,750	Income EST.
						$6,250	Income		
SW Minn	$1,175 per acre price		42.6	Acres	$6,350 per acre			$270,213	Value
	Cash rent	$60	$2,553	Rent		$166	Rent	$105,787	Income EST.
					$7,064	Income			
					Average of total value				$235,904
					Average of total income				$105,235

Table 4: Sow farm investment in sow shares

	2400 sow center		
$50,000	Initial investment		
$400	Per sow		
125	Shares to sell weaned piglets		
$1,800	Current share value in the same business with facilities inventory and other investments		
Current value	$225,000		
Pig per profit annually	$3	$215,625	After 23 years
Annual income at 25 psy	$9,375	Annual	

invested in land, in sow shares, or in a packer. I have to believe hindsight is very telling in that the meat packers have enjoyed significant gains. While there is no guarantee in performance in any of the examples, the stock market has been very good to the meat packers. Sow shares have provided income greater than the land, and while the land has appreciated significantly, it is without saying that $50,000 would not have bought enough land even in 1996 to make much of a start on any income production. My point in this exercise is even if you are employed with little opportunity to invest in pigs where you are today, our industry has carried significant gains over the last two decades. Choosing a pathway forward can be intimidating. Jumping into the business in order to gain experience along the way, may prove to bring the best lessons about our business.

Comparison of owning land, vs owning sow shares...or what if you bought shares of Tyson, Smithfield, or Hormel. (Table 2, 3 and 4).

Opportunities for pig ownership may change with consolidation. Opportunities for investment in pig barns or simply stock ownership will likely continue. Industry production opportunities evolve slowly. I have found just being prepared and waiting patiently for the right time with the right partners has been the best plan for me. Independent producers even as large of 50,000 pigs sold annually, now question their place at the negotiating table. Veterinarians skilled at production negotiations and networking will continue to be involved with these groups. Additionally, veterinarians with a desire to remain in production must align with groups or packers to join in these opportunities. Consolidation in production could limit these opportunities in the future. My advice for choosing this pathway for experience, never go into a production investment without understanding the very real possibility that additional capital is likely to be required during certain periods. Always understand the numbers for projections and performance expectation and remain close to the outcomes. Finally, know the business you are in. Whether that be genetics, weaned pigs, semen production, or simply barn ownership. Whatever your investment, learn what makes that investment solid. Experience is a good teacher, but remember the test comes first followed by the lesson.

Conclusion

- Why I see things the way I do? Business ownership and a strong curiosity for biological systems
- What may be coming in Swine Medicine? Multiple diverse opportunities and intense competition
- What you can do about? Read, Study, Network and Pray

References

1. Vlaskovits, Patrick. Henry Ford, Innovation, and That "Faster Horse" Quote. Harvard Business Review. August 29, 2011.
2. AVMA. 2016 AVMA Report on Veterinary Markets. 2016; 54
3. American Association of Swine Veterinarians. Salary Survey Results. 2017.
4. Waddell, John T. Built to last: 50 years of AASV. AASV. 2019; 13-31
5. Dicks, Michael, et al. Veterinary practice consolidation demystified. AVMA. 2019.
6. Apiam Animal Health. https://www.apiam.com.au/management-team/.
7. Firkins, Larry. University of Illinois Executive Veterinary Program. https://vetmed.illinois.edu/evp/.
8. National Pork Board. VCPR Definition. PQA Plus Training Manual. 2019.
9. Profeta, Louis. The Patient in Room 9 Thinks He is God. 2010.
10. Main, Rodger. Personal Communication. Iowa State University Veterinary Diagnostic Laboratory.
11. AASV E-Letter. Job Openings Section.
12. AVMA. Resource Tools. https://www.avma.org/resources-tools.
13. Vinson, Ralph. The 1990's, the Last Decade of Private Swine Practice? AASV Howard Dunne Lecture. 1986.
14. Johnson, Clayton. Personal Communication.
15. Felder, Richard. Learning Styles and Strategies Handout. North Carolina State University.
16. Wagner, Rodd. Stop Using Employee Friendships to Measure Engagement. Harvard Business Review. August 7, 2015.
17. American Association of Lab Animal Sciences. https://www.aalas.org/education/aalas-learning-library. Accessed November 2019.

Effect of two PRRS MLV doses compared to a single dose vaccination program on the wild-type virus shedding and mortality of growing pigs from endemic sources

C. Moura[1], DVM; R. Philips[2], DVM; G. Silva[1], DVM, MS, PhD; D. Linhares[1], DVM, MBA, PhD
[1]Veterinary Diagnostic and Production Animal Medicine Department, Iowa State University, Ames, Iowa;
[2]Boehringer Ingelheim Animal Health, Duluth, Georgia

Introduction

Porcine reproductive and respiratory syndrome virus (PRRSV) continues to cause significant economic impact in the swine industry.[1] It has been demonstrated that PRRS modified-live virus (MLV) vaccination, with a full dose, significantly reduces clinical consequences of PRRSV challenge compared to non-vaccinates.[2-8] Aiming to reduce clinical signs and shedding, some veterinarians have recommended vaccination of growing pigs with two doses of attenuated vaccines. In fact, it has previously been described that the application of two PRRSV MLV vaccinations was effective in reducing the spread of wild-type virus in growing pig populations.[9-13] However, there is limited information about the effect of the double PRRSV MLV vaccination on the productivity of growing pigs, compared to the single dose protocol. Objectives of this study were 1) to evaluate the levels of PRRSV shedding over time from growing pigs vaccinated with two doses of PRRSV MLV vaccine, compared to matched groups vaccinated with a single dose of the same vaccine; 2) to describe performance of pig groups vaccinated with two doses of PRRSV MLV vaccine, compared to single dose vaccinated groups; 3) to assess the effect of frequency of wild-type virus detection and diversity of wild-type virus detection on mortality of groups vaccinated with two doses of PRRSV MLV vaccine, compared to single dose vaccinated groups.

Methods

This randomized field study was conducted from fall/2017 to summer/2019. Eighty-one growing pig groups, were conveniently selected from seven PRRSV-stable[14] and two -unstable[14] sow farms in US, were randomly allocated, by a coin toss, into two groups and were placed at high pig dense areas (Table 1). Group 1 received one full IM dose of Ingelvac PRRS® MLV vaccine around weaning or processing age. Group 2 received the same vaccination as group one, with an additional dose three to four weeks after the first.

Six cotton ropes were spatially distributed in each group for oral fluid sampling[15] every three weeks, totaling 8 sample collections over time. Samples were sent to ISU-VDL for PRRSV rRT-PCR testing and every positive sample was submitted for PRRSV ORF-5 sequencing. Production records from wean-to-finish phase were collected from all groups. Number of positive and negative samples for ORF-5 sequencing, at each of the eight collection points, was recorded for all pig groups. Binomial regression models were built to compare levels of PRRSV shedding, at each sampling point, for the two vaccination groups. Poisson regression models were performed to asses the effect of the vaccination program on mortality. Additionally, the frequency of virus detection and the virus diversity of each group were both included as interaction terms with vaccination group. Virus diversity was defined based on a cut-off of at least 99% of homology similarities between virus. Analysis were conducted separately for PRRSV-stable and -unstable scenarios.

Results

Study included 136,482 pigs, 634 sample collections and 3,646 oral fluid samples. There were 33 single and 33 double dose groups on PRRSV-stable scenario, while there were 6 single and 9 double dose groups on PRRSV-unstable scenario.

Table 1: Baseline production for each vaccination treatment group

Vaccination treatment	1 dose group Mean (σ)	2 dose group Mean (σ)
Weight started (lb)	14.97 (1.92)	15.07 (1.89)
Average days on feed	160 (7.20)	159 (6.40)
Pigs started	68,512 (1173.93)	67,970 (978.19)
Groups started	39	42

Among PRRSV-stable flows, the double dose groups had a significant reduction on levels of PRRSV shedding at weeks 7 ($P = 0.0027$), 10 ($P = 0.0003$) and 19 ($P = 0.0155$), compared to the single dose groups. The single dose groups had significant reduction on levels of PRRSV shedding only at week 16 ($P = 0.0216$), compared to the double dose groups. However, for the PRRSV-unstable scenario, the double dose groups had a significant reduction on levels of PRRSV shedding only at week 7 ($P = 0.0167$), compared to the single dose groups (Figure 1).

Single dose groups, from PRRSV-stable scenario, had significantly higher mortality (mean 5.93%) when compared to the double dose groups (5.34%; $P = 0.0002$) with rate ratio 1.11 (95% CI: 1.05 –1.17) (Figure 2). Moreover, the effect of the double dose program to reduce mortality increased at higher levels of PRRSV challenge, which was measured by the frequency of wild-type detection and diversity of wild-type detection in the study groups (Figures 3 and 4).

Single dose group, from PRRSV-unstable scenario, had significantly higher mortality (mean 19.45%) when compared to the 2-dose group (14.63%; $P < 0.0001$) with rate ratio 1.33 (95%CI: 1.22–1.44) (Figure 2). Moreover, the effect of the double dose program to reduce mortality increased at higher levels of PRRSV

Figure 1: Binomial regression model for wild-type PRRSV shedding comparison between vaccination treatment groups, from PRRSV-stable (Panel A) and PRRSV-unstable (Panel B) breeding herds.

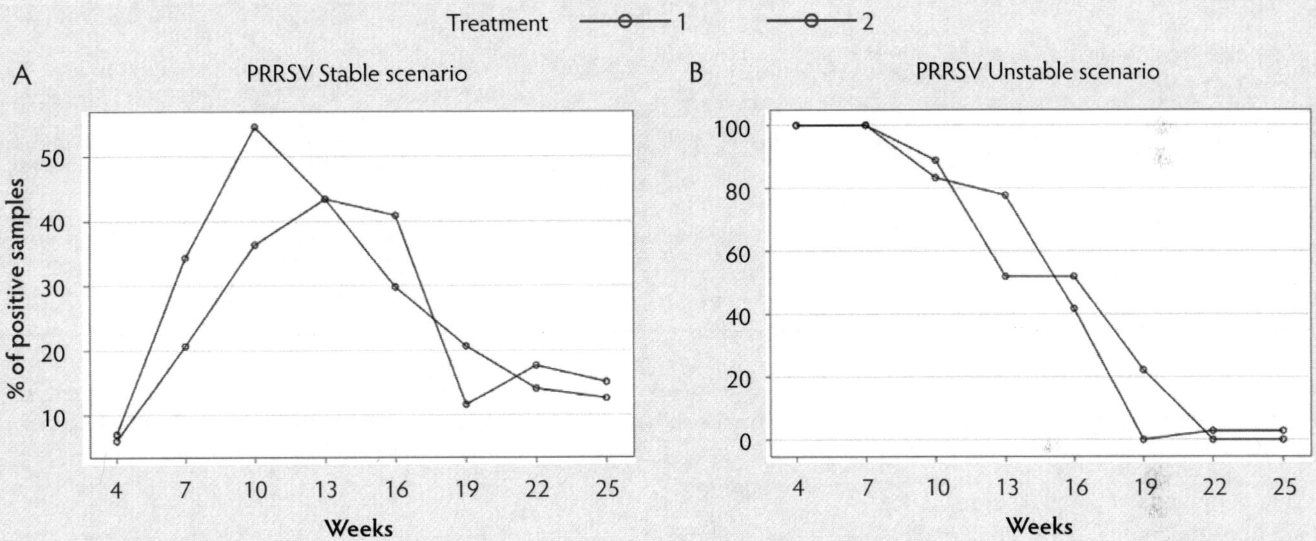

Figure 2: Poisson regression model for mortality comparison between vaccination treatment groups, from PRRSV-stable (Panel A) and PRRSV-unstable (Panel B) breeding herds.

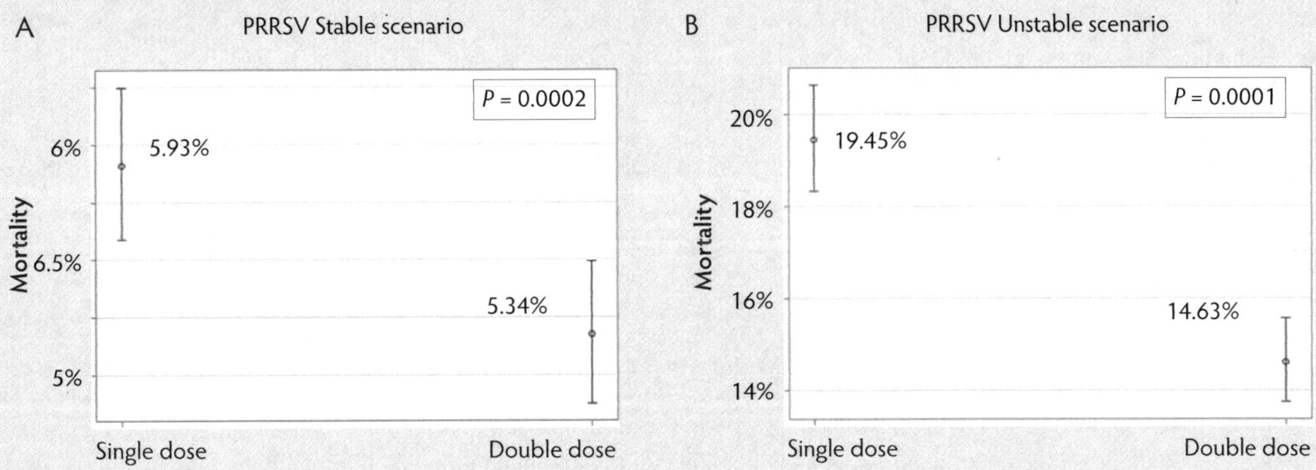

challenge, which was measured by the frequency of wild-type detection and diversity of wild-type detection in the study groups (Figures 3 and 4).

Conclusion

Under study contitions, there was no statistical difference on the levels of PRRSV shedding from vaccination groups at PRRSV-unstable scenario. However, for the PRRSV-stable scenario, there was a reduction of PRRSV shedding on double dose groups at weeks 7, 10 and 19. Double dose group had significantly lower mortality rate than single dose groups, for PRRSV-stable and -unstable scenarios. The efficacy of the double dose protocol to reduce mortality was higher when frequency of wild-type virus detection and wild-type virus diversity increased.

References

1. Holtkamp D, Kliebenstein J, Neumann E, et al. Assessment of the economic impact of porcine reproductive and respiratory syndrome virus on United States pork producers. *Journal of Swine Health and Production.* 2013;21(2):72-84.

2. Pileri E, Mateu E. Review on the transmission porcine reproductive and respiratory syndrome virus between pigs and farms and impact on vaccination. *Vet Res.* 2016;47(1):108.

3. Linhares DC, Cano JP, Wetzell T, Nerem J, Torremorell M, Dee SA. Effect of modified-live porcine reproductive and respiratory syndrome virus (PRRSv) vaccine on the shedding of wild-type virus from an infected population of growing pigs. *Vaccine.* 2012;30(2):407-413.

4. Opriessnig T, Pallares FJ, Nilubol D, et al. Genomic homology of ORF 5 gene sequence between modified live vaccine virus and porcine reproductive and respiratory syndrome virus challenge isolates is not predictive of vaccine efficacy. *Journal of Swine Health and Production.* 2005;13(5):246-253.

5. Zuckermann FA, Garcia EA, Luque ID, et al. Assessment of the efficacy of commercial porcine reproductive and respiratory syndrome virus (PRRSV) vaccines based on measurement of serologic response, frequency of gamma-IFN-producing cells and virological parameters of protection upon challenge. *Vet Microbiol.* 2007;123(1-3):69-85.

6. Cano G, Cavalcanti MO, Orveillon FX, et al. Production results from piglets vaccinated in a field study in Spain with a Type 1 Porcine Respiratory and Reproductive virus modified live vaccine. *Porcine Health Manag.* 2016;2:22.

7. Lyoo KS, Choi JY, Hahn TW, Park KT, Kim HK. Effect of vaccination with a modified live porcine reproductive and respiratory syndrome virus vaccine on growth performance in fattening pigs under field conditions. *J Vet Med Sci.* 2016;78(9):1533-1536.

8. Kristensen CS, Kvisgaard LK, Pawlowski M, et al. Efficacy and safety of simultaneous vaccination with two modified live virus vaccines against porcine reproductive and respiratory syndrome virus types 1 and 2 in pigs. *Vaccine.* 2018;36(2):227-236.

9. Dee SA, Philips R. Using vaccination and unidirectional pig flow to control PRRSV porcine reproductive and respiratory syndrome virus transmission. *Swine Health and Production.* 1998;6(1):21-25.

10. Dee S, Philips R. Using vaccination and unidirectional pig flow to control PRRSV transmission. *Swine Health and Production.* 1998;6(1):21-25.

11. Philips RC, Dee SA, Aasv. Evaluation of mass vaccination and unidirectional flow for the elimination of PRRS virus from grow/finish populations. *American Association of Swine Veterinarians, Proceedings.* 2002:165-166.

12. Philips R DS. Elimination of the PRRS virus from a farrow to finish single site farm using mass vaccination and unidirectional pig flow strategy. Paper presented at: American Association of Swine Practitioners1999; St. Louis.

13. Philips RC DS. Evaluation of mass vaccination and unidirectional flow for elimination of PRRS. Paper presented at: 4th International Symposium on Emerging and Re-emerging Pig Diseases2003; Rome.

14. Holtkamp D, Polson D, Torremorell M. Terminology for classifying swine herds by porcine reproductive and respiratory syndrome virus status. *Journal of Swine Health and Production.* 2011;19(1):44-56.

15. Rotolo ML, Sun Y, Wang C, et al. Sampling guidelines for oral fluid-based surveys of group-housed animals. *Vet Microbiol.* 2017;209:20-29.

Figure 3: Poisson regression model for mortality comparison between vaccination treatment groups, at different levels of PRRSV wild-type detection frequencies, from PRRSV-stable (Panel A) and PRRSV-unstable (Panel B) breeding herds.

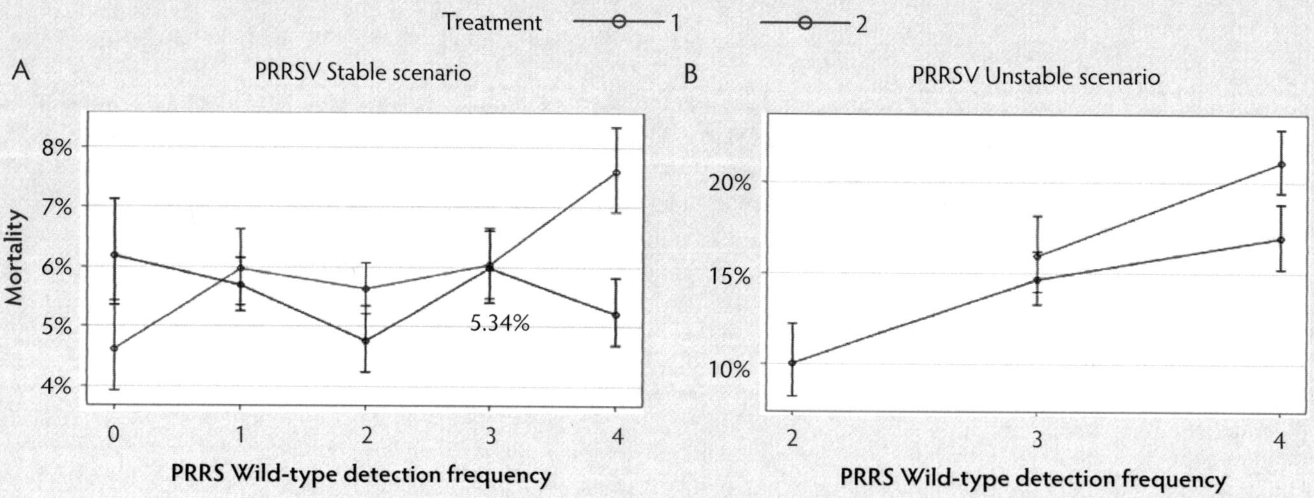

Figure 4: Poisson regression model for mortality comparison between vaccination treatment groups, at different levels of PRRSV wild-type detection diversities, from PRRSV-stable (Panel A) and PRRSV-unstable (Panel B) breeding herds.

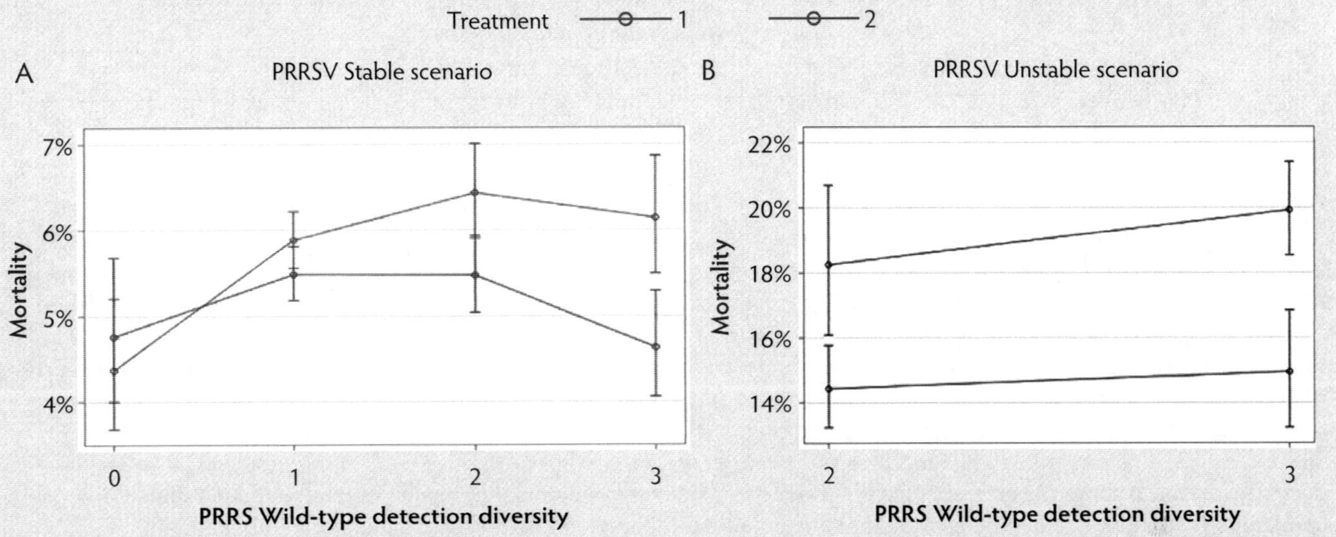

Fecal samples contain detectable PRRSV antibodies

A. Henao-Diaz[1], DVM, MS; G. Wang[2], DVM, PhD; J. Ji[3], MS, PhD; L Giménez-Lirola[1], PhD; J. Zimmerman[1], DVM, PhD
[1]Veterinary Diagnostic and Production Animal Medicine, Iowa State University, Ames, Iowa;
[2]Harbin Veterinary Research Institute, Chinese Academy of Agricultural Sciences, Harbin, China;
[3]Department of Statistics, Iowa State University, Ames, Iowa

Introduction

Detection of porcine reproductive and respiratory syndrome virus (PRRSV) antibody in aggregate specimens, e.g., oral fluids (OFs), processing fluids, has been an active area of research in recent years. Aggregate specimens provide significant advantages for surveillance at the herd level, but testing individual animals is still necessary under specific circumstances. Serum is the historical specimen for individual animal antibody monitoring, but fecal samples (FS) do not require animal handling and are easily collected by a single person. The presence of antibody in feces (coproantibody) against specific pathogens has been described in humans and animals, e.g., sheep, mice, dogs, and primates. Likewise, in swine, coproantibody against African swine fever virus, classical swine fever virus, hepatitis E virus, and porcine epidemic diarrhea virus has been documented.[1,2] Herein, we describe the detection of PRRSV coproantibody in fecal samples from pigs of known PRRSV infection status.

Materials and methods

Twelve individually-housed pigs were vaccinated with a modified live virus vaccine (Ingelvac® PRRS MLV) and sampled from -5 to 42 days post vaccination (DPV). A total of 132 serum and 572 oral fluids were tested for antibody using commercial specimen-specific PRRSV IgG ELISAs (IDEXX OF Ab test, IDEXX PRRS X3 Ab Test, IDEXX Laboratories, Inc). A total of 573 fecal samples were collected and then assayed for PRRSV IgG antibody by diluting samples 1:1 with ELISA kit diluent containing 1500 ppm chitosan (Figure 1) and then testing 200 μL of sample on the PRRSV OF ELISA using the procedure described by the manufacturer.

Results and discussion

Using a cutoff of $S/P \geq 0.4$, the first positive serum-ELISA and OF-ELISA samples were detected on DPV 8. Using a cutoff of $S/P \geq 0.1$, PRRSV antibody was detected in fecal samples on DPV 10 (Figure 2). Thus, this pilot study demonstrated that PRRSV antibody is present in swine feces and that coproantibody ontogeny mirrors serum and oral fluid antibody kinetics (Figure 2). Future research will focus on the development of an assay specifically for fecal samples and creating a practical approach for collecting samples to test adult animals (sows or boars) in PRRSV surveillance programs.

Acknowledgments

The technical support of Dr. Ronaldo Magtoto is gratefully acknowledged. PRRSV ELISA kits were provided by IDEXX Laboratories, Inc. Henao-Diaz received a scholarship through the program CONACYT Mexico, Becas al extranjero 2017.

References

1. Nieto-Pelegrín, et al. First detection of antibodies against African swine fever virus in faeces samples. *Transbound Emerg Dis* 2015; 62: 594–602.

2. Seo S won, et al. Detection of antibodies against classical swine fever virus in fecal samples from wild boar. *Vet Microbiol* 2012 ;161: 218–221.

® registered trademark of Boehringer Ingelheim Vetmedica GmbH.

Figure 1: Fecal sample clarification process. A. Left to right: undiluted fecal sample, fecal sample:kit sample diluent with coagulant (1:1), vortexed sample, and clarified fecal sample. B. Clarified fecal sample supernatant ready for ELISA testing.

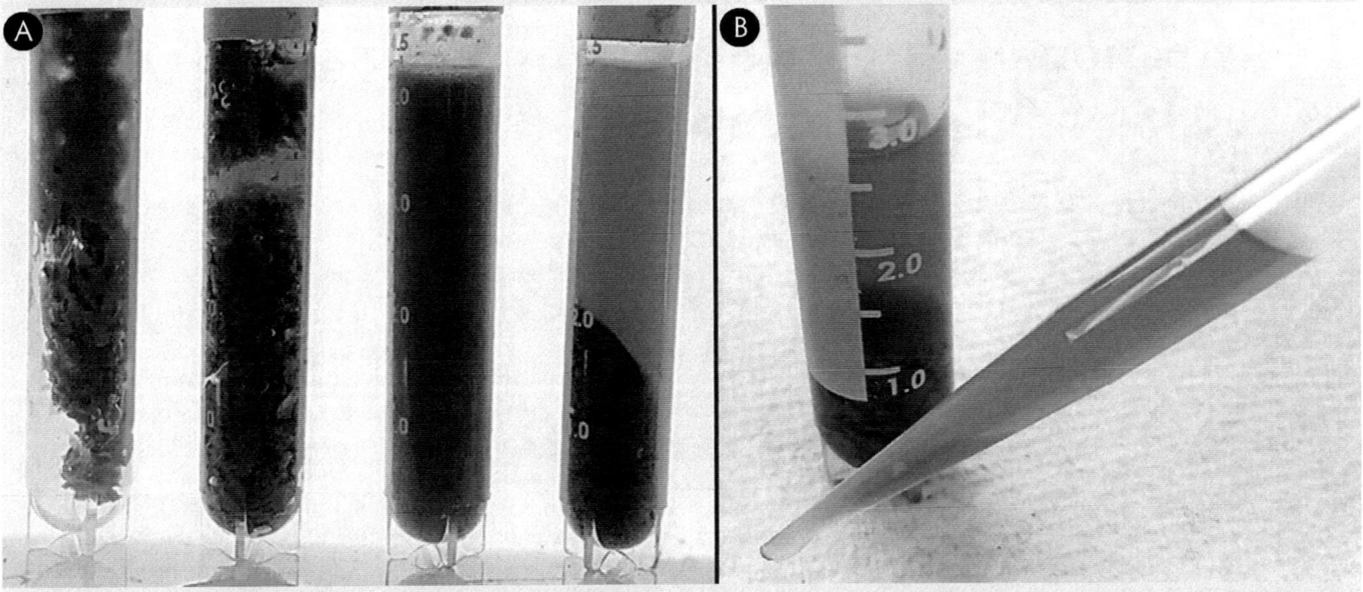

Figure 2: PRRSV ELISA antibody detection (12 pigs mean S/P) in oral fluids, serum, and fecal samples by day post vaccination.

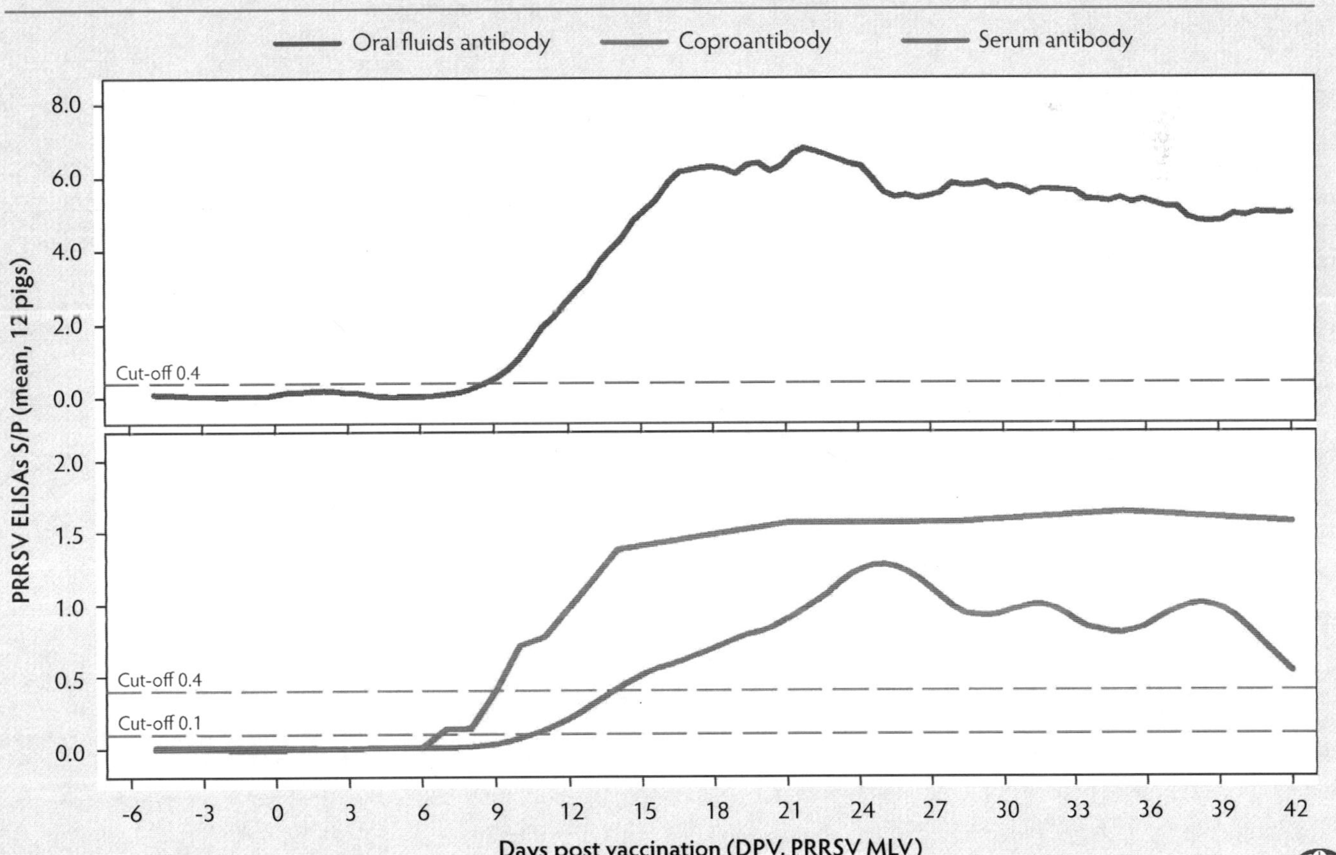

Use of an adapted commercial serum antibody enzyme-linked immunosorbent assay (ELISA) for the detection of anti-PRRSV antibody isotypes in processing fluid specimens

Will A. Lopez, DVM; Daniel C. L. Linhares, DVM, MBA, PhD; Jeff J. Zimmerman, DVM, PhD DACVPM; Ronaldo L. Magtoto, PhD; Luis Gimenez-Lirola, MS, PhD
Veterinary Diagnostic and Production Animal Medicine, Iowa State University, Ames Iowa

Introduction

Our initial work showed that PRRSV IgG antibody could be detected in processing fluids using the same PRRSV ELISAs used for serum or oral fluid samples.[1] Thus, IgG testing of processing fluids could be a cost-effective way to confirm the continued freedom from PRRSV infection over time. The problem we then addressed was how to adapt this approach to processing fluids collected in PRRSV-infected or vaccinated sow herds. We know that maternal (colostral) IgG cannot be differentiated from the IgG produced by the piglets in response to infection. However, infected piglets also produce PRRSV-specific IgM and IgA. Therefore, the objective of this study was to develop and evaluate IgM- and/or IgA-specific ELISA as a means to detect active infection in the face of IgG maternal antibodies.

Materials and methods

Serologic assays currently used to detect PRRSV IgG antibodies in swine serum (i.e., PRRS X3 Ab Test, IDEXX Laboratories, Inc., Westbrook, ME) were optimized for the detection of different antibody isotypes (IgM, IgA, IgG, IgM + IgA) in processing fluids. Processing fluid samples from PRRSV-positive breed-to-wean herds were used to optimize the PRRS ELISA for antibody detection. Then processing fluid samples from PRRSV-naïve breed-to-wean herds were used to establish the test cut-off, i.e., the point that best discriminated positive and negative samples. Processing fluid samples were also tested for PRRSV RNA by RT-qPCR to confirm their positive or negative status.

Results

The IgG results obtained with the modified kit showed improved discrimination between positive and negative processing fluid samples compared to the original kit. Work-in-progress has shown promising results with the PRRS IgA ELISA for processing fluids, with additional optimization to be done for IgM detection. Establishing cut-offs and evaluating the diagnostic performance of the isotype-specific antibody processing fluid PRRS ELISAs is ongoing.

Conclusions

In particular, results from this study suggest that IgG testing of processing fluid samples could be highly useful for monitoring breeding herds that are supposed to be naïve for PRRSV. This offers a more affordable alternative to PCR testing and could help increase the number of samples tested and the frequency of monitoring, thus, improving PRRSV surveillance protocols in breeding herds. Additional research may succeed in adapting this technology to the detection of IgM and/or IgA antibody in processing fluids from PRRSV-infected or vaccinated herds.

Reference

1. Lopez WA, Angulo J, Zimmerman JJ, Linhares DCL. Porcine reproductive and respiratory syndrome monitoring in breeding herds using processing fluids. *J Swine Health Prod.* 2018;26(3).

Increasing the functionality of your processing fluid toolbox beyond PRRSV monitoring: PCV2, PEDV, and PDCoV

G. Trevisan[1], DVM, MBA; J. Angulo[2], DVM; C. Johnson[3], DVM; M. Jensen[2], DVM; V. Utrera[2], DVM, MS, PhD; T. Marostica[1], DVM; P. Gauger[1], BS, DVM, MS, PhD; D.C.L. Linhares[1], DVM, MBA, PhD
[1]Veterinary Diagnostic and Production Animal Medicine, Iowa State University, Ames, Iowa;
[2]Zoetis, Parsippany, New Jersey; [3]Carthage Veterinary Service, Carthage, Illinois

Processing fluid (PF) samples are a practical, time-efficient, and affordable diagnostic tool to monitor porcine reproductive and respiratory syndrome virus (PRRSV) in the newborn piglet population at processing time.[1,2] Detection of porcine circovirus type 2 (PCV2) in PF has also been reported.[3-5] The first part of this study had the purpose to use PF to screen the newborn population for PCV2, classify batches according to PCV2 status, and associate detection of PCV2 in the newborn population with downstream performance. The second part investigated PF to detect porcine epidemic diarrhea virus (PEDV) and porcine deltacoronavirus (PDCoV).

For the PCV2 study, PF samples were collected from 4 sow farms, by room and day of processing, stored at -20°C and tested by pools of 5:1 for PCV2 by PCR according to rooms that generated a wean-to-finish (W2F) cohort group. Two farms were naïve for PRRSV (ELISA-negative),[6] and 2 farms were PRRS-positive. Placement batches were classified as positive or negative for PCV2 according to PCR results. All piglets were vaccinated for PCV2 at weaning. Association of PF PCR result for PCV2 and key W2F performance indicators including mortality, culls, total losses (mortality + culls) was investigated using a Chi-square test. Moreover, a generalized linear model was used to compare average age daily gain (ADG), feed efficiency (FE), and average days to market (AVM) between groups testing positive versus negative for PCV2.

In addition, weekly PF samples were tested by RT-qPCR for PEDV and PDCoV following two outbreaks of PEDV and one outbreak of PDCoV, each occurring at different time points. Day zero for each of the 3 outbreaks was defined as the day of recognition of clinical signs in the farm and detection of a positive result by RT-qPCR in feces. Stored PF was tested in pools of 5:1 by the room, day and week of collection and tested for a period 2 weeks prior and four weeks after the outbreaks.

An intermittent pattern of PCV2 detection in sow farms was observed. Only one farm obtained negative results for PCV2 for a period longer than 4 weeks. A total of 47 W2F cohort groups were generated, from which 10 were removed from the analysis for being PRRSV RT-qPCR-positive on PF. From the remaining 37 placements, 32 tested PCR-positive and 5 PCR-negative for PCV2. No PCV2-negative groups were detected from PRRS-positive farms. Groups negative for PCV2 at processing time had 1.29% (CI 0.86-1.73) higher mortality, 0.87% (CI 0.55-1.19) higher culls, and 2.14% (CI 1.67-2.66) higher total loss when compared with the positive groups. All comparisons had a Chi-square P-value < 0.0001. No differences were detected for ADG, FE, AVM between PCV2-positive vs. PCV2-negative groups. Additionally, no differences were encountered for W2T cohort groups coming from ELISA negative vs. PRRS-positive farms having PF negative result for PRRSV on RT-qPCR.

Regarding the PDCoV outbreak investigation, PCR results were positive 6 days before the outbreak and continued to test positive up to 17 days after outbreak onset. Regarding the PEDV investigation, no detection of PEDV on PF was encountered prior to day zero, or up to 4 days after. In this outbreak, the PF samples tested positive by PEDV RT-qPCR for up to 32 days post-outbreak. On the second PEDV outbreak, no detection in PF was encountered between 36 days prior to outbreak until 17 days post outbreak.

This work demonstrates that using PF is useful to monitor PCV2 status in the newborn population at processing time. Additionally, PF PCR results allowed classifying placement batches according to PCV2 status. Batches classified as PCV2-negative had higher mortality, culls and total losses when compared with batches having a PCV2- positive result on PF. Further studies are needed to clarify the association of PCV2-negative closeouts having worse KPI indicators. Also, PF can be utilized as a practical tool to monitor PDCoV activity on the farm, but not for PEDV. Previous work demonstrated that PDCoV causes viremia 2 to 5 days post-infection[7] suggesting RT-qPCR positive PF may be from viremic piglets considering PF is a serosanguineous fluid. PEDV was not detected in PF samples before the outbreak and further studies still needed to understand why this happened. For PEDV monitoring, fecal samples continue to be the optimal choice.

References

1. Lopez WA, Angulo J, Zimmerman JJ, Linhares DCL. Porcine reproductive and respiratory syndrome monitoring in breeding herds using processing fluids. *Journal of Swine Health and Production.* 2018;26(3):146-150.

2. Trevisan G, Jablonski E, Angulo J, Lopez WA, Linhares DCL. Use of processing fluid samples for longitudinal monitoring of PRRS virus in herds undergoing virus elimination. *Porcine Health Manag.* 2019;5:18.

3. Lopez W, Linhares D. Processing fluids, blood serum, and tail blood swabs to detect PRRSV RNA and PCV2 DNA by PCR-Based assays. Paper presented at: 2017 ISU James D. McKean Swine Disease Conference, 2017; Ames, IA.

4. Castevens K, Spronk E, Garbes N, Trevisan G. Use of processing fluids to monitor sows for porcine circovirus type 2 and determine parity influence on viral detection. Paper presented at: 50th Annual Meeting of the American Association of Swine Veterinarians, 2019; Orlando.

5. Perkins S, Davenport J, Jabloinski E, Amodie D, Trevisan G. Comparison of processing fluid and placental umbilical cord serum to detect porcine circovirus type 2 in sow farms. Paper presented at: 50th Annual Meeting of the American Association of Swine Veterinarians, 2019; Orlando.

6. Holtkamp DJ, Polson DD, Torremorell M. Terminology for classifying swine herds by porcine reproductive and respiratory syndrome virus status. *Journal of Swine Health and Production.* 2011;19(1):44-56.

7. Vitosh-Sillman S, Loy JD, Brodersen B, Kelling C, Doster A, Topliff C, Nelson E, Bai J, Schirtzinger E, Poulsen E, Meadors B, Anderson J, Hause B, Anderson G, Hesse R. Experimental infection of conventional nursing pigs and their dams with Porcine deltacoronavirus. *J Vet Diagn Invest.* 2016;28(5):486-497.

Estimating the sensitivity of two sample types for detection of *Mycoplasma hyopneumoniae* early and late after experimental and natural infection

A. Sponheim[1,2], DVM; J. Alvarez[3,4], DVM, PhD; E. Fano[2], DVM, MS, PhD; E. Schmaling[2]; S. Dee[5], DVM, MS, PhD, DACVM; D. Hanson[5]; T. Wetzell[2], DVM; M. Pieters[1], DVM, PhD
[1]Department of Veterinary Population Medicine, College of Veterinary Medicine, University of Minnesota, St. Paul, Minnnesota; [2]Boehringer Ingelheim Animal Health USA Inc., Duluth, Georgia; [3]Centro de Vigilancia Sanitaria Veterinaria VISAVET, Universidad Complutense, Madrid, Spain; [4]Departamento de Sanidad Animal, Facultad de Veterinaria, Universidad Complutense, Madrid, Spain; [5]Pipestone Applied Research, Pipestone, Minnesota

Introduction

Detection of *Mycoplasma hyopneumoniae* infection in live pigs is a critical component to measure the success of disease control or elimination strategies. However, *in vivo* diagnosis of *M hyopneumoniae* is difficult and the imperfect sensitivity of diagnostic tools, which is expected to vary over the course of infection, has been considered as one of the main challenges.[1] No studies estimating the sensitivity of the two sample types yielding the highest *in vivo* detection rates, laryngeal swabs and deep tracheal catheters, collected early and late after infection following a known exposure time, are available. Furthermore, no studies comparing sample types in experimentally and naturally infected pigs in a large population under controlled field conditions are available. Therefore, the objective of this study was to estimate the sensitivity of laryngeal swabs and deep tracheal catheters for detection of *M hyopneumoniae* early and late after infection in experimentally and naturally infected pigs.

Materials and methods

A subset of 90 pigs tested prior to experimental inoculation were confirmed negative for *M hyopneumoniae* using real-time PCR on laryngeal swabs and deep tracheal catheters. The following day, 320 8-week old seeder pigs were intra-tracheally inoculated with *M hyopneumoniae* 232 and immediately placed with 1,920 contact pigs to achieve a 1:6 seeder-to-contact ratio. A subset of seeders and contacts were longitudinally sampled at 7, 28, 97, and 113 days post-inoculation (dpi) and at 28, 56, 84, and 113 days post-exposure (dpe) respectively using laryngeal swabs followed by deep tracheal catheters. Samples were tested for *M hyopneumoniae* by a species-specific real-time PCR.

The number of positive pigs and the Ct value for each sample type were compared at each sampling using McNemar exact test and Wilcoxon signed rank test for paired samples in seeders and contact pigs using R.[2] The sensitivity obtained for each sample type in seeders at each sampling was calculated as the percentage of positive seeders from the total of seeders using EpiTools along with the Clopper-Pearson (exact) confidence limits.[3] Bayesian modeling was used to estimate the sensitivity of both sample types in the absence of a reference test in contacts since the true status (infected/not infected) could not be determined. A two dependent tests – one population model was fitted at each sampling.[4] To ensure model identifiability, the analyses incorporated prior information regarding the test sensitivities, while a weakly informative prior was used for the prevalence, and a perfect specificity was assumed as previously reported in the literature.[5,6] Beta distributions for the sensitivity of each sample type were formulated on BetaBuster, assuming that the most likely value was the point estimate for the same sample type in seeder pigs, and a 95% confidence that values were above the lowest bound of the 95% confidence interval estimated for seeder pigs (LS = mode 43%, 95% > 15% and DTC = mode 83%, 95% > 57%).[7] A uniform (0,1) distribution was used as the prior for the prevalence. The influence of the priors on the posterior estimates was assessed in a sensitivity analysis consisting in replacing alternatively the informative priors for the sensitivity of each test and the prevalence by uniform (0,1) distributions. The analyses were carried out in WinBUGS.[8] Three chains with different starting values were run for 50,000 iterations after burn-out of the initial 5,000 iterations and estimates were obtained after thinning the chains by collecting one in 10 consecutive samples to remove autocorrelation. Model convergence was checked visually by assessing the mixing of the chains and more formally using the Gelman-Rubin diagnostic.[9]

Pigs were observed at each collection time point for detection of cough, suggestive of *M hyopneumoniae* infection.[10] Coughing was evaluated on a group basis.[11]

Results and discussion

Mean detection of *M hyopneumoniae* by real-time PCR for seeders and contacts is presented in Table 1. Positive real-time PCR results were obtained in seeders for both sample types (when collected) at all sampling days throughout the study. The first real-time PCR positive result in contacts was detected at 28 dpe in deep tracheal catheter, and additional positive results in both sample types were obtained at the three subsequent samplings. In seeders at all time points, and contacts beginning at 56 dpe, deep tracheal catheters yielded a significantly ($P < 0.001$) higher proportion of real-time PCR positive results compared with laryngeal swabs; mean Ct values in positive deep tracheal catheters were also significantly ($P < 0.001$) lower than those from positive laryngeal swabs (data not shown). The contact pig population was still coughing at 113 dpe, while the seeder pig population was not.

The mean sensitivity of laryngeal swabs and deep tracheal catheters for *M hyopneumoniae* detection by real-time PCR at each sampling in seeders is presented in Table 2, along with the posterior estimates for sensitivity for contacts. Sensitivity prior to and after 100 dpe were classified as early and late infection, respectively, based on prior research where coughing in seeder pigs and direct contact pigs ceased between 63 and 91 dpi/dpe and coughing in indirect contact pigs ceased between 119 and 155 dpe.[11] For both seeders and contacts, the sensitivity of deep tracheal catheters was higher than the one obtained in laryngeal swabs at all samplings.

Sensitivity posterior estimates for contacts largely agreed with priors. Conditional dependence between laryngeal swabs and deep tracheal catheters was low, although variable depending on the sampling date. Conditional dependency between test results was assumed in the models given the relatively close proximity between sampling sites, although model results suggested that dependency was in fact low at all sampling dates. Still, using a simpler model without the dependency terms did not lead to change in the posterior estimates. No differences in the posterior estimates (changes below 5%) were obtained when the informative priors were replaced with non-informative distributions in the sensitivity analysis.

Under the conditions of this study, the proportion of real-time PCR positive deep tracheal catheters was higher than laryngeal swabs at all samplings and deep tracheal catheters were a more

Table 1: DNA detection[1] (%) of seeders and contacts, early and late after infection with *Mycoplasma hyopneumoniae**

Days post inoculation (dpi) of *seeders*	Laryngeal swab	Deep tracheal catheter
7	44[a]	80[b]
28		94
97	57[a]	86[b]
113	27[a]	71[b]
Days post exposure (dpe) of *contacts*		
28	0	1
56	9[a]	31[b]
84	40[a]	79[b]
113	53[a]	90[b]

[a,b] Different superscript letters within rows denote statistically significant differences (*P*-value < 0.001).
[1] Real-time PCR. Samples with Ct values < 38 were considered positive.
* All data is available in another publication.[12]

Table 2: Mean sensitivity of the two sample types of seeder and contact *Mycoplasma hyopneumoniae* detection by real time PCR.*

Days post inoculation (dpi) of seeders	Laryngeal swab	Deep tracheal catheter
7	0.44[a]	0.80[b]
28[1]		0.94[b]
97	0.57[a]	0.86[b]
113	0.27[a]	0.71[b]
Days post exposure (dpe) of contacts		
56	0.30	0.85
84	0.46	0.89
113	0.59	0.91

Clear: early infection. Shaded: late infection.
[a,b] Different superscript letters within rows denote statistically significant differences (*P* value < 0.001).
[1] 31 seeders sampled.
* All data is available in another publication.[12]

sensitive sample than laryngeal swabs. The sensitivity of both sample types varied over time and by exposure method, and these factors should be considered when designing diagnostic strategies.

References

1. Sibila et al. Current perspectives on the diagnosis and epidemiology of *Mycoplasma hyopneumoniae* infection. *Vet J.* 2009;181:221-231.

2. R Development Core Team. R: A language and environment for statistical computing [Computer software]. R Foundation for Statistical Computing. Vienna, Austria. 2008. Available at: `http://www.r-project.org/`.

3. Sergeant, ESG. EpiTools epidemiological calculators. Ausvet Pty Ltd. 2018. Available at: `http://epitools.ausvet.com.au`.

4. Branscum et al. Estimation of diagnostic-test sensitivity and specificity through Bayesian modeling. *Prev Vet Med.* 2005;68(2-4):145-163.

5. Strait et al. Real-time PCR assays to address genetic diversity among strains of *Mycoplasma hyopneumoniae. J. Clin. Microbiol.* 2008;37:620-627.

6. Fablet et al. Estimation of the sensitivity of four sampling methods for *Mycoplasma hyopneumoniae* detection in live pigs using a Bayesian approach. *Vet Microbiol.* 2010;143:238-245.

7. Su. BetaBuster. 2019. Available at: `https://cadms.vetmed.ucdavis.edu/diagnostic/software`.

8. Lunn et al. WinBUGS – a Bayesian modeling framework: concepts, structure and extensibility. *Stat. Comput.* 2000;10:325-337.

9. Gelman et al. Inference from iterative simulation using multiple sequences. *Stat. Sci.* 1992;7:457-511.

10. Nathues et al. Value of the clinical examination in diagnosing enzootic pneumonia in fattening pigs. *Vet. J.* 2012;193:443-447.

11. Fano et al. Dynamics and persistence of *Mycoplasma hyopneumoniae* infection in pigs. *Can. J. Vet. Res.* 2005;69:223-228.

12. Sponheim et al. Comparison of the sensitivity of laryngeal swabs and deep tracheal catheters for detection of *Mycoplasma hyopneumoniae* in experimentally and naturally infected pigs early and late after infection. Vet. Microbiol. In press.

Development of cost-effective surveillance protocols to minimize the risk of *Mycoplasma hyopneumoniae* introduction to sow farms

A. M. Betlach[1,2], DVM; R. Valeris-Chacin[3], DVM, MS; R. S. Singer[3], DVM, MPVM, PhD; A. Rovira[4], DVM, MS, PhD; M. Allerson[5], DVM, PhD; M. Pieters[1], DVM, PhD

[1]Department of Veterinary Population Medicine, College of Veterinary Medicine, University of Minnesota, St. Paul, Minnesota; [2]Swine Vet Center, St. Peter, Minnesota; [3]Department of Veterinary and Biomedical Sciences, College of Veterinary Medicine, University of Minnesota, St. Paul, Minnesota; [4]Veterinary Diagnostic Laboratory, University of Minnesota, St. Paul, Minnesota; [5]Holden Farms Inc., Northfield, Minnesota

Introduction

*Mycoplasma hyopneumoniae (*M hyo), the etiologic agent of enzootic pneumonia in swine,[1,2] continues to be a prevalent and economically important, respiratory pathogen.[3] Over the years, advancements in diagnostic and control strategies have been achieved through an improved understanding of M hyo ecology and epidemiology.[4] However, outbreaks of M hyo in sow farms continue to occur, in which the source of infection is often not identified.[5] The utilization of replacement gilts is commonly practiced for genetic advancement, parity maintenance, and sow farm productivity. Nevertheless, replacement gilts are perceived as a risk for pathogen entry. For M hyo surveillance, serum and/or oral fluid samples are commonly collected from incoming replacement gilts for the detection of antibodies or genetic material, respectively. However, diagnostic limitations exist, particularly due to their lack of sensitivity during acute stages of infection, which leads to diagnostic uncertainty.[6] Therefore, the objective of this study was to develop cost-effective surveillance protocols to minimize the risk of M hyo introduction into naïve sow farms using simulation modeling with the incorporation of epidemiological parameters for natural infection.

Materials and methods

Thirty-two, 5-week old gilts were randomly selected from a M hyo and PRRSV negative herd. Gilts were transported to an isolation unit at two different time points, namely for establishment of M hyo natural infection and assessment of transmission. Two gilts were experimentally inoculated with M hyo and housed with an age-matched naïve gilt for 24 days to obtain natural infection in the contact naïve gilt. Post-infection confirmation, the naturally infected gilt was relocated and housed with 29 age-matched naïve gilts for 8 weeks to obtain an initial 3% prevalence in the population. Individual tracheal and serum samples were obtained on 0, 1, 2, 4, 6, and 8 weeks post-exposure (i.e. wpe), along with the collection of an oral fluid sample from the group. At 8 wpe, all gilts were humanely euthanized and bronchial swabs were collected. Macroscopic lung lesion scoring was also performed. The M hyo natural transmission rate (β) was estimated using a Bayesian logistic regression model (assuming an SI dynamic model). In addition, differences in time-to-detection based on sample type were evaluated using a Cox Proportional Hazard model and log rank test. Statistical modeling was conducted using R v3.2. Effectiveness of different surveillance protocols for M hyo detection was evaluated using simulation-based modeling while incorporating natural transmission coefficient and diagnostic sensitivities.

Results and discussion

At 8wpe, 27% (8/29) of the naïve contact gilts became infected from the naturally infected gilt ($\beta = 0.33$). Using Idexx ELISA, *M hyopneumoniae* antibodies were initially detected in the naturally infected gilt at 6 weeks post-infection. In addition, 1 contact gilt became seropositive at 8 wpe. Oral fluids were negative for *M hyopneumoniae* detection at all samplings during the investigation, regardless of the presence of infected gilts in the group. At all samplings, *M hyopneumoniae* was detected by PCR in tracheal samples from the naturally infected seeder. At 6 and 8wpe, 3% (1/29) and 17% (5/29) of the contact gilts were identified as *M hyopneumoniae* positive by PCR. Furthermore, 27% (8/29) of the gilts were positive for *M hyopneumoniae* via bronchial swabs. Overall, these results showed that use of pathogen-specific diagnostic methods is key in *M hyopneumoniae* detection, especially considering acute stages of infection. Moreover, simulation based-modeling was useful to identify the ideal testing strategy to maximize the use of resources while minimizing the risk of pathogen introduction.

References

1. Mare CJ, Switzer WP. New Species: *Mycoplasma hyopneumoniae*; a causative agent of virus pig pneumonia. Vet Med Small Anim Clin, 1965;60:841-846.

2. Goodwin RF et al. Production of enzootic pneumonia in pigs with mycoplasma. Vet Rec, 1965; 77:1247-1249.

3. Pieters MG, Maes D. Mycoplasmosis. In Diseases of Swine. Hoboken, NJ, USA: John Wily & Sons, Inc, 2019;863-883.

4. Maes D et al. Update on *Mycoplasma hyopneumoniae* infections in pigs: Knowledge gaps for improved disease control. Transbound Emerg Dis, 2018;1-15.

5. Anderson A et al. Use of molecular characterization tools to investigate *Mycoplasma hyopneumoniae* outbreaks. Proceedings from the American Association of Swine Veterinarians Annual Meeting, Denver, CO. 2017; 60-62.

6. Pieters M et al. Comparison of sample types and diagnostic methods for in vivo detection of *Mycoplasma hyopneumoniae* during early stages of infection. Vet Microbiol, 2017;203:103-109.

Impact of using nurse sows or custom-made vaccines in the occurrence of influenza A virus in pigs prior to weaning

Jorge Garrido Mantilla[1]; Marie Culhane[1]; Jeremy Pittman[2]; Juan Sanhueza[1]; Julio Alvarez[3]; Matt Allerson[4]; Montserrat Torremorell[1]

[1]College of Veterinary Medicine, University of Minnesota, St. Paul, Minnesota; [2]Smithfield Hog Production – North Region, Waverly, Virginia; [3]VISAVET Health Surveillance Center, Universidad Complutense, Madrid, Spain; [4]Holden farms, Northfield, Minnesota

Introduction

Vaccination is the most common intervention to control influenza A virus (IAV) infections in pigs. However, the high genetic and antigenic diversity of IAV often reduces the effectiveness of vaccine control programs. Both commercially licensed and custom made vaccines can be effective at reducing IAV prevalence at weaning but their effectiveness depends in part on the antigenic matching of the vaccine virus with the wild-type virus circulating in the pigs. In this study, we evaluated a company driven strategy to reduce levels of IAV at weaning by regularly updating the custom made vaccines with circulating IAV strains obtained from the system sow farms identified as part of their surveillance system.

Methods

This case report included data collected for 2.5 years (July 2016-January 2019) from 35 farrow-to-wean farms belonging to an integrated production system in North Carolina. Vaccines were administered seasonally (fall and spring) to all dams in the farm at once (mass vaccination) and/or pre-farrowing. IAV presence of the sow farms was determined by monthly IAV rRT-PCR testing of nasal wipes collected from litters within 1 week of weaning. The genetic sequences obtained from the IAV isolates from the positive sow farm samples were compared to the IAV isolates of the vaccine used on those sow farms. Specifically, the similarity (e.g. homology) between the hemagglutinin (HA) and neuraminidase (NA) sequences of the circulating IAV strain and the vaccine IAV strain was determined by pairwise alignment of the translated protein sequences and key amino acid comparison.

Results

Custom-made vaccines were updated 4 times throughout the period of this study. Out of all 1,071 sampling events, 141 were positive yielding 66 HA complete sequences. Fifty-four H1 (22 H1-δ, 28 H1-γ, and 4 H1-npdm clades) and 14 H3 (12 IV-A and 2 IV-B clusters) circulating IAV strains were identified. The overall HA amino acid similarity ranged from 95% to 99%. However, similarity of the HA antigenic sites ranged from 53% to 99%. The proportion of positive sampling events decreased with time from a peak of 54.5% in July 2016 to 2.6% in January 2019 (Figure 1).

Discussion

In this case report, we documented the decrease of IAV infections at weaning in an integrated production system that uses a systematic approach to monitor, select, and update IAV strains in their custom vaccines. Having a long-term approach to the control of IAV is effective but requires the on-going characterization of IAV strains and frequent vaccine strain updates.

Introduction

Influenza A virus transmission from nurse sows (NS) to adopted pigs has been already established through an experimental model. However, the importance of these results obtained experimentally in the field have not been evaluated. Use of NS to adopt fall behind piglets is a common practice in swine farms to increase productivity. NS are usually selected at weaning when in endemic farms piglets may test IAV positive. We hypothesize that if the skin of NS is contaminated with nasal and oral secretions that contain viable IAV, NS could be a source of IAV infection to the newly adopted litters, perpetuating the infection in farrowing rooms. In this study, we evaluated the impact of nurse sows on the transmission of IAV under field conditions.

Methods

A cohort study to test IAV status in nurse and control sows and their corresponding litters was carried out. We enrolled 184 sows (92 NS and 92 controls) selected from 3 farrow to wean farms with a history of IAV infection. Nurse sows were selected at weaning and transferred to rooms containing younger pigs to be adopted. Once NS were enrolled, a control sow that was rearing her own litter located in the same room where the NS was located was also enrolled in the study. NS and controls were sampled using udder skin wipes (UW) and oral swabs (OS) at the time of enrollment and again at weaning. Six pigs from each litter were sampled randomly by collection of OS at enrollment, 2 days post-enrollment (2DPE), 4 days post-enrollment (4DPE), at 14 days of lactation (14DL) and at weaning (WA). All samples were tested for IAV by rRT-PCR. Oral swabs collected from pigs were tested in pools of 3 (2 pools per litter). Litters were categorized as positive when at least one pool was IAV rRT-PCR positive.

Results

At enrollment, none (0/86) of the sows tested IAV positive by rRT-PCR in the control group compared to 3.37% (3/89) in the nurse sow group (positive NS were excluded from the analysis). At weaning, 17.72% (14/79) of sows in the control group and 15.09% (11/73) in the nurse sow group tested IAV positive by rRT-PCR. 75% of the UW collected from nurse sows were positive at enrollment compared to 23% of the control sows. The odds of litters testing IAV positive were higher in litters adopted by NS compared to controls for 2 DPE, 4 DPE and 14 DL sampling dates, while no difference was observed at weaning. Overall, the IAV prevalence at weaning for all three farms in the study, as assessed by testing OS collected from piglets, was quite high (Table 1).

Conclusions and discussion

Litters adopted by nurse sows were more likely to become IAV-infected shortly after adoption, but at weaning, there were no differences in IAV status between litters of nurse and controls sows. This study supports the role of nurse sows in the transmission and persistence of IAV during the pre-weaning period under field conditions.

Figure 1: Shows proportion of positive sampling events by collection of nasal wipes or udder skin wipes by month for the whole system (35 farms) using IAV custom-made vaccines in sows.

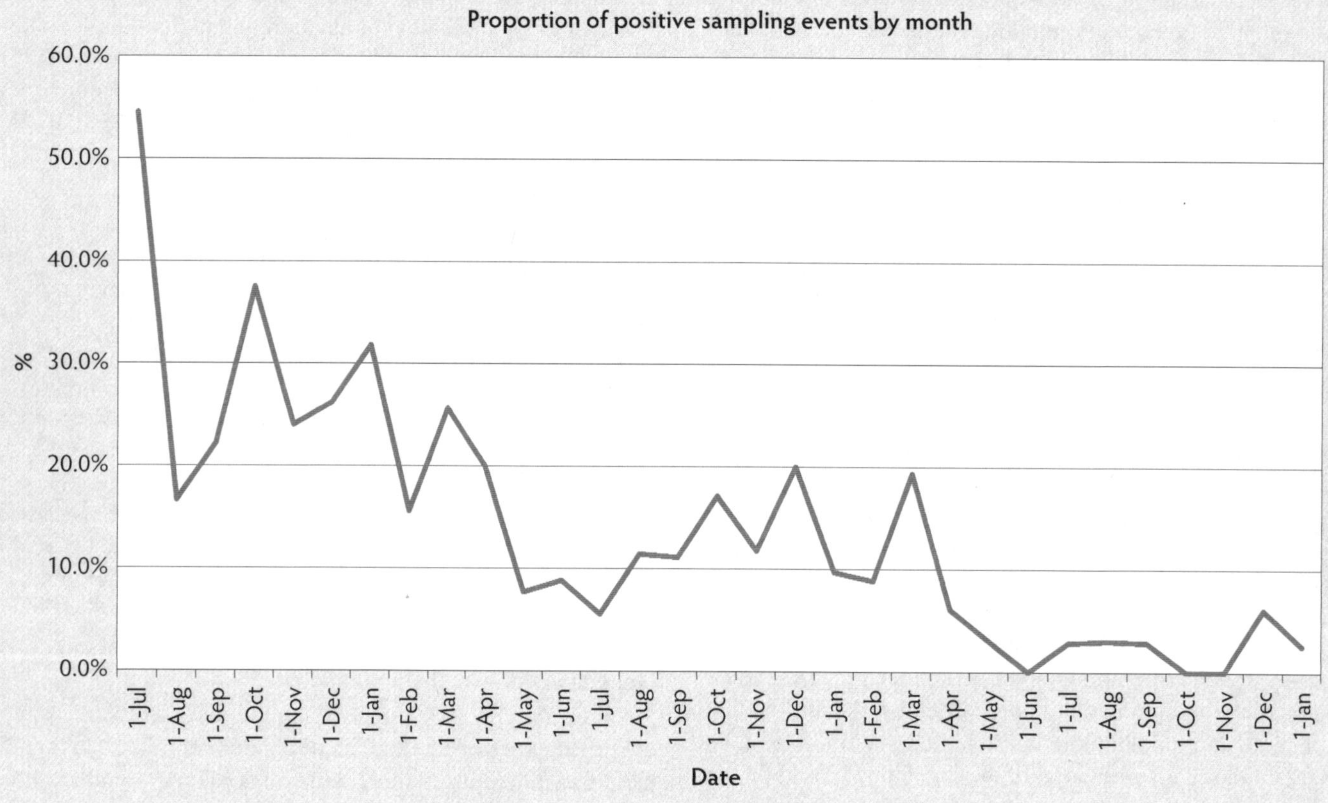

Table 1: Percent IAV rRT-PCR oral swab positive piglets at each time point from enrollment, while nursing, and at weaning, by group and by farm. Numbers in columns with different superscripts were statistically different ($P < 0.05$)

	0 DPE	2 DPE	4 DPE	14 DL	At wean
Control	12.09[a]	14.89[a]	22.99[a]	46.75[a]	63.95[a]
NS	12.22[a]	30.23[b]	36.90[a]	59.38[a]	56.41[a]
Farm 1					38.9
Farm 2					95.8
Farm 3					51.6

Persistent atypical porcine pestivirus (APPV) infection in gilts

Alexandra Buckley[1]; Bailey Arruda[2]; Shollie Falkenberg[1]; Laura Miller[1]; Kelly Lager[1]
[1]National Animal Disease Center, US Department of Agriculture, Ames, Iowa; [2]College of Veterinary Medicine, Iowa State University, Ames, Iowa

Introduction

Atypical porcine pestivirus (APPV) is a recently discovered virus that can be sporadically detected in swine. It has been associated with Type AII (viral related) congenital tremors (CT) in newborn pigs and APPV was shown to cause CT by direct inoculation of fetal amniotic vesicles with APPV-positive serum. Much is still unknown about the pathogenesis and epidemiology of this virus, especially the long-term consequences of CT in gilts. This report characterizes: 1) the duration of viremia and antibody response in gilts born APPV positive, 2) the transmission potential of APPV, and 3) if persistently infected gilts produce APPV and CT positive piglets.

Materials and methods

Nine gilts were selected from a commercial grow/finish barn in Iowa based on originating from a CT litter at birth and the persistent presence of APPV in serum throughout the finishing phase. These gilts were obtained from the commercial farm at market weight and transported to the National Animal Disease Center (NADC) for further study. Gilts were bled monthly and oral fluids were collected prior to breeding and during gestation. The gilts were synced and bred via artificial insemination with APPV negative semen. One gilt was necropsied for tissue collection to evaluate viral distribution at the time of breeding. While housed at the NADC, three separate groups of naïve contact swine were allowed fence-line contact with the gilts to study transmission potential. Contact pig(s) were allowed nose-to-nose contact for approximately 1 month and then removed. Blood, oral, and rectal swabs were collected at weekly intervals for the month during the transmission period as well as for approximately 1 month after the contact period and removal. Post farrowing, piglets were observed daily for approximately the first week for tremors. They were bled and swabbed at farrowing and at weekly intervals until necropsy. Gilts were euthanized between 1 and 2 weeks after farrowing and issues were collected. Two to three piglets were euthanized both at birth and between 5 and 6 weeks from each litter. Again a full set of tissues were collected. Rectal swabs, oral swabs, oral fluids and serum will be tested for APPV RNA by PCR and the antibody response determined by ELISA. Tissue will be tested for APPV RNA by *in situ* hybridization.

Results

Viremia was detected at multiple time points in the nine gilts for five months after birth. While viremia was not consistently detected at each sample time point, oral fluids and oral swabs were consistently positive for APPV by PCR testing. Transmission of the virus was successful in the first group (2 pigs); first pig positive for APPV in at least one sample at 25 days post contact (dpc), second pig did not consistently sample positive in serum and swabs until 53 dpc. The second group (1 boar) was placed in contact about 2 months after the first group was removed during the gestation period of the gilts. It became APPV positive by PCR. The third group of contact pigs (2 pigs) was exposed to the gilts at the end of gestation. The first pig became positive at 35 dpc, while the second pig had yet to have a positive sample by 56 dpc. Piglets born to persistently infected APPV sows were negative for APPV in serum at birth and did not demonstrate visible tremors. Oral swabs from some piglets in two litters were positive for APPV by PCR on the day of birth, but all gilts had positive oral swabs on the day of farrowing as well. Serum and oral swabs collected at 4 weeks of age were all negative for APPV. *In situ* hybridization and ELISA results for gilts, contact pigs, and piglets are pending.

Conclusion

Pigs born APPV positive with and without CT tested positive by PCR for APPV for an extended period of time in both serum and oral fluid samples. These animals have been demonstrated to transmit virus to naïve contact animals. Furthermore, the virus can be found in a wide variety of tissues and sample types. Understanding infection dynamics of APPV and applicable sample types for monitoring infection following exposure can aide control measures for this virus in addition to reducing Type AII CT due to APPV.

Efficacy of an inactivated Seneca Valley virus vaccine in pregnant sows

Alexandra Buckley; Kelly Lager
National Animal Disease Center, US Department of Agriculture, Ames, Iowa

Introduction

Senecavirus A (SVA), commonly known as Seneca Valley virus (SVV), is a causative agent of vesicular disease in swine that is clinically indistinguishable from foot-and-mouth disease (FMD). Due to the severe consequences of FMD entering a FMD-free country, a foreign animal disease investigation must be performed every time a vesicular lesion is observed. These investigations have economic ramifications especially for the state and federal government. A large portion of the investigations have been centered at sow slaughter facilities, with some facilities requiring investigations weekly. Having an effective SVV vaccine would be a valuable tool for the swine industry to improve swine health and reduce disruptions related to foreign animal disease investigations.

Materials and methods

Part 1: Pregnant sows were randomly placed into 4 groups: Group 1 (vax + challenge, n = 9), Group 2 (vax + sham, n = 3), Group 3 (no vax + challenge, n = 5) and Group 4 (no vax + sham, n = 3). Around 70 and 90 days of gestation Groups 1 and 2 were vaccinated intramuscularly (IM) with 2 mL of whole-virus inactivated SVV (2015) with an oil-in-water emulsion adjuvant. Sows were intranasally challenged around 104 days of gestation (0 dpc) with 5 mL of a 2018 SVV isolate (10^7 $TCID_{50}$/mL, challenge) or media (sham). Sows were rectal swabbed on 0, 1, 2, 3, 5, 7, 14, 21, and 28 days post challenge (dpc) and checked for vesicular lesions. Sows were bled on -35, -14, 0, 4, 7, 14, and 28 dpc in addition to the day of farrowing.

Part 2: At farrowing, colostrum was collected and piglets were bled prior to suckling. Piglets from Groups 1 and 3 were bled and swabbed around day 7, 21, and necropsy. Piglets (3-6 days old) from Groups 2 and 4 were challenged orally with 2 mL of SVV (10^7 $TCID_{50}$/mL, challenge). Piglets were bled and swabbed on 0, 4, 7, 14, 21 dpc. Swabs, serum, and colostrum were tested for SVV RNA by PCR and the antibody response was determined by virus neutralization (VN) assay.

Results

None of the vaccinated sows (Group 1) developed clinical signs of vesicular disease after challenge and some sows that were not vaccinated (Group 3) did develop vesicular lesions after challenge with a 2018 SVA isolate. In addition, only a mild transient diarrhea was observed in some piglets challenged with SVA, otherwise no specific clinical disease was recognized in Groups 2 and 4 piglets. PCR and VN results are pending as the second round of sows recently farrowed. Previous research has demonstrated immunity in sows given this vaccine and challenged with a 2015 SVV isolate (10^6 $TCID_{50}$/mL). Unpublished work has also demonstrated efficacy of this vaccine in growing pigs against homologous challenge. We hypothesize the whole-virus inactivated SVV vaccine will provide protection against clinical disease, prevent viremia, and reduce or eliminate shedding of virus in feces in sows after challenge with a 2018 SVV isolate. We also hypothesize that vaccinated sows will pass SVV immunity on to their offspring and piglets challenged with a 2018 SVV isolate will be protected.

Conclusion

An effective vaccine against SVA could have a positive impact on welfare in the swine industry and reduce the economic burden of foreign animal disease investigations.

Pseudorabies virus (PRV) antibody detection in serum and oral fluid specimens

T.-Y. Cheng[1], DVM, MS; A. Buckley[2], DVM, MS; A. van Geelen[2], DVM, PhD; K. Lager[2], DVM, PhD; A. Henao-Díaz[1], DVM, MS; K. Poonsuk[1], DVM, PhD; P. Piñeyro[1], DVM, PhD; J. Ji[1], PhD; C. Wang[1], PhD; D. Baum[1], DVM, PhD; R. Main[1], DVM, PhD; J. Zimmerman[1], DVM, PhD; L. Giménez-Lirola[1], PhD
[1]Iowa State University, Ames, Iowa; [2]Agricultural Research Service (ARS), United States Department of Agriculture (USDA), Ames, Iowa

Introduction

Pseudorabies virus (PRV) can be eradicated from domestic swine using marker vaccines and accompanying ELISA(s). However, PRV circulates in free-ranging feral swine populations on all continents, with documentation of occasional spillover into domestic herds. This on-going threat to PRV-free swine populations justifies efforts to improve PRV surveillance. Thus, the objective of this study was to evaluate the detection of PRV antibodies in swine oral fluid.

Materials and methods

Oral fluid and serum samples were obtained from 12- to 16-week-old pigs allocated to 4 groups (10 pigs per group): 1) negative control (NC), 2) wild-type PRV inoculation (PRV 3CR Ossabaw), 3) PRV vaccination (MLV) (Ingelvac® Aujeszky MLV), and 4) PRV vaccination followed by PRV inoculation (PRV 3CR Ossabaw) at 21 days post vaccination. Depending on the group, serum and oral fluid samples were collected from individual animals for up to 49 days. Serum and oral fluid samples were tested for PRV antibody using a PRV whole virus indirect IgG ELISA; oral fluid samples were tested using a modified protocol. ROC curve and repeatability analyses were performed to evaluate assay performance.

Results and discussion

Using a serum PRV whole virus indirect IgG ELISA (IDEXX Laboratories, Inc., Westbrook, ME) adapted to the oral fluid matrix, PRV antibody was detected in oral fluid samples from treatment groups PRV, MLV, and MLV-PRV in a pattern similar to serum (Figure 1). Vaccination alone produced a low oral fluid antibody response (groups 3 and 4), but a strong anamnestic response was observed following challenge with wild-type virus (group 2). Analyses of the oral fluid PRV indirect ELISA results showed good binary diagnostic performance (area under ROC curve = 93%) and excellent assay repeatability (intra-class correlation coefficient = 99.3%). The demonstrable presence of PRV antibodies in swine oral fluids suggests its possible use in PRV disease surveillance.

Acknowledgments

PRV indirect ELISA kits [ADV(S)] were provided by IDEXX Laboratories, Westbrook, ME.

This research was funded by Swine Health Information Center (17-201; Ames, IA).

Figure 1: Serum and oral fluid antibody responses ($\bar{x}$, standard error) based on ELISA testing (ADV (S) *i*ELISA Kit, IDEXX Laboratories, Inc., Westbrook, ME). Day 7: intramuscular administration of modified live virus vaccine (Ingelvac® Aujeszky MLV, Boehringer Ingelheim, Ridgefield, CT). Day 28: intranasal administration of PRV isolate 3CR Ossabaw ($1 \times 10^{3.5}$ $TCID_{50}$ per pig).

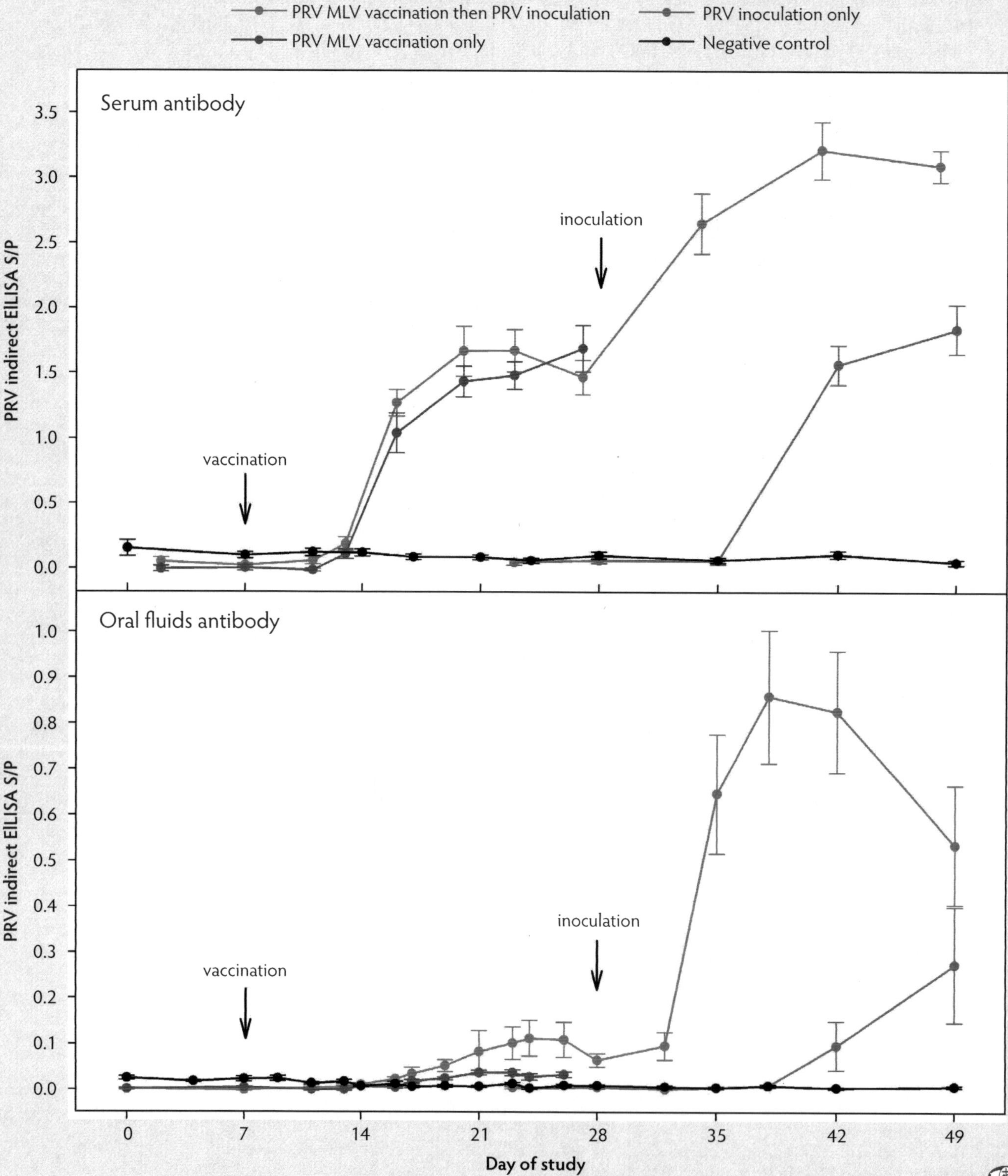

Susceptibility of attenuated *Salmonella* vaccines intended for swine to zinc oxide *in vitro*

E. Burrough[1], DVM, PhD, DACVP; L. Burrough[2]; Y. Yin[2], MD, MPA; L. Dai[2], PhD; O. Sahin[1,2], DVM, MS, PhD, DACVM
[1]Veterinary Diagnostic and Production Animal Medicine, Iowa State University, Ames, Iowa;
[2]Veterinary Microbiology and Preventive Medicine, Iowa State University, Ames, Iowa

Introduction

Salmonellosis in swine has traditionally been associated with either *Salmonella enterica* serovar Choleraesuis or Typhimurium.[1] Contamination of pork with zoonotic strains of *Salmonella*, like serovar Typhimurium, is a major concern for the swine industry and thus there is a need to establish control measure for these bacteria on farm.[2] Vaccination for *Salmonella* has been used on swine farms historically; however, inactivated vaccines often give mixed results due to the intracellular nature of infection and antigenic diversity.[3] More recently, attenuated live vaccines administered orally have been used with a higher degree of success likely due to the stimulation of cell-mediated immunity.[3,4]

There are currently two commercially available vaccines for pigs in the United States containing attenuated strains of *S* Choleraesuis alone (Argus® SC/ST, Merck Animal Health) or in combination with *S.* Typhimurium (Enterisol® *Salmonella* T/C, Boehringer Ingelheim). These products are labeled for administration as a live oral vaccine to pigs 2 - 3 weeks of age and older, which is around the time of weaning in current commercial practices (approximately 3 weeks of age). Weaning stress often manifests as post-weaning diarrhea (PWD) that is commonly associated with *Escherichia coli* infection.[5] Interventions to mitigate post-weaning diarrhea include in-feed therapeutic antibiotics, organic acids, probiotics, prebiotics, and pharmaceutical concentrations of heavy metal micronutrients, specifically copper and zinc.[6,7]

In the United States, pharmacological doses of zinc [in the form of zinc oxide (ZnO)] are fed to pigs in efforts to improve growth and reduce PWD.[8,9] Common protocols include feeding 2,000 to 3,000 ppm of Zn for the first few weeks post-weaning.[10] As pharmacologic concentrations of Zn may be in the feed at the same time live *Salmonella* vaccines are being administered, there is the potential that these vaccine strains may be inhibited and result in suboptimal response to vaccination.

Accordingly, the objective of this study was to assess the susceptibility of the two commercially available, attenuated live *Salmonella* vaccines marketed for swine in the United States to ZnO in vitro using an agar dilution technique. The hypothesis of this study being that commercial attenuated oral vaccine strains, prepared from historical isolates, will be susceptible to pharmacologic concentrations of Zn in the form of ZnO.

Materials and methods

Vaccine isolates

Two commercially available vaccines (Argus® SC/ST, Merck Animal Health; Enterisol® *Salmonella* T/C, Boehringer Ingelheim) were purchased from a retail supplier and reconstituted per manufacturer instructions. Reconstituted preparations were then immediately plated onto blood agar and Hektoen enteric (HE) selective agar media for isolation. Inoculated plates were incubated overnight at 37°C. Five well-isolated colonies were selected from each vaccine preparation for further characterization. As the combination vaccine (Enterisol®) contains two different serovars, a total of ten colonies were selected with five each of two different color phenotypes on HE (green colonies for Choleraesuis and black colonies for Typhimurium). Isolates were confirmed as *Salmonella* by MALDI-TOF and serogrouping was performed to ensure the different HE phenotypes matched the expected serogroups (Group C1 and Group B for Choleraesuis and Typhimurium, respectively). Pure cultures of each isolate were prepared in glycerol stock solution and frozen at -80°C until further use.

Media preparation

Mueller Hinton (MH) agar (Oxoid) was prepared according to the manufacturer instructions and combined with decreasing amounts of a 100 mg/mL solution of ZnO nanoparticles (30 nm average size; Inframat, Manchester, CT, USA) suspended in ddH2O to produce media with ZnO concentrations of 6, 3, 1.5, 0.75, 0.375, 0.187, and 0.093 mg/mL. These media were autoclaved and cooled before pouring into empty Petri dishes. Constant stirring was used as the media cooled to keep the ZnO particles evenly suspended in the media.

Determination of the minimum inhibitory concentration of ZnO

A fresh agar-grown culture of each vaccine isolate was suspended in MH broth (Oxoid) and adjusted to OD600 = 0.01, and then 1 µl of each suspension was spotted in triplicate onto MH agar plates containing increasing concentrations of ZnO nanoparticles (0, 0.093, 0.187, 0.375, 0.75, 1.5, 3, 6 mg/mL). The final inoculum concentration was approximately 10^4 CFU per spot. A 1 µl spot of MH broth alone was also added to each plate as a control. Two or three experimental replicates were performed on each isolate to determine the minimum inhibitory concentration (MIC).

The MIC was considered the minimum concentration at which at least one of the experimental replicates showed no growth (on all triplicate spots) on the agar after incubation at 37°C for 24 h.

Results

Results of MIC testing for all vaccine isolates are summarized in Table 1. The median MIC of the five Choleraesuis isolates from the Argus® vaccine was 0.375 mg/mL (range 0.093 – 0.75 mg/mL. The median MIC of the five Choleraesuis isolates from the Enterisol® vaccine was 0.75 mg/mL (range 0.75 – 1.5 mg/mL) while all five Typhimurium isolates from this vaccine had MICs of 0.75 mg/mL.

Discussion

Pathogenic intestinal bacteria must acquire metal ions, including zinc, from the intestinal contents while also preventing potentially toxic intracellular accumulation of these metals. *Salmonella* Typhimurium requires the zinc transporter ZnuABC for homeostasis and survival in zinc deficient environments and can utilize this high-affinity transporter to counter the zinc-sequestering effects of calprotectin during intestinal inflammation.[11,12] While zinc is important for *Salmonella* survival in the intestine, growth inhibition has been reported for *Salmonella* spp. and other enteric pathogens on MH agar containing high concentrations of zinc sulfate (1-2 mg/mL) in vitro.[13] Agar dilutions of ZnO nanoparticles have also been used previously to determine the MIC for various enteric bacteria including *Campylobacter* spp., *Escherichia coli*, and *Salmonella enterica* serovar Enteritidis.[14,15] In these studies, it was determined that ZnO is bactericidal and proposed mechanisms include, but are not limited to, direct damage of bacterial cell membranes and the production of reactive oxygen species.[14,16]

Salmonellosis in swine has been associated with multiple *Salmonella* serovars, particularly Choleraesuis and Typhimurium, and more recently serovar 4,[5],12:i:-.[1,17] Further characterization of the recently emergent *S.* 4,5,12:i:- has revealed these isolates are often multidrug resistant and resistant to heavy metal micronutrients.[18,19] Given that zinc, in the form of ZnO, is commonly included at 2,000 – 3,000 ppm in commercial weaned pig diets to reduce diarrhea,[20,21] it is interesting that such feeding for prolonged periods may actually impair the intestinal immune response to *S.* Typhimurium.[22] Furthermore, there is a strong association between antimicrobial resistance and decreased susceptibility to heavy metal micronutrients in *Salmonella* isolates from swine, swine feed, and barn floors.[6] Accordingly, there is concern that feeding high concentrations of heavy metals, such as zinc and copper, may lead to co-selection of strains harboring

Table 1: Susceptibility (MIC) of *Salmonella* isolates to ZnO nanoparticles as determined by the agar dilution test

			ZnO nanoparticle concentration							
	Serotype	Source	0 mg/mL	0.093 mg/mL	0.187 mg/mL	0.375 mg/mL	0.75 mg/mL	1.5 mg/mL	3 mg/mL	6 mg/mL
Isolate #1	Choleraesuis	Vac #1	2/2*	1/2	0/2	0/2	0/2	0/2	0/2	0/2
Isolate #2	Choleraesuis	Vac #1	3/3	3/3	2/3	1/3	0/3	0/3	0/3	0/3
Isolate #3	Choleraesuis	Vac #1	3/3	3/3	3/3	2/3	0/3	0/3	0/3	0/3
Isolate #4	Choleraesuis	Vac #1	3/3	3/3	3/3	3/3	2/3	0/3	0/3	0/3
Isolate #5	Choleraesuis	Vac #1	2/2	2/2	2/2	2/2	0/2	0/2	0/2	0/2
Isolate #6	Choleraesuis	Vac #2	2/2	2/2	2/2	2/2	1/2	0/2	0/2	0/2
Isolate #7	Choleraesuis	Vac #2	2/2	2/2	2/2	2/2	1/2	0/2	0/2	0/2
Isolate #8	Choleraesuis	Vac #2	3/3	3/3	3/3	3/3	3/3	2/3	0/3	0/3
Isolate #9	Choleraesuis	Vac #2	3/3	3/3	3/3	3/3	3/3	0/3	0/3	0/3
Isolate #10	Choleraesuis	Vac #2	3/3	3/3	3/3	3/3	0/3	0/3	0/3	0/3
Isolate #11	Typhimurium	Vac #2	3/3	3/3	3/3	3/3	0/3	0/3	0/3	0/3
Isolate #12	Typhimurium	Vac #2	3/3	3/3	3/3	3/3	0/3	0/3	0/3	0/3
Isolate #13	Typhimurium	Vac #2	2/2	2/2	2/2	2/2	0/2	0/2	0/2	0/2
Isolate #14	Typhimurium	Vac #2	2/2	2/2	2/2	2/2	0/2	0/2	0/2	0/2
Isolate #15	Typhimurium	Vac #2	2/2	2/2	2/2	2/2	0/2	0/2	0/2	0/2

Vac #1 = Argus® SC/ST, Merck Animal Health
Vac #2 = Enterisol® *Salmonella* T/C, Boehringer Ingelheim
* Results reflect the number of experimental replicates (each having 3 spots) with growth / the number of replicates performed.
Shaded cells indicate the MIC of ZnO nanoparticles for each isolate, where lack of bacterial growth (in all 3 spots) was observed in at least one of the experimental replicates.

antimicrobial resistance genes which has potential implications on public health.[19]

The objective of this study was to determine the in vitro susceptibility to ZnO nanoparticles of attenuated live vaccine strains that might be orally administered to pigs during the same period as high concentration of ZnO feeding. In total, 13 of the 15 *Salmonella* vaccine isolates were inhibited on MH agar containing 750 ppm ZnO or less, only a single isolate had any growth on MH agar containing 1,500 ppm ZnO, and no isolate from either vaccine was able to grow in the presence of 3,000 ppm ZnO or greater. These data suggest both vaccines may have suboptimal performance when administered concurrently with pharmacological concentrations of ZnO (2,000 – 3,000 ppm) in the feed. In vivo evaluation of piglet response to orally administered, attenuated live *Salmonella* vaccines while pharmacological Zn is present in the feed appears warranted.

Funding

The authors received no external financial support for the research described or authorship of this proceedings.

References

1. Griffith, R.W., Carlson, S.A., Krull, A.C., 2019. Salmonellosis, In: Zimmerman, J.J., Karriker, L.A., Ramirez, A., Schwartz, K.J., Stevenson, G.W., Zhang, J. (Eds.) *Diseases of Swine*. Wiley-Blackwell, Hoboken, NJ, pp. 912 - 925.

2. Hauser, E., Tietze, E., Helmuth, R., Junker, E., Blank, K., Prager, R., Rabsch, W., Appel, B., Fruth, A., Malorny, B., 2010. Pork contaminated with *Salmonella enterica* serovar 4,[5],12:i:-, an emerging health risk for humans. *Appl. Environ. Microbiol.* 76, 4601-4610.

3. Wales, A.D., Davies, R.H., 2017. *Salmonella* vaccination in pigs: A review. *Zoonoses Public Health* 64, 1-13.

4. Gradassi, M., Pesciaroli, M., Martinelli, N., Ruggeri, J., Petrucci, P., Hassan, W.H., Raffatellu, M., Scaglione, F.E., Ammendola, S., Battistoni, A., Alborali, G.L., Pasquali, P., 2013. Attenuated *Salmonella enterica* serovar Typhimurium lacking the ZnuABC transporter: an efficacious orally-administered mucosal vaccine against salmonellosis in pigs. *Vaccine* 31, 3695-3701.

5. Fairbrother, J.M., Nadeau, É., Gyles, C.L., 2005. *Escherichia coli* in postweaning diarrhea in pigs: an update on bacterial types, pathogenesis, and prevention strategies. *Anim. Health Res. Rev.* 6, 17-39.

6. Medardus, J.J., Molla, B.Z., Nicol, M., Morrow, W.M., Rajala-Schultz, P.J., Kazwala, R., Gebreyes, W.A., 2014. In-feed use of heavy metal micronutrients in U S swine production systems and its role in persistence of multidrug-resistant *Salmonellae*. *Appl. Environ. Microbiol.* 80, 2317-2325.

7. Rhouma, M., Fairbrother, J.M., Beaudry, F., Letellier, A., 2017. Post weaning diarrhea in pigs: risk factors and non-colistin-based control strategies. *Acta Vet. Scand.* 59, 31.

8. Carlson, M.S., Hill, G.M., Link, J.E., 1999. Early- and traditionally weaned nursery pigs benefit from phase-feeding pharmacological concentrations of zinc oxide: effect on metallothionein and mineral concentrations. *J. Anim. Sci.* 77, 1199-1207.

9. Hill, G.M., Cromwell, G.L., Crenshaw, T.D., Dove, C.R., Ewan, R.C., Knabe, D.A., Lewis, A.J., Libal, G.W., Mahan, D.C., Shurson, G.C., Southern, L.L., Veum, T.L., 2000. Growth promotion effects and plasma changes from feeding high dietary concentrations of zinc and copper to weanling pigs (regional study). *J. Anim. Sci.* 78, 1010-1016.

10. Burrough, E.R., De Mille, C., Gabler, N.K., 2019. Zinc overload in weaned pigs: tissue accumulation, pathology, and growth impacts. *J. Vet. Diagn. Invest.* 31, 537-545.

11. Ammendola, S., Pasquali, P., Pistoia, C., Petrucci, P., Petrarca, P., Rotilio, G., Battistoni, A., 2007. High-affinity Zn^{2+} uptake system ZnuABC is required for bacterial zinc homeostasis in intracellular environments and contributes to the virulence of *Salmonella enterica*. *Infect. Immun.* 75, 5867-5876.

12. Liu, J.Z., Jellbauer, S., Poe, A.J., Ton, V., Pesciaroli, M., Kehl-Fie, T.E., Restrepo, N.A., Hosking, M.P., Edwards, R.A., Battistoni, A., Pasquali, P., Lane, T.E., Chazin, W.J., Vogl, T., Roth, J., Skaar, E.P., Raffatellu, M., 2012. Zinc sequestration by the neutrophil protein calprotectin enhances *Salmonella* growth in the inflamed gut. *Cell Host Microbe* 11, 227-239.

13. Surjawidjaja, J.E., Hidayat, A., Lesmana, M., 2004. Growth inhibition of enteric pathogens by zinc sulfate: an in vitro study. *Med. Princ. Pract.* 13, 286-289.

14. Xie, Y., He, Y., Irwin, P.L., Jin, T., Shi, X., 2011. Antibacterial activity and mechanism of action of zinc oxide nanoparticles against *Campylobacter jejuni*. *Appl. Environ. Microbiol.* 77, 2325-2331.

15. Graham, L.L., Feero, S.E., 2019. The *Campylobacter fetus* S layer provides resistance to photoactivated zinc oxide nanoparticles. *Can. J. Microbiol.* 65, 450-460.

16. Sirelkhatim, A., Mahmud, S., Seeni, A., Kaus, N.H.M., Ann, L.C., Bakhori, S.K.M., Hasan, H., Mohamad, D., 2015. Review on zinc oxide nanoparticles: antibacterial activity and toxicity mechanism. *Nanomicro. Lett.* 7, 219-242.

17. Arruda, B.L., Burrough, E.R., Schwartz, K.J., 2019. *Salmonella enterica* I 4,[5],12:i:- associated with lesions typical of swine enteric salmonellosis. *Emerg. Infect. Dis.* 25, 1377-1379.

18. Mourão, J., Novais, C., Machado, J., Peixe, L., Antunes, P., 2015. Metal tolerance in emerging clinically relevant multidrug-resistant *Salmonella enterica* serotype 4,[5],12:i:- clones circulating in Europe. *Int. J. Antimicrob. Agents* 45, 610-616.

19. Figueiredo, R., Card, R.M., Nunez-Garcia, J., Mendonça, N., da Silva, G.J., Anjum, M.F., 2019. Multidrug-resistant *Salmonella enterica* isolated from food animal and foodstuff may also be less susceptible to heavy metals. *Foodborne Pathog. Dis.* 16, 166-172.

20. Jensen-Waern, M., Melin, L., Lindberg, R., Johannisson, A., Petersson, L., Wallgren, P., 1998. Dietary zinc oxide in weaned pigs - effects on performance, tissue concentrations, morphology, neutrophil functions and faecal microflora. *Res. Vet. Sci.* 64, 225-231.

21. Milani, N.C., Sbardella, M., Ikeda, N.Y., Arno, A., Mascarenhas, B.C., Miyada, V.S., 2017. Dietary zinc oxide nanoparticles as growth promoter for weanling pigs. *Anim. Feed Sci. Technol.* 227, 13-23.

22. Schulte, J.N., Brockmann, G.A., Kreuzer-Redmer, S., 2016. Feeding a high dosage of zinc oxide affects suppressor of cytokine gene expression in *Salmonella* Typhimurium infected piglets. *Vet. Immunol. Immunopathol.* 178, 10-13.

Candidate virulence-associated genes identified by genome analysis of *Streptococcus suis* strains from the United States

April A. Estrada[1]; Douglas G. Marthaler[2]; Connie Gebhart[1]
[1]College of Veterinary Medicine, University of Minnesota, Minneapolis, Minnesota;
[2]College of Veterinary Medicine, Kansas State University, Manhattan, Kansas

Introduction

Streptococcus suis is a significant economic and welfare concern in the swine industry. *S suis* is a Gram-positive bacterium and the species contains both pathogenic and commensal strains. Pathogenic *S suis* strains are associated with systemic and neurological disease (meningitis, arthritis, endocarditis/epicarditis, septicemia) in growing pigs[1,2] while commensal strains normally reside in the upper respiratory tracts of pigs, which commonly serve as carriers of the bacterium.[1,3,4] *S suis* can also be an opportunistic pathogen associated with co-infections with other bacterial and viral pathogens.[2,5]

Comprehensive characterization of contemporary strains of *S suis* in the United States is lacking, hindering our understanding of pathogenic strains associated with systemic and neurological disease. Furthermore, this deficiency impacts the effectiveness of prevention, treatment, and control strategies for *S suis*. In a previous study, we characterized 208 *S suis* isolates collected between 2014 to 2017 across the United States by serotyping and multilocus sequence typing (MLST).[6] The predominant serotype and sequence types (ST) (listed in decreasing order) were 1/2, 7, 2 and ST28, ST94, ST1, ST108, respectively. These 208 *S suis* isolates were also characterized into pathotypes (pathogenic [n = 139], possibly opportunistic [n = 47], and commensal [n = 22]) using an approach similar to previously published methods.[7,8] "Pathogenic" isolates were obtained from systemic tissues of pigs with signs of neurological or systemic disease (meningitis, arthritis, epicarditis, or septicemia). "Possibly opportunistic" isolates were from lung tissues of pigs without signs of neurological or systemic disease. "Commensal" isolates were from laryngeal, tonsil, or nasal samples retrieved from farms with no known history or current control methods for *S suis* disease. Investigation of associations between subtype and pathotype classifications identified multiple subtypes that corresponded to the pathogenic pathotype. These subtypes were serotypes 1, 1/2, 2, 7, 14, and 23 and ST1, ST13, ST25, ST28, ST29, ST94, ST108, ST117, ST225, ST373, ST961, and ST977, of which serotype 1/2 and ST28 were the most predominant US subtypes identified in the study.

Although the same study concluded that ST appeared to be a stronger predictor of pathotype than serotype,[6] there were limitations of that study due to the large amount of genetic diversity of the *S suis* strains. Therefore, alternative subtyping techniques with higher discriminatory power are needed to better differentiate pathogenic from commensal strains. Pan-genome analysis is one technique for characterizing strain-level diversity of bacterial species as it describes shared (core) and unique (accessory) traits within a species.[9] The accessory genome contains genes not shared by all the isolates (including strain-specific genes) and is where the diversity among isolates can be observed, and novel genes can be discovered. In a recent United Kingdom study, genome-wide association studies were used to identify genetic markers in the *S suis* accessory genome for differentiating isolates into invasive disease-associated and non-disease associated groups.[10] That pathotyping tool, which consisted of a generalized linear model and a multiplex-PCR assay, identified three genetic markers (*S suis* species-specific, invasive disease-associated, and non-disease associated markers) for subtyping *S suis* isolates from England and Wales. Furthermore, the pathotyping tool demonstrated performance metrics (e.g. sensitivity, specificity, negative predictive value, and positive predictive value) similar to serotyping, MLST, virulence-associated genes, minimum core genome sequence typing for predicting the clinical significance of *S suis* isolates. We applied pan-genome analysis to the same 208 US *S suis* isolates[6] to identify accessory genes that corresponded to the pathogenic pathotype and thus may serve as novel candidate virulence-associated genes (VAGs) of *S suis*. In addition, the identification of candidate VAGs may elucidate a VAG panel for the prediction of pathogenicity of *S suis* isolates in the United States.

Materials and methods

The same 208 *S suis* genomes used in the previous serotype and MLST characterization study[6] were assembled into draft genomes to further conduct pan-genome analysis. Genome assemblies were generated using MEGAHIT and polished using Pilon. Only contigs > 500 bp were kept for annotation by Prokka to predict coding sequences. The pan-genome was annotated using Roary using a 90% blastp identity cut-off to define clusters of genes. Individual genes were tested by chi-square and Fisher's exact test to determine associations with pathotype. Genes lacking a significant (P-value < 0.05) association with pathotype were removed, and the data were further filtered by removing genes detected in less than 50% (69/139) of isolates within the pathogenic pathotype. The remaining genes were analyzed using the LASSO (Least Absolute Shrinkage and Selection Operator) shrinkage regression model, which reduces coefficients to zero and gradually eliminating genes from the model. The pathotype was entered as the dependent variable (response) in the logistic regression model, with focus on the pathogenic pathotype as the indicator of pathogenicity. The LASSO analysis was performed using the R statistical package glmnet, and the best lambda penalty value was used to determine the fewest number of genes. The functions of proteins encoded by the candidate VAGs were predicted to explore relevance to *S suis* pathogenicity. Genes lacking annotation by the Prokka program were compared against the NCBI non-redundant protein database using blastn/blastx, with a maximum expectation value of 10^{-6} to obtain more significant hits and reduce random background noise.

Results

In this study, a pan-genome approach was utilized to determine candidate VAGs of *S suis* that could be used to differentiate pathogenic and commensal strains. The average *S suis* genome length, GC content, and number of predicted coding sequences for the 208 US isolates were 2.2 Mb, 41%, and 2,113 coding sequences, respectively. These values are consistent with the characteristics of *S suis* genomes deposited in the GenBank/EMBL/DDBJ database. Furthermore, *S suis* illustrated an open pan-genome as the total number of genes in the pan-genome increased with each additional genome, indicating a potential to discover novel genes with the sequencing of more *S suis* strains. A total of 12,395 genes were identified in the pan-genome and 1,034 of them were core genes as they were present in all or nearly all genomes (at least 206/208) that were sequenced.

However, we focused on accessory genes (n = 554) to investigate associations between candidate VAGs and pathotype by using LASSO (a shrinkage regression model) to identify the fewest number of genes that could be included in a panel to differentiate between isolates classified as pathogenic and commensal. The LASSO associated 27 pan-genes with the pathogenic pathotype. Three pan-genes (pan-gene2859, pan-gene3448, and pan-gene4779) were further investigated for use in a panel predictive of pathotype based on high LASSO coefficients, presence among pathogenic isolates, and absence in commensal isolates.

We investigated the genes individually and as combinations of genes to explore predictivity of the pathogenic pathotype (Table 1). As individual genes, over 75% of the isolates positive for each gene were classified as the pathogenic pathotype: pan-gene2859 (77.0%, 137/178), pan-gene3448 (78.6%, 132/168), and pan-gene4779 (77.9%, 134/172). However, the combination/profile of pan-genes 2859, 3448, and 4779 was observed in 78.8% (164/208) of *S. suis* isolates, from which 80.5% (132/164) were classified as the pathogenic pathotype. These results can be interpreted as the probability that an isolate is pathogenic given the profile of pan-genes 2859, 3448, and 4779, which is slightly greater than that obtained for each individual gene. Furthermore, the profile of pan-genes 2859, 3448, and 4779 contained the lowest percentage of commensal (1.8%, 3/164) and possibly opportunistic (17.7%, 29/164) isolates. However, the high number of possibly opportunistic isolates, which are isolates not associated with systemic and neurological disease but rather associated with respiratory disease, that are positive for all three pan-genes still contributed to a lower the probability that an isolate is pathogenic given the profile.

As expected, the *S suis* isolates with the candidate VAG profile (pan-genes 2859, 3448, and 4779) were predominantly characterized as serotype 1/2 and ST28, which were previously identified as potential subtype indicators of pathogenicity (Table 2).[6] Other subtypes that possessed the candidate VAG profile included serotypes 1, 2, 7, 14, and 23, and ST1, ST13, ST25, ST29, ST94, ST108, ST255, ST373, ST961, and ST977, which were also previously identified as subtypes corresponding to the pathogenic pathotype.[6] Only a single subtype (ST117) failed to possess the pan-genes 2859, 3448, and 4779 profile.

Both pan-gene2859 and pan-gene4779 were annotated by Prokka as hypothetical proteins or proteins uncharacterized *in vivo*. Pan-gene2859 had several BLAST hits to a putative uncharacterized conserved secreted protein in *Streptococcus pneumoniae* (e-value = 5e-19, 28% identity). This similarity to a protein in *S pneumoniae*, another meningeal pathogen, suggests this protein could play a role in pathogenicity. Pan-gene4779 could not be further characterized by searching the NCBI non-redundant protein database. High scoring BLAST hits in this database pertained only to other hypothetical proteins in *S suis* strains. Pan-gene3448 was annotated as a S26 family signal peptidase, which is involved in the processing of newly synthesized secreted proteins.

Discussion

This study demonstrated the use of a pan-genome approach to identify novel candidate VAGs of strains in the United States and further investigated the utility of a VAG panel as a typing method for *S suis*. This approach elucidated various genes capable of differentiating among pathotypes and a novel panel of three candidate VAGs for the prediction of pathogenicity. Protein function prediction of these genes revealed a putative secreted protein, S26 family signal peptidase, and two hypothetical proteins. However, further research is needed to characterize these candidate VAGs and to better understand their roles in *S suis* pathogenicity. Using this novel VAG panel, we were able to detect the prevalent *S suis* subtypes in the United States including serotype 1/2 and ST28, which are associated with systemic and neurological disease (classified as the pathogenic pathotype).[6] These findings suggest pan-genome analysis may enhance strain selection efforts for preventative strategies, such as vaccine development, by presenting a highly discriminatory typing technique for *S suis*. Our study may also inform new diagnostic approaches and novel gene targets for virulence studies. More specifically, it will contribute to the development of knock-out studies targeting these candidate VAGs to further investigate the pathogenesis of *S suis* infection.

Table 1: Numbers and proportions of *S suis* isolates of each pathotype that are positive for pan-genes and pan-gene panels representing the candidate virulence-associated genes, pan-genes 2859, 3448, and 4779

Pan-gene or pan-gene panels	Pathogenic (n = 139)		Possibly opportunistic (n = 47)		Commensal (n = 22)		Total no. of isolates containing the pan-gene or pan-gene panel
	No.	Proportion	No.	Proportion	No.	Proportion	
pan-gene2859	137	0.770	34	0.191	7	0.039	178
pan-gene3448	132	0.786	31	0.185	5	0.030	168
pan-gene4779	134	0.779	34	0.198	4	0.023	172
pan-gene2859, pan-gene3448	132	0.800	29	0.176	4	0.024	165
pan-gene2859, pan-gene4779	134	0.798	31	0.185	3	0.018	168
pan-gene3448, pan-gene4779	132	0.795	30	0.181	4	0.024	166
pan-gene2859, pan-gene3448, pan-gene4779	132	0.805	29	0.177	3	0.018	164

Table 2: Pathogenic subtypes of *S suis* possessing the candidate virulence-associated gene profile of pan-genes 2859, 3448, and 4779

Pathogenic subtypes*		No. of isolates positive for the pan-gene profile	No. of isolates negative for the pan-gene profile
Serotypes	1	11	0
	1/2	54	0
	2	17	0
	7	21	2
	14	5	0
	23	10	0
Sequence types	1	17	0
	13	5	0
	25	2	0
	28	52	0
	29	2	0
	94	18	0
	108	17	0
	117	0	2
	225	3	0
	373	4	1
	961	10	0
	977	9	0

* Identified as potential subtype indicators of pathogenicity in our previous study.[6]

References

1. Clifton-Hadley FA. 1984. Studies of *Streptococcus suis* type 2 infection in pigs. Vet Res Commun 8:217–227.

2. Reams RY, Glickman LT, Harrington DD, Thacker HL, Bowersock TL. 1994. *Streptococcus suis* infection in swine: a retrospective study of 256 cases. Part II. Clinical signs, gross and microscopic lesions, and coexisting microorganisms. J Vet Diagn Invest 6:326–334.

3. Clifton-Hadley FA, Alexander TJ. 1980. The carrier site and carrier rate of *Streptococcus suis* type II in pigs. Vet Rec 107:40–41.

4. Arends JP, Hartwig N, Rudolphy M, Zanen HC. 1984. Carrier rate of *Streptococcus suis* capsular type 2 in palatine tonsils of slaughtered pigs. J Clin Microbiol 20:945–947.

5. Galina L, Collins JE, Pijoan C. 1992. Porcine *Streptococcus suis* in Minnesota. J Vet Diagn Invest 4:195–196.

6. Estrada AA, Gottschalk M, Rossow S, Rendahl A, Gebhart C, Marthaler DG. 2019. Serotype and genotype (Multilocus Sequence Type) of *Streptococcus suis* isolates from the United States serve as predictors of pathotype. J Clin Microbiol 57.

7. Higgins R, Gottschalk M, Beaudoin M, Rawluk SA. 1992. Distribution of *Streptococcus suis* capsular types in Quebec and western Canada. Can Vet J Rev Veterinaire Can 33:27–30.

8. Weinert LA, Chaudhuri RR, Wang J, Peters SE, Corander J, Jombart T, Baig A, Howell KJ, Vehkala M, Välimäki N, Harris D, Chieu TTB, Van Vinh Chau N, Campbell J, Schultsz C, Parkhill J, Bentley SD, Langford PR, Rycroft AN, Wren BW, Farrar J, Baker S, Hoa NT, Holden MTG, Tucker AW, Maskell DJ. 2015. Genomic signatures of human and animal disease in the zoonotic pathogen *Streptococcus suis*. Nat Commun 6:6740–6748.

9. Medini D, Donati C, Tettelin H, Masignani V, Rappuoli R. 2005. The microbial pan-genome. Curr Opin Genet Dev 15:589–594.

10. Wileman TM, Weinert LA, Howell KJ, Wang J, Peters SE, Williamson SM, Wells JM, Langford PR, Rycroft AN, Wren BW, Maskell DJ, Tucker AW. 2019. Pathotyping the zoonotic pathogen *Streptococcus suis*: Novel genetic markers to differentiate invasive disease-associated isolates from non-disease-associated isolates from England and Wales. J Clin Microbiol 57.

Importance of capsular immunity in protection against *Glaesserella parasuis*

Samantha J. Hau[1,2]; Susan L. Brockmeier[1]
[1]Virus and Prion Research Unit, NADC, ARS, USDA, Ames, Iowa;
[2]Oak Ridge Institute for Science and Education, Oak Ridge, Tennessee

Background

Glaesserella (*Haemophilus*) *parasuis* is a Gram negative bacterium in the Pasteurellaceae family. It is the causative agent of Glässer's disease in swine, a disease characterized by fibrinopurulent polyserositis, arthritis, and meningitis, and a bacterial contributor to porcine respiratory disease complex.[1] *G parasuis* is an early colonizer of the swine respiratory tract and a member of the respiratory microbiota of healthy animals. Isolates of *G parasuis* are characterized by capsular polysaccharide into 15 serovars (SV) through serology or PCR analysis of the capsule locus.[2,3] Capsule is an important virulence factor of *G parasuis*, though isolate SV does not correlate exactly with disease potential.[4,5] *G parasuis* control has been challenging in production settings due to the ubiquitous distribution of *G parasuis*, rapid onset of disease, and the lack of broadly effective vaccines. The difficulty generating broadly effective vaccines is attributed, in part, to strain and serovar specific immunity associated with bacterin vaccination, which results in minimal protection against heterologous isolates. To better understand the role of capsular immunity and assess the capacity of protein based immunity to provide protection against challenge, we evaluated an avirulent, capsule-deficient mutant of the SV5 strain HS069 (HS069Δcap) as a whole cell bacterin and compared its efficacy against homologous and heterologous challenge with a bacterin derived from the wild type HS069.

Methods

To assess the ability of protein based immunity to provide broad protection against *G parasuis*, pigs were vaccinated twice with HS069Δcap, the wild type HS069, or adjuvant only at a 21 day interval. Serum samples were assessed for antibody titers at days 0, 21, and 42. On day 42 of the study, animals were challenged with the homologous SV5 strain (HS069), or one of the following heterologous strains: 12939 (SV1), 2170B (SV4), Nagasaki (SV5), or MN-H (SV13). Animals were evaluated over a 14 day period for clinical signs of Glässer's disease and euthanized when severe disease presented. Samples were collected to confirm systemic distribution of *G parasuis*. Hydrophobicity was assessed by combining 1.2 mL of bacterial suspension with 0.1 mL xylenes. The mixture was incubated for 10 minutes at 37°C and vortexed for 2 minutes. After resting for 15 minutes, the change in OD_{400} was measured as an indication of hydrophobicity. More hydrophobic isolates move into the xylenes phase and the OD_{400} decreases. Hydrophobicity correlates with reduced capsule production and increased outer membrane exposure. Capsule was also visualized using transmission electron microscopy after performing a capsule stain with ferritin.

Results

Vaccination with HS069Δcap bacterin and the wild type HS069 bacterin generated strong antibody titers to whole cell sonicate of all challenge strains and boost vaccination resulted in an amnestic antibody responses (Figure 1). Animals vaccinated with HS069Δcap were protected against challenge with the homologous strain (HS069) (Figure 2A) as well as the heterologous strains 12939 (Figure 2B), 2170B (Figure 2C), and MN-H (Figure 2D); however, animals challenged with the heterologous SV5 strain Nagasaki showed no improvement in survival, though median survival time was extended (Figure 2E). Comparatively, vaccination with the wild type HS069 provided protection against all challenge strains, including Nagasaki (Figure 2A-E). To understand more about the importance of capsular immunity in protection against Nagasaki, a hydrophobicity assay was utilized to determine difference in capsule production. Nagasaki was significantly more hydrophilic than HS069 ($P = 0.03$), therefore produced more capsule than HS069 (Figure 3). Transmission

Figure 1: Serum antibody titers to HS069. Antibody titers to the wild type HS069 increased after vaccination and showed an amnestic response for both vaccine groups. This was similarly seen for all challenge strains and the mutant HS069Δcap (data not shown).

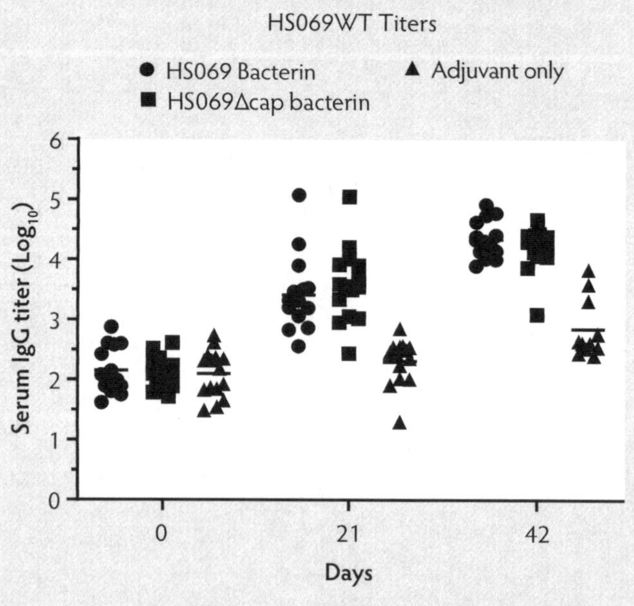

Figure 2: Challenge survival curves. Survival of HS069 bacterin vaccinated (circle), HS069Δcap bacterin vaccinated (square), and adjuvant only vaccinated (triangle) against challenge with wild type HS069 (A), 12939 (B), 2170B (C), MN-H (D), and Nagasaki (E).

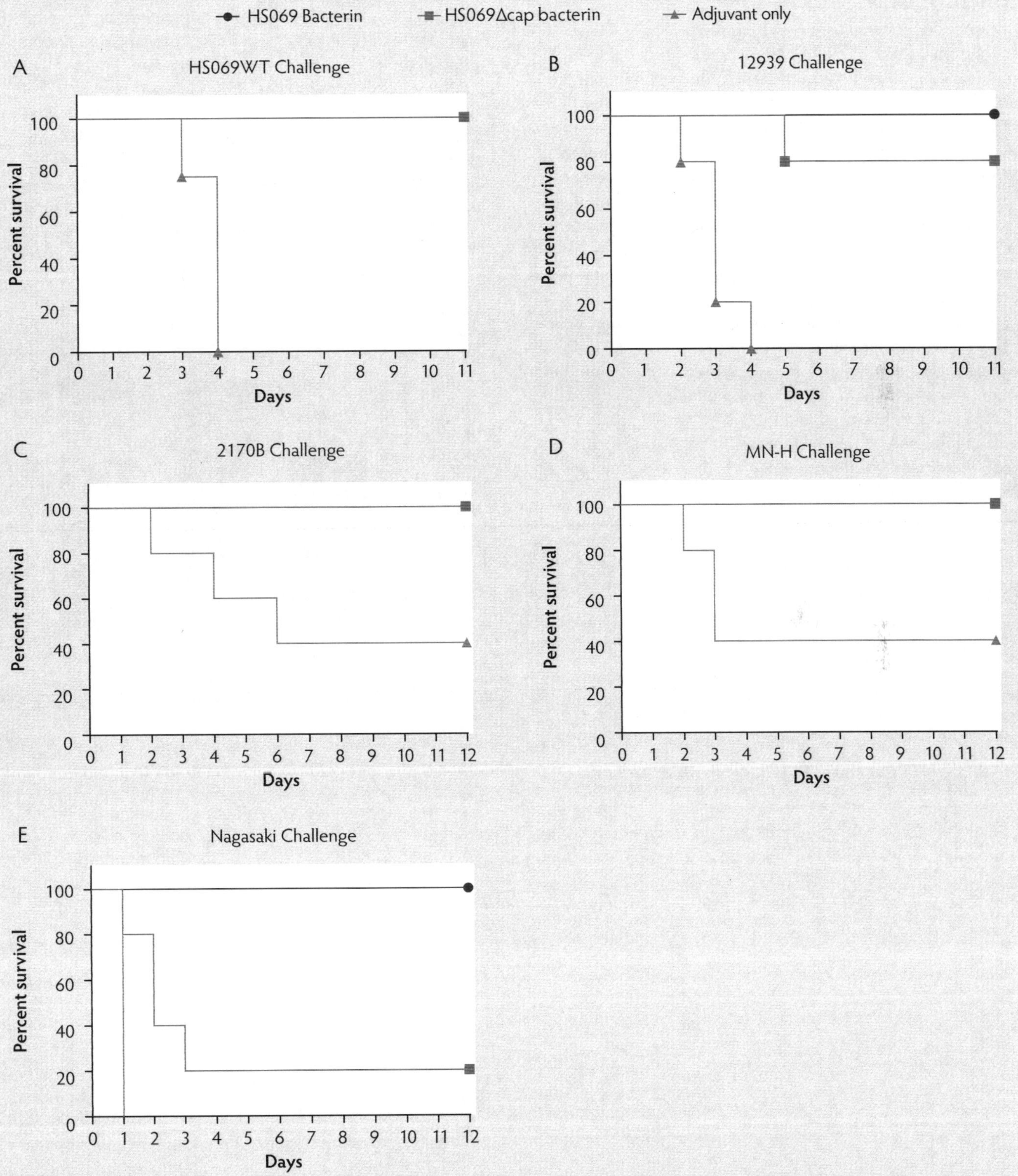

electron microscopy also revealed increased capsule on the cell surface and a more uniform organization of capsule in Nagasaki as compared to HS069.

Conclusions

Differences in protection between bacterins generated from HS069Δcap and the wild type HS069 against challenge with Nagasaki indicate the importance of capsular directed immunity to protection against this *G parasuis* isolate. For the other challenge isolates in this study, capsular immunity was not essential to prevent the development of disease. Assessment of capsular polysaccharide indicated differences in the quantity and distribution of capsule between Nagasaki and HS069, which may have resulted in the differences in protection between the bacterins after challenge with Nagasaki. The greater amount of capsule produced by Nagasaki may necessitate immunity directed at capsule to prevent systemic disease. This provides important information for *G parasuis* vaccine development. Capsular immunity is only essential for some isolates of *G parasuis*, while protein directed immunity can provide protection against less encapsulated strains. Strain selection is important when generating bacterin vaccines, as some strains will generate better heterologous protection through protein directed antibodies. However, knowledge of the strains circulating within the herd is important when determining whether capsular or protein based immunity will be effective.

References

1. Aragon V, et al. Chapter 55 Glasser's Disease. In *Diseases of Swine*, edited by JJ Zimmerman, LA Karriker, A Ramirez, KJ Schwartz and GW Stevenson, 760-69. West Sussex, UK: John Wiley & Sons, Inc, 2012.

2. Kielstein P and VJ Rapp-Gabrielson. Designation of 15 serovars of *Haemophilus parasuis* on the basis of immunodiffusion using heat-stable antigen extracts. *J Clin Microbiol.* 1992;30: 862-5.

3. Howell KJ, et al. Development of a multiplex PCR assay for rapid molecular serotyping of *Haemophilus parasuis*. *J Clin Microbiol.* 2015; 53: 3812-21.

4. Wang X, et al. Polysaccharide biosynthesis protein CapD is a novel pathogenicity-associated determinant of *Haemophilus parasuis* involved in serum-resistance ability. *Vet Microbiol.* 2013; 164: 184-9.

5. Oliveira S, et al. Characterization of the diversity of *Haemophilus parasuis* field isolates by use of serotyping and genotyping. *Am J Vet Res.* 2003; 64: 435-42.

Figure 3: Hydrophobicity assay to assess capsule production. Results are reported as a percent of the absorbance at OD_{400} of the initial suspension. Because capsule is a hydrophilic substance, a greater portion of the suspension remains in the aqueous phase when the strain is encapsulated, resulting in a higher percent absorbance post-incubation. Nagasaki was found to be significantly more hydrophilic than HS069. Because the outer membrane is hydrophobic, the unencapsulated HS069Δcap moved into the hydrophobic xylenes phase, resulting in a lower percent absorbance post-incubation. Significant differences in hydrophobicity are identified by different letters ($P < 0.05$).

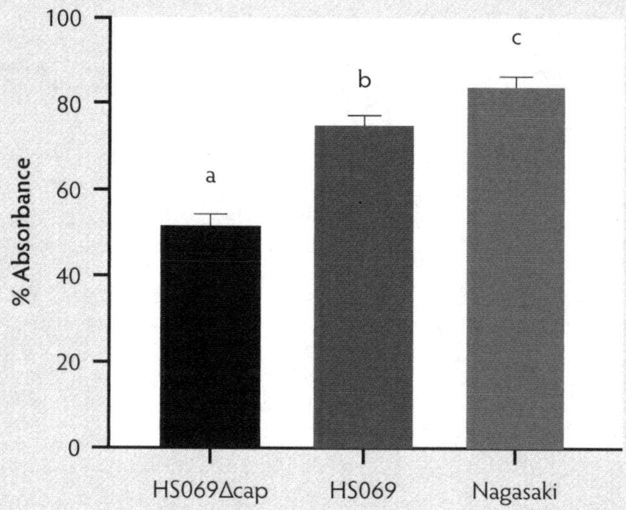

Scoring lesions in slaughtered pigs through artificial intelligence technology: the first extensive investigation

A. R. Trachtman[1], DVM; L. Bergamini[2], BEng; A. Palazzi[2], BDEng; A. Porrello[2], BDEng; A. Capobianco Dondona[3], DVM, MS; E. Del Negro[3], BEng; S. Calderara[2], BDEng, PhD; G. Marruchella[1], DVM, MS
[1]University of Teramo, Faculty of Veterinary Medicine, Loc. Piano d'Accio, Teramo, Italy;
[2]AImageLab, University of Modena and Reggio Emilia, Via Vivarelli, Modena, Italy;
[3]Farm4Trade s.r.l.,Via Marino Turchi, Chieti, Italy

Introduction

The slaughterhouse is widely recognized as a key check point for assessing the health status of farm animals, as well as the effectiveness of strategies implemented to prevent or treat disease conditions. In this respect, a number of scoring methods have been developed to estimate the economic impact of diseases in slaughtered pigs, the greatest attention being paid to the so-called porcine respiratory disease complex. As a general rule, the ideal scoring method should be simple, fast, easily standardisable, providing suitable data for statistical analysis. Methods currently available to score pneumonia and pleurisy at slaughter do well fit such requirements. Despite this, the collection of such data is sill very expensive and time-consuming, thus making difficult, if not impossible, a systematic recording.[1,2]

Artificial intelligence is a discipline aiming to develop intelligent agents, i.e. machines that can perceive the environment and take action to maximize their success with regard to a defined target.[3] Over the last few decades, a number of attempts have been made to apply artificial intelligence ("machine learning") to human health and medicine, mostly in the field of diagnostic imaging .[4]

To the best of our knowledge, artificial intelligence has been never applied to the identification and quantification of gross lesion in animals. Therefore, the present study aimed to train an artificial intelligence-based system, in order to automatically score pleurisy in slaughtered pigs. The automation of lesion scoring would be very helpful to enable the systematic inspection of slaughtered livestock, positively influencing the herd management, the animal welfare and the profitability of farming.

Materials and methods

As the first step, a total of 5.252 half-carcasses were inspected and photographed with smartphone cameras under field conditions. Pleurisy was then scored by two skilled veterinarians, using a simplified version of the "pleurisy evaluation on the parietal pleura" (PEPP) method[2], as follows: absence of pleurisy = 0 point; pleurisy affecting the cranial chest wall (five intercostal spaces) = 1 point; pleurisy affecting the caudal chest wall (all the remaining intercostal spaces) = 2 points; pleurisy affecting both the cranial and the caudal chest wall = 3 points. All the scores were agreed by the veterinarians and recorded on a Microsoft Excel spreadsheet.

Thereafter, all pictures were annotated ("segmented") using an ad hoc open-source tool (Labelme), the following features being systematically labelled: the silhouette of the half-carcass, the vertebral bodies, the diaphragm, the cranial chest wall, the caudal chest wall, lesions and artefacts. Segmented pictures were used to train different "convolutional neural networks" (CNNs; deep learning methods, currently representing the state of the art in the field of artificial intelligence), both available in scientific literature[5,6] and de-novo developed. A rule-based classifier was finally employed to convert the segmentation masks, as provided by the trained CNNs, into a score consistent with the PEPP method.

The performances of the trained CNNs were compared with those provided by the veterinarians, the latter being considered as the gold standard. In addition, the accuracy of the trained CNNs was evaluated by an independent test set of 200 images.

Results and discussion

Overall, the de-novo developed and trained CNN showed the best performances, being able to effectively discriminate between healthy and diseased half-carcasses (accuracy = 95%). More in detail, it effectively identified healthy (accuracy = 98%) and severely affected (accuracy = 100%) half-carcasses. On the other hand, the same trained CNN provided accuracy values ranging between 68% and 78% for less severe pleurisy; such values were mainly influenced by the less accurate evaluation of small lesions located at the edge between the cranial and caudal chest walls.

Artificial intelligence-based methods are very topical, intriguing and virtually able to revolutionize most human activities, including those in the field of biomedical sciences. The present study represents the first application of such methods to the assessment of pathological findings in veterinary medicine. Results indicate that the trained CNN is able to effectively recognize the different anatomical districts of the carcasses, as well as to discriminate health from diseased chest walls. We consider that our data encourage further investigations, in order to extend the same technology to the scoring of other disease conditions, enzootic pneumonia in primis. This could allow a fast, cheap and systematic way of recording lesions in slaughtered pigs, thus providing, in real-time, an enormous amount of valuable information to better manage the health of livestock.

References

1. Luppi A., Merialdi G. Lesioni al macello. In: Martelli P. *Le Patologie del Maiale.* 1° ed. Le Point Veterinaire Italie; 199-218. 2013.

2. Di Provvido A., Trachtman A. R., Farina E., Odinzov Vaintrub M., Fragassi G., Vignola G., Marruchella G. Pleurisy evaluation on the parietal pleura: an alternative scoring method in slaughtered pigs. *J Swine Health Prod.* 2019; 27(6):312-316.

3. Poole D., Alan M., Randy G. Computational Intelligence and Knowledge. In: *Computational intelligence: a logical approach.* Oxford University Press; 1-22. 1998.

4. Kononenko I. Machine learning for medical diagnosis: history, state of the art and perspective. *Artif Intell Med.* 2001; 23:89-109.

5. He K., Zhang X., Ren S., Sun J. Deep residual learning for image recognition. *Proceedings of the IEEE conference on computer vision and pattern recognition.* 770–778. 2016.

6. Deng J., Dong W., Socher R., Li LJ., Li K., Fei-Fei L. Imagenet: A large-scale hierarchical image database. *Proceedings of the IEEE conference on computer vision and pattern recognition.* 248–255. 2009.

Effects of duration of ultraviolet-C radiation exposure on porcine reproductive and respiratory syndrome virus

A. Riley[1], BS; S. Tousignant[2], DVM; L. Dalquist[3], DVM

[1]University of Minnesota College of Veterinary Medicine, St. Paul, Minnesota; [2]Boehringer Ingelheim Animal Health USA Inc., Duluth, Georgia; [3]Christensen Farms, Sleepy Eye, Minnesota

Introduction

Ultraviolet-C (UV-C) irradiation has been used as a disinfectant as early as 1910.[1] The DNA and RNA of pathogens is disrupted by the ultraviolet rays, which disables the pathogens from multiplying and causing diseases.[2] UV-C has been proven to inactivate porcine reproductive and respiratory syndrome virus (PRRSV) after ten minutes of exposure.[3] Therefore, the objective of this study was to validate appropriate UV-C exposure time in a UV-C Chamber as a means of inactivating PRRSV on various surfaces.

Materials and methods

Aluminum soda cans and plastic sandwich containers were used in this study. Each item had 10 centimeter sampling squares drawn in three locations (top, side, and front) and then pre-inoculation swabbed (pre-swabs) with a synthetic swab (Puritan, Guilford, ME) and Modified Eagle Medium (MEM) culture media (Life Technologies Corporation, Grand Island, NY). The items were then inoculated with two milliliters of commercially available modified live PRRSV vaccine in the three locations. The vaccine was evenly distributed across the sampling square, and set aside to dry for twenty minutes. The item was then placed in a commercially available UV-C Chamber (Once Inc. Bioshift® Pass-Through UV-C Chamber, Plymouth, MN) and exposed to UV-C radiation for ten minutes. Control samples were placed in the UV-C Chamber for 10 minutes without the UV-C bulbs being operated. The sampling squares were then swabbed (post-swabs) as described above. The study was replicated five times, and the process was repeated using an exposure period of 5 minutes.

All swabs were immediately labeled and stored in an ultra-low (-80° C) freezer until being sent to the University of Minnesota College of Veterinary Medicine Diagnostic Lab to be tested. The pre-swabs were pooled in groups of five and tested by polymerase chain reaction (PCR), and the post-swabs were tested by virus isolation (VI), batched by UV-C exposure duration. The results were recorded and analyzed.

Results

Table 1 presents the test results for PCR and VI on pre- and post-swabs. All of the pre-inoculation control swabs were negative by PCR for PRRSV. All of the UV-C radiated post-swabs were negative by PRRSV VI at both five and ten minutes. One out of five non-UV-C radiated post-swabs on the sandwich container were negative by PRRSV VI at ten-minute exposure time, and zero were negative at five-minute exposure time. Four out of five of the non-UV-C radiated post-swabs on the aluminum soda can were negative on the ten-minute exposure time, and one out of five were negative on the five-minute exposure time.

Discussion and conclusions

Five minutes of UV-C radiation was determined to be enough time to kill an extremely high level of a modified live PRRSV vaccine. This was an important finding as workers have found that the longer their food is under UV-C radiation, the worse it tastes. Future research may include repeating the study with a field strain of PRRSV, measuring the effect of objects shadowing each other in the UV-C Chamber and how age of the UV-C bulbs effect the virucidal capacity of the chamber.

References

1. Hijnen, W.A.M., Beerendonk, E.F., Medema, G.J., Inactivation credit of UV radiation for viruses, bacteria and protozoan (oo)cysts in water: A review, Water Research, Volume 40, Issue 1, 2006, Pages 3-22.

2. Stephan, A. B., 2017 Reduction in Swine Pathogen Numbers by a UVC Germicidal Chamber., Once Inc., National Hog Farmer 2017.10.31.

3. Dee, S., Otake, S., Deen, J., 2011. An evaluation of ultraviolet light (UV254) as a means to inactivate porcine reproductive and respiratory syndrome virus on common farm surfaces and materials. Vet. Microbiol. 150, 96–99. doi:10.1016/j.vetmic.2011.01.014.

Table 1: Polymerase chain reaction (pre-swabs) and virus isolation (post-swabs) test results

		5 minutes						10 minutes					
		Pop can			Sandwich container			Pop can			Sandwich container		
		pre-swab	post-swab		pre-swab	post-swab		pre-swab	post-swab		pre-swab	post-swab	
New bulb	trial 1	Negative	Negative	trial 1	Negative	Negative	trial 1	Negative	Negative	trial 1	Negative	Negative	
	trial 2	Negative	Negative	trial 2	Negative	Negative	trial 2	Negative	Negative	trial 2	Negative	Negative	
	trial 3	Negative	Negative	trial 3	Negative	Negative	trial 3	Negative	Negative	trial 3	Negative	Negative	
	trial 4	Negative	Negative	trial 4	Negative	Negative	trial 4	Negative	Negative	trial 4	Negative	Negative	
	trial 5	Negative	Negative	trial 5	Negative	Negative	trial 5	Negative	Negative	trial 5	Negative	Negative	
		Pop can			Sandwich container			Pop can			Sandwich container		
		pre-swab	post-swab		pre-swab	post-swab		pre-swab	post-swab		pre-swab	post-swab	
Control (no bulb)	trial 1	Negative	Positive	trial 1	Negative	Positive	trial 1	Negative	Positive	trial 1	Negative	Positive	
	trial 2	Negative	Positive	trial 2	Negative	Positive	trial 2	Negative	Negative	trial 2	Negative	Positive	
	trial 3	Negative	Negative	trial 3	Negative	Positive	trial 3	Negative	Negative	trial 3	Negative	Positive	
	trial 4	Negative	Positive	trial 4	Negative	Positive	trial 4	Negative	Negative	trial 4	Negative	Positive	
	trial 5	Negative	Positive	trial 5	Negative	Positive	trial 5	Negative	Negative	trial 5	Negative	Negative	

A comparison of alternative media for *Mycoplasma hyopneumoniae* aerosol exposure

K. Hewitt, BS[1]; M. Hensch[2], DVM; A. Maschhoff[2], DVM
[1]Iowa State University, Ames, Iowa; [2]The Maschhoffs, Carlyle, Illinois

Introduction

Mycoplasma hyopneumoniae (Mhp) causes significant economic losses to the swine industry.[1] Effective planned Mhp exposure of replacement breeding animals aids Mhp elimination protocols. Friis medium (FM) is the current gold standard to dilute lung homogenate (LH) in Mhp exposure protocols.[2] Less costly media are available, but their effect on Mhp viability is unknown. The goal of this study was to evaluate alternative media used in Mhp LH preparation on aerosol exposure efficacy.

Materials and methods

Four barns were used for the trial, each having 10 pens housing 125 pigs/pen. The finishing site selected was part of a porcine reproductive and respiratory syndrome virus (PRRSV) and Mhp negative flow. At -14 days post-infection (DPI), tracheobronchial (TB) swabs from 30 pigs chosen randomly were confirmed Mhp-negative by polymerase chain reaction (PCR). There were no observed clinical signs indicating the presence of other respiratory pathogens prior to the start of the study.

Four media were evaluated: Friis medium (FM), phosphate buffered saline (PBS), purchased saline (PS), and homemade saline (HS). Each barn was assigned a single medium. For HS preparation, 37 grams of table salt was dissolved into 1 gallon of distilled water. Two routes of administration (ROA) were evaluated per medium: intra-tracheal (IT) inoculation and aerosol exposure (AE) (8 total treatments).

All exposures were completed at 0 DPI (pig age approximately 22 weeks). Inoculum was prepared using a ratio of 80 mL of known-Mhp-positive LH diluted into one gallon of medium. After mixing, inoculum used for IT exposure was stored on ice packs until administered; inoculum used for AE was stored at room temp (approximately 80 degrees F) for 60 minutes before use. Ten pigs in each barn's southwest-most pen (40 total) were ear tagged and inoculated via IT administration of 10 mL diluted chilled inoculum delivered via a rubber urethral catheter (positive controls). Pigs in the northeast-most pen were exposed using 1 gallon of aerosolized inoculum mist delivered via two Hurricane foggers per pen over 15-20 minutes.

At 4 DPI, 30 random pigs from each AE pen (120 total) were ear tagged for repeat sampling. TB swabs were collected individually from tagged pigs at 4 DPI (all groups, N = 160) and 14 DPI (AE only, N = 120) using a sterile swab attached to a modified post-cervical artificial insemination catheter. Swabs were transported in 5 mL falcon tubes containing 1 mL PBS then tested for Mhp in pools of 5 by PCR at Iowa State University Diagnostic Laboratory (ISU-VDL).

At 57 DPI, lungs from a subset of tagged pigs were examined post-harvest at a packing plant.

Results

By 4 DPI, Mhp was detected by PCR from all groups, regardless of ROA (Table 1); 4 DPI IT samples were 100% PCR Mhp-positive when tested individually; these were not retested at 14 DPI.

At 14 DPI, 100% of pooled samples from each aerosol group were PCR positive. All samples were tested individually and every sample from each media type was confirmed PCR positive.

At harvest, 87.5% of 96 examined lungs showed gross lesions of pneumonia, predominately in the anterior-ventral lobes.

In this study, all media and ROA conferred rapid and complete Mhp colonization within 14 DPI, indicating Mhp viability was not adversely affected by media type or ROA. Storage for 60 minutes did not hamper efficacy inoculum used for aerosol ROA. More rapid colonization was seen with IT ROA versus aerosol. Homemade saline proved to be as effective as the other media as a diluent for LH. HS advantages include cost savings and ease of formulation in most veterinary clinics.

References

1. Maes D., Segales J., Mynes T., Silbila M., Pieters M., Hasebrouck F., Control of *Mycoplasma hyopneumoniae* infections in pigs. *Rev Vet Microbiol.* 2008; 126:297-309

2. Robbins RC, Betlach AM, Mondragon-Evans MR, et al. Development of a herd-specific lung homogenate for exposure to *Mycoplasma hyopneumoniae* under field conditions. *J Swine Health Prod.* 2019;27(4):221–227.

Table 1: Detection of *Mycoplasma hyopneumoniae* by PCR in tracheobronchial swabs on days 4 and 14 post-inoculation

ROA and media	Price per gallon	4 DPI				14 DPI			
		Pooled		Individual		Pooled		Individual	
		Positive/ Total	Average Ct	Positive/ Total	Average Ct	Positive/ total	Average Ct	Positive/ total	Average Ct
Intra-tracheal									
Friis medium media	$366.16	2/2	27.9	10/10	29.5				
Phosphate buffered saline	$143.85	2/2	26.1	10/10	26.9				
Purchased saline	$49.21	2/2	23.3	10/10	27.6				
Homemade saline	$0.89	2/2	23.3	10/10	24.6				
Aerosol									
Friis medium media	$366.16	3/6	32.6			6/6	26.4	30/30	27.6
Phosphate buffered saline	$143.85	6/6	31.2			6/6	23.6	30/30	23.4
Purchased saline	$49.21	5/6	32.7			6/6	24.4	30/30	23.5
Homemade saline	$0.89	5/6	28.6			6/6	22.8	30/30	23.3

Tracheobronchial swabs were not collected for intra-tracheal inoculated pigs on 14 DPI because all swabs were individually PCR positive on 4 DPI.

Impact on piglet behavior after use of flunixin meglumine at time of castration

K. L. Kalbian, BS; M. Ceballos, PhD; M. Pierdon, VMD, DACAW
New Bolton Center, University of Pennsylvania, Kennett Square, Pennsylvania

Introduction

Piglets undergo castration, a painful procedure, but producers have limited options for pain mitigation. Injectable products labelled for piglets for the reduction of pain are not available in the United States. Flunixin meglumine is a non-steroidal anti-inflammatory drug that is labelled for the treatment of pyrexia in swine associated with respiratory disease. It has recently received a label in its transdermal form for the treatment of pain related to foot rot in cattle. Drugs labelled for a species are given priority when determining if off label use is acceptable. Despite that, flunixin meglumine is not well studied for pain relief in piglets. Time and efficiency are concerns when it comes to castration in a farm setting. Administration of the flunixin meglumine at the time of castration reduces the amount of time spent handling piglets and the number of times the piglets have to be handled, a stressful process. Several published studies have been conducted on the use of flunixin meglumine for castration pain relief. Two were conducted in Germany[1,2] and one in the US[3] Post-castration cortisol was lower in pigs given flunixin starting at 30 minutes post-castration as well as a reduction in post-castration pain related behaviors.[1] A reduction in post-castration pain behavior and decreased cortisol with flunixin was also reported independent of whether it was given 30 minutes prior to or at the time of castration.[3] The objective of this study is to determine if intramuscular flunixin meglumine administered at the time of castration reduces the occurrence of pain related behaviors in piglets.

Materials and methods

From June until August 2019, 50 litters of piglets were castrated at 7-10 days of age. Males were randomly assigned to receive an intramuscular injection of flunixin meglumine at 2.2 mg/kg (CF) or sterile water (CC). Female piglets received the control injection and sham castration (S). Following intramuscular injection, the piglets were castrated or sham castrated. The piglets were videotaped for 1 hour before and 24 hours following castration. At 10 different time intervals (-60, 0, 30, 60, 90, 120, 180, 480, 720, 1080 minutes) over a 30-minute period, the behavior of 110 piglets in 10 litters (CC = 30, CF = 31, S = 49) was coded every 30 seconds. Postures recorded include lying, standing, walking, non-nutritive suckling and suckling. Point behaviors were coded including tail wagging, trembling, playing, and scratching. Coding was done using Noldus Observer v.11.4. Those coding video and administering treatments were blinded to treatment. STATA v.15 was used for statistical analysis. All data was non-normal. Kruskal Wallis test was used to determine the effect of treatment on different behaviors at different time points. Bonferroni correction was used for multiple comparisons. Results are reported as median and interquartile range.

Results

Immediately after castration CC pigs showed less tail wagging (0.01+/-0.03) and CF pigs showed less tail wagging (0.03+/- 0.07) than S pigs (0.07+/-0.10) ($P < .001$). At that same time point CC pigs showed less walking (0.03+/-.05) and CF pigs showed less walking (0.03+/- 0.09) than S pigs (0.09+/-0.08) ($P < 0.001$). CF pigs showed more tail wagging at time 180 (0.02+/-0.08) compared to S pigs (0.0+/-0.02) ($P < 0.04$) and time 1080 (0.03+/-0.08) compared to S pigs (0.0+/-0.02) ($P < 0.01$). CF pigs showed more walking at time 180 (0.03+/-0.02) compared to S pigs (0.01+/-0.05) ($P < 0.05$). At time 180 CF pigs showed increased standing (0.15+/-0.20) compared to S pigs (0.04+/-0.13) ($P < 0.05$) as well as at time 480 (0.19+/-0.27) compared to S pigs (0.13+/-0.17) ($P < 0.05$). CF pigs showed reduced lying in a group (0.47+/-0.58) compared to S pigs (0.89+/-0.55) at time 180 ($P < .05$).

Discussion

In the period monitored immediately post castration the behavioral effects of castration were decreased tail wagging and walking in the castrated groups compared to sham pigs. Flunixin meglumine altered behaviors previously associated with castration pain such as social cohesion[3] and tail wagging[1] but in a negative way with increased tail wagging and less time sleeping in the group at certain times. Increased standing and increased walking in the flunixin treated pigs is non-specific in relation to pain but the flunixin treated pigs were not different from those pigs that were not given pain relief. Based on the timepoints studied, and behavioral measures included, flunixin meglumine did not have a positive impact on pain behaviors associated with castration. This could be because the dose of flunixin needed to provide pain relief and peak serum concentrations of flunixin in piglets is not well elucidated. Behavioral repertoire in young pigs is limited and physiological factors, such as cortisol levels would also be useful in interpretation of these changes in behaviors.

References

1. Langhoff, R., Zoels, S., Barz, A., Palzer, A., Ritzmann, M., & Heinritzi, K. 2008. [Investigation about the use of analgesics for the reduction of castration-induced pain in suckling piglets]. Berliner und Munchener Tierarztliche Wochenschrift, 122(9-10), 325-332.

2. Reiner, G., Schollasch, F., Hillen, S., Willems, H., Piechotta, M., & Failing, K. 2012. [Effects of Meloxicam and Flunixin on pain, stress and discomfort in male piglets during and after surgical castration]. Berliner und Munchener Tierarztliche Wochenschrift, 10(7-8), 305-314.

3. Sutherland, M. A., Davis, B. L., Brooks, T. A., & Coetzee, J. F. 2012. The physiological and behavioral response of pigs castrated with and without anesthesia or analgesia. Journal of animal science, 90(7), 2211-2221.

Evaluation of location and fogger type on disinfectant surface area coverage of supplies entering a commercial sow farm

K. Leuck[1], BS; C. LeFevre[2], DVM; M. Jansen[3], DVM; D. Amodie[3], MS

[1]University of Illinois College of Veterinary Medicine, Urbana, Illinois; [2]Carthage Veterinary Services, Carthage, Illinois; [3]Zoetis, Parsippany, New Jersey

Introduction

Biosecurity is essential to limiting the entry of pathogens into swine farms. Pathogens have the potential to enter farms via multiple routes including humans, animals, feedstuffs and new supplies. Chemical decontamination of new supplies through the use of aerosolized disinfectants is a common practice on commercial sow farms and provides a crucial opportunity to prevent the introduction of disease. The objective of this study was to determine the ability of current industry standard fumigation room practices to thoroughly cover all surfaces of supplies entering a site through the evaluation of percent surface area coverage.

Materials and methods

This study took place on a commercial sow farm with a 1.8 × 2.1 × 2.4m fumigation room. Four new, white 0.3 × 0.2 × 0.08m cardboard boxes were used for each trial to mimic boxes of supplies entering a site and served as a measurable surface for disinfectant to be applied. Boxes were evenly spaced on the top and middle shelves of a 1.5 × 0.6 × 1.5m metal wire shelving unit with three shelves. The shelving unit was located on the north side of the room. The room also contained a 0.6m ventilation fan on the east side of the room to exhaust the fumigation room after the disinfectant was applied. To visualize what percent surface area was covered by disinfectant, purple clothing dye (Rit, Nokoma Products LLC) was added to an accelerated hydrogen peroxide disinfectant (Intervention®, Virox Animal Health) at 1:16 dilution according to manufacturer's recommendation. Clothing dye was added at a 1:20 ratio of dye to disinfectant solution. Two brands of foggers, the Fogmaster® Jr. 5330 (Fogmaster Corporation) and the Hurricane Ultra II (Curtis Dyna Fog, Ltd), as well as a manual pump sprayer (Chapin International Inc.) were used to apply the disinfectant solution.

Accelerated hydrogen peroxide disinfectants work by producing destructive hydroxyl free radicals that attack membrane lipids, DNA and other essential cell components.[1] At recommended 1:16 dilution, Intervention® claims to be effective against: bacteria, antibiotic resistant bacteria, viruses, fungi, mold and mildew with a 5 minute contact time and is effective against Foot and Mouth Disease virus with a 10 minute contact time.[2] Applying disinfectants to supplies entering the farm was achieved with automated fogging machines or manual application. The Fogmaster® Jr. 5330 is an automated fogger that emits particles at an adjustable size of 15-40 microns at a rate of 0-120mL per minute.[3] The fogger has a stationary neck piece and disinfectant was emitted at a fixed angle from the unit. In this study, the Fogmaster® Jr. 5330 flow rate was set to the maximum setting indicated by a gradient on the adjustable nozzle. The Hurricane Ultra II is an automated fogger that emits particles at a size of 5.5 microns at an adjustable rate of 0-283 mL per minute.[4] It additionally has a tilt locking neck feature which allows the angle at which the disinfectant can be sprayed to be adjusted. The flow rate was set to the highest setting of 283 mL per minute. Manual spay was achieved using the Chapin 20003 3-Gallon Lawn and Garden Sprayer which emits a flow rate of 1,514-1,894 mL per minute.[5]

Four treatment groups were evaluated: T01-single Fogmaster® Jr. 5330; T02-single Hurricane Ultra II; T03- two Hurricane Ultra II's; T04- manual sprayer. Fogging units for T01-03 were placed in location A and/or location B. Fogger location A was a shelf in the southwest corner of the fumigation room, 0.58m from the ceiling. Fogger location B was the floor in the southeast corner of the fumigation room. Foggers for T01 and T02 were placed in location A and foggers for T03 were placed in both locations A and B. (Figure 1). The sprayer used in T04 was operated manually. At location A the Hurricane fogger used in T02 and T03 was pointed downwards at 80°. At location B the Hurricane fogger used in T03 was pointed upwards at 130°.

Each treatment group was evaluated in a separate trial with four new boxes for each trial. Each trial was replicated three times for every treatment group. Foggers in treatment groups T01, T02 and T03 were run for 10 minutes, followed by the ventilation fan for 10 minutes, boxes were then removed from the fumigation room and set to dry for another 10 minutes before images were taken of each side. The manual sprayer operated in treatment group T04 was sprayed on all surfaces of the boxes until visual coverage was achieved. On average, each box took 1 minute to cover with disinfectant; totaling 4 minutes per trial. T04 boxes were placed on the shelving unit in the same orientation as the automatic fogging trials and were not manipulated during the spraying process to expose different surfaces. Once each of the four boxes was sprayed using the manual pump sprayer, the ventilation fan was run for 10 minutes, the boxes were then removed from the fumigation room and allowed to dry for another 10 minutes before images were taken of each side. Contact time for the disinfectant was maintained for 10 minutes via the ventilation period before being manipulated and brought into the farm for each treatment group.

Figure 1: Orientation of foggers in the fumigation room. Foggers in T01 and T02 were placed in location A and foggers in T03 were placed in location A and B.

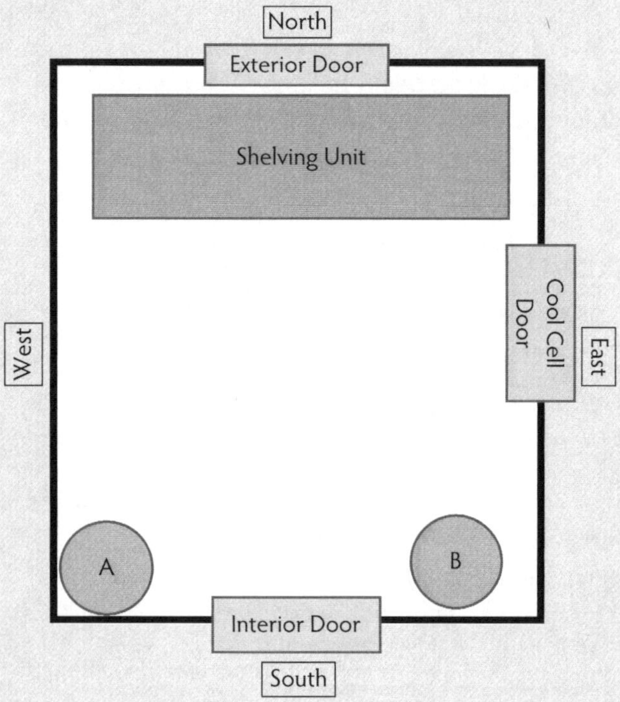

Figure 2: Average total % box coverage.

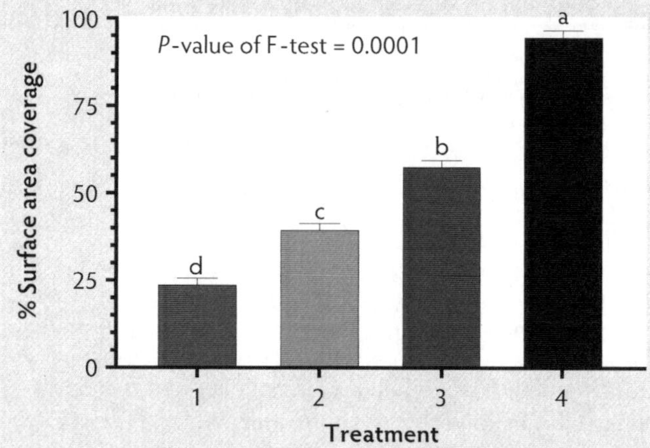

Visual surface area coverage was evaluated using National Institute of Health's Image J Software (version 2.0.0) on all surfaces of the boxes. Images of each side of each box were taken at a consistent height with a black contrasting background and were uploaded in the imaging software to be evaluated using the same parameters of hue, saturation and brightness to evaluate the surface area coverage.

Using the SAS Proc Mixed Procedure (SAS 9.4, Cary, NC), total box coverage was analyzed with a model that considered the fixed effect of fogger location and the random effect of the residual error. Least Squares Means (LSMeans) were calculated for each fogger location and compared by the two-sided student's t-test at the 5% level of significance.

Results

The average percent coverage for T01, T02, T03 and T04 was 24.02%, 39.88%, 57.84% and 95.08%, respectively (Figure 2). Statistical analyses indicated that T04 covered significantly more surface area than all other treatments ($P = 0.0001$), followed by T03, T02 and T01 which covered significantly less than all other treatments.

Discussion

Proper disinfection requires physical contact of disinfectant agents with all surfaces of supplies entering a site. No matter the location of either automated fogger, complete surface area coverage was not achieved. Incomplete coverage leaves the potential for pathogen entry which could result in disease outbreak. While some numerical differences in coverage were noted depending on the location of the fogger(s), manually spraying supplies achieved the highest coverage and may be the most reliable method of supply disinfection until further improvements in fumigation room practices are made. When applying disinfectant with a manual sprayer in a practice setting, while dye will not be used to assure surfaces are visually covered, they should appear saturated to visualize coverage.

Evaluation of current standard fumigation room practices were also considered in this study to look for additional improvements that can be recommended to swine producers to improve the effectiveness of their biosecurity procedures. Recommendations include the following: 1. Establish clean and dirty lines for the exterior entrance of the fumigation room. This can be further improved by utilizing a physical barrier to establish clean and dirty lines such as the placement of the shelving unit for placement of supplies. 2. Include a checklist for supplies that are entering the farm to enhance communication about fogging procedures and to help trace disease outbreak sources. 3. Remove outer packaging that supplies arrive in to allow disinfectant solution to come into contact with the actual supplies that are entering the farm rather than their packaging. Studies show that complex substrates such as carboard are ineffectively disinfected no matter orientation of the substrate in the fogging chamber or disinfectant solution used.[6] 4. Space out boxes on the shelving unit to minimize contact and allow disinfectant solutions to reach all surfaces. 5. Train individuals to identify empty fogger reservoirs and measure proper dilutions to maintain fogging machine reservoirs. 6. Do not use the fogging chamber for storage. Fogging chambers that are cluttered may prevent the disinfectant emitted from the fogging unit from reaching the new supplies to allow physical contact and disinfection to occur.

References

1. Center for Disease Control. Chemical Disinfectants. 2008. Available at: https://www.cdc.gov/infectioncontrol/guidelines/disinfection/disinfection-methods/chemical.html Accessed 21 October 2019.

2. Virox Animal Health. Intervention. 2017. Available at: https://www.viroxfarmanimal.com/wp-content/uploads/2018/11/7561_Virox_Intervention-CON-RefSheet_06.pdf Accessed 21 October 2019.

3. Fogmaster. Fogmaster® Jr. 5330. 2013. Available at: https://www.fogmaster.com/hand_held_5330.htm Accessed 21 October 2019.

4. Curtis Dyna Fog. Hurricane Ultra II. 2018. Available at: http://www.dynafog.com/wp-content/uploads/2018/06/Hurricane-Ultra-II-Product-Data-Sheet.pdf Accessed 21 October 2019.

5. Chapin. Chapin 20003 3-Gallon Lawn and Garden Sprayer. 2019. Available at: https://chapinmfg.com/Product/slug/chapin-20003-3-gallon-lawn-and-garden-sprayer Accessed 21 October 2019.

6. Kettelkamp et al. Impact of item orientation on the efficacy of supply decontamination with aerosolized chemical disinfectants. AASV. 2019.

Assessing sampling strategies for optimal detection of influenza A virus in endemically infected pigs post weaning

M. Neiley[1]; B. Blair[1]; J. Lowe[1,2]

[1]University of Illinois College of Veterinary Medicine, Urbana, Illinois; [2] Lowe Consulting Ltd., Mahomet, Illinois

Introduction

Influenza A virus (IAV) continues to circulate endemically in swine populations in the United States. H1N1, H1N2, and H3N2 are recognized as the three predominant subtypes circulating in the North American populations.[1] Characterized by 8 RNA gene segments, the virus maintains the unique ability to reassort with other strains and undergo antigenic drift.[2] In endemically infected populations, co-circulation of multiple variants of one virus has been reported in a single batch of pigs.[3] As a result, new strains of the virus are continually emerging throughout swine populations. Introduction of new variants not only complicates control efforts but may also lead to emergence of an antigenically unique virus. The high morbidity associated with IAV results in reduced economic performance within swine populations. IAV has been identified as a major contributor to porcine respiratory disease complex (PRDC) and when compounded with other pathogens, losses can exceed $10 per pig.[4,5] Given the economic losses incurred with endemic IAV and the rapid emergence of new strains in the endemically infected swine populations, active surveillance is crucial to finding and implementing effective control strategies.

Efforts to improve surveillance have increased in the last decade but a consistent and dependable system has not been established.[6] Viral isolation can be difficult because without proper storage and transport media, the enveloped virus is easily desiccated.[7] Despite detection via real time RT-PCR, lack of viral isolation and subsequent genome sequencing adds to the ambiguity of the current circulating strains. Current recommendations for detecting IAV in swine herds are to collect epithelial cells and nasal secretions using rayon tipped nasal swabs and performing M gene real time RT-PCR. Nasal swabs are effective at detecting virus as well as providing viable virus for isolation.[8] However, collection of nasal swabs is invasive, labor intensive and requires a large number of samples.[8] Other major ante mortem sampling methods include nasal wipes, oral fluids and air sampling but an understanding of their relative effectiveness is lacking for IAV detection and isolation. Nasal wipes are less invasive than nasal swabs and are therefore less labor intensive but may not provide the same virus yield as nasal swabs.[8] Pen based sampling techniques like oral fluids and air sampling have gained popularity because they are non-invasive and can detect virus with fewer samples. Oral fluids are sensitive for the detection of viral pathogens like IAV however, viral isolation is poor making this strategy less desirable for surveillance.[9,10] Air sampling can also screen for IAV in the environment but has limited evidence for virus isolation.[11]

In addition to variable success with sampling protocols, clinical disease in pigs may be transient. Current guidelines recommend sampling acutely ill, febrile pigs with nasal discharge and cough to get the greatest yield of virus.[2] However, febrile illness signs may not be apparent until several days post infection, after viral load has decreased, and subclinical disease may go completely undetected.[12] Thus, only sampling clinically ill animals will not provide accurate measures of prevalence or identification of all circulating strains in an endemically infected herd.[8,13] Recognizing clinical signs earlier may be beneficial for detecting the virus at peak shedding and new "outbreaks" of IAV. Technological advancements, like cough count monitors in barns, allow producers to recognize clinical respiratory disease earlier and without relying on personnel to monitor the herd for symptoms. Cough count monitors such as SOMOS (Sound Talks Inc) have been shown to detect cough sound signatures from pigs infected with respiratory pathogens like PRRS, *Mycoplasma* spp. and IAV.[14]

Determining the prevalence of circulating IAV strains in US swine herds requires active surveillance strategies. Active surveillance is difficult to implement due to the limitations of sampling techniques and the technical limitations of sampling a statistically relevant number of pigs.[9] Increasing the sensitivity of detection methods without increasing labor intensity will be beneficial in expanding IAV surveillance efforts. This study looks to determine the optimal sampling technique for IAV surveillance in endemically infected pigs.

Materials and methods

Pigs from 0-7 days post weaning from a standard commercial finishing site equipped with SOMO (Sound talks Inc.) recording devices were used in the study. Pigs arrived from a single sow farm known to be endemically infected with IAV. Two barns on the finishing site were selected and 60 pigs from each barn were enrolled in the study. The groups were identified as Pen A and Pen B. Pen A pigs arrived at the finishing site 5 days prior to pigs housed in Pen B. 60 pigs in each pen were individually identified with ear tags and were housed with other pigs not individually enrolled in the study. Pigs were managed according the system's normal protocol.

Twelve samplings, conducted on Monday, Wednesday and Friday each week over 4 weeks were performed. Two forms of pen based/barn-based sampling techniques were employed, oral fluids collection and air sampling. All pigs within the pen were included in group sampling. Oral fluid samples were collected by hanging cotton ropes on a dividing panel within each pen for 30 minutes. Once collected, the oral fluids deposited within the rope were extracted and placed into sterile 50 mL tubes. Samples were then aliquoted into two separate 2mL fractions, one diluted 1:1 with RNAlater (Thermo Fisher Scientific) (OFRL) and one without a diluent (OFNE) for submission. A cyclonic filter collection device (ACD-200 BOBCAT Air Sampler- Drexel, MO) was used to collect an air sample (AS) from the same location in each barn for 30 minutes during each sampling. Once the filtration run was complete the filter was removed and rinsed using the provided filter reagent into a sterile cup.

Individual samples were collected of the identified pigs within each pen. Nasal wipes (NW) were collected using a previously described technique.[8] Briefly, a sterile 5.08cm × 5.08cm cotton gauze square was wiped across the pigs snout, ensuring contact was made with the external nares. Once collected, the gauze squares were placed into a 15mL Eppendorf tube containing 5 mL of viral transport medium composed of Brain Heart Infusion Broth with Penicillin G sodium salt and streptomycin sulfate. These samples were then vortexed and aliquoted into 1 mL samples for submission. Nasal swabs (NS) were collected using the standard described method by introducing a sterile rayon micro-tipped swab into the nasal cavity of the pig. Swabs were then placed in 1 mL of RNAlater (Thermo Fisher Scientific) in a sterile 2 mL tube. All biologic samples were transported on ice. Once processed, samples were stored at -20° C until all samples could be submitted to the St. Jude's Children's Research Hospital's infectious disease laboratory to undergo standard RT-PCR analysis for the M gene.

All samples yielding RT-PCR cycle threshold (CT) values > 38 were considered negative for IAV. Samples that yielded CT values < 38 were considered positive for IAV. Daily respiratory distress index (RDI) as recorded by the SOMO recording devices for each barn were also reported. RDI or cough count score is calculated based on an algorithm that assigns a linear score to the cough sound signatures collected during the time period of interest.

Results and discussion

All sampling methods detected the presences of IAV on multiple days during the study. NW and NS however were the only techniques to detect the presences of virus each day. Mean daily prevalence for NW was 32.6% and 48.3% for Pen A and Pen B respectively (Table 1). In comparison, NS mean daily prevalence for Pen A and Pen B was 28% and 33% (Table 1). AS were found positive for 33% of the days within Pen A as compared to 30% of the days within pen B. Samples were positive 50% and 58.3% of the time for OFNE and 58% and 25% of the time for OFRL for pen A and B respectively (Table 1).

RDI was compared to the calculated daily prevalence in each pen for each of the individual sampling methods. RDI was observed to be highly variable and episodic. A relationship between RDI and NS can be observed as they move in concert over the first 21 days of the study (Fig. 1). After this time, RDI begins to decline while NS prevalence increases, ending the previously observed relationship. No statistical relationship however could be found between RDI and NS. NW prevalence was not associated with changes in RDI. If it is assumed that an increased RDI is associated with the presence of clinical disease, NW appears to find virus even in the absence of clinical signs. There was no association between AS, OFRL, or OFNE positive results and RDI scores, NW or NS daily prevalence.

Under these conditions, NW detected higher mean daily prevalence of IAV compared to NS. The lowest prevalence measured by NW within the study was 6.7% compared to a prevalence of 1.7% measured by NS. Meaning NW had a higher likelihood of detecting positive animals. Assuming the measured minimum prevalence, 42 NW samples would be required to achieve a 95% confidence in the detection of at least 1 positive sample, while 160 NS samples would be required to achieve the same level of confidence. Within this study both NS and NW would have detected a positive with 95% confidence on 10 of 12 sampling days.

While nasal swabs are thought to be the gold standard for virus detection and subsequent viral isolation our study reveals that nasal wipes may actually prove more sensitive within populations endemically infected with IAV. Pen based collection methods, although validated in the literature as a screening tool, resulted in unreliable detection of IAV under these conditions. While further research is needed to compare the quality of virus collected by both techniques for further sequencing, nasal wipes prove to be the best sampling technique to identify the presence of disease with the fewest number of samples. These data suggest that the optimal strategy when sampling for IAV within endemically infected populations is to collect 45 nasal wipes to confidently identify the presence of the pathogen.

References

1. Webby, R. J., Rossow, K., Erickson, G., Sims, Y. & Webster, R. Multiple lineages of antigenically and genetically diverse influenza A virus co-circulate in the United States swine population. 103, 67–73 (2004).

2. USDA/APHIS. Influenza A Virus in Swine Surveillance. (2018).

3. Simon-grifé, M. et al. Swine influenza virus infection dynamics in two pig farms ; results of a longitudinal assessment. Vet. Res. 43, 24 (2012).

4. Haden, C., Painter, T., Fangman, T., Holtkamp, D. Assessing production parameters and economic impact of swine influenza, PRRS and *Mycoplasma hyopneumoniae* on finishing pigs in a large production system. Proc AASV (2012).

5. Rose, N. et al. Dynamics of influenza A virus infections in permanently infected pig farms : evidence of recurrent infections, circulation of several swine influenza viruses and reassortment events. 1–14 (2013). doi:10.1186/1297-9716-44-72.

6. Vincent, A., Awada, L., Brown, I., Chen, H., Claes, F., Dauphin, G., ... Ciacci-Zanella, J. (2014). Review of Influenza A Virus in Swine Worldwide: A Call for Increased Surveillance and Research. *Zoonoses and Public Health, 61*(1), 4–17. https://doi.org/10.1111/zph.12049.

7. Spickler, A.R. 2016. *Swine Influenza*. Retrieved from http://www.cfsph.iastate.edu/DiseaseInfo/factsheets.php.

Table 1: Results of IAV sampling by sample day and pen. NS-nasal swabs, NW-nasal wipes, AS-air samples, OFNE-oral fluids without transport media, OFRL- oral fluids in RNAlater, RDI-respiratory distress index

	Pen A						Pen B					
Study Day	**NS**	**NW**	**AS**	**OFNE**	**OFRL**	**RDI**	**NS**	**NW**	**AS**	**OFNE**	**OFRL**	**RDI**
0	20.0%	23.7%	+	-	+	0.48	85.0%	50.0%	+	+	-	0.80
1						0.24						0.55
2	36.7%	15.3%	-	-	+	0.06	20.0%	51.7%	-	+	+	0.12
3						0						0.24
4						0						0.09
5	11.7%	6.70%	+	+	-	1.06	1.70%	6.70%	-	-	-	0.87
6						1.96						1.51
7	26.7%	8.30%	-	-	-	2.2	1.70%	16.7%	-	+	-	0.63
8						2.14						0.92
9	30.0%	33.3%	+	+	-	2.72	41.7%	40.0%	+	+	+	0.77
10						1.41						0.96
11						0.43						0.16
12	1.70%	41.7%	-	-	-	0.44	1.70%	56.7%	-	-	-	0.28
13						0.57						0.45
14	26.7%	63.3%	-	-	-	2.94	5.00%	66.7%	-	+	-	1.47
15						1.81						0.41
16	21.7%	50.0%	+	+	+	1.75	35.0%	73.3%	+	-	-	0.94
17						1.83						0.61
18						2.09						0.68
19	30.0%	13.3%	-	+	+	0.88	40.0%	46.7%	-	+	-	0.28
20						0.57						0.23
21	43.3%	61.7%	+	+	+	1.22	41.7%	58.3%		-	-	0.08
22						0.83						0.15
23	55.0%	11.7%	-	+	+	0.87	61.7%	23.3%		-	-	0.18
24						0.97						0.38
25						0.81						0.06
26	33.3%	61.7%	+	-	+	0.23	70.0%	90.0%	-	+	+	0.08
27						0.58						0.18
Mean	**28.1%**	**32.6%**	**33.3%**	**50.0%**	**58.3%**	**1.21**	**33.8%**	**48.3%**	**30.0%**	**58.3%**	**25.0%**	**0.42**
Min	**1.70%**	**6.70%**				**0**	**1.70%**	**6.70%**				**0.06**
Max	**55.0%**	**63.3%**				**2.94**	**55.0%**	**90.0%**				**1.51**

8. Nolting, J. M., Szablewski, C. M., Edwards, J. L., Nelson, S. W. & Bowman, A. S. Nasal Wipes for Influenza A Virus Detection and Isolation from Swine. 1-5 (2015). doi:10.3791/53313.

9. Decorte, I., Steensels, M., Lambrecht, B., Cay, A. B. & De, N. Detection and Isolation of Swine Influenza A Virus in Spiked Oral Fluid and Samples from Individually Housed , Experimentally Infected Pigs : Potential Role of Porcine Oral Fluid in Active Influenza A Virus Surveillance in Swine. 1–15 (2015). doi:10.1371/journal.pone.0139586.

10. Romagosa, A., Gramer, M. & Joo, H. S. Sensitivity of oral fluids for detecting influenza A virus in populations of vaccinated and non-vaccinated pigs. 110–118 (2011). doi:10.1111/j.1750-2659.2011.00276.x.

11. Corzo, C. A., Culhane, M., Dee, S. & Morrison, R. B. Airborne Detection and Quantification of Swine Influenza A Virus in Air Samples Collected Inside , Outside and Downwind from Swine Barns. 8, 1–7 (2013).

12. Janke, B. H. Diagnosis of swine influenza. J. Swine Heal. Prod. 8, 79–84 (2000).

13. Kyriakis, C., Zhang, M., Wolf, S., Jones, L. & Shim, B.-S. Molecular epidemiology of swine influenza A viruses in the Southeastern United States, highlights regional differences in circulating strains. Vet. Microbiol. 211, 174–179 (2017).

14. Polson, D. D., Playter, S., Berckmans, D., Stoffel, A., & Quinn, B. (2018). Precision livestock farming (PLF) for pig health and production : Sound as a diagnostic sample, 3–6.

Figure 1: The mean daily prevalence for nasal swabs (NS) in Pen A plotted over time, compared to the respiratory disease index (RDI) scores for each day.

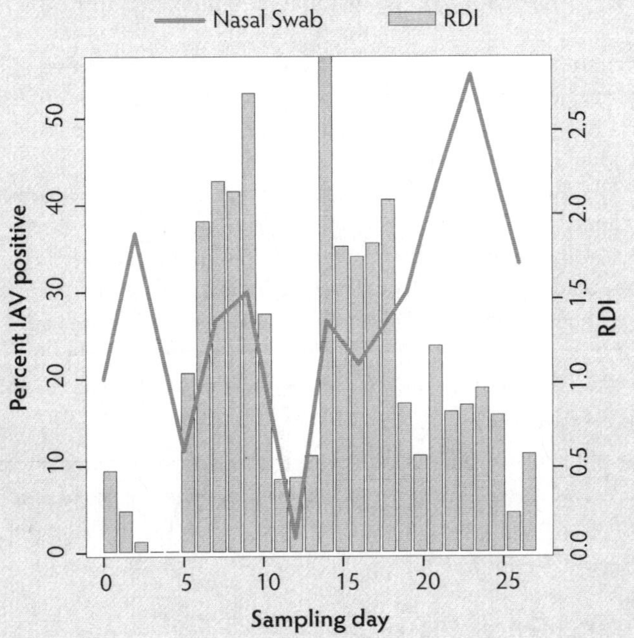

A spatial and temporal analysis of cull sow movements

M. Hermes[1]; B. Blair[1]; J. Lowe[1,2]

[1]University of Illinois College of Veterinary Medicine, Urbana, Illinois [2]Lowe Consulting Ltd., Mahomet, Illinois

Introduction

Assuming a cull rate approaching 50% within the US swine industry more the 3.2million cull sows are expected to enter the marketing channel annually. This vast number of cull sows in conjunction with the limited number of federally inspected cull plants (FIP), leads to co-mingling of sows from varying sources as they prepare to enter these plants. Unlike that of the lean hog market, where greater than 95% of animals move direct farm to plant, the cull sow marketing network is unique as the large majority of sows, greater than 90%, are sold to an intermediary collection point (CP) prior to being moved to slaughter. This structure within the cull sow market exist for many reasons, such as a propensity of some FIP's for a certain type of sow or as a convenient route for sow farms to sell cull sows as the nearest FIP may be many states away. An intermediary CP allows for both to occur, as many collection points are located in sow dense regions and allow sows to be sorted by weight and type creating full loads to move to the FIPs. While benefits of the current market structure exist, drawbacks exist as well. The consolidation of cull sows within a region into a singular CP and then again into one of the 17 FIP in the US creates an extreme amount co-mingling. This co-mingling poses a threat to the US swine industry, as diseases harbored across many different farms are now present at a few select locations.

While the structure of the cull marketing channel allows for disease introduction risk, the complexity of the system does as well. Sows within the cull marketing system are transported at least 2 times. They are culled from the source farm (SF), transported to an intermediate CP, and then move again to a FIP. While most sows most likely follow this trajectory, previous work hypothesized[1] that up to 14% of all cull sows make an additional stop, as they move between different buying stations. This hypothesized extended period of time may allow for the replication of disease within the channel, creating a reservoir population. This allows the opportunity for disease to spread within the marketing channel, posing a risk as trucks move animals from farm to CP. During the porcine epidemic diarrhea virus (PEDV) US outbreak of 2014, studies[2] showed that trailers could become contaminated with PEDV while unloading at a slaughter facility.[2] Other studies,[3] have shown that the exposure of naïve pigs on a trailer contaminated with $\geq 10^3$ $TCID_{50}$/mL of porcine respiratory and reproductive syndrome (PRRS) for 2h led to the infection of the animals. From studies such as these the risk that animal movements from a farm to a highly comingled point may pose extravagant risk if lapses animal movement related biosecurity exist. Lapses in areas such as truck washes or during the loading or unloading of animals at the farm or the CP, all create the opportunity to carry disease back to a system or individual farm.

Both the structure and complexity of the cull sow marketing channel make understanding how and why sows move through the channel of outmost importance. Previous work[1] observed the national scope of the cull sow market with sows from 21 states, 297 premises entering 1 plant over a course of 1 week. The combination of the systems structure, national scope and complexity creates an opportunity for the formation of unmanaged reservoir population, making every contact within marketing channel one of high risk. Because of this in light of the threat of foreign animal disease, further knowledge of this market must be obtained to mitigate risk and preserve business continuity in the event of a disease outbreak.

The objective of this study was to create the most comprehensive cull sow data set to date. Then, use the data generated to describe both the spatial and temporal movement of cull sows within the US marketing channel.

Material and methods

Data was collected with the help of the USDA/APHIS Brucellosis Laboratory. Premise ID tags submitted in association with brucellosis surveillance samples were collected. Premise ID's collected represented 7 cull slaughter plants: Tyson (Hillshire/Dean), Newbern, TN; Bob Evan Farms, Xenia, OH; Johnsonville Sausage, Watertown, WI; Johnsonville Sausage (Banner Creek), Holton, KS; F.B. Purnell Sausage, Simsonville, KY; Caughman's, Lexington, SC; J&B, Pipestone, MN. Tags were collected 1 week each month for a total of 6 months: 5/18, 6/18, 7/18, 2/19, 3/19, and 4/19. Sow ID, Premise ID, state, plant and kill date was recorded for each tag collected. Latitude and longitude, for each premise, was identified through a public database, and then found on Google Maps for location confirmation. Premises belonging to participating production systems were distributed and cull dates of individual sows identified.

Total time spent in the market channel was defined as kill date less cull date. Distance traveled from source farm to FIP, and the time spent in the marketing channel was calculated for each sow. The average straight line distance traveled to each plant, the number of states represented, and the number of premises represented was calculated for each plant within the study. Relationships between plant capacity, season, price, and the previously mention variables were explored, through the use of correlation. All data analysis and data visualization was performed in R statistical software.[4]

Results and discussion

In total 17493 individual sow tags were collected representing ~8.4% of the national kill, from 7 cull sow plants responsible

33% of the national kill. This means that data was collected of ~25% of sows entering one of the 7 plants during the study timeframe. Sows traveled a mean distance of 534km +/- 363km (Figure 1) representing 32 states and 1211 unique premise. 59.7%, 33.4%, and 6.9% of sows from a single premise traveled to 1, 2, or 3+ plants respectively. 16.9% of individual sows have reported cull dates. This showed that 66.1% of all sows were in the market channel for ≤ = 3days. 24.3% for 4-5 days, and 9.6% for > 5days with a mean of 3.08 days and a maximum of 40days spent in the channel.

Of the 7 plants represented the average distance traveled by a sow varied by plant, with sows entering J&B traveling the least 189.4km and sow entering Tyson the most at 736km (Table 1). 13 states were represented within a state on average, with a maximum of 29 states and a minimum of 1 state represented within a singular plant. On average 220 premise IDs were collected from a cull plant in the study, with a maximum of 483 and a minimum of 25 unique IDs being observed. No correlations could be found between plant capacity, season, price and number of states, distances, or premises.

Figure 1: The distribution of distances sows traveled to arrive a cull sow plant.

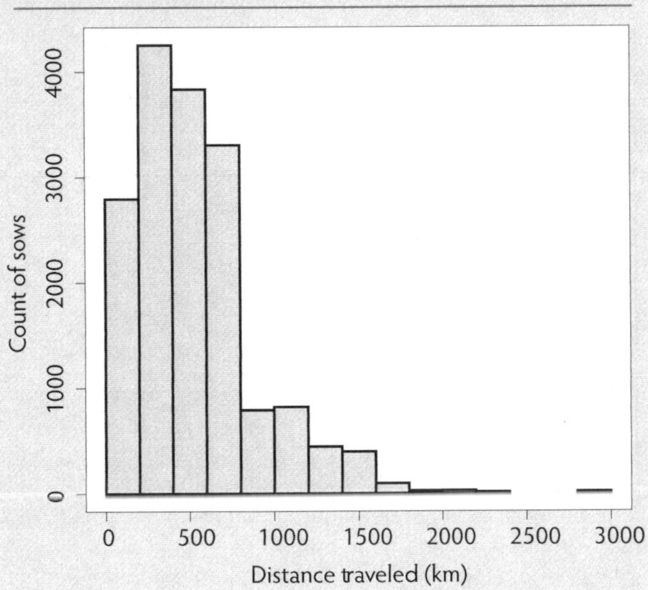

Although this data is not a complete view of the cull marketing channel, we believe this to be the most comprehensive view of the US cull sow marketing channel ever performed, with nearly 18000 sows identified over 6 week study. This work supports the findings of previous work completed[1] showing the national scope of the cull sow market. With the average plant collecting cull sows from 13 states and over 220 premises, the opportunity to spread disease, in a sporadic nature, across the US exist. With 32 states and over 1200 farms represented within the 7 plants enrolled within the study, makes apparent the amount of comingling that must occur at the plant or before arriving at the FIP. This raises concerns regarding the risk of disease movement within the cull sow marketing channel or the movement of disease back to farms or systems. Because of the national nature of the cull sow market and the consolidation that must occur prior to arrival at FIP, complex indirect contact between farms in many states. Because of these indirect contacts, the risk of disease introduction and the possible source of disease introduction can be hard to pinpoint.

This study also confirms the hypothesis that a small but significant percentage of sows remain in the cull sow marketing network for an extended period of time.[1] With an average of 3.08days spent within the marketing channel and 9.6% of all sows remaining longer than 5 days, disease not only has the opportunity to move within the network as mentioned previously but has the opportunity to replicate within cull sow over these extended periods of time. The replication of disease within this unmanaged population of animals produces the opportunity for the formation of a reservoir of disease. This reservoir can complicate disease trace back efforts from routine surveillance within FIPs as both the time and pathogen load exist within the channel for animals to become infected while in transit to the FIP. In the case of a foreign animal disease outbreak, this reservoir may hinder eradication efforts of not properly managed as every sow farm in the US would have indirect contact to the cull sow marketing network.

Table 1: Average distance, number of unique states, and number of unique premises represented at a cull sow plant.

Plant	Average distance (km)	# States	# Premises
Banner Creek	438.4	16	264
Bob Evans	548.8	7	119
Coughman	291.5	1	25
J & B	189.4	4	31
Johnsonville	526.6	10	179
F. B. Purnell	484.2	24	440
Tyson	736.0	29	483

This spatial and temporal description of the US cull sow marketing network provides the evidence to suggest that the possibility exist for both the spread of disease and the propagation of disease within the cull sow marketing channel. Information collected during this study and more like it must be utilized to generate a further understanding of the complex relationships that exist regarding the movement of animals within this network. Without a better understanding of the dynamic nature of this system, the impacts of events such as a stop movement order on a state or population may not be fully understood. Without more complete understanding of the effects such an event may have on the US cull sow marketing network, efficiently and quickly controlling the spread of a disease may prove impossible.

References

1. Blair, B. & Lowe, J. Describing the cull sow market network in the US : A pilot project. Prev. Vet. Med. 162, 107–109 (2019).

2. Lowe J., Gauger P., Harmon K., Zhang J., Connor J., Yeske P., Loula T., Levis I., Dufresne L., Main, R. Role of Transportation in Spread of Porcine Epidemic Diarrhea United States. *Emerg. Infect.* Dis. 2014;20(5):872-874.

3. Dee, S. A., Deen, J., Otake, S. & Pijoan, C. An experimental model to evaluate the role of transport vehicles as a source of transmission of porcine reproductive and respiratory syndrome virus to susceptible pigs *Can J Vet Res* 2004;128–133.

4. R Core Team, 2017. R: A language and environment for statistical computing. R Foundation for Statistical Computing, Vienna, Austria. `https://www.R-project.org/`.

A fluid dynamics approach for the evaluation of chemical disinfection via fogger

R. Schultz[1]; B. Blair[1]; J. Lowe[1,2]
[1]University of Illinois College of Veterinary Medicine, Urbana, Illinois; [2]Lowe Consulting Ltd., Mahomet, Illinois

Introduction

Biosecurity remains an important pillar within the swine industry in the prevention of infectious disease introduction into swine populations. Many routes have been implicated as means of pathogen entry, including the entry of animals, people, supplies or equipment. Strategies have been developed to help farms to mitigate this, such as Danish entry or isolation facilities, and while most techniques have proved beneficial, methods commonly used to deal with supply/equipment entry such as UVC or chemical disinfection, currently lack the scientific evidence proving their effectiveness.

Chemical disinfection, a method typically used to deal with large supply or equipment entries into a farm, have used varying classes of disinfectants such as potentiated hydrogen peroxide or quaternary ammonias. Both possess the ability to deactivate pathogens when applied directly to the infected surfaces at a high enough concentration and long enough duration of contact.[1-4] However, questions still remain about delivery techniques employed on farms to apply these disinfectants and the amount of surface coverage it creates.[5] One common technique utilizes a commercial fogger to aerosolize disinfectant within a room in an attempt to contact all sides of the supplies. However some research exist that demonstrated that disinfection of a surface by use of a fogger is dependent on the orientation of the surface in relation to the fogger.[5] This study raises the question, how effective foggers are at contacting all sides of a box? Boxes within the fogging chambers are often stacked or placed side by side as to increase the number of items within the chamber possibly decreasing the contact made by the disinfectants to the sides of the boxes. As previously discussed it is imperative that contact is made on all parts of the item when using these aqueous chemical disinfectants if the goal is to prevent the entry of contaminated items into farms. The objective of this project is to develop a computer simulated model to evaluate the delivery of chemical disinfectant by a commercial fogger through the use of computational fluid dynamics (CFD), allowing for the evaluation of fogging chambers under many different constraints and scenarios.

Materials and methods

3-d models were created of the disinfection chambers, which was made up of 4 components: the chamber, a wire shelving unit, cardboard box, and Hurricane Ultra II fogger. Each component was drawn to scale by use of the computer-aided design (CAD) software Autodesk Fusion 360. 3 disinfection chambers of varying sizes were drawn: a 1.52 × 1.52m, 1.5 × 3.0m and 1.5 × 6.1m all 2.4m in height. A shelving unit measuring 0.61m × 1.22m × 1.83m was place at the center of each chamber. 1 box measuring 0.36m × 0.36m × 0.36m, was placed in the middle of the center shelf. The Hurricane Ultra II fogger was as drawn to scale and placed on the floor in the corner of the chamber. The fogger was aimed directly at the center of the box. The full configuration of the one of the chambers can be seen in Figure 1. Once the models were completed they were uploaded into Autodesk CFD 2019, to undergo simulation. Boundary conditions of the 3 inlets, the fogger nozzles, were set so velocity at the inlet was equal to 6007ft/min as measured by a wind vane anemometer. Results of the model were reported as the sheer stress generated on each box surface. It was assumed that sheer acted as fair representation of contact as contact must occur to produce sheer. These results were captured as a color-coded image of the amount of sheer generated on each side, with dark blue being no sheer to red being the most sheer. 6 images were captured for each simulation representing each side of the box. Simulations were run under steady state conditions as to decrease the amount of computing power needed to generate results. Steady state conditions allow for a conservative look at the amount of contact generated by a fogger, as it continues to run until no further change in sheer is generated.

Once all simulations were complete the results generated were subjected to image analysis using the colorsys library within the python programming environment.[6] The goal of this analysis was to determine the percent coverage of the box. The way in which this was performed was by capturing the RGB color code associated with each pixel within the image. Pixels matching the color code of the color associated with 0 sheer stress (0, 0, 153) were considered to not have made contact with the disinfectant. Results were reported as the percent coverage of a side or the percentage of pixels not equal to 0.

To validate the CFD models a live simulation of a 1.4 × 1.4 × 2.4m room with one centralized box was compared to a CFD model of the same specifications. The room was set up with the box in the center of the room, the fogger was placed in the corner of the room and set to directly hit the center of the single box which was placed on a wire box to simulate the wire shelving unit. The fogger was run for a total of ten minutes. Coverages during the live simulations were calculated by applying a highlighter solution through the fogger. After the trial was run, pictures were captured under black light to be used to calculate the coverage on each side of the box. Images were taken of each side of the box, all from the same distance, and these images underwent RGB analysis similar

Figure 1: CAD drawing of a 1.52 × 1.52 × 2.44m fogging chamber used within the model.

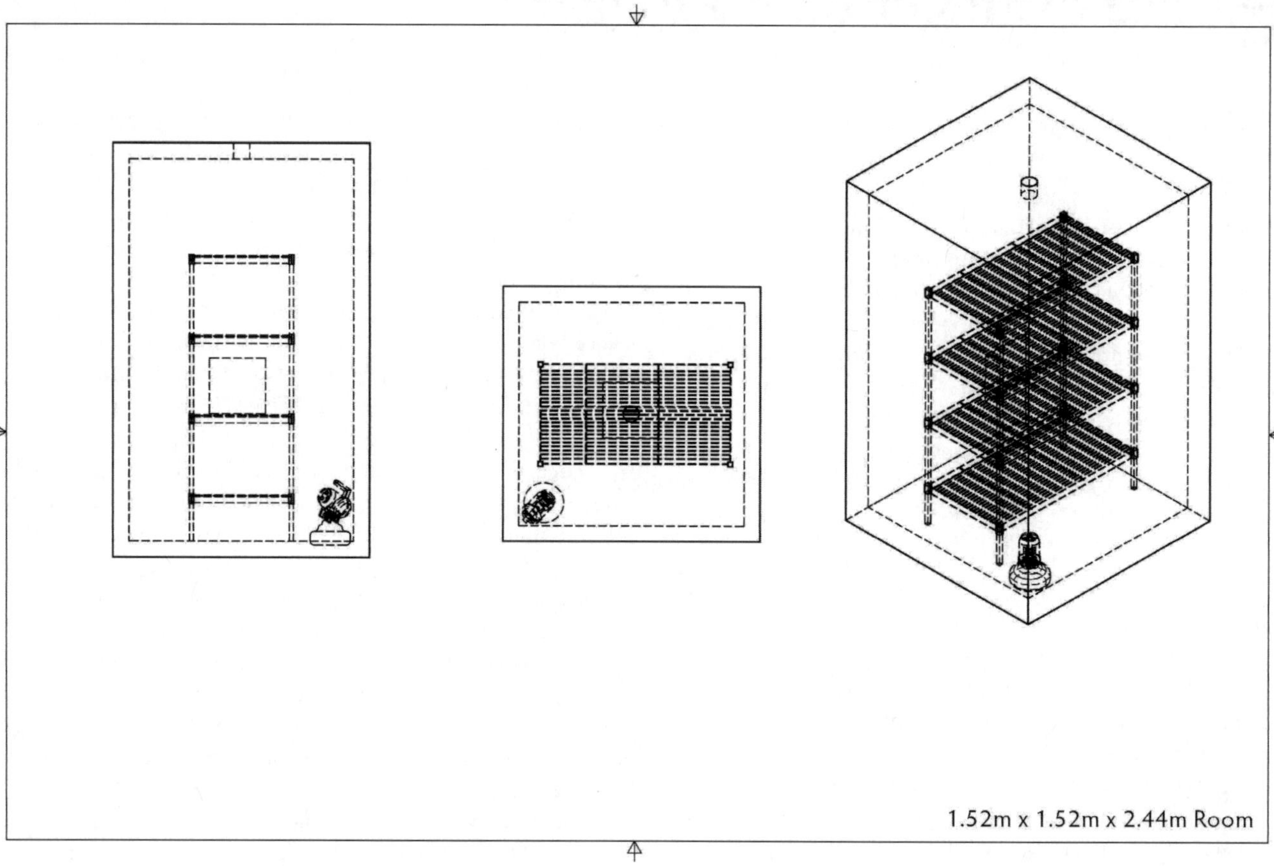

to that previously described to determine the proportion of pixels covered by highlighter compared to the portion of the box that was not covered by the highlighter solution. 3 differing cut points were used to determine if a pixel was to be considered covered in highlighter: 15, 20, and 25% of the maximum green value. Each were then compared to the proportion of each representative surface undergoing sheer in the CFD model for a comparative analysis.

Results and discussion

Model validation (Fig. 2) revealed the correlation between the coverage of live simulations and CFD model ranged from 0.986-0.992. When the cut value is set to equal 15% of the maximum possible green value for each pixel the correlation value was equal to .992, with the mean coverage by the live simulation equal to 77.5% and the model equal to 75.1%. From this it can be concluded that the model and the live simulations conclude approximately the same results and acts as a sufficient means to assess the performance of a fogging chamber.

Total coverage, as calculated by the model, varied across the 3 configurations ranging from 0.1%-83.1% (Table 1). The box in the 1.5 × 1.5m chamber is seen to have the most coverage (Fig. 3) with a maximum coverage of 100% on the left side of the box nearest the fogger, while only 54% coverage was achieved on the back of the box. The total coverage within the 1.5 × 3.0m chamber was increasingly worse (Fig. 4). A maximum coverage of 31% achieved on the front of the box. While only 3% of the back of the box came into contact with disinfectant. Finally, within the largest room (Fig. 5), 1.5 × 6.1m, coverage is nearly non-existent with maximum coverage observed on the front of the box at 0.38% and while all other side saw 0% coverage. Overall, the front side of the box is consistently seen to have the most coverage while the back of the box is seen to have the least amount of coverage. This concludes that because the fogger is directly pointed at the front side of the box and this side sees the most coverage, the disinfectant acts more like a high velocity mist, undergoing directional motion, than that of a fog dissipating throughout the chamber. Because of this delivery of a disinfectant to a surface is strongly influenced by the orientation of the surface to the fogger.

The results of this experiment reveal the delivery of disinfectant via fogger to the surface of boxes proves grossly inadequate. Even under the most conservative analytical techniques and scenarios the maximum total coverage of a box is only seen to reach 83.1%.

Figure 2: Model agreement with live simulation. Images show the comparison of the front of the box. Table compares calculated coverage at different cut values of green.

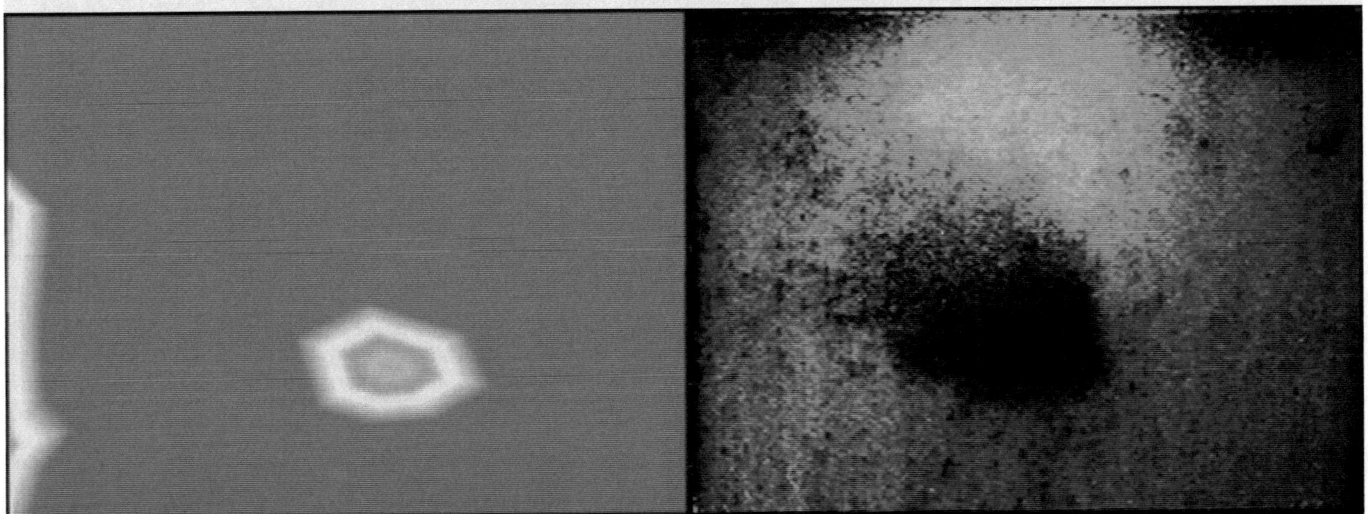

	Model	Sim G=0.15	Sim G=0.2	Sim G=0.25
Front	99.4	96.8	91.1	87.6
Top	93	95.4	92	88.9
Bottom	71.9	77	72.7	67.3
Right	67.2	66.3	61.9	57
Left	87.4	89	87.2	85.2
Back	31.7	40.2	33.3	29.5

Table 1: Percentage of a side covered by disinfectant as calculated by a CFD model.

Chamber size	Front	Near	Far	Top	Bottom	Back	Total
1.5 × 1.5m	88%	100%	88%	81%	88%	54%	83%
1.5 × 3.0m	31%	23%	4%	10%	31%	3%	17%
1.5 × 6.1m	0.4%	0%	0%	0%	0%	0%	0.1%

Large amounts of variation exist across all sides of the boxes and overall there is not enough contact generated between the disinfectant and the box to provide adequate disinfection when delivered under these conditions. It is also observed that as the room becomes larger, the amount of contact created decreases drastically. It can be speculated that because behavior of the disinfectant during delivery, like that of a high velocity mist, it is likely that gravity overcomes the momentum of the particles as they move through the chamber, causing them to fall short of the box. Due to the strong negative effect of both surface orientation and distance, we believe that the conservative approach described to measure the contact generated by a fogger on a box proves that under the constraints presented, foggers serve as an inadequate method to deliver disinfectant to surfaces. We also believe that the CFD modeling techniques defined serves as an adequate method to test the theoretical coverage generated by delivery of disinfectant via gaseous or aqueous solutions. At this time, the data presented within this research would suggest that disinfection of items by use of a fogger within a chamber should be strongly discourage and alternative methods should be explored.

Figure 3: Model generated image of coverage of a box within a 1.5 × 1.5m chamber.

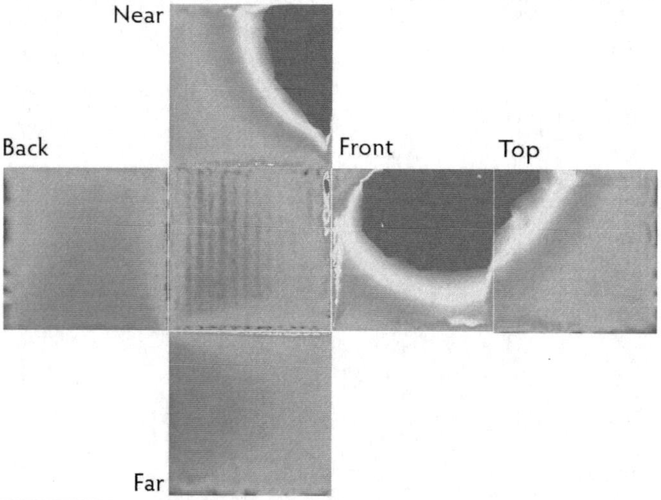

Figure 4: Model generated image of coverage of a box within a 1.5 × 3.0m chamber.

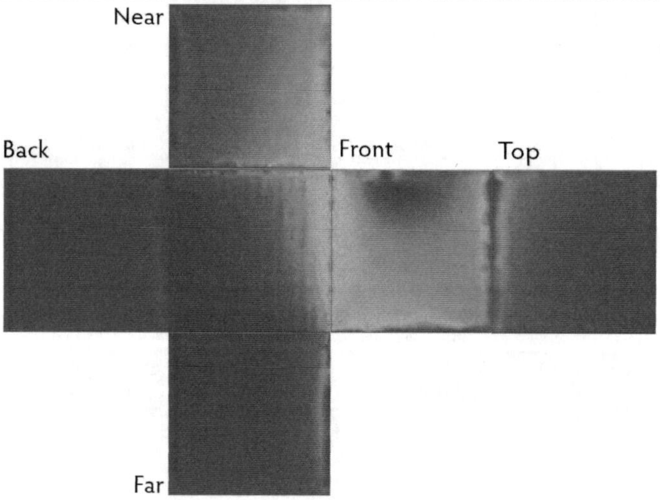

Figure 5: Model generated image of coverage of a box within a 1.5 × 6.1m chamber.

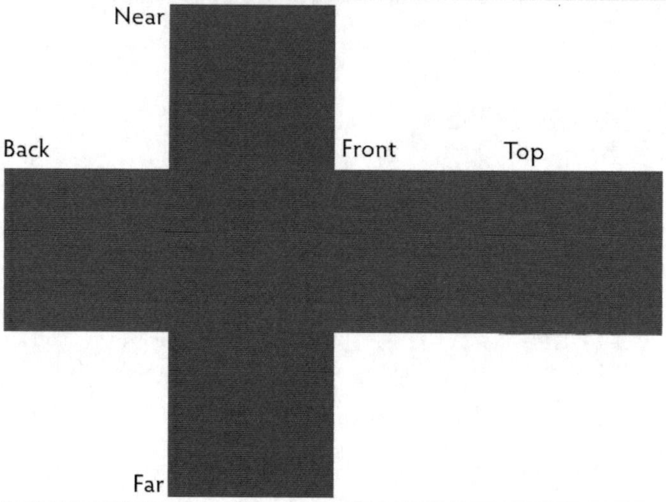

References

1. French, G. L. *et al.* Tackling contamination of the hospital environment by methicillin-resistant Staphylococcus aureus (MRSA): a comparison between conventional terminal cleaning and hydrogen peroxide vapour decontamination. 31–37 (2004). doi:10.1016/j.jhin.2004.03.006

2. Gerba, C. P. Quaternary Ammonium Biocides : Efficacy in Application. 81, 464–469 (2015).

3. Jimenez, L. & Chiang, M. Virucidal activity of a quaternary ammonium compound disinfectant against feline calicivirus : A surrogate for norovirus. 19–21 doi:10.1016/j.ajic.2005.11.009

4. Lineback, C. B. *et al.* Hydrogen peroxide and sodium hypochlorite disinfectants are more effective against Staphylococcus aureus and Pseudomonas aeruginosa biofilms than quaternary ammonium compounds. 5, 1–7 (2018).

5. Kettelkamp, E., Kordas, J., Blair, B., & Lowe, J. (2019). Impact of item orientation on the efficacy of supply decontamination with aerosolized chemical disinfectants. In *Impact of item orientation on the efficacy of supply decontamination with aerosolized chemical disinfectants.* American Association of Swine Veterinarians.

6. van Rossum, G. Python Tutorial. (2016).

Causes of lameness in sows euthanized for lameness

N. Fanzone, BA; J. B. Engiles, VMD, DACVP; K. Wulster, VMD, DACVR; M. Pierdon, VMD, DACAW
New Bolton Center, University of Pennsylvania, Kennett Square, Pennsylvania

Introduction

Sow lameness is a welfare concern and a major economic concern for pork producers. The nature of loose housing makes early detection, assessment, and treatment of lameness difficult. Advanced lameness in sows often leads to involuntary removal, stunting productivity and incurring cost to producers. The prevalence of lameness in sows housed in pens varies by study and is frequently over 20%. A survey of Belgian farms measured lameness prior to parturition and found prevalence up to 23.1%.[1] Another study found a 26.2% prevalence of lameness[2] and another, 22.9% prevalence.[3] The cost of lameness is associated with early removal from the herd and decreased productivity. Anil et al. found that lame sows have a 1.710 times higher risk of removal from the herd within 350 days of lameness assessment, compared with the risk of removal for a non-lame sow. They also reported a lower number of piglets born alive for lame compared to non-lame sows.[4] In order to undertake effective prevention or treatment strategies, it is important to understand the underlying pathological basis of lameness. The objective of this study was to use radiography, gross necropsy, and histopathology to determine the cause of lameness in sows euthanized for lameness in herds using pen gestation.

Materials and methods

Farms agreed to participate and notify us when sows were being euthanized for lameness. Herd veterinarian or barn staff clinically confirmed the lameness in the affected limb or limbs. Sows were euthanized at the farm and all four limbs were removed by disarticulating the coxofemoral joint and excising the scapulae, including intact glenohumeral joints. Limbs were transported to the University of Pennsylvania New Bolton Center where they underwent radiography, gross evaluation, bacteriology, and histopathology. Digital radiographs of distal limbs including carpi and tarsi were taken prior to gross dissection of all appendicular joints and surrounding soft tissues including dewclaws and hooves. Distal interphalangeal (DIP) joints were scored as having a lesion or normal. Gross evaluation of soft tissues, appendicular joints and bones, including bandsaw subsections were performed for all regions of limbs, and representative lesions were fixed in 10% formalin prior to histo-processing, including, when applicable, bone decalcification in an aerated 15% formic acid solution. Lesions were categorized as septic by presence of fibrinosuppurative exudates affecting soft tissues, synovial structures and or bone; osteochondrosis (OC)/osteochondritis dissecans (OCD) by presence of characteristic lesions at predilection sites;[4] or degenerative osteoarthritis (OA) based on characteristic gross lesions affecting cartilage, synovium and subchondral bone. Statistics were done using Stata IC v.15. All data was non-normal. Significance for odds ratios was calculated using fisher's exact test. Mann Whitney U test was used to calculate the difference in front compared to hind lameness.

Results

Limbs from 18 sows (5 herds) from parity 0-6 were collected. Median parity was 1+/-1. Hind limbs were more likely to be affected (17/36) than front (8/36) ($P < 0.05$). The odds of a distal limb lesion being in the lame limb were higher than non-lame limbs (OR = 12.1; CI 3.3-47.6; $P < 0.001$) whereas the odds of a proximal limb lesion being in the lame limb were not different than non-lame limbs (OR = 2.1; CI 0.70-6.3; $P < 0.05$). Odds of a lame limb having a septic process were higher than a non-lame limb (OR = 65.7; CI 11.1-631.9) ($P < .001$) whereas odds of a lame limb having OCD were not different than non-lame limbs (OR = 0.91; CI 0.28-2.9) ($P < 0.05$). Clinical lameness was attributed to OCD in 1/18 animals while lameness in the remaining 17/18 animals was attributed to septic processes. Culture recovered mixed growth. *Trueperella pyogenes* was isolated in 10 of 13 animals (76.9%). Radiography of DIP joints, using microscopic and gross analysis for confirmation, showed 66.7% sensitivity and a 99.1% specificity with a 6.3% incidence of lesions in the DIP joint.

Discussion

Septic lesions were identified in lame sows more commonly than OC, OCD, or OA and was more often correlated with the clinically lame limb resulting in their euthanasia. The high occurrence of lameness in the distal limb and cultures showing mixed bacterial growth and high frequency of *T pyogenes*, an opportunistic pathogen, suggests these lesions may have a traumatic origin rather than a systemic bacterial infection. The frequency and type of lesions identified in these sows differ from previous reports[5,6] and indicates the need for further investigation of their etiopathogenesis, earlier detection methods such as radiology and either prevention or early antibiotic treatment.

References

1. Pluym, L., Van Nuffel, A., Dewulf, J., Cools, A., Vangroenweghe, F., Van Hoorebeke, S. and Maes, D., 2011. Prevalence and risk factors of claw lesions and lameness in pregnant sows in two types of group housing. Veterinarni Medicina, 56(3), pp.101-109.

2. Anil, S.S., Anil, L., Deen, J., Baidoo, S.K. and Walker, R.D., 2007. Factors associated with claw lesions in gestating sows. Journal of Swine Health and Production, 15(2), pp.78-83.

3. Zurbrigg, K. and Blackwell, T., 2006. Injuries, lameness, and cleanliness of sows in four group-housing gestation facilities in Ontario. Journal of Swine Health and Production, 14(4), pp.202-206.

3. Anil, S.S., Anil, L., Deen, J., Baidoo, S.K. and Walker, R.D., 2009. Effect of lameness on sow longevity. Journal of the American Veterinary Medical Association 235(6), pp.734-738.

4. Jensen, T.B. and Nils, T., 2009. Causes of and predisposing risk factors for leg disorders in growing-finishing pigs. CAB Rev Perspect Agric Vet Sci Nutr Nat Res 4(4), pp.1-8.

5. Dewey, C.E., Friendship, R.M. and Wilson, M.R., 1993. Clinical and postmortem examination of sows culled for lameness. The Canadian Veterinary Journal, 34(9), p.555.

6. Heinonen, M., Oravainen, J., Orro, T., Seppä-Lassila, L., Ala-Kurikka, E., Virolainen, J., Tast, A. and Peltoniemi, O.A.T., 2006. Lameness and fertility of sows and gilts in randomly selected loose-housed herds in Finland. Veterinary Record, 159(12), pp.383-387.

Lung homogenate optimization for successful *Mycoplasma hyopneumoniae* exposure in gilts during acclimation

M. McMahon[1]; B. Garcia Morante[1], DVM, MS, PhD; A. Betlach[1,2], DVM; C. De Abreu[1]; R. Robbins[3], DVM, PhD; P. Yeske[2], DVM, MS; M. Pieters[1], DVM, PhD

[1]Department of Veterinary Population Medicine, University of Minnesota, St. Paul, Minnesota; [2]Swine Vet Center, St. Peter, Minnesota; [3]Seabord Foods, Guymon, Oklahoma

Problem statement and objectives

Mycoplasma hyopneumoniae causes chronic respiratory disease in swine and is known to economically impact herds worldwide.[1] To control *M hyopneumoniae* in sow herds, gilt acclimation has become a common practice in the United States by use of exposure of seeder animals, tracheal inoculation or aerosol. Gilt acclimation aids to achieve uniform exposure of female pigs prior to entry to the breeding herd, or to create a "Day 0" for disease elimination purposes. One method for gilt acclimation includes the use of herd-specific lung homogenate for exposure.[2-4] However, the characterization, preparation, and handling of lung homogenate has been minimally evaluated, which poses risks for exposure failure.[5] The aims of this study were to determine the optimal conditions for preparation of lung homogenate for gilt exposure to *M hyopneumoniae* by 1) evaluating bacterial load in different anatomical lung sections and 2) the effect of lung homogenate homogeneity on *M hyopneumoniae* detection by real-time PCR.

Materials and methods

A graphical representation of the study experimental design is shown in Figure 1. All procedures and protocols were approved by Institutional Animal Care and Use Committee of the University of Minnesota. In order to evaluate bacterial load in different anatomical lung sections, a total of nine lung donor pigs were selected from three different farms (n = 3/farm). Donor pigs were selected based on Ct values from deep tracheal catheters (DTC) tested for *M hyopneumoniae* by real-time PCR. Pigs were categorized into the following Ct value groups: low ≤ 24 Ct, medium 25-30 Ct, and high 31-39 Ct. Selected pigs were humanely euthanized, followed by observation of lung lesion scores and collection of bronchial swabs from each lung lobe. For each lobe, tissue was blended at a 70:30 proportion of tissue and Friis medium using a Ninja® Blender (© 2019 SharkNinja Operating LLC) to create lung lobe-specific homogenate samples. Real-time PCR for *M hyopneumoniae* detection was performed on bronchial swabs and lung lobe homogenates (5 replicates each).

In order to evaluate the effect of lung homogenate homogeneity on *M hyopneumoniae* detection, lungs were collected from two donor pigs (n = 1/farm) as previously described, and blended whole at a 70 tissue: 30 Friis ratio. The resulting lung homogenate was used to obtain 50:50 and 30:70 dilutions. All three lung homogenate dilutions were further diluted at a 1:10 ratio and treated as follows: i) No filtration, ii) Gross filtration using a nylon stocking and iii) Fine filtration (100µm). Filtered and diluted products were tested for *M hyopneumoniae* via real-time PCR (5 replicates each).

Statistical analysis was performed using ANOVA to compare Ct values of lung homogenates from different lung lobes and different preparation conditions. Additionally, the correlation between Ct values from bronchial swabs, lung lobe homogenates and lung lobe lesion scores were evaluated using Pearson coefficient.

Results

Bacterial load in different anatomical lung sections

Lesions were observed on all evaluated lungs. All lobe-specific bronchial swabs and 85% of lung homogenate samples were positive for *M hyopneumoniae*, regardless of evident lung lesions. Mean Ct values in lung homogenate were significantly lower in low DTC Ct pigs compared to medium and high DTC Ct pigs ($P < 0.0001$), along with less variation of lung homogenate Ct values across all lobes for low DTC Ct pigs. All lung homogenates were positive from pigs with low DTC Ct values, in contrast to pigs with medium and high DTC Ct values, where there were both negative and positive lung homogenates.

Effect of lung homogenate homogeneity on *M hyopneumoniae* detection by real-time PCR

Dilution of the lung homogenate generated conflicting results among samples, without a clear indication of its effect on *M hyopneumoniae* detection by real-time PCR. Regarding filtration methods, both sets of samples showed a within pig significant difference in lung homogenate Ct values for the lung homogenates that were diluted at a 70:30 ratio and filtered either using a nylon stocking or fine filtration (100 µm) ($P \leq 0.046$). The type of filtration, namely gross using a nylon stocking vs fine (100 µm), did not have a significant effect.

Discussion and conclusion

Based on the conditions of this study, the detection of *M hyopneumoniae* in DTC was useful when selecting a donor pig for lung homogenate preparation. Selecting a pig with a low DTC Ct

Figure 1: Study experimental design.

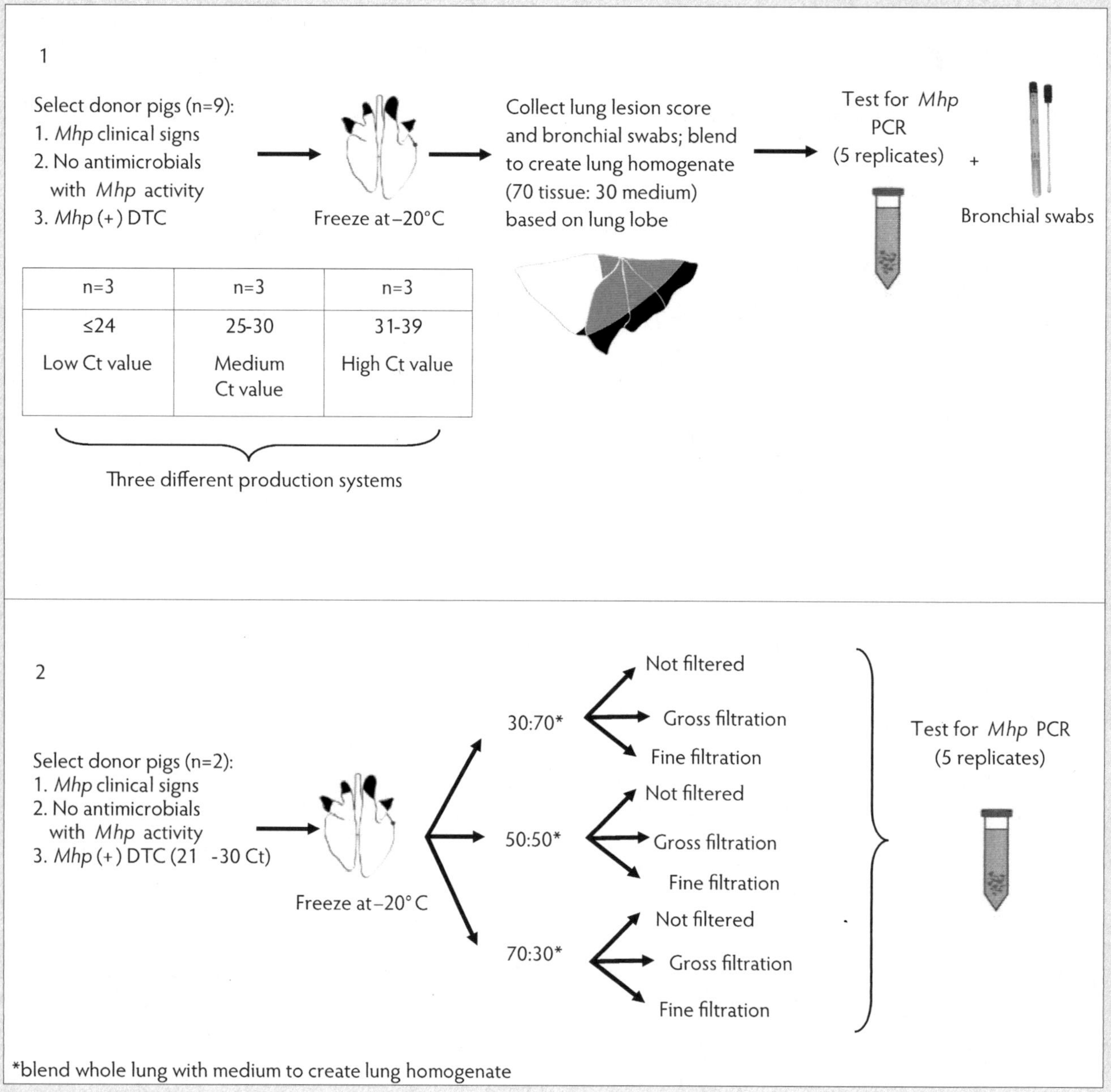

Mhp: Mycoplasma hyopneumoniae

DTC: Deep tracheal catheter

value provided a relatively consistent load of *M hyopneumoniae* throughout the whole lung, as well as a higher bacterial load. In addition, lung lesion scores were not a suitable method for selecting a donor pig, as *M hyopneumoniae* was detected in lobes without any lung lesions and vice versa.

These preliminary results suggested that dilution could have an effect on lung homogenate Ct value when filtration is applied. The lowest dilution resulted in significantly lower lung homogenate Ct values when filtration was applied compared to no filtration. However, type of filtration did not appear to have a significant effect.

It is important to note that this is pilot study and a greater sample size may be needed for definitive results. A larger sample size would yield a more accurate analysis of the trends observed.

In conclusion, these results suggested the selection of donor pigs with low DTC Ct values, as they correlated with greater and consistent bacterial load in the lung homogenate. Our results indicate that all lung tissue could be employed for lung homogenate preparation, regardless of the presence/absence of lesions. If performing a 70:30 dilution, it may be useful to filter the lung homogenate, as for other dilutions filtration may not be necessary. Results from this study can be readily applied in the field to increase the likelihood of successful gilt exposure programs for *M hyopneumoniae* disease control.

This project was funded by the National Pork Board and the Pork Checkoff.

References

1. Goodwin RFW, Pomeroy AP, Whittlestone P. Production of enzootic pneumonia in pigs with a mycoplasma. Vet Rec. 1965;77:1247-1249.

2. Fano, E., Payne, B., 2015. *Mycoplasma hyopneumoniae* gilt acclimation and sow herd stability: essentials to the systematic control approach. In: AASV Annual Meeting. Orlando, Florida. pp. 175–178.

3. Centeno, N., Chévez, J., Fano, E., 2016. Mexican swine industry on *Mycoplasma hyopneumoniae* gilts acclimatation. In: Proceedings of the 24th IPVS Congress. Dublin, Ireland, 31, 2013.

4. Garza-Moreno, L., Segalés, J., Pieters, M., Romagosa, A., & Sibila, M. (2018). Acclimation strategies in gilts to control *Mycoplasma hyopneumoniae* infection. Veterinary microbiology, 219, 23-29.

5. Robbins RC, Betlach AM, Mondragon-Evans MR, et al. Development of a herd-specific lung homogenate for exposure to *Mycoplasma hyopneumoniae* under field conditions. J Swine Health Prod. 2019;27 (4):221–227.

Effect of time and temperature on decontamination of supplies entering a farm

J. Kordas[1]; B. Blair[1]; J. Lowe[1,2]
[1]College of Veterinary Medicine, University of Illinois, Urbana, Illinois; [2]Lowe Consulting Ltd., Mahomet, Illinois

Introduction

Entry of equipment and supplies pose threats of disease introduction to swine farms. Although practices have been established to handle the admittance of such items, questions regarding the efficacy of complete decontamination by use of these techniques still remain. The use of fogging rooms and UV chambers have traditionally been used to pass both personal and supply items into farms. However, recent research has begun to question the effect of both technologies on pathogens commonly found on farm, especially in complex situations where virus may be shielded by organic material or other supplies.

The efficacy of traditional fogging chambers for decontamination on farms has begun to come into question in recent research. While it is not in question that chemical disinfectants are effective against swine pathogens, disinfection using standard industry foggers demonstrated only partial efficacy in one recent study due to gravitational accumulation of the disinfectant as well as the complexity of the surface to which it was applied.[1] The implication being that foggers do not fully aerosolize the chemicals but generate condensation that retains the physical properties of a liquid, inhibiting its ability to fully permeate the objects within the chamber and leaves areas untouched by the disinfectants.

Additionally, UV disinfection of personal item decontamination for entry onto swine farms has raised mixed reviews as to its practical application on farms. Another study suggests that UVC chamber decontamination may not be homogenous and do not fully inactivate surface pathogens on items entering facilities.[2] Furthering this idea, a 2019 publication concluded that operant conditions had significant impact on the efficacy of such chambers, due to shadowing potential and differing item placement, full decontamination using UVC on salmonella inoculated items was never achieved at times up to 40 minutes.[3]

Other methods of decontamination have since been investigated or proposed, such as fumigation or time/temperature treatment. Disinfection of pathogens by time/temperature treatment has been researched within feed ingredients and on contaminated livestock trailers. Temperatures of 71° C for 10 minutes as well as 20°C for 7 days were adequate in deactivating porcine epidemic diarrhea virus (PEDV) in contaminated fecal samples.[4] Additionally, it was found that the half-lives of viruses within certain feed ingredients were dependent on both the virus and specific ingredient, with soybean meal, and lysine hydrochloride harboring viable virus the longest at up to 26.6 days.[5] While this data is indicative of the possible use of temperature and time to neutralize infectious agents, no research exists on its use for the decontamination of supplies entering a farm.

Materials and methods

The intent of this study was to investigate the use of time/temperature treatment on the decontamination of Salmonella and porcine respiratory and reproductive syndrome virus (PRRSV) on a surface. Salmonella and PRRSV were selected as indicator organisms for salmonella's resistance to environmental degradation and PRRSV's importance to the swine industry. Inoculation of brain heart infusion broth (BHIB), which served as a nutrient medium for each pathogen, was performed by adding one dose of Enterisol® Salmonella T/C. Ingelvac PRRSV MLV vaccine was reconstituted in the BHIB, rather than the provided sterile diluent. 1mL, the equivalent of one vaccine dose, of each inoculated solution was then placed on a smooth plastic surface. Once contaminated, all surfaces were placed in low humidity (< 20%) rooms at the three different temperatures: 4° C, 18° C, and 37°C. Six replicates of each treatment group were performed. Infected surfaces were swabbed each day for 5 days (PRRSV) or 7 days (salmonella) to monitor for pathogen degradation collecting a total of 216 samples. Surfaces were swabbed using a cotton tipped applicator in a repeatable, systematic fashion to ensure sampling of the entire surface each day.

Once collected, salmonella samples were directly plated from the swabs onto salmonella selective XLD agar and incubated for 24 hours after which time colony counts were performed. Directly plated salmonella vaccine served as a positive control, while a clean plastic surface, subject to each time and temperature treatment served as negative controls for this portion of the study. PRRSV samples were collected, placed in 1ml of sterile water and stored at -20° C until further processing and cell culture could take place.

Once all PRRSV samples were collected, 500uL of sample was used to infect MARC-145 cells cultured to near 100% confluency in 24-well plates. Complete Roswell Park Memorial Institute medium + 5% Fetal Bovine Serum was added as nutrient media to each well to ensure no cell damage would occur due to lack of nutrients, Samples were incubated at 37° C with 5% carbon dioxide and cytopathic effect of each well was assessed at 3 and 6 days post infection. Vaccine reconstituted in BHIB and not subject to time or temperature treatments served as the positive control for this portion of the study while clean surfaces, subject to the same time/temperature treatments as the samples served as the negative controls. All positive controls yielded cytopathic effect while all negative controls yielded no change in cellular structure.

Results and discussion

Complete decontamination of salmonella on of surface was achieved by use of time/temperature treatment when subjected to 37° C for three or more days. Colony count was reduced to one third of the original inoculate after a single day incubation period then fell to 0 on day 3. While impacts were made to salmonella growth at other temperatures complete decontamination of all samples was not achieved during the 7 day study period. At 4° C colony count remained consistent until day 7 where a 50% reduction in colony growth was seen. All 7 day colony counts for the samples held at 18° C did not fall below 83% of the original inoculate (Table 1). Decontamination of PRRSV, a quite fragile pathogen as compared to salmonella, was attained by day 5 at 18°C and within 2 days at 37° C (Table 1). While some deactivation of PRRSV was observed at 4° C by day 4, complete decontamination was not accomplished throughout the five day study.

The results of this study show that the use of time and temperature may be a viable option to decontaminate items upon entry onto farms. While not all items, such as vaccines, should be exposed to the most effective high temperatures, placement of an items at cooler temperatures for a longer period may prove efficacious. Additionally, there are plans for exploring the effect of humidity on the decontamination process. It is inferred that higher humidity levels may increase the time necessary for pathogen degradation, as the gold standard for growth of many bacterial species in an incubation chamber is high humidity at 37° C. Compiling these additional data points as well as expanding the range of temperature/time treatments may allow for the development of a dose response curve to aid in implementing the most effective temperature/time treatment for a specific environment. While further research exploring the effect of temperature/time on different surfaces and pathogens such as herpesvirus and influenza virus must still be completed, this research would suggest that when exposed to 37° C for greater than two days pathogens will become deactivated, while longer times, up to or exceeding seven days, may be necessary for decontamination at lower temperatures.

Additionally, treatment of objects by temperature/time is not inhibited by the orientation or composition of objects like other commonly used techniques, such as fogging. All surfaces within the environment are subject to equal treatment based on time and temperature. With limited chemical use, there are reduced human health safety concerns associated with this proposed method for decontamination. For its lack of safety concerns, ease of use, and efficacy as described by this research, the use of temperature/time exposure must be a consideration when designing biosecurity for the decontamination of items upon entry into a farm.

Table 1: Percent of positive samples after time/temperature treatment

Temp. (C°)	Time (days)	% Salmonella positive	% PRRSV positive
4	1	100	100
4	2	100	100
4	3	100	100
4	4	100	83.3
4	5	100	33.3
4	6	100	
4	7	50	
18	1	100	100
18	2	100	83.3
18	3	100	33.3
18	4	83.3	16.7
18	5	100	0
18	6	83.3	
18	7	100	
37	1	33.3	100
37	2	33.3	0
37	3	0	0
37	4	0	0
37	5	0	0
37	6	0	
37	7	0	

References

1. Kettelkamp, E., Kordas, J., Blair, B., & Lowe, J. Impact of item orientation on the efficacy of supply decontamination with aerosolized chemical disinfectants. Proc AASV. 2019:11-114.

2. Rieland, K., Yang, M., & Torremorell, M. (2019). Evaluation of Efficacy of Ultraviolet Germicidal Chambers in Swine Farms. Swine Disease Eradication Center. Retrieved from https://vetmed.umn.edu/sites/vetmed.umn.edu/files/shmp_20181119.48_evaluation_of_ultraviolet_germicidal_chambers-sciencepage.pdf.

3. Kordas, J., Kettelkamp, E., Blair, B., Lowe, J. Implications of operating conditions in UV light decontamination chambers for items entering swine facilities. Proc AASV 2019: 275-276.

4. Thomas, Paul R.; Karriker, Locke A.; Ramirez, Alejandro; Zhang, Jianqiang; Ellingson, Joshua S.; Crawford, Kimberly K.; Bates, Jessica L.; Hammen, Kristin J.; and Holtkamp, Derald J., "Evaluation of time and temperature sufficient to inactivate porcine epidemic diarrhea virus in swine feces on metal surfaces" (2015). Veterinary Diagnostic and Production Animal Medicine Publications. 120. https://lib.dr.iastate.edu/vdpam_pubs/120.

5. Dee, S. A., Bauermann, F. V., Niederwerder, M. C., Singrey, A., Clement, T., Lima, M. D., ... Diel, D. G. (2019). Correction: Survival of viral pathogens in animal feed ingredients under transboundary shipping models. *Plos One*, *14*(3). doi: 10.1371/journal.pone.0214529.

Screening the vulnerability of porcine reproductive and respiratory syndrome virus (PRRSV) introduction in breeding herds using a short survey

Broc W. Mauch[1]; Gustavo Silva[1]; Gustavo Machado[2]; Derald Holtkamp[1]; Daniel Linhares[1]
[1]Iowa State University College of Veterinary Medicine, Ames, Iowa;
[2] North Carolina State University College of Veterinary Medicine

Introduction

Porcine reproductive and respiratory syndrome (PRRS) continues to plague the swine industry with 20-30% of herds breaking annually. These outbreaks cost the swine industry and its producers $664 million every year. Swine producers continue to offer efforts to increase biosecurity with the intention to provide the best welfare for their pigs and produce a wholesome product for consumers. However, there are still gaps in biosecurity measures when considering prevention of PRRS outbreaks. The industry needs the ability to better predict operation risk of a PRRS outbreak as well as to evaluate farm specific biosecurity. The objective of this study is to measure and benchmark the relationship between key biosecurity aspects and PRRS outbreaks in breeding herds, while validating a short biosecurity screening survey.

Materials and methods

Production systems across the US were invited to enroll at least 5 breeding herds provided that the system would share the AASV or Morrison Swine Health Monitoring Program (MSHMP) weekly status for PRRS for each farm dating back to 2014. A survey of 44 biosecurity measures were completed for each breeding herd. Survey questions were presented in multiple choice and short answer format. Content of the questions revolved around herd demographics, PRRS outbreak history, frequency of risk events, surrounding swine density, transport biosecurity practices, carcass disposal, people movement and other risk events. The answers to these questions were obtained from the production systems' herd veterinarians, farm managers, and herd health management. Three machine learning algorithms were used to identify key biosecurity factors and practices associated with PRRS outbreaks. The most accurate, based on ability to explain variability between farms in the frequency of reported PRRS outbreaks, was selected. Moreover, positive predictive value (PPV) from the models were utilized as a biosecurity score. We investigated the correlation between PPV and the reported frequency of outbreaks.

Results and discussion

A total of 13 production systems were enrolled in the study (188 sow herds), representing 15 states. The best machine learning algorithm predicted PRRS occurrences with an accuracy of 78.5%. The 44 biosecurity measures were ranked according to their contribution to the model prediction. The first four most important variables were devotion to breeding genetic replacements, number of swine premises within a three mile radius, number of breeding females on the premises, and distance to the nearest public road. The probability that a herd had reported an outbreak were higher in farms that had raised breeding animals as genetic replacements. The higher the number of swine facilities within a three mile radius and the closer the farm was to a public road, the more likely the facility would be expected to break with PRRS. Finally, the PPVs were correlated with the observed presences of PRRS outbreaks, $r = 0.72$, P-value < 0.01, with the observed number of outbreaks (x-axis) compared with the model's predicted probabilities of a new PRRS case (Figure 1). The majority of farms who had reported no outbreaks (red) in the past five years had low (< 50%) probability of having a PRRS outbreak while a larger number of PRRS cases were expected for farm who had reported at least one outbreak (blue), with few exceptions.

Figure 1: Outbreak history vs. positive predictive value.

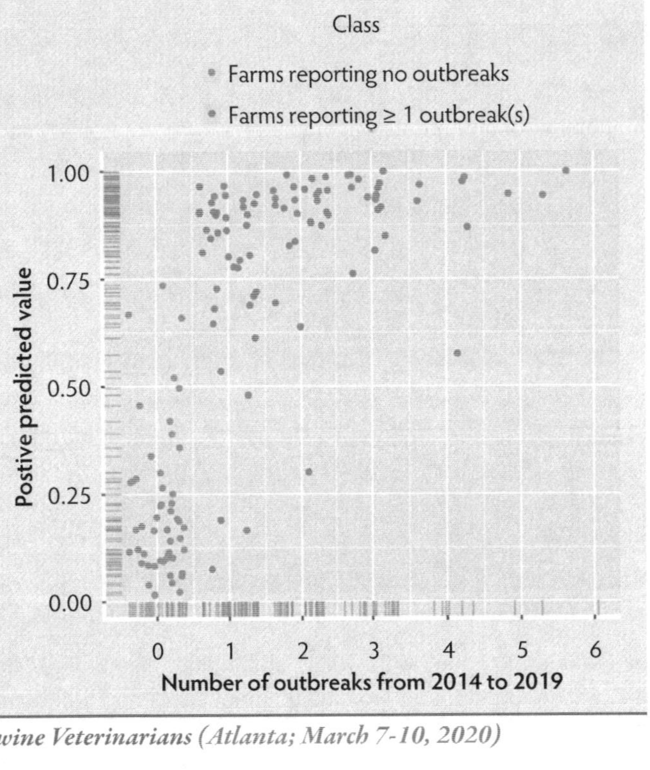

Conclusion

The analysis offers a flexible, shortened approach to screen breeding herd's PRRS biosecurity vulnerability. This study highlights the value of using data to build upon our understanding of biosecurity risk in an operation.

Evaluating the dilution effect for detection of *Mycoplasma hyopneumoniae* in processing fluids by PCR

G. Fry[1], BS; A. Sponheim[2], DVM; D. Murray[3], DVM; L. Johnson[3], DVM; C. Vilalta[4], DVM, PhD; M. Pieters[4], DVM, PhD

[1]Purdue University, West Lafayette, Indiana; [2]Boehringer Ingelheim Animal Health USA Inc., Duluth, Georgia; [3]New Fashion Pork, Jackson, Minnesota; [4] University of Minnesota, St. Paul, Minnesota

Introduction

Mycoplasma hyopneumoniae (*M hyopneumoniae*) can be challenging to detect especially in known positive, low prevalence herds.[1] Current diagnostic strategies are labor intensive and costly. Processing fluids (PF) are used for PRRSV and are a novel population-based method for detection of *M hyopneumoniae* by PCR.[2,3] *Mycoplasma hypopneumoniae* colonizes the respiratory tract of pigs, although it has been isolated outside of the respiratory tract from organs such as the spleen and liver.[4-7] *Mycoplasma hyopneumoniae* has been detected in PF in seven known positive breeding herds within one production system. However, detection of *M hyopneumoniae* in PF is inconsistent and the origin of *M hyopneumoniae* in PF remains unknown.[8,9] Field data suggests detection in low prevalence herds could be impacted by a dilution effect, resulting in suspect results. Assessing detection of *M hyopneumoniae* in pooled litter samples could optimize cost and increase diagnostic confidence of the sample type to correctly classifying herds. Therefore, the goal of this study was to characterize the dilution effect on detection of *M hyopneumoniae* in PF by PCR.

Materials and methods

PF were prepared by placing tails and testicles of pigs collected during processing directly in a Ziploc® bag. The tissues were refrigerated for 24 hours, then only the serosanguinous exudate was poured off and submitted for diagnostic testing.[10,11] Negative PF were collected from a *M hyopneumoniae* naïve herd and confirmed by PCR. A field isolate was diluted with negative PF to a Ct value of 31.23 to represent one positive litter based on data from field samples of a low prevalence, endemic herd.[9] The positive litter was diluted 5, 10, 20, 30, 40 and 50 times to mimic the dilution effect of the different number of litters. Two separate preparations were made, one unconcentrated, representing normally prepared PF, and the other concentrated. Per each dilution, 500µl aliquots were prepared for the unconcentrated preparation (n = 10) and 1000 µl aliquots for the concentrated preparation (n = 10). All aliquots were frozen immediately after preparation. Vacuum centrifugation was used to reduce the volume of concentrated preparations by 50%. All aliquots were submitted to the University of Minnesota's Veterinary Diagnostic Lab for *M hyopneumoniae* PCR. A cutoff Ct of ≤35.99 was considered positive and ≥ 36 to < 40 was considered suspect. Pool sensitivity was calculated by dividing any positive aliquots by the total number of aliquots and the 95% CI was calculated using R.

Results

Ct values for all preparations are shown in Table 1. *Mycoplasma hyopneumoniae* PCR results were as follows: Unconcentrated aliquots, dilutions of 5, 10 and 20 were 10/10 positive (100% pool sensitivity; 95%CI: 69%-100%). Dilution of 30 were 9/10 positive (90% [55%-100%]). Dilution of 40 were 8/10 positive (80% [44%-97%]) and dilution of 50 were 4/10 positive (40% [12%-74%]). Concentrated aliquots, dilutions of 5, 10, 20, and 30 were 10/10 positive (100% [69%-100%]). Dilution of 40 were 8/10 positive (80% [44%-97%]) and dilution of 50 were 5/10 positive (50% [19%-81%]).

Discussion and conclusions

Concentrated preparations exhibited greater detection in dilutions of 30 and 50 versus unconcentrated. Pool sensitivity of all other dilutions was similar. There was no loss in pool sensitivity in unconcentrated samples in dilutions of 5, 10, 20 and in concentrated samples in dilutions of 5, 10, 20, 30. Every sample tested suspect or positive. Suspect results increased with dilution size. Concentration may be beneficial when sequencing a sample with a Ct value slightly greater than 30. Based on the results of this study, PF samples for detection of *M hyopneumoniae* via PCR should be limited to pools of 30 litters or less considering the risk and cost of missing detection. If a suspect is detected, the population should be retested in smaller pool sizes. Current monitoring in field conditions has yielded many suspect results. This study provides one possible explanation for the cause of suspect results. It is important to note that this study used an initial litter based Ct value of 31.23 as representative of field application and based on preliminary data; however, this value may vary within herds. Additionally, a small sample size resulted in a wide 95% CI. An extension of this study would be continuous monitoring via PF of positive farms and farms eradicating *M hyopneumoniae*. Potential future applications could be using PF as a surveillance method for *M hyopneumoniae* negative herds. However, currently it is recommended to collect deep tracheal catheter samples from the breeding herd and test for *M hyopneumoniae* PCR in addition to PF

Table 1: PCR Cycle threshold values for the different dilutions of the concentrated and unconcentrated PF. Values ≥ the threshold of 36 (shaded in grey) are considered suspects.

	Aliquot	Initial Ct value	Litter pool size					
			1:5	1:10	1:20	1:30	1:40	1:50
Unconcentrated PF	1	31.23	32.25	33.26	34.48	35.42	36.49	35.59
	2	31.23	33.01	33.40	34.35	35.72	37.60	35.98
	3	31.23	31.71	33.13	34.83	35.64	35.02	36.73
	4	31.23	31.84	32.71	34.40	35.29	35.42	36.71
	5	31.23	31.39	33.32	35.03	35.12	35.31	36.87
	6	31.23	32.30	33.28	35.03	35.20	35.51	37.04
	7	31.23	32.28	33.69	34.07	35.89	34.85	35.94
	8	31.23	32.47	33.45	34.83	35.42	34.99	35.52
	9	31.23	32.42	32.97	35.06	36.01	35.00	36.58
	10	31.23	32.39	33.00	35.07	35.96	34.47	36.22
Concentrated PF	1	31.23	33.19	34.13	33.68	34.77	35.12	35.85
	2	31.23	33.57	33.53	34.34	34.87	35.55	35.29
	3	31.23	33.16	34.16	34.10	34.48	35.18	36.11
	4	31.23	33.89	33.35	34.26	35.07	35.48	35.79
	5	31.23	32.93	33.70	34.31	34.13	35.27	35.17
	6	31.23	32.08	33.30	34.58	34.29	35.23	36.07
	7	31.23	32.32	33.33	34.30	34.89	35.12	36.86
	8	31.23	32.77	33.17	33.67	35.41	35.23	36.58
	9	31.23	33.09	33.32	34.43	35.59	36.26	36.36
	10	31.23	32.64	33.21	34.04	34.73	36.82	35.63

for confirmation.[1] This is due to a lack of understanding of both the origin of *M hyopneumoniae* in PF and variation in detection between strains. Further work is needed to determine if *M hyopneumoniae* in PF is due to environmental contamination or if the bacterium is present in non-respiratory tissues. Ongoing *M hyopneumoniae* monitoring of breeding herds via PF in smaller litter pools could be adopted to help understand the detection of *M hyopneumoniae* in PF.

References

1. Sponheim A, Fano E, Polson D, et al. A diagnostic approach for detection of *Mycoplasma hyopneumoniae* in low prevalence scenarios. European Symposium of Porcine Health Management Proceedings. 2018; 306.

2. Lopez W, Angulo J, Zimmerman J, et al. Porcine reproductive and respiratory syndrome monitoring in breeding herds using processing fluids. *J Swine Health Prod.* 2018;26(3):146-150.

3. Vilalta C, Sanhueza J, Alvarez J, et al. Use of processing fluids and serum samples to characterize porcine reproductive and respiratory syndrome virus dynamics in 3 day-old pigs. *Vet. Microbiol.* 2018; 225:149-156.

4. Friis, NF. *M. suipneumoniae* and *M. flocculare* in comparative pathogenicity studies. *Acta. Vet. Scand.* 1974; 15: 507–518.

5. Le Carrou J, Laurentie M, Kobisch M, et al. Persistence of *Mycoplasma hyopneumoniae* in experimentally infected pigs after Marbofloxacin treatment and detection of mutations in the parC gene. *Antimicrob. Agents Chemother.* 2006; 50:1959-1966.

6. Marois C, Le Carrou J, Kobisch M, et al. Isolation of *Mycoplasma hyopneumoniae* from different sampling sites in experimentally infected and contact SPF piglets. *Vet Microbiol.* 2006; 120:96-104.

7. Woolley L, Fell S, Gonsalves J, et al. Evaluation of clinical, histological and immunological changes and qPCR detection of *Mycoplasma hyopneumoniae* in tissues during the early stages of mycoplasmal pneumonia in pigs after experimental challenge with two field isolates. *Vet Microbiol.* 2012; 161:186-195.

8. Sponheim A et al. Detection of *Mycoplasma hyopneumoniae* and *Mycoplasma hyorhinis* by PCR in processing fluids. Presented at: University of Minnesota's Allen D. Leman Swine Conference; 2018 September 15-18; Saint Paul, Minnesota, USA.

9. Vilalta C, Sanhueza J, Murray D, et al. Detection of *Mycoplasma hyopneumoniae* in piglet processing fluids. Veterinary Record. 2019; 185: 510.

10. Baker J, Murray D, Sponheim A, et al. Utilizing piglet-processing fluids to detect PRRSV by PCR in a low-prevalence population. American Association of Swine Veterinarians; 2018.

11. Vilalta C, Baker J, Sanhueza J, et al. Effect of litter aggregation and pooling on detection of porcine reproductive and respiratory virus in piglet processing fluids. *J Vet Diagn Invest.* 2019; 31(4):625-628.

Evaluation of foggers and stir fans to improve coverage of fogged disinfectants used in supply entry rooms

Jamie Madigan[1], BS; Shamus Brown[2], DVM; Derald Holtkamp[3], DVM; Nathan Schaefer[4], DVM
[1]North Carolina State University, Raleigh, North Carolina; [2]Iowa Select Farms, Iowa Falls, Iowa;
[3]Iowa State University, Ames, Iowa; [4]Boehringer Ingelheim Animal Health USA Inc., Duluth, Georgia

Introduction

Supply entry is a known risk for disease introduction, which may be mitigated by fogging supplies with disinfectants as they enter disinfectant and decontamination (D&D) rooms. Previous studies have shown that though fogging the disinfectant can help mitigate the risk of disease entry, complete coverage of supplies is challenging and difficult to monitor. The objectives of this pilot study were to determine if 1) a water soluble dye is a viable tool to evaluate disinfection coverage and 2) which, if any, combination of foggers and stir fans improved disinfectant application coverage.

Materials and methods

Sprayola® dye was chosen as the dying agent as it was practical for on-farm evaluation and cost effective. A Colorimeter® app on an iPhone was used to measure surface area coverage using saturation levels on the on the HSV (hue, saturation, value) scale.

A model D&D room measuring 12' × 12' × 8' was used for the study. Two wire shelving units were placed in the middle of the room, parallel to each other. Three 12" × 12" × 12" Delphi boxes were equally spaced on the top and middle shelf of each shelving unit, for a total of 12 boxes. The center Delphi boxes on each middle shelf was used for coverage measurements, recorded on all six sides of the boxes, which were assigned a letter. A stir fan was placed perpendicular to each shelving unit, and angled upward toward the boxes. Two foggers (Dyna-Fog® Hurricane Ultra II) were placed on the ground in opposite corners of the room at 45 degree angles toward the shelves. Each fogger contained a mixture of 8 ounce Intervention®, 6 ounce SPRAYOLA, and 1 gallon water.

Four different protocols were evaluated: 1) one fogger, 2) one fogger plus two stir fans, 3) two foggers, and 4) two foggers plus two stir fans. Each protocol was replicated five times, and all boxes were rinsed, bleached, and dried between replicates. Foggers ran for 5 minutes, then the D&D room sat for 10 minutes before entry. When stir fans were used, they were turned on at the 3 and 5 minute interval for 15 seconds. Before and after each replicate, the center boxes were removed, each side was visually evaluated, and photographed using the COLORIMETER app. Differences in surface coverage between protocols were assessed using one-way ANOVA to compare saturation values by surface using JMP. Pairwise comparisons between each protocol were carried out using the Tukey-Kramer adjustment for multiple comparisons.

Results

The least squares mean saturation values for each protocol and box surface are shown in Table 1. The protocol with two foggers with two stir fans had the most surface area coverage, followed by two foggers, one fogger with two stir fans, and one fogger.

Conclusion

This project demonstrated that the dying agent SPRAYOLA mixed with disinfectant could be an effective tool to visually assess supplies entered through a D&D room.

The results suggest that increasing the number of foggers and the use of fans increased surface area coverage. Even though coverage was relatively poor on surfaces A (bottom), D and F (sides facing other boxes), this pilot study shows promise in the use of air circulation and increased number of foggers in improving total surface area coverage.

Table 1: Statistical difference between protocols by box side coverage based on saturation levels from HSV scale*

	Box side					
Protocol	A (Bottom)	B (Top)	C	D	E	F
One fogger	3.2^{A}	47.1^{A}	$4.1^{A,B}$	4.3^{A}	4.2^{A}	3.3^{A}
One fogger, two fans	3^{A}	56.4^{A}	3.1^{A}	3.3^{A}	6.7^{A}	7.7^{B}
Two foggers	$3.3^{A,B}$	79.8^{B}	22.3^{C}	3.3^{A}	4.6^{A}	3.5^{A}
Two foggers, two fans	4.7^{B}	75.8^{B}	$20.2^{B,C}$	3.3^{A}	31.4^{B}	4.8^{A}

^{A,B,C} Different superscripts within column indicate statistically significant differences between protocols ($P < 0.05$).

* HSV scale is 0 to 100, with 100 indicating complete saturation of the surface with the primary color.

Field evaluation of a liquid *Saccharomyces cerevisiae* fermentation prototype on lactation performance in sows

S. McCallister[1]; K. Robertson[2], DVM, DACVPM; D. Petry[2], MBA, PhD; C. Sexton[3], DVM; B. Meisheid[2]
[1]North Carolina State University, Raleigh, North Carolina; [2]Diamond V Mills, Cedar Rapids, Iowa;
[3]Livestock Veterinary Services, Kinston, North Carolina

Introduction

Producing a robust pig that is equipped to withstand the stress of weaning is critical to starting pigs well in the nursery. A unique *Saccharomyces cerevisiae* fermentation product (Original XPC™; Diamond V Mills, Cedar Rapids, Iowa, USA) delivered to sows via the feed throughout gestation and lactation has been shown to increase litter weight and decrease pre-wean mortality of pigs.[1] The purpose of this field study was to evaluate the effects of this technology on lactation performance of sows when a liquid *S cerevisiae* fermentation prototype was delivered orally in lactation only.

Materials and methods

A batch of 353 sows due to farrow in June 2019 in an 1,800-sow farm in eastern North Carolina was enrolled in the study. Prior to the study, water lines were cleaned in 10 farrowing rooms according to the protocol in Figure 1. The study was designed as a non-blinded experimental trial with the experimental unit being the farrowing room. The sows were assigned to farrowing rooms according to the usual management on the farm.

The rooms were assigned to one of two treatments using a non-random method based on ease and safety of administering product through water medicators. Sows in control rooms (CTL) received fresh drinking water without additives. Sows in the treatment rooms (SCP) were administered 15mL per sow per day of a liquid *S cerevisiae* fermentation prototype (Diamond V Mills, Cedar Rapids, Iowa, USA) in the drinking water via water medicators. The daily dose of SCP for the treated rooms was mixed in a stock solution with water so that treatment was delivered over a 6- to 8-hour period. The *S cerevisiae* fermentation prototype used in this study is known to contain a similar fermentation metabolite profile to Original XPC and was chosen due to the ease of administering product to individual rooms through the drinking water when compared to administering product to individual rooms through the feed.

Cross-fostering was completed by 24 hours after birth and all sows were given an opportunity to nurse 14 or 15 pigs. The number of pigs given to nurse and litter weight were recorded 24 hours after birth and again 24 hours prior to weaning. Three litters were randomly chosen within each room using a random number generator. Individual weights were recorded for pigs in these litters. Data were analyzed using a mixed model in SAS with room as the experimental unit. The model included treatment and parity as fixed effects with number of pigs given a chance to nurse and number of days given to nurse as covariates.

Results and discussion

A total of 273 sows were included in data analysis; 149 in the treatment group and 124 in the control group. Sows were excluded from data analysis when cross-fostering occurred after 24 hours and when pigs were weaned early and comingled in farrowing crates to make space for sows farrowing late in the batch. A significant increase in litter wean weight was observed for sows administered SCP compared to sows in the CTL group. There was also a trend toward lower pre-wean mortality for sows in the SCP group as well as an increase in the number of pigs weaned per sow when compared to sows in the CTL group. These results are summarized in Table 1.

Figure 1: Water line cleaning protocol.

Subject: Proxy-Clean Protocols for Swine Producers

Suggested Water Line Cleaning Program with Animals Present

For cleaning the lines while animals are housed create stock solutions as described below and run the program up to 10 days depending on degree of fouling.

Pump Setting	Water Quantity	Day 1&2	Day 3&4	Day 5&6	Day 7-10
		Proxy-Clean	Proxy-Clean	Proxy-Clean	Proxy-Clean
1:128	5 Gallons	8 oz.	12 oz.	16 oz.	20 oz.

A summary of 21-day adjusted individual pig weights is provided in Table 2. Wean weights were adjusted to 21 days using the method reported by Cutshaw et al.[2] Differences in average 21-day wean weight were not significant, leading the authors to attribute increased litter wean weights observed to be primarily due to reduction in pre-wean mortality. For individual 21-day wean weights, homogeneity of dispersion was analyzed using a chi-square test. A significant difference ($P < 0.05$) was found with pigs from SCP sows having a narrower dispersion compared to the CTL 21-day pig weights. This measure indicates less variation in 21-day adjusted weaning weights of pigs from dams in the SCP group vs. those sows in the CTL group.

Reducing mortality in US commercial pork production is a priority in the industry and an ongoing topic of research interest.[3] The *S cerevisiae* fermentation prototype tested in this study decreased pre-wean mortality thereby increasing total litter wean weight when administered to sows only in lactation. This reduces the cost of product when compared to administering in both lactation and gestation. Assuming 2.3 lactation cycles per year and a weaned pig value of $35, an increase of one pig weaned per sow per year would increase income by $35.00 per sow. The liquid *S cerevisiae* fermentation product has a yearly cost of $3.27 per sow, therefore profitability was increased in this study by $31.73 per sow annually with a return on the investment in product of 9.7:1.

References

1. Petry et al. The effects of Diamond V Original XPCTM on weaning performance when fed to sows during gestation and lactation. AASV 2019; 235.

2. Cutshaw et al. Relationships among sow productivity traits within purebred and crossbred litters. Livestock Science, Vol. 170, December 2014; 193-202.

3. Multiple authors. Improving Pig Survivability project©2018; Retrieved from `https://piglivability.org/about-the-project`.

Table 1: Sow performance on piglets in lactation

Treatment group	Litter wean weight	Pre-wean mortality %	Pigs weaned per litter
CTL	181.6	11.2	12.41
SCP	193.5	8.3	12.85
Difference	+11.9	-2.9	+0.44
P	0.013	0.104	0.084

Table 2: Number and percentage of pigs weaned by 21-day adjusted weight category

	Ave 21d weight (lbs.)	Coefficient of variation[a]	< 10 lbs.		10-11.9 lbs.		12-13.9 lbs.		14-15.9 lbs.		≥ 16 lbs.	
			No.	%	No.	%	No.	%	No.	%	No.	%
CTL	15.2	22.9	12	8%	8	5%	29	19%	34	23%	67	45%
SCP	15.0	19.5	9	5%	16	9%	32	19%	49	29%	66	38%

[a] Coefficient of variation ($P < 0.05$)

Genetic analysis of swine and human influenza A virus subtype H3N2 from 2015-2018 in the United States

D. Kinkade[1]; B. Blair[1]; J. Lowe[1,2]

[1]University of Illinois College of Veterinary Medicine, Urbana, Illinois; [2]Lowe Consulting Ltd., Mahomet, Illinois

Introduction

Influenza A virus (IAV) is a potentially devastating respiratory disease that has the potential to have major human and swine health implications. Therefore, IAV surveillance has become an increasingly vital aspect of both swine and human health in recent years. This is due in large part to the fact that novel influenza A virus (IAV) strains can be achieved in a wide variety of ways, making IAV a readily adaptable and highly mutagenic virus. The swine industry is a perfect example of this, because IAV not only has the ability to be transmitted to swine hosts from human or avian sources, it has also been shown to be capable of transmitting from swine to other species such as humans which is of particular interest. The highly variable nature and ability to cross species of Influenza A virus is due to its ability to readily undergo antigenic mutations (drift) and genetic reassortment due to the exchange of genes between two or more IAVs. These characteristics result in a rapidly evolving and extremely complex virus that demands attention from both a public safety and economic perspective.[1]

These unique characteristics of influenza A viruses described above also make the virus a prime candidate for a pandemic. In 2009, the United State and Mexico suffered from precisely that. A H1N1 IAV was transmitted from human to human across both countries. Upon genetic characterization and analysis of this virus, this H1N1 subtype was of swine origin and contained a novel reassortment of the RNA segments that make up its genetic profile. While there has been a significant increase in the quantity of gene sequencing of IAV, most of these sequences are of human or avian origin. The relatively limited amount of sequencing on swine origin viruses made it difficult to determine exactly how, where, and when this virus was able to become pandemic.[2]

Previous research has shown that the H3N2 subtype has an exceptionally high potential for cross species transmission given its more recent triple reassortant nature.[3] Influenza A virus's genome is comprised of eight RNA segments. Due to reassortment, this H3N2 triple reassortant strain contains RNA segments that originate from swine, avian, and human hosts. This influenza strain is one of the most prominent seen in swine populations to date. The prototype sequence commonly used to represent this strain of influenza is A/Swine/Texas/4199-2/98.

Previous studies have demonstrated that humans in regular contact with swine are at a much higher risk of contracting variant influenza strains. Studies show reassortant H3N2 subtype of swine origin has the capabilities of infecting humans and has done so in recent years. The humans that are most prone to becoming infected with these viruses have notably increased contact with swine in one form or another whether that be contact at a county fair or living on a pig farm.[4] While there is some information available about IAV being transmitted from swine to humans, there is limited information available on transmission of human-origin influenza virus to swine. The use of phylogenetics has the potential to elucidate not only how the virus is changing over time, but how it is moving between different species. This study uses phylogenetic techniques to determine the frequency and direction of H3N2 influenza A spillover events between human and swine hosts from 2015-2018.

Materials and methods

Hemagglutinin (HA) sequences from H3N2 viruses found in human and swine hosts from July 2015 – July 2018 were obtained from the NCBI's Influenza Virus Database. Two reference strains, the H3N2 triple reassortant strain commonly found in swine (A/Swine/Texas/4199-2/98) and the seasonal H3N2 virus most common in humans (A/Michigan/15/2014) were obtained as reference host genomes for each respective host. In total, 7,610 influenza sequences were collected for analysis in this study. 6,754 of those sequences were from human hosts and 856 of those sequences were from swine hosts. The sequences were named given the nomenclature system embedded in the NCBI influenza database and were not changed. All sequence data was downloaded as a FASTA file from the NCBI database for analysis.

The FASTA file containing all of the HA RNA segment sequences were aligned for further analysis using the MUSCLE alignment function within the MEGA-X software program. Evolutionary rates were determined using the Bayesian MCMC (Markov Chain Monte Carlo) method implemented in BEAUTI, a part of the BEAST 1.10.4 software package, under strict and then relaxed clock conditions and an uncorrelated log normal rate distribution model. A strict clock model assumes that every branch of the phylogenetic tree evolves according to the same evolutionary rate, while an uncorrelated relaxed fit model allows each branch of the phylogenetic tree to have its own evolutionary rate.[5] A Bayes factor using marginal likelihoods implemented in Beast selected the best-fitting models from what BEAUTI generated. MCMC chains for each year were then run for 100 million generations with sampling every 10,000th generation. The resulting trees were summarized in a maximum clade credibility tree after performing a 10% burn-in using the TreeAnnotator program that is also a part of the BEAST 1.10.4 software package. The output trees were then reviewed and analyzed using FigTree 1.4.4.

A total of six phylogenetic trees: one strict and one relaxed fit tree for the 2015-2016 influenza season (July 2015 – July 2016), one strict and one relaxed fit tree for the 2016-2017 influenza season (July 2016 – July 2017), and one strict and one relaxed fit tree for the 2017-2018 influenza season (July 2017 – July 2018) were generated. All HA sequences which originated from swine hosts were highlighted within each tree to easily differentiate them from human host influenza strains. Human origin viruses located in a swine origin clade were considered a swine to human spillover event and swine origin viruses located in a human origin clade were considered human to swine spillover events.

Results and discussion

Both the strict and relaxed clock models yielded the same results. Therefore, only the strict fitting tree for each influenza season will be described. For the 2015-2016 influenza season (July 2015- July 2016), there were no spillover events documented in either direction: all swine strains were located in swine origin clades and all human strains were located in human origin clades. For the 2016-2017 influenza season (July 2016 – July 2017), there were two spillover events that were both human like HA which was identified in swine hosts. One swine case (A/swine/North Carolina/A02216473/2016) was identified in the first spillover event and two cases were observed in the second spillover event (A/swine/Oklahoma/A02216476/2017 and A/swine/Oklahoma/A02216474/2017). For the 2017-2018 influenza season (July 2017- July 2018), there were also two spillover events, both were human like HA was found in swine hosts. One spillover event was contained to a single swine case (A/swine/Illinois/A02155577/2018) and the other spillover event involved four swine cases (A/swine/Arkansas/A01678528/2017, A/swine/Oklahoma/A01678529/2017, A/swine/Arkansas/A01678570/2018 and A/swine/Arkansas/A01678568/2018). Therefore, over the course of three years, four isolated spillover events, each with human like H3 genes found in swine, involving eight swine cases were recorded. No swine origin influenza strains were found in human hosts. These results demonstrate that the current H3s are more likely to move from human hosts to swine hosts which is contrary to conventional wisdom.

While our data set may have been limited to the sequences that have been submitted to NCBI's influenza database, we were still able to clearly identify four isolated events of influenza moving from human to swine hosts that has not been found in prior studies. This is in direct contrast to what one might expect given previous reports of IAV subtype H3N2 infecting humans in direct contact with swine, especially since no cases if swine like HA were identified in human hosts. While this does not mean that swine origin H3N2 is not transmissible from swine to human hosts, it suggests that it is a less common occurrence than the infection of swine with human origin H3N2.

Further research for past and future H3 strains and exploration of additional HA types and genes would be build a more complete understanding of the frequency of cross species transmission. IAV is rapidly evolving and the utilization of advanced phylogenetic approaches such as this offers a promising method of tracking and monitoring its transmission between species.

References

1. Chiapponi C, Ebranati E, Pariani E, et al. Genetic analysis of human and swine influenza A viruses isolated in Northern Italy during 2010-2015. Zoonoses Public Health. 2018;65(1):114-123. doi:10.1111/zph.12378

2. Escalera-Zamudio M, Cobián-Güemes G, de los Dolores Soto-del Río M, et al. Characterization of an influenza A virus in Mexican swine that is related to the A/H1N1/2009 pandemic clade. Virology. 2012;433(1):176-182. doi:10.1016/J.VIROL.2012.08.003

3. Ma W, Lager KM, Lekcharoensuk P, et al. Viral reassortment and transmission after co-infection of pigs with classical H1N1 and triple-reassortant H3N2 swine influenza viruses. J Gen Virol. 2010;91(9):2314-2321. doi:10.1099/vir.0.021402-0

4. Epperson S, Jhung M, Richards S, et al. Human infections with influenza A(H3N2) variant virus in the United States, 2011-2012. Clin Infect Dis. 2013;57(SUPPL.1). doi:10.1093/cid/cit272

5. A. J. Drummond, S. Y. W. Ho, M. J. Phillips and A. Rambaut (2006) Relaxed phylogenetics and dating with confidence. PLoS Biology 4(5): e88.

Evaluation of reported PCV2 sequences for evidence of recombination

Meggan Bandrick[1], DVM PhD; Prerak Desai[2], PhD; Gonzalo Rincon[2], PhD; Dennis Foss[1], DVM PhD
[1]Zoetis, Global Biologics Research, Veterinary Medical Research and Development, Kalamazoo Michigan;
[2]Zoetis, Global Genetics Research, Veterinary Medical Research and Development, Kalamazoo Michigan

Introduction

Porcine circovirus (PCV) is a nonenveloped, single-stranded DNA virus. PCV viruses have similar genomic structures and encode two major structural proteins, the capsid (encoded by ORF2) and the replicase (encoded by ORF1). Today, six PCV2 genotypes have been described[1] and another two including potential recombinant forms[2] have been proposed. Genotypes are delineated by a *P*-distance of 0.035[3] and proposed to have a maximum intra-genotype *P*-distance of 13%.[2] PCV2 genotypes can be further organized by cluster.[4] A striking feature of PCV2 is its high rate of evolution. Despite being a DNA virus, PCV2 has an evolutionary rate similar to RNA viruses.[5] In addition to the high rate of point mutations, PCV2 viruses recombine.[6-8] Recombinant viruses have been found world-wide and are fit enough to compete with parental strains.[2,9] In fact, approximately 20-35% of PCV2 viruses are recombinants.[2,9] Co-infection of PCV2 viruses is common[8] and this allows multiple PCV2 viruses to interact within a single cell and share genetic material. The objective of the study was to evaluate reported PCV2 sequences for evidence of recombination and to understand the immune coverage offered by Fostera Gold PCV to those viruses.

Materials and methods

A total of 4500 publicly available (as of June 2017) PCV2, ORF2 nucleotide sequences were downloaded from GenBank. The nucleotide sequences were aligned using MAFFT[10] and a maximum likelihood tree was constructed using FastTree.[11] BAPS (Bayesian analysis of Population Structure)[12] was used to cluster the strains and perform admixture analysis to identify field strains with possible recombination events. The de novo identified sequence clusters were then mapped to known sequence clusters as defined by Xiao et. al.[13] Within each cluster identified by using BAPS, PCA (principal component analysis) for the DNA distance matrix was used to identify the minimum number of PCs (K) that explain 95% of the variance in the distance matrix. The cluster was further partitioned into k sub-clusters using K means clustering. A strain nearest to the sub-cluster medoid was selected as a representative for the sub-cluster. Strains representing recombinant forms were included within the analysis and across cluster (with the exception of cluster 11, PCV2e as recombinant viruses were not found within the cluster).

A subset of 160 sequences from global locations and representing the diversity of PCV2 were selected for further analysis.[14] These 160 sequences were evaluated for T cell epitope content [sum of putative class I and class II T cell epitopes restricted by Swine Leukocyte Antigen (SLA)] along with sequences from three commercial PCV2 vaccines [chimeric PCV1-PCV2a (cPCV2a) monovalent, baculovirus-expressed PCV2a capsid protein monovalent, and bivalent cPCV2cPCV2b]. We then compared the putative epitopes of each vaccine to itself and against the 160 representative PCV2 strains to calculate a relatedness score based on T cell epitopes shared between pairs of sequences (shared EpiCC score). We also calculated average EpiCC scores for the comparisons of each vaccine against all the PCV2 representative strains. The average EpiCC score is an estimation of the vaccine T cell epitope "coverage" for the complete set of strains, which may be used to identify the vaccine that shares, on average, more T cell epitope content with the circulating strains and has the best potential to confer the broadest cross-reactive cell-mediated immunity and protection.

Results

A total of 4500 PCV2 sequences were obtained from GenBank for analysis. Of these, 3463 were identified with a specific date. Most sequences were deposited in GenBank between 2005 and 2016. The percent of each genotype reported from 2002 to 2016 changed markedly over time. PCV2a and PCV2b were more commonly reported before 2011, but since 2014, PCV2d has predominated, both globally (Figure 1a) and in the US (Figure 1b).

The PCV2 sequences were further characterized into clusters within each genotype. These clusters show the gene flow over time into two super clusters, PCV2b/d and PCV2a (Figure 2).

Of 1037 full genome sequences from 41 countries between 1980 and 2017, 19.4% were recombinants. In general, most of these recombinant PCV2 viruses contained sequences primarily from two genotype clusters, although many contained multiple genotype-derived sequence. Most recombinants included portions of the most prevalent genotypes, PCV2a, 2b and 2d. Cluster 2 included the greatest number of recombinant viruses. PCV2e included the least number of recombinants; PCV2e also had a low prevalence, limited geography, and limited sequence diversity.

Of the subset of 160 representative strains, 40 of 160 (25%) were recombinant. These 160 sequences were also evaluated for T cell epitope content, along with sequences from commercial PCV2 vaccines.[14] The recombinant strains identified in that analysis are highlighted in Figure 3. This radar plot shows the relative T cell epitope coverage (EpiCC score) of the recombinant strains (indicated with blue lines). The bivalent PCV2aPCV2b vaccine had,

Figure 1: PCV2 epidemiology. Percent of each genotype of PCV2 deposited in GenBank between 2002 and 2016, globally (A) or in the US (B).

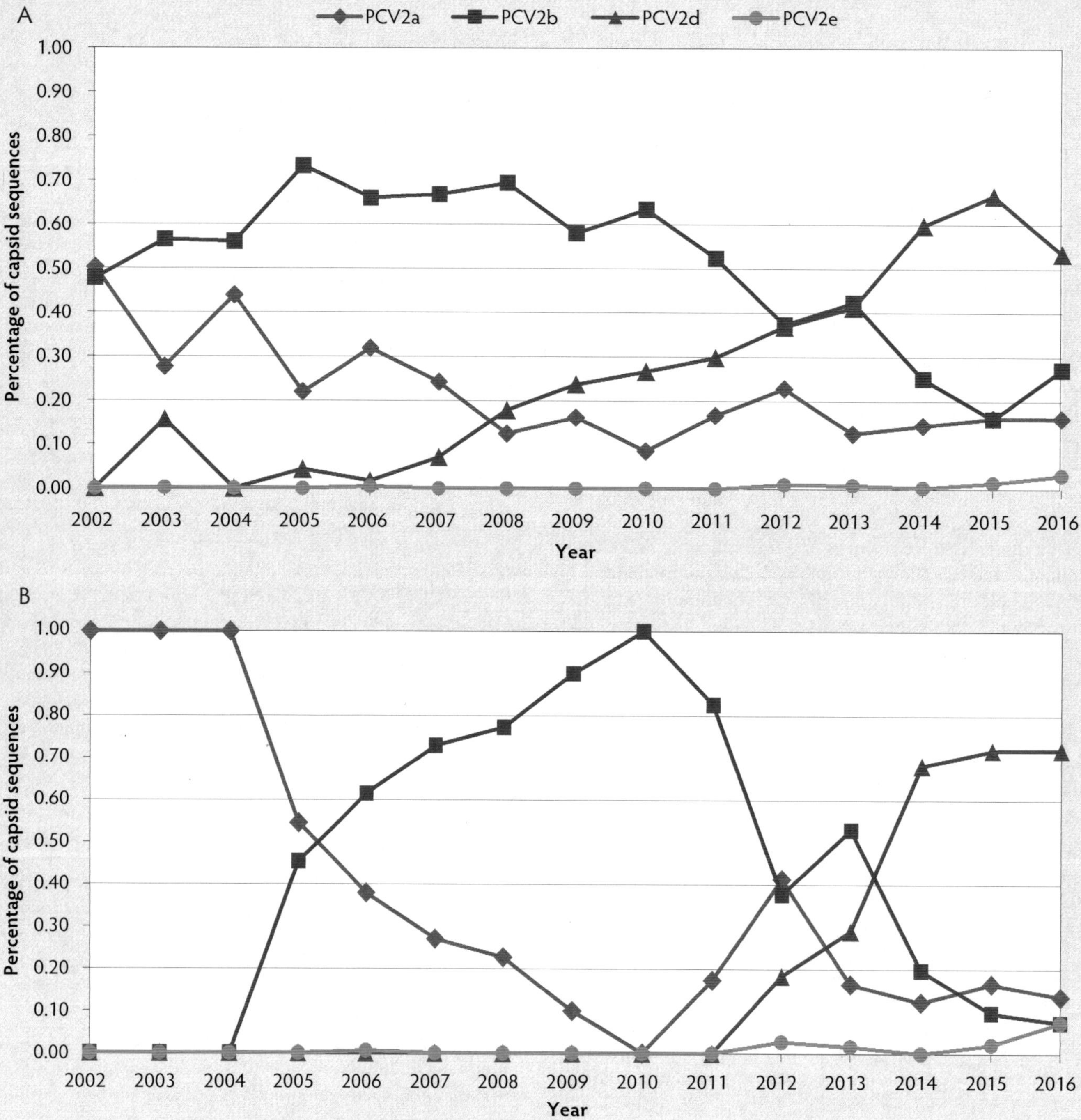

Figure 2: Gene flow of PCV2 genotype, showing the main sequence clusters of PCV2b/d and PCV2a. Clusters 8 and 6 have been circled as this is where the PCV2b and PCV2a vaccine strains within Fostera Gold PCV MH cluster.

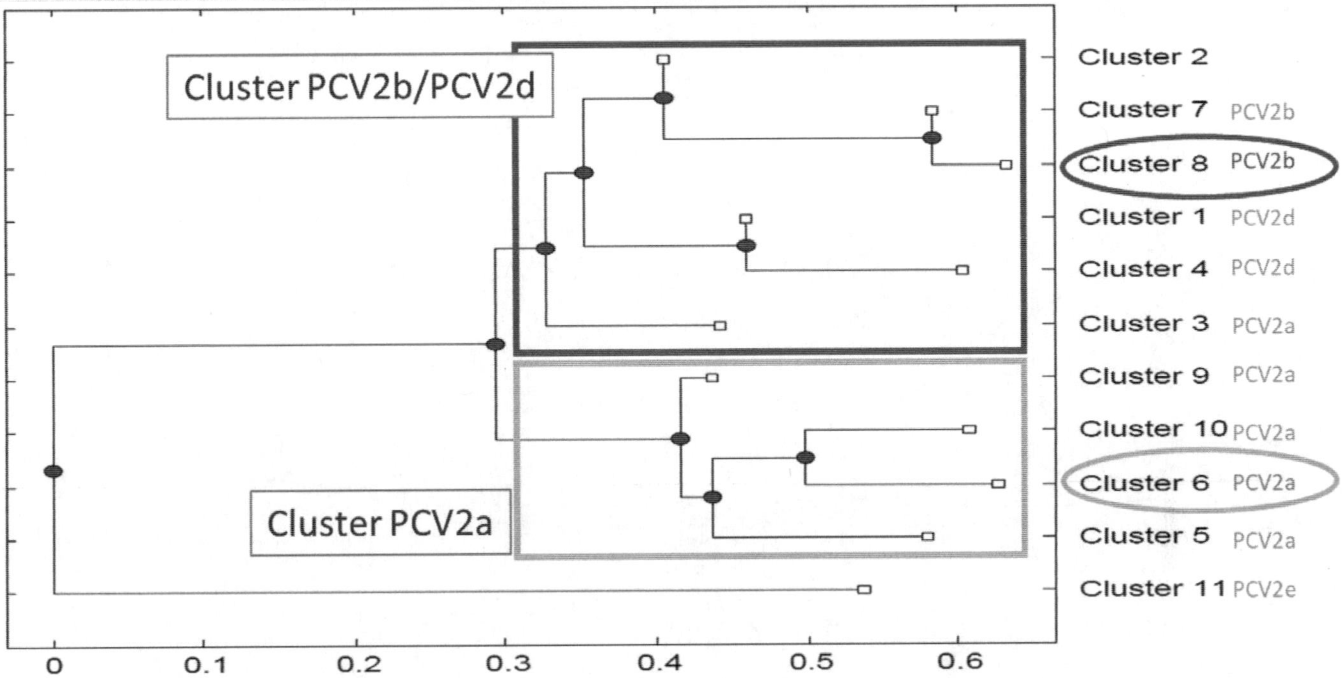

on average, the greatest T cell epitope relatedness among vaccine and field virus epitopes (highest average EpiCC score). Further, the recombinant strains, including recombinant strains from the US, had more T cell epitopes in common with the bivalent PCV2aPCV2b vaccine compared to the monovalent cPCV2a or monovalent baculovirus-expressed PCV2a capsid protein vaccines.

Conclusion

The diversity in PCV2 is driven by ongoing genetic change including recombination. The ability of PCV2 viruses to recombine and the preponderance of recombinant viruses agrees with previous reserach. Others have shown that recombinant PCV2 viruses are able to induce disease.[15,16] Therefore, PCV2 vaccines should protect against recombinant viruses. Our analysis shows that the bivalent PCV2 vaccine, containing PCV2a and PCV2b, offers the best potential to confer the broadest cross-reactive cell-mediated immunity and protection to this diverse virus. The predicted broad coverage by the bivalent PCV2 vaccine is due to the representation from both superclusters of the virus, namely the PCV2a group and the PCV2b/PCV2d group. Further, the pig's immune system may be best educated by the bivalent vaccine in terms of protection against diverse PCV2 field viruses including recombinants that contain bits of PCV2a or PCV2b or PCV2d, because of the epitope overlap in the vaccine and field viruses. Taken together, the bivalent vaccine offers the best potential for broad PCV2 protection.

References

1. Bao F et al. Retrospective study of porcine circovirus type 2 infection reveals a novel genotype PCV2f. Transbound Emerg Dis. 2018. 65(2):432-440.

2. Franzo G and Segales J. Porcine circovirus 2 (PCV-2) genotype update and proposal of a new genotyping methodology. 2018. PLoS ONE 13 (12): e0208585. https://doi.org/10.1371/journal.pone.0208585.

3. Segales J et al. PCV-2 genotype definition and nomenclature. Vet Rec. 2018. 62(26):867-868.

4. Olvera A, et al. Molecular evolution of porcine circovirus type 2 genomes: phylogeny and clonality. Virology. 2007. 357(2):175-185.

5. Firth C et al. Insights into the Evolutionary History of an Emerging Livestock Pathogen: Porcine Circovirus 2. J Virol. 2009.83(24)12813-12821.

6. Hesse R et al. Evidence for recombination between PCV2a and PCV2b in the field. Virus Research. 2008.132(1–2):201–207.

7. Lefebvre DJ et al. Recombination of two porcine circovirus type 2 strains. Arch Virol 2009.154, 875-879.

8. Gerber PF et al. Association of concurrent porcine circovirus (PCV) 2a and 2b infection with PCV associated disease in vaccinated pigs. Res Vet Sci. 2013;95(2):775-781.

9. Franzo G et al. Phylodynamic analysis of porcine circovirus type 2 reveals global waves of emerging genotypes and the circulation of recombinant forms. Mol Phylogen Evol 2016.100:269-280.

10. Katoh K, Standley DM. MAFFT multiple sequence alignment software version 7: improvements in performance and usability. Mol Biol Evol 2013.30:772-780.

11. Price MN et al. FastTree 2--approximately maximum-likelihood trees for large alignments. 2010. PLoS One 5:e9490.

12. Cheng L et al. Hierarchical and spatially explicit clustering of DNA sequences with BAPS software. Mol Biol Evol 2013. 30:1224-1228.

13. Xiao CT et al. Global molecular genetic analysis of porcine circovirus type 2 (PCV2) sequences confirms the presence of four main PCV2 genotypes and reveals a rapid increase of PCV2d. J Gen Virol 2015. 96:1830-1841.

Figure 3: Recombinant viruses applied to the Epivax plot from the global strain analysis. Orange and grey circles represent the EpiCC score of a bivalent PCV2a/PCV2b vaccine (Fostera Gold PCV) and a baculovirus-expressed PCV2a capsid protein monovalent, respectively. Recombinant strains are indicated with blue lines. Data on file Study Report B820Z-US-18-789.

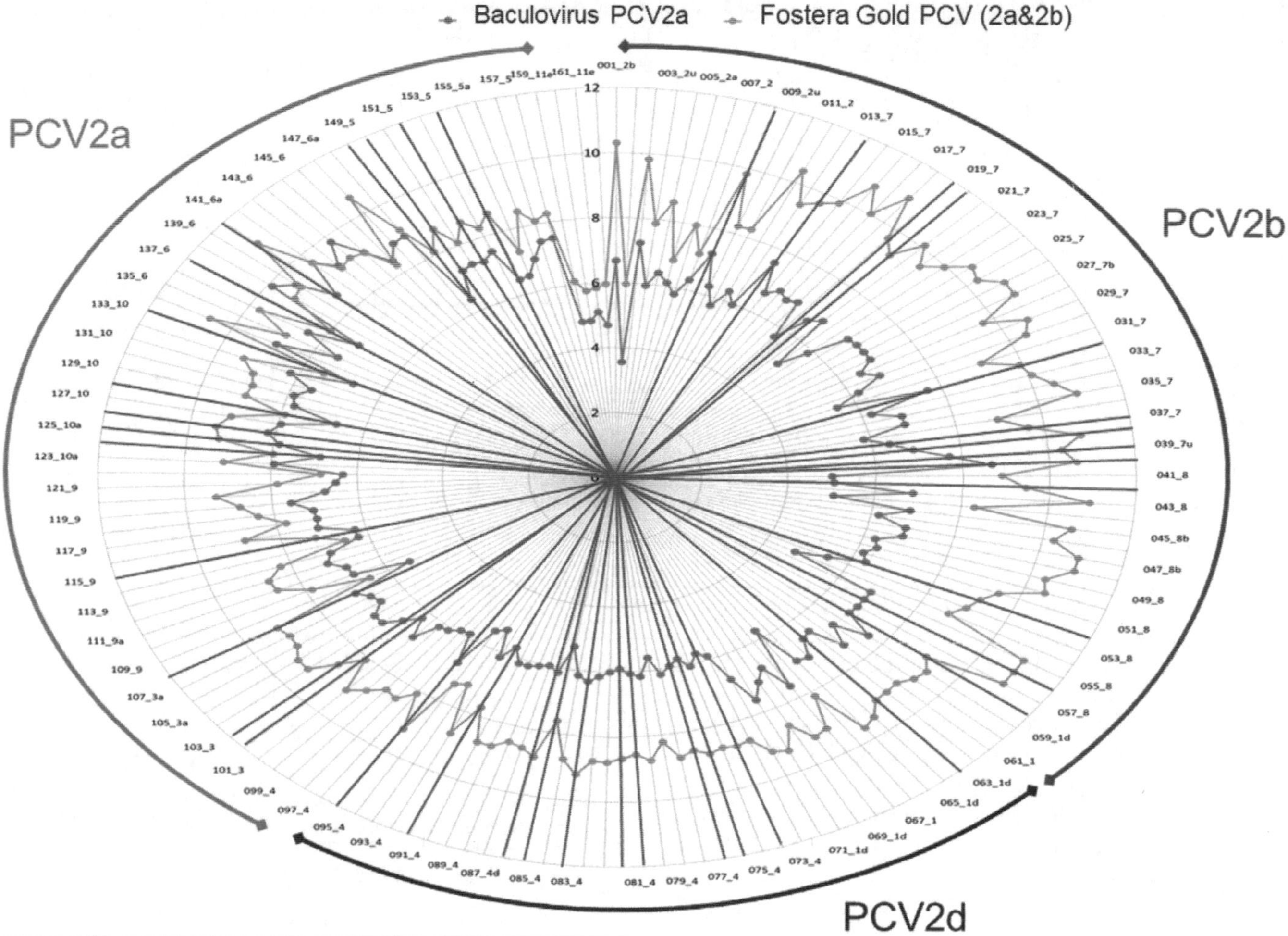

14. Zoetis, study on file B820Z-US-18-789.

15. Neira V, et al. Genetic analysis of porcine circovirus type 2.

16. Huang Y et al. Evidence for different patterns of natural inter-genotype recombination between two PCV2 parental strains in the field. Virus Res. 2013.175(1):78-86.

A heterologous prime boost mass vaccination protocol for IAV-S control

Bob Thompson[1], DVM, MS; Marie Culhane[2], DVM, PhD, DVM, Larry Coleman[3], DVM; Micah Jansen[2]; Lucina Galina Pantoja[2], DVM, PhD

[1]Independent consultant, Franklin, Kentucky; [2]University of Minnesota, Saint Paul, Minnesota; [3]Vet Care and Consultation, Broken Bow, Nebraska; [4]Zoetis, Parsippany, New Jersey

Introduction

Elimination of influenza A virus in swine (IAV-S) can be achieved in breed to wean herds[1] but many challenges remain. The herd veterinarian attempted influenza eliminations at a multiplier herd using a whole herd, double mass vaccination using either a homologous autogenous (100% similarity, 1st attempt) or commercial vaccines (98.6% similarity, 2nd attempt), but elimination was not achieved or maintained. Recently, the advantages of heterologous prime boost vaccination have been described, and under experimental conditions, enhanced HI antibody responses and reduced viral shedding were reported.[2] This study was set up to evaluate the implementation of a heterologous prime boost vaccination with a goal of consistently weaning IAV-S negative pigs at a commercial herd.

Materials and methods

The evaluation was conducted in a 11,000-sow multiplier farm with an internal GDU for gilt replacements. The resident IAV-S was a H3N2 cluster IVA virus 98.6% similar to FluSure XP® (Zoetis). The virus was not causing clinical disease at the sow farm, but the farm was weaning positive pigs that impacted customer herds who were mixing pigs from multiple sources and reporting clinical disease and secondary bacteria. Replacement gilts received FluSure XP® at 6 weeks post wean, as well as 25 and 28 weeks of age pre-breeding. Biosecurity protocols allowed only semen to enter the farm. Similarly, the protocol did not allow sick employees to enter the site and they were required to record body temperatures daily ≤ 100° F. Additionally, all weaned pigs and cull sows were shuttled out of sow farm, all supplies entered through a decontamination chamber, and feed was supplied by a dedicated feed mill. At week 13, the whole herd was primed with an autogenous vaccine matching the resident strain (Sequivity™, Merck) and on week 22, the herd was boosted with FluSure XP. From weeks 26-31, a modified McRebel was implemented allowing cross fostering of pigs < 24 hrs. of age, creating fall back litters ≤ 5 days of age, and only allowing newly farrowed sows to be used as nurse sows and segregating farrowing staff. At week 32 fall back, litters ≤ 11 days were allowed. The IAV-S status was monitored by PCR from udder wipes from 10 litters of ready-to-wean pigs. To determine a baseline, PCR testing took place at week 26 (4 weeks post-booster), and then weekly from week 30-39 (Table 1).

Results and discussion

The PCR results showed that 30% of the litters were positive (3/10) at week 26. Ten weeks post booster and four weeks post implementation of McRebel all PCR tests were negative except on week 37 which resulted in 2/10 positive samples with high Cts. From that point on, all PCR tests were negative up until week 40. Weaning negative IAV-S pigs appears to be possible using a heterologous prime boost and following stringent biosecurity and management protocols. Success requires an integrated approach using an improved protocol of mass vaccination and minimizing the risk of transmission. Now that negative pigs can be weaned, the focus has been shifted to creating an antibody decay profile from replacement gilts to determine optimal time for vaccination. The protocol described here can be improved upon based on the findings of this field evaluation, i.e., the gap between the prime and the boost can be shortened and the on-site holding nursery where commercial pigs are held prior to sale can be depopulated. This protocol can be customized to control herd specific strains from other herds.

References

1. Thompson et al. Elimination of influenza A virus in multiple breed to wean herds. IPVS. 2016;595.

2. Culhane et al. Efficacy of prime-boost vaccination protocols in pigs challenged with influenza A viruses. AASV. 2019;164.

Table 1: Calendar of activities following a heterologous prime boost mass vaccination for swine influenza. Interventions and IAV-S PCR testing dates and results are shown.

Wk.	**13**	**14**	**15**	**16**	**17**	**18**	**19**	**20**	**21**	**22**
Intervention	Prime mass vaccination									Boost mass vaccination
Wk.	**23**	**24**	**25**	**26**	**27**	**28**	**29**	**30**	**31**	**32**
Intervention				Modified McRebel, fall back litters ≤ 5 days of age allowed						
				Only newly farrowed nurse sows						
				PCR*				PCR*		
Results				3/10				0/9**	0/10	0/9***
Wk.	**33**	**34**	**35**	**36**	**37**	**38**	**39**	**40**	**41**	**42**
Intervention	Fall back litters ≤ 11 days allowed									
	PCR									
Results	0/10	0/10	0/10	0/10	2/10	0/10	0/10	0/10		

* PCR results expressed as (#positive/total samples)
** One tube leaked
*** One tube with not enough sample for testing.

Progress towards improved *Mycoplasma hyopneumoniae* surveillance

Marisa Rotolo[2]; Seth Krantz[3]; Dapeng Hu[1]; Chong Wang[1]; Thaire Marostica[1]; Alexandra Henao-Diaz[1]; Eduardo Fano[4]; Dale Polson[4]; Deanne Hemker[2]; Edgar Tapia[5]; Silvia Zimmerman[5]; AW Tucker[2,6]; J. P Cano[2]; Jeffrey Zimmerman[1]; M. J Clavijo[1,2]

[1]Department of Veterinary Diagnostic and Production Animal Medicine, College of Veterinary Medicine, Iowa State University, Ames, Iowa; [2]Pig Improvement Company, Hendersonville, Tennessee; [3]Tosh Farms, Henry, Tennessee; [4]Boehringer Ingelheim, St. Joseph, Missouri; [5]IDEXX Laboratories, Westbrook, Maine; [6]University of Cambridge, Cambridge, United Kingdom

Introduction

The cost of *Mycoplasma hyopneumoniae* (MHP) ($375-400 million annually) has led the industry towards MHP elimination, but the slow rate of MHP transmission, the complexity of swine production systems, and the uncertainty regarding the best approach for MHP surveillance has hindered the development of effective control/elimination programs. Therefore, the purpose of this study was to compare the MHP surveillance (probability of detection) using tracheal swab (TS), serum (S), and oral fluid (OF) samples.

Materials and methods

This study was conducted in one room (46 pens, 28 pigs per pen) of a wean-to-finish site sourced with 21-day-old MHP-negative barrows. At 9 weeks of age, 10 seeder pigs in a pen located in the center of the room were inoculated with MHP lung homogenate, leaving the rest of the pigs in the pen as contact pigs. Beginning at the time of inoculation and every two weeks thereafter (7 total samplings), one OF, one TS, and 4 S samples were collected from each of the 45 contact pens in the room. For individual animal samples (TS, S), pigs were selected at random at each sampling. Samples were tested for DNA (TS and OF) or antibody (serum). The diagnostic data was then modeled using a piecewise exponential survival model to determine the probability of detection by sample size, test performance, and prevalence.

Results and discussion

All samples collected prior to inoculation were MHP DNA and antibody negative. All seeder pigs (n = 10) and contact pigs (n = 18) in the inoculated pen were confirmed positive by TS PCR and S ELISA by 7 DPI and 28 DPI, respectively. For contact pens, the first positive S ELISA sample occurred at 14 DPI, 28 DPI for TS PCR and 42 DPI for OF PCR. By the last sampling, 100% of the TS PCRs and S ELISAs were positive and 69% of the OF samples were positive. Preliminary results from the model produced similar estimates between DNA and antibody detection for individual samples.

The pigs' delayed response to MHP exposure renders traditional surveillance methods ineffective and mandates the development of a new approach to MHP surveillance. Probability of detection estimates showed that TS and S performed similarly. Oral fluids did not provide a high probability of detection at smaller sample sizes, but the results justify further evaluation in surveillance. Overall, this study suggested that the optimal MHP surveillance design may require a combination of diagnostic specimen types. For all sample types, fixed spatial sampling proved to be as good (or better) than random sampling. MHP remains challenging, however our research shows that the right combination of sampling tools can improve detection.

Activated medium chain fatty acids: a multifaceted tool

Benjamin Bass[1], PhD; Stacie Crowder[2], PhD; Murali Raghavendra Rao[2], PhD; Annie Lerner[3], PhD; Jason Woodworth[3], PhD; Scott Dee[4], DVM, PhD; Terry Weeden[2], PhD; Brenda DeRodas[2], PhD; Peter Karnezos[1], PhD

[1]Land O'Lakes, Arden Hills, Minnesota; [2]Purina Animal Nutrition, Gray Summit, Missouri; [3]Kansas State University, Department of Animal Sciences and Industry, Manhattan, Kansas; [4]Pipestone Applied Research, Pipestone Veterinary Services, Pipestone, Minnesota

Introduction

The swine industry is changing. There is a desire to reduce the amount of antibiotics being used in livestock production. Additionally, the potential for virus transmission in feed[1,2] has prompted the search for natural mitigation products. Activated medium chain fatty acids (MCFA) are a tool that can be used, in different combinations and concentrations, to reduce or remove in-feed antimicrobials, for in-feed pathogen mitigation, and to optimize performance. OutPace® FA is designed with patented, activated medium chain fatty acids to optimize pig performance. Dominnate™ contains a blend of carefully researched, activated medium chain fatty acids designed to support biosecurity protocols.

Materials and methods

Performance

A meta-analysis was conducted which included nine individual trials representing data for the complete nursery period (13 to 54 lbs. body weight), as well as the late nursery only (33 to 59 lbs. body weight). A total of five studies were included for the complete nursery period and four trials were included in the late nursery analysis. Average daily gain, average daily feed intake, and feed conversion were compared between a control diet, and control diet containing OutPace® FA (5 lbs./ton Phase 1 and 2, 2.5 lbs./ton Phase 3). Pigs were fed a budget of 5.0 lbs. of Phase 1, 13.0 lbs. of Phase 2, and Phase 3 diets for the remainder of the study.

Nonmedicated feed

A study was conducted at a commercial wean-to-finish research facility in north central Iowa. Weaned pigs were individually weighed and placed in pens (22 pigs/pen, 11 pens/treatment) and pens were randomly assigned to receive either a medicated diet (Mecadox) or an unmedicated diet containing OutPace® FA (Table 1). Pigs were provided commercial nursery diets and fed a budget of 5.0 lbs. of Phase 1 diets, 13.0 lbs. of Phase 2 diets, and Phase 3 diets through the end of the 42-d study. Body weight and feed intake were recorded to determine average daily gain, average daily feed intake, and feed conversion.

Table 1: Nonmedicated feed study treatments

	OutPace	Medicated
Phase 1	20 lbs./ton OutPace	50 g/ton Mecadox
Phase 2	20 lbs./ton OutPace	50 g/ton Mecadox
Phase 3	10 lbs./ton OutPace	25 g/ton Mecadox

Feed mitigation

To determine efficacy of an MCFA blend to mitigate PEDV in feed, a corn and soybean meal-based diet was mixed, and treatments were added to equal aliquots of feed. Treatments consisted of 1) Negative control (no additive, not inoculated with PEDV); 2) Positive control (no additive, inoculated with PEDV); 3) As 2 + 0.3% commercial formaldehyde (Sal CURB, Kemin, Des Moines, IA); 4) As 2 + 0.8% Dominnate™ (Dominnate, PMI, Arden Hills, MN) equivalent. Following 24 h incubation, all treatments, except the Negative control, were inoculated with PEDV and incubated for 24 h at room temperature. Following incubation, PBS was added, and samples were incubated an additional 24 h. The PBS was collected and used to determine presence of virus following treatment using qPCR, and to conduct bioassays on pigs to determine infectivity.

In a separate study, 300 pigs (33 lb. body weight; 100 pigs/treatment) were used to evaluate the efficacy of Dominnate™ as a feed mitigant. Complete feed was mixed to either contain no mitigant, or to contain 0.8% Dominnate™ equivalent. Feed was placed in individual feed bins and each bin was assigned to receive 1 of 3 challenge treatments. Treatments consisted of 1) Porcine epidemic diarrhea virus (PEDV); 2) porcine respiratory and reproductive syndrome virus; and 3) Seneca Valley A. Viral challenge was delivered via a 1.0 lb. ice block, consisting of either 100 mL Seneca Valley A (5 logs TCID50/mL, Ct = 20.72), 100 mL PRRSV 174 (5 logs TCID50/mL, Ct = 21.38), or 100 mL PEDV (5 logs TCID50/mL, Ct = 24.25) balanced with 154 mL MEM. Blocks were frozen at -80C and placed into each feed bin on days 0 and 6 of the study on top of the feed. As the blocks melted liquid permeated the feed, which was augured into feeders in the room for pigs to consume via natural feeding behavior. Inoculated feed was delivered by auger to pens for 15 days. The percentage of infected pigs was determined by viral recovery using rectal swabs (n = 30 pigs/treatment), and disease was determined by observation of clinical signs (n = 6 pens/treatment).

Results

Performance

When fed from 33 to 59 lbs. of body weight (late nursery), pigs provided OutPace® FA had 5.0% higher average daily gain (1.47 vs. 1.40 lbs./d), consumed 1.9% more feed per day (2.18 vs. 2.14 lbs./d), and were 3.2% more efficient than pigs fed the control diet (Figures 1, 2, and 3). Pigs that were provided OutPace® FA from 13 to 54 lbs. body weight (complete nursery) had a 6.1% increase in average daily gain (1.05 vs. 0.99 lbs./d), a 2.2% higher average daily feed intake (1.37 vs. 1.34 lbs./day) and were 3.7% more efficient than pigs fed control diets (Figures 1, 2, and 3).

Nonmedicated feed

There were no differences in body weight at any time during the nursery period between pigs provided medicated diets, or unmedicated diets (Table 2). Additionally, early and overall ADG, ADFI, and feed conversion were not different between the treatments (Table 3). However, feeding unmedicated diets with OutPace® FA provided a cost savings compared to medicated diets in this study (Table 4).

Feed mitigation

When added to feed, Dominnate™ reduced detectable PEDV compared to the positive control and commercial formaldehyde (Table 5). Additionally, while positive control pigs became infected, Dominnate™, as well as commercial formaldehyde, was able to mitigate the infectivity of PEDV in feed. In the second

Figure 1: Effect of OutPace® FA on average daily gain. Overall nursery (5 trials; growth from 13 to 54 lbs.). Late nursery (4 trials; growth from 33 to 59 lbs.).

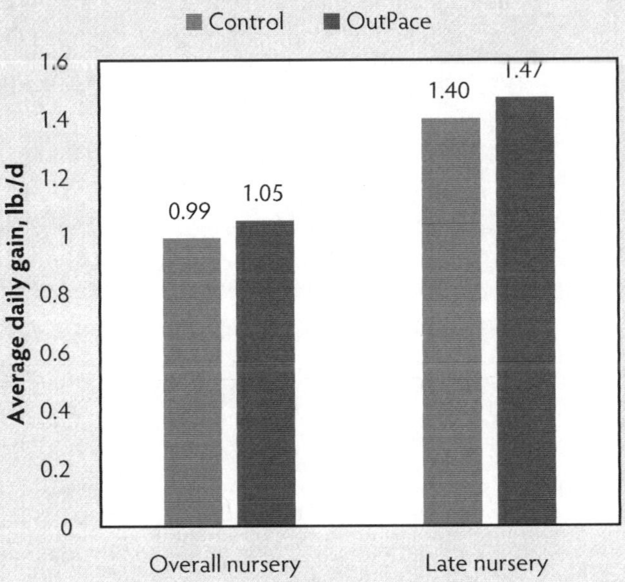

Figure 2: Effect of OutPace® FA on average daily feed intake. Overall nursery (5 trials; growth from 13 to 54 lbs.). Late nursery (4 trials; growth from 33 to 59 lbs.).

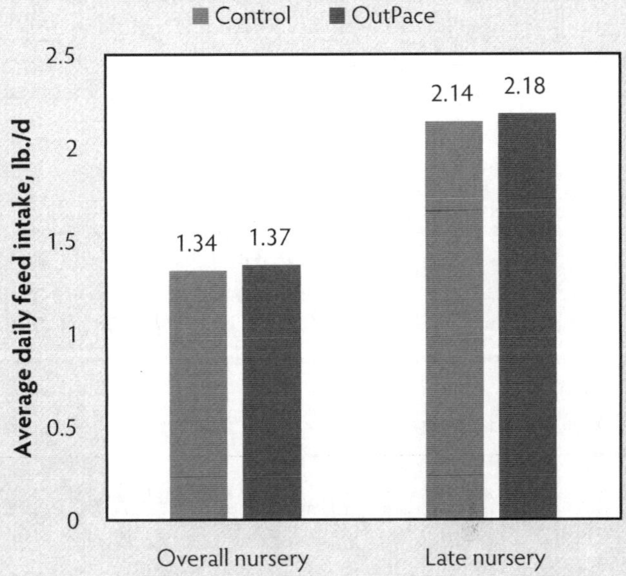

Figure 3: Effect of OutPace® FA on feed conversion. Overall nursery (5 trials; growth from 13 to 54 lbs.). Late nursery (4 trials; growth from 33 to 59 lbs.).

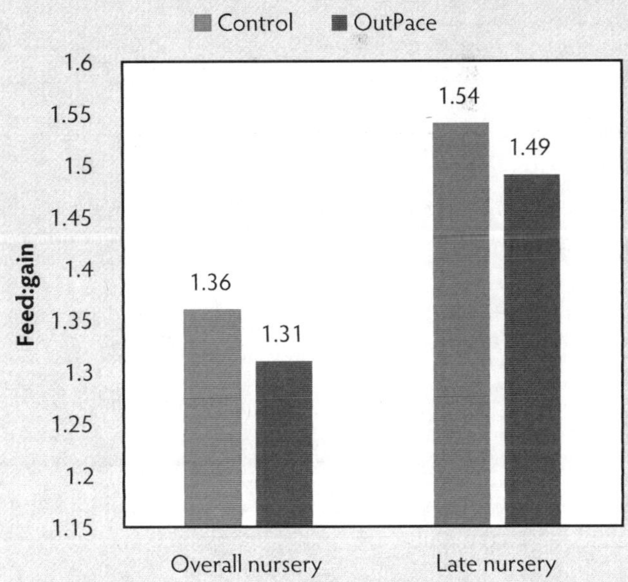

study, 100% of control-fed pigs tested were positive for PRRSV, 70% were positive for PEDV, and 20% were positive for Seneca Valley A compared to 0% for PRRSV, 23% for PEDV, and 0% for Seneca Valley A in the Dominnate™-fed pigs (Table 6). Additionally, providing Dominnate™ in the feed reduced the incidence of clinical signs for each virus when compared to control-fed pigs. Finally, the overall average daily gain was higher in challenged pigs that were provided diets containing Dominnate™ compared to control-fed pigs.

Discussion

Medium chain fatty acids are a useful tool that can be employed in swine diets to accomplish a variety of outcomes. Providing OutPace® FA during the nursery period improved gain, intake, and feed efficiency. In the absence of in-feed antimicrobials, an elevated inclusion of OutPace® FA provided similar performance as the medicated feed and provided a favorable economic return. When the focus is enhancing feed biosecurity, the MCFA blend present in Dominnate™ was able to reduce the amount of PEDV RNA recovered in treated feed samples and prevented, or reduced, the transmission of PEDV through the feed. Collectively, these studies highlight the effectiveness of activated MCFAs in swine production.

References

1. Dee et al. An evaluation of contaminated complete feed as a vehicle for porcine epidemic diarrhea virus infection of naïve pigs following consumption via natural feeding behavior. *BMC Vet Res*. 2014. 10:176.

2. Dee et al. An evaluation of porcine epidemic diarrhea virus survival in individual feed ingredients in the presence or absence of a liquid antimicrobial. *J Porcine Health Manage*. 2015. 1:9.

Table 2: Body weight (lbs.) of pigs fed either a medicated or unmedicated diet

	OutPace	Medicated	*P*-value
D 0	12.1	12.1	0.90
D 7	15.4	15.1	0.62
D 21	27.7	27.3	0.82
D 42	55.9	56.1	0.94

Table 3: Nursery performance of pigs fed either a medicated or unmedicated diet

	OutPace	Medicated	*P*-value
D 0 to 7			
ADG, lb./d	0.481	0.416	0.30
ADFI, lb./d	0.487	0.460	0.58
F:G	1.10	1.12	0.87
D 0 – 42			
ADG, lb./d	1.02	1.01	0.90
ADFI, lb./d	1.43	1.43	0.94
F:G	1.40	1.41	0.27
Livability, %	96.3	93.4	0.28

Table 4: Economics of OutPace in unmedicated diets

	OutPace	Medicated
Cost/lb. gain	$0.1898	$0.2010
Cost 40 lb. gain	$7.59	$8.04
Savings/pig	$0.45	-

Table 5: Effect of chemical mitigant used to treat swine feed contaminated with PEDV

	Viral RNA in feed	Bioassay results					Viral RNA in cecum
Treatment[1]	Feed Ct[2]	-2 dpi[3]	0 dpi	3 dpi	5dpi	7 dpi	
Negative control	> 36	---[4]	---	---	---	---	-
Positive control	28.0[b]	---	---	+--	++-	+--	+
Formaldehyde	29.2[b]	---	---	---	---	---	-
Dominnate™	32.2[a]	---	---	---	---	---	-

1 Negative control (NC) = non-chemically treated corn-soybean meal-based feed not inoculated with PEDV; Positive control (PC) = NC inoculated with PEDV; Formaldehyde = PC + 0.3% Sal CURB (Kemin Industries, Des Moines, IA); Dominnate™ = PC + 0.8% Dominnate equivalent (PMI, Arden Hills, MN).
2 A cycle threshold (Ct) > 36 was considered negative for presence of PEDV RNA. Lower Ct value equates to more PEDV RNA in the sample.
3 dpi = Day post-inoculation.
4 A (-) indicates a PEDV negative fecal swab from one pig where a (+) indicates a PEDV positive fecal swab.
ab Means with differing superscripts within column differ ($P < 0.05$).

Table 6: Effect of Dominnate™ on diagnostics, clinical signs, and performance[1]

	Infection (virus recovery)			Disease (clinical signs)			Gain
	PEDV[2]	PRRSV	SVA	PEDV	PRRSV	SVA	ADG, lbs./d
Control[3]	77%	100%	20%	100%	100%	100%	0.93[b]
Dominnate™	23%	0%	0%	33%	0%	0%	1.56[a]

1 Inoculated feed was delivered by auger to pens for 15 days. The percentage of infected pigs was determined by viral recovery (n = 30 pigs/trt), and disease was determined by observation of clinical signs within a pen (n = 6 pens/trt).
2 PEDV = porcine epidemic diarrhea virus; PRRSV = porcine reproductive and respiratory Syndrome virus; SVA = Seneca Valley A virus
3 Control = complete feed with no mitigant; Dominnate™ = Control + 0.8% Dominnate™ equivalent (PMI, Arden Hills, MN)
ab Means with differing superscripts within column differ ($P < 0.05$).

Estimation of water-soluble zinc amino acid complex supplementation on growth performance and immune response of nursery pigs

P. Aparachita[1]; S. D. Carter[1]; C. V. Cooper[1]; J. A. Harshman[1]; A. M. Sawyer[1]; T. J. Scaff[1]; Z. J. Rambo[2]; T. L. Ward[2]
[1]Oklahoma State University, Stillwater, Oklahoma; [2]Zinpro Corporation, Eden Prairie, Minnesota

Introduction

Zinc is used for supporting growth and boosting the immune function of weaned pigs during the nursery period. Conventional administration of zinc is limited to dietary intake whereas supplementing zinc via drinking water is now available. We conducted two experiments evaluating the effects of titrated levels of a water-soluble zinc amino acid complex on growth performance and immune response of nursery pigs. Based on the results, we were able to estimate the optimal dose response of zinc using non-linear statistical models.

Materials and methods

Two experiments utilizing 280 crossbred weaned pigs were allotted to one of four water zinc treatments (10 pigs/pen; 7 pens/treatment) in a randomized complete block design. The titrated levels of water-soluble zinc amino acid complex (TruCare, Zinpro Corporation, Eden Prairie, MN) were 0, 20, 40, 80, and 0, 40, 80, 160 mg Zn/L water for Experiment 1 and 2, respectively. All pigs were fed common corn-soybean meal based diets in 4 phases: phase 1 (added 2,500 mg Zn/kg as ZnO; d 0-7), phase 2 (added 1,750 mg Zn/kg as ZnO; d 7-14), phase 3 and phase 4 (added 200 of Cu/kg as $CuSO_4$; d 14-23 and d 23-42, respectively). Pen was the experimental unit for ADFI, ADG, and G:F. On d 23, all pigs were challenged with an intramuscular injection of lipopolysaccharide (LPS O55:B5; 12 μg/kg BW). Blood was collected from two pigs per pen (n = 14 per treatment) for analysis of serum tumor necrosis factor-alpha (TNF-α) and serum zinc at h 0 (baseline), h 3, and h 12 (post-LPS). Data from each experiment were analyzed to estimate the optimal dose to improve ADG, G:F, serum Zn, and reduce serum TNF-α. Data were analyzed using SAS version 9.4 statistical program.[1] PROC NLIN and PROC NLMIXED were used to estimate initial parameters of non-linear segmented models[2] in order to plot the best fit of zinc dose using quadratic broken line model with a random component.[3]

Results and discussion

Combining data from the common treatments in both experiments (0, 40, and 80 mg Zn/L) demonstrated that pigs drinking TruCare had improved ADG (linear, $P = 0.03$) and G:F (linear, $P < 0.001$) from d 0-42 (Figure 1). Three hours following LPS, the febrile response was mitigated for pigs consuming TruCare with a decrease in serum TNF-α (linear, $P = 0.04$). Furthermore, the depression in serum zinc following LPS was mitigated with pigs consuming TruCare having a linear increase in serum zinc ($P = 0.04$). Using all treatments across both experiments, the estimated optimal doses of TruCare (0, 20, 40, 80, and 160 mg Zn/L) for the ideal response in ADG, G:F, TNF-α and serum zinc were 74, 70, 40, and 160 mg Zn/L, respectively (Figure 2). In conclusion, supplementing TruCare appears to alleviate the hypozincemia condition induced by inflammation. Additionally, our data suggest that the optimum dose of TruCare in drinking water is 40-80 mg Zn/L to enhance growth performance and mitigate febrile responses in newly weaned pigs.

References

1. Copyright© 2013, SAS Institute Inc., Cary, NC, USA.

2. Fadel, J. G. (2004). Technical Note: Estimating parameters of nonlinear segmented models. *Journal of dairy science, 87*(1), 169-173.

3. Robbins, K. R., Saxton, A. M., & Southern, L. L. (2006). Estimation of nutrient requirements using broken-line regression analysis. *Journal of animal science, 84* (suppl_13), E155-E165.

Figure 1: Combining results of (A) ADG, (B) G:F, (C) serum TNF-α, and (D) serum zinc from Experiment 1 and 2 for treatment 0, 40, and 80 mg Zn/L water of TruCare. Linear increases in ADG and G:F were observed from d 0-42 ($P < 0.05$), while a linear decrease in serum TNF-α ($P = 0.04$) and a linear increase in serum zinc ($P = 0.04$) were observed across hours of LPS challenge on d 23.

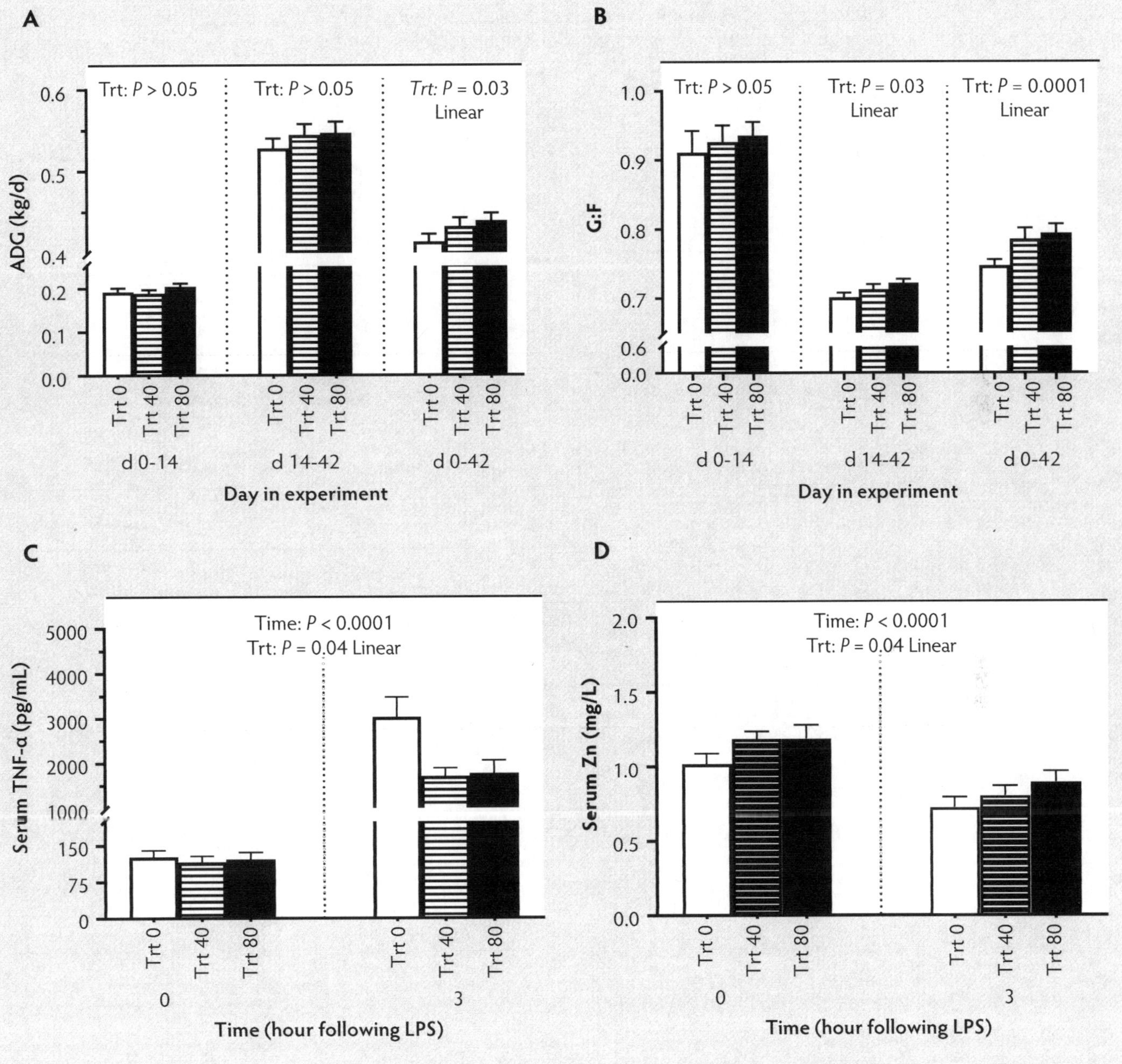

Figure 2: Fitted regression lines for (A) ADG, (B) G:F, (C) serum TNF-α, and (D) serum zinc using a quadratic broken line model with experiment as a random component. Vertical dotted line indicates the best estimated dose of TruCare.

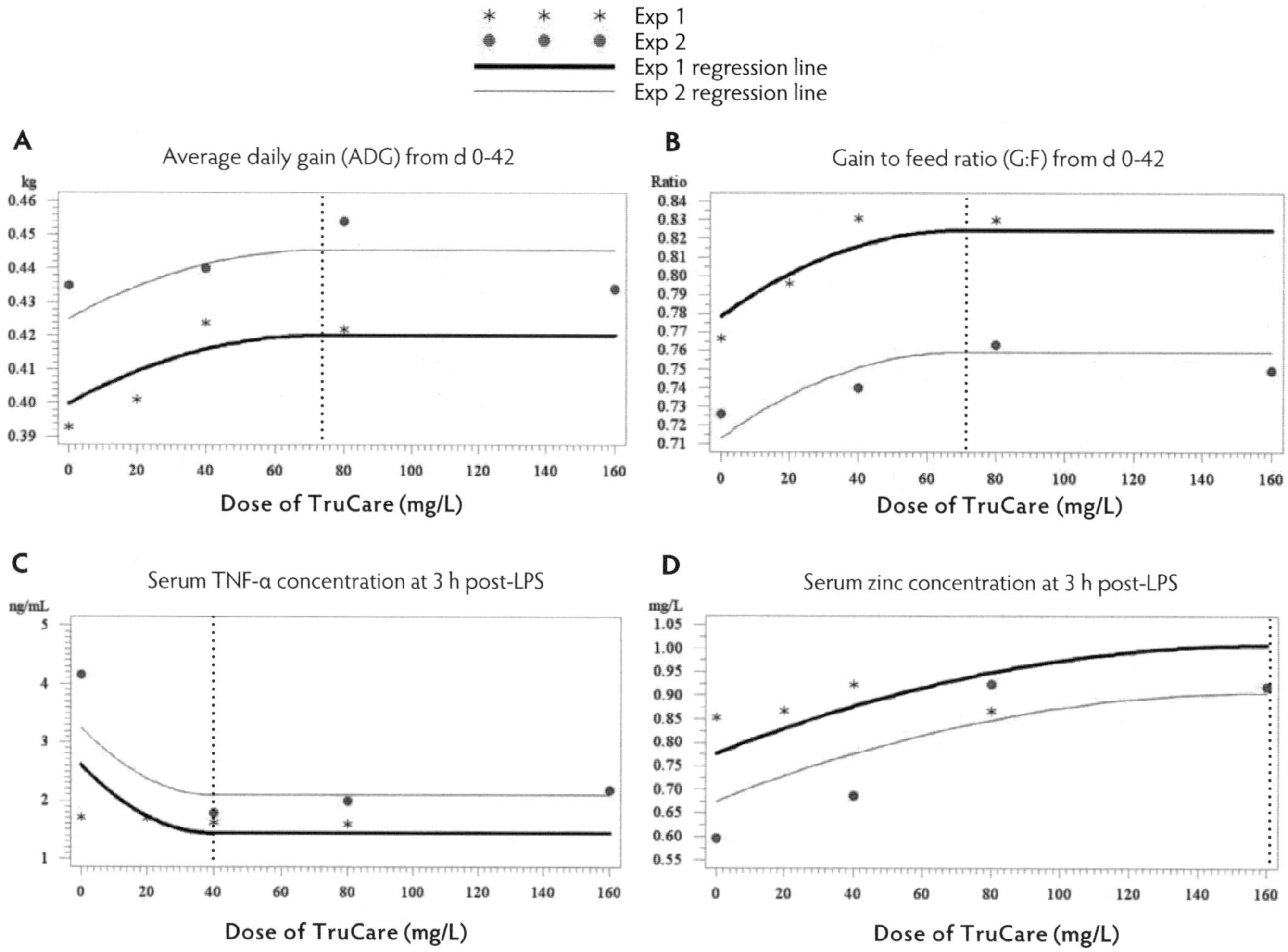

Efficacy of Vigilex™ feed ingredient in reducing PEDV contamination risk in complete feed and impact of PEDV transmission and nursery performance in pigs infected via non-feed vectors

Kari Saddoris-Clemons, PhD and Sara Ebarb, MS
Provimi, Cargill Animal Nutrition and Health, Brookville, Ohio

Introduction

The US experienced a porcine epidemic diarrhea virus (PEDV) epidemic in 2014 and the virus still remains circulating in 2019. Feed was shown to be a potential vector for PEDV transmission.[1] Further work has indicated that PEDV can survive in a simulated transboundary shipping model in ingredients, such as conventional and organic soybean meal, choline, lysine, and vitamin D.[2] PEDV is viable in complete feeds for up to 45 days and for at least 30 days in DDGS, meat and bone meal, red blood cells, Lysine-HCl, DL-Methionine, choice white grease, and choline chloride.[3] PEDV survived for the longest in soybean meal, where virus was viable for as long as 180-days post-infection.[3] Due to the potential for virus to remain infectious in complete feed and feed ingredients, interest exists to include a mitigant in feed to reduce the chance of feed being a vector for disease transmission. Provimi's Vigilex™ feed ingredient, a proprietary oil blend, was developed to reduce the risk of PEDV contamination in feed. The objective of the first study was to determine the ability of Vigilex to reduce PEDV contamination risk in complete feed. The objective of the second study was to determine the impact of Vigilex on PEDV transmission and nursery pig performance in pigs infected through a non-feed vector.

Materials and methods

Both studies were conducted at BSL-2 accredited facilities.

Study 1 (in-vitro and bioassay)

The study consisted of two parts, an *in vitro* and an *in vivo* bioassay portion. In the *in-vitro* portion, the ability of Vigilex™ feed ingredient to reduce the risk of PEDV infection in complete feed following a 24-hour incubation period was evaluated. For the in-vitro portion, Vigilex was added to complete sow feed at a rate of 0.4%. Complete feed with and without 0.4% Vigilex was shipped to a BSL-2 laboratory where samples were stored a 4°C. A US PEDV prototype strain cell culture isolate USA/IN19338/2013 was isolated and propagated in Vero cells. Virus stock utilized was harvested at passage 7 and the titer was confirmed to be 10^5 $TCID_{50}$/mL. All work was conducted at 4°C in order to provide optimal growth conditions for PEDV. Three, 90 g subsamples of feed (3 replications/trt) were collected from each complete feed sample. Complete feed without the Vigilex™ feed ingredient was used for both the strict non-infected control and the infected control treatment. Virus stock solution was diluted 1:10 and 10 mL of the diluted stock solution was added to 90 g of feed and mixed for 5 minutes. The strict non-infected control was inoculated with 10 mL of phosphate buffered saline (**PBS**). Immediately following the mixing period, the time 0 (**T0**) sample was collected from each treatment group and another sample was collected 24 h post-virus inoculation (T24). At the designated time points, a 20 g subsample was collected from each replicate and suspended in 80 mL of PBS and stored at 4°C overnight. Following the rest period, the mixture was vigorously mixed and a 5 mL aliquot from the supernatant was obtained and stored at -80°C for qPCR analysis. An additional 50 mL aliquot of the supernatant was collected at T24 and stored at -80°C for use in the subsequent swine bioassay. The feed suspension was evaluated by use of a PEDV N-gene–based RT-qPCR assay.[4]

The swine bioassay was conducted following the in-vitro study. Pigs (n = 20) were obtained from a commercial swine farm and confirmed fecal PCR negative for PEDV and Delta Coronavirus and serum negative for PEDV antibodies upon arrival. Pigs (10 days of age) were weighed, blocked by body weight, and randomly assigned to one of the two treatment groups. Treatments consisted of the supernatant collected from the infected control sample and the infected feed + 0.4% Vigilex sample obtained from T24 of the in-vitro assay. Treatments were housed in separate rooms within the BSL-2 facility to prevent transmission from one treatment group to another. Pigs were housed across two pens with 5 pigs/pen within each room. Following a 4-day acclimation period to a common dry diet, inoculation material was thawed on ice at room temperature and each of the 3, T24 time point replicates were pooled, and gently mixed immediately prior to administration. All pigs were orally gavaged with 10 mL of the appropriate treatment material. Pigs were monitored for clinical signs during the 10-day study period. Individual fecal swabs were collected on D0, 7, and 10 post-inoculation for PEDV PCR testing and serum was collected on D0 and 10 post-inoculation for detection of PEDV antibodies. Pigs were necropsied on D10 and tissue samples were collected from the ileum and submitted for PEDV immunohistochemistry (**IHC**) testing. All diagnostics were conducted at Iowa State University Veterinary Diagnostic Laboratory.

Study 2

To determine the effect of the Vigilex™ feed ingredient on PEDV transmission and pig performance in pigs infected by a non-feed vector a seeder pig model was utilized. Upon arrival to the study facility, eighty (n = 80), 10-day old pigs were blocked by body weight and randomly assigned to one of two treatments within block. Animals were housed in separate rooms to prevent potential of PEDV transmission between treatment groups. Pigs were acclimated to the facility and a common dry diet for 2 days. Dietary treatments were initiated on Day 3 post-weaning. Dietary treatments consisted of 1) Standard commercial diet with Denagard®/CTC and 2) Standard commercial diet with Denagard®/CTC + 0.4% Vigilex. Dietary treatments were fed from study Day 0-21. On D7, two pigs from each pen were randomly selected as seeder pigs. Each seeder pig was orally gavaged with 10 mL of an inoculate containing 10^4 $TCID_{50}$/mL of PEDV (2013 strain isolated in North Carolina). Individual fecal swabs were collected on D7, 9, 11, 13, 15, 17, 19, and 21 and analyzed for PEDV by PCR at Iowa State University Veterinary Diagnostic Laboratory. In addition, daily observations were conducted, and fecal scores were assigned to each pig. Pigs were weighed on D0, 6, and 21. Feed added was recorded and feed remaining was weighed on D21 to determine feed disappearance.

Results and discussion

Study 1

The strict negative control samples remained PEDV negative at both T0 and T24. Table 1 shows PEDV levels present in feed samples at both timepoints in the *in-vitro* study. The PC treatment averaged 2.6×10^6 genomic copies/ml of PEDV at T0 (5 min post-mixing) and was reduced to 1.46×10^6 genomic copies of PEDV at 24 hr post-inoculation. Complete feed treated with 0.4% Vigilex averaged 1.63×10^6 genomic copies/ml at T0 and was reduced to 8.92×10^4 genomic copies/ml. The baseline to calculate virus reduction over 24 hr was selected to be the control T0 due to previous work indicating a consistent immediate reduction in virus copies at the initial timepoint in feed samples containing the Vigilex™ feed ingredient. Over the 24-hr period PEDV copies declined 0.24 logs in the control feed sample and declined ($P < 0.01$) 1.46 logs in the feed sample containing Vigilex. This indicates that Vigilex starts to work to reduce PEDV levels in feed on contact and continues to work over time. Under laboratory conditions, the Vigilex™ feed ingredient was able to reduce virus concentration in feed by greater than 1 log. However, viral DNA was still present in the sample. The bioassay was conducted in pigs using the material collected at the 24-hr period to determine if the virus remaining in the feed was live, infectious virus.

Table 2 shows the results of the bioassay. All pigs (10/10) administered the T24 PEDV infected control sample were confirmed PEDV positive by fecal swab on D7 and 3/10 pigs remained positive on D10. In addition, clinical signs of watery diarrhea were observed from D4-D8 post-inoculation and 5/10 were serum ELISA positive for PEDV on D10. Pigs can seroconvert to be PEDV ELISA positive between D10-D21 days post-exposure. All intestinal samples were IHC negative for PEDV on Day 10 post-inoculation. This result was not surprising as PEDV IHC positives are usually observed in the first 4 days post-infection. Pigs administered the T24 PEDV infected + Vigilex sample were negative for PEDV by fecal swab on D7 and D10 post-inoculation. No clinical signs of disease were observed during the study period, and all pigs were serum ELISA negative for PEDV, and all intestinal samples were IHC negative for PEDV. This indicates that the PEDV detected by PCR in the Control sample, 24-h post-inoculation contained infectious virus that was capable of causing clinical disease in 14-day old pigs. The PEDV detected in the feed sample treated with the Vigilex™ feed ingredient at 24h post-inoculation by PCR was inactivated virus that was not capable of inducing clinical disease in 14-day old pigs.

Table 1: The effect of Vigilex™ feed ingredient on PEDV concentration in complete feed following 24 hr of virus exposure

Treatment	Time 0 h Ct-value	Time 24 h Ct-value	Genomic copies/ml at T0	Genomic copies/ml at T24	Log reduction vs control T0 over 24 h
Negative control	NA	NA	NA	NA	NA
Positive control	26.4	27.1	2.6×10^6	1.46×10^6	0.25
Vigilex™ feed ingredient	27.6	31.6	1.63×10^6	8.92×10^4	1.46

Table 2: The number of pigs fecal PCR positive for PEDV during the swine bioassay

	Number of pigs fecal PCR positive for PEDV		
Treatment	0 DPI	7 DPI	10 DPI
Infected control	0/10	10/10	3/10
Infected +Vigilex™ feed ingredient	0/10	0/10	0/10

Study 2

The seeder pig model utilized in the transmission study was successful at infecting 80 and 90% of pigs chosen as seeder pigs by 1-day post-infection, in the Control and Vigilex treatment groups, respectively (Table 3). By day 3 post-challenge, 93 and 90% of the non-seeder pigs were fecal PCR positive for PEDV in the control and Vigilex treatments, respectively. By D5 post-infection, all pigs were 100% positive. The load of fecal viral shedding was between 8-9 logs of virus for all treatment groups within 3 days post-inculcation of the seeder pigs (Table 4). As expected seeder pigs were shedding more virus on day 1 post-infection; however, there were no meaningful differences between treatment groups. The performance results are shown in Table 5. Final body weight was significantly ($P < 0.05$) greater in pigs fed the Vigilex™ feed ingredient on both D6 and D21. Overall ADG and ADFI tended ($P < 0.10$) to be greater for pigs fed Vigilex compared to pigs fed the Control diet. The benefits of Vigilex when used in diets at the level to reduce the risk of PEDV transmission in feed does not help reduce the risk of PEDV transmission or reduce viral shedding when infection occurs through a non-feed vector. However, the Vigilex™ feed ingredient does appear to provide a performance benefit of improved feed intake and gain, and this benefit is maintained during a PEDV outbreak.

In conclusion, the Vigilex™ feed ingredient, when included in complete feeds at 0.4%, reduced the risk of PEDV transmission in feed within 24 h of exposure. Vigilex was not able to reduce the risk of transmission or reduce fecal shedding when virus infection occurs via a non-feed vector, but did improve the performance of pigs following a PEDV infection, when included at a rate of 0.4% in complete nursery feeds.

Acknowledgments

A special thanks to Dr. Zhang from Iowa State University for conducting the in-vitro portion of the feed assay and providing the PEDV virus stock.

References

1. Dee, S., T. Clement, A. Schelkoph, J. Nerem, D. Knudsen, J. Christopher-Hennings, and E. Nelson. An evaluation of contaminated complete feed as a vehicle for porcine epidemic diarrhea virus infection of naïve pigs following consumption via natural feeding behaviour: proof of concept. BMC Veterinary Research 2014, 10:176.

Table 3: The percentage of pigs fecal PCR positive for PEDV following inoculation of the seeder pigs

Day post-challenge	CON	CON-SEEDER	Vigilex™	Vigilex™ -SEEDER
1 DPC	6.7%[a]	80%[b]	3.3%[a]	90%[b]
3 DPC	93%	100%	90%	90%
5 DPC	100%	100%	100%	100%
7 DPC	100%	100%	100%	100%
9 DPC	100%	88.9%	100%	100%
11 DPC	83.3%	88.9%	80%	60%
13 DPC	50.%	44.4%	56.7%	70%
14 DPC	66.7%	44.4%	60%	70%

a,b Rows with different superscripts differ, $P < 0.05$.

Table 4: Fecal shedding of PEDV (logs of virus) in positive pigs following inoculation of the seeder pigs

Day post-challenge	CON	CON-SEEDER	Vigilex™	Vigilex™SEEDER
1 DPC	5.76[ab]	7.05[bc]	5.35[a]	8.44[c]
3 DPC	8.25[a]	9.06[ab]	8.87[b]	8.97[ab]
5 DPC	8.63[b]	6.28[a]	8.55[b]	6.39[a]
7 DPC	7.06[b]	4.94[a]	6.89[b]	4.83[a]
9 DPC	4.80	4.75	4.72	5.48
11 DPC	5.61	6.75	5.43	5.34
13 DPC	5.55	4.48	5.12	6.06
14 DPC	0.62	0.70	0.52	0.85

a,b Rows with different superscripts differ, $P < 0.05$.

Table 5: The effect of Vigilex™ feed ingredient on pig performance during a non-feed exposure to PEDV in nursery pigs[1]

Variable	Control	Vigilex™	SE	*P*-value
Initial BW (D0) lb	7.52	7.57	0.628	0.803
Pre-Challenge BW (D6), lb	9.57[a]	10.12[b]	0.721	0.036
Final BW (D21), lb	18.04[a]	19.01[b]	0.913	0.011
ADG (D0-6), lb/d	0.342	0.425	0.040	0.193
ADG (D6-D21), lb/d	0.538	0.593	0.022	0.157
ADG (D0-D21), lb/d	0.483	0.545	0.018	0.071
ADFI (D0-D21), lb/d	0.672	0.716	0.037	0.074

[1] PEDV challenge was conducted on study day 7.
[a,b] Means within a row with different superscripts are different ($P < 0.05$)

2. Dee., S. F. Bauermann, M. Neiderwerder, A. Singrey, T. Clement, M. deLima, et al. 2018. Survival of viral pathogens in animal feed ingredients under transboundary shipping models. PLoS ONE. 13(3):e0194509.

3. Dee, S., C. Neil, T. Clement, A. Singrey, J. Christopher-Hennings, and E. Nelson. 2015. An evaluation of porcine epidemic diarrhea virus survival in individual feed ingredients in the presence or absence of a liquid antimicrobial. Porcine Health Management. 1:9.

4. Schumacher, L., J.C. Woodworth, C. K., Jones, Q. Chen, J. Zhang, et al. 2015. Evaluation of the minimum infectious dose of porcine epidemic diarrhea virus in virus-inoculated feed. AJVR. 77(1):1108-14.

Porcine circovirus type 3: update on field experiences with Sequivity™ vaccination for this emerging pathogen

Jack Creel[1], DVM; Brian Schantz[2], DVM; Greg Armbruster[1], DVM, MBA; Justin Cagle[1], DVM; Kimberly Crawford[1], DVM, MS; Kelly Greiner[1], DVM; Megan Inskeep[1], DVM; Doug King[1], DVM; James Lehman[1], DVM, MS Jeremy Maurer[1], VMD; Brett O'Brien[1], VMD, MS; Abigail Redalen[1], DVM; Channing Sebo[1], DVM; Brad Thacker[1], DVM, PhD;
[1]Merck Animal Health, DeSoto, Kansas; [2]Diversified Swine Management, Laurel, Nebraska

Introduction

Porcine circovirus type 3 (PCV3) was first reported in the US in 2015.[1,2] Several conditions have been attributed to PCV3 including porcine dermatitis and nephropathy syndrome (PDNS), multi-systemic inflammation, myocarditis, congenital tremors, and reproductive failure including abortions, increased mummified fetuses and non-viable piglets, reduced litter size and delayed return to estrus. In 2018, customers began to request PCV3 vaccine produced by Merck Animal Health's Sequivity™ Technology RNA-Particle (RP) platform.[3] Sequivity Technology utilizes an alphavirus replicon vector system to produce vaccines that contain a pathogen's gene of interest (GOI) that encodes for a protective antigen. For PCV3, the gene of interest is ORF2 which encodes for the capsid protein of the virus.

This paper provides an update of field experiences as Merck Animal Health works with customers to implement a PCV3 prescription vaccination program in their swine herds. These customer interactions include preliminary data from an on-going survey of veterinarians that have prescribed PCV3 vaccines, our normal day-to-day contacts with customers and a case study of a sow herd where diagnostic testing was done prior to and after whole herd vaccination.

Reasons for vaccination

Survey findings to-date indicated that increased mummified and stillborn fetuses were the most common reproductive reasons for implementing vaccination. Reduced conception and/or farrowing rates were also reported as reasons for vaccination. Growing pig performance and weaned pig quality were reported as well. Diagnostic findings of positive PCR results in processing fluids, tissues, sow and pig blood and oral fluids were also reported as reasons for implementing vaccination. Fetal death resulting in mummified or stillborn fetuses that contain high amounts of PCV3 (low PCR cycle time or CT values) provide the clearest link of PCV3 infection with disease. The link between PCV3 infection and clinically apparent disease in growing pigs is less clear.

Vaccination strategies

To date, PCV3 sequence analysis of PCV3 field isolates has found minor amino acid homology differences. Accordingly, customers implementing a prescription based vaccination program have requested the same ORF 2 construct. Veterinarians typically prescribe the vaccine for use in breeding animals. Typically, vaccination is implemented by performing a mass vaccination of sow herd including developing gilts, twice at a 3-4 week interval. Follow up vaccinations are done by periodic whole herd boosters or by production phase (weaning, post pregnancy examination or prior to farrowing). In some herds, no additional vaccinations are done following the mass vaccinations although in some of these herds, vaccination of gilts twice prior to breeding is continued. A few herds vaccinate replacement gilts at younger, growing ages.

Monitoring strategies

Diagnosis of PCV3 is mostly limited to testing a variety of sample types by PCR. To-date, serological assay development efforts have not resulted in a test that is useful in the field. Identification of PCV3 in tissues by in situ hybridization (ISH) has been reported. Increasingly, processing fluids are used to determine sow herd PCV3 status and level of infection prior to and after implementing a vaccination program. We have encountered several situations where processing fluids contained a relatively high level of PCV3 based on PCR testing (low CT values) prior to vaccination and were then completely free of PCV3 (high CT values) following vaccination. The time from vaccination to negative processing fluids has been around 4 months in several situations. In one particular case, a pig vaccine study was initiated approximately 4 months after mass vaccination of the sow herd. Pig vaccination was performed at processing and weaning. Processing fluids collected by litter were all negative. Repeated sampling of the pigs (sera) and their pens (oral fluids) from weaning to market yielded PCR negative results at all timepoints. Accordingly, neither vertical or lateral transmission of PCV3 was detected following whole sow herd vaccination.

Reported benefits

From the survey, the reported production benefits from prescription vaccination programs in sow herds included improvement in weaned pig quality, decreased mummified and still born pigs, improvement in conception and farrowing rates and increased liveborn. In some herds, it was too early to determine any benefit and in others no impact was observed. In our day-to-day discussions with customers, determining the impact of vaccination is often difficult as the clinical presentation of PCV3 can be vague and the magnitude of the problem is much less than other disease such as PRRS (sows and pigs) or PCV2 (pigs). In any sow herd, fertility and litter size can be impacted by other factors besides PCV3, especially PRRS outbreaks which can impact herd performance for several months.

The clearest indication of the benefits of vaccination are related to decreasing the level of infection in the sow herd based on processing fluid testing as discussed above. In most cases, this observation is based on before and after vaccination comparisons. There is one report of reduced levels of PCV3 in processing fluids in sows vaccinated during mid-gestation compared to non-vaccinated sows within the same farms and breeding groups.[4] Accordingly, the benefit of vaccination depends on the degree that PCV3 field infection is impacting production, which might vary from herd to herd. We also have little understanding of the long-term consequences of PCV3 infection in individual sows. There is circumstantial data that suggests viremia can last for several months but no data to indicate when PCV3 infection is cleared and the subsequent duration of immunity. Furthermore, the lack of an accurate serological test impedes the needed laboratory studies and field investigations to address these questions.

Sow farm case study

This case increasingly appears to be representative of what we are observing in other sow herds that have implemented a prescription vaccination program. A 5,000 sow herd and the accompanying downstream pig flow first experienced what was diagnosed as PCV3 related disease in late January of 2019. PCV3 was detected in several cases of respiratory disease in nursery age pigs. The pigs were also infected with PRRSV and/or swine influenza virus. In these cases, the PCV3 PCR CT values ranged from 25-27. At the same time, a diagnosis of PCV3 myocarditis based on ISH was made in a separate case of a five-week-old pig where no other pathogens (PCV2, swine influenza virus, PRRSV, porcine parainfluenza virus) were detected. The CT value in this case was 27.1. In early February, PCV3 was detected by PCR in mummified fetuses (CT = 19.1), stillborn fetuses (CT = 19.0) and processing fluids (CT = 17.1). These findings led to vaccination of the sow herd and incoming gilts.

The sow herd was mass vaccinated in early April (week 14) and early May (week 18). Processing fluids were then collected weekly starting towards the end of May (week 21) with collection by parity (gilts vs. sows) starting the first week of June (week 23). The PCV3 PCR results were analyzed using statistical process control methods (Statistix 10, Analytical Software, Tallahassee, FL).

From weeks 23 to 29, the CTs of processing fluids from gilt litters were lower (more virus) than from sow litters. From weeks 30-33, the CTs were similar in gilts compared to sows but the values were trending up (declining virus levels). From week 34 to 40, samples were either negative or had CT values within 1.2 CTs of the negative cut-off. Accordingly, the time interval between the second mass vaccination and processing fluids becoming nearly negative was 16 weeks.

In week 45 (early November), a survey of downstream pigs at 5, 10 and 15 weeks of age (WOA) was performed. Serum (20 pigs per age group, tested in pools of 5) and oral fluids (2 pens per age group) were tested for PCV3 by PCR. All serum pools were PCR negative for PCV3. Oral fluids samples at 5 WOA had CTs of 34.0 and 34.6; 10 WOA – both negative; and at 15 WOA – CTs of 31.1 and 36.0. The virus levels in these samples were much lower than what was detected prior to sow herd vaccination.

With regard to reproduction, stillborn, mummy and total dead born rates appear to have started declining the first week of September (week 36) but it is still too early to determine if this trend will continue.

Concluding remarks

Understanding the disease ramifications of PCV3 infection and the role of vaccination for controlling PCV3 is definitely a "work in progress." Advances are being made but much work needs to be done. The field experiences reported here strongly suggest that vaccine could impact the level of infection in sow herds, which in turn may provide production benefits. However, the magnitude of the benefits provided by vaccination is still unknown at this time and may vary from herd to herd.

References

1. Palinski R, Pineyro P, Shang P, et al. 2016. A novel porcine circovirus distantly related to know circoviruses is associated with porcine dermatitis and nephropathy syndrome and reproductive failure. *J Virol.* 2017; 91(1):e01879-16. Doi: 10.1128/JVI.01879-16.

2. Phan TG, Giannitti F, Rossow S, et al. 2016. Detection of a novel circovirus in pigs with cardiac and multi-systemic inflammation. *Virol J.* 2016; 13:184. doi: 10.1186/s12985-016-0642-z.

3. Thacker, B. Armbruster, G., Byers, E., Cagle, J., Crawford, K., Creel, J., Lehman, J., Mogler, M., Roerink, R., Sebo, C., Strait E., Wei, H. Porcine circovirus type 3: Field experiences with a potentially emerging pathogen. In: Proc. 50th Annual Meeting of the American Association of Swine Veterinarians, Orlando, Florida, 2019, pp. 245-246.

4. Vennekotter, A., Byers, E., Thacker, B., Cagle, J. Evaluation of PCV3/2d prevalence in placental umbilical cord blood and processing fluids and reproductive outcomes following mass vaccination with an RNA-particle platform PCV3/2d vaccine. In: Proc. 50th Annual Meeting of the American Association of Swine Veterinarians, Orlando, Florida, 2019, pp. 89-92.

Field experiences with PCV2 unstable sow herds and downstream pig flows

James Lehman[1], DVM, MS; Keith Aljets[2], DVM; Nicolas Rippel[2], DVM; Grant Allison[3], DVM; Greg Armbruster[1], DVM, MBA; Justin Cagle[1], DVM; Kimberly Crawford[1], DVM, MS; Jack Creel[1], DVM; Robyn Fleck[1], DVM; Kelly Greiner[1], DVM; Megan Inskeep[1], DVM; Doug King[1], DVM; Jeremy Maurer[1], VMD; Brett O'Brien[1], VMD, MS; Abigail Redalen[1], DVM; Joe Roder[1], DVM, PhD; Channing Sebo[1], DVM; Brad Thacker[1], DVM, PhD

[1]Merck Animal Health, DeSoto, Kansas; [2]Veterinary Medical Center, Williamsburg, Iowa; [3]Walcott Veterinary Clinic, Walcott, Iowa

Introduction

The role of sow herd instability for swine pathogens is well recognized although knowledge gaps still exist with regard to: 1) understanding disease and performance impacts in the breeding herd and downstream growing pigs; 2) diagnostic strategies to accurately assess and monitor sow herd pathogen status; 3) understanding the contributing factors that destabilize a sow herd; 4) intervention methods to stabilize the sow herd; and 5) control methods to promote consistent stability over time. Furthermore, each pathogen has distinct features that limit a "one size fits all" approach for achieving stability over time. The focus of this paper is porcine circovirus type 2 (PCV2). The overall objective of this paper is to increase the awareness of swine veterinarians to the potential role that sow herd instability can play in effectively controlling PCV2.

PCV2 most commonly causes disease in grower phase pigs (10-16 weeks of age). Early infection of pigs can occur if the virus is transmitted from the dam in utero, via colostrum or via direct contact prior to weaning. PCV2 can also cause reproductive disease but in most cases, the impact on reproductive performance is relatively insignificant. Porcine circovirus associated disease (PCVAD) is still occasionally observed in late finishing pigs if the onset of infection is delayed. Investigation of these cases most often points to off-label use and/or lack of compliance with vaccination protocols as the main causes of vaccine failure.

Our knowledge of PCV2 instability in sow herds has been derived from diagnostic investigations of cases of PCVAD in vaccinated pigs reported to Merck Animal Health by producers and herd veterinarians. All adverse event (safety or efficacy) complaints are investigated with the goal of identifying the most likely contributing causes. Our rule-outs for failure of PCV2 vaccine efficacy include: 1) poor vaccination protocol compliance; 2) off-label use: 3) infection with PCV2 prior to vaccination; 4) maternal antibody interference; and 5) immunosuppressive diseases such as PRRSV at the time of vaccination. Investigations sometimes include PCV2 sequencing although the emergence of new strains has not been incriminated in vaccine failure cases to-date. Assessing sow herd PCV2 status has become a routine component of PCV2 vaccine efficacy complaint investigations.

Disease and performance impacts

A case report of PCV2 instability in a sow herd occurred in 2012 and has been previously reported.[1] In this case, PCVAD was observed in offspring in the late nursery phase. These pigs had been vaccinated for PCV2 at processing and weaning. There were no clinical signs of reproductive loss in the sow herd. Testing of colostrum for PCV2 by PCR found high PCV2 prevalence and high virus levels while testing of other herds within the system found negative colostrum samples and their downstream flows were not experiencing PCVAD. We also observed high maternally-derived antibody (MDA) levels in the unstable herd, presumably due to stimulation of the sow's immune system due to the apparent active infection. Our hypothesis was that a high percentage of piglets were already infected at the time of vaccination and that this, combined with the high maternal antibody levels at vaccination, may have contributed to the vaccine failure.

Subsequently, we have investigated other similar cases where vaccine protocol compliance has been adequate but mortality rates in vaccinated pigs approached levels observed in 2006 before vaccines became available. Co-factors, especially PRRSV, can play a role in the overall severity of the problem but some affected herds are free of PRRSV.

Diagnostic approaches and challenges

The main emphasis in investigating cases of suspected sow herd instability is to determine the timing of infection in the offspring along with assessing MDA levels. Blood sampling of pigs for detecting viremia by PCR has been done at various ages prior to the onset of clinical disease (weekly intervals back to birth). However in our experience, blood sampling of young pigs is not very sensitive in that the lack of viremia does not mean the pigs are not infected. If any viremia is detected, the level of infection is considered quite high. PCR testing of sow serum is likewise often unrewarding. In both pigs and sows where viremia is not detectable, we suspect the infection is still present in tissues but is not at a high enough level to appear in the blood. In a previous vaccination/challenge study, vaccination prevented viremia but tissue infection was still detectable.[2] It is also likely that circulating PCV2 antibodies (passive in pigs or active in sows) are able to block the ability of the PCR assay to detect PCV2 in serum. Because of the lack of sensitivity for testing the pigs and sows for viremia, the focus for understanding sow herd instability is on the sow (whole colostrum containing whey and cellular fractions), fetal blood (PUCS), pre-suckle pig blood or processing fluids. Other than pre-suckle blood, the risk of environmental contamination of the samples needs to be addressed during sample collection and interpretation of the results.

Factors contributing to PCV2 instability in sow herds

Unstable sow herds have one or both of two common factors. We have observed that within large production systems, affected sow farms often had common gilt sources. This led to investigating the status of replacement gilts and we found that gilts were PCV2 viremic upon introduction into the sow herd. Since many gilt sources are relatively high health, it was perceived that the PCV2 vaccination program did not need to be as robust as used for of commercial, conventional health status pigs. Consequently, clinical disease is prevented but the gilts are viremic and shedding PCV2 when they are moved to the sow farm. These infected, shedding gilts then contribute to increased vertical transmission of PCV2 from dams to their offspring.

The second potential contributing factor is the emergence of PCV2d (Chinese mutant) that was first reported in the United States in 2012 and became the predominate genotype starting in 2014.[3] There are two potential ways that PCV2d may have changed the overall pattern of PCV2 viremia in sow herds. The first is that the aggregate immunity of the unstable sow herd had developed from prior vaccination with PCV2a-based vaccines and/or previous exposure to PCV2a and PCV2b field viruses. It is possible that the aggregate immunity in the sow herd did not completely prevent PCV2d infection, allowing for virus circulation and vertical transmission to the offspring. The second way is that PCV2d may be more pathogenic compared to previous genotypes (PCV2a and PCV2b). The ability of PCV2d to quickly become the dominant virus in the US swine population suggests that PCV2d is more "fit" and the severity of clinical disease would be expectedly greater, especially in subpopulations of inadequately immunized pigs. Several studies have demonstrated that current commercial PCV2 vaccines are effective against PCV2d challenge.[4-6] However, regardless of the vaccine or vaccination protocol used, some animals will not be become fully immunized. If the degree and duration of viremia in these non-immunized pigs is greater with PCV2d compared to previous variants, then the rate of clinically apparent PCVAD will be greater. In addition, it is possible that PCV2d will be more readily found in sows and newborn piglets in unstable sow herds.

Intervention strategies

When diagnostic efforts indicate that sow herd instability is the cause of PCV2 disease in vaccinated pigs, sow herd vaccination becomes necessary. Stabilization of the sow herd may be accomplished by mass vaccinating the sow herd twice at a four week interval. Resolution of downstream PCV2 and cessation of detectable shedding in sows (colostrum samples are no longer PCV2 positive by PCR), regardless of PCV2 genotype, occurs usually within 3-5 months.

Thorough review of the PCV2 status of growing age replacement gilts and the PCV2 vaccination strategy for these gilts is also necessary. Shifting to a vaccination program that provides a high level of protection based on the absence of viremia the entire life of the gilt is recommended. Continual monitoring of gilts for PCV2 viremia is also recommended.

For pigs, we have not been able to alter the impact of vertical transmission by lowering the age at first vaccination. As discussed above, many times PCV2 antibody levels in the sows are high resulting in more potential for MDA interference of vaccine efficacy, which is more likely to occur with processing and weaning vaccination protocols. Higher MDA levels can occur following sow herd vaccination as well so monitoring MDA levels in two- to three-week-old pigs by IFA is often done for several months post sow herd vaccination. It is not uncommon for problem herds to move their PCV2 vaccination protocol to later time points ranging between three and eight weeks of age.

Control strategies

Once the sow herd has stabilized, follow up vaccinations include boostering previously vaccinated gilts at selection and sows at each parity (pre-breeding or mid gestation) or on a calendar basis (1-2 times per year). Recently, some herds have started to use prescription vaccine containing a PCV2d construct. The vaccine is often administered in combinations with other antigens, as prescribed by the Attending Veterinarian, which is allowed under the prescription vaccine guidelines.

In some cases, a sow herd vaccination "maintenance program" is started as a preventative measure in herds that are stable, especially in large production systems where some of the herds are unstable and if the gilt source is common.

Regardless, the impact of PCV2 sow herd instability justifies monitoring replacement gilts for PCV2 viremia and periodic assessment of sows (via colostrum) for PCV2 infection and in some cases, monitoring the pigs for PCV2 infection and MDA levels.

Concluding remarks

Control of PCV2 infections remains critical for successful swine production. This virus is still capable of causing severe economic losses of up to $20-30 per pig in affected groups. Although vaccination of pigs has been highly successful, there are situations where vaccination protocols need to be revisited due to sow herd instability. Sow herd instability can be a critical underlying factor that needs to be addressed in order to gain control of PCV2 in the downstream pigs.

References

1. Aljets, K., Rippel, N., Brinkman, M., Lehman, J., Thacker, B. Diagnostic and control strategies for managing PCV2 in the breeding herd. In: Proc. 46th Annual Meeting of the American Association of Swine Veterinarians, Orlando, Florida, 2015, pp. 393-394.

2. Thacker, B, Kniffen, T, Hesse, R, Madson, D, Strait, E Laboratory Studies on Circumvent® PCV M: Combination *Mycoplasma hyopneumoniae* & Porcine Circovirus Type 2 Vaccine In: Proc. 43rd Annual Meeting of the American Association of Swine Veterinarians, Denver, Colorado, 2012, pp. 123-126.

3. Xioa, C., Halbur, P., Opriessnig, T. Complete genome sequence of a novel porcine circovirus type 2b variant present in cases of vaccine failures in the United States. J. Virol 86: 12469, 2012.

4. Opriessnig, T., Gerber, P., Xiao, C., Halbur, P., Matzinger, S., Meng, X. Commercial PCV2a-based vaccines are effective in protecting naturally PCV2b-infected finisher pigs against experimental challenge with a 2012 mutant PCV2. Vaccine 32: 4342-4348, 2014.

5. Strait, E., Suddith, A., Allen, M., Bijlsma, J., Sanders, E., Lathouwers-van Schijndel, J., Hesse, H. Experimental PCV2d challenge of pigs vaccinated with prototype vaccines containing ORF2 from PCV2a, PCV2b or PCV2d. In: Proc. 48th Annual Meeting of the American Association of Swine Veterinarians, Denver, Colorado, 2017, pp. 189-190.

6. Thacker, B., Schlueter, R., Creel, J., Madson, D., Hesse, R. Circumvent® PCV-M G2: PCV2d challenge study and field production data update. In: Proc. 47th Annual Meeting of the American Association of Swine Veterinarians, New Orleans, Louisiana, 2016, pp.151-153.

Updates on influenza A vaccination using the SEQUIVITY™ technology

Channing Sebo[1], DVM; Pravina Kitikoon[1], DVM, PhD; Tara Donovan[2], DVM; Chandra Morgan[1], BS; Susan Knetter[1], PhD; Heather Dempsey[1], BS; Kimberly Crawford[1], DVM, MS; Mark Mogler[1], PhD; Brad Thacker[1], DVM, PhD; Erin Strait[1], DVM, PhD
[1]Merck Animal Health, De Soto, Kansas; [2]Hanor, Spring Green, Wisconsin

Introduction

Influenza A virus in swine (IAV-S) is a complicated and costly pathogen for the swine industry. Controlling this virus is a major challenge due to the vast diversity of the hemagglutinin (HA) proteins present among IAV-S circulating in the swine population. IAV-S undergoes both antigenic shift and drift leading to this extensive diversity.[1] Successful vaccination strategies are dependent on detailed sequence analysis in order to match the HA antigens in the vaccine to the strains present within the farms.[2,3] The SEQUIVITY™ Dashboard developed by Merck Animal Health is a sequence analysis tool that is the cornerstone for assisting customers with vaccination design. The Dashboard is used to track the movement and diversity of IAV-S throughout the United States and Canada, which further enables identifying emerging and evolving strains.[4]

Evaluating vaccination success within a farm is often heavily dependent on subjective clinical observations. Merck Animal Health is committed to assisting customers with vaccine evaluations that rely on laboratory testing along with clinical observations. An example of successful IAV-S vaccination that was previously reported demonstrated a reduction in clinical signs, a rise in hemagglutination inhibition (HI) titers, and a reduction in virus levels in weaning age piglets based on udder wipes and nasal swabs.[4] A continuation of this sow farm monitoring is described below.

Key antigenic sites are an important aspect of IAV-S sequence analysis, and incorporation of key antigenic site matching into vaccine design increases the likelihood of a successful vaccination response. An example of field evidence that validates the utility of key antigenic site analysis for H1N2 Delta 1a is presented.

SEQUIVITY™ Dashboard IAV-S trends

The SEQUIVITY™ Dashboard has over 12,000 IAV-S sequences dating back to 2003, with over 1,800 sequences available for analysis from the past year. Examining the trends of new and emerging strains allows our customers to make informed strain selections in vaccine design. The SEQUIVITY™ Dashboard has been tracking IAV-S sequence data to highlight strains that are making a regional impact, moving from farm to farm, and identifying which strains pose the most risk for introduction. This approach allows customers to move away from a farm-by-farm vaccine design approach to a more proactive strategy, based on regional sequence data.

An IAV-S analysis was performed looking at the trends over the past two years as shown in Table 1. The analysis compares two time periods; the first from December 2017 through December 2018 and the second from December 2018 through December 2019. Important changes identified include a decline of over 8% in both H1 gamma and H1 delta 1a strains. Strains with a rise in prevalence were H1 alpha and H1 pandemic strains of over 5% each, with the most increase found in H1 delta 2 rising from 8.6% to 20.2%. In our experience over the past several years, these findings indicate significant changes in IAV-S circulating field strains that in turn require corresponding alterations in vaccine design.

Long term sow farm vaccination strategy update

Previously, we reported clinical and laboratory findings from a sow herd that was mass vaccinated twice at a 4 weeks interval with prescription IAV-S vaccine.[4] Sample collection for laboratory testing was done before vaccine implementation, 1 month after the second mass vaccination, and 4 months after pre-farrow vaccination began. It was found that antibody titers measured via HI assay using a homologous H1N2 virus were significantly higher after vaccination implementation than before vaccination. The observed boostering of these low titers demonstrated the value of using a prescription vaccine in order to increase the level of herd immunity. Positive response to vaccination was further demonstrated by the decline in percentage of PCR positive udder wipes at each sampling time point. Prior to vaccination, 36.8% of litters tested positive by PCR for IAV-S via udder wipes, while all sampled litters tested negative at the last sampling 7 months after vaccination began. Nasal swab testing aligned and supported the udder wipe findings, as 36.7% of nasal swabs from weaned pigs were positive prior to vaccination intervention, and dropping to 0% by the time of the last sampling.

This herd has continually been monitored to measure the long-term benefits from IAV-S vaccination due to IAV-S sow farm stability. The farm has continued to use a combination of a pre-farrow vaccination with intermittent mass vaccinations and is routinely tested for strain consistency. No new IAV-S strains have been identified on this farm. The farm was sampled again approximately 13 months after vaccination implementation (Steady State 1) and again at 21 months post-implementation (Steady State 2). In order to evaluate post-vaccination serological responses and the farm's

Table 1: IAV-S strain trends over time

Strain	Dec. 2017-Dec. 2018 (% of total sequences)	Dec. 2018-Dec. 2019 (% of total sequences)	Change
H3 cluster 4	10.8%	11.4%	0.6%
H3 human-like	19.6%	16.5%	-3.1%
H1 alpha	6.1%	11.8%	5.7%
H1 beta	0.8%	0.3%	-0.5%
H1 pandemic	5.2%	10.3%	5.1%
H1 gamma	30.6%	22.5%	-8.1%
H1 delta 1a	14.6%	5.7%	-8.9%
H1 delta 1b	3.7%	1.3%	-2.4%
H1 delta 2	8.6%	20.2%	11.6%

ongoing status, serum was obtained from 2-5 day-old pigs (average 3 days of age), at each of these two timepoints. Udder wipes at weaning were also collected at the 21-month post-implementation timepoint.

The serum collected was tested by HI assay using a homologous H1N2 virus to the vaccine virus as previously reported.[4] Results of all sampling time points are shown in Table 2. HI titers remained elevated at the 7 and 13 month samplings. The second steady state samples at 21 months are currently being tested and will be available for presentation.

Additional udder wipe samples were collected just prior to weaning at approximately 3 weeks of age and 21 months post-vaccination (Steady State 2). These samples were all negative. Results of testing at all timepoints are displayed in Table 2. Prior to vaccination, 36.8% of the litters were positive at weaning, this dropped to 12.5% after 2 mass vaccinations, and have remained negative since 7 months after the implementing the vaccination program.

In addition to the HI response and udder wipe data, we sought to further examine an alternative way to measure the status of the farm and the long-term effects of IAV-S vaccination. A nucleoprotein (NP) ELISA (IDEXX Swine Influenza Virus Ab test,, IDEXX Laboratories, Westbrook, ME) was performed on retained pre-vaccination and 7-month post-vaccinationserum samples, as well as the 13 and 21 month samplings. The NP ELISA measures antibody against the nucleoprotein of IAV-S. The nucleoprotein is conserved such that all strains contain the same antigen, therefore the test is not strain specific. NP antibodies present in the serum can be generated from exposure to circulating field virus or from vaccination with a whole virus vaccine. IAV-S vaccines do not contain nucleoprotein antigen, only HA. Since no whole virus IAV-S vaccines have been used on this farm for several years, any detectable NP antibodies would be due to exposure from circulating field virus within the farm.

Serum was collected at 3 days of age as an indirect measure of the maternal antibody transfer of the NP antibodies from the sow. The percent NP ELISA positive piglets are shown in Table 2. Prior to IAV-S vaccination, 100% of the piglets were NP positive and 85.5% were positive after the second mass plus pre-farrow vaccinations (4 months). At the 13 and 21 month sample times, the percent positives dropped to 12.5% and 33.3% respectively.

These laboratory diagnostic findings correlate with the decline in clinical signs on this farm over the course of the last two years according to the farm staff and attending veterinarian. Piglets are consistently weaned negative for IAV-S virus and clinical signs are minimal within the nursery.

Key site field application

Key sites or antigenic epitopes have been studied in IAV-S with published data on specific amino acid positions for H3 and H1 delta viruses.[5-7] The SEQUIVITY™ Dashboard evaluates these sites every time sequences are analyzed for vaccine design. Key sites are areas within the HA protein of IAV-S that have been found to be antigenically important, with even a seemingly minor change of a single amino acid having the potential for creating a lack of cross-protection from one strain to another. The need for field validation of these key sites is important to improve vaccination analysis from simply percent identity (homology) of nucleotides and/or amino acids.

A 500 head sow farm had documented a single H1N2 delta 1a strain on the farm for 3+ years with clinical signs present in both farrowing and the nursery. A IAV-S vaccination was implemented in the sow farm (two mass vaccinations with a subsequent pre-farrow vaccination) and the farm clinically responded with decreased clinical signs present in farrowing and nursery after implementation. Approximately 5 months post vaccine implementation, clinical signs within the farm began to rise and sequences were obtained. The farm was continually monitored and sequenced over the next 8 months. Evaluating the key sites of this H1 delta 1a virus on the SEQUIVITY™ Dashboard identified a key site mutation at site 140 from the vaccine/original virus. An

Table 2: Long term sow farm monitoring: HI titer to H1N2 virus, udder wipe PCR and NP ELISA results

	Pre-vaccination	Post-2x mass vaccination	Post-mass and pre-farrow vaccination	Steady state monitoring 1	Steady state monitoring 2
Time from 1st vaccination:	**0 months**	**3 months**	**7 months**	**13 months**	**21 months**
HI titer to H1N2 (geometric mean) @ 3 DOA	75.9[a]	187.0[b,c]	276.9[c]	159.5[b]	*Result pending
Udder wipes (no. and % PCR positive litters) @ 3 WOA	36.8%[a] (7/19)	12.5%[a,b] (2/16)	0.0%[b] (0/18)	*Not performed	0.0%[b] (0/30)
NP ELISA % positives @ 3 DOA	100%[a] (17/17)	*Not performed	85.5%[a] (118/138)	12.5%[b] (2/17)	33.3%[b] (10/30)

[a,b] Values with different superscripts within a row were significantly different ($P < 0.05$).

amino acid change from a lysine (K) to Arginine (R) was noted. This mutation was documented in multiple case submissions. This new virus with the K140R mutation was isolated from clinical samples and used via hemagglutination inhibition (HI) assay to further understand the lack of apparent protection.

The new virus had a 98.4% amino acid homology to the vaccine strain. Antisera against an isolate homologous to the vaccine was available from three different pigs and was tested by HI assay using either the H1N2 delta 1a homologous virus or the new isolate with the K140R mutation. The results are shown in Table 3. Pigs A, B and C developed titers as expected to the homologous virus with titers ranging from 1:80 to 1:320. When the antiserum was tested against the new isolate with the key site mutation, titers ranged from 1:10 to 1:20, representing 8-16 fold lower titers. This loss of HI titer cross-reactivity was found even though the virus was 98.4% homologous to the original vaccine virus. Considering these results, the farm plans to implement a vaccine change to include the new strain with the K140R mutation.

Summary

The SEQUIVITY™ Dashboard has become an important tool in monitoring for IAV-S diversity in the field. The Dashboard enables tracking new and emerging strains, identifying area dominant strains, and determining trends in the prevalence of existing strains. The most notable trends found were declines in the prevalence of H1 gamma and delta 1a viruses and increasing prevalence of H1 alpha, H1 pandemic and H1 delta 2 strains.

It is important to continually monitor farms for new viruses and measure response to vaccination overtime. Demonstration of the positive response to vaccination within the sow farm was depicted through a maintained HI titer response as well as through udder wipes remaining PCR negative at weaning, 21 months post program implementation. The NP ELISA findings at 13 and 21 months, in contrast to results pre-vaccination and at 7 months, suggests the circulation of field virus has diminished substantially. Further study is needed to confirm the utility of the NP ELSIA for monitoring sow herds for field virus circulation and to monitor long term effectiveness of a herd's IAV-S vaccination strategy.

We demonstrated the utility of key site evaluation for vaccine strain selection. A change at site 140 in a H1 delta 1a virus with 98.4% homology to the parent and vaccine strain yielded a 8-16 fold lower HI titers. Further research is needed to determine if the single mutation is the direct driver of the reduced cross reaction, or if additional changes could influence the decline. Further studies are also needed to validate the key sites within other IAV-S H1 and H3 clades, especially those that are commonly found in the field and that are often used in prescription vaccines.

References

1. Vincent AL et al. Swine Influenza Viruses: A North American Perspective. *Adv V Res.* 2008: 127-154.

2. Vander Veen RL et al. Alphavirus replicon vaccines. *Ani Health Res Rev.* 2012: 1-9.

3. Tamura S et al. Mechanisms of Broad Cross-Protection Provided by Influenza Virus Infection and Their Application to Vaccines. *Jpn J Infect Dis.* 2005: 195-207.

4. Sebo et al. Influenza A vaccination using the SEQUIVITY™ Technology. AASV. 2019.

5. Lewis N et al. Substitutions near the Hemagglutinin Receptor-Binding Site Determine the Antigenic Evolution of Influenza A H3N2 Viruses in US Swine. Journal of Virology. 2014: 88: 4752-4763.

6. Absente E et al. The molecular determinants of antibody recognition and antigenic drift in the H3 hemagglutinin of swine influenza A virus. Journal of Virology. 2016. 90: 8266 –8280.

7. Rajao D et al. Antigenic and genetic evolution of contemporary swine H1 influenza viruses in the United States. Virology. 2018. 518: 45-54.

Table 3: HI titer response to homologous 140K isolate vs. heterologous 140R new farm isolate

Pig ID	Homologous isolate (140K)	Farm isolate (140R)	Titer difference
A	320	20	16 - fold
B	80	10	8 - fold
C	160	10	16 - fold

Field evaluation of a liquid *Saccharomyces cerevisiae* fermentation prototype on breeding performance in sows

S. McCallister[1]; K. Robertson[2], DVM, DACVPM; D. Petry[2], MBA, PhD; C. Sexton[3], DVM; B. Meisheid[2]
[1]North Carolina State University, Raleigh, North Carolina; [2]Diamond V Mills, Cedar Rapids, Iowa;
[3]Livestock Veterinary Services, Kinston, North Carolina

Introduction

Several studies have evaluated the lactation performance of sows fed a unique *Saccharomyces cerevisiae* fermentation product (Original XPC™; Diamond V Mills, Cedar Rapids, Iowa, USA) throughout gestation and lactation. In these studies, a trend toward reduction in wean-to-first-service interval has been observed.[1-3] The purpose of this field study was to evaluate the effects of this technology on breeding performance of sows when a liquid *S cerevisiae* fermentation prototype was delivered orally in lactation only.

Materials and methods

A batch of 353 sows due to farrow in June 2019 in an 1,800-sow farm in eastern North Carolina was enrolled in the study. Prior to the study, water lines were cleaned in 10 farrowing rooms according to the protocol in Figure 1. Sows were assigned to farrowing rooms according to the usual management on the farm.

Farrowing rooms were assigned to one of two treatments using a non-random method based on ease and safety of administering product through water medicators. Sows in control rooms (CTL) received fresh drinking water without additives. Sows in the treatment rooms (SCP) were administered 15mL per sow per day of a liquid *S cerevisiae* fermentation prototype (Diamond V Mills, Cedar Rapids, Iowa, USA) in the drinking water via water medicators. The daily dose of SCP for the treated rooms was mixed in a stock solution with water so that treatment was delivered over a 6- to 8-hour period. The SCP used in this study is known to contain a similar fermentation metabolite profile to Original XPC and was chosen due to the ease of administering product to individual rooms through the drinking water when compared to administering product to individual rooms through the feed.

Wean-to-first-service interval, 36- and 56-day conception rate, and farrowing rate were recorded from herd production records in the parity after sows received SCP or CTL treatments in lactation.[5] Data were analyzed using a mixed model in SAS with treatment and parity as fixed effects, and number of pigs nursed and days nursed as covariates. Conception rates will be analyzed using a binomial distribution.

Results and discussion

Results on lactation performance are presented elsewhere in these proceedings.[4] A total of 290 sows were included in data analysis; 132 in the CTL group and 158 in the SCP group. A numerical decrease in wean-to-first-service interval was observed in the SCP group (7.6 days) when compared to the CTL group (9.2 days). Additional statistical analysis, 36-day pregnancy data, 56-day pregnancy data, and farrowing rate data were not available at the time of publication but will be presented in Atlanta at the AASV Annual Meeting, March 8, 2020.

Sows in this study were weaned and bred during mid-July in North Carolina in the southeast United States. The average ambient temperature was 83° F during lactation with 17 days reaching 95° F or higher. In this geography in the summer season elevated wean-to-first-service intervals are routinely observed.

Failure to observe statistical significance in the wean-first-service days between SCP and CTL groups can be attributed to the extreme variation observed in this production parameter in the current field study. In previous studies cited in which trends for decreased wean-first-service interval were observed, Original XPC was fed throughout gestation and lactation.[1-3] Additional studies with larger numbers of sows will be needed to determine the impact of the *S cerevisiae* fermentation prototype on breeding productivity when provided to sows during lactation only vs. throughout gestation and lactation.

References

1. Petry et al. The effects of Diamond V Original XPCTM on weaning performance when fed to sows during gestation and lactation. AASV 2019; 235.

2. Diamond V internal communication. Swine Research Summary: Effect of yeast culture on the reproductive performance of fifth parity sows. SW07, 1993.

3. Kim, et al. Effect of supplementing *Saccharomyces cerevisiae* fermentation product in sow diets on reproductive performance in a commercial environment. Can. J. Sci. 90: 229-232.

4. McCallister et al. Field evaluation of a liquid *Saccharomyces cerevisiae* fermentation prototype on lactation performance in sows. AASV 2020.

5. PigKnows™, LLC. Greely, CO.

Figure 1: Water line cleaning protocol.

Subject: Proxy-Clean Protocols for Swine Producers

Suggested Water Line Cleaning Program with Animals Present

For cleaning the lines while animals are housed create stock solutions as described below and run the program up to 10 days depending on degree of fouling.

Pump Setting	Water Quantity	Day 1&2 Proxy-Clean	Day 3&4 Proxy-Clean	Day 5&6 Proxy-Clean	Day 7-10 Proxy-Clean
1:128	5 Gallons	8 oz.	12 oz.	16 oz.	20 oz.

A multifaceted approach to selection for post weaning pig survival

M. Romoser, MS; D. Grossi, DVM, PhD; M. Duggan, MS
Fast Genetics, Saskatoon, Saskatchewan

Introduction

Mortality in the post-weaning phases is one of the most profit limiting factors in commercial pork production. With increased mortality, those that die represent significant investment losses, increased cost of production per animal marketed, and lost opportunity to sell a full-value market hog.[1] While environment, disease, and management effects on pig survivability are significant,[2] heritability is in the low to moderate range.[3] The aim of the current work is to determine if selecting for pig survivability from birth through the finishing phase, is possible using pedigree and genotype information from nucleus herd progeny. Identifying animals who genetically are more likely to survive to market weights, proposes substantial incentives both economically and for animal well-being.

Materials and methods

The study was conducted at a terminal sire genetic nucleus herd. Sires (n = 652) of progeny born from 1/1/2006 to 1/1/2019 were evaluated based on their offspring's likelihood to survive from birth to market. All animals in the genetic nucleus farm were put on test for the purpose of collecting growth performance phenotypes. The proportion of a sire's progeny that completed performance testing to the total pigs born for that sire was used to calculate the Survivability Ratio (SR). A bivariate analysis was used to calculate the correlations between SR and other estimated breeding value in the sire index. A mixed linear model was used to determine the Survivability EBV (SEBV) with fixed effects of SR~ year-season (YS) and random effect of animal.

Results and discussion

Of the 652 sires included, maximum and minimum number of offspring born for each sire were 131 and 2 respectively. Similarly, the maximum number of offspring that completed performance testing was 88 with the minimum being 0 offspring completing the testing period. After performance testing, a moderately high proportion of the variability in SR was explained by SEBV ($R^2 = 0.50$) (Figure 1). Phenotypic correlations for SR were low for total feed intake (-0.02), age (0.12), back fat (-0.06), loin depth (0.10), average daily gain (-0.10), and intramuscular fat (-0.04) (Table 1).

The results from the current work suggests a SR phenotype may serve as a potential predictor for a sire's SEBV. Further investigation is warranted to test the genetic correlations with survivability, and its value as an index in selecting pigs with improved fitness and livability.

References

1. Maes D, Larriestra AJ, Deen J, Morrison RB. A retrospective study of mortality in grow-finishing pigs of a multi-site production system. J Swine Health Prod. 2001; 9:267–274.

2. Mehling S, Henao-Diaz A, Maurer J, et al. Mortality Patterns in a Commercial Wean-To Finish Swine Production System. Vet Sci. 2019;6(2):49. Published 2019 May 29. doi:10.3390/vetsci6020049.

3. Hermesch S., Li L., Doeschl-Wilson A. B., Gilbert H. (2015) Selection for productivity and robustness traits in pigs. Animal Production Science 55, 1437-1447. https://doi.org/10.1071/AN15275.

Figure 1: Relationship of Survivability Ratio (SR) with Survivability Estimated Breeding Value (SEBV). A moderately high positive correlation exists between SR and SEBV, with an R^2 value of 0.50.

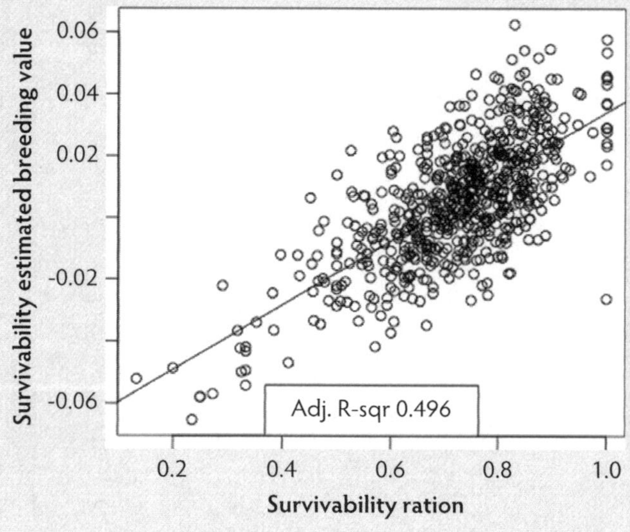

Table 1: Phenotypic correlations with survivability ratio

Trait	r_p
Total feed intake	-0.02
Age	0.12
Back fat	-0.06
Loin depth	0.10
Average daily gain	-0.10

ProVent® ECL, a unique bacillus-based direct-fed microbial (DFM), as a tool to improve wean-to-first service interval distribution and sow productivity

T. Specht, DVM; M. Heimerl; B. Heimerl
Heimerl Farms, LTD, Johnstown, Ohio

Introduction

A strategic evaluation of a sow herds' reproductive fitness can identify opportunities for improved profitability in a swine production system. Although many metrics can be used to evaluate sow productivity, the distribution of wean-to-first service interval (WTFSI) within a sow herd has proven to be a sensitive lead indicator of overall reproductive performance. Factors influencing WTFSI are many, including lactation feed intake, lactation length and bacterial and viral disease pressure.

Within production systems, downstream losses in performance that are associated with health challenges can originate from health or management issues at the sow farm. Recent studies have shown evidence of pathogenic microorganisms being vertically transmitted from the sow to her offspring and subsequent carry over into the nursery. Consequently, interventions for managing enteric health at the sow farm should be considered.[1]

A 2,400 head crated breed to wean facility (PIC multiplication) that was built in 2004 and located in southern Ohio was evaluated to monitor key sow productivity data during a 22-week period, from January 1, 2019 through June 1, 2019. In the fall of 2018, pre-wean scours were observed. Gaining control of the scours and improving health of the system was of continued priority. A diagnostic review was executed, focusing on pre-wean diarrhea. Previous consultation with United Animal Health™ on a different sow unit within this same system included the use of PathKinex,™ a comprehensive microbial surveillance tool. PathKinex™ results identified and quantified toxin genes present and determined the compatible use of ProVent® ECL, a Direct Fed Microbial (DFM) as a potential intervention.

Past success with ProVent® ECL at the previous sow unit drove the decision to also implement ProVent® ECL in both, gestation and lactation diets at this unit to help reduce the incidence of pre-wean scours and improve overall enteric health. Product was implemented during the 2nd week of November 2018, however, a six week pre-load period is recommended to allow for the bacillus strains to fully colonize in the animal. Therefore, data collection began Wk. 1- January 2019 and was compared to historical data captured through MetaFarms data system.

At this farm where ProVent® ECL was to be evaluated, fecal swabs were collected from scouring piglets and sent to the Iowa State University Veterinary Diagnostic Lab in May of 2018. Results indicated a strong Rotavirus Group C (CT – 14.2 -27.2) presence. Also, on culture hemolytic *E coli* and *C perfringens* were isolated and grown. Based on these results, a Rotavirus vaccine was initiated week 35 in 2018 as a management tool along with strategies to improve the overall hygiene at the sow unit.

Although ProVent® ECL was used to manage pre-weaning enteric challenges, further understanding of this DFM's impact on WTFSI distribution, subsequent total born and farrowing rate were evaluated.

Materials and methods

ProVent® ECL was implemented in lactation and gestation diets to reduce clinical scours associated with *E coli* and *Clostridium*. ProVent® ECL is a 6-strain, bacillus-based product developed by United Animal Health™. Bacillus strains within this product were isolated from, and, identified as unique to high performing animals through the strain selection process. Selected bacillus strains were found to colonize in the intestine of pigs. A secondary selection method was used to identify additional properties unique to each candidate strain, and their influence on certain bacteria including *E coli, C perf.* and *C diff.* Previous studies with the bacillus strains in Provent® ECL have shown to increase feed intake.[1]

ProVent® ECL was included in lactation and gestation diets at a level of 1 pound/ton of complete feed beginning the 2nd week of November 2018. A six-week loading period was allowed for colonization of the bacillus and modulation of the microbiota in the sow herd.

Key production metrics of pre-weaning mortality and then days to first service, conception rates, farrowing rates and number born alive were monitored via MetaFarms data starting Week 1, 2019. This data was compared to historical data collected from 2017 and 2018. Although, initial focus of evaluation was pre-wean mortality, an expanded review of other parameters associated with sow reproductive performance was performed. Data was evaluated for the sow herd's wean-to-first-service interval distribution, along with number born and farrowing rate, comparing each data set on a quarterly basis to minimize seasonal impact.

Results and discussion

Pre-weaning mortality

Prior to ProVent® ECL implementation, average PWM% during Q1-Q3 of 2017 was 17.90% and during Q1-Q3 of 2018,

PWM% was 17.30% (Figure 1). ProVent® ECL was implemented the 2nd week of November 2018, providing for the six-week bacillus loading period through the final week of December 2018. The average PWM% was reduced to 15.50% during Q1-Q3 of 2019 (Figure 1) ($P < 0.01$).

Wean to first service interval

Weekly production records for weaned sow groups indicated a reduction in weaned sows returning to estrus past five days. A closer evaluation showed significant improvements in wean-to-first service interval. Compared to 2018 (prior to ProVent® ECL implementation), 2.1% more sows were bred between days 3-7 post wean in 2019 (post ProVent® ECL implementation) (Figure 2, $P < 0.0001$). Total born increased in 2019, with sows bred during days 3-5 post wean also having higher total born compared to day 6-9. Previous research has shown that a shorter WTFS interval is associated with a higher total number born in subsequent litters.[2]

Economic models were created to quantify overall improvements recognized in WTFS interval, total born, farrowing rate and prewean mortality, with an additional $23.23 captured per mated female (Figure 3). Economic improvements should be evaluated on a per system basis due to individualized values placed on each parameter.

Percent sows bred in the narrow window of 3-5 days post weaning may also serve as a lead indicator for early disease detection in the sow herd. During this evaluation, H1N1 influenza was detected within this sow unit. Consequently, a drop in the number of sows bred days 3-5 post weaning was recognized (Figure 4). Possible reasons for this may be a reduction in lactation feed intake during disease pressure.

Management practices

Multiple management considerations were also made to improve the overall performance and health of this sow unit. Specific management protocols implemented and followed include offering fresh feed daily with feeders being cleaned out and ensuring water quantity and quality availability. Environmental protocols included proper washing, disinfecting and drying of farrowing rooms prior to subsequent groups loading. Farm employees were coached to get all lactating sows up twice daily to promote feed and water intake and to monitor for health and treatment needs in a timely manner. A renewed focus on Day 1 pig care was also implemented. Consequently, a focus on management practices and implementation of ProVent® ECL resulted in improved performance and economic outcomes. This positive impact is similar to previously reported results where ProVent® ECL was utilized in a sow unit in combination with improved sanitation to improve pre-weaning mortality.[3]

Figure 1: Pre-weaning mortality by quarter. Direct fed microbial (Provent® ECL) would have only influenced 2019 as sow feeding started mid November 2018.

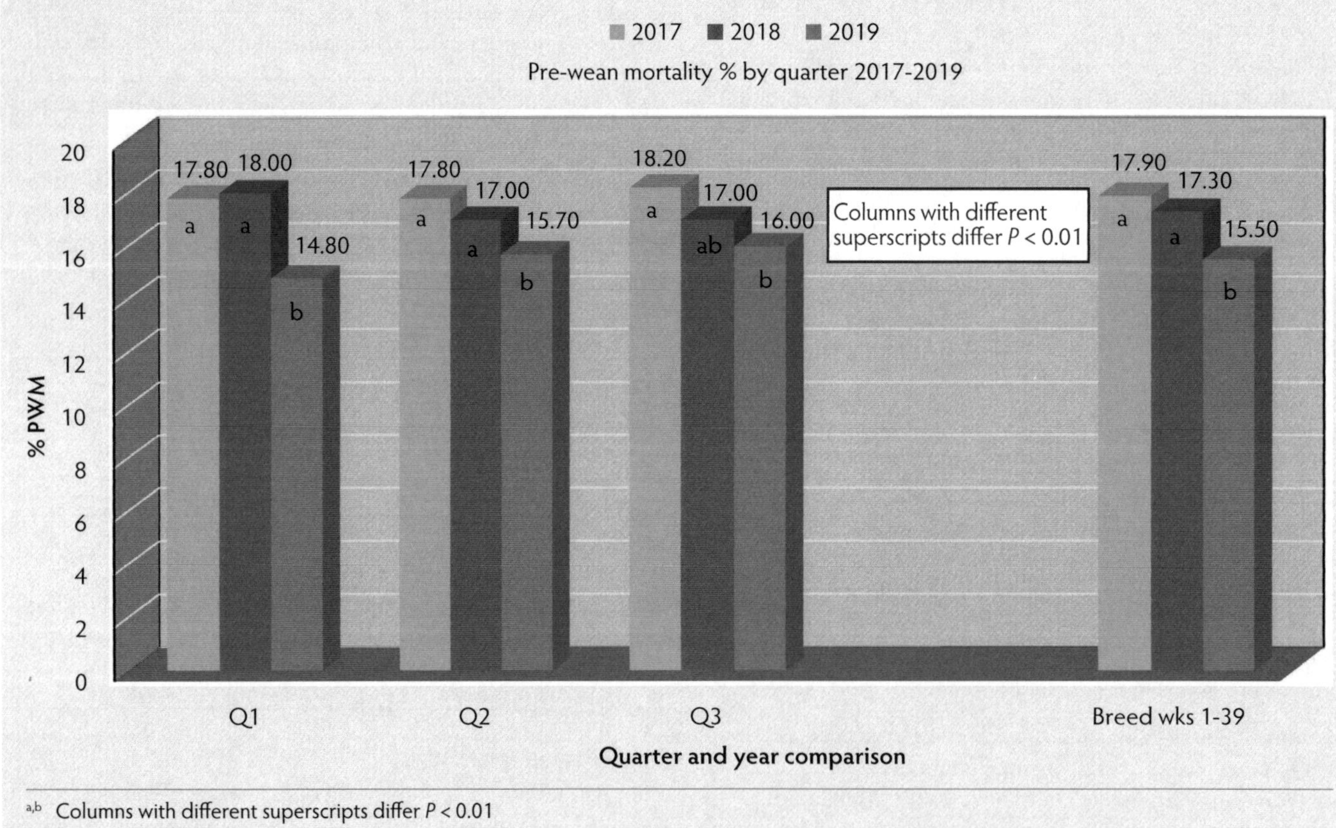

a,b Columns with different superscripts differ $P < 0.01$

Figure 2: Total born and percent of sows by day of first insemination post wean for 2018 (blue) and 2019 (red). 2019 represents sows fed a direct fed microbial (Provent® ECL).

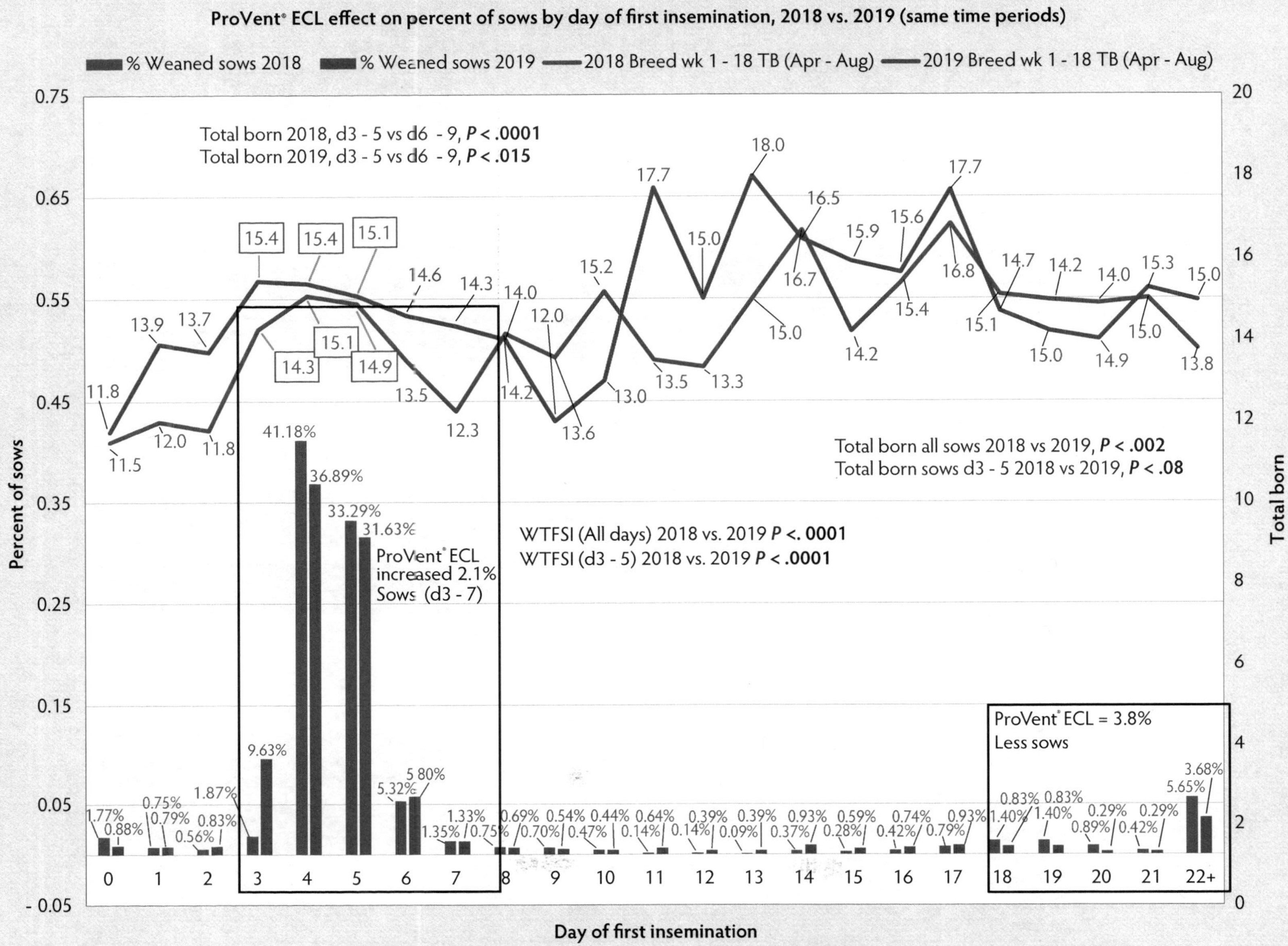

In summary, the distribution of WTFS interval is a key metric in overall reproductive efficiency and can be positively influenced by several factors, including the introduction of a unique bacillus based direct fed microbial in sow diets. While pre-weaning mortality was the initial focus for using ProVent® ECL, improvements in piglet survival, WTFS interval and total born were observed. Focusing on increasing sows in heat within 5 days post weaning can be a lead indicator for sow fertility and disease monitoring and can be used as a common score card evaluation on all farms.

References

1. Augspurger et al 2018. The effect of a bacillus subtilis-based direct -fed-microbial feed additive on the lactation performance of sows and their offspring. J. Anim. Sci. 96(Suppl. 2.):187-188(Abstr.)

2. M. Rix, R. Ketchem & V. Duttlinger. Why is wean-to-service interval critical to your farm? National Hog Farmer, Nov 29,2017.

3. Farkas, A. New tools for scour diagnosis and control...But don't forget sanitation is king! National Hog Farmer. Oct. 26, 2018.

Figure 3: Calculated return per mated sow in 2018 (blue) and 2019 (Provent® ECL influence, red) by measured metric.

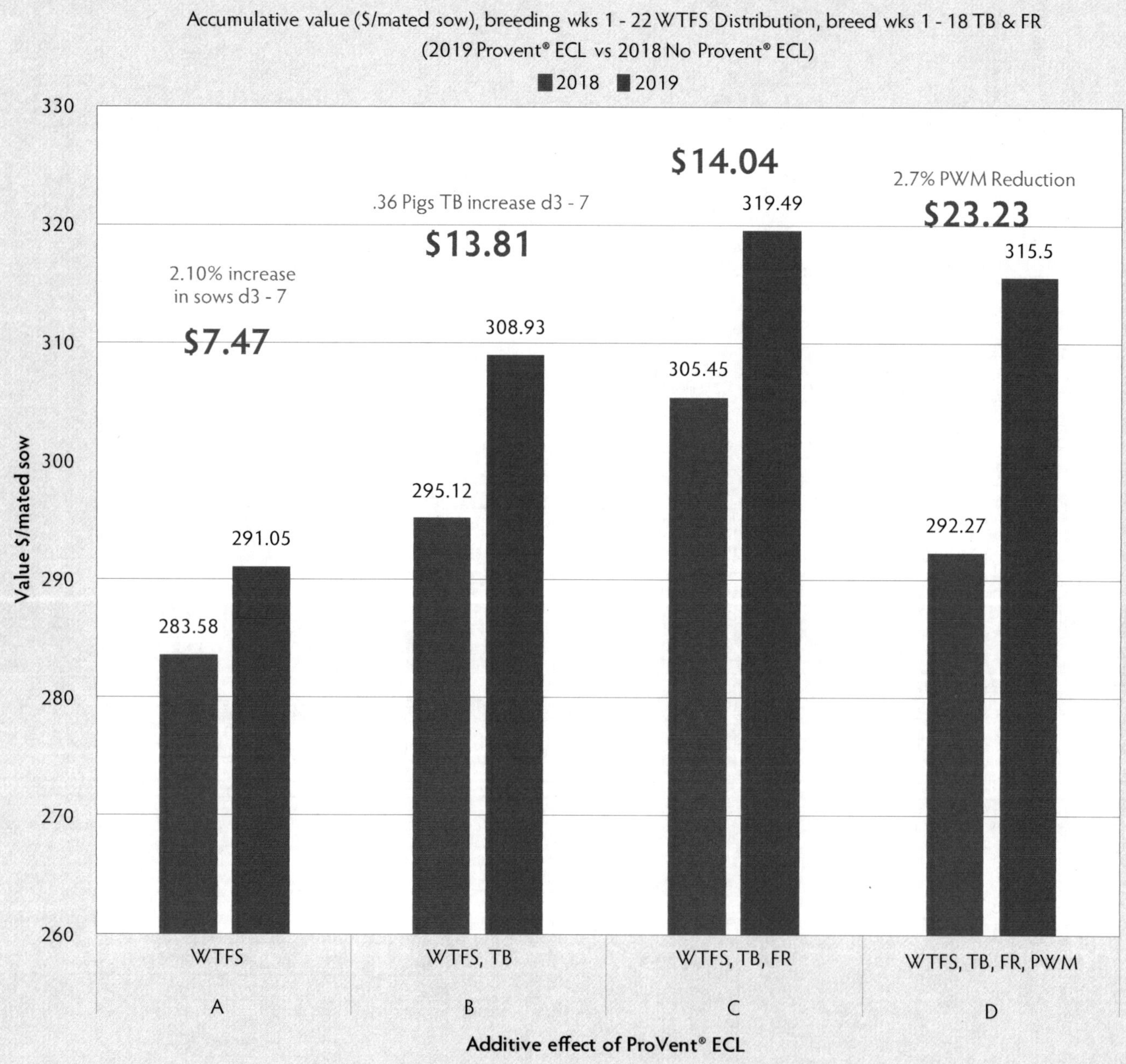

Figure 4: Percent of parity 1 females bred 3-5 days post weaning in 2018 (blue) and 2019 (influence of Provent®ECL, red). Suggests this metric may be a sensitive indicator to health status.

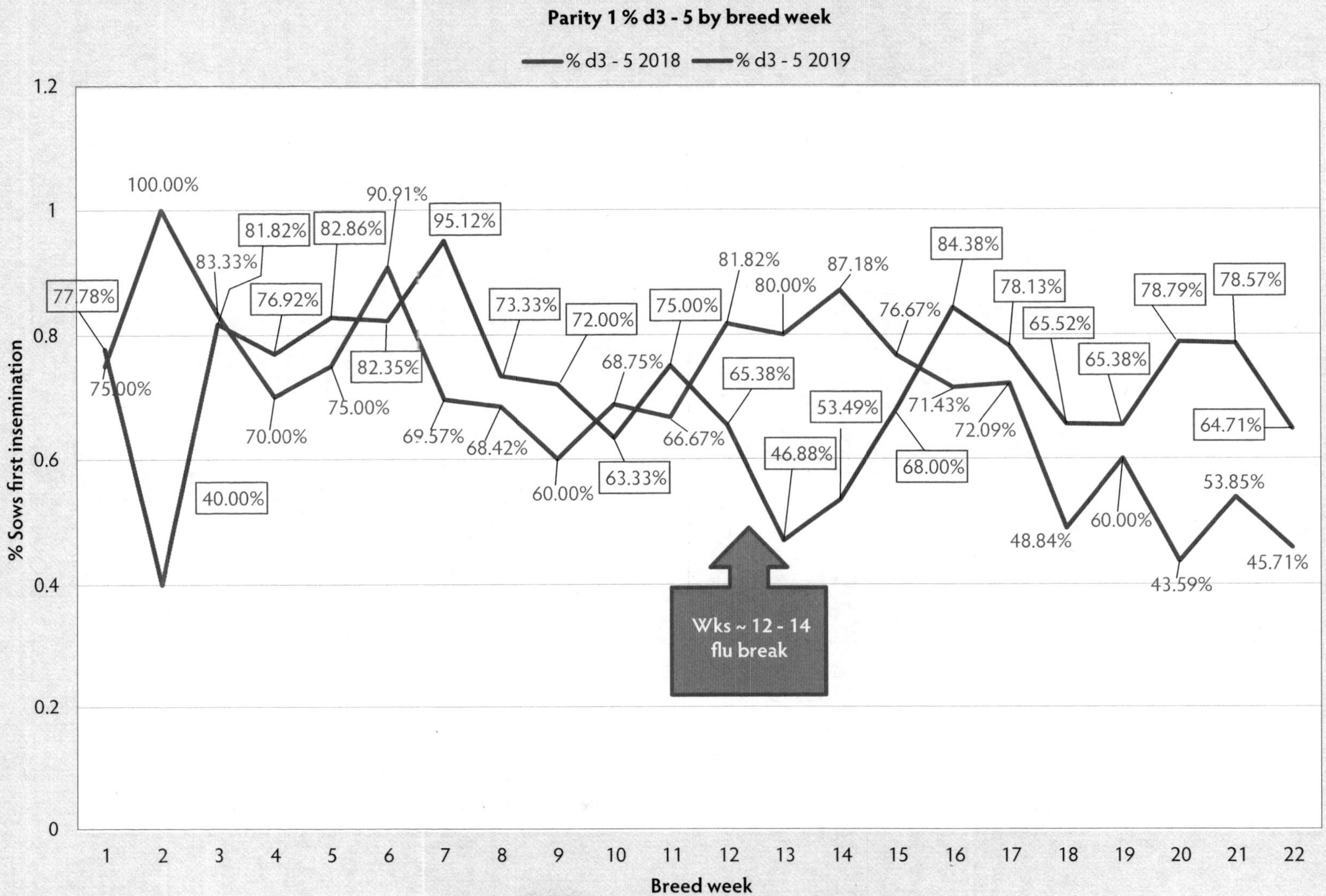

Benefits of feeding M-Mobilize™, a novel clay-based product with *Bacillus* strains, in the presence and absence of toxin contaminated diets

A. Woodward, PhD
United Animal Health, Sheridan, Indiana

Introduction

Mycotoxins, produced by mold and present on a variety of grain products used in livestock feed, negatively impact production performance and reduce profitability.[1,2] In the swine industry, three main toxins of concern are deoxynivalenol (DON), fumonisin (FUM), and zearalenone (ZEA).[3] While all three toxins cause liver damage, the structure and bioavailability of each mycotoxin play significantly different roles in reducing production parameters across phases.

Many products which assist in mycotoxin-challenge scenarios are available; however, few offer multi-coverage against the deleterious impacts of mycotoxins in all production phases. M-Mobilize,™ with its combination of a patented, specialized hydrated sodium-calcium aluminosilicate and novel, *Bacillus subtilis* strains, offers benefits in the presence of multiple mycotoxins. Furthermore, the combination of specialized ingredients enhances performance and increases profitability even in the absence of a mycotoxin challenge.

Materials and methods

Two experimental scenarios were tested to determine the impact of M-Mobilize™ in diets (1) with a mycotoxin challenge, and (2) without a mycotoxin challenge. In scenario 1, a meta-analysis was performed on 10 trials (n = 3,120, where 8 of the 10 trials showed a statistically significant response, $P < 0.10$) across various swine production phases with pigs fed a low toxin, positive control, a high toxin, negative control, or the high toxin, negative control diet plus M-Mobilize™ (Table 1). Toxins tested included DON, FUM and ZEA in nursery, grow/finish, and gilt production phases. Performance parameters of growth were measured for trials including DON. Morphologic parameters for gut health were measured for FUM, and both growth and metabolic parameters of toxin uptake were measured in ZEA trials.

In scenario 2, a meta-analysis was performed on 5 trials (n = 1,260, where 3 of the 5 trials showed statistically significant response, $P < 0.05$) across various swine production phases with pigs fed a non-toxin diet control or the non-toxin diet plus M-Mobilize™ (Table 2). Both nursery and grow/finish production phases were included.

In both scenarios, animals were placed in pens, divided into replicates and randomly assigned a treatment depending on body weight for each trial. Treatment diets were formulated to meet or exceed common industry recommendations for maximum performance by production phase, and animals were fed their treatment

Table 1: Trials included in the meta-analysis for scenario 1, with individual trial performance and statistical significance. Overall meta-analysis showed a 53.4% ($P < 0.05$) improvement in growth when M-Mobilize™ was added with various levels of different toxins.

Trial	Toxin	High toxin level, ppm	Days	Control ADG, lbs./d	High toxin level ADG, lbs./d	High toxin level + M-Mobilize™ ADG, lbs./d	SEM	Recovery, %	*P*-value
1	DON	4.0	14	1.98	1.69	1.75	0.03	20.7	< 0.001
2	DON	3.5	18	1.13	0.87	0.92	0.08	19.2	< 0.001
3	DON	1.0	42	0.79	0.77	0.81	0.03	200.0	0.265
4	DON	3.0	21	1.79	1.59	1.87	0.01	140.0	0.081
5	DON	5.0	22	1.35	1.17	1.12	0.05	(27.8)	< 0.001
6	DON	3.0	15	1.31	1.22	1.30	0.05	89.4	0.115
7	DON	3.0	42	2.07	1.94	1.99	0.02	36.5	< 0.001
8	DON	5.0	42	2.63	2.15	2.27	0.04	25.5	< 0.001
9	ZEA	1.2	14	2.59	1.87	2.06	0.12	26.4	< 0.001
10	ZEA	0.7	16	2.24	1.70	2.03	0.12	61.1	0.053

Table 2: Trials included in the meta-analysis for scenario 2, with individual trial performance and statistical significance. Overall meta-analysis showed a 5.6% ($P < 0.05$) improvement in growth when M-Mobilize™ was added to control diets in various non-mycotoxin scenarios.

Trial	Scenario	Days	Control ADG, lbs./d	Control + M-Mobilize™ ADG, lbs./d	SEM	Difference, %	P-value
1	Enterically healthy	14	0.36	0.41	0.02	5.1	0.136
2	Enterically healthy	21	1.32	1.36	0.02	3.0	0.050
3	Enterically healthy	28	2.65	2.80	0.05	5.8	0.050
4	Stressed, non-toxic	13	0.25	0.31	0.03	24.0	0.012
5	Stressed, non-toxic	21	0.47	0.46	0.03	(2.1)	0.958

diets for the duration of each experiment. Pen weights and feed intakes were recorded per production phase and utilized to calculate average daily gain (ADG) and average daily feed intake (ADFI); these metrics were used to calculate feed efficiency.

Results and discussion

M-Mobilize™ affects growth performance during a vomitoxin challenge

Growing pigs fed high DON (3 ppm added DON) diets tended to decrease ($P = 0.081$) ADG by 11% compared to pigs fed low DON diets, leading to a 4.3 lb. decreased ($P = 0.081$) end body weight after 21 d. Inclusion of M-Mobilize™ in the diet recovered all of lost the ADG, allowing for 5.9 lbs. increased ($P = 0.031$) end body weight compared to non-supplemented high DON diets.

In the nursery, pigs fed high DON (3 ppm added DON) diets had a tendency for decreased ($P = 0.079$) end body weight by 1.8 lbs. compared to pigs fed low DON diets after only 15 d on experimental diets. This decreased growth was caused by an 8% numerical decrease ($P = 0.115$) in ADG. Inclusion of M-Mobilize™ in high DON diets recovered 89% of lost ADG, leading to a 1.3 lb. heavier ($P = 0.069$) end body weight compared to non-supplemented high DON diets.

These results demonstrate the effectiveness of M-Mobilize™ in restoring growth performance lost during times when a high DON challenge is present in the diet.

M-Mobilize™ effects on intestinal integrity during a fumonisin challenge

Diets containing high FUM can damage the intestinal epithelium, allowing for colonization of pathogenic bacteria.[4] Compared to low FUM diets, nursery pigs fed high FUM (25 ppm) diets had 25% reduced villi height 14-d after challenge. However, animals receiving M-Mobilize™ in high FUM diets had a 75% increase in villi height and a more intact intestinal lining compared to those pigs fed non-supplemented high FUM diets.

Data shows that when faced with a high FUM challenge scenario, M-Mobilize™ has the ability to sustain the intestinal lining, which can reduce pathogen colonization.

M-Mobilize™ impact on zearalenone uptake and metabolism

Gilts challenged with a high ZEA (675 ppb) diet had 24% decreased ($P = 0.016$) ADG and 19% decreased ($P = 0.004$) ADFI, leading to a decrease ($P = 0.031$) of 7.7 lbs. end body weight after 16 d on treatment. Supplementation with M-Mobilize™ recovered 61% of growth and 50% of feed intake, creating a 4.7 lb. heavier ($P = 0.081$) pig at experiment end.

Zearalenol is produced by metabolism of zearalenone in the liver; therefore, it serves as a good indicator of zearalenone uptake from the intestine.[5] Pigs fed M-Mobilize™ with a high ZEA diet had a 29% increase ($P = 0.199$) in fecal zearalenone combined with a 32% decreased ($P = 0.110$) fecal zearalenol compared to non-supplemented gilts.

The change in concentrations of fecal zearalenone and fecal zearalenol indicate that M-Mobilize™ supports a reduction in uptake and metabolism of mycotoxins, thus leading to a healthy pig overall.

M-Mobilize™ affects growth during low mycotoxin challenges

Nursery pigs fed diets supplemented with M-Mobilize™ in diets devoid of mycotoxin challenges demonstrated a 5% increase ($P = 0.136$) in ADG when diets were supplemented from weaning to 14-d post-wean. Additionally, pigs receiving M-Mobilize™ during the late nursery period (42 – 63 d old) increased ($P = 0.165$) end body weight 0.8 lbs., brought on by a 3% increase ($P = 0.050$) in ADG, after only 21 d.

In stress scenarios where pigs were deprived of feed and water, supplementation of M-Mobilize™ increased ($P = 0.012$) ADG up to 24% over non-supplemented, stressed pigs.

In 8 out of 10 high mycotoxin diets, M-Mobilize™ statistically ($P < 0.10$) improved growth performance in a variety of production scenarios, while when no toxin was present, M-Mobilize™

statistically ($P < 0.05$) improved performance in 3 out of 5 trials. The ability of M-Mobilize™ to elicit a performance response in both the presence and absence of mycotoxin and stress challenges indicates that M-Mobilize™ provides low risk insurance during all production phases.

References

1. Bryden WL. Mycotoxin contamination of the feed supply chain: Implications for animal productivity and feed security. *Animal Feed Sci and Tech.* 2012;173:134.

2. Shull LR and Cheeke PR. Effects of synthetic and natural toxicants on livestock. *J Anim Sci.* 1983;57:330.

3. Diekman MA et al. Mycotoxins and Swine Performance. *Pork Industry Handbook.* 1992;E2420.

4. Oswald IP et al. Mycotoxin fumonisin B1 increases intestinal colonization by pathogenic *Escherichia coli* in pigs. *App Env Micro.* 2003;69:5870.

5. Olsen M et al. Reduction of zearalenone to zearalenol in female rat liver by 3α-hydroxysteroid dehydrogenase. *Acta Pharma Toxic.* 1981;48:157.

The effects of sulfated polysaccharides from macro-algae on gut health and immunity

P. Olsen, DVM; J. A. Erceg, MS; M. Berri, Msc; C. Rousseaux, PhD;
P. Nyvall Collen, PhD; M. Garcia Suarez, DVM

Introduction

"Gut Health" is an increasingly popular buzzword aimed to combine nutritional efficiency and immunity into one, with the goal of making animals able to better face daily stress and challenges. Recent research has highlighted the potential of in-feed marine macroalgal polysaccharides as a reliable solution to strengthen animal defenses.[1]

The cell wall of marine macroalgae is mainly composed of water-soluble sulfated polysaccharides which present a complex structure.[2] The molecular weight and several structural specificities, including branched conformation, presence of several sugars and more particularly rare ones (xylose and rhamnose) and sulfation, confer to algal polysaccharides unique biological activities. Among others, modulation of the immune response and reinforcement of the gut barrier function are being investigated.[3,4]

Innovative technology and marine bioactive ingredient extraction know-how can lead to the development of new in-feed products. The present work aimed at evaluating the capacity of two different marine macroalgae extracts (MSP®) to stimulate the gene expression of different proteins involved in gut health and immunity.

Materials and methods

Study 1

A crude extract (MSP®$_{IMMUNITY}$) containing algal sulfated polysaccharides was prepared from the green algae *Ulva armoricana*. The ability of MSP®$_{IMMUNITY}$ to stimulate the expression of immune response mediators was evaluated using an *in vitro* system of porcine differentiated intestinal epithelial cells IPEC-1. Three doses (1, 0.1 and 0.01 mg/mL) were tested with *E coli* O111:B4 LPS as positive control and untreated cells as a negative control. MSP®$_{IMMUNITY}$ was also tested on a human embryonic kidney cell line, HEK293, transfected either to express none (null) or one specific immune receptor (TLR4/MD2/CD14, TLR2, TLR5, TLR9, NOD1 and NOD2), in order to identify which TLR or NLR receptor was involved. Expression of interleukin (IL-8) with ELISA was used as a marker of stimulation.

Data are expressed as the mean value ± S.E.M. of triplicate assays and were analyzed using the Kruskal-Wallis test followed by Bonferroni-Dunn post-test group comparison tests of means using the GraphPad (GraphPad Prism version 4.00 for Windows; GraphPad Software, USA). Statistical differences between the various treatments were considered significant when *P* values were < 0.01.

Study 2

The effects of a red algae extract (MSP®$_{BARRIER}$) from *Solieria chordalis* on intestinal cell processes were identified using *in vitro* cell models HT-29 MTX and Caco-2 cells for the evaluation of mucin and tight junction proteins respectively. This was done under normal physiological and inflammatory conditions (using *E coli* O111:B4 LPS in the cell culture medium), for 1 and 4 hours of incubation of MSP®$_{BARRIER}$ at the dose of 1mg/mL in the cell culture medium. Gene expression was measured for several membrane-bound (MUC4) and gel-forming (MUC1, MUC2, MUC5AC) mucin proteins, and for scaffolding (ZO-1, ZO-2) and transmembrane (claudin-2) tight junction proteins; all using qRT-PCR. Results were considered significant when the gene expression was ≥ 2 compared to the control.

Results and discussions

Study 1

RT-qPCR assay showed that MSP®$_{IMMUNITY}$, at the dose of 1 mg/mL, induces an increased expression of several cytokines by IPEC-1 cells in comparison with untreated cells (Table 1). The relative expression of IL-8 (Fc = 313.53 ± 47.5), CCL20 (Fc = 159.44 ± 42.5), TNFα (Fc= 82.63 ± 15.8), IL-1α (Fc= 22.96 ± 3.2) were significantly ($P < 0.01$) increased compared to the control (Figure 1 and Figure 2, respectively). Furthermore, MSP®$_{IMMUNITY}$ increased significantly ($P < 0.01$) the relative expression of IL-6 (Fc = 30.58 ± 7), IL-1β (Fc = 4.92 ± 1.6), and TGFβ (Fc = 4.83 ± 0.7 Fc) mRNA (Table 1). However, MSP®$_{IMMUNITY}$ did not produce any significant change on the relative gene expression of IL-10 and IL-12 cytokines, and CCL25 and CCL28 chemokines (Table 1). When tested with HEK293 cells on several membrane receptors, MSP®$_{IMMUNITY}$ stimulated the expression of immune factors via the activation of TLR4 and TLR2 membrane receptors.

These results show that MSP®$_{IMMUNITY}$ has the capacity to stimulate, *in vitro*, the expression of cytokines involved in the activation, recruitment and migration of lymphocytes, as well as dendritic cells, to modulate the innate and adaptive response. The stimulation of anti-inflammatory immune mediators (PPAR-γ) was also observed. This stimulation seems to come from the activation of TLR4 and TLR2 receptors. However, this route of activation of immune mediators may not be exclusive.

Study 2

The addition of MSP®$_{BARRIER}$ upregulated gene expression regarding mucin secretion. Tight junction protein expression ZO-1

Table 1: Main functions of target mediators and the influence of 1 mg/mL of $MSP_{IMMUNITY}$ on their gene expression

Mediator	Main functions	Degree of expression compared to control
TNFα	Phagocytosis and chemo-attraction of neutrophils	82.6 ± 15.8**
IL-1α	Proliferation of CD4+ T-cells and fibroblasts, proliferation and maturation of B-cells	229.6 ± 3.2**
IL-8	Recruitment of neutrophils, phagocytosis	313.5 ± 47.5**
CCL20	Recruitment of lymphocytes and dendritic cells, antimicrobial activity	159.4 ± 42.5**
IL-6	Differentiation of cytotoxic T-cells, proliferation and maturation of B-cells and excretion of IgA.	30.6 ± 7.0**
IL-1β	Proliferation, differentiation, apoptosis, expression of adhesion molecules and chemokines	4.9 ± 1.6**
IL-12p40	Production of IFN-γ, differentiation of Th1 cells, activation of NK cells and development of cytotoxic T- cells	3.9 ± 0.66**
TGF β	Differentiation of B-cells, switch IgM/IgA, induction of regulatory T-cells	4.8 ± 0.66**
PPARγ	Transcription factor with anti-inflammatory function, inhibition of TNFα and IL1β production	3.7 ± 0.78**
IL-12p35	Production of IFN-γ, differentiation of Th1 cells, activation of NK cells and development of cytotoxic T-cells	4.2 ± 0.63
IL-10	Differentiation and production of IgA+ plasma cells, control of inflammatory response, stimulation of regulatory T-cells	2.5 ± 0.61
CCL25	Migration and homing of T-cells and IgA+ plasma cells	6.2 ± 2.5
CCL28	Migration and homing of T-cells and IgA+ plasma cells	3.2 ± 1.1

** signifies a value of ($P < 0.01$)

and CLDN2 were upregulated by a factor of 3 and 7 respectively (Figure 3). Similar results were seen with MUC2 and MUC4 expression which were more than doubled when a dose of $MSP®_{BARRIER}$ was introduced (Figure 4).This suggests that overall upregulation of scaffolding and transmembrane proteins, both necessary for optimal tight junction protein function, was increased due to the inclusion of $MSP®_{BARRIER}$. There were no significant differences in regulation of MUC1, MUC5AC, and ZO-2. These results indicate that $MSP®_{BARRIER}$ may have the ability to reinforce the intestinal barrier in both normal physiological and inflamed conditions by upregulating the gene expression for production of both mucins and tight junction proteins.

Conclusion

In conclusion $MSP®_{IMMUNITY}$ and $MSP®_{BARRIER}$ extracts both showed positive effects *in vitro* on the immune system and the intestinal epithelial barrier respectively. Furthermore, these two extracts have the potential to be fed simultaneously to increase overall animal gut health. More research is needed to determine the interaction between these extracts and their effects *in vivo*.

References

1. Barcelo A., J. Claustre, F. Moro, J-A. Chayvialle, J-C. Cuber, P. Plaisancié. 2000. Mucin secretion is modulated by luminal factors in the isolated vascularly perfused rat colon. Gut. 46: 218-224

Wijesekaraa I., R. Pangestutia, S-K. Kim. 2010. Biological activities and potential health benefits of sulfated polysaccharides derived from marine algae. Carbohyd Polym 84:14–21.

2. Wijesekaraa I., R. Pangestutia, S-K. Kim. 2010. Biological activities and potential health benefits of sulfated polysaccharides derived from marine algae. Carbohyd Polym. 84:14–21. 3. M. Lahaye, A. Robic. 2007. Structure and functional properties of Ulvan, a polysaccharide from green seaweeds, Biomacromolecules. 8:1765–1774.

4. M. Berri, M. Olivier, S. Holbert, J. Dupont, H. Demais, M. Le Goff, P. Collen. 2017. Ulvan from Ulva amoricana (Chlorophyta) activates PI3K/ Akt signaling pathway via TLR4 to induce intestinal cytokine production. Algal Research. 28: 39-47.

Figure 1: Relative gene expression of IL-8 and CCL20 by intestinal epithelial cells.

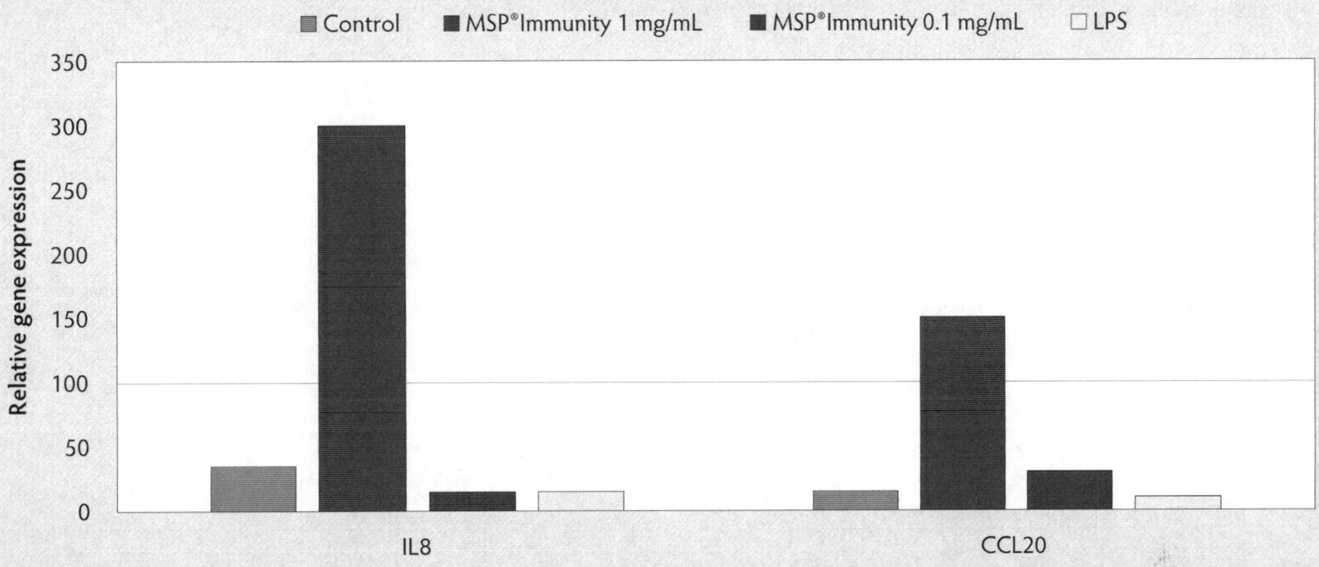

Figure 2: Relative gene expression of IL1α and TNFα by intestinal epithelial cells.

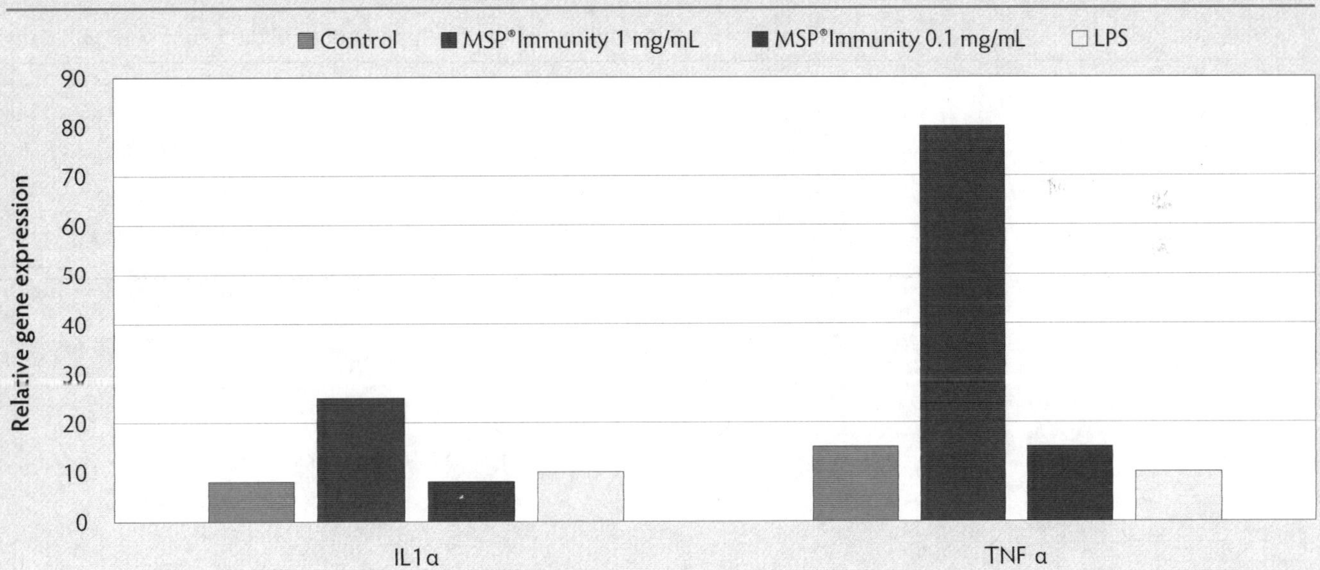

Figure 3: mRNA relative expression of ZO-1 and CLDN2 regarding tight junction protein reinforcement.

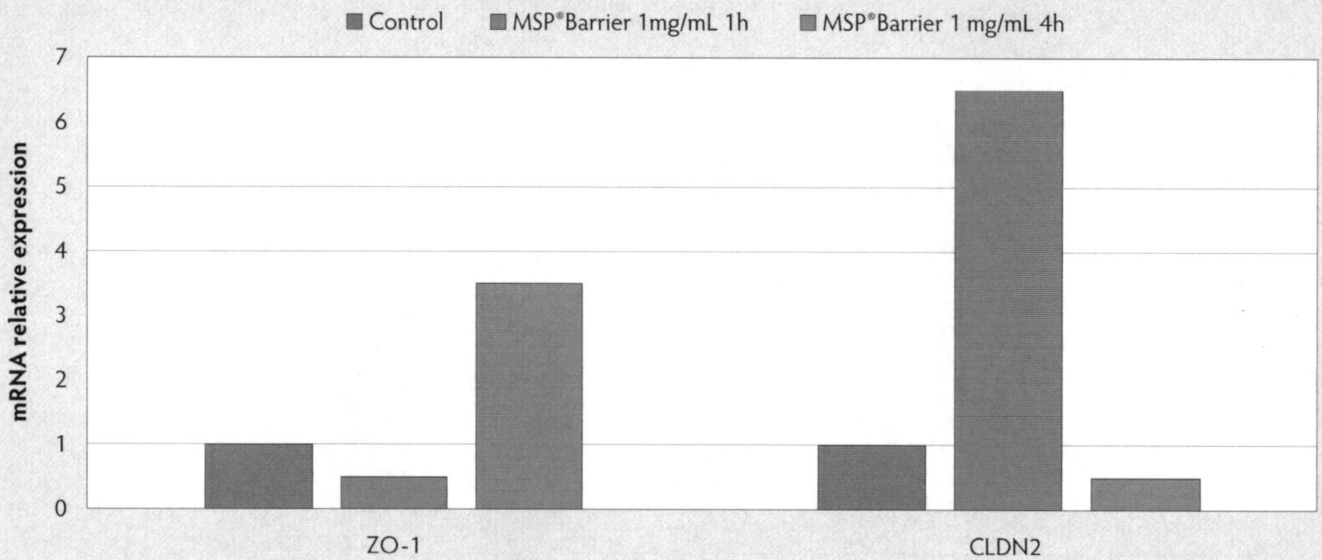

Figure 4: mRNA relative expression of MUC2 and MUC4 regarding increases in mucin production.

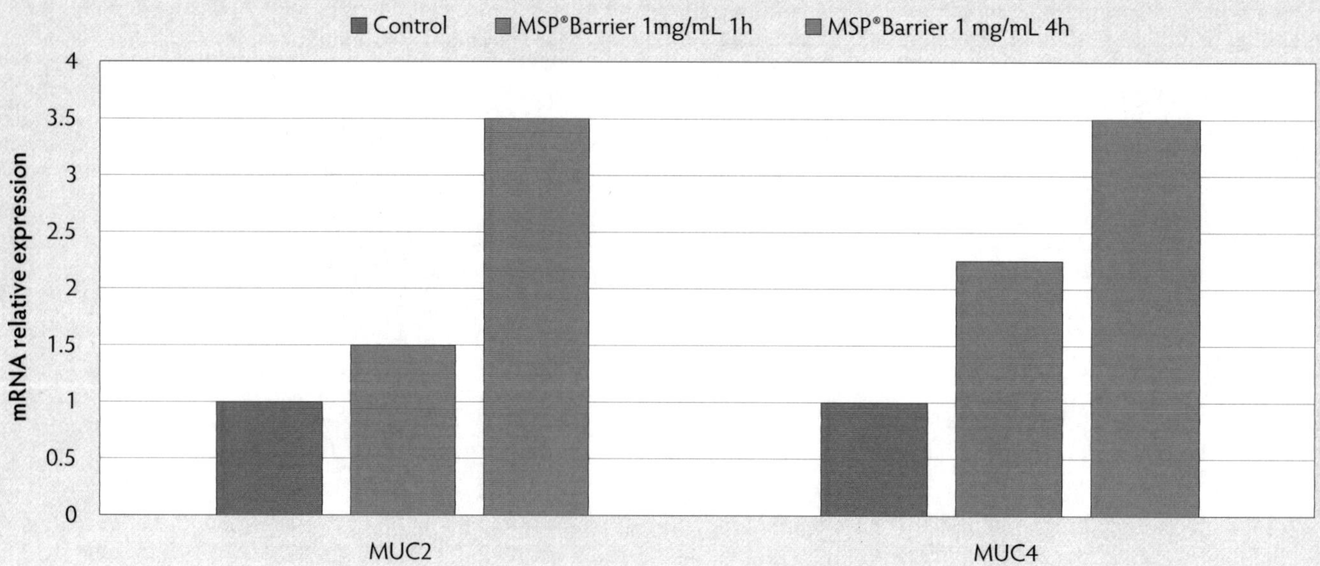

Detection of *Mycoplasma hyopneumoniae* in oral fluids?

Kent Doolittle[1]; Wendy Witbeck[1]; Silvia Zimmerman[1]; Jeff Zimmerman[2]; Ana Paula Silva[2]; Henrique Meiroz-De-Souza-Almeida[3]; Rachel Derscheid[2]; Bailey Arruda[2]
[1]IDEXX; [2]Iowa State University; [3]FCAV- São Paulo State University (UNESP)- Jaboticabal-SP, Brazil

Introduction

Mycoplasma hyopneumoniae (MHP) is the etiologic agent of enzootic pneumonia (EP), a chronic respiratory disease in pigs.[1] MHP is highly prevalent worldwide and causes significant financial loss to the swine industry.[2]

Prevention, control, and/or elimination of swine pathogens is based on understanding patterns of infection. For most pathogens, the industry relies on oral fluids for routine surveillance/detection. The advantages of oral fluids (reduced labor, lower cost, and higher probability of detection) are well-established. However, the detection of MHP in oral fluids has been problematic and many producers monitor MHP by PCR testing of tracheal swab or laryngeal swab samples from individual pigs.

Nevertheless, diagnostic technologies continue to evolve. In the pilot study described herein, the detection of MHP in swine oral fluids was evaluated using a newly released commercial MHP PCR (IDEXX RealPCR M hyo).

Materials and methods

Pigs (n = 50) were allocated to 5 inoculation groups (n = 10 per group): 1) negative control (Friis medium); 2) *M flocculare*; 3) *M hyorhinis*; 4) *M hyosynoviae*; and 5) MHP. Within each group, pigs were housed in 5 pens, i.e., 2 pigs per pen. Pen-based oral fluid samples were collected daily from day post inoculation (DPI) 0 to 56.

At the termination of the study, all oral fluids from the MHP group (n = 280) were analyzed by Method 1 (IDEXX Mag Beads extraction and RealPCR M hyo, Ct cut off < 40) and Method 2 (MagMAX™ Pathogen RNA/DNA kit and published PCR[3], Ct cut off < 37). In addition, weekly oral fluid samples from groups 1, 2, 3, and 4 were analyzed by Method 1 (n = 180).

Results

All oral fluids (n = 180) from groups 1, 2, 3, and 4 were negative by Method 1, i.e., no false positive results were observed. Results from testing oral fluid samples from the MHP group are reported in Table 1.

Table 1: MHP PCR positive oral fluid samples by pen

	Method 1	Method 2
Pen A	11+/56 (20%)	2+/56 (4%)
Pen B	10+/56 (18%)	0+/56 (0%)
Pen C	7+/56 (13%)	1+/56 (2%)
Pen D	23+/56 (41%)	9+/56 (14%)
Pen E	10+/56 (18%)	4+/56 (7%)
Total Positive	**61**	**16**

Conclusion

Results of paired testing revealed differences among testing protocols in terms of MHP DNA detection. Method 1 returned 61 positives vs 16 from Method 2. Of the 16 samples reported positive by Method 2, only 2 samples were not detected by Method 1 (Pen A, day 36 and Pen C, day 26). It was also noted that 31 of the Method 1 positives were above 37 Ct; however, the effect of extraction protocol on nucleic acid detection is known to be significant.[4] Additional research is necessary, but MHP DNA detection technology continues to improve and oral fluids may ultimately be a useful sample for routine surveillance/detection of MHP.

References

1. Maes D, Sibila M, Kuhnert P, et al. Update on *Mycoplasma hyopneumonia* infections in pigs: knowledge gaps for improved disease control. *Transbound Emerg Dis.* 2018;65(Suppl.1):110–124.

2. Holst S, Yeske P, Pieters M. Elimination of *Mycoplasma hyopneumoniae* from breed-to-wean farms: A review of current protocols with emphasis on herd closure and medication. *J Swine Health Prod.* 2015;23(6):321–330.

3. Strait EL, Madsen ML, Minion FC, Christopher-Hennings J, Dammen M, Jones KR, et al. Real-time PCR assays to address genetic diversity among strains of *Mycoplasma hyopneumoniae*. *J Clin Microbiology*. 2008;46: 2491–2498.

4. Chittick WA, Stensland WR, Prickett JR, et al. 2011. Comparison of RNA extraction and RT-PCR methods for the detection of PRRSV in porcine oral fluid specimens. *J Vet Diagn Invest*. 23:248-253.

Successful *Mycoplasma hyopneumoniae* elimination using Tilmicosin (Pulmotil AC®) water medication with a shortened herd closure

A. Bents[1], DVM; R. Evelsizer[2], DVM; M. Pieters[3], DVM, PhD; J. M. Hammer[2], DVM
[1]Hubbard Feeds, Mankato, Minnesota; [2]Elanco Animal Health, Greenfield, Indiana;
[3]University of Minnesota, St. Paul, Minnesota

Introduction

Mycoplasma hyopneumoniae (*M hyopneumoniae*) is a pathogen affecting swine worldwide. Respiratory disease caused by *M hyopneumoniae* leads to significant economic losses due to decreased pig performance. Partial control of *M hyopneumoniae* infections can be achieved using various strategies; however, the most effective means of control is disease elimination. Herd elimination of *M hyopneumoniae* has been demonstrated using a variety of methods.[1] Previous *in vitro* studies indicate that Tilmicosin (Pulmotil AC®) is effective against *M hyopneumoniae*.[2] Therefore, this trial was designed to perform *M hyopneumoniae* elimination using Tilmicosin in combination with a shortened herd closure.

Materials and methods

A 3,000 sow herd with a history of respiratory disease (porcine reproductive and respiratory syndrome virus [PRRSV] and *M hyopneumoniae* positive) and associated production losses in finishers embarked on a *M hyopneumoniae* elimination program.

The farm had an off-site isolation facility (gilt development unit (GDU), housing a 70 day supply of replacement gilts. A previous protocol described a 240-day closure (exposure using lung homogenate to *M hyopneumoniae* negative gilt entry) period.[1] Although completely stocked with unexposed gilts and a fully stocked exposed GDU, only a 180 day closure could be achieved in this project due to management constraints of maintaining breeding targets.

To develop the lung homogenate, gilts showing clinical symptoms in the sow unit were sampled via laryngeal swabs and tested with a PCR[3] to select candidates for lung harvest. The gilts housed separately in the GDU were then individually inoculated with lung homogenate. Three weeks post-inoculation (PI), the inoculated GDU gilts exhibited moderate clinical respiratory disease, all of which were then treated via with tulathromycin (Draxxin®) according to label indications.

Proportioner medicators delivered 1 g Pulmotil AC® per sow daily in farrowing, and 0.5 g per sow daily in gestation for 21 days. Tulathromycin was administered to any coughing animals in the sow herd or GDU during the elimination period and Pulmotil AC® was administered to eliminate *M hyopneumoniae* shedding. All piglets received tulathromycin at weaning for the duration of the elimination project. Pulmotil AC® was continued in farrowing until the conclusion of the 240-day closure period.

Post-elimination laryngeal swabs were collected at 4-week intervals on selected pools of animals in the herd until samples were negative by real-time PCR. When the herd was *M hyopneumoniae* PCR negative, whole-herd *M hyopneumoniae* vaccination quarterly was implemented to maintain herd immunity.

A timeline of events in the elimination protocol applied at this farm is presented in Table 1.

Results and discussion

By three weeks PI, 4/4 laryngeal swab pools collected from the youngest gilts were positive for *M hyopneumoniae* with a Ct value range from 25-30. By day 113, all 8 pools of animals in the sow herd were negative. Following this negative test, inoculated and now PCR negative gilts from the GDU were moved to the sow farm. Sampling continued and at day 240, 40 random samples from throughout the sow herd were all negative.

Starting with pigs weaned approximately 90 days PI, or approximately 70 days after starting Pulmotil AC® to the entire herd, clinical respiratory symptoms were notably reduced. Due to the closure and elimination program, antibiotic costs were reduced by 25%, and mortality reduced by 50% from pre-elimination rates. By these parameters, the elimination was deemed successful. Following surveillance sampling, the herd has remained *M hyopneumoniae* negative for over 18 months, despite its location in a swine dense region, and experiencing a subsequent PRRSV break.

Based on these results, Pulmotil AC® administered in water to the sow herd coupled with piglet injection with tulathromycin at weaning can be used for *M hyopneumoniae* elimination in farrow-to-wean herds.

References

1. Holst, S, P Yeske, and M Pieters. 2015. Elimination of *Mycoplasma hyopneumoniae* from breed-to-wean farms: A review of current protocols with emphasis on herd closure and medication. J Swine Health Prod. 23:321-330.

2. Wu, C C, T R Shryock, T L Lin, and M F Veenhuizen. 1997. Testing antimicrobial susceptibility against *Mycoplasma hyopneumoniae* in vitro. J Swine Health Prod. 5:227-330.

3. Pieters, M, J Daniels, and A Rovira. 2017. Comparison of sample types and diagnostic methods for in vivo detection of *Mycoplasma hyopneumoniae* during early stages of infection. Vet Microbiol. 203:103-109.

Table 1: Timeline of events in the *M hyopneumoniae* elimination process.

M hyo elimination timeline	Event
D -14	New gilts enter GDU
D -14	Gilts with clinical signs identified in sow herd for lung harvest to make lung homogenate inoculum
D0	Lung homogenate exposure to gilts in GDU
D18	Inoculated gilts 100% positive for M hyo by PCR; first dose of Tulathromycin administered to gilts in GDU; entire sow herd started on Pulmotil AC®. Piglets receiving Tulathromycin at weaning.
D45	GDU and gilts in sow farm all found to be positive for M hyo by PCR; second dose of Tulathromycin administered to all gilts in GDU
D70	GDU gilts detected positive for M hyo by PCR
D99	GDU gilts with some positive and some negative M hyo PCR pools
D113	All swab pools negative on PCR for M hyo
D120	All gilts in inoculated and treated GDU moved into sow farm
D124	New M hyo negative gilts purchased and moved into GDU; vaccinated with commercial M hyo vaccine (not inoculated with homogenate)
D133	One suspect pool detected in sow herd by PCR
D164	PCR testing for M hyo revealed one weakly positive pool
D180	Herd re-opened. Vaccinated but non-inoculated gilts moved into sow herd
D240	All samples negative for *M hyopneumoniae* detection

Prevacent® PRRS: determining safety

Christopher L. Puls[1], PhD; Patrick Hoffmann[1], DVM; Ryan Saltzman[2], DVM; Jessica Risser[1], DVM; R. Evelsizer[1], DVM; J. Mark Hammer[1], DVM
[1]Elanco Animal Health, Greenfield, Indiana; [2]Veterinary Resources, Inc., Ames, Iowa

Background and objectives

Porcine reproductive and respiratory syndrome (PRRS) is a significant endemic disease in most swine producing areas of the world. The economic losses can be substantial and are estimated at $1 billion in the USA alone annually.[1] Vaccination and management strategies are partially effective in controlling this endemic disease, but reoccurrence is common. The Domestic Swine Diseases Monitoring Program reports vaccine or vaccine-like strains in nearly 25% of samples evaluated.[2] It is known that Modified Live Vaccines (MLV) have the potential to shed and transmit (i.e. spread) within populations of pigs.[3] Veterinarians and producers are concerned about the shedding potential of MLV PRRS vaccines as controlling the number of PRRS isolates circulating in a pig population should be an objective in PRRS control.

Prevacent® PRRS is a relatively new MLV Lineage 1 PRRS vaccine, which licensing data indicates may have a relatively low shedding and spreading potential. Prevacent's shedding and transmission study demonstrated low transmission to unvaccinated commingled sentinel pigs when housed with IM vaccinated cohorts.[4] Over the course of the 21 day study, only one sentinel pig tested positive on nasal swabs taken approximately 3 days apart. This pig was nasal swab positive once and was not detected as viremic during the study. At necropsy, no PRRS virus was detected in tissues of sentinel animals. However, all vaccinated pigs were viremic for the duration of the 21 day study. Also, all vaccinated pigs were intermittently nasal swab positive (Figure 1) with 4 of 9 vaccinated pigs nasal swab positive on day 21. At necropsy (day 21), all vaccinated pigs were also broncho-alveolar lavage fluid positive. These findings, if validated, could be valuable to swine practitioners in the field at controlling transmission of vaccine virus.

To further define the potential of shedding and transmission, a longer duration study was performed to assess the shed and spread potential of Prevacent PRRS. The objective of the presented study was to assess the shedding and transmission profile of two licensed PRRS MLV vaccines.

Materials and methods

Forty weaned pigs, 17 to 18 days of age and weighing 12.0 ± 1.28 lbs, were randomly assigned by weight and gender to 4 equal treatment groups housed in two separate rooms of a Biosecurity Level 2 facility. Each room had two pens of 10 pigs consisting of 5 vaccinated and 5 unvaccinated sentinel pigs per pen. Vaccine was administered according to label directions. All unvaccinated pigs were mock vaccinates with 1 mL of sterile water. Treatment groups were: TG01 - unvaccinated sentinels housed with TG02, TG02 – pigs vaccinated according to the label (2 mL IM) with

Figure 1: Quantitative PCR results for viremia and nasal swabs of pigs vaccinated with Prevacent PRRS.

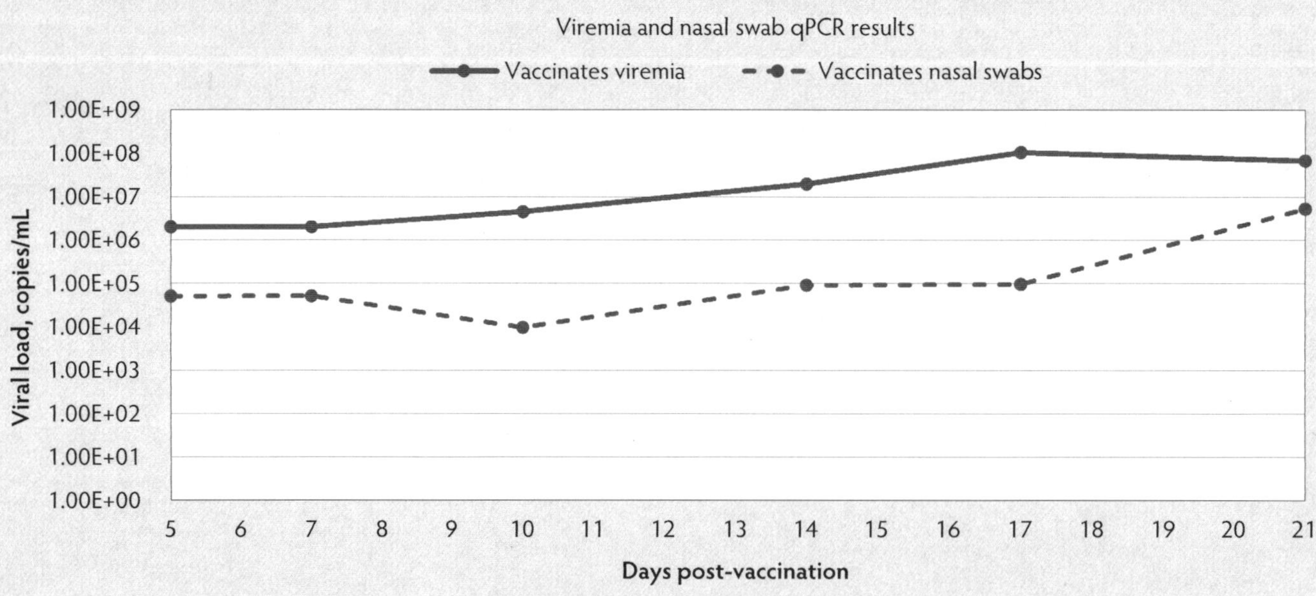

Ingelvac® PRRS MLV, TG03 – unvaccinated sentinels housed with TG04, and TG04 – pigs vaccinated according to the label (1 mL IM) with Prevacent PRRS. Blood samples and nasal swabs were collected once per week (days 0, 7, 14, 21, 28, 35, 42, 49, 56, 63, and 70 of study). Quantitative PCRs (qRT-PCR) were performed to assess viral load in each sample as well as IDEXX ELISA PRRS X3 to assess humoral responses. ISUVDL performed all testing. All blood and nasal swab qRT-PCR were tested immediately to determine the status of the sentinel pigs. Periodic ORF-5 sequencing was performed to confirm positive results were that of the respective MLV PRRS strains. As to evaluate the primary vaccine shedding and transmission potential and minimize infected sentinel pigs from amplifying virus, the following removal criterion was developed: sentinel pigs were removed from the study and humanely euthanized if viremic (PCR CT value < 37) on any sampling or if two consecutive positive nasal swabs (CT value< 30) were identified. Shedding of vaccine virus was defined as any vaccinated pig testing nasal swab PCR positive during the study. Transmission of the vaccine was defined as any sentinel pig becoming viremic or testing nasal swab PCR positive during any point of the study. Additionally, oral fluids were collected from each pen on days 0, 14, 28, 42, 56, and 70 of study. Body temperature was measured on days 0, 1, 3, 7, and 10. Pigs were also weighed weekly during the study. This study protocol was reviewed and approved by Elanco IACUC prior to initiation of the study.

Results

All (10 of 10) sentinel pigs housed with pigs vaccinated with Inglevac PRRS MLV became viremic by day 42 of study and removed. Nine of 10 sentinel pigs housed with pigs vaccinated with Prevacent PRRS became viremic and were removed during the study. One sentinel pig remained in the Prevacent PRRS room throughout the study and never tested nasal swab positive or became viremic. When comparing the CT values at the time of removal, sentinels housed with pigs vaccinated with Prevacent PRRS had a mean CT value of 32.0 vs. 24.4 for sentinels housed with pigs vaccinated with Ingelvac PRRS MLV. Serum CT values were lower ($P < 0.05$) for pigs vaccinated with Inglevac PRRS MLV than those vaccinated with Prevacent on days 14, 21, 28, and 35 of study (Figure 2; Table 1). All (10 of 10) pigs vaccinated with Ingelvac PRRS MLV tested nasal swab positive at least once during the study, whereas only 7 of 10 pigs vaccinated with Prevacent PRRS tested nasal swab positive during the study. Peak shedding of Prevacent PRRS occurred on day 7 of study, when 4 of 10 vaccinated pigs were nasal swab PCR positive with mean CT values of 35.4 and a low of 30.4, and on day 21, when 4 of 10 vaccinated pigs were nasal swab PCR positive with mean CT values of 35.8 and a low of 33.0. All Prevacent PRRS vaccinated pigs were negative (CT value > 37) from day 35 to the end of study (day 70). Similarly, peak shedding of Inglevac occurred on day 7 and day 21 of study. On day 7, 9 of 10 pigs vaccinated with Ingelvac had nasal swab PCR positives, with a mean CT value of 31.3 and a low of 25.6. On day 21, 7 of 10 pigs vaccinated with Ingelvac had nasal swab PCR positives, with a mean CT value of 34.5 and a low of 31.1. Pigs vaccinated with Prevacent PRRS had significantly higher ($P < 0.05$) nasal swab CT values on day 7 than pigs vaccinated with Ingelvac PRRS MLV (Figure 3; Table 1). Duration of shedding in

Figure 2: Serum CT values by week.

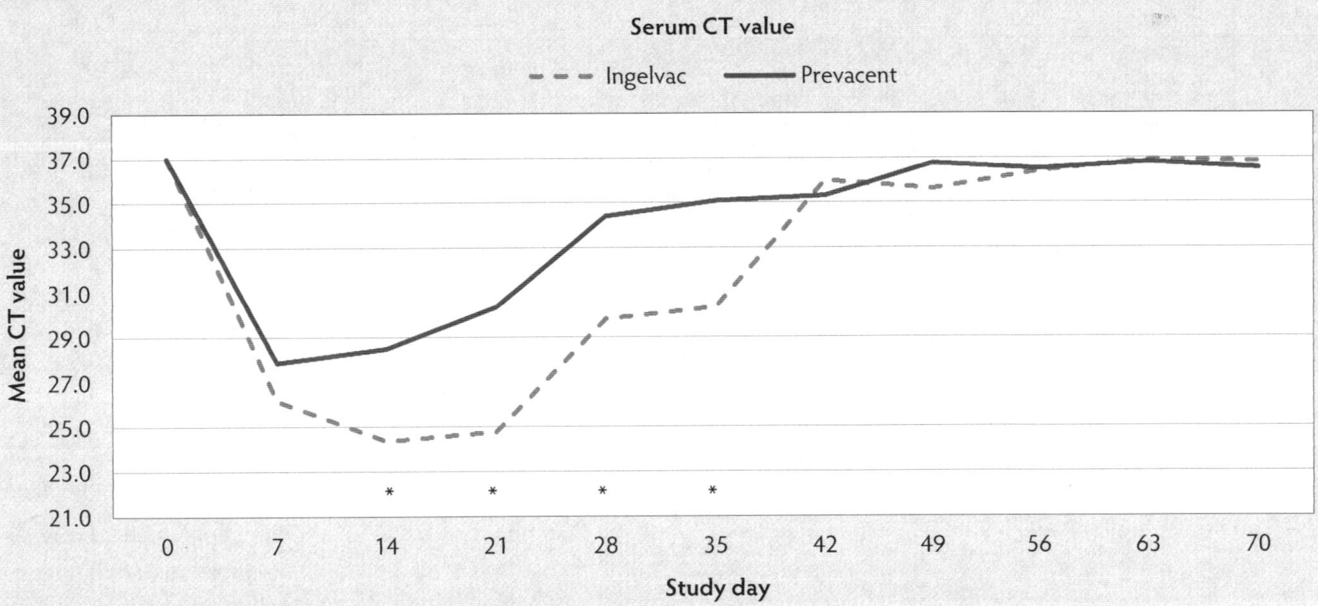

* Denotes difference ($P < 0.05$).

Table 1: Results of shedding and transmission of commercial vaccine for viremia and nasal swabs

		Viremia				Nasal swabs			
Study day	Treatment	No. pigs qPCR positive	LSMean CT	Log$_{10}$	Estimated RNA, copies/mL	No. pigs qPCR positive	LSMean CT	Log$_{10}$	Estimated RNA, copies/mL
0	TG01	0/10	37.0^{a}	0	0.00	0/10	37.0^{a}	0	0.00
	TG02	0/10	37.0^{a}	0	0.00	0/10	37.0^{a}	0	0.00
	TG03	0/10	37.0^{a}	0	0.00	0/10	37.0^{a}	0	0.00
	TG04	0/10	37.0^{a}	0	0.00	0/10	37.0^{a}	0	0.00
7	TG01	0/10	37.0^{a}	0	0.00	0/10	37.0^{a}	0	0.00
	TG02	9/10	26.2^{b}	5.9	737,223.90	9/10	31.3^{c}	4.4	22,928.80
	TG03	1/10	36.5^{a}	0.4	1.80	0/10	37.0^{a}	0	0.00
	TG04	10/10	27.9^{b}	5.7	466,550.90	4/10	35.4^{b}	1.7	46.30
14	TG01	1/10	35.6^{a}	0.7	4.20	2/10	36.4^{a}	0.8	5.00
	TG02	10/10	24.4^{c}	6.7	4,891,025.30	3/10	35.6^{a}	1.3	19.00
	TG03	1/10	36.2^{a}	0.5	2.40	0/10	37.0^{a}	0	0.00
	TG04	10/10	28.5^{b}	5.5	310,097.70	1/10	36.7^{a}	0.4	1.40
21	TG01	1/9	35.5^{a}	0.8	5.00	0/9	37.0^{a}	0	0.00
	TG02	10/10	24.8^{c}	6.6	3,697,429.30	7/10	34.5^{b}	2.8	656.50
	TG03	4/9	34.6^{a}	2.0	105.20	1/9	36.9^{a}	0.4	1.30
	TG04	10/10	30.4^{b}	5.0	89,124.10	4/10	35.8^{ab}	1.5	33.80
28	TG01	3/8	32.6^{ab}	2.4	267.20	4/8	35.1^{ab}	2.1	116.90
	TG02	10/10	29.8^{b}	5.1	127,848.80	2/10	36.6^{a}	0.7	4.40
	TG03	3/5	32.9a^{b}	3.0	980.70	2/5	33.4^{b}	2.2	172.50
	TG04	6/10	34.4^{a}	2.6	363.50	2/10	36.6^{a}	0.7	4.20
35	TG01	2/5	31.1^{b}	2.7	524.30	2/5	32.8^{b}	2.2	168.30
	TG02	8/10	30.4^{b}	3.9	8,244.20	3/10	35.7^{a}	1.1	12.20
	TG03	0/2	37.0^{a}	0	0.00	0/2	37.0^{a}	0	0.00
	TG04	5/10	35.1^{a}	1.8	60.50	0/10	37.0^{a}	0	0.00
42	TG01	3/3	25.9^{b}	5.7	525,532.00	2/3	31.8^{b}	3.2	1,489.00
	TG02	6/10	36.0^{a}	1.8	57.90	0/10	37.0^{a}	0	0.00
	TG03	1/2	36.9^{a}	1.3	17.40	0/2	37.0^{a}	0	0.00
	TG04	5/10	35.3^{a}	1.7	51.70	0/10	37.0^{a}	0	0.00
49	TG01	.	.	.	.	.	.	.	.
	TG02	5/10	35.6^{a}	1.6	42.10	2/10	35.9^{a}	0.8	5.60
	TG03	0/1	37.0^{a}	0	0.00	0/1	37.0^{a}	0	0.00
	TG04	2/10	36.8^{a}	0.6	2.70	0/10	37.0^{a}	0	0.00

[a,b,c] Within day, means with different superscript are different ($P < 0.05$).

Table 1 continued on next page

Table 1: Continued

Study day	Treatment	Viremia: No. pigs qPCR positive	Viremia: LSMean CT	Viremia: Log_{10}	Viremia: Estimated RNA, copies/mL	Nasal swabs: No. pigs qPCR positive	Nasal swabs: LSMean CT	Nasal swabs: Log_{10}	Nasal swabs: Estimated RNA, copies/mL
56	TG01	.	.	.	.	.	.	.	.
	TG02	2/10	36.4[a]	0.7	3.60	1/10	37.0[a]	0.3	0.80
	TG03	0/1	37.0[a]	0	0.00	0/1	37.0[a]	0	0.00
	TG04	2/10	36.5[a]	0.6	3.30	0/10	37.0[a]	0	0.00
63	TG01	.	.	.	.	.	.	.	.
	TG02	1/10	36.9[a]	0.3	0.90	0/10	37.0[a]	0	0.00
	TG03	0/1	37.0[a]	0	0.00	0/1	37.0[a]	0	0.00
	TG04	1/10	36.8[a]	0.3	1.00	0/10	37.0[a]	0	0.00
70	TG01	.	.	.	.	.	.	.	.
	TG02	1/10	36.8[a]	0.3	1.00	0/10	37.0[a]	0	0.00
	TG03	0/1	37.0[a]	0	0.00	0/1	37.0[a]	0	0.00
	TG04	1/10	36.5[a]	0.4	1.40	0/10	37.0[a]	0	0.00

[a,b,c] Within day, means with different superscript are different ($P < 0.05$).

oral fluids was less in the Prevacent PRRS room than the Ingelvac PRRS MLV room. Oral fluids collections were only positive on day 28 for the Prevacent vaccinated room whereas at least 1 of the 2 pen samples collected from the Ingelvac vaccinated room were positive on day 14, 28, 42, and 56 of study. Body temperature and weight gain were not significantly ($P > 0.05$) different during the study, although pigs vaccinated with Prevacent PRRS were 2.9 lb heavier at day 70 than pigs vaccinated with Ingelvac PRRS MLV.

Discussion and conclusion

The statistically lower serum CT values on days 14, 21, 28, and 35 indicate higher viremia (viral load) in Inglevac PRRS MLV vaccinated pigs compared to Prevacent PRRS vaccinated pigs. Nasal swabs and oral fluids were sporadically positive throughout the study for both vaccinated groups, but shedding both within the animal and within the environment appear to be less for Prevacent vaccinated pigs than those vaccinated with Ingelvac. Nonetheless, both vaccines do shed and transmit vaccine virus to sentinel pigs, but the amount and duration appear different. It is important to note that although this study was conducted for a longer duration than the initial licensing study, it was still carried out in a small number of animals in a highly controlled setting. Larger groups of vaccinated and sentinel pigs need to be evaluated to better understand the transmission of Prevacent PRRS to unvaccinated sentinels or in large populations of vaccinated pigs. Furthermore, all sentinels were removed because of viremia and not positive nasal swabs. Perhaps a change in removal criteria or a replacement strategy would have shown different results. Likewise, although viremia at the time of removal is drastically less in sentinels housed with pigs vaccinated with Prevacent, it remains unclear what the performance and viremia would look had sentinels been allowed to remain as part of the study population. Further evaluations of Prevacent shedding and transmission are warranted.

References

1. Holtkamp DJ. Assessment of the economic impact of porcine reproductive and respiratory syndrome virus on United States pork producers. J Swine Health and Prod. 2013. 21:72.

2. Swine Health Information Center Domestic Swine Disease Monitoring Program Report `https://fieldepi.research.cvm.iastate.edu/domestic-swine-disease-monitoring-program/` Accessed Nov. 12, 2019

3. Renukaradhya, GJ et al. Live porcine reproductive and respiratory syndrome virus vaccines: Current status and future direction. Vaccine. 2015. 33:4069.

4. Elanco. Data on file.

Figure 3: Nasal swab CT values by week.

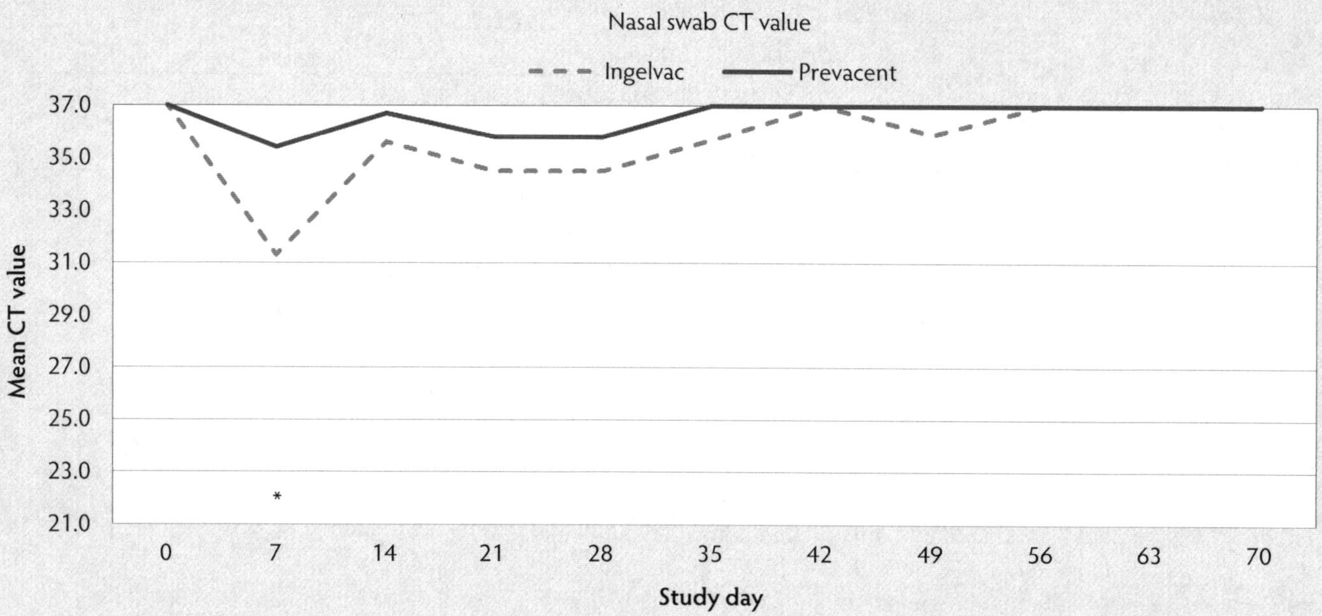

* Denotes difference ($P < 0.05$)

Utilizing LeeO™ individual animal identification system in commercial research

Paul Yeske[1], DVM, MS; J. Tyler Holck[2], DVM, MS, MBA
[1]Swine Vet Center, St Peter, Minnesota; [2]Feed His People, LLC, Gilbert, Iowa

Introduction

The LeeO™ individual animal identification system has been in use in The Netherlands for several years. Westfort Meat Products is a pork processor that purchases 100% of its live pork from Dutch farmers. Westfort utilizes the LeeO radio frequency identification (RFID) tags to track pigs from birth to slaughter, providing individual tracking from farrowing crate all the way to the final carcass scale and quality assessment in their pork plant. This birth to market information provides valuable information on reproductive as well as grow/finish production and is available real-time via the LeeO cloud-based software. Prairie Systems (Spencer, Iowa) is launching the LeeO system in the US with targeted swine market segments of 1) genetic suppliers and internal multiplication, 2) commercial research, and 3) commercial production.

The Swine Vet Center is a swine exclusive veterinary consulting practice which services swine producers with clinical and data services as well as commercial research. The LeeO system was utilized in a Swine Vet Center PRRS research trial to assess the efficiency in utilizing individual RFID technology to gather data and monitor pigs in a wean-to-finish study vaccinated and control animals.

Study design and use of LeeO

A 2400 head commercial research barn with two rooms (east and west rooms with separate airspace) comprised of eight large (150 head) pens each was utilized for this trial. The commercial pigs were sourced from 2 high health (PRRS, Mycoplasma and APP negative) sow farms and weaned/placed one week prior to the start of the trial. A total of 1016 pigs were selected and individually tagged, registered into LeeO, and weighed on October 31, 2019. The vaccine trial compared 480 saline injected pigs located across 4 pens in the east room and 480 PRRS vaccinated pigs along with 56 non-vaccinated pen mates to the PRRS vaccinated pigs all located in 4 pens in the west room. 30 pigs in each treatment group (60 total) were selected for blood sampling, tagged with a different colored tag, and assigned to a diagnostic sampling subgroup within LeeO.

The steps to initiate the trial and the LeeO individual animal data gathering process were:

1. Eligible pigs were gathered on one side of the large pen and pigs were picked up individually.
2. Pigs were tagged with a LeeO UHF (ultra-high frequency) ear tag
3. The UHF tag was then read with a LeeO reader to register the pig into LeeO
4. The UHF tag was then read a second time, gender recorded, and the pig placed on the LeeO scale
5. The pig was then administered either saline (east room) or the PRRS vaccine (west room) and placed on the other side of the pen

This process started with a pace of 2-3 pigs processed per minute for the first two pens and became more efficient with the final two pens processing 5-6 pigs per minute (Figure 1). The streamlining of the pig processing with LeeO was attributed to more effective handling of the pigs and equipment and fine-tuning of the scale settings to minimize bottlenecks while still maintaining quality data collection (avoiding translocation and wrong IDs being read).

The data from the day one activities was also available real-time, with weight distribution by pen (Figure 2) assessed during breaks from the pig processing.

Additional data sampling occurred with individual weights collected at approximately 9, and 21 weeks post-vaccination. In addition, blood samples were collected from the 60 pigs assigned to the diagnostic sampling group on days 0, 4, and 8 post-vaccination along with weeks 3, 5, 9, 11, 13, 17, and 21 post-vaccination using the tags to locate the pigs for testing.

Conclusion

The use of the LeeO system provided an accurate and efficient method to identify individual pigs and conduct the required processes associated with this commercial research trial, including individual animal weights and diagnostic sampling. Advantages of the LeeO system over traditional tags or other RFID tags include the automatic linkage with scales and pen locations as well as real-time access to the data. Additional observations and data will be shared at the 2020 AASV annual convention as the commercial trial is concluded.

Figure 1: Individual animal processing efficiency using LeeO™ system.

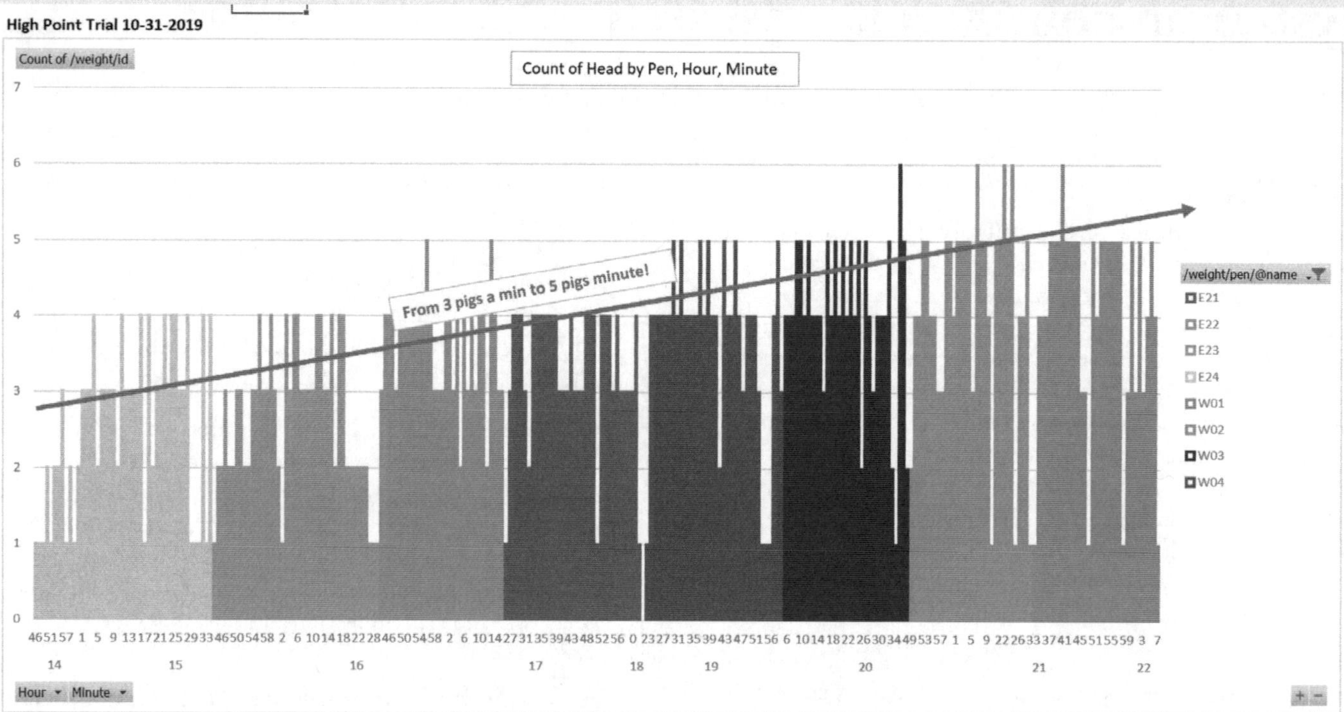

Figure 2: Distribution of individual animal weights at processing using LeeO™ system.

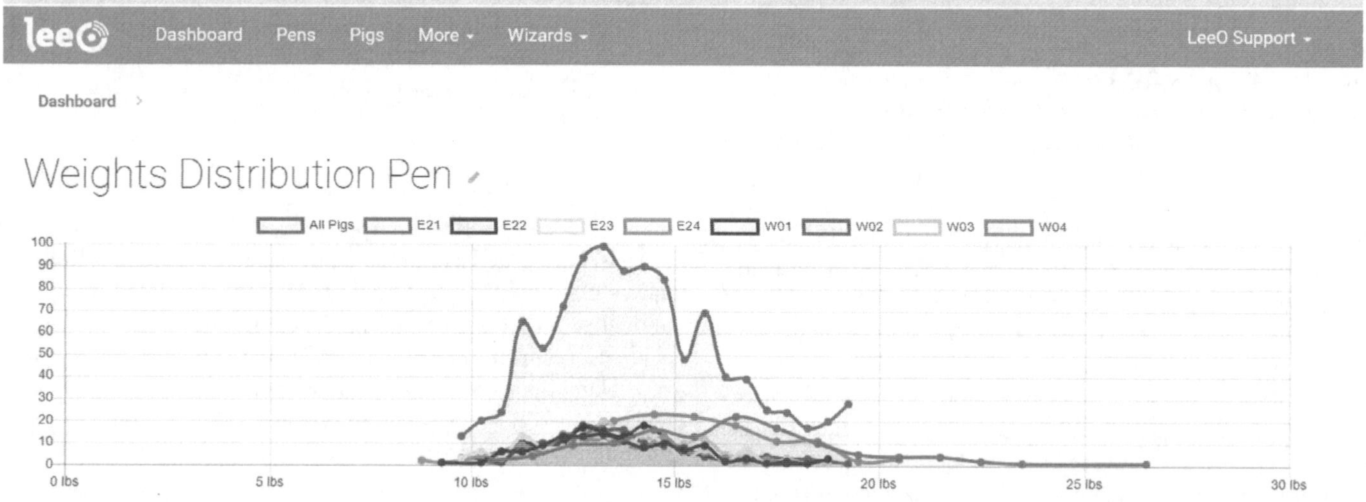

VFD inspections update: education to enforcement

J. Tyler Holck, DVM, MS, MBA
Feed His People, LLC

VFD regulations and compliance activities

In January 2017, the US Food and Drug Administration's (FDA) Center for Veterinary Medicine (CVM) and animal drug manufacturers completed the voluntary transition of all medically important antimicrobial drugs used in or on animal feed from over-the-counter (OTC) to Veterinary Feed Directive (VFD) marketing status, under Guidance for Industry (GFI) #213.[1] FDA employed a phased-in compliance strategy for the VFD final rule and GFI #213 implementation efforts. Under this strategy, FDA's initial focus was to educate affected stakeholders (e.g., producers, veterinarians, feed mills, and retail establishments) on the new VFD regulation before taking enforcement action. FDA responded to various individual questions and challenges as they arose as part of this initial education phase. FDA collected stakeholder feedback from veterinarians, producers, feed manufacturers and other segments of the animal agriculture industry.

Along with publishing the VFD final rule in June 2015, CVM developed a framework for inspections of distributors, veterinarians, and producers involved in the VFD process (Appendix I).[2] Inspection activities were carried out by FDA's Office of Regulatory Affairs (ORA) and under contract by participating state feed regulatory programs. During fiscal years 2016 and 2017, inspections focused primarily on providing education to producers, veterinarians, and VFD medicated feed distributors (e.g., feed mills and retailers of VFD medicated feed) as part of the phased-in compliance strategy. In fiscal year 2018, FDA transitioned toward an inspectional approach focused on compliance with the VFD requirements and documenting violations when found.

In March 2019, the FDA announced the availability of an updated guidance for industry (GFI) #120 titled "Veterinary Feed Directive Regulation Questions and Answers."[3] The revised draft guidance added 53 new VFD questions and responses — and revised 14 — covering a variety of topics. The Q&A document began with general VFD questions related to the rule itself, drugs, recordkeeping, etc., then continued with questions focused on the specific stakeholders, namely veterinarians, distributors (feed mills) and clients (producers, owners, animal caretakers).

FDA is now expanding review of VFD medicated feed and VFD-related records into routine Current Good Manufacturing Practice (CGMP) inspections of medicated feed mills, including licensed medicated feed mills, and to Drug Residue Investigations.

Enforcement strategy

During the initial educational phase, investigators were encouraged to discuss deficiencies that were not significant to public health (e.g., minor recordkeeping discrepancies) with the involved producer, veterinarian, or VFD distributor and obtain voluntary corrective action.

FDA issued one Warning Letter (WL) in fiscal 2018 based on a medicated feed CGMP and VFD inspection. The feed mill adulterated and misbranded VFD feed by distributing VFD feed to other distributors without first receiving an acknowledgment letter, in addition to adulterating and misbranding medicated and non-medicated feed for other reasons.

CVM will continue to expand the comprehensive VFD compliance strategy and intends to continue monitoring compliance of all parties to ensure the VFD regulation is followed. Although FDA will continue to provide education, enforcement strategies will also be used when voluntary compliance is not achieved.

Any questions about the VFD regulation may be directed to `AskCVM@fda.hhs.gov`.

Lessons learned from inspections

A summary of VFD inspections was released by the FDA in August of 2019, summarizing VFD inspections conducted in Fiscal Years 2016 - 2018.[4] Annual rates of compliance were reported for each of the three years assessed in the report. Overall, approximately 91% of the VFD final inspections were classified as no action indicated (NAI), and feedback showed that affected parties were generally aware of and in compliance with the VFD final rule. The majority of inspected VFD feed distributors notified FDA of their intent to distribute VFD feeds. As of August 2019, there are over 9,600 facilities in all 50 states, Puerto Rico, and Canada that have notified FDA of their intent to distribute VFD medicated feed, a substantial increase over the approximately 1,400 facilities at the end of 2015.

The majority of VFD feed distributors who manufacture feed that were inspected followed labeling and recordkeeping requirements showing that VFD feeds were generally manufactured in compliance with the VFD final rule (Appendix II).

Drug inventory or production records showed the correct amount of drug was added to the feed for the VFD reviewed ranged from 90.3% to 96.7% during the three years summarized in this report and labels and formulas matched the VFD reviewed in 87.2 to 96.6% of instances. However, some VFD feed labels were missing the VFD caution statement. "Caution: Federal law restricts medicated feed containing this veterinary feed directive (VFD) drug to use by or on the order of a licensed veterinarian."

All inspected veterinarians were licensed in the state where the VFD feed authorized on the VFD order(s) was being fed or were operating through reciprocity or a similar type of program in 100 % of occurrences. However, investigators/inspectors found occurrences where the veterinarian's written or electronic signature appeared to be missing from the VFD due to technical issues. A majority of VFDs contained caution and warning statements matching the indication for use. However, some VFDs were either missing withdrawal time, special instructions, and/or cautionary statements, or included caution or warning statements for combination VFD drugs when the combination was not ordered.

Regarding the producer/client, VFD feeds were generally fed according to the instructions on the feed label and the VFD.

Compliance to the expiration date on the VFD feed being fed was 75 to 100%. Compliance to the VFD being fed to authorized animals was 90 to 100 %, for the duration identified on the VFD was 89.3% to 100%, and with special instructions in 91.3% to 100% of the occurrences.

Electronic/digital VFD solutions

Paper and word processing methods of writing VFDs and scripts involve a large amount of time, paperwork and resources to ensure clear communication, accuracy of drug label requirements, completeness of information and compliance with regulations. The FDA has approved the use of electronic VFDs (eVFDs) instead of paper or hard copy VFDs, as long as the electronic VFDs meet the requirements of 21 CFR part 11. An eVFD must be signed by the veterinarian, which can be done with a Part 11 compliant digital signature.

The industry standard VFD software is provided by GVL (GlobalVetLink), which has been available since 2005. Over 300,000 VFDs have been written in GVL since January 2017. Of those VFD's, approximately one-third, or over 100,000 have been written for swine (Figure 1).

With the GVL platform, veterinarians, producers, and feed distributors gain access to their VFDs in a secure, CFR 21 Part 11 compliant database, thus fulfilling the two-year storage requirements in Guidance 120. The GVL system has been scrutinized by a third party auditor who specializes in Part 11 compliance, and found GVL VFDs to be in compliance. Feed distributors receive copies of VFDs from veterinarians through GVL, and have the option to upload digital copies of paper, emailed or faxed VFDs in their account. GVL's VFD system also allows distributors to communicate directly with veterinarians and request VFDs for customers, and it notifies all parties as VFDs near expiration for easier renewal.

In addition, the GVL VFD system is integrated with Prairie Systems' Feed Allocation System (FAS) software. By integrating with Prairie Systems using the industry approved standard, the actual feed order process will not change, but simply be enhanced as FAS will automatically check for a valid VFD in the GVL database. If no VFD exists to enable the order to be completed, a request for an VFD will be automatically generated on behalf of the producer and sent to GVL. This request will contain all the necessary information required, such as requested medication, animal group details, farm location, and mill information, to initiate a request of the VFD to the veterinarian identified with a VCPR for that producer. The veterinarian will then review the request and either complete the VFD or decline the request. The

Figure 1: GVL digital VFDs.

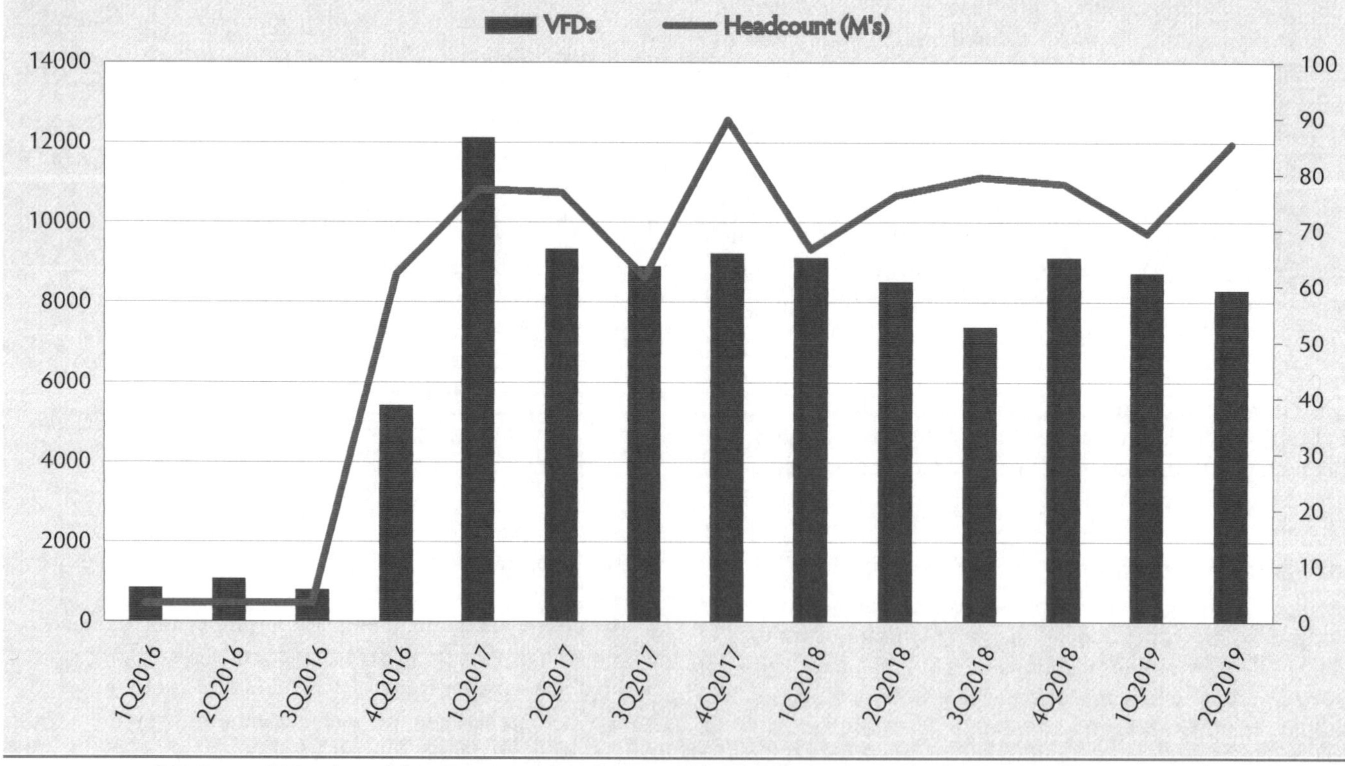

integration of two third party systems working collaboratively to ensure each and every feed order is valid prior to reaching the manufacturing partner is an industry first.

Summary

The implementation of the new VFD rule has in general been very smooth for the feed distributors, veterinarians, and producers supporting the swine industry, thanks in part to the GVL eVFD solution enabling compliance to the new regulations. The FDA has stated that the initial educational phase of inspections has come to a close and they will now address any compliance infractions with enforcement. It is important for the swine industry to maintain its focus and discipline to ensure continued compliance with the VFD regulations.

References

1. CVM GFI #213 New Animal Drugs and New Animal Drug Combination Products Administered in or on Medicated Feed or Drinking Water of Food-Producing Animals. 2013. Available at: https://www.fda.gov/regulatory-information/search-fda-guidance-documents/cvm-gfi-213-new-animal-drugs-and-new-animal-drug-combination-products-administered-or-medicated-feed. Accessed 28 October 2019.

2. Federal Register Veterinary Feed Directive. 2015. Available at: https://www.federalregister.gov/documents/2015/06/03/2015-13393/veterinary-feed-directive. Accessed 28 October 2019.

3. Small Entity Compliance Guide CVM GFI #120 Veterinary Feed Directive Regulation Questions and Answers. 2019. Available at: https://www.fda.gov/regulatory-information/search-fda-guidance-documents/cvm-gfi-120-veterinary-feed-directive-regulation-questions-and-answers. Accessed 28 October 2019.

4. FDA Publishes Assessment of Compliance with Veterinary Feed Directive Final Rule. 2019. Available at: https://www.fda.gov/animal-veterinary/cvm-updates/fda-publishes-assessment-compliance-veterinary-feed-directive-final-rule. Accessed 28 October 2019.

Appendix I – VFD inspection example questions

Using the database of distributor notifications, FDA field staff will begin inspections at VFD distributors with additional follow-up at veterinarians and producers. At the VFD distributor, an investigator may examine three randomly selected VFD forms. From these, they will pick one form to follow back to the vet and to the producer. The investigators will be looking for the required information on the VFD order (see previous "VFD Required Information" section).

Specific example questions for each party may include:

Distributor questions:

1. Did the distributor notify FDA of the intent to distribute VFD feeds? (Verify VFD Distributor Notification Listing on fda.gov.)
2. Does the distributor keep copies of VFD orders for at least 2 years, if distributing VFD feed to the end user?
3. If the operation being inspected distributes VFD feed to other distributors and not to the end user, does this operation keep copies of acknowledgement letters for at least 2 years from the date of last shipment under the acknowledgement letter?

VFD distributor questions:

1. Does the feed label contain the VFD Caution statement?
2. Do the drug inventory or production records show the correct amount of drug added?
3. Do the labels and formulas match the VFD orders?

Veterinarian questions:

- Does the veterinarian have a valid license in those states where VFD feed is being fed?
- If No, can the veterinarian provide any evidence that he or she is complying with applicable state veterinary licensing and practice requirements in the states in which the veterinarian does not have a license?
- Does the veterinarian know that either the state or federal requirements for veterinary client patient relationship (VCPR) apply in each state?
- Can the veterinarian show any medical record(s) for the client's animals named on the VFD?
- Does the veterinarian keep copies of VFD orders for at least 2 years?

Producer (client) questions:

- Does the client keep copies of VFD orders for at least 2 years?
- Did the client feed the VFD feed to the authorized number of animals on the VFD order?
- Did the client feed the VFD feed for the identified duration on the VFD order?
- Did the client stop feeding the VFD feed prior to the expiration date on the VFD order?
- Did the client follow the withdrawal period for the VFD feed, if any?
- Did the client follow any special instructions or caution statements on the VFD order, if any?
- If a combination VFD feed was fed, was its use consistent with the affirmation statement on the VFD order?

- Does the client have labels for VFD feeds? If Yes,
- Does the feed label contain the VFD Caution statement?
- Did the drug level on the label match the drug level on the VFD form?
- Is the drug level and indication on the VFD form consistent with the approval?

Appendix II -VFD required information

This information is required on a lawful VFD:

- Veterinarian's name, address, and telephone number;
- Client's name, business or home address, and telephone number;
- Premises at which the animals specified in the VFD are located;
- Date of VFD issuance;
- Expiration date of the VFD;
- Name of the VFD drug(s);
- Species and production class of animals to be fed the VFD feed;
- Approximate number of animals to be fed the VFD feed by the expiration date of the VFD;
- Indication for which the VFD is issued;
- Level of VFD drug in the feed and duration of use;
- Withdrawal time, special instructions, and cautionary statements necessary for use of the drug in conformance with the approval;
- Number of reorders (refills) authorized, if permitted by the drug approval, conditional approval, or index listing;
- Statement: "Use of feed containing this veterinary feed directive (VFD) drug in a manner other than as directed on the labeling (extralabel use), is not permitted";
- An affirmation of intent for combination VFD drugs as described in 21 CFR 558.6(b)(6); and
- Veterinarian's electronic or written signature.

You may also see the following optional information on the VFD:

- A more specific description of the location of the animals (for example, by site, pen, barn, stall, tank, or other descriptor the veterinarian deems appropriate);
- The approximate age range of the animals;
- The approximate weight range of the animals; and
- Any other information the veterinarian deems appropriate to identify the animals at issue.

Also, VFD drug labeling and advertising must prominently and conspicuously display the VFD caution statement.

Diagnostic trends of swine nursery systemic pathogens

Eduardo Fano[1], DVM, MS, PhD; Ana Paula Silva[2], DVM; Bailey Arruda[2], DVM, PhD; Jeffrey Zimmerman[2], DVM, PhD; Rodger Main[3], DVM, PhD; Maria Jose Clavijo[2], DVM, PhD
[1]Boehringer Ingelheim Animal Health USA Inc., Duluth, Georgia; [2]Iowa State University, VDL, Ames, Iowa

Introduction

Swine systemic bacterial diseases continue to challenge pork production by reducing productivity, compromising welfare and leading to an over-dependence on antimicrobials. Legislative pressures for reduction of antimicrobial use in animal agriculture will likely lead to increased disease occurrence in systems that rely heavily on antimicrobials for control. *Mycoplasma hyorhinis* and *Haemophilus parasuis* are detected annually in over 50% of polyserositis cases and *Streptococcus suis* has raised again as a significant pathogen. These systemic pathogens present a significant collective threat to young pigs, occur across a relatively well-defined production period (nursery and early finishing stages), and symptoms of the diseases are often mistaken for each other. It is critical to refocus the attention on these pathogens and begin to close important knowledge gaps in order to develop improved control and prevention strategies.

Mycoplasma hyorhinis (*M hyorhinis*)

M hyorhinis is an increasingly prevalent disease challenge for swine veterinarians and producers. It is considered a commonly occurring microorganism in swine populations—it colonizes the ciliated upper respiratory tract and may not show any clinical signs. However, under certain conditions (stresses like nursery comingling or infection with other pathogens), *M hyorhinis* can move from being an upper respiratory commensal to a systemic pathogen, causing outbreaks of polyserositis and arthritis. Piglets of nursery age are the most susceptible to clinical disease caused by systemic infection. With a good understanding of how *M hyorhinis* causes disease, diagnostics can be used to identify it in clinically affected pigs, confirming the presence of the systemic disease.

Haemophilus parasuis (*H parasuis*)

H parasuis continues to challenge swine producers causing fibrinous polyserositis, meningitis and arthritis, syndrome commonly known as Glässer's Disease. It is considered one of the most prevalent bacterial infections during the nursery stage, and regular medication to control this disease is an emerging concern.

Streptococcus suis (*S suis*)

For decades, *S suis* has challenged producers causing systemic disease pre and post weaning. However, for the last couple of years, several US practitioners have reported an increase in *S suis* associated-disease in the field. A survey performed at the VDL at ISU showed that the number of cases with *S suis* has increased. While this increase can be due to more awareness and improved ability to diagnose *S suis*, it can also be due to an increased disease challenge. In fact, *S suis* was recently considered the number one bacterial agent that poses the most risk for the US swine industry.

Diagnostic trends (2014-2019, preliminary results)

The main goal of this study was to describe diagnostic trends of the main nursery systemic pathogens: *M hyorhinis, S suis, H parasuis*. Frequency of detection of these swine systemic bacteria was investigated in systemic cases (samples) submitted to the Iowa State University Veterinary Diagnostic Laboratory (Ames, Iowa) between 2014 and 2019. Between 2014 and 2019, *M hyorhinis* was detected in over 50% of systemic samples every year. An increase of the frequency of detection was observed in 2018 and 2019 (over 60%). Between 2015 and 2019, *H parasuis* was detected consistently in over 60 % of systemic samples every year, observing an increase in 2018. Between 2015 and 2019, *S suis* was detected (culture) in 36 % of all type of samples in the last 6 years, observing an increase of submissions and detection frequency in 2018 and 2019. Definitive/final results of this study will be presented during the AASV 2020, including all information by year and type of sample and a detailed discussion by pathogen.

Farmera™: a new generation of farm management

Jens Kjaer[1], DVM; Kellie Cicconi-Hogan[1], PhD; Kaylee Hillinger[1];
Dale Polson[1], DVM, PhD; Bernd Grosse-Liesner[2], DVM
[1]Boehringer Ingelheim Animal Health USA Inc., Duluth, Georgia;
[2]Boehringer Ingelheim Vetmedica, Ingelheim, Germany

Introduction: what is FARMERA?

In swine production and health management today, relevant information is aggregated and interpreted primarily through manual and analog methods, such as unrelated spreadsheets and handwritten barn sheets. There is a clear need for technology that can quickly and efficiently gather, disseminate and ultimately generate predictability of information for a given farm. While some technology does currently exist, existing products and services fail to address the need for simplicity and real time access of health-related information collected with digital tools. Additionally, these products do not address, assess or related the relationship of that data to the performance of the farm.

A new and innovative tool by Boehringer Ingelheim Animal Health USA Inc., FARMERA, addresses this gap by efficiently identifying health issues significantly earlier than today's typical farm, recommending appropriate actions or interventions and showing the economic value thereof.[1,2] Boehringer Ingelheim's initial focus has been on providing real time health related information combined with farm performance parameters to producers. Our target personnel are the farm or system supervisor and the barn caregiver, followed by the veterinarian tending to the farm. We collect data on multiple variables e.g. temperature, humidity, water, feed from multiple sources and subsequently present them on a single mobile platform, an application called FARMERA.

A key feature of the app is inventory management, treatment input and mortality assessment of all of the supervisor's barns at a glance, providing a clear snapshot of the overall 'health of the system'. This also allows the supervisor to maximize their time by prioritizing the farms with the greatest need for attention and care. Additionally, the platform has a unique feature in the integration of data from SoundTalks®, a cough monitoring system. SOUNDTALKS devises and provides notifications and alerts for critical situations, such as strong deviations from the previously assessed baseline, which may indicate significant respiratory events.

Benefits to the swine producer

For a given swine production system, the aggregation and collection of all the data in one place provides streamlined, transparent and timely availability of information on the environment and animal health in the barn. Furthermore, the collection of the data electronically has now allowed swine production systems to eliminate classical barn sheets. This is valuable to the producer, as this has traditionally been a cumbersome and error-prone element of the swine production process.

For production supervisors, the FARMERA app also provides a way to monitor the activities of contract growers, as it gathers data that has been entered into the application daily - and signifies if data is missing. In case mistakes are made, it is relatively easy for the supervisors to make corrections.

For attending veterinarians, it is very helpful to have historical data on temperature, humidity, water consumption and feed consumption, as this often allows the veterinarians to correlate changes in these variables with treatments and mortalities to disease outbreaks and other events.

Materials and methods: the application development approach

The development of the application has been an iterative process that is heavily influenced by the user. This was done through repeated user interviews by dedicated user experience (UX) designers. The UX designers sought input from the end user on how to make the application as user friendly as possible, as well as what key elements of the application were missing or needed, by interviewing caregivers, supervisors and vets at an interval of 4 weeks. Between the interviews, the software development team would conduct 2 development sprints and seek input from the end-users in between, so that when they returned to the users after 4 weeks, new and further developed tools would be presented that had been directly influenced or requested by the user.

Results: current and ongoing impacts

At the time of writing this abstract (November 15th), FARMERA is installed in almost 46 barns in 4 major systems in the USA. The system provides alerts of any deviations from the norm, as well as historical and real-time data including graphical representations of the information. Furthermore, the notifications will direct the supervisor and the veterinarian to the barns with the biggest health problems. The SOUNDTALKS monitors are also providing early warnings on cough activity and cough index. As the cough monitors measure and report cough 24 hours a day, the cough monitor is able to identify increased cough activities 2-5 days before humans,[3] and the caregiver is able to initiate further diagnostic measure and treatment faster than the veterinarian or caregiver alone.

Discussion: future opportunities

In addition to sensor technologies in the barns, the FARMERA app will allow the veterinarian and/or the CEO of the systems to have an up-to-date overview of health and performance via dedicated dashboards that are a constantly updated feed. This will soon include diagnostic data from the different flows directly available on a dashboard. Furthermore, we are expecting that machine learning will eventually identify risk factors for disease outbreaks that will provide information used to prevent disease outbreaks.

Additionally, we expect the early cough identification to foster onsite diagnostic tests, such as Mobinostics™, allowing for pen-side confirmation of the diagnosis and potentially lead to optimal treatment within hours.

References

1. Polson D et al. Classification of cough patterns in growing pigs using continuous sound monitoring and an algorithm-based respiratory distress index. ESPHM. 2018.

2. Krantz S et al. Experiences with the SOMO+ respiratory distress monitor to evaluate spatial spread of *M hyopneumoniae* within in a barn. AASV. 2019.

3. Polson D et al. Precision livestock farming (PLF) for pig health and production: Sound as a diagnostic sample. AASV. 2018.

Role of vaccine-derived immunity for heterologous protection against PRRS: what we have learned

R. Philips, DVM
Boehringer Ingelheim Animal Health USA Inc., Duluth, Georgia

Introduction

Controlling PRRS requires a systematic approach that is multifactorial and employs complementary components for PRRS control such as immunity, biosecurity, pig-flow management, and diagnostic testing and monitoring. These complementary components are utilized in protocols that are targeted at optimizing population immunity and minimizing exposure and transmission by phase of production. We have learned that population based vaccination protocols can provide protective immunity that can mitigate the consequences of PRRSV infection, thus reducing the clinical impact of disease and subsequently improving health and performance – a direct benefit of vaccine-derived immunity.[1-3] Additionally, we have learned from both controlled and field-based studies, that vaccine-derived immunity can reduce wild-type PRRSV shedding within and from vaccinated populations, decreasing the risk of virus exposure – an indirect benefit of vaccine-derived immunity.[1,2,4-6]

Heterologous protection: direct benefit of vaccine-derived immunity

The ever-expanding diversity of PRRSV field strains frequently evokes the question: "Will currently available commercial vaccines protect against the contemporary and relevant wild-type PRRSV strains that I have in my system?" Effective vaccines need to provide heterologous protection. Over the years, Ingelvac PRRS® MLV has been shown to provide heterologous protection against both Type 1 and Type 2 PRRSV strains.[7,8] The INGELVAC PRRS MLV vaccine has demonstrated heterologous protection against genetically diverse PRRSV isolates, including strains currently circulating in the US industry. Patterson et al. summarized 21 independent studies that evaluated the INGELVAC PRRS MLV vaccines following challenge with eight genetically diverse PRRSV isolates where vaccinated pigs had significantly reduced lung lesions compared to non-vaccinated controls.[9] More recently, INGELVAC PRRS MLV was shown to protect against the currently circulating and particularly virulent heterologous PRRSV lineage 1 field strains RFLP 1-7-4 and 1-3-4 (Tables 1 and 2, respectively).[10] Vaccination reduced clinical signs of PRRSV infection, which led to improved health and average daily weight gain (ADWG) performance. The studies reported add to the accumulating data that demonstrate the ability of INGELVAC PRRS MLV to protect against relevant PRRSV infection. In these latest studies, heterologous protection was observed by vaccination against two new and particular virulent isolates responsible for a growing number of infections in the US, providing mitigation of clinical disease and subsequently significant improvement of health and ADWG.

Reduction in viral shedding frequency and duration: indirect benefit of vaccine-derived immunity

In addition to reducing clinical disease, it has been observed that INGELVAC PRRS MLV vaccine-derived immunity may result in reducing the aerosol shedding of wild-type PRRSV. A study that compared non-vaccinated pigs to those vaccinated with INGELVAC PRRS MLV, found that following inoculation with wild-type PRRSV, the frequency of detection of wild-type PRRS RNA was significantly less in the vaccinated group ($P < 0.0001$).[6] In addition, the duration of aerosol detection of wild-type PRRS RNA was shorter for the vaccinated group compared to the non-vaccinated group (Table 3). Another study explored a therapeutic model to determine the impact of INGELVAC PRRS MLV vaccine-derived immunity on the reduction of shedding in infected pigs.[5] In this study, 10% of the pigs from each of 2 rooms/groups were inoculated with wild-type PRRSV to mimic a field-based PRRS virus exposure. At 8 and 36 days post-inoculation, all pigs in the vaccine group were vaccinated with INGELVAC PRRS MLV, while the control group received saline only. Air sampling was performed daily to assess frequency and duration of aerosol shedding. Compared with the non-vaccinated control group, the vaccinated group demonstrated a significant reduction in the frequency of PRRSV detection in air samples, a reduction in the duration of PRRSV detection in air samples and a significant reduction in the cumulative PRRSV shed in the air (Table 4 and Figure 1). These research findings demonstrate that vaccine-derived immunity can reduce the amount of PRRSV shed within and from vaccinated farms, thereby reducing the risk of transmission and potential impact of the virus within and between swine farms and populations in swine-dense-areas.

Effect of PRRSV challenge dose in vaccinated pigs

Is there a PRRSV challenge dose where vaccination prevents the consequences of infection? A study was designed to evaluate the effect of PRRSV challenge dose in INGELVAC PRRS MLV vaccinated pigs, and biological and clinical responses to various doses of virulent challenge.[11] This controlled, randomized, blinded study involved pigs who were either vaccinated with 2 mL of INGELVAC PRRS MLV on Day 0 or left unvaccinated. On Day 28, pigs were challenged with virulent PRRSV isolate (dose range 1 to 4 log_{10} 50% TCID/mL). Viremia levels, pyrexia, ADWG and clinical signs were assessed. In the study, vaccinated groups showed

Table 1: Efficacy of INGELVAC PRRS MLV against PRRS 1-7-4 challenge. Lung lesions (median percent) and post-challenge average daily weight gain (ADWG)

Treatment	Lung lesions (median %)	ADWG in lbs
Ingelvac PRRS MLV	8.4[a]	0.61[a]
Competitive vaccine	12.9[a]	0.49[a]
Placebo (non-vaccinated challenge control)	25.4[b]	0.24[b]

[a] Significantly different from the placebo at $P \leq 0.05$.

Table 2: Efficacy of INGELVAC PRRS MLV against PRRS 1-3-4 challenge. Lung lesions (median percent) and post-challenge average daily weight gain (ADWG)

Treatment	Lung lesions (median %)	ADWG in lbs
Ingelvac PRRS MLV	14.9[a]	+0.317[a]
Competitive vaccine	27.4[a]	+0.052[c]
Placebo (non-vaccinated challenge control)	58.3[b]	-0.25[b]

Different superscripts indicate significant ($P \leq 0.05$) differences.

Table 3: Detection of PRRSV in air by PCR

Parameter (days)	Non-vaccinated control group	Vaccinated group
Frequency post-vaccination	0/28 days[a]	5/28 days[b]
Duration post-vaccination	0	6
Frequency post-inoculation	21/118 days[a]	4/118[b]
Duration post-inoculation	36 days	6 days

Rows with different superscripts differ, $P < 0.05$.

Table 4: Frequency of air-PRRSV qPCR positive days

	Challenge-control group	Challenge-vaccinated group
Number of positive days	31	17*
Duration (last positive day)	70	45

* Statistical significance between groups at alpha level of 0.05.

higher ADWG at all doses of heterologous PRRSV challenge compared with non-vaccinated groups, demonstrating that vaccination reduced the negative impact of PRRSV challenge on ADWG. At lower challenge doses (≤ 2 logs), vaccinated groups performed similarly to the non-challenged control group (Table 5). At all challenge dose levels, vaccinated groups had a significant decrease in days pyrexic compared to non-vaccinated groups, and at PRRSV challenge doses of ≤ 2 logs, the post-challenge days pyrexic of vaccinated groups were similar to the non-challenged group (Table 6). Vaccinated groups demonstrated reduced post-challenge viremia at all challenge levels compared to non-vaccinated groups and at lower challenge doses (≤ 2 logs), vaccinated pigs had similar levels of viremia as the vaccinated/non-challenged group. The study concluded that at all challenge doses, vaccine-derived immunity mitigated the consequences of PRRSV infection by reducing post-challenge viremia, leading to a reduction in clinical signs, and improvement in health and ADWG performance. At a challenge dose of ≤ 2 logs, the consequences of challenge in vaccinated pigs were similar to non-challenged pigs.

Figure 1: Cumulative days of PRRSV RNA detected in air samples by RT-PCR.

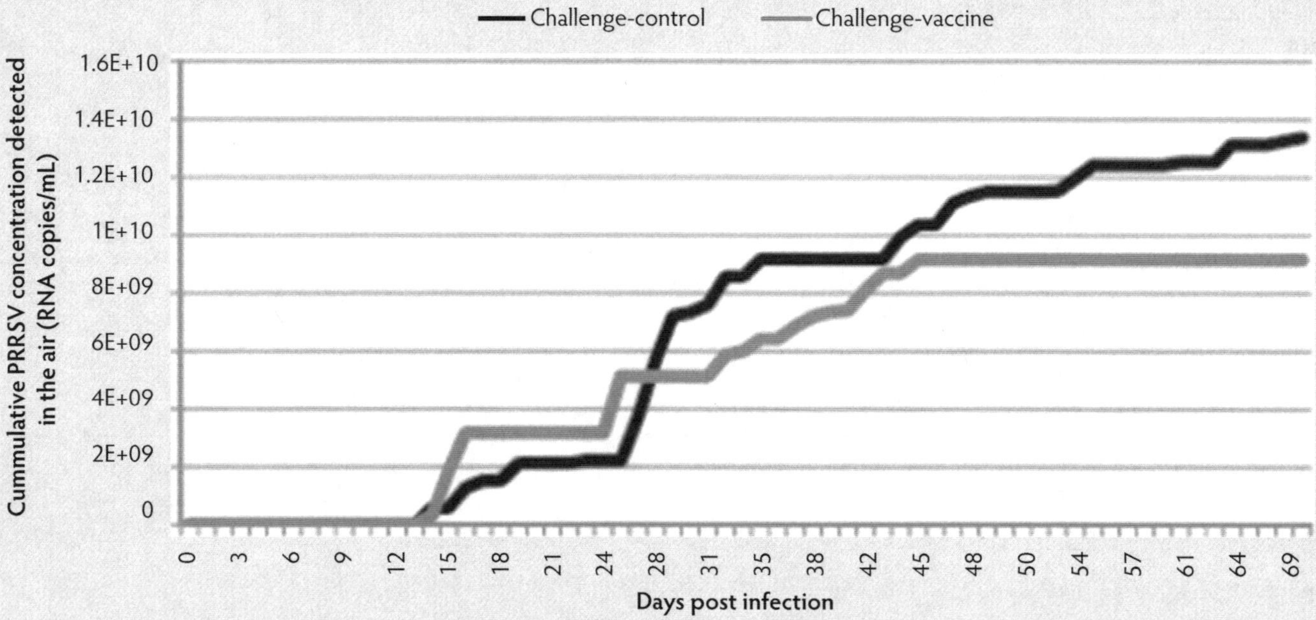

Table 5: Average daily weight gain in lbs. (ADWG)

Treatment group	4 log challenge	3 log challenge	2 log challenge	1 log challenge	No-challenge
INGELVAC PRRS MLV	1.4[1]	1.29[1]	1.70[1]	1.64[1]	1.67
Challenge control (non-vaccinated)	1.18	1.06	1.15	1.23	-

[1] Statistically significant difference ($P < 0.05$) in ADWG between INGELVAC PRRS MLV and Challenge control groups based on model prediction.

Table 6: Mean number days pyrexic post-challenge

Treatment group	4 log challenge	3 log challenge	2 log challenge	1 log challenge	No-challenge
Ingelvac PRRS® MLV	4.4[1]	4.2[1]	1.0[1]	1.4[1]	1.8
Challenge control (non-vaccinated)	11.2	8.8	10.0	6.0	-

[1] Statistically significant difference ($P < 0.05$) in number days pyrexic between Ingelvac PRRS® MLV and Challenge control groups based on model prediction.

Summary

Together, the research outlined confirms that the direct benefits of vaccine-derived immunity can optimize population immunity to help improve pig health and performance. The research also demonstrates the indirect benefit of vaccine-derived immunity to reduce exposure/transmission by reducing wild-type virus shedding and the level of virus challenge within and from vaccinated populations. The effect of PRRSV challenge dose demonstrated that vaccinated pigs exposed to moderate to low levels of virus did not show any significantly different clinical signs from non-challenged pigs. These findings support the importance of implementation of vaccine in a systematic and coordinated methodology along with the other complimentary components for PRRS control, maintaining immunity in system- and area/regional-based PRRS control programs. In addition to supporting health and performance, the data supports the conclusion that vaccine-derived immunity can also reduce the level of viral shedding within and from vaccinated populations, potentially to the point that PRRSV exposure may become inconsequential.

References

1. Polson, D., et al. 2008. Improved growing pig performance in a large production system applying an intensive management and vaccination protocol. *IPVS Proceedings.*

2. Angulo, J., et al. 2012. Growth performance improvement and mortality reductions derived from a PRRS large-scale control project in the United States. *IPVS Proceedings.*

3. Oropeza-Munoz, A., et al. 2012. A methodical approach to PRRS management, a success story of production improvements. *IPVS Proceedings.*

4. Cano, JP., et al. 2007. Impact of a modified-live porcine reproductive and respiratory syndrome virus vaccine intervention on a population of pigs infected with a heterologous isolate. *Vaccine.* 25 (2007) 4382-4391.

5. Linhares, D., et al. 2012. Effect of modified-live porcine reproductive and respiratory syndrome virus (PRRSv) vaccine on the shedding of wild-type virus from an infected population of growing pigs. *Vaccine.* 30 (2012) 407-413.

6. Dee, S., et al. 2014. Reduction of wild-type PRRS virus shedding in aerosol of growing pigs by modified-live virus vaccination at weaning. *AASV Proceedings.*

7. Opriessnig, T., et al. 2005. Genomic homology of ORF 5 gene sequence between modified live vaccine virus and porcine reproductive and respiratory syndrome virus challenge isolates is not predictive of vaccine efficacy. *J Swine Health and Prod.* 13:246-253.

8. Charerntantanakul, W., 2012. Porcine reproductive and respiratory syndrome virus vaccines: Immunogenicity, efficacy and safety aspects. *World J Virology.* 1:23-30.

9. Patterson, A., et al. 2013. Modified live PRRSV vaccination is efficacious following challenge with eight genetically diverse PRRSV isolates. *A. D. Leman Swine Conference Proceedings.*

10. Patterson, A., et al. 2019. Efficacy of Ingelvac PRRS® modified live virus vaccine against heterologous porcine reproductive and respiratory syndrome virus challenges. *J Vaccines Vaccin 10:407.*

11. Haiwick, G., et al. 2018. Examination of viremia and clinical signs after challenge with a heterologous PRRSV strain in PRRS Type 2 MLV vaccinated pigs: A challenge-dose study. *PLoS ONE* 13(12): e0209784.

Comparison of two water-soluble chlortetracyclines, with and without agitation

D. Nolan[1], DVM; K. Brown[2]; K. Ewing[2]; M. Hensch[2], DVM; C. Shull[2], PhD; C. Sparks[1], PhD; M. Wilcox[1]
[1]Huvepharma, Inc., Peachtree City, Georgia; [2]The Maschhoffs, LLC, Carlyle, Illinois

Introduction

Water soluble antimicrobials have been used for over a half a century in swine production for the treatment and control of disease. Swine veterinarians have a renewed interest in applicable information on the use of water soluble products following the implementation of the FDA GFI #213[1] requiring their involvement in the prescribing of products that had previously been over-the-counter.

The renewed interest has brought up even more basic questions than how the antibiotic works, such as, is it even being delivered to the animal? In the initial use of the water soluble products stock tanks were the primary method for delivery to the animal. Today we use a stock solutions to be siphoned or pumped into the water drinking line and delivered via pipes and a wide variety of drinkers. So both initial amount of water to dissolve the product and delivery are considerably different.

A study deploying a 2 × 2 factorial arrangement of treatments was designed to evaluate two commercially available competing water soluble chlortetracyclines (Chloronex® and Pennchlor 64®),with a goal of determining the concentration of active ingredient in the stock solution, the water that pigs drink and in pig serum. While visually there is a difference between the two products in presentation in the package and then both time into solution and residual sediment, regardless of the use agitation, it is not known if this is chemically or biologically significant.

A previous poster at the 2014 AASV meeting evaluating Pennchlor 64 demonstrated a large numerical increase in the amount of chlortetracycline reaching the drinker when agitation was used.[2] Therefore, this trial focused on the amount of medication in the stock solution, at the drinker, and in the pig (plasma levels); with and without agitation after 24 and 48 hours of medication. Samples were analyzed for chlortetracycline (CTC) and 4-epi chlortetracycline (epi-CTC) a microbiologically active degradation product of chlortetracycline. Total chlortetracycline (total CTC) is a mathematically derived combination of CTC and epi-CTC.

Materials and methods

Four commercial wean to market farms in the Midwest USA were selected. Medication and equipment used were ordered by the producer and received onto farms through their normal distribution channels. Fours barns or rooms per site were randomly assigned to a treatment with a dedicated medicator serving each barn/room.

Treatments were Chloronex® and Pennchlor 64® (Chl and Pen). With agitation (ChlA and PenA), and without agitation (ChlN and PenN). Agitation treatments were done by use of 120v Little Giant® submersible fountain pump. Agitation, when used, was done continuously for the duration of the study. Stock solution was mixed at the label dose of 5 packets 25.6oz. (102.4g active) of medication to each 4 gallons of stock solution and dispensed at 1oz. per gallon (1:128) to achieve 10mg/lb of body weight. Non agitation treatments were stirred for 20 seconds after the final packet was added to the solution.

Medicators were validated and confirmed to be dispensing 1 oz. stock solution per gallon of drinking water, before and after the trial.

Five pigs per barn/room were randomly selected and tagged for serial sample collection and five pigs per barn/room were randomly selected at each unique sampling period. Plasma samples were taken from pigs at zero hours (pre-medication), 24 hours, and 48 hours after onset of treatment. New stock solutions were mixed at time zero and 24 hours with stock solution samples taken at zero hours (15 minutes) and 24 hours. Drinker water samples were taken at zero hours (pre-medication), 24 hours and 48 hours. Stock solution and drinker water samples were frozen immediately. Blood was collected in EDTA tubes and spun and plasma separated prior to freezing. Water and plasma samples were shipped by courier on dry ice for analysis at Iowa State University Veterinary Diagnostic Chemistry Lab. Water samples were tested by HPLC and plasma samples were tested by mass spectrometry.

Results and discussion (Table 1)

Agitation

While there were some minor differences in concentration in the stock solution and at the drinker due to mixing method, there were no differences ($P > 0.05$) between treatments in the plasma associated with agitation. The values that demonstrated an increase ($P < 0.05$) when agitation was used were the 24 hr CTC and 48hr epi-CTC and total CTC in the stock solution and 24 and 48 hour epi-CTC at the drinkers. There was an interaction between CTC product and mixing method for epi-CTC level at the drinker at the 24 and 48 hour period, with the agitation pump increasing epi-CTC in the Pen product compared to hand mixing whereas there was no difference in mixing method for the Chl product.

Table 1: Summary results

	Product			Mixing			*P*-values		
	Chl	Pen	SEM	Hand mixed	Agitation pump	SEM	Product	Mixing	Product × mixing
Stock solution, g/oz									
0 Hour (fresh stock)									
CTC	0.75[a]	0.54[b]	0.059	0.62	0.67	0.059	0.002	0.31	0.92
Epi-CTC	0.04[b]	0.17[a]	0.018	0.10	0.11	0.018	< 0.001	0.68	0.54
Total	0.79	0.71	0.072	0.72	0.78	0.072	0.20	0.33	0.89
24 Hour (old stock)									
CTC	0.65[a]	0.48[b]	0.037	0.50[b]	0.63[a]	0.037	0.01	0.04	0.58
Epi-CTC	0.09[b]	0.24[a]	0.036	0.15	0.18	0.036	0.001	0.35	0.09
Total	0.75	0.72	0.057	0.65	0.81	0.057	0.69	0.06	0.69
24 Hour (fresh stock)									
CTC	0.70[a]	0.52[b]	0.051	0.61	0.60	0.051	< 0.001	0.69	0.28
Epi-CTC	0.05[b]	0.17[a]	0.019	0.11	0.11	0.019	< 0.001	0.75	0.56
Total	0.74	0.68	0.064	0.72	0.71	0.064	0.09	0.64	0.24
48 Hour (old stock)									
CTC	0.71[a]	0.47[b]	0.056	0.55	0.64	0.056	0.001	0.09	0.95
Epi-CTC	0.07[b]	0.25[a]	0.029	0.13[b]	0.19[a]	0.029	< 0.001	0.02	0.12
Total	0.78	0.72	0.079	0.67[b]	0.83[a]	0.079	0.27	0.02	0.57
Drinker samples, g/gal									
0 Hour									
CTC	0.63	0.52	0.141	0.55	0.61	0.141	0.11	0.37	0.08
Epi-CTC	0.02	0.02	0.014	0.02	0.03	0.014	0.86	0.41	0.27
Total	0.66	0.55	0.073	0.57	0.64	0.073	0.11	0.32	0.11
24 Hour									
CTC	0.71[a]	0.61[b]	0.144	0.65	0.68	0.144	0.04	0.54	0.64
Epi-CTC	0.05[b]	0.13[a]	0.056	0.05[b]	0.13[a]	0.056	< 0.001	< 0.001	0.001
Mixing									
Hand mixed	0.04[b]	0.05[b]	0.123	-	-	-	-	-	-
Agitation pump	0.07[b]	0.20[a]	-	-	-	-	-	-	-
Total	0.76	0.74	0.097	0.70[b]	0.81[a]	0.097	0.62	0.03	0.10
48 Hour									
CTC	0.75	0.65	0.094	0.69	0.71	0.094	0.20	0.79	0.68
Epi-CTC	0.07[b]	0.14[a]	0.030	0.06[b]	0.15[a]	0.030	0.001	< 0.001	0.03
Mixing									
Hand mixed	0.04[b]	0.07[b]	0.130	-	-	-	-	-	-
Agitation pump	0.09[b]	0.22[a]	-	-	-	-	-	-	-
Total	0.81	0.80	0.120	0.75	0.86	0.120	0.83	0.13	0.35

Table 1: Continued

	Product			Mixing			*P*-values		
	Chl	Pen	SEM	Hand mixed	Agitation pump	SEM	Product	Mixing	Product × mixing
Plasma, ppb (ng/mL)									
0 Hour	5.4	4.7	2.97	4.8	5.3	2.97	0.55	0.66	0.92
24 Hour									
CTC	558.5[a]	409.3[b]	49.11	496.9	470.9	49.11	< 0.001	0.51	0.89
Epi-CTC	126.5[a]	91.1[b]	13.44	115.1	102.6	13.44	< 0.001	0.14	0.79
Total	685.0[a]	500.6[b]	61.46	612.1	573.5	61.46	< 0.001	0.42	0.94
48 Hour									
CTC	568.7[a]	441.4[b]	59.51	502.6	507.5	59.51	< 0.001	0.89	0.06
Epi-CTC	134.0[a]	101.5[b]	18.67	121.6	113.9	18.67	< 0.001	0.31	0.66
Total	702.8[a]	542.9[b]	76.36	624.3	621.5	76.36	< 0.001	0.95	0.10

Product comparison

There were differences between products in CTC and epi-CTC concentration in the stock solution at all time periods. Compared to Pen, Chl had higher concentrations of CTC (Chl .65 to .75 vs. Pen .47 to .54 g/oz.) and lower epi-CTC (Chl .05 to .09 vs. Pen .17 to .25 g/oz.), but similar total CTC (Chl .74 to .79 vs. Pen .68 to .72 g/oz.). Water drinker measures at 0 hour (~15 minutes after beginning the barn on medication) were lower than expected, likely due to the procedure and residual fresh water in the lines. The protocol called for flushing of lines and the last drinker until the medication was present then sampling. At 24 hours Chl had higher CTC concentration than Pen at the drinkers (0.71 vs 0.61 g/gallon; $P = 0.04$). There were no differences in CTC concentration at the drinker between treatments at 48 hours. At 24 and 48 hrs, Chl had lower epi-CTC (0.05 vs. 0.13 and 0.07 vs. 0.14 g/gallon, respectively; $P < 0.05$). There were no differences between products in total CTC concentration at the drinkers at any of the time periods. At 24 and 48 hours on water chlortetracycline treatment, plasma levels for Chl compared to Pen were higher for CTC (559 vs. 409 and 569 vs. 441 ng/mL, respectively; $P < 0.001$), epi-CTC (127 vs. 91 and 134 vs. 102 ng/mL, respectively; $P < 0.001$), and Total CTC (685 vs. 501 and 703 vs. 543, respectively; $P < 0.001$).

Discussion

While total CTC in the stock solution was similar between products, Chl had significantly more pure CTC, whereas Pen had significantly more epi-CTC. This trend was also present at the drinkers. This difference may explain the difference in plasma CTC levels. Chl had significantly more CTC, epi-CTC and total CTC levels in plasma; however, the clinical implications are unknown. If the CTC plasma levels were at or over the target MIC level for the disease to be treated, the implications of higher CTC levels would likely be mute. However, if the 33% higher total CTC plasma levels achieved with the Chl helped to reach or surpass the target MIC it could make a significant difference in the clinical outcome. The higher levels achieved with the Chl could also have potential value in the treated animals that have lower than average water intake because of disease or other factors. The increased levels achieved with Chl could potentially play a further role in reducing a bacterial population from returning post treatment or even developing antibiotic resistance from sub-therapeutic treatment.

® Huvepharma and Chloronex are registered trademarks of Huvepharma EEOD

® Pennchlor 64 is a registered trademark of Pharmgate Animal Health

® Little Giant is a registered trademark of Franklin Electric Co., Inc.

References

1. FDA GFI#213. Available at: https://www.fda.gov/media/83488/download. Accessed Nov. 15, 2019.

2. Girad, T'Lee et al. Determining the effect of agitation of chlortetracycline stock solutions on chlortetracycline concentrations at different areas of water delivery systems. Poster, AASV 2017.

Prevalence and consequences of sow anemia

G. Almond[1], DVM, PhD; K. Castevens[1], BS; E. Noblett[1], BS, MS; J. Ferreira[1], DVM, DVSc; T. Gillespie[2], DVM, DABVP; C. Olsen[3], DVM, MS; J.P. Nielsen[4], DVM, PhD
[1]College of Veterinary Medicine, North Carolina State University, Raleigh, North Carolina; [2]Performance Health PC, Battle Ground, Indiana; [3]Pharmacosmos Inc, Watchung, New Jersey; [4]Royal Veterinary College, Copenhagen, Denmark

Introduction

The iron demands and hemoglobin (Hb) concentrations of sows are typically ignored in pork production. Sows with Hb levels less than 10 g/dL are at risk for an increased occurrence of stillbirths.[1] The objectives of this study were to evaluate Hb concentrations in different parities of sows at various stages of reproduction, and to determine the correlation between sow Hb concentrations and the numbers of stillborn piglets.

Materials and methods

Study 1: The initial cross-sectional study included 2683 sows from 11 different farms (2400-4200 sows/farm) in two states (North Carolina and Indiana). Approximately 250 blood samples were collected from 10 sows/parity/stage on each farm. Stages were defined as early, mid, and late gestation, and early and late lactation. Parity groups were 0, 1, 2, 3, and ≥ 4. The Hb concentrations were measured on the farm using a HemoCue Hb 201+.™ Blood samples were taken from the ear veins of sows. Data was analyzed using analysis of variance with state, farm, parity, and stage as the independent variables. Means were compared with Tukey's HSD test.

Study 2: The second study was conducted on 5 farms (1700-4000 sows/farm) and included 50-150 sows/farm. Sows were from varying parities. Blood samples were collected in late gestation (>110 days) and after farrowing for Hb determinations. Litter demographics were recorded for each sow. Lung flotation tests were performed to distinguish post-partum deceased piglets and intra-partum or pre-partum deaths. Pearson correlations and regression models were used to assess relationships among the various parameters.

Results

Study 1: The Hb concentrations differed ($P < .05$) between North Carolina (9.7 ± .03 g/dL) and Indiana (10.5 ± .05 g/dL). Also, Hb concentrations differed among reproductive stages (Figure 1) with mid-gestation having the highest ($P < .05$) concentrations (10.7 ± .06 g/dL), early lactation with the lowest concentrations (9.1 ± .05 g/dL), and concentrations rising over late lactation (9.4 ± .06 g/dL) and early gestation (10.6 ± .06 g/dL). This trend was consistent among parities and farms. There also was an effect of parity, with Hb concentrations decreasing with increasing parity.

Study 2: Post-farrowing Hb concentrations were less ($P < .05$) than the late gestation concentrations. Parity > 4 sows had lower ($P < .05$) post-farrowing Hb concentrations than sows in other parities. Total number of stillborn pigs was 1.08 + 0.07/litter and pigs born alive was 13.3 + 0.16/litter. Late gestation Hb concentrations were negatively correlated ($P < .01$) to the number of stillborn pigs, and the number of pre-partum stillborn pigs.

The regression model predicted that for each one unit increase in a sows late gestation Hb, the count of stillborn piglets per litter will change by 0.82× after adjusting for confounding in the model ($P = 0.015$). For example, if one sow has a late gestation Hb of 10 g/dL and had 2 stillborn piglets, and another a sow had Hb of 11 g/dL, the model would predict her to have 2*0.82 = 1.64 stillborn piglets after adjusting for confounding factors.

Discussion

This method of Hb assessment provided a simple and inexpensive method of Hb evaluation on commercial farms. The difference between states was surprising; however, it was suspected that sow diets were a significant factor. If 10 g/dL is used as an indicator of anemia, then a large proportion of sows were anemic; thereby, revealing potential issues with anemia in sows, particularly in lactation. Additional studies are necessary to evaluate the influence of Hb status on reproductive performance.

Based on the results, Hb concentrations are associated with the occurrence of stillborn piglets, particularly, pre-partum stillbirths. The precise mechanism to explain the relationship between sow anemia and stillbirths is speculative; however, iron deficiency may contribute to impaired uterine contractions at farrowing. The low Hb concentrations in late gestation are likely due, at least in part, to transfer of iron from the dam to the fetuses through the maternal uteroferrin - transferrin – ferritin pathway.[2] Additional studies are required to devise methods to reverse the anemia in older parity sows.

References

1. Bharttarai S, Framstad T, Nielsen JP. Stillbirths in relation to sow hematological parameters at farrowing: A cohort study. *J Swine Health Prod.* 2018;26:215-222.

2. Ducsay CA, Buhl WC, et al. Role of uteroferrin in placental iron transport: Effect of maternal iron treatment on fetal iron and uteroferrin content and neonatal hemoglobin. *J Anim Sci.* 1984;59:1303-1308.

Figure 1: Hemoglobin concentrations (mean ± SEM) in sows at different stages of reproduction and different parities. Note the low concentrations indicative of anemia (Hb < 10 g/dL) in higher parity sows during late gestation, and all parities in lactation.

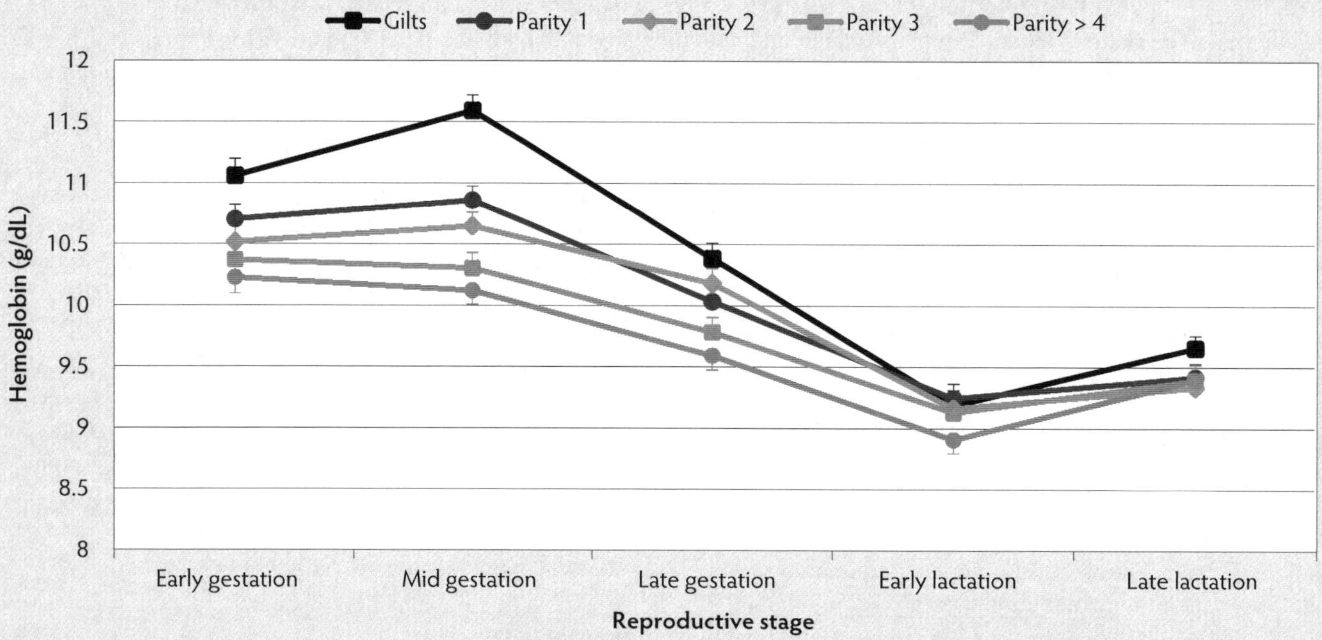

Field trial results: a second dose of injectable iron to maximize growth

Chris Olsen, DVM, MS
Pharmacosmos Inc. Watchung, New Jersey

Introduction

The profitability of swine production is dependent on many factors including, herd health, efficiency of production, growth performance, the price of market pigs, and many others. While the price and conditions of the swine market in the United States cannot be controlled, it is important that producers optimize as many other aspects of their operation as possible. This allows them to remain competitive and provides the best opportunity for profitability. One area where efficiency can be optimized is through wean-to-finish growth performance. Reducing the number of days on feed to reach a target weight or increasing liveweight at marketing can have profound impacts on the bottom line.

While there are many things that can influence wean-to-finish growth performance, several recent studies have demonstrated that optimizing the dose and timing of injectable iron administration during the suckling period can have positive impacts on wean-to-finish growth.[1-5] With minimal input costs, this has proven to result in a significant return on investment for the producer.[6]

Knowing this information, a field study was conducted at the request of a farrow-to-finish producer in North Carolina with 25,000 sows. Specifically, the producer was interested in exploring the impact of a second 200 mg dose of Uniferon® on post-weaning growth performance.

Materials and methods

Baby pigs were enrolled into the trial on a single 2,500 sow farm within the producer's system in North Carolina. 1,686 baby pigs were allotted to one of two treatment groups at 2-4 days of age, which is when processing (castration, iron injections) normally occurs within the system. Within a litter, two baby pigs were visually weight-matched, weighed and assigned to a treatment group. Group A (n = 841), received a single 200 mg dose of Uniferon® at the time of enrollment. Group B (n = 845) received 200 mg of Uniferon® at the time of enrollment and again at 6-10 days of age. All baby pigs were individually weighed at enrollment, weaning (25 ± 1 days) and at the end of the nursery stage (42 days on feed). Blood hemoglobin was measured in 100 baby pigs per treatment group at the time of weaning.

Results and discussion

At weaning, 89% of the baby pigs in Group B, which received two doses of Uniferon®, had an optimal blood hemoglobin concentration of ≥ 110 g/L compared to 20% in the single-dose Group A (Table 1). Pigs in Group B weighed 0.5 lbs more at weaning and 1.0 lbs more at the end of nursery compared to Group A (Table 2). The weight gain improvements observed in this study were very similar to that reported previously (Fredericks, Mahan) and therefore the producer felt comfortable ending the trial after nursery weights were collected.

Based on these results, a second injection of Uniferon® has been implemented across the entire system of 25,000 sows. Prior to this change, no routine baby pig handling events occurred between processing and weaning. Because no convenient timing existed within the current protocols, it was decided to time the second dose of iron based on best-practice recommendations. A targeted range of 8-10 days of age was selected based on the knowledge that the fastest-growing baby pigs outgrow their supply of iron from the first dose by day 12-14,[7] and to ensure adequate time between the two doses. Since making this change, the producer continues to be satisfied with the results and improved pig growth despite requiring an extra handling event in the farrowing house.

Conclusion

This study reflects the findings of many previous studies which have shown that optimal dosing of Uniferon® is research-proven to boost hemoglobin levels, average daily gain and your return on investment.[1,3,6]

® Registered trademark of Pharmacosmos Inc.

Table 1: Mean and distribution of blood hemoglobin concentration at weaning

Treatment group	Deficient (< 90 g/L)		Sub-clinical (90-110 g/L)		Optimal (≥ 110 g/L)		Overall mean Hb
	%	mean	%	mean	%	mean	
A (200 mg)	33%	79.4	47%	99.2	20%	118.7	96.7
B (400 mg)	2.0%	72.0	9.0%	101.7	89%	129.5	125.9

Table 2: Live weights by treatment group

Treatment	Number enrolled	Weight (lbs)		
		Enrollment (2-4 days of age)	Weaning (25 days of age)	Nursery (42 days on feed)
A (200 mg)	841	3.7	13.8	51.0
B (400 mg)	845	3.8	14.3	52.0

References

1. Bhattarai, S. et al. Association between hematological status at weaning and weight gain post-weaning in piglets. *J Livestock Sci.* 2015;182:64.

2. Dunnigan, S. et al. The effects of hemoglobin levels on growth rates in pigs. AASV. 2013;305.

3. Fredericks, L., et al. Evaluation of the impact of iron dosage on post-weaning weight gain, and mortality. AASV. 2018;315.

4. Mahan, D. et al. Effect of injected and dietary iron in young pigs on blood hematology and postnatal pig growth performance. *J Anim Sci.* 2011;89:4068.

5. Perri, A. et al. An investigation of iron deficiency and anemia in piglets and the effect of iron status at weaning on post-weaning performance. *J Swine Health Prod* 2016;**24**:10.

6. Olsen, C. The economics of iron deficiency anemia on US swine production: An annual impact of 46-335 million US dollars. AASV. 2019;351.

7. Unpublished field data.

Is perception everything when it comes to disinfectants?

Jeff Odle
Lanxess Corporation, Actives and Disinfectants, Pittsburgh, Pennsylvania

Introduction

Perception can be defined as a mental impression. Right or wrong, it is what we believe to be true in our mind. Rick Riordan once quoted, "Humans see what we want to see." A perfect example of this would be the famous analogy "Is the glass half full or half empty?" It depends on the person and how they interpret what they see. Another quote by an anonymous author states "Perception and reality are two different things." Unfortunately, on many occasions this is proven to be true.

The reality in today's global swine production industry is that African swine fever virus (ASFV) is a game changer. The devastating impact of ASFV on swine production across Eastern Europe and Asia has fundamentally changed how swine producing companies operate. The virus is present in over 50 countries. Losses to the global swine herd are expected to be as high as 25%, according to analysts, forecasting which is supported by the OIE. There is now an intense focus on biosecurity in trying to contain and eradicate the virus and exclude it from reaching unaffected areas. The perception some disinfectant manufacturers in the USA would like growers to believe is that their products are effective against viruses such as ASFV and FMD, even though they have not been tested using standard microbiological protocols approved by EPA against those particular pathogens. The objective of this paper is to address some of the perceptions, or better yet the misperceptions and misrepresentations, around disinfectant chemistries being promoted in the industry which could potentially lead to an outbreak in the US.

Disinfectant efficacy

Products making disinfection claims must be registered and approved by the Environmental Protection Agency (EPA). There are very few exceptions and are reserved primarily for products containing natural ingredients. This helps protect the environment and the consumer from hazardous products or products with unproven efficacy. Marketing literature is required to reflect only claims that have been reviewed and approved by EPA. Any claims in marketing that are not approved on the master label are considered misbranding and subject to enforcement actions by EPA. The master label for each EPA registered antimicrobial can be found on the Pesticide Product and Label System database at `https://iaspub.epa.gov/apex/pesticides/f?p=PPLS:1`.

However, knowing that EPA rarely takes enforcement action when it comes to marketing communications, some disinfectant manufacturers make claims where no proof of efficacy has been reviewed by EPA. This leaves growers with the expectation that the disinfectant works, when in fact they're risking a potential outbreak by using an untested and unapproved product. In order to make claims of efficacy against OIE listed pathogens, independent study data needs to be provided to EPA during the submittal process for review and approval. Unfortunately, some marketing claims can be intentionally or unintentionally misleading. For example, one disinfectant manufacturer distributed hard copy literature at AASV 2019 implying their glutaraldehyde/quaternary ammonium based product sold in the UK was effective against African Swine Fever virus. This particular product was not registered nor commercially available in the US. Makes one wonder why a disinfectant producer would market at a US conference a product not approved for use in the US. Was the intent to lead the audience to believe all glut/quat products are effective against ASFV, including their glut/quat sold in the US, which is a completely different formulation than the product sold in the UK? Did the audience come away with the perception that the glut/quat sold in the US was effective against ASFV?

This same disinfectant manufacturer handed out marketing literature dated from 2017, at the Allen D. Leman Swine Conference in September 2019. This communication stated their glut/quat product was effective at inactivating foot-and-mouth disease and cited an efficacy suspension study conducted in the UK. The EPA requires a use dilution test for efficacy. EPA rarely approves test results from international laboratories especially when the methodology used is not an approved protocol that EPA supports. The suspension test conducted in the UK is generally easier to pass than the use dilution test. In the use dilution test, the pathogen is applied and dried onto a carrier whereas the suspension test is exactly as the name implies – the pathogen is held in suspension where it can be attacked from all sides by the antimicrobial. The actual glut/quat product label on file with EPA, dated October 2018, does not approve of a FMD claim. So why did the marketing communication say the product was effective against FMD, especially since FMD is an OIE listed pathogen and requires a data review by EPA? This is considered misbranding by EPA and the sale of misbranded products is illegal under the Federal Insecticide, Fungicide, and Rodenticide Act.

Chlorine-based, dry powder boot baths have become popular in livestock production facilities. Most of these are unregistered products being marketed as absorbents or deodorizers. As long as no claims for pathogen control are made, this is legal. Or is it? The main purpose of a boot bath in swine and poultry production is for pathogen control. One steps into a boot bath, not with the intent to deodorize the boots but to disinfect the boots. Having chlorine chemistry, as the base of the powder, is intended to give the impression the powder is effective against the pathogens of concern. It is well documented that chlorine in solution can be an effective disinfectant, but concentration and contact time are critically important. The key to that statement is 'chlorine in solution' not 'chlorine as a dry powder'. The perception is that the dry powder boot baths contain chlorine, therefore they must be effective disinfectants. As demonstrated in Figure 1 below, a laboratory study[1] was conducted comparing the efficacy of two chlorine based dry powders, Virkon™ S Disinfectant and Virucide as a powder and Virkon™ S at the proper use dilution in a 1% solution as specified in the use directions on the product label. Only the Virkon™ S dilution delivered the 3.75 log reduction required for efficacy of a disinfectant by EPA.

Figure 1: Efficacy of dry powder products vs Virkon™ S 1% solution.

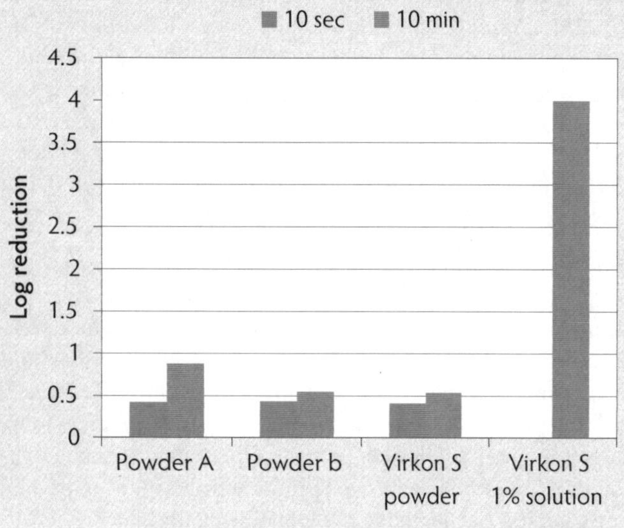

Conclusion

The ASFV is a highly complex, multi-layered enveloped virus, which in most cases enveloped viruses are easy to kill. Due to the pH stability (3.9-11.5) ASFv acts more like a non-enveloped virus and thus much more resilient to some disinfectant chemistries. Because it is highly virulent, there are very few laboratories capable of handling or testing the virus safely. In order for the US to keep ASFV out of the country, ports of entry and growers will need to depend on disinfectant products with proven efficacy against the virus. Each individual should ask to see the supporting study data to prove efficacy and they should check the PPLS database from EPA that shows that data has been reviewed and approved as scientifically valid.

Reference

1. The Efficacy of Virkon™ S in a Competitive Comparison with Dry Powdered Products against *Salmonella seftenberg* in a Simulated Foot Pan Test with Feces by Dr. Carl Erkenbrecher and Sherrill Nurnberg. Chemours CDR-178-16.

Fat-soluble vitamin update: an overview of results with liquid and injectable fat-soluble vitamins

R. L. Stuart, PhD
Stuart Products, Inc., Bedford, Texas

Introduction

Fat-soluble vitamins A, D, E are vital nutrients for all swine. These vitamins are routinely added to swine feeds at or above National Research Council (NRC) recommended levels. Husbandry practices, improved genetics and replacement of feed-grade antibiotics have dramatically increased the compulsion for supplementing higher levels of fat-soluble vitamins. Furthermore, nutritional needs for fat-soluble vitamins has not kept pace with strives in genetic production. As the number of newborn pigs increases, the importance of fat-soluble vitamin status of colostrum and milk also increases.

Field observations due to fat-soluble vitamin insufficiencies include lack of utilization of feed-grade vitamin E-acetate in nursery pigs and, low vitamin E and D status in finishing swine. Gestating sows injected with fat-soluble vitamins pre-partum have shorter delivery times, reduction in stillborns, and reduction in prolapses.

Supplementation via drinking water and/or injection is utilized to strategically enhance fat-soluble vitamin status during critical times. The primary emphasis for NRC dietary requirements is minimum daily requirements for animal gain and feed conversion. Besides animal gain, research has shown that under certain conditions, higher levels of fat-soluble vitamins impact other factors besides daily gain, primarily the immune system and reproduction .[1-5]

Previous research has demonstrated that either water-supplemented or injected fat-soluble vitamins dramatically enhances serum levels,[6-8] and improvements in antioxidant status[8]. For vitamin E, serum C_{max} is attained very quickly post-injection and also diminishes within 24 hours post-injection, while serum 25-OH-D_3 status is maintained for a much longer time compared to either vitamin E or vitamin A.[6]

The purpose of this paper is to review observations that measured the fate of injected and/or orally administered fat-soluble vitamins during two critical times – gestating gilts/sows and newly-weaned pigs, and offer recommended optimal serum levels for fat-soluble vitamins.

Importance of bioavailability of fat-soluble vitamins

A unique issue with fat-soluble vitamins is bioavailability of oral and injectable products. Just because a label states the potency of a vitamin does not mean that the vitamin is bioavailable, especially fat-soluble vitamins. University experiments and field demonstrations have shown that a 5 mL-injection of VITAL E®-Repro (Stuart Products, Inc.) reduces farrowing time approximately 5 min/pig.[9] Instead of utilizing VITAL E-Repro, competitive formulations have been evaluated instead of VITAL E -Repro. The competitive A, D and E products all have the same identical preparation that contains an equal amount of vitamin E, lesser quantity and different source of vitamin A, and lesser quantity of vitamin D compared to VITAL E-Repro.

To determine serum fat-soluble vitamin status between the two products, six gestating sows of similar parity and breed were utilized in a study to measure plasma concentrations of fat-soluble vitamins.[10] Three sows were injected with either 5 mL VITAL E®-Repro or 5 mL Vitamin E+AD (Durvet, Inc.). Table 1 shows levels of vitamins injected for each treatment. Plasma samples were collected prior to injection (d 0) and d 1, 3 and 7 post-injection. Plasma was analyzed for 25-OHD_3, retinol, retinyl palmitate (source of vitamin A in VITAL E-Repro), retinyl propionate (source of vitamin A in Vitamin E+AD), and alpha-tocopherol (Table 1). As expected, VITAL E-Repro-injected sows had dramatically higher 25-OH D_3 on days 1, 3, and 7 post-injection ($P < 0.0001$) compared to the competitive product. Although small, the plasma retinol levels were numerically higher in the VITAL E-Repro-injected sows ($P < 0.09$). Due to the different sources and quantity of vitamin A in the two products, retinyl palmitate levels were dramatically higher in VITAL E-Repro injected sows.[10] When retinyl propionate contained in Vitamin E+AD was injected, none was found for any plasma sampling days. The lack of retinyl propionate in serum has been confirmed in other studies in pigs and calves. For vitamin E status, even though an equal amount of vitamin E was injected[1] (500 I.U.), VITAL E-Repro-injected sows had 211% higher plasma status on day 1 ($P < 0.003$); and on day 3 the difference was[6] 1% higher ($P < 0.021$) (Table 1). This experiment and other experiments demonstrate the importance of bioavailability and the superiority of palmitate-form of vitamin A activity in injectable fat-soluble vitamin products.

In another study to determine serum and tissue distribution of vitamins,[11] newly-weaned pigs were injected with VITAL E-A+D to measure serum levels of vitamins E, A and D and tissue levels of vitamins A and E. Treatments were non-injected control (no vitamins) and intramuscular injection of 3 mL VITAL E-A+D having retinyl palmitate (300,000 IU), d-alpha-tocopherol (900 IU), and vitamin D (30,000 IU) per pig. All pigs were bled at d 0, 3, 7, and 14 post-injection for E, A,

Table 1: Relative bioavailability of VITAL E-Repro compared to Vitamin E+AD in gestating sows

Vitamin	VITAL E-Repro	Vitamin E+AD	SEM	*P*-values	TRT
25-OHD$_3$, ng/mL					
d 0	31.87	33.73	2.664	0.629	0.0007
d 1	53.43	34.83	2.727	0.0001	
d 3	79.77	35.93	2.639	< .0001	
d 7	76.63	34.83	2.202	< .0001	
Retinol, µg/mL					
d 0	0.15	0.13	0.011	0.220	0.0606
d 1	0.15	0.12	0.011	0.098	
d 3	0.18	0.12	0.012	0.006	
d 7	0.16	0.13	0.009	0.061	
Retinyl palmitate, µg/mL					
d 0	0.05	0.03	0.009	0.094	0.0155
d 1	9.72	0.05	2.184	0.009	
d 3	9.21	0.03	1.446	0.001	
d 7	0.06	0.04	0.006	0.043	
Retinyl propionate, µg/mL					
d 0	ND	ND*	-	-	-
d 1	ND	ND	-	-	
d 3	ND	ND	-	-	
d 7	ND	ND	-	-	
Alpha-tocopherol µg/mL					
d 0	0.73	0.80	0.153	0.764	0.0135
d 1	9.23	2.97	1.211	0.003	
d 3	3.43	2.13	0.346	0.021	
d 7	1.30	1.37	0.158	0.771	

* ND = not detected.

and D analysis. Tissue samples were collected in three pigs per treatment pre-injection and day 7 and 14 post-injection. Tissues were analyzed for vitamin A and vitamin E. Plasma 25-OH D3 was dramatically higher (3.44 vs. 71.00 ng/mL at d 3 post-injection) and retinol (0.09 vs. 0.18 µg/mL at d 3 post-injection) concentrations increased from d 3 to 14 post-injection ($P < 0.05$) (Table 2). Plasma retinyl palmitate was detected only in the vitamin treatment at d 3 and 7 post-injection (115.51 and 4.97 µg/mL, respectively). Concentrations of liver retinol (0.63 vs. 19.10 µg/g) and liver retinyl palmitate (43.33 vs. 199.13µg/g) at d 7 post-injection also were increased on d 14 post-injection ($P < 0.05$) whereas neither were detected in other tissues. The alpha-tocopherol concentrations increased in plasma ($P < 0.05$) at d 3 (0.99 vs. 7.14 µg/mL) and d 7 post-injection and in liver (4.30 vs. 11.17 µg/g; $P < 0.10$), heart (4.60 vs. 24.80 µg/g; $P<0.10$), and muscle (3.07 vs. 8.30 µg/g; $P < 0.05$) at d 7 post-injection. Brain was the only tissue that did not have an increase in vitamin E status (Table 2). This experiment revealed that a fat-soluble vitamin injection in pigs increased plasma and liver status of alpha-tocopherol, retinol, and retinyl palmitate, and plasma 25-hydroxycholecalciferol concentrations. Alpha tocopherol was distributed to all tissues examined but retinol and retinyl palmitate were found only in the liver. Although plasma vitamin E levels decreased, tissue levels of vitamin E and A increased in tissues except brain (Table 3). This experiment demonstrated that even though there were post-injection reductions in serum levels, there was a corresponding increase in various tissue levels, especially for vitamin E.

Optimum serum fat-soluble vitamin levels

The NRC does not provide optimum serum/plasma reference levels for fat-soluble vitamins in swine. The most critical times for optimum fat-soluble vitamin status are pre-farrow and pre-breeding for gilts and sows, and post-weaning in pigs. Table 4 presents the optimum levels of fat-soluble vitamins for various classes of

Table 2: Effects of VITAL E-A+D injection on serum vitamin levels in weaned-pigs (Jang, 2018)[7]

Vitamin	Control	VITAL E-A+D	SEM	*P*-value
25-OHD3 (ng/mL)				
d 0	4.33	4.01	0.432	.058
d 3	3.44	71.00	6.979	< .001
d 7	4.79	36.35	3.256	< .001
d 14	7.24	15.50	1.220	< .001
Retinol (μg/mL)				
d 0	0.11	0.08	0.007	0.005
d 3	0.09	0.18	0.010	< .001
d 7	0.10	0.13	0.008	0.03
d 14	0.14	0.17	0.006	0.002
Retinyl Palmitate (μg/mL)				
d 0	ND	ND		
d 3	ND	115.51	13.693	
d 7	ND	4.97	1.089	
d 14	ND	ND		
Alpha-tocopherol (μg/mL)				
d 0	1.69	1.64	0.119	0.77
d 3	.99	17.14	1.188	< .001
d 7	.80	2.35	0.137	< .001
d 14	.66	0.99	0.275	0.41

Table 3: Tissue vitamin E levels in VITAL E-A+D injected weaned pigs

Tissue	Control	Vital E A+D	SEM	*P*-values
Brain, μg/g				
d 7	8.20	7.57	1.333	0.14
d 14[1]	7.40	8.33	0.566	0.32
Heart, μg/g				
d 7	4.60[e]	24.80[d]	3.975	0.06
d 14	2.17[b]	4.57[a]	0.594	0.047
Lung, μg/g				
d 7	2.17	18.80	5.905	0.13
d 14	9.60	10.93	1.717	0.31
Muscle, μg/g				
d 7	3.07[c]	8.30[a]	0.655	0.01
d 14	2.80	3.90	0.734	0.23

[a-c] $P < 0.05$;
[d-e] $P < 0.10$.
[1] One sample was removed from the control treatment.

swine. Most levels are dramatically higher than reference values from veterinary diagnostic laboratories and typically cannot be attained by using NRC recommended levels.

Published research has demonstrated that for optimum innate immune responses, serum values need to be higher than typically found when NRC levels are fed.[12] Due to reduced feed intake in gestating sows, the suggested daily vitamin E intake is 400 I.U. per day which is approximately four-times the NRC recommended daily level.

Field observations from veterinarians

Veterinarians have reported that sow and newly-weaned pig performance has been improved by enhancing fat-soluble vitamin status. A reduction in mortality of nursery pigs has also been observed which is what Pallaras et al., 2002 reported.[13] In gilts and sows, a reduction in delivery time, reduction in stillborns especially older sows, and reduction in prolapses have been reported.

Product recommendations

Newborns

VITAL E®-Newborn, 0.5 mL (250 I.U. vitamin E, 25,000 I.U. vitamin A and 25,000 I.U. vitamin D)

Nursery

EMCELLE® TOCOPHEROL (400 I.U. vitamin E per gallon drinking water)

EMCELLE® E-D3 Liquid (400 I.U. vitamin E and 24,000 I.U. vitamin D per gallon drinking water)

Grow/finish

EMCELLE® E-D3 Liquid (200 I.U. vitamin E and 12,000 I.U. vitamin D per gallon drinking water)

EMCELLE® D3 Liquid (20,000 I.U. vitamin D per gallon drinking water)

Gilts and sows

VITAL E®-Repro, 5 mL (1,500 I.U. vitamin E, 1,000,000 I.U. vitamin A, and 500,000 I.U. vitamin D)

Administer 1-2 weeks pre-partum and again at weaning.

Conclusions

Of the three fat-soluble vitamins, vitamin E appears to be the most deficient in all post-weaning phases of swine production, while vitamin D supplementation may also be critical in the nursing pig due to low levels of vitamin D in milk. Reduction in mulberry heart has been observed when fat-soluble vitamin status in post-weaned pigs is above 4 µg/mL. For newly-weaned pigs, the bioavailability of vitamin E is more critical than inclusion level. The need to supplement with a micellized, biologically-available source of vitamins E, A and D as found in EMCELLE TOCOPHEROL, EMCELLE ED_3 Liquid, or VITAL E-injectables is critically important to be assured that fat-soluble vitamin levels are optimized.

Injecting gilts and sows prior to farrowing with VITAL E-Repro, reduces farrowing time and reduces stillborns in older sows. Reduced innate immunity are areas affected by inadequate fat-soluble vitamin intake. In general, the NRC recommended levels for fat-soluble vitamins fat-soluble vitamins are too low, especially vitamin E for the gestating sow and biologically available sources for the nursery pig.

References

1. Hidiroglou, M., T. R. Batra, E. R. Farnworth, and F. Markham. 1995. Effect of vitamin E supplementation on immune status and α-tocopherol in plasma of piglets. Reprod. Nutr. Dev. 35:443–450.

2. Jensen, M., C. Fossum, C., M. Ederroth, R.V.J. Hakkarainen, R.V.J. 1988. The effect of vitamin E in the cell-mediated immune response in pigs. J. Vet. Med. B. 1988. 35:549–555.

3. Lee, G. Y. and S. N. Han. 2018. The Role of Vitamin E in Immunity. Nutrients. 10: 161

4. Saul, L, I. Mair, A. Ivens, P. Brown, et al. 2019. 1,25-Dihydroxyvitamin D3 Restrains CD4 T Cell Priming Ability of CD11c Dendritic Cells by Upregulating Expression of CD31. *Frontiers in Immunology*. 10: 3389.

5. Britt, J. 2014. Vitamin A injection improves embryonic development resulting in uniformity of piglet size at birth for Gilts and Sows. News to Use February, 2014. Stuart Products, Inc.

6. Specht, T.A., D.C. Mahan, N.D. Fastinger and R.L. Stuart. 2003. Effects of d-alpha-tocopherol alcohol or acetate in water soluble or emulsified form to the drinking water of weanling pig. J. Anim. Sci. 81 (Suppl. 2): p 88 (Abstract No. 356).

7. Jang, Y.D., M.D. Lindemann, H.J. Monegue, and R.L. Stuart. 2014. The effects of fat-soluble vitamin administration on plasma vitamin status of nursing pigs differ when provided by oral administration or injection. Asian-Australas J Anim Sci. 5:674-82.

8. Copeland, K. R., D. B. Scales, G. M. Hill, J. E. Link, K. C. Turner, C. J. Rozeboom, and R. L. Stuart. 2017. Effects of vitamins E and D on performance and antioxidant enzymes in nursery pigs. Midwest ASAS Meeting, March 13, 2017.

Table 4: Optimum serum fat-soluble vitamin levels for swine

Class	Retinol (ng/mL)	25-OH-D3 (ng/mL)	α-tocopherol (µg/mL)
Neonates (1-4 days)	≥ 100	≥ 10	≥ 2.00
Nursing (14-21 days)	≥ 200	≥ 15	≥ 6.00
Post-weaning (35-60 days)	≥ 300	≥ 20	≥ 4.00
Grow/finish	≥ 400	≥ 40	≥ 3.00
Gestating gilts/sows	≥ 500	≥ 60	≥ 4.00
Lactating sows	≥ 500	≥ 70	≥ 4.00

9. Becera, Roel. 2017. Improving sow productivity and nursery pig performance by injecting sows with d-alpha-tocopherol, retinyl palmitate, and cholecalciferol. Master's Thesis. North Carolina State University.

10. Jang, Y. D., 2019. Unpublished data.

11. Jang, Y. D., M. J. Rotering, and R. L. Stuart. 2018. Distribution of injected fat-soluble vitamins in plasma and tissues of nursery pigs. Leman Swine Conference, St. Paul, MN.

12. Jensen, S. K. and C. Lauridsen. 2017. Fat Soluble Vitamins - availability and recommendations for highly prolific sows and their progeny. 78th Minnesota Nutr. Conf. Sept. 20-21, 2017. pp 165-175.

13. Pallares, F.J., M. J. Yaeger, B. H. Janke, G. Fernandez and P.G. Halbur. 2002. Vitamin E and selenium concentrations in livers of pigs diagnosed with mulberry heart disease. J Vet Diagn. Invest 14:412–14.

Integrating environmental data from Maximus controls with sow production data

Tom Stein, DVM, MS, PhD
Maximus Systems & Maximum Ag Technologies, Saint-Bruno, Quebec, Canada and Jenison, Michigan

As pork producers look to the future, they have important decisions to make now about how to get ready for emerging smart technologies. Which barn control platform to adopt will be one of the critical success factors for future-proofing your production system. Producers will want to have a single system that can connect to or integrate with these new technologies. It will be smart barn control systems – "brains in the barn" – that become the single management and control platform for integrating all these new in-the-barn technologies.

Which management software platform to adopt is another critical success factor for the same reasons. Producers will want a single software platform that seamlessly works with a smart control system because otherwise it will be too hard to manage. Brains in the barn will communicate in real-time and in both directions with brains in the office to spearhead the next evolution in pork production management.

Maximus controllers in sow farms continuously generate hourly and daily environmental, feed, and water data. This includes inside barn high and low temperatures, humidity, and water usage. There is a lot of interest in the effects of heat stress on sows and nutritional technologies that might mitigate these effects. In this presentation, we will show how it is possible to create and use reports that integrate the environmental and barn data coming from Maximus controllers with sow performance data.

The highest priority is to know what is happening with the environment, including water consumption, against what you programmed for the room or building. We have found a dashboard approach works best. It should include visual data from real-time observations collected by the control. Managers should be able to see quickly out-of-process events and be able to drill-down into the details to find out what's wrong and intervene immediately.

We also use a real-time alarm management system which creates a history of all alarms as well as when the alarm was acknowledged and when the room or building returned to normal (Figure 1).

For analytics that pull together both environmental and production data, we use a framework that incorporates the concepts of "exposure" and outcomes. Examples of "exposures" are: average temperature, high temperature, low temperature, net temperature difference, humidity, CO_2 levels, outside temperature, outside-inside temperature differential, and water disappearance.

We define exposures that occur over periods of time. Examples include exposure during Service Week, exposure during particular gestation weeks, exposure at farrowing or by lactation days or weeks, or exposure for weaned sow cohorts just after weaning or during the days within the wean-to-first-service interval.

Outcomes include conception rate, pregnancy rates at 35, 72, and 105 days of gestation, farrowing rate, total born, live born, mummies, and stillborn. Other outcomes include culling, death, and non-productive days.

Producers are starting to ask us to create other types of analysis. Figures 2 and 3 are examples of questions producers have asked.

Figure 1: Alarm management and communication system example.

Maximus Alarms
Hyland Pork

Interval: Last 1 Day(s)

Creation Time	Description	Type
Site : Elm Creek		
Building : Elm Creek 3		
11/12/2019 23:44	Room B Room B Feed maximum run time	Critical
Building : Elm Creek 2		
11/12/2019 16:45	Alarm has been acknowledged.	Normal
11/12/2019 16:44	Room D Room D Feed maximum run time	Critical
11/12/2019 16:16	Room C Room C Feed maximum run time back to normal.	Normal
11/12/2019 15:26	Alarm has been acknowledged.	Normal
11/12/2019 11:22	Module 15 back to normal.	Normal
11/12/2019 05:46	Module 15 Room D Fan 3 Low current	Critical
Building : Elm Creek 1		
11/12/2019 11:25	Average temp. Room C has returned within the normal range. Current value = 9.9°C.	Normal
11/12/2019 10:59	Alarm has been acknowledged.	Normal
11/12/2019 10:44	Average temp. Room A has returned within the normal range. Current value = 8.8°C.	Normal
11/12/2019 10:37	Alarm has been acknowledged.	Normal
11/12/2019 10:29	Module 2 Room A Fan 4 High current	Critical
11/12/2019 10:28	Average temp. Room C is under the critical low limit of 9.7°C. Current value = 2.9°C.	Critical
11/12/2019 10:28	Average temp. Room A is under the critical low limit of 8.5°C. Current value = 6.7°C.	Critical
11/12/2019 10:28	Average temp. Room D is under the critical low limit of 9.7°C. Current value = 1.9°C.	Critical
11/12/2019 08:32	Alarm has been acknowledged.	Normal
Building : Elm Creek 4		
11/12/2019 10:02	Room D Room D Feed maximum run time back to normal.	Normal
11/12/2019 10:02	Room A Room A Feed maximum run time back to normal.	Normal
11/12/2019 10:02	Alarm has been acknowledged.	Normal
11/12/2019 08:44	Room A Room A Feed maximum run time	Critical
11/12/2019 03:02	Room D Room D Feed maximum run time	Critical
11/12/2019 00:44	Room D Room D Feed maximum run time	Critical

Figure 2: Inside barn temperature and conception rate.

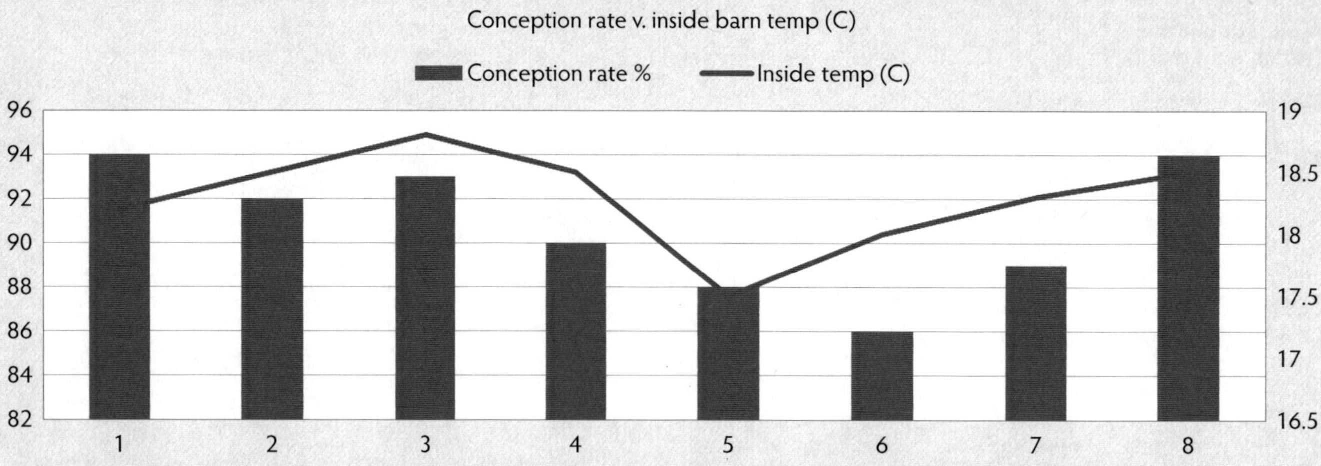

Figure 3: Relationship between barn temperature, feed use, & abortion.

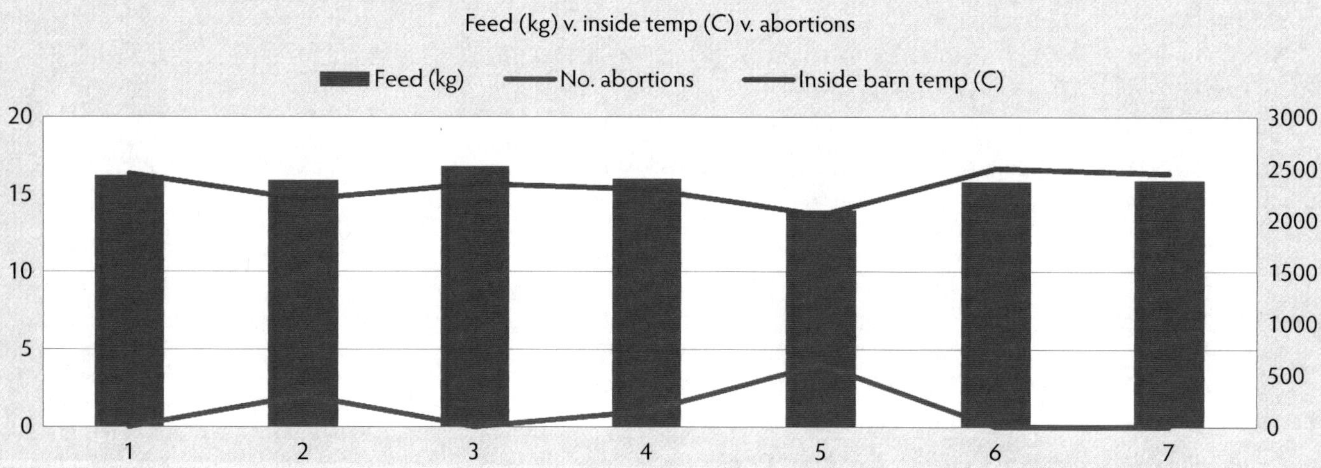

A killed *Streptococcus suis* vaccine formulated in Amplivac™ elicits maternal immunity capable of protecting piglets

Scott Dee[1]; Joel Nerem[1]; Dan Hanson[1]; Roy Edler[1]; Ethan Schmaling[2]; Kelli Peters[2]; Keith Wilson[2]; Thomas Wetzell[2]; Kellie Hogan[2]; Paulraj Lawrence[2]

[1]Pipestone Applied Research, Pipestone, Minnesota; [2]Boehringer Ingelheim Animal Health USA Inc., Duluth, Georgia

Introduction

Streptococcus suis (*S suis*) is a gram positive bacterium that causes septicemia, bronchopneumonia, arthritis, and fatal meningitis in young piglets. Piglets are usually infected through exposure to asymptomatic sows. Thirty-five different serotypes of *S suis* have been identified,[1] however, recent work has suggested that only 29 unique serotypes exist.[2,3] In addition, many isolates remain non-typeable using traditional methods. Of the 35 serotypes described, serotype 2 is most commonly associated with disease, but serotypes 1, 1/2, 3, 4, 5, 7, 8, 9 and 16 are also frequently isolated from diseased swine.[1] Recently *S suis* infections have become more problematic in pre and post wean pigs. Currently, there are no commercial *S suis* vaccines available in the USA. The objective of this study was to test the effect of an autogenous vaccine used in sows and piglets that were experimentally challenged with a homologous serotype.

Materials and methods

A pig flow in the Midwest region of the USA that had been experiencing significant mortalities from *Streptococcus suis* type 1 was selected for this study. The vaccine used in the trial was a *S suis* serotype 1 manufactured at Newport Laboratories, Inc. autogenous vaccine using AMPLIVAC adjuvant. A 1 mL, intramuscular injection was administered to piglets and/or sows. Treatment groups were; A) piglets vaccinated at processing and weaning from sows vaccinated at 5 weeks and 3 weeks prefarrow, B) Unvaccinated piglets from sows vaccinated at 5 and 3 weeks prefarrowing C) piglets vaccinated at processing and weaning from unvaccinated sows, and D) non–vaccinate control group. Treatments groups were assigned by litter, and all parities were equally represented among all treatment groups. A total of 36 piglets were enrolled per treatment group. No antibiotics or acidifiers were given at weaning or in the nursery feed. Two weeks after weaning, all pigs received a challenge dose of a homologous *S suis* type 1 at log 1×10^9 log via intramuscular and intraperitoneal routes, and all mortalities were recorded and tissues collected for confirmation of *S suis* infection for 14 days post challenge (Figure 1). Clinical scoring of piglets post-challenge was performed on a scale of 1-4, depending on the severity of disease. A score of 0 was considered normal; 1- mild lameness, still mobile; 2- severe lameness and no motivation to walk; 3-paddling and CNS signs, mostly euthanized; 4- found dead. A cumulative clinical score was calculated for each treatment group.

Results and discussion

Table 1 shows the mortality and cumulative clinical scores. The challenge model was successful, as shown by the mortality in the non-vaccinated control treatment group. Under the conditions of this trial, the *S suis* serotype 1, Newport Laboratories, Inc. autogenous vaccine using AMPLIVAC adjuvant provided a level of protection to the piglets when used as prefarrow shot only (group B) or in combination as piglet dose (group A) (Table 1). Mortality rates were lower in the vaccinated treatment groups versus the non-vaccinated controls. Clinical scores were higher in the non-vaccinated controls versus the vaccinated treatment groups. This trial supports the use of autogenous vaccine as a valuable tool in managing the effects of *S suis* related health and production challenges within populations of pigs. The importance of developing a vaccine with a homologous strain of *Streptococcus suis* from the target population is critical. Well executed diagnostics, as well as a good history of the clinical presentation and long term serotype surveillance are cornerstones in developing and maintaining a successful autogenous vaccine program.

References

1. Goyette-Desjardins G et al. *Streptococcus suis*, an important pig pathogen and emerging zoonotic agent—an update on the worldwide distribution based on serotyping and sequence typing. *Emerg. Microbes Infect.* 2014;3:e45.

2. Gottschalk et al. Susceptibility of *Streptococcus suis* to penicillin. *J Vet Diag Invest.* 1991;3:170-172.

3. Food and Drug Administration. Veterinary Feed Directive Fact Sheet.2017. Available at www.fda.gov/animalveterinary/developmentapprovalprocess/ucm449019.htm. Accessed 20 October 2017.

Figure 1: Timeline of events during the study.

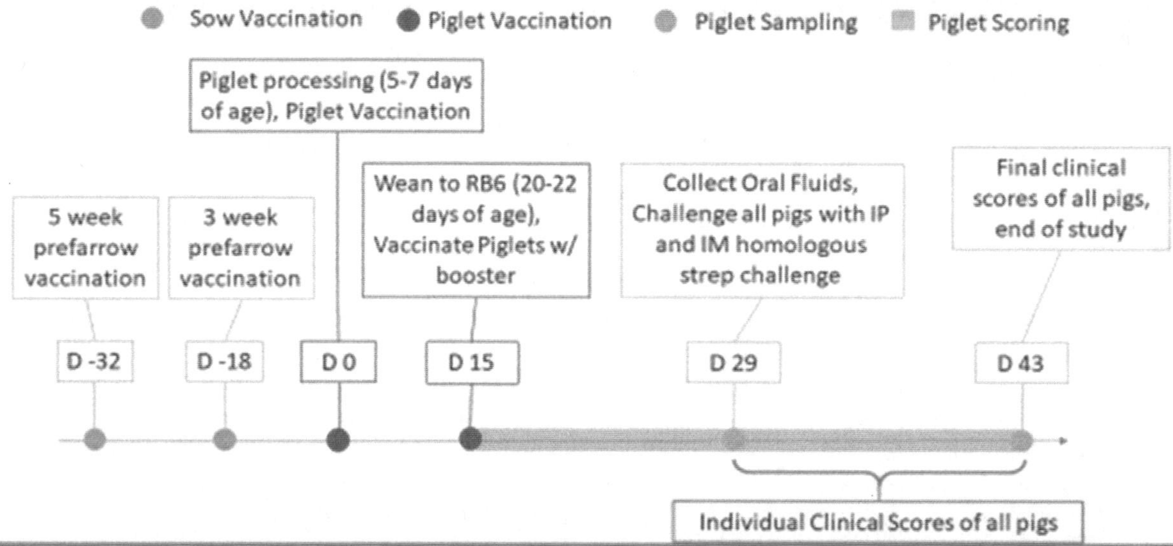

Table 1: Mortality and clinical scores

	Treatment			
	A	B	C	D
	Newport	Newport	Newport	NVC
	(Sow + pig)	(Sow)	(Pig)	
Number of pigs on test	36	36	36	36
Total mortality	9	7	15	23
Cumulative *S suis* clinical score observation rate	266/504	268/504	356/504	424/504

Genetic selection can be used as a viable tool to improve natural robustness to PRRS challenge

J. R. Dunkelberger[1], PhD; P. K. Mathur[2], PhD; E. Little[3], MSc; D. Hanson[3], BA;
J. M. Eggert[1], PhD; S. Dee[3], DVM, PhD; E. F. Knol[2], PhD
[1]Topigs Norsvin USA, Burnsville, Minnesota; [2]Topigs Norsvin Research Center B.V., Beuningen, The Netherlands;
[3]Pipestone Applied Research, Pipestone, Minnesota

Introduction

PRRS remains the most economically devastating disease of the US swine industry. Various control strategies can be used to help prevent or mitigate the impact of disease following infection; however, existing control methods are only partially effective. Therefore, the need for alternative, or complementary, control strategies remains. One such alternative is to use genetic selection to breed pigs for improved host response to PRRS. The feasibility of this approach is supported by evidence of naturally-occurring genetic variation in reproductive performance under PRRS challenge for purebred and crossbred sows[1-3] and growth performance under PRRS challenge for maternal by-products.[4-11] However, the extent of genetic variation in host response to PRRS for commercial finishing pigs, either between or within genetic sources, has not yet been investigated. Therefore, the objectives of this study were to: 1.) evaluate performance differences between Topigs Norsvin-sired finishers vs. finishers sired by boars from a competing genetic source; and 2.) to quantify genetic variation in host response to PRRS challenge for Topigs Norsvin-sired finishing pigs.

Materials and methods

Commercial finishers sired by a Topigs Norsvin terminal sire line (n = 730 TN Tempo × TN70) and competitor line (n = 729 competitor boar x TN70) were farrowed at a commercial sow farm. At weaning, pigs were transported to a commercial research facility and placed into pens by genetic group. The research barn was filled in 5 wean groups over the course of 1.5 weeks and each pig was given a minimum 4-day acclimation period prior to vaccination with a modified live PRRSV vaccine. Four weeks later, each pig was experimentally inoculated with 10^5 $TCID_{50}$ of the 1-7-4 PRRSV isolate and followed until marketing.

Phenotypes recorded throughout the study included body weight and feed intake at each diet change (recorded on a pen basis), used to calculate wean-to-finish ADG and FCR. Individual phenotypes were recorded for: body weight at 0, 13, and 42 days post-infection (dpi) (TN Tempo sired-pigs only), market weight and hot carcass weight (used to calculate carcass yield), days on feed, BF, loin depth, lean meat percentage, full value status, and mortality from 0 dpi to market. The effect of genetic group (TN Tempo-sired vs. competitor-sired) on each trait was analyzed using a mixed linear model in SAS, except for percent full value pigs and mortality, which were analyzed using a LOGIT model.

Genetic analyses were conducted for the TN Tempo-sired pigs only. Tissue collected from TN Tempo-sired pigs at birth was used for genotyping with the Illumina 50K single nucleotide polymorphism chip. Genotypes were then used to construct a genomic relationship matrix to account for relatedness between individuals for genetic parameter estimation of mortality and ADG from: 0 to 13 dpi, 13 to 42 dpi, and 0 dpi to market. ADG was analyzed by phase as a continuous variable using animal models with ASReml 4.0. A similar approach was used for mortality, as well as analysis of mortality using a LOGIT model.

Results and discussion

Multiple pathogens (in addition to PRRSV) were detected post-challenge, including, but not limited to: *Haemophilus parasuis, Actinobacillus suis, Streptococcus suis, Salmonella,* and *Pasteurella multocida.*

Results of performance differences between genetic lines showed that pigs from the TN Tempo-sired group had significantly greater ADG (1.70 vs. 1.58 lbs./d; $P < 0.0001$), lower FCR on a live-weight basis (2.39 vs. 2.45; $P = 0.0003$) and a carcass basis (3.22 vs. 3.31; $P = 0.0003$), fewer days on feed (158 vs. 164 d; $P < 0.0001$), higher carcass yield (74.2% vs. 74.0%; $P = 0.04$), fewer mortalities (13% vs. 22%; $P < 0.0001$), and more full-value pigs (84.6% vs. 74.2%; $P < 0.0001$) than the competitor line following PRRS challenge. No evidence of a difference in loin depth (2.50 vs. 2.54 in.; $P = 0.20$), BF (0.68 vs. 0.67 in.; $P = 0.45$), or lean meat percentage (55.9% vs. 55.9%; $P = 0.93$) was detected between genetic groups.

Results of genetic analyses showed that both growth rate and mortality post-challenge were heritable traits. Heritability estimates for ADG were 0.27 (0.05), 0.16 (0.05), and 0.26 (0.05) when analyzed from 0 to 13 dpi, 13 to 42 dpi, and 0 dpi to market, respectively. Heritability estimates for mortality were 0.10 (0.07) and 0.10 (0.04), when analyzed using a LOGIT model, or on the observed scale, respectively. Additional results of genetic parameter estimation provided evidence of substantial genetic variation (0.38) in mortality post-challenge. Together, the observed genetic variation and heritability estimate for mortality indicate the possibility of reducing the incidence of mortality under challenge by 5.7% after a single generation of single-trait selection.

In conclusion, results show that the TN Tempo-sired pigs were naturally more robust to the multifactorial PRRS challenge than pigs sired by boars from a competing genetic source. Genetic parameter estimates for growth rate and mortality post-challenge provide evidence of naturally-occurring genetic variation in robustness to challenge. The extent of the genetic variation detected in this study demonstrates an opportunity to further enhance the performance and survivability of the TN Tempo line under disease challenge via genetic selection.

References

1. Lewis CRG et al. Genetic parameters for performance traits in commercial sows estimated before and after an outbreak of porcine reproductive and respiratory syndrome. *Journal of Animal Science*. 2009; 87.

2. Rashidi et al. Variation among sows in response to porcine reproductive and respiratory syndrome. *Journal of Animal Science*. 2014; 92:1.

3. Serão NVL et al. Genetic analysis of reproductive traits and antibody response in a PRRS outbreak herd. *Journal of Animal Science*. 2014; 92:7.

4. Boddicker NJ et al. Evidence for a major QTL associated with host response to porcine reproductive and respiratory syndrome virus challenge. *Journal of Animal Science*. 2012; 90:6.

5. Boddicker NJ et al. Validation and further characterization of a major quantitative trait locus associated with host response to experimental infection with porcine reproductive and respiratory syndrome virus. *Animal Genetics*. 2013; 45:1.

6. Boddicker NJ et al. Genome-wide association and genomic prediction for host response to porcine reproductive and respiratory syndrome virus infection. *Genetics Selection Evolution*. 2014;46:1.

7. Hess AS et al. Comparison of host genetic factors influencing pig response to infection with two North American isolates of porcine reproductive and respiratory syndrome virus. *Genetics Selection Evolution*. 2016:48;43.

8. Dunkelberger JR et al. Effect of a major quantitative trait locus for porcine reproductive and respiratory syndrome (PRRS) resistance on response to coinfection with prrs virus and porcine circovirus type 2b (PCV2b) in commercial pigs, with or without prior vaccination for PRRS. *Journal of Animal Science*. 2017a; 95.

9. Dunkelberger JR et al. Genomic regions associated with host response to porcine reproductive and respiratory syndrome vaccination and co-infection in nursery pigs. *BMC Genomics*. 2017b;18:865.

10. Waide EH et al. Genomic prediction of piglet response to infection with one of two porcine reproductive and respiratory syndrome virus isolates. *Genetics Selection Evolution*. 2018;50:3.

11. Putz AM et al. Novel Resilience Phenotypes Using Feed Intake Data From a Natural Disease Challenge Model in Wean-to-Finish Pigs. *Frontiers in Genetics*. 2019;9.

Recent findings of the impact of *Lawsonia intracellularis* challenge on gut health

F. Leite, DVM, MS, PhD
Boehringer Ingelheim Animal Health USA Inc., Duluth, Georgia

Introduction

Our knowledge of what contributes to a healthy gastro-intestinal tract in the pig is increasing, as is the understanding of how pathogens can compromise this health. A growing body of evidence is finding gut health as a factor that must be prioritized to allow the pig to achieve its full genetic potential as well as be less susceptible to not only enteric but also respiratory disease.[1,2] Having tools to promote a healthy gut is very important to achieve maximum efficiency and overall health on the farm.

Lawsonia infection impairs intestinal structure and function

Research has led to interesting findings on how Ileitis or porcine proliferative enteropathy (PPE) caused by the bacterial pathogen *Lawsonia intracellularis*, can compromise gut health. The hallmark lesion of PPE is the thickening of the intestinal wall which can be present primarily in the ileum, but may occur in any portion of the intestinal tract.[3,4] The intestine is lined with cells called enterocytes. These cells are responsible for the absorption of the nutrients that will lead to feed efficiency and daily weight gain. *L intracellularis* is an obligate intracellular pathogen that replicates inside of enterocytes primarily present in the crypts structures of the intestine leading to their increase and thickening of tissue.[3,4] Research has found that receptors to key nutrients like amino acids, glucose and lipids are decreased in enterocytes infected by *Lawsonia*.[5] Importantly also, infection leads to a decrease in goblet cells which are responsible for producing the mucus that separates bacteria and other microorganisms from the enterocyte lining of the gut.[3,4] Research from the University of Edinburgh confirmed this and found that infected pigs have a decrease of the important protective mucus layer.[6]

Recently, we have found that not only are the crypts enlarged and goblet cells decreased with infection, but also villi height can be significantly decreased.[7] Again, demonstrating another significant change infection can lead to in intestinal structure. A decrease in villi height and enlargement of crypts due to hyperplasia has been found as a general feature of many intestinal inflammatory conditions.[7,8] We have also analyzed total gene expression (transcriptome) of infected tissue in a model of subclinical infection in which animals did not present many clinical signs of diarrhea. We found that several pathways and genes associated with exacerbated cell proliferation and inflammation were activated, especially in animals with more severe lesions and at peak stage of infection.[7] These lesions and responses are far from that of a normal functioning and healthy intestinal tract.

Understanding that *Lawsonia* can lead to significant lesions in the gut we evaluated if supplementing pigs with an organic zinc form could enhance the pigs response to infection.[9] Zinc is known to have anti-inflammatory properties and is crucial for optimal innate and adaptive immune function.[10] Interestingly, we found that supplementation of animals with an organic zinc form can lead to a decrease in lesion severity.[9] Recently, it was found that this zinc supplement when coupled with an oral live vaccine led to a reduction in mortality in animals challenged with *L intracellularis*.[11]

Lawsonia alters the gut microbiome and response to other pathogens

Not only can *Lawsonia* infection directly impact performance, but it may also increase susceptibility to other pathogens. The pig gut is colonized by an array of microorganisms known collectively as the microbiome (or microbiota). The bacteria in this community can have different relationships amongst each other as well as with the pig host. A healthy gut is one with a bacterial community that allows for proper intestinal function and promotes colonization resistance to pathogens, which ultimately leads to optimal production performance. *Lawsonia* seems to lead to the opposite microbiome as that which would be desired.

An epidemiologic study conducted in France found that *Lawsonia* infection was a risk factor for increased *Salmonella* shedding in growing-finishing pigs.[12] With this in mind, research was conducted to investigate if prevention of *Lawsonia* with an oral live vaccine could impact *Salmonella* shedding.

It was also investigated if changes in the microbiome (bacterial community) in the gut could be involved in this response. To test this, four treatment groups with one set of control non-infected animals were compared. The infected groups were challenged with either *S* Typhimurium alone or with *L intracellularis*, with and without *L intracellularis* vaccination.

Vaccination of the co-infected pigs significantly decreased their fecal shedding of *Salmonella* at seven days after the *Salmonella* Typhimurium challenge. At this stage significantly fewer pigs in the vaccinated/co-infected group, just 38% compared to 100% of pigs in the other challenged groups, were actively shedding *Salmonella* and this trend persisted at later time-points. At seven days post *Salmonella* challenge, co-infected pigs shed around 100 times more *Salmonella* than the group co-infected that had received vaccine. The effect of vaccination on shedding of *Salmonella* was only present in co-infected animals suggesting that it was mediated by the effect of the vaccine on *Lawsonia* infection.[13]

When the microbiome response of animals at the seven-day time point was analyzed, the co-infected vaccinated group had a different bacterial community than those of the co-infected group that had not received vaccine. Among several differences, *Clostridium butyricum* was much more abundant in the group that had received the vaccine as compared to those in the co-infected non-vaccinated group.[13]

Clostridia species produce short-chain fatty acids by fermenting dietary fiber. These fatty acids include butyrate, which *C butyricum* is known for producing. Butyrate has anti-inflammatory properties and also can down-regulate the expression of virulence genes in *Salmonella* which allow its invasion of the gut.[14,15]

In this and other ways, then, a deleted population of butyrate-producing Clostridia bacteria could promote *Salmonella* infection. Since butyrate and other changes in the microbiome could potentially have an effect on more *Salmonella* serovars that share similar virulence factors, we predicted that *Lawsonia* vaccination could have an impact on other serovars as well when co-infecting a pig with *Lawsonia*.[13]

Interestingly, a recent study from University of Veterinary Medicine Hannover in Germany found that when comparing tylosin treatment to vaccination with an oral live *Lawsonia* vaccine, in animals co-infected with *Lawsonia* and *Salmonella* Derby, the vaccinated group had significantly fewer animals shedding *Salmonella* Derby in feces and significantly fewer pigs with positive cultures from ileocecal lymph nodes. These findings are important, as they suggest vaccination may help control *Salmonella* and potentially improve food safety.[16]

To investigate the relationship of *Lawsonia* and *Salmonella* Typhimurium further, we used an *in vitro* model to test how *Lawsonia* and *Salmonella* would behave alone or when co-infecting enterocytes.[17] *Salmonella* is known to induce an inflammatory response when it infects enterocytes and this response is important to enhance its infection.[18]

We found that *Lawsonia* had an additive effect on the induction of inflammatory markers (IL-8 and TNF) that contribute to the pathology of *Salmonella* Typhimurium in co-infected enterocytes.[17] This suggests that the host response to infection, in addition to changes in the microbiome, may be another way in which these two pathogens may favor each other.[17]

Recently, in a different *ex vivo* model, researchers at Iowa State University have found that animals co-infected with *Lawsonia* and *Mycoplasma hyopneumoniae* had a significant 60% increase in the invasion of *Salmonella* Typhimurium in intestinal tissue.[19] This has further demonstrated that *Lawsonia* can promote *Salmonella* infection, which could be influenced by *M hyopneumoniae* as well.

Adding to the spectrum of pathogens *Lawsonia* may enhance, researchers from University of Minas Gerais in Brazil have presented results showing a synergistic effect in pigs co-infected with *Lawsonia* and *Brachyspira hyodysenteriae*. Co-infected pigs presented more severe clinical signs and microscopic lesions as compared to each singly infected group.[20]

Conclusion

Our knowledge is increasing of how *Lawsonia* can affect the gut by disturbing its function, structure, microbial community composition and interaction with other pathogens. With this, it is important to prevent this costly disease.

References

1. Moeser A. et al., Weaning stress and gastrointestinal barrier development: Implications for lifelong gut health in pigs. Anim Nutr. 2017 Dec;3(4):313-321.
2. Niederwerder MC. Role of the microbiome in swine respiratory disease. Vet Microbiol. 2017 Sep;209:97-106.
3. Lawson GHK, Gebhart CJ. 2000. Proliferative enteropathy. J Comp Pathol. 122:77–100.
4. Vannucci et al. Proliferative Enteropathy, Diseases of Swine, 11th edition, 2019.
5. Vannucci FA et al. Laser microdissection coupled with RNA-seq analysis of porcine enterocytes infected with an obligate intracellular pathogen (*Lawsonia intracellularis*). BMC Genomics. 2013 Jun 24;14:421.
6. Bengtsson RJ et al., *Lawsonia intracellularis* infection of intestinal crypt cells is associated with specific depletion of secreted MUC2 in goblet cells. Vet Immunol Immunopathol. 2015 Nov 15;168(1-2):61-7.
7. Leite F et al., A Cell Proliferation and Inflammatory Signature Is Induced by *Lawsonia intracellularis* Infection in Swine. MBio. 2019 Jan 29;10(1).
8. Erben U et al., A guide to histomorphological evaluation of intestinal inflammation in mouse models. Int J Clin Exp Pathol. 2014. 7:4557– 4576.
9. Leite F et al., The effects of zinc amino acid complex supplementation on the porcine host response to *Lawsonia intracellularis* infection. Vet Res. 2018 Sep 10;49(1):88.
10. Bonaventura P et al., Zinc and its role in immunity and inflammation. Autoimmun Rev. 2015. 14:277–285
11. Mueller et al., Effect of feeding Availa-Zn and/or vaccine on growing pig performance and mortality during *Lawsonia intracellularis* challenge. Leman Swine Conference Proceedings 2018.
12. Beloeil PA et al., Risk factors for Salmonella enterica subsp. enterica shedding by market-age pigs in French farrow-to-finish herds. Prev Vet Med. 2004 Apr 30;63 (1-2):103-20.
13. Leite F. et al. Vaccination Against *Lawsonia intracellularis* Decreases Shedding of Salmonella enterica serovar Typhimurium in Co-Infected Pigs and Alters the Gut Microbiome. Sci Rep. 2018 Feb 12;8(1):2857.
14. Gantois, I. et al. Butyrate Specifically Down-Regulates Salmonella Pathogenicity Island 1. Gene Expression. 72, 946–949 (2006).
15. Blander, JM et al. Regulation of inflammation by microbiota interactions with the host. Nat. Immunol. 18, 851–860 (2017).
16. Visscher et al. Spread of an Experimental *Salmonella Derby* Infection in Antibiotic-Treated or *Lawsonia intracellularis* Vaccinated Piglets. Animals (Basel). 2018 Nov 12;8(11).
17. Leite F et al., The effects of *Lawsonia intracellularis*, *Salmonella* enterica serovar *Typhimurium* and co-infection on IL-8 and TNFα expression in IPEC-J2 cells. Vet Microbiol. 2019 Apr;231:76-79.
18. Kim HB, Isaacson RE; *Salmonella* in Swine: Microbiota Interactions. Annu Rev Anim Biosci. 2017 Feb 8;5:43-63.
19. Helm ET et al., *Mycoplasma hyopneumoniae-Lawsonia intracellularis* dual challenge modulates intestinal integrity and function. J Anim Sci. 2019 May 30;97(6):2376-2384.
20. Daniel, AGS et al., Synergic effect of *Brachyspira hyodysenteriae* and *Lawsonia intracellularis* Coinfection. Leman Swine Conference Proceedings 2018.

New labor-free fly control ClariFly 267 (diflubenzuron) achieves 99% reduction in housefly (*Musca domestica*) populations in swine facilities

Tom Pastor, BS; Gene Spellman, BS
Central Life Sciences, Schaumburg, Illinois

Introduction

Despite extensive investments in biosecurity measures and protocols, farms still have unexplainable cases of viral diseases. Fly control is now recommended as an important plank of swine biosecurity protocols and disease risk reduction.[1-3] Flies can carry and transmit important swine disease viruses such as porcine reproductive and respiratory virus (PRRSV),[1] porcine epidemic diarrhea virus (PEDV),[2] Senecavirus A (SVA)[3] and now African swine fever virus (ASFV).[4] Flies can carry infectious bacteria,[5] and antibiotic-resistant bacteria.[6] Flies can travel up to four miles.[7] Fly control in swine till now has been labor-intensive, sporadically effective, non-coordinated, using multiple products with harsh chemicals. A search for new products and new technology for fly control was conducted. ClariFly 267 (diflubenzuron) insect growth regulator larvicide was developed and approved for swine. This research shows ClariFly 267 in-feed larvicide achieved a 99% reduction in housefly populations in swine finishing facilities.

Materials and methods

A 2,400 head finisher in the Eastern Corn Belt was selected. The facility was a curtain-sided, tunnel ventilated double wide finisher with two rooms of 600 head on each side of the curtain. ClariFly 267 (diflubenzuron) in-feed larvicide was incorporated within the typical ration at the rate of 0.20 mg ClariFly 267 per kg of body weight per day, for the five-week study. Feed samples were collected at the start and mid-point of the study. The feed samples were assayed for proper levels of ClariFly 267. To count flies and monitor the fly population reduction, two Starbar EZ Trap sticky fly traps were hung in each room. Traps were placed approximately seven feet above the floor. Fans and air vents were avoided for trap placement to avoid air movement interfering with fly activity. Traps were hung for 1 week and replaced with new, unused traps for the next week of the study. Weekly trap collection, fly counting and new trap placement was repeated throughout the five week study.

Results

House Fly Counts are shown in Table 1. After five weeks on ClariFly 267 in-feed larvicide, the fly populations, as represented by the flies caught and counted on the traps collected weekly during the study, were reduced more than 99% by the end of week five versus the end of week one, showing that ClariFly 267 was very effective in reducing fly populations. Orderly placing weekly traps next to each other offers a visually striking picture of the reduction in fly population achieved with ClariFly 267 (Figure 1).

Discussion

ClariFly 267 in-feed larvicide proved extremely effective in reducing housefly populations in this study, achieving a 99% reduction in fly counts in each room. ClariFly 267 in-feed larvicide offers fly control options in a labor-free fashion, since ClariFly 267 is mixed in the swine ration. The pigs do the work, and are effective in delivering ClariFly 267 where it is needed-in the manure. ClariFly 267 is a chitin-synthesis inhibitor. Chitin is a key component of the fly's exoskeleton. Inhibiting chitin synthesis blocks further development of the fly in the larval stage. ClariFly 267 does not kill adult flies. Adult flies are prevented from developing from earlier larval stages. Existing adult flies eventually die off and are not replaced when ClariFly 267 is fed.

Table 1: Fly count results

	House fly count			
Date	**Room 1A**	**Room 1B**	**Room 2A**	**Room 2B**
Week 1 (7/14/2016)	1080	2100	916	636
Week 2 (7/21/2016)	1200	1086	1900	1850
Week 3 (7/28/2016)	972	873	1150	1265
Week 4 (8/4/2016)	56	48	92	71
Week 5 (8/11/2016)	4	10	8	3

Figure 1: Visual lineup of traps from Room 2B.

It is important to show patience when feeding ClariFly 267, as adult flies are still visible during the initial feeding stages. In room 2B, week 3 fly counts (1150) were higher than week 1 fly counts (916), and then dropped tremendously in weeks 4 and 5. Conclusions on ClariFly effectiveness would have been faulty in early feeding stages. ClariFly 267 offers a convenient and effective option to achieve fly control in swine biosecurity protocols. Fly control also provided an improved working environment for pigs and employees, as reported by the study cooperator. More research is needed in the area of improved working environment and animal welfare to quantify the benefits resulting from improved fly control.

References

1. Otake S, Dee SA et al. Transmission of porcine reproductive and respiratory syndrome virus by houseflies, (Musca domestica Linnaeus). *Vet Rec* 2003. 152: 73-76.

2. Allison G, et al. PEDV positive bioassay reveals houseflies (Musca domestica) can transmit infectious PEDV to pigs. *Proceedings 50th AASV 2019.*

3. Allison G, et al. PEDV, PRRSV, and Seneca Valley virus PCR positive flies (Musca domestica). What now? *Proceedings 49th AASV 2018.* 394-396.

4. Olesen, A, et al. Infection of pigs with African swine fever virus via ingestion of stable flies (Stomoxys calcitrans) *Transboundary Emerging Diseases 2018,* 65: 1152-1157.

5. Khamesipour F, et al. A systematic review of human pathogens carried by the housefly (Musca domestica L.). *BMC Public Health 2018,* 18:1049.

6. Zurek L, et al. Insects Represent a Link between Food Animal Farms and the Urban Environment for Antibiotic Resistance Traits. *Appl Environ Microbiol. 2014 Jun;* 80(12): 3562–3567.

7. Nazni WA, Luke H et al. Determination of the flight range and dispersal of the house fly, Musca domestica (L.) using mark release recapture technique. *Trop Biomed 2005 Jun;*22(1):53-61.

Impacts of copper supplementation on performance during a natural porcine reproductive and respiratory syndrome virus (PRRSV) and influenza A disease outbreak in a large-scale commercial research facility

F. Sandberg[1], PhD; L. Ochoa[1], PhD; B. Lawrence[2], PhD; K. Soltwedel[1], PhD; M. Bible[1], PhD
[1]Furst-McNess Company, Freeport, Illinois; [2]Novus International, St Charles, Missouri

Introduction

PRRSV has been estimated to cost the US swine industry $664 million per year.[1] Costs arise from compromised performance in multiple stages of production (sow, nursery and finishing). Influenza A can also influence performance to a significant extent. Late finishing death loss and poor performance can be very costly due to accumulated value of the animal, and large feed consumption at this stage. Pathogen challenges reduce performance through anorexia, increased energy and amino acid requirements and through repartitioning of nutrient and energy intake towards protective and immune mechanisms.[2] Mineral nutrition during pathogen challenges is complex, and trace-minerals such as copper play an important role in many metabolic functions, enzyme systems and immune functions. Copper deficiency can influence both innate and acquired immunity including the number of antibody-producing cells, microbicidal activity of macrophages, cytolytic activity of natural killer cells, T cell function and cytokine expression of the immune system.[3] Copper is a common feed additive for growth promotion when added above basal requirement levels. The addition of copper to a diet can come from in-organic sources such as copper sulfate ($CuSO_4$) and copper chloride ($Cu_2[OH]_3CL$), or, from organic sources where elemental copper is bound to some type of ligand e.g. copper methionine hydroxy analogue chelate $Cu(HMTBa)_2$ and several types of Cu amino acid complexes. The type and level of copper fed influences the level of accumulation of Cu in key mineral reservoirs in the animal such as the liver where it has been shown that pigs fed 170 ppm of copper from $Cu(HMTBa)_2$ had 30.5 ppm Cu in liver tissue, as compared to 18.37 ppm for pigs fed 170 ppm Cu from $CuSO_4$ and 6.83 ppm for pigs fed a basal requirement level of 6 ppm Cu from $CuSO_4$.[4] During disease challenges such large differences in key reservoirs may play an important role in how immunity and other physiological mechanisms are expressed to overcome a disease challenge. However, the extent to which performance is influenced by Cu supplementation in a large-scale live production environment during a joint PRRSV and influenza A disease challenge has to our knowledge not been previously reported. The first objective of this study is to report the performance of finishing pigs during a naturally occurring PRRS and influenza A disease exposure where pigs were supplemented with Cu from either $CuSO_4$, $Cu_2[OH]_3CL$ or $Cu(HMTBa)_2$. The second objective was to attempt to quantify an estimate of the financial cost of that co-infection in late finishing pigs.

Materials and methods

A feeding trial was conducted with 1,312 PIC 359 pigs averaging 10.5 lb and stocked at 28 to 30 pigs per pen with 10 replicates per treatment in a commercial research facility. Pigs were allocated by environment, sex and body weight to evaluate the impact of 3 copper (Cu) supplementation and source of Cu treatments on performance of pigs: 1) 150 ppm of Cu from $Cu_2[OH]_3CL$; 2) basal 12 ppm Cu from $CuSO_4$; and 3) 50 ppm of Cu from $Cu(HMTBa)_2$. Data reported here is for the period of Day 84 (Approx 130 lb LW) to Day 176 (285 lb LW) on trial. Following sickness behavior reported by farm staff, oral fluids collected on d 126 of the study indicated pigs were positive for PRRSV (1-4-2) with US Ct 32.8 and 32.4, and at day 136 oral fluids indicated pigs were positive for influenza A with US Ct 35.8, 28 and 35.4 conducted by Iowa State Veterinary Diagnostic labs. Pigs were marketed in 4 cuts, with 8 pigs harvested from each pen for cuts 1,2,3 and remaining pigs on cut 4. All data were analyzed using the MIXED procedure of SAS (SAS Institute, Inc., Cray, NC) with a randomized complete block design. Pen served as the experimental unit for growth performance and health status, and the pig served as the experimental unit for carcass characteristics. Differences among treatments were considered significant with P-value < 0.05, and a trend if P-value > 0.05 and < 0.10.

Results and discussion

At Day 84 of the trial there was no difference in BW ($P = 0.56$) and pigs were approximately 130 lb. At market time (day 176) pigs that were fed either 150 ppm Cu as $Cu_2[OH]_3CL$ (286.4 lb) or 50 ppm Cu as $Cu(HMTBa)_2$ (287.2 lb) were not different, but both of these tended ($P < 0.10$) to be greater than 12 ppm of Cu as $CuSO_4$ (281.0 lb). In Table 1 are the values for ADG for the different measurement periods and these was numerically greater for both supplemental levels of Cu as compared to $CuSO_4$. Notably, the growth rates fell from day 126-147 to day 147-176 by 40%, 55% and 45% for 150 ppm Cu as $Cu_2[OH]_3CL$, 12 ppm of Cu as $CuSO_4$ and 50 ppm Cu as $Cu(HMTBa)_2$, respectively where the positive diagnosis was made for PRRSV and influenza A coinfection.

Viral challenges are often associated with strong fever responses that are associated with a significant increase in energy requirements.[2] In this trial, as shown in Table 2, feed efficiency increased as compared to the previous phase by 33%, 80% and 31% for 150 ppm Cu as $Cu_2[OH]_3CL$, 12 ppm of Cu as $CuSO_4$ and 50 ppm Cu as $Cu(HMTBa)_2$, respectively. During the challenge period (147-176) CuSO4 had a 33% greater feed efficiency as compared to both the other supplemental copper levels. The results of this trial would indicate that higher levels of copper play a potentially important role in maintaining energetic efficiency in pigs during a viral disease challenge.

The second objective of this trial was to try and estimate the financial cost of a coinfection of PRRSV 1-4-2 and influenza A in late finishing pigs, and the key financial responses are summarized in Table 3. There was a trend for 12 ppm Cu as $CuSO_4$ to have a higher cost/lb than either 150 ppm Cu as $Cu_2[OH]_3CL$ or 50 ppm as Cu as $Cu(HMTBa)_2$. The period of day 147-176 represents the last 20 lb of gain from approximately 260 to 280 lb. The relative feed cost per pig using the determined cost/lb of gain from this trial standardized to 20 lb of gain are $8.72, $11.52 and $9.06 for 150 ppm Cu as $Cu_2[OH]_3CL$, 12 ppm of Cu as $CuSO_4$ and 50 ppm Cu as $Cu(HMTBa)_2$, respectively. Thus, during a late finishing health challenge, only supplementing basal levels of 12 ppm Cu in this trial increased feed cost per pig by $2.63. The treatment cost was also the lowest for 150 ppm Cu as $Cu_2[OH]_3CL$ and highest for 12 ppm of Cu as $CuSO_4$ by $9.70 per pen (or $0.32 per pig) with 50 ppm Cu as $Cu(HMTBa)_2$ being intermediate.

Mortality from day 147 to 176 was greatly affected by Cu level and source, with 50 ppm Cu as $Cu(HMTBa)_2$ being 3.35% lower than 12 ppm of Cu as $CuSO_4$ and 2.38% lower than 150 ppm Cu as $Cu_2[OH]_3CL$ ($P < 0.05$). Assuming a group of 1000 pigs the pig losses would equate to 34, 44 and 10 pigs lost for 150 ppm Cu as Cu2[OH]3CL, 12 ppm of Cu as $CuSO_4$ and 50 ppm Cu as $Cu(HMTBa)_2$, respectively. Taking an estimated $120 investment

Table 1: Average daily gain (lb/d) of pigs fed different sources and levels of Cu in late finishing prior to, and during exposure to PRRSV (1-4-2) and influenza A virus

Phase (d)	$Cu_2[OH]_3CL$	$CuSO_4$	$Cu(HMTBa)_2$	*P*-value
84-105	2.35	2.35	2.39	NS
105-126	2.06	2.05	2.04	NS
126-147	1.93	1.95	1.89	NS
147-176	1.16	0.88	1.03	0.16

Table 2: Feed efficiency of pigs fed different sources and levels of Cu in late finishing prior to, and during exposure to PRRSV (1-4-2) and influenza A virus

Phase (d)	$Cu^2[OH]^3CL$	$CuSO^4$	$Cu(HMTBa)_2$	*P*-value
84-105	2.37	2.40	2.41	NS
105-126	2.79	2.81	2.84	NS
126-147	3.27	3.22	3.31	NS
147-176	4.34[b]	5.79[a]	4.35[b]	< 0.05

[a,b] Superscripts differ at $P < 0.05$.

Table 3: Mortality, cost/lb, treatments and treatment cost during period from day 147-176 of the experiment

Parameter	$Cu_2[OH]_3CL$	$CuSO_4$	$Cu(HMTBa)_2$	*P*-value
Cost/lb of gain, $	0.436[y]	0.576[x]	0.453[y]	< 0.10
Percent mortality, %	3.38[a]	4.35[a]	1.00[b]	< 0.05
Pigs treated	7.1	9.5	8.7	NS
Treatment cost, $	15.1[y]	24.8[x]	20.7[x,y]	< 0.10

[x,y] Superscripts differ at $P > 0.10$ and $P < 0.05$.
[a,b] $P < 0.05$.

per animal to that phase, and an estimated loss in revenue of $130 per pig, the net margin cost to the overall group of pigs is $250 per pig that dies. Therefore, the reduction in net margin of all remaining pigs in group would be $8.8, $11.50 and $2.53 for 150 ppm Cu as $Cu_2[OH]_3CL$, 12 ppm of Cu as $CuSO_4$ and 50 ppm Cu as $Cu(HMTBa)_2$, respectively.

Implications

Further research is warranted to explore feeding high levels of chelated Cu to pigs and sows during times of PRRSV or influenza A challenges, and the conclusion from the data presented here is to use higher levels of organic Cu (in the form of 50 ppm Cu as $Cu(HMTBa)_2$) in PRRSV active flows to reduce the financial cost of the disease, and also to consider using higher levels of Cu during time of year where influenza A is present.

References

1. Holtkamp DJ et al. Assessment of the economic impact of porcine reproductive and respiratory syndrome virus on United States pork producers. *J Swine Health and Prod.* 2013;21:72-84.
2. Sandberg FB et al. The effects of pathogen challenges on the performance of naïve and immune animals: the problem of prediction. *Animal* 2007; 1:67-86.
3. Bonham M et al. The immune system as a physiological indicator of marginal copper status? *British J of Nutrition.* 2002;87:393-403.
4. Zhao J et al. Effects of a chelated copper as growth promoter on performance and carcass traits in pigs. *Asian Australas J Anim Sci.* 2014;27:965-973.

Theory and practice of cleaning and disinfection in swine biosecurity

J. A. Ramirez, MS, PhD
Virox Animal Health, Oakville, Ontario, Canada

Introduction

Much has been written about the chemical disinfection of premises as a key component of biosecurity in swine farming and transport. Many of the studies on chemical disinfection focus on the influence of practical in-use conditions on the efficacy of the disinfectant, namely; soil load (amount and type of soils), temperature, concentration and contact times. Much less has been written about other key aspects of cleaning and disinfection (C&D) that end up having a tremendous impact on the overall quality of hygiene outcomes (and therefore pig productivity metrics). Examples of these variables are proper cleaning prior to disinfection and proper equipment and technique in maximizing worker productivity and process efficiency. In this paper we relay our field experience over the past 18 months with a novel systems approach to cleaning and disinfection in swine operations. We challenge existing conventions with respect to C&D in swine operations that unnecessarily hamper productivity and efficiency, leveraging recent advances in disinfection technology.

Best practice in cleaning & disinfection - theory

The C&D process in swine production can be summarized in seven discrete steps:[1]

1. Dry cleaning

In general, this refers to ensuring gross soils (feces, bedding, feed, crusts, etc) are removed in order to decrease the soil load present for the wet-cleaning and/or disinfection steps. It typically involves the use of shovels and brushes to shear or scrape-off large particulate matter before the use of water. The dismantling of equipment and other components should be done here in preparation for the next steps.

2. Soaking with a diluted detergent cleaner

The next step is to soak all surfaces with a diluted detergent product - this step introduces chemical agents that are surface active that help dislodge soils attached to surfaces through detersive action. These agents further help solubilize and hold in suspension the soils so that they are readily removed with a water rinse. A foamed detergent phase is most effective as it allows for longer contact time on vertical surfaces, and makes it easy to visually discern application.

3. High-pressure cleaning

A high-pressure water rinse follows the detergent soak, pressure in the order of between 750-1,800 psi (50-120 atm) is preferred - this adds mechanical action (and temperature if hot water is used) to the chemical action of the detergent pre-soak to further loosen soils and enable their more effective removal.

4. Drying

Drying after cleaning is recommended to avoid dilution of the disinfectant by remaining rinse water.

5. Disinfection

Disinfectant at the proper concentration to mitigate the infectious agents of concern should be applied, preferably at a pressure about an order of magnitude lower than that of the high pressure rinse, anywhere between 60-200 psi. The key should be to ensure all surfaces are treated with the correct concentration of the disinfectant.

6. Drying

Drying is key given disinfectants are typically irritant or corrosive to skin and eyes and are orally toxic in the case of accidental residue ingestion.

7. Validation

Periodic validation of the C&D process through a structured visual and microbiological sampling audit helps identify areas for process improvement and training deficiencies.

Actual practice in cleaning & disinfection

It is interesting how much corner-cutting occurs in C&D in actual production systems. Since the impact of disease outbreak or subclinical infections due to herd exposure to pathogenic agents is well understood and documented, it is no surprise that awareness for the disinfection step is highest. However, the effects of pig exposure to non-infectious antigens (such as those present in an unclean environment with feces and dust) are also economically important as documented by van Heugten et al.,[2] amongst others. This in itself should hint at the importance of not just disinfection, but cleaning as an effective means of reducing background pressure from non-infectious antigens present in soils, dust and feces. Cargil & Benhazi[3] showed that there was a 14.4% improvement in average daily weight gain for weaners and 8.1% for finishers when the premises were cleaned+disinfected vs. disinfected alone.

In our experience with large scale top 40 powerhouse integrators in the United States, very few sites observe the recommended

C&D process described above. Most farms that have slatted floors do not perform the initial gross-soil-removal, dry cleaning step; reasoning that the slots on the floor prevent the accumulation of gross soils. Adherence to the dry-cleaning step is universal in the C&D of hog trailers, where excess amounts of bedding and feces accumulate with each transport. Detergent soaking is much less prevalent, with an estimated rate of compliance of about 40%. Many producers and transporters rely on high pressure or high volume water treatment as the first step in the C&D process. Many view presoaking with a detergent as an extra step that is not really necessary if the power washing step is properly performed.

After the power-washing step, compliance with the drying step prior to disinfectant application is virtually zero. Drying is a time-consuming step whereby the premises, equipment and workers remain idle and hence is widely viewed as a process in inefficiency that can be easily avoided. The result is that disinfectant is applied unto surfaces that are still wet, in many cases with substantial pools of rinse water that naturally dilute the disinfectant, possibly affecting its efficacy. It is instructive to review the dependency of disinfection efficacy on concentration. Since the pioneering papers by Chick[4] and Watson[5], it is widely accepted that for a given level of reduction in the viable population of an infectious agent, the dependency between contact time and disinfectant concentration is given by an exponential expression of the type:[6]

$$\frac{t_2}{t_1} = \left[\frac{C_1}{C_2}\right]^{\eta}$$

Where η is the concentration exponent or dilution factor, and varies with the type of disinfectant.[6,7] It is clear from this expression that the higher the concentration exponent; η, for a given disinfectant, the more dramatic the dilution effect will be. For quaternary ammonium compounds (a commonly used disinfectant referred to as a quat), the concentration exponent reported[7] is between 0.8 and 2.5. For another commonly used disinfectant, glutaraldehyde, the reported[6] exponent is 1.0. This means that in the worst case, for a quat disinfectant, dilution by a pool of standing rinse water to 50% of target concentration would require a contact time of almost 1 hour (versus the label-specified 10 minutes) to ensure disinfection. For the case of the glutaraldehyde, while not as dramatic, a 50% dilution still requires a contact time of 20 minutes; that is, 2 times that specified on the label.

The application of disinfectant is almost exclusively done following a prescribed "dosage" of disinfectant based on a specified rate of application, typically about 1 L for every 4 m^2 or about 1 gal per 150 sqft.[1] This is acceptable as long as it can be assured that the in-use concentration being dispensed is the correct one for efficacy. In the author's experience, virtually no farm manager or operating crew properly understand how in-line chemical injection systems work, and as a consequence, farms are operating with incorrect disinfectant concentrations. The prevalence in the field of manual-adjustable in-line injectors and improperly selected dilution tips on fixed injectors virtually guarantee this. The gravity of the situation is exacerbated considering the discussion on dilution effects just discussed above. The second obvious, yet unaddressed issue is the difficulty of "eyeballing" a 1 gal per 150 sqft application under the pressures of limited time and the uncomfortable working conditions in a barn.

Under the time pressures in a farrowing environment, sows are moved in many times before the environment is fully dry. Less time-sensitive operations like nurseries and finishing barns typically can afford several hours to let the environment dry. Transport trailers that are subject to "baking" dry naturally during that process, and even those that are not "baked" generally dry on their way to animal pick-up due to forced convection currents from the highway drive. Unfortunately, because of the nature of the farrowing process, it is there where the risk of harm to the animals from residual disinfectant is greatest. The sow has to lie on her side for considerable time in the crate. The newborns have very sensitive skin and because of their lower body mass are more vulnerable to oral ingestion of residual chemical. Hence, it is most critical that the proper disinfectant drying time be allotted for in the farrowing rooms.

In the author's experience, some farms and a majority of transport operators perform some type of visual inspection after C&D, many have experimented generally unsuccessfully with ATP bioluminescence as a validation technique for cleaning, while others have used RT-PCR selectively to detect surface presence of key viruses like PEDV or PRRS.

Improvements to recommended and actual practice - field experience

In light of the preceding gaps in C&D practice in the field, the following measures have been proposed by the author and adopted by a growing number of farms and transport operators in North America:

Adoption of a detergent soak step

The use of a detergent ultimately improves crew efficiency in the overall washing process. Hurnick[8] has shown that the use of a presoak detergent will reduce the overall wash time for a finishing pen by 46.5% if cold water is used. Even when presoak and cold water rinse is used, it still takes 30% less time to clean the barn as compared to hot water pressure washing alone. The reason is that it is more difficult to remove soils by using mechanical action (pressure) alone, the detersive and emulsifying action of a detergent allow the user to power-wash the area in a continuous mode without constantly having to stop to dislodge recalcitrant soils.

Furthermore, the use of a detergent actually can reduce the infectious agent load by as a much as 2 logarithms (99%) as has been widely reported in the literature[9,10] as well as help prevent the formation of biolfilms[9]. Improved cleaning by using a detergent, and preferably hot water in the power rinse will also reduce or eliminate the potential for interference with the disinfectant. The quenching effects of soil on the activity of disinfectants has been widely documented and studied.[11]

It is important to execute proper technique in the application and rinsing of a presoak detergent. The technique of foam-up/rinse-down should be followed. Foaming from the bottom in an upwards direction helps avoid overuse/wastage of detergent and ensures all spots are treated, as detergent rundown when foaming from the top prevents clear demarcation of treated vs. non-treated areas. Rinsing from the top in a downward direction (rinse-down) reduces water consumption by taking advantage of rinse water downward-runoff from higher elevations.

An additional benefit of a detergent presoak is gained if a detergent product with biocidal properties - a cleaner/disinfectant- is employed in the detergent presoak step. In this case, the reductions in infectious agent described above are further supplemented by additional reduction by action of the detergent-disinfectant product itself. Products that include label instructions for use as a one-step cleaner/disinfectant are generally suitable for use as an effective biocidal-detergent presoak.

Mitigating dilution effects when applying disinfectant to wet surfaces

The effect of dilution on the disinfectant was explained in terms of the concentration exponent, η, of the disinfectant. Clearly, in order to minimize dilution effects, disinfectants with low η are preferred. The η coefficients for different biocidal actives are presented in Table 1.[6]

Disinfectants that are more tolerant to dilution effects (e.g. low η) are preferred whenever there is little time to allow for drying after the power wash rinse. Hydrogen peroxide-based disinfectants, with an η of lower than 1 are highly preferred.

Application of disinfectant - overcoming the key failure mode

As discussed above, the disinfectant application method is perhaps the weakest point in the overall C&D process. Too much room for human judgement (and error) exists, which is exacerbated by the generally poor training and high turnover of personnel in cleaning crews. The following measures however, have been adopted successfully over the past 18 months by many forward-thinking production and transport companies:

Table 1: Concentration exponent, η, of common disinfectants

Biocide	η
Hydrogen peroxide	0.5
Glutaraldehyde	1.0
Quaternary Ammoniums (Quats)	0.8-2.5
Phenolics	4.0-9.9
Aliphatic alcohols	6.0-12.7

1. Adoption of adequate in-line dilution equipment. Venturi chemical injectors fitted with the proper internal dilution orifice tip (which makes it inaccessible to the operator) should be employed. The injector should be properly calibrated (this means that it has been fitted with the correct dilution orifice tip internally) by the manager or a designated individual with the proper training. A chemical disinfectant whose in-use concentration can be measured by a simple means, such as colored reactive paper strips, is preferred.
2. Use of a foaming gun/nozzle and a chemical disinfectant capable of producing high-cling, viscous foam. There are several reasons that make foamed application of the disinfectant desirable. Foam allows more efficient coverage of the surfaces with less dispensed volume of liquid, due to the higher specific volume of the foam phase vs. a liquid. Secondly, the lower density of a foam phase results in lower gravity drag force from vertical and inverted surfaces, allowing for longer time of contact for disinfection. Third, the foam phase itself, having a high interfacial concentration of surfactant, hampers the evaporation rate of the disinfectant through the well known *monolayer resistance* effect described in colloid science.[12] This again results in extended contact time for disinfection.
3. A more immediate and intuitive benefit comes from the obvious visual cue for coverage that comes from a foamed solution. The benefits of switching from a clear visual judgement of surface coverage instead of an estimate of coverage from clear, non-foaming liquid disinfectants such as those prevalent in farm and trailer disinfection today are quite dramatic. We have measured the difference in different environments in the field and present them in Table 2. The better coverage provided by the foam phase and the better visual indication of coverage results in a 76% and 7-fold reduction in application times for farrowing and finishing barns respectively, with the reduced product wastage resulting in an overall reduction of cost-in-use of disinfectant of 45% and 38% respectively, in spite of the lower in-use dilution as compared to prevalent disinfectants in use today.

Leveraging additional efficiencies by choice of disinfectant

A powerful tool for potentiating the biocidal action of a disinfectant is utilization at higher temperatures. As demonstrated long ago by Chick,[4] the biocidal process follows first order kinetics on the concentration of target microorganisms, n, as;

$$-\frac{dn}{dt} = Kn$$

where, n is the number of viable cells per unit volume, K is a kinetic constant and t is time. Given the kinetic constant, K, is described by the same Arrhenius type expression used for chemical reactions;[4] it is not difficult to arrive at the following relationship describing the variation of biocidal effect with temperature:[13]

$$\frac{t_1}{t_2} = \Theta^{(T_2 - T_1)}$$

Table 2: Field data for current practice with prevalent quat-glutaraldehyde or phenolic disinfectant with existing application equipment/practice vs. calibrated in-line dilution equipment, high foaming gun/wand and high foaming hydrogen peroxide-based cleaner-disinfectant

	Farrowing farm		Finishing barn	
	Current products used @ 1:256	**Intervention® used @ 1:64**	**Current products used @ 1:256**	**Intervention® used @ 1:64**
Coverage	4.6 gal per crate	0.64 gal per crate	1 gal = 30 sqft	1 gal = 281 sqft
Product use	2.3 oz. per crate	1.3 oz per crate	1 gal : 8,000 ft²	1 gal :17,980 ft²
Cost	$0.56 per crate	$0.30 per crate	$0.39/100 ft²	$0.17/100 ft²
Time	55 s per crate	13 s per crate	39 s per 100 ft²	4-5 s per 100 ft²

where, t_1 and t_2 are the times for disinfection to a given level, T_1 and T_2 are the corresponding temperatures and Θ is a chemical-biocide dependent constant.[13] It is customary to express Θ, in terms of its value over a 10° difference, the Θ_{10}; in which case it varies between 1.5 for aldehydes, to up to 50 for aliphatic alcohols. The Θ_{10} for a formulated disinfectant based on hydrogen peroxide and anionic surfactants such as Intervention® is around ~2;[14] that is, the biocidal activity can be doubled with every 10° C increase in temperature. Thus, using dilution water at 30° C (86° F) could reduce the contact time from 5 min to 2.5 min, while at 40°C (104°F) it could be just over 1 minute.

Of course, this applies to other disinfectants as well. For example, a mixture of quat (Θ_{10} = 3.5) and glutaraldehyde (Θ_{10} = 1.5) would likely have a Θ_{10} ~ 2. The problem here lies with the intrinsic respiratory toxicity and skin and eye corrosivity for these chemicals, which is exacerbated by higher temperatures by the same underlying principle that accelerates biocidal activity. Thus, in practice, utilization of these chemicals at elevated temperature is not only difficult, but quite dangerous as well.

Another area where process efficiencies can be gained in C&D is in the drying period after disinfection. It was mentioned above that the reason for this drying period is to avoid exposure of the animals to the biocidal chemical agents - which in most cases are irritants or corrosive to skin and/or eyes, or respiratory tract, and/or orally toxic. Naturally, one way of avoiding the down time associated with disinfectant drying is to use disinfectant agents that are not toxic; particularly agents that are formally classified as non-irritant or corrosive to skin, eyes or the respiratory tract and are not considered orally toxic.

Conclusion

Our field experience with many large integrators across North America has revealed a number of deficiencies in the C&D process that can be easily corrected by proper attention to equipment selection and improved training. Understanding and refreshing some fundamental concepts related to C&D helps the individual responsible for animal health or biosecurity implement certain trivial improvements to the C&D program. This knowledge couple with new, recently available chemical technologies can dramatically improve process flows, reduce cost and improve hygiene and productivity outcomes and metrics.

References

1. Dewulf, J. and F. van Immerseel, Biosecurity in animal production and veterinary medicine, Acco Leuven, Den Haag, 2018.
2. Van Heugten, E., Spears, J. W. and M. T. Coffey, The effect of dietary protein on performance and immune system response in weakling pigs subjected to an inflammatory challenge, J. Anim. Sci., 72, 2661:2669.
3. Cargil, C. and T. Benhazi, The Importance of cleaning in All IN ALL OUT management systems, 1998 Proceedings 15th IPVS Congress, Birmingham, England, July 5-9, (3), 3.
4. Chick, H. An Investigation into the Laws of Disinfection, J. Hyg., 1908, 8(1), 92:158.
5. Watson, H. E., A note on the variation of the rate of disinfection with the change in the concentration of the disinfectant, J. Hyg., 1908, 8(4):536:542.
6. Maillard, J.Y., Testing the effectiveness of disinfectants and sanitizers, Handbook of Hygiene Control in the Food Industry, Woodhead Pub. Series in Food Science, Technology & Nutrition, 2005, 641:671.
7. Reichart, O. Modeling the destruction of Escherichia coli on the base of reaction kinetics, Int. J. food Microb., 1994, 23,449:465.
8. Hurnik, D. Investigation into optimal washing and disinfection techniques for pig pens, 2005, Proceedings of the 5th London Swine Conference, London, Ontario, April 6-7.
9. Fotheringham, V.J.C., Disinfection of livestock production premises, Rev. Sci Tech. Off. int. Epiz., 1995, 14(1), 191:205.
10. Maertens, H., De Reu, K., Van Weyenberg, S., Van Coillie, E., Meyer, E., Van Immerseel, F., Vandenbroucke, V., Vanrobaeys, M. and J. Dewulf, Evaluation of the hygienogram scores and related data obtained after cleaning and disinfecting poultry houses in Flanders during the period 2007 to 2014, Poultry Sci., 2017, 97, 620:627.
11. Lambert, R.J.W. and M.D. Johnson, The effect of interfering substances on the disinfection process: a mathematical model, J. Applied Microb., 2001, 91, 548:555.
12. Farajzadeh, R., Kristen, R. and P.L.J. Zitha, Foam film permeability: Theory and experiment, Advanc. in Colloid & Interf. Sci., 2008, 137, 27:44.
13. Bean, H. S. (1967) Types and characteristics of disinfectants, J. Appl., Bact., 1967, 30(1), 6:16.
14. Personal Communication, 2002, Virox Technologies Inc. Oakville, ON, Canada.

Field experiences using Circo/MycoGard® in the US and beyond

F. Chamba[1], DVM, MS, PhD, SHM-EVP; C. Francisco[1], DVM, MS, SHM-EVP; C. Smith[1], DVM; A. Hintz[1], DVM; D. Rosener[1], DVM; P. Canning[2], DVM, PhD; G. Greaves[2], DVM, MS
[1]Pharmgate Animal Health, Wilmington, North Carolina;
[2]South West Ontario Veterinary Services, Stratford, Ontario, Canada

Introduction

Circo/MycoGard® is the recently introduced PCV2/Mhp combination vaccine from Pharmgate Animal Health. Circo/MycoGard® is a PCV2b capsid (ORF2) vaccine produced as virus-like particles (VLP) in a baculovirus expression system. Circo/MycoGard® has been in the market since 2017 and several studies have been completed in the field. Currently, small-scale experimental studies cannot be used to infer vaccine efficacy in the field given the variability of exposures and infection dynamics (infection pressure) among flows as well as differences in environment, facilities and management conditions. Moreover, the different pathogens (PRRSV, Mhp, PCV2, IAV and endemic bacteria) co-infecting pigs in the same populations makes it more critical to conduct vaccine comparative studies under field situations. This way, results and conclusions can be inferable to similar flows or populations across the US industry as well as in other countries. Pharmgate is driven to help farmers produce more pork with fewer resources and secure pork supply for the near future. In this matter, Pharmgate is driven to conduct field vaccine research so we can provide applicable information to our customers in the US and worldwide. In this publication, we summarized all the recent studies that have been conducted with Circo/MycoGard®.

Materials and methods

A total of 7 field studies (9,949 pigs) have been conducted so far and summarized in this manuscript. All studies were conducted in commercial and conventional grow-finish facilities in different production companies and locations in the US (studies 1-6) and Canada (study 7). All studies have been conducted between 2017 and 2019. In addition, almost all studies were done comparing current commercially available PCV2/Mhp combination vaccines except study 4. All vaccines were administered as single-dose vaccines around weaning age. Studies 1 and 2 were field challenge studies and the rest were field natural exposure/infection studies for both PCV2 and Mhp. Pigs, pens or closeouts (barns) were the experimental units and randomized according to each study as described in Table 1. Study 7 was a flow-based design and study 3 was a pen-based approach.

Furthermore, study 2 had PRRSV as a co-challenge pathogen. Challenge material for both pathogens were administered individually through IM and/or IN routes in studies 1 and 2. Challenge materials consist of culture supernatants (PCV2 and PRRSV) or lung homogenates (Mhp). In study 2, pigs were challenge ~4 weeks after vaccination/weaning, and 7 (Mhp) and 11 (PCV2d) weeks after vaccination/weaning in study 1. Almost all studies had a duration from wean-to-market except studies 5 and 6. Finally, all studies measured key performance indicators such as survivability and ADG. Studies 3 and 7 also included gross margin (USD) and live weight (lb) per ton of feed, and FCR as key production indicators. Study design details are summarized in Table 1.

Data analysis was performed using R statistical software.[1] Statistical analysis was adjusted by start weight, sex, block (barn location in the pen-based study only) and days-on-feed (flow-based study only) effects. Gross margin (USD) and live weight (lb) per ton of feed, survivability (dead or alive), FCR and ADG were the outcomes for logistic or linear generalized linear models in the R package lme4.[2] Vaccine group was the main predictor in those models. Means were estimated using the R package lsmeans.[3] Gross margin (USD) per ton of feed was calculated for each pen (study 3) or closeout (study 7) based on a feed price of 300 and 263 USD per ton in studies 3 and 7 respectively; and 0.60 and 0.76 USD per pound of live hog weight in studies 3 and 7 accordingly.

Results

Tables 2-4 summarized all the performance indicators evaluated in 9,949 pigs from the 7 field studies. Overall, key production indicators such as gross margin (USD) and live weight (lb) per ton of feed, and FCR suggested that Circo/MycoGard® performed similarly to the other PCV2/Mhp vaccines used in those studies (Table 2). In study 7, Vac C was the control group and in study 3, non-vaccinated controls performed similarly to vaccinated groups. Additionally, Circo/MycoGard® pig survivability was similar to the other vaccines except in study 3 where Circo/MycoGard® pig survivability was suboptimal compared to other vaccines and better than the non-vaccinated controls (Table 3). Depending on the severity of the infections, non-vaccinated control pig survivability was similar or worse than vaccinated controls. PRRSV co-infection increased the severity of PCV2/Mhp infections as appreciated in the lower pig survivability in the non-vaccinated controls in studies 2 and 4 (Table 3).

Overall, Circo/MycoGard® ADG was similar to the other vaccines except in study 3 where Circo/MycoGard® ADG was

Table 1: Characteristics of field Circo/MycoGard® studies

Characteristic	Study 1	Study 2	Study 3	Study 4	Study 5	Study 6	Study 7
Exposure	Challenge	Challenge	Natural	Natural	Natural	Natural	Natural
Pathogens	PCV2d Mhp232	PCV2d PRRSV184	PCV2 Mhp	PCV2d Mhp PRRSV	PCV2 Mhp	PCV2d	PCV2 Mhp
No. units	128 pigs	2500 pigs	40 pens 880 pigs	160 pigs	100 pigs	150 pigs	10 closeouts 6031 pigs
Randomization	Weight, litter, pen	pen	barn	pen	pen	pen	flow
Vac. age, days	21	21	20	10	21	21	33
Duration, days	141	132	165	159	57	28	175
Facility	WTF	WTF	WTF	NF	N	N	NF
Treatments	CMG Control Vacc A Vacc B	CMG Control Vacc A Vacc B Vacc C	CMG Control Vacc A Vacc B	CMG Control	CMG Control Vacc A Vacc B	CMG Control Vacc A Vacc B Vacc C	CMG Vac C
Outcomes	Survivability ADG	Survivability ADG	GM/TF LW/TF Survivability FCR ADG	Survivability ADG	Survivability ADG	Survivability ADG	GM/TF LW/TF Survivability FCR ADG

GM/TF = Gross margin (USD) / ton of feed
LW/TF = Live weight (lb) / ton of feed

Table 2: Pig feed efficiency in field Circo/MycoGard® studies

Study	Indicator	Circo/MycoGard®	Non-vaccinated	Vac A	Vac B	Vac C	*P* - value
3	Gross margin (USD) / ton feed	147	152	146	143	-	0.23
	Live weight (lb) / ton feed	744	753	744	738	-	0.23
	FCR	2.85	2.82	2.85	2.86	-	0.57
7	Gross margin (USD) / ton feed	115	-	-	-	111	0.66
	Live weight (lb) / ton feed	879	-	-	-	869	0.66
	FCR	2.65	-	-	-	2.67	0.73

suboptimal compared to the other vaccines and better than the non-vaccinated controls (Table 4). Depending on the severity of the infections, ADG in non-vaccinated controls was similar or worse than vaccinated pigs. Similar to the pig survivability results, ADG of non-vaccinated controls was worse than vaccinated pigs in studies with PRRSV co-infections. Circo/MycoGard® performed similarly to other vaccines in multiple field studies with different infection dynamics, designs and conditions.

Discussion and conclusions

Field studies are critically important for vaccine evaluations and applicability of those results across similar populations or flows in the industry. Vaccines are key for producers and veterinarians to improve control and eradication methods of swine pathogens and hence, secure pork supply for the near future. In this publication, we summarized all Circo/MycoGard® field studies that Pharmgate has conducted so far and concluded that overall, Circo/MycoGard® performed similarly to other vaccines in several studies with different infection dynamics, designs and flow characteristics.

Given the different approaches in study designs, our results also highlighted the importance of a more comprehensive understanding and interpretation of vaccine field study results. Variation of infection dynamics among groups within the same

Table 3: Pig survivability in field Circo/MycoGard® studies

Study	Circo/MycoGard®	Non-vaccinated	Vac A	Vac B	Vac C	*P* - value
1	90.8	96.9	90.9	90.9	-	0.68
2	85.4[b]	56.7[c]	95.4[a]	95.3[a]	94.1[a]	<.01
3	80.5	78.2	81.1	85.7	-	0.98
4	87.3[a]	64.3[b]	-	-	-	<.01
5	93.2	96.2	96.8	96.7	-	0.90
6	97.6	100	100	100	100	0.49
7	92.5	-	-	-	94.1	0.93

a,b,c Superscripts within a row differ statistically ($P < 0.05$)

Table 4: Pig daily weight gain in field Circo/MycoGard® studies

Study	Circo/MycoGard®	Non-vaccinated	Vac A	Vac B	Vac C	*P* - value
1	1.72	1.70	1.71	1.71	-	0.89
2	1.75[b]	1.64[c]	1.81[a]	1.80[a]	1.80[a]	<.01
3	1.32	1.31	1.29	1.33	-	0.11
4	1.33[a]	1.23[b]	-	-	-	0.03
5	1.63	1.62	1.52	1.66	-	0.19
6	0.72	0.72	0.73	0.74	0.74	0.97
7	1.68	-	-	-	1.71	0.32

a,b,c Superscripts within a row differ statistically ($P < 0.05$)

flow affects the outcomes of vaccine field trials. In natural infection studies, generally non-vaccinated controls performed similarly to vaccinated groups unless there are severe PCV2, Mhp and PRRSV co-infections. Given the variabilities on infection dynamics within and between flows, we should not expect non-vaccinated controls to perform worse than vaccinated groups in flows where infections are mild or stable (high health flows). Perhaps, a better approach in those high health flows is to design trials where the current vaccine is the positive vaccinated control, and the test group is the new vaccine being tested. In one of our studies, we showed that this type of studies can be executed on a flow-based where one weaned group is vaccinated with one vaccine and the next weekly weaned group is vaccinated with other vaccine. This weekly intercalating enrollment continues until the desired number of groups (closeouts) is reached. This new epidemiological approach allows to adjust better for the variability of infections over time.

On the other hand, given that non-vaccinated controls do not perform worse than vaccinated groups in most of natural infection vaccine field studies, administering challenge materials to produce severe disease is common. Challenge studies are a repeatable creation of severe disease that allows a complete, consistent and adequate vaccine evaluation, especially when PRRSV is administered. However, there can be ethical and pig welfare concerns when conducting these types of studies in commercial facilities with few biocontainment measures. Additionally, morbidities and mortalities in challenge studies are higher than what is generally observed in the field. In sum, study design, herd immunity and infection dynamics are critical aspects to consider when designing and interpreting PCV2/Mhp vaccine field studies.

References

1. R: A language and environment for statistical computing v. 3.4.2 (R Foundation for Statistical Computing, Vienna, Austria, 2017).

2. Bates, D., Mächler, M., Bolker, B. & Walker, S. Fitting linear mixed-effects models using lme4. *arXiv preprint arXiv:1406.5823* (2014).

3. Lenth, R. V. Least-Squares Means: The R Package lsmeans. *Journal of Statistical Software* **69**, doi:10.18637/jss.v069.i01 (2016).

Controlling an outbreak of *Mycoplasma hyopneumoniae* and reducing vertical shedding in a naive gilt population using Aivlosin® 17% (tylvalosin) Medicated Premix

T. Faulkner[1], MS P. Canning[1], DVM, PhD; J. Mora[2], DVM, PhD; G. Wideman[1], DVM; D. MacDougald[1], DVM; B. Maxwell[3], F Chamba[3], DVM, MS, PhD, SHM-EVP.
[1]South West Ontario Veterinary Services, Stratford, ON, Canada;
[2]ECO Animal Health Ltd., London, UK; [3]Pharmgate Animal Health, Wilmington, NC, USA

Introduction

Mycoplasma hyopneumoniae (Mhp) is known to be a primary agent causing pneumonia in swine, and it is known to have an adverse economic impact on swine production by causing respiratory disease, delayed growth and anorexia.[1,2] In addition, infection with Mhp can cause an increased susceptibility to secondary infections leading to other complications within the herd.[3] Controlling disease caused by Mhp can be accomplished by strict management practices such as the use of vaccination programs and medications. This case study examined the use of Aivlosin® 17% Medicated Premix for the control of an Mhp outbreak in a sow farm of 2200 females. In the spring of 2018, a farrow-to-wean farm was undergoing a repopulation when the gilt multiplier broke with Mhp. In July 2018, a diagnostic investigation revealed that the received gilts were infected with Mhp and experiencing clinical signs. The growing pigs in this system are raised from birth without the use of antibiotics, making the impact of operating as a Mhp positive unstable site very costly, especially in finishing. Therefore, a treatment plan was used to reduce shedding of the pathogen from the sow herd to weaned piglets and to pursue an Mhp elimination.* The treatment plan involved the implementation of Aivlosin® 17% Medicated Premix (ECO Animal Health Ltd.) for gestating and lactating gilts. The goal of this case study was to gather information regarding the prevalence of Mhp in the gilt population and to assess the efficacy of the control program in preventing clinical disease and shedding into the nursery and finisher flows.

Materials and methods

Five cohorts of gilts were sampled to estimate Mhp prevalence. The gilts in each of these groups were sampled using tracheobronchial swabs (TBS) pre- and post-treatment plan implementation. Between 30 and 43 gilts were sampled pre-treatment and the same gilts were sampled again post-treatment. The treatment plan involved administering Aivlosin® 17% Medicated Premix to gilts at a dose of 5mg tylvalosin/kg BW/day for 7 days pre-farrow in gestation housing, and for another 7 days on entry into the farrowing rooms (Treatment A). A one-off whole herd (gilts only) pulse of Aivlosin® 17% Medicated Premix was administered at a dose of 5mg tylvalosin/kg BW/day for 10 days (Treatment B). Cohort 1A and Cohort 2A received Treatment A; Cohort 3AB, Cohort 4AB and Cohort 5AB were exposed to treatment B at the same time and after that, they received treatment A as they came due to farrow. Each cohort was part of one weekly breeding group and about 30 days passed between the enrolment of each cohort. Samples were sent to Animal Health Laboratory (Guelph, Ontario) where Mhp PCR testing on individual swabs was performed.

Concurrent to the gilt treatment, once per month, one group of weaned pigs from the sow farm was sampled in the nursery and tracked throughout finishing for coughing score and prevalence of Mhp. At each nursery site, 100 TBS were collected within 7 days of entry and tested for Mhp using PCR in pools of 3. Oral fluids were also taken at the end of the nursery period and tested for Mhp by PCR. Between 4 and 16 oral fluid samples were taken at each site depending on the size of the nursery. At the corresponding finishing sites, coughing scores and oral fluids were collected at 4, 8, and 12 weeks post-placement for Mhp PCR. Between 5 and 14 oral fluid samples were collected at each of the three sampling periods at each finishing site. Coughing indexes were calculated using the Boehringer Ingelheim Cough Index Calculator application. Blood samples and TBS were not collected in the finishing phase. All TBS and oral fluid samples were submitted to the Veterinary Diagnostic Laboratory at the Iowa State University for Mhp PCR.

Results and discussion

For each cohort of gilts, the percentage of Mhp PCR positive gilts were estimated pre- and post-administration of the Aivlosin® treatment. The percentage of gilts positive for Mhp within each cohort were as follows: Cohort 1A: 53% positive pre-treatment & 22% positive post-treatment, Cohort 2A: 57% positive pre-treatment & 15% positive post-treatment, Cohort 3AB: 26% positive pre-treatment & 1% positive post treatment, Cohort 4AB: 1% positive pre-treatment & 0% positive post-treatment, Cohort 5AB: 12% positive pre-treatment & 0% positive post-treatment. The ct values of the gilts in the first three cohorts before treatment ranged from 32.9 – 40.0. After treatment, the ct values ranged from 33.59 – 40.00. The last two cohorts did not have any gilts that tested positive for Mhp post-treatment.

* Sow herd medication is allowed under this program

Apart from the first nursery group, which tested 43% of TBS Mhp PCR positive at ~7 days post-placement, all subsequent nursery groups sampled with TBS tested PCR negative for Mhp (Table 1). Of the 7 finishing sites that had oral fluids (OF) tested for Mhp by PCR, three sites were consistently negative at all three testing time points. The first finishing site was the only site that tested positive at all three sampling time points. However, the last sampling of the first site showed a low percentage of oral fluids that tested Mhp PCR positive. The last two finishing sites sampled were negative at all sampling time points (Table 1). Cough score indexes were very low at each finishing site, which indicated an absence of respiratory complications. The implementation of the Mhp control program was an effective way to reduce shedding from sows to piglets. This was visible through the reduction of Mhp positive gilts post- Aivlosin® treatment as well as through a decline in Mhp prevalence at nursery and finishing sites based on the TBS and oral fluid sampling that we did. Finally, given that we did not collect any serum and/or other swabs such as TBS from finishing pigs to further confirm low Mhp prevalence or elimination, our results should be interpreted cautiously, especially when extrapolating our medication program results to a Mhp elimination strategy.

® Registered trademark of ECO Animal Health Ltd., London, United Kingdom

References

1. Maes, D., Segales, J., Meyns, T., Sibila, M., Pieters, M., & Haesebrouck, F. (2008). Control of *Mycoplasma hyopneumoniae* infections in pigs. *Veterinary microbiology*, *126*(4), 297-309.

2. Villarreal, Iris, et al. "Effect of vaccination of pigs against experimental infection with high and low virulence *Mycoplasma hyopneumoniae* strains." *Vaccine* 29.9 (2011): 1731-1735.

3. Thacker, Eileen L., et al. "*Mycoplasma hyopneumoniae* potentiation of porcine reproductive and respiratory syndrome virus-induced pneumonia." Journal of clinical microbiology 37.3 (1999): 620-627.

Table 1: *Mycoplasma hyopneumoniae* prevalence results from nine weaned groups from the affected sow farm.* Group 1 was enrolled after the start of Treatment A of the gilts at the sow farm.

	Nursery period		Finisher period					
			4 weeks		8 weeks		12 weeks	
Groups	% of TBS Mhp PCR positive at < 7 days post-placement in nursery†	% of OF Mhp PCR positive at 6 weeks post-placement in nursery‡	% of OF Mhp PCR positive	Cough index¶	% of OF Mhp PCR positive	Cough index	% of OF Mhp PCR positive	Cough index
Group 1	43	NA	-	-	-	-	-	-
Group 2	0	6	50	0.34	80	1.24	20	0.18
Group 3	NA	0	0	0	0	-	0	0.04
Group 4	NA	0	0	0.14	30	0.04	50	0.20
Group 5	0	50	0	-	40	0.62	0	-
Group 6	NA	0	NA	NA	NA	NA	NA	NA
Group 7	0	0	0	0.08	0	-	60	0.38
Group 8	0	0	-	-	0	0.08	0	-
Group 9	NA	0	-	-	0	0.02	0	0.06

* Once per month, one group of weaned pigs from the affected sow farm were sampled and subsequently tracked throughout finishing for cough score and prevalence of Mhp. At each nursery site, 100 tracheobronchial swabs were collected within 7 days upon entry into the nursery and tested for Mhp using PCR in pools of 3. Oral fluids were also taken at the end of the nursery period and tested for Mhp with PCR. At the corresponding finishing sites, cough scores and oral fluids were collected at 4, 8, and 12 weeks into the finisher. MHP PCR testing was completed on the oral fluids.

† % of TBS Mhp PCR Positive = (the number of Mhp positive pools ÷ total number of pools tested) x 100.

‡ % of OF Mhp PCR Positive = (the number of Mhp positive oral fluids ÷ total number of oral fluids tested) x 100.

¶ Coughing was measured at the finishing sites using the Boehringer Ingelheim Cough Index app at 4, 8, and 12 weeks. A CI greater than 2.5 indicates respiratory disease in the room tested.

\- No data collected at this time point.

NA No data was required to be collected at this time point.

Mhp = Mycoplasma hyopneumoniae; PCR = polymerase chain reaction; TBS = tracheobronchial swab; OF = oral fluid; CI = cough index

Evaluation of the response to PRRSGard® administration in weaned pigs

C. Smith[1], DVM; F. Chamba[1], DVM, MS, PhD, SHM-EVP; J. Pittman[2], DVM, MS, DABVP; G. Rawal[3], DVM, MS; J. Zhang[3], MD, MS, PhD; C. Francisco[1], DVM, MS, SHM-EVP

[1]Pharmgate Animal Health, Wilmington, North Carolina; [2] Smithfield Hog Production-North Region, Waverly, Virginia; [3]Iowa State University, Ames, Iowa

Introduction

Porcine reproductive and respiratory syndrome virus (PRRSV) continues to cause significant economic and performance losses in swine production in the United States.[1] Multiple practices have been employed in attempts to control the disease, including modified-live virus (MLV) vaccines to reduce clinical signs, viremia and lung lesions, which improve health and performance in pigs. There are several different PRRSV MLV products commercially available to producers. The choice between products may be influenced by cost, efficacy, decreased performance or setback directly following vaccination (drag), and the ability of the vaccine to spread to non-target or non-vaccinated pig populations.

PRRSGard® is a unique PRRSV MLV vaccine with a chimeric virus composed from a proprietary, highly attenuated backbone and structural proteins from a highly virulent isolate (MN184) within ORF5 lineage 1.[2] Additionally, a 23-nucleotide insert has been positioned between ORF1b and ORF2 to be used as a genetic marker for differential rRT-PCR diagnostic testing. The objective of this report is to summarize the immunological response, shedding profile, and production performance drag of weaned pigs vaccinated with PRRSGard®.

Materials and methods

Weaned pigs approximately 3 weeks old and known to be naïve for PRRSV were sourced from multiple sow farms. Eight hundred fifty-three mixed breed pigs were sorted by sow farm source and randomly allocated into 35 of 44 total pens in a commercial, tunnel ventilated, wean-to-finish barn following the farm's standard operating procedures. The day after placement, pens were assigned to various trials based upon location within the barn and air flow in order to limit potential vaccine virus spread to other trial pigs. The assignments are illustrated in Figure 1. Pigs in pens not associated with a trial were not vaccinated with PRRSGard® or sham vaccine. The farm's standard operating procedures included a "pull program" which involved leaving several hospital pens empty at placement with multiple fixed time identification and removal points of the slowest growing pigs throughout the barn. These pigs were used to populate the hospital pens. Any pigs removed from their original pen and placed in hospital pens were immediately removed from their respective trials and no further information was collected.

Trial 1: Performance drag

Twenty-eight total pens were selected and assigned to either vaccinated or sham-vaccinated control groups. All pigs within each pen were individually identified by ear-tag, weighed, and vaccinated with the appropriate product. The vaccinated group received 1ml of PRRSGard® intramuscularly and the sham-vaccinated control group received 1ml of vaccine diluent intramuscularly. Six oral fluid samples were taken on day 0 and 48 post-vaccination in the control pens, and tested by a commercial PRRSV rRT-PCR to ensure no virus circulation within the controls. Pigs were individually weighed again 48 days post-vaccination. Average daily gain (ADG) and survivability were estimated utilizing generalized linear and logistic regression models in the R statistical software[3] packages lme4[4] and lsmeans.[5] The analysis was adjusted by start weight and block (barn location) effects. A P-value of <0.05 was used to indicate statistical significance.

Figure 1: Barn diagram with trial and treatment assignments.

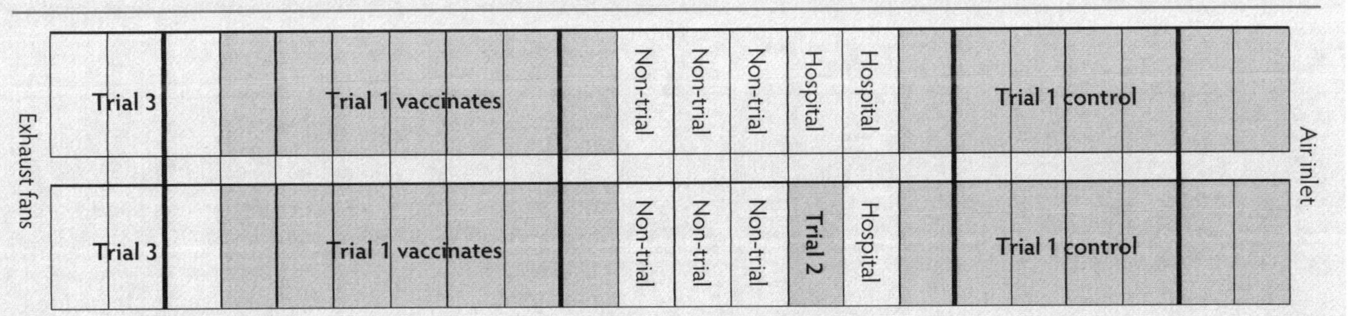

Trial 2: Immune response characterization

Twenty-five pigs in one pen were vaccinated with 1ml of PRRSGard® intramuscularly. Serum samples were collected from each pig on days 0, 7, 14, 21, 28, 35, and 41 post-vaccination. Serum samples were tested by a commercial PRRSV rRT-PCR at day 0. All samples were tested by the PRRSGard® specific rRT-PCR utilizing the integrated genetic marker, a commercial PRRSV ELISA and a PRRSV174 serum neutralization (SN) assay. The PRRSGard® specific rRT-PCR was developed and performed by Dr. Jianqiang Zhang and Dr. Gaurav Rawal at the Iowa State University Veterinary Diagnostic Laboratory. The PRRSGard® specific rRT-PCR did not cross react with many PRRSV field and vaccine strains as well as other common swine pathogens. The analytical sensitivity based on ct values was similar to a current commercial PRRSV rRT-PCR. Plotting of those results was done using the R package ggpubr.[6]

Trial 3: Shed and spread

Six pens containing a total of 144 pigs were selected and pigs within each pen were individually identified by ear tag and randomly assigned to the vaccinated or sham-vaccinated groups. All six individual pens were composed of 50% vaccinated and 50% control pigs. The vaccinated group received 1ml of PRRSGard® intramuscularly and the sham-vaccinated control group received 1ml of vaccine diluent intramuscularly. Serum samples were collected from all pigs at vaccination on day 0 and again on day 41. Day 0 samples were tested by a commercial rRT-PCR. All samples were tested by a commercial PRRSV ELISA and the PRRSGard® specific rRT-PCR. A sample with an S/P ratio of ≥0.4 was considered ELISA positive in all trials. A sample with a ct value of < 40 was considered rRT-PCR positive in all trials. The basic reproduction number (R_0) of PRRSGard® spread during those 6 weeks was estimated using the attack rate method in the R package R0.[7]

Results

Trial 1: Performance drag

Oral fluid samples tested PRRSV negative by the commercial rRT-PCR at days 0 and 48 post-vaccination. There was not enough evidence of differences in ADG and survivability between the PRRSGard® and control groups. The descriptive and statistical analysis of performance parameters is summarized in Table 1.

Trial 2: Immune response characterization

All weaned pigs were confirmed to be PRRSV negative by rRT-PCR and ELISA prior to vaccination. Two pigs were removed from the trial between days 14 and 21 per the farm's standard operating procedure to identify the slowest growing pigs and placed them in a hospital pen. Viremia was detected in sera of 19/25, 22/25, 19/23, 21/23, 20/23 and 19/23 at 7, 14, 21, 28, 35, and 41 days post-vaccination, respectively. Viremia begin dropping off 35 days post-vaccination. The mean rRT-PCR cycle threshold (ct) values over time are illustrated in Figure 2. As expected, ELISA results were negative until the 2nd sampling event (14 days post-vaccination) with a markedly increase in the percentage of positives in subsequent sample events. The ELISA results are summarized in Figure 3. Serum neutralization assay results were unavailable at the time of paper submission but they will be summarized in the accompanying presentation.

Trial 3: Shed and spread

All weaned pigs were confirmed to be PRRSV negative by rRT-PCR and ELISA prior to vaccination. At 41 days post-vaccination, 57/71 (80%) of the negative control pigs tested PRRSGard® rRT-PCR negative. Only one of the negative control pigs tested positive in the ELISA with an S/P ratio of 0.454. The basic reproduction number (R_0) for viremic pigs and serological positive pigs was 1.1 and 1.0, respectively. Our results are indicative of the limited transmission of PRRSGard® to contact-naïve pigs during those 6 weeks of evaluation.

Discussion

PRRSV continues to be a problem for many swine producers. Modified-live vaccines remain one of the best tools to control the clinical signs of disease associated with PRRSV. Traditional attenuation methods rely on altering field viruses, so they are hopefully less virulent, but similar enough to stimulate a protective immune response to subsequent natural infection. This balance between altering a virus, while also keeping it similar, may often be difficult to achieve. PRRSGard® takes a unique approach by the application of a chimera. The replication machinery in ORF1a and ORF1b are from a proprietary strain that has been well attenuated. However, the structural proteins (ORF3-7) are from a highly virulent isolate (MN184) that was not modified using traditional attenuation methods. The purpose of the chimera pro-

Table 1: PRRSGard® performance drag summary

Group	No. pigs	ADG (lb/day)	95% CI	Survivability %	95% CI
PRRSGard®	345	1.17	1.15 – 1.19	96.7	94.1 – 98.1
Control	339	1.19	1.17 – 1.21	96.5	93.8 – 98.0
Difference	-	0.02	-	0.02	-
P-value	-	0.14	-	0.87	-

Figure 2: PRRSGard® viremia over time.

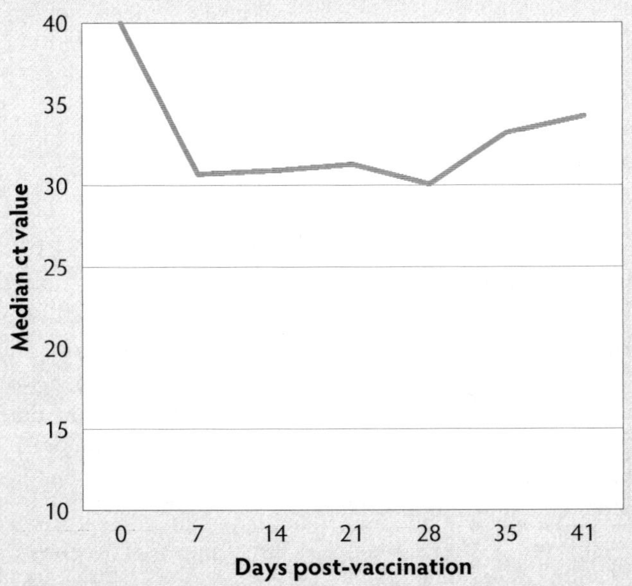

Figure 3: ELISA positive pigs over time.

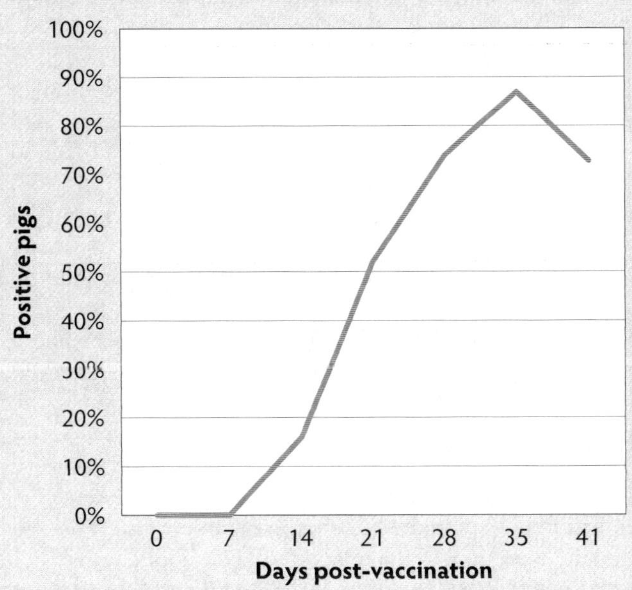

duction is a vaccine virus that is highly infectious and stimulates a strong, relevant immune response but is lowly transmissible and does not negatively impact performance through signs similar to disease expression.

Vaccination of weaning age pigs with PRRSGard® did not result in reduction of performance as measured by average daily weight gain and survivability when compared to sham-vaccinated pigs. This limited impact of PRRSGard® on performance is important in high health flows where pigs need to be vaccinated given the high risk of lateral introductions. Additionally, PRRSGard® induced high levels of replication seven days post-vaccination and a subsequent immune response two weeks later. Finally, PRRSGard® spreads slowly within naïve populations with direct contact and following vaccination of 50% of the animals. This last point is important because it suggests that there was limited exposure of non-target animals to vaccine virus under field conditions. Even more, a PRRSV vaccine with limited transmission is important when considering the risk of area spread, especially transmission to naïve sow farms. A PRRSV vaccine with limited transmission and performance drag can be applied in other production situations that warrant further exploration. In summary, these trials begin characterizing the benefits of PRRSGard® as a new tool to aid in the control of PRRSV in the US and worldwide.

References

1. Holtkamp D. et al. Assessment of the economic impact of porcine reproductive and respiratory syndrome virus on the United States pork producers. *Swine Health and Production.* 2013. 21:72-84.
2. Chamba F. et al. Experimental safety and efficacy of a unique MLV PRRSV vaccine: PRRSGard®. *AASV proceedings.* 2018. 170-174.
3. R: A language and environment for statistical computing v. 3.4.2 (R Foundation for Statistical Computing, Vienna, Austria, 2017).
4. Bates, D., Mächler, M., Bolker, B. & Walker, S. Fitting linear mixed-effects models using lme4. arXiv preprint arXiv:1406.5823 (2014).
5. Lenth, R. V. Least-Squares Means: The R Package lsmeans. Journal of Statistical Software 69, doi:10.18637/jss.v069.i01 (2016).
6. ggpubr: "ggplot2" based publication ready plots. R package version 0.1.7 (2018).
7. R0: Estimation of R0 and Real-Time Reproduction Number from Epidemics v. R package version 1.2-6 (2015).

Beyond bone: the importance of vitamin D for immune function in swine

S. Hough, DVM; J. Bergstrom, PhD; C. Bradley, PhD; J. Hahn, PhD
DSM Nutritional Products

Introduction

Vitamin D has long been recognized as essential to the skeletal system by maintaining calcium and phosphorus homeostasis.[1] Newer evidence acknowledges its significant role in supporting immune function as well. The presence of vitamin D receptors and D3-metabolizing enzymes in almost all the cells of the immune system, including monocytes, macrophages, dendritic cells and activated T and B cells, suggests a potential role of $1,25(OH)_2D_3$ in immune function.[2] There are also vitamin D receptors in enteroctyes, myocytes, mammary gland cells and many various endocrine cells, making it a very versatile and essential micronutrient.[3] Circulating $1,25(OH)D_3$ can be taken up by macrophages, neutrophils or epithelial cells at locations exposed to the external environment and can be converted with extra-renal 1-alpha-hydroxylase to $1,25(OH)_2D_3$.[4] This allows vitamin D to act locally within or around immune cells. Epidemiological studies also link inadequate vitamin D levels to higher susceptibility of immune-mediated disorders in humans.[2] This suggests that the vitamin plays a role in supporting immune function. Here we will outline the positive effects Vitamin D has on immune function and how that correlates to overall pig health and production.

Immunomodulatory effects of vitamin D

Intestinal epithelial cells are the first line of defense against pathogenic antigens present in the lumen of the gut. Vitamin D helps with the physical barrier by upregulating genes that encode proteins required for tight junctions (occluding), gap junctions (connexion) and adherens (E-cadherin) in humans.[5] Vitamin D has been associated with attenuating the negative effects of rotavirus challenge in pigs by potentially promoting intestinal integrity. *In vitro* studies have shown the benefit of vitamin D when porcine intestinal cells were challenged with a culture-adapted rotavirus strain from Ohio State University.[6] The porcine intestinal epithelial cell line (IPEC-J2) model showed increased viability of intestinal cells and decrease in viral quantity with Vitamin D treatment.[6] The IPEC-J2 cells expressed high baseline levels of 1-alpha-hydroxylase. Porcine rotavirus (PRV) infection, alone, resulted in a significant increase in 1-alpha-hydroxylase mRNA in these cells, which augmented the production of active vitamin D.[7] In *in-vivo* studies, both Zhao et al., (2014) and Tian et al., (2016) have shown that rotavirus-challenged pigs given 5,000 IU/kg of vitamin D had improved average daily gain, feed intake and fecal consistency scores when compared to control animals and those given only 200 IU/kg of vitamin D.[8,9] Dietary supplementation with 5,000 IU vitamin D in challenged pigs was associated with improved villus height when compared to challenged pigs fed 200 IU/kg. These studies demonstrate the potential benefit of vitamin D in the presence of a common enteric pathogen of swine.

Hy-D® preparation

Vitamin D can be acquired through the diet and endogenously by exposure to UVB light that triggers vitamin D synthesis in the body. Typically, both routes are biologically inert and require multiple phases of hydroxylation in the body for activation. Hy-D®, a source of 25-OH-D3 for feed or water supplementation, is a DSM product that provides an efficient form of the vitamin that doesn't require the initial liver hydroxylation step.

In Swine, Hy-D® has been associated with improved structural soundness in gilts[10] and an increase in number of liveborn and weaned piglets in both sows and gilts.[11,12] In growing animals, Hy-D® has been shown to reduce osteochondrosis lesions.[13] In a Canadian study completed in 2013, weaned pigs that were supplemented using Hy-D® had higher serum levels of $25(OH)D_3$, which correlated to an increased number of blood leukocytes, particularly lymphocytes and granulocytes.[14] After two weeks of supplementing typical levels of Vitamin D_3 in the feed, an average of 40.4 ng/mL of $25(OH)D_3$ was found in the serum of those animals that received 2,000 IU/kg from Hy-D® with 1,500 IU/kg vitamin D_3, compared to 11.2 ng/mL of $25(OH)D_3$ for those that received 3,500 IU/kg vitamin D_3 and 7.87 ng/mL for the control group that received 1,500 IU/kg vitamin D_3. This suggests there is improved uptake of $25(OH)D_3$ *via* the intestinal mucosa when compared to the traditional vitamin D_3 preparations, and highlights Hy-D® as a superior dietary supplement for weaned pigs. Hy-D also improved the capacity of leukocyte subsets to survive and upregulated their functional antimicrobial responses compared to control.[16] Improved phagocytosis in this cell line could possibly influence the response to many common viruses of swine (PRRS, PCV2, etc.). It has also been shown that dendritic cells are also affected as exposure to vitamin D directly affects the expression of genes involved in metabolic pathways, such as glycolysis, which results in the induction of antioxidant proteins.[15] This relationship helps prevent damage to important cellular components possibly caused free radicals and shows additional support of the innate immune system.

Historical studies have indicated low Vitamin D levels in the serum of commercial nursery pigs, which may suggest inadequate support of the immune system.[16] The post-weaning phase is a critical, developmental period that is key to securing health and optimum performance in the future. Enteric and respiratory challenges are more likely during the early nursery period, and most pigs are at a disadvantage with respect to their vitamin D status, as many pigs are born with inadequate serum levels of vitamin D.[17] Recent data from several studies show there is an improvement in serum vitamin D status when HyD® is administered through either feed or water. Results from one such study in France showed a marked difference in serum levels when nursery pigs were fed 50µg 25(OH)D3/kg HyD® for a total of 48 days versus 2,000 IU D3/kg (Figure 1).[18] Hy-D® is the preferred form for feed or water application because its absorption is less affected by enteric insults, such as infection or associated malabsorption, and is minimally dependent on the presence of dietary fat for absorption. The liver plays a critical role in functional homeostasis by detoxifying the body of various metabolites through portal vein filtration, regulating glycogen stores and processing minerals.[19] The hepatic system is typically very busy during weaning due to increased cortisol release, feed change, and exposure to new pathogens. The liver can suffer collateral damage from viral, bacterial or parasitic infection; and its efficiencies can become taxed. Hy-D® by-passes the liver hydroxylation step and provides a more effective form of the vitamin, which is important for young pigs. While Hy-D® is the preferred form of vitamin D activity to provide in the feed, the water-soluble option provides veterinarians with an alternative, accessible solution to support the immune function of nursery pigs during maternal antibody decay, disease challenge, and personal adaptive immune development.

Conclusion

It has been shown that vitamin D may support nutrient absorption during an enteric challenge and help prevent an overactive immune response, and therefore helping the animal conserve its reserves. Common infections stress the animal resources making various bodily systems less efficient. Vitamin D was shown to help pigs to better withstand such challenges. Hy-D® can be added to the list of available tools that can be used to help combat post-weaning threats.

Figure 1: Serum 25(OH)D3 levels from pigs fed either 2,000 IU D3/kg or 50 µg 25(OH)D3/kg from HyD®.

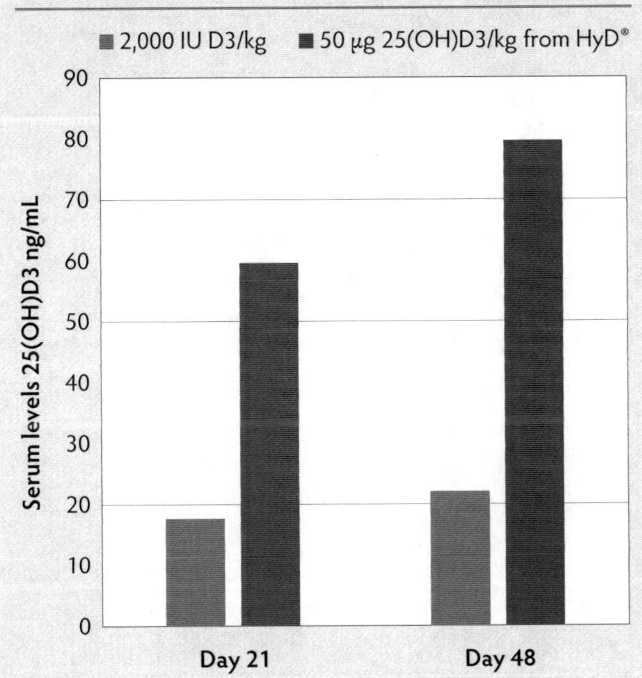

References

1. DeLuca, H.F. (2004). Overview of general physiologic features and functions of vitamin D. *American Journal of Clinical Nutrition.* 80, 1689-1696.

2. Baeke, F., T. Takiishi, H. Korf, C. Gysemans, and C. Mathieu. (2010). Vitamin D: modulator of the immune system. *Current Opinion in Pharmacology, 10*(4), 482–496.

3. Combs, G.F. (2008) The Vitamins: Fundamental Aspects in Nutrition and Health - 3rd ed. ISBN: 9780128029657

4. Kamen, D. and V. Tangpricha. (2010). Vitamin D and molecular actions on the immune system: modulation of innate and autoimmunity. *Journal of Molecular Medicine.* 88, 441-450.

5. Schwalfenberg, G. (2011) A review of the critical role of Vitamin D in the functioning of the immune system and the clinical implications of vitamin D deficiency. *Molecular Nutrition and Food Research.* 55, 96-108.

6. Zhao, Y., Z. Ran, Q. Jiang, N. Hu, B. Yu, L. Zhu, L. Shen, S. Zhang, L. Chen, H. Chen, J. Jiang and S. Chen. (2019). Vitamin D Alleviates Rotavirus Infection through a Microrna-155-5p Mediated Regulation of the TBK1-IRF3 Signaling Pathway In Vivo and In Vitro. *International Journal of Molecular Sciences.* 20, 3562.

7. Zhao, Y., B. Yu, X. Mao, J. He, Z. Huang, P, Zheng, J. Yu, G. Han, X. Liang and D. Chen. (2015). Effect of 25-hydroxyvitamin D_3 on rotavirus replication and gene expressions of RIG-I signaling molecule in porcine rotavirus-infected IPEC-J2 cells. *Archives of Animal Nutrition.* 69(3), 227-235.

8. Zhao, Y., B. Yu, X. Mao, J. He, Z. Huang, P, Zheng, J. Yu, G. Han, X. Liang and D. Chen. (2014). Dietary vitamin D supplementation attenuates immune response of pigs challenged with rotavirus potentially through the retinoic acid-inducible gene I signaling pathway. *British Journal of Nutrition.* 112, 381-389.

9. Tian, G., X Laing, D. Chen, Z. Mao, J. Yu, P. Zheng, J. He, Z. Huang and B. Yu. (2016). Vitamin D_3 supplementation alleviates rotavirus infection in pigs and IPEC-J2 cells via regulating the autophagy signaling pathway. *Journal of Steroid Biochemistry and Molecular Biology.* 163, 157-163.

10. Brana, D., J. Gabriel-Landon, J. Cervantes-Lopez and J.A. Cuaron. (2012) Nonruminant nutrition: Vitamins and Minerals. *Journal of Animal Science.* 98(Supp 1), 114.

11. Planchenault, D., B. Renouf, A. Meuter and J.M. Thoby. (2018) Effect of dietary 25-hydroxycholecalciferol and cholecalciferol on the performance of gestation and lactating sows and their litter. *Journees Recherche Porcine.* 50, 143-144.

12. Coffey, J.D., E.A. Hines, J.D. Starkey, C.W. Starkey and T.K. Chung. (2012) Feeding 25-hydroxycholecalciferol improves gilt reproductive performance and fetal vitamin D status. *Journal of Animal Science.* 90, 3783-3788.

13. Sugiyama, T., T.K Chung, H. Yonekura, E. Azem and T. Hayakawa. (2013) Effects of 25-hydroxycholecalciferol on the development of osteochondrosis in swine. *Journal of Animal Science.* 84, 341-349.

14. Konowalchuk, J. D., A. M. Riegea, M. D. Kiemele, D. C. Ayres and D. R. Barreda. (2013) Modulation of weanling pig cellular immunity in response to diet supplementation with 25-hydroxyvitamin D3. *Veterinary Immunology and Immunopathology.* 155(2), 57-66.

15. Vanherwegen, A.S., C. Gysemans, C. Mathieu. (2017). Vitamin D endocrinology on the cross-road between immunity and Metabolism. *Molecular and Cellular Endocrinology*, 453, 52-67.

16. Arnold, J., D. Madson, S. Ensley, J. Goff, C. Sparks, G. Stevenson, T. Crenshaw, C. Wang and R. Horst. (2015). Survey of serum vitamin D status across stages of swine production and evaluation of supplemental bulk vitamin D premixes used in swine diets. *Journal of Swine Health and Production.* 23 (1), 28-34

17. Goff, J.P., R.L Horst and E.T. Littledike. (1984) Effect of sow vitamin D status at parturition on the vitamin D status of neonatal pigs. *Journal of Nutrition.* 114, 163-169.

18. Meuter, A., J. Thoby, and J. Lichou. (2016) Modulation de la reponse immunitaire de porcelets sevres par une supplementation en 25-hydroxycholecalciferol. *Journees Recherche Porcine - 48th meeting proceedings.* Paris, France, 347-348.

19. Talwani, R., B.L. Gilliam and C. Howell. (2011). Infectious diseases and the liver. *Clinics in Liver Disease.* 15(1), 111-130.

Differences in number of pigs completing the nursery phase when using Swine Awake™ from NutriQuest® in PRRSV-positive pigs

Deborah M. Murray[1], DVM; Tyler J. Holck[2], DVM, MS, MBA; Sarah A. Weiland[3], MS
[1]New Fashion Pork, Jackson, Minnesota; [2]Feed His People, LLC, Gilbert, Iowa; [3]NutriQuest, Mason City, Iowa

Introduction

Porcine reproductive and respiratory syndrome virus (PRRSV) affects not only sow herd performance, but also the survivability and the downstream performance of pigs weaned from infected herds. While the offspring may be introduced to the virus during lactation, PRRSV continues to impact piglet performance through the nursery phase due to seroconversion and continued virus introduction to new pen or barn mates. In a controlled PRRSV challenge study in newly weaned pigs, pigs given Swine Awake exhibited improved average daily gain (ADG) and feed conversion, reduced microscopic lung lesion scores, and an enhanced adaptive immune response by reduced neutrophil count, increased lymphocyte count, and increased IL-1β and IL-12 at 21 dpi.[1,2] In commercial nursery flows, the number of pigs completing the nursery phase was increased by 1.9% in pigs provided NutriQuest Swine Awake.[2]

A field observation was performed to observe the impact of NutriQuest Swine Awake on the number of pigs completing the nursery phase in pigs weaned from a PRRSV positive sow flow.

Materials and methods

Pigs utilized in this study were weaned from two PRRSV unstable filtered sow farms run as one flow with parity segregation. The farms had a previous PRRSV break in December of 2017, following which the herd was closed and taken PRRSV negative. The herd was then re-opened and broke in January of 2019. It was suspected that the PRRSV strain of the 2019 break had been in the farm during the previous herd break. PRRSV status was confirmed with processing fluid and serum. Pigs were double stocked into wean to finish barns starting at approximately 2200 head and split-out at the end of the nursery period, half of which remained in the original site and the other half moved to a finishing barn. Observations were made by barn, as Swine Awake is a water-soluble product the entire barn was either provided the product or not. Comparisons were made in a before and after manner. Results were collected at split-out, which occurred at an average of 7.8 and 8.0 weeks after weaning for non-Swine Awake and Swine Awake barns, respectively. Nursery pigs weaned from these sow farms had been consistently seroconverting 1-2 weeks prior to nursery split out. Denagard®/CTC® was used for two weeks in the middle of the nursery period in all barns. In treated barns, Swine Awake™ was administered via water for 21 consecutive days at an estimated dosage of 5 mg/pig/day. Closeouts provided results for the number of pigs completing the nursery phase as a percentage of total pigs placed. Data were analyzed using a two-sample student t-test in SAS with barn as the experimental unit.

Results and discussion

In barns given Swine Awake the percent of pigs completing the nursery phase was 97.7% vs. 93.5% for barns not provided with Swine Awake ($P = 0.004$; Figure 1). This is a higher change in percent of pigs completing the nursery phase than has been reported in previous field trials with Swine Awake. Previous PRRSV challenge studies have shown pigs provided Swine Awake exhibit enhanced ADG and feed conversion.[1,2] Due to management practices on farm, growth parameters including ADG, average daily feed intake (ADFI) and feed conversion could not be collected. Given the difference in percent pigs completing the nursery phase it is expected that these parameters would be improved with the inclusion of Swine Awake.

Conclusions

In barns provided with Swine Awake a significant increase in the proportion of pigs completing the nursery phase was observed. With an average of 2213 pigs placed, this increase resulted in an average of 93 more pigs per barn.

References

1. Song, R., T. Holck, S.M. Sinn, R.E. Musser, and K.G. Friesen, 2018. A novel, non-antimicrobial protein improves growth performance and immune responses of pigs experimentally infected with porcine reproductive and respiratory syndrome virus. *J. Anim. Sci.* 96, suppl.2 (abstract): 127.

2. Sugg, E., R. Jones, T. Holck, S. Heintzelman, S.A. Weiland. 2018. The impact of Swine Awake® on commercial nursery flows from a PRRSV positive source. *J. Anim. Sci.* 97, suppl.2. (abstract): 136.

Figure 1: Impact of Swine Awake on pigs sold as percentage of pigs placed.[a]

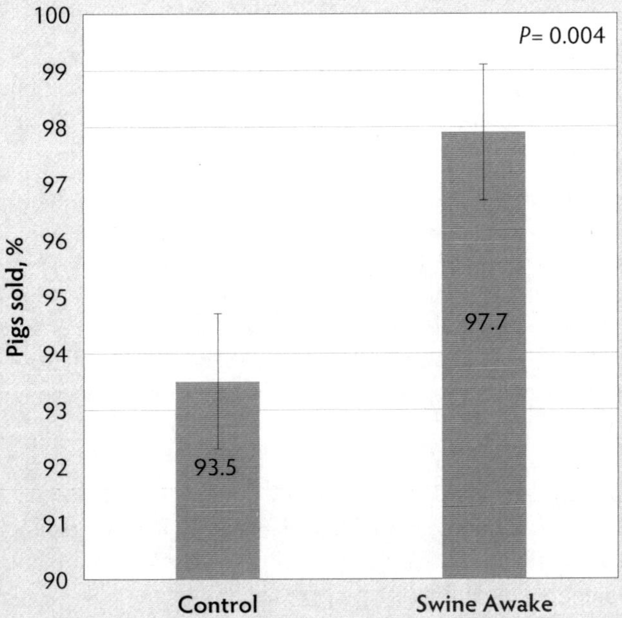

[a] Average pigs placed = 2213

Development and validation of PRRS biosecurity survey for breeding herds

D. Torrents, DVM, PhD; L. Nodar, DVM; D. Angelats, DVM; J. Miranda, DVM
HIPRA, Amer, Spain

Introduction

Porcine reproductive and respiratory syndrome (PRRS) is a major health, production and financial problem in swine production worldwide. Biosecurity is one of the key factors to prevent and control PRRS. Assessment of farm's biosecurity level through a systematic and complete survey should be included in any PRRS control program. The aim of this study was to build and validate a PRRS biosecurity survey in commercial breeding herds.

Material and methods

PRRS biosecurity survey was developed by a meta-analysis of papers previously published about epidemiology and control of PRRSV.[1,2] The total number of questions included in the survey was 68. Fifteen PRRS experts through a Delphi approach discussed the weight of each group of questions. From the total, 23 questions referred to internal risk factors (management, immunization and general characteristics that weighted 33 %) and 45 to external risks (semen, replacement animals, transport, location and other measures that weighted 67%) (Figure 1).

Once the survey was developed and in order to validate it in commercial breeding farms we used 35 PRRSv positive breeding herds enrolled in a PRRS monitoring project. The monitoring project was designed based on the AASV guidelines which consisted of monthly individual blood testing of 30 piglets at weaning. Serum from those piglets were pooled (5 pools in total) and tested for PRRSV by RT-PCR. Farms were classified as Positive Unstable and Positive Stable. The survey was filled out once in all the breeding farms. An ANOVA test was used to evaluate the association between PRRS status and PRRS risk scoring.

Results

The results of the survey showed that the total risk score in Positive Stable herds was statistically significantly lower than in Positive Unstable herds (1482 ± 148 vs. 1945 ± 317, *P*-value < 0.001). Moreover, the internal and external risk score was lower in Positive Stable herds than in Positive Unstable (Internal: 380 ± 91 Positive Stable vs. 481 ± 121 Positive Unstable, *P*-value < 0.05; External: 1102 ± 204 Positive Stable vs. 1464 ± 264 Positive Unstable, *P*-value < 0.05).

Discussion and conclusion

In this case, there is a clear relation between higher risk scoring in the PRRS biosecurity survey and PRRS instability. So, those farms with worse biosecurity measures are more unstable to PRRS than those farms with better biosecurity measures. Biosecurity scoring provided by the PRRS biosecurity survey may be useful to assess vulnerabilities on biosecurity protocols in order to reduce the frequency of PRRS outbreaks and may help producers and veterinarians prioritize investments in improving biosecurity practices over time.

References

1. Hermann JR, Muñoz-Zanzi CA, Roof MB, Burkhart K, Zimmerman JJ. Probability of porcine reproductive and respiratory syndrome (PRRS) virus infection as a function of exposure route and dose. Vet Microbiol. 2005, 110:7-16.

2. Pileri E, Mateu E. Review on the transmission porcine reproductive and respiratory syndrome virus between pigs and farms and impact on vaccination. Vet Res. 2016, 47:108.

Figure 1: Weight of external and internal risks.

From depopulation to repopulation: a Chinese farm underwent African swine fever virus (ASFV) outbreak and survives with ASFV

Linan Weng
Thermo Fisher Scientific

Pork consumption in China accounts for 50% of the world pork consumption. In China, the per capita consumption was approximately 40 kilograms. After the 2018 outbreak of African swine fever virus (ASFV), inventory decreased by 70%.

Many changes of the biosecurity measures and concepts of Chinese pig farms have been implemented since the occurrence of ASFV. The main reason for the spread of ASFV so quickly in China was attributed by the weakness of biosecurity systems.

The pig farmers in China realized that environmental sample testing was even more important than animal specimen testing. The farms that want to restock, are required to take many different environmental samples before they can introduce new animals onto the farm. China is currently investing many resources to develop ASFV vaccine. The launch of the vaccine will motivate farmers to restock.

From the national level, ASFV caused the improvement of industrial layout and production. ASFV revealed the weakness of biosecurity system in most Chinese farms. From now on, the industry will pay more attention to management and biosecurity. There will be more scaled farms instead of backyard.

Recently pig market price continues to increase, so the China government is providing subsidies to encourage pig farms to expand production. Some farms have been rehabilitated successfully. Some farms are producing well although ASFV is circulating in the farms. More research institutions have joined the R&D of vaccine program. ASFV impacted the beef price as well. Meat price of beef and lamb is increasing as well. China is also encourages production of alternatives for protein supplement, such as poultry and dairy.

Reducing piglet crushing with the application of the SmartGuard® system

Seth Krantz[1], DVM; Matthew Rooda[2]; Abraham Espinoza[2]
[1]Tosh Pork, LLC, Henry, Tennesee; [2]Swine Tech, Cedar Rapids, Iowa

Introduction

Significant mortality has been attributed to piglets being laid on in the first few days post-partum. In the farm in which this study was conducted, laid on pigs constitute 64% of the recorded pre-weaning mortality events. SwineTech has created technology to address this common animal care issue. SwineTech is an ag-tech company that offers pork producers unique tools that enable a more efficient farrowing operation. Their intelligent SmartGuard® device used voice recognition to provide aid to piglets in distress, along with sensors that control each microclimate and the health and well-being of each sow and her piglets. This trial will review a large field evaluation of the technology within a 3,200 sow farm.

Materials and methods

231 sows and their litters were enrolled in this study. 120 of the sows were assigned to the SmartGuard® treatment group while 111 were assigned to the control group which was left unguarded. Both treatments were represented within farrowing rooms and treatment groups were separated between rows so that all sows and litters in a particular row were among the same treatment group. Due to the length of time covered during this trial, treatments were repeated in rooms but during subsequent turns through a room, treatment was alternated by row as to minimize any environmental or location effects. Parity was evenly distributed between treatments. Any cross fostering that occurred was done within the first 24 hours post-partum and cross fostering may have occurred between treatment groups.

Sows within the SmartGuard® treatment group were fitted with the SmartGuard® wearable after being placed in the farrowing stall (Figure 1). The wearable was automatically paired with the control unit which was located on the creep divider panel. Microphones within the control unit detect distress squeals and respond by stimuli to the sow to encourage her to stand. The initial stimulus is vibration and if no response occurs, the stimulus progresses to a gentle TENZ impulse.

All sows within both treatment groups were monitored during farrowing and day one pig care was performed per the farm's standard operating procedure. This includes recording times of piglet births and any interventions performed. Midwifery only occurred during the normal daytime working hours as this farm does not have extended hour monitoring.

Number of pigs born alive and mortality due to crushing during the first three days post-partum were recorded and evaluated between treatment groups.

Results and discussion

Born alive between the treatment groups was no different, giving no advantage to crushing mortality in either group. When comparing the number of pigs lost due to crushing within the first three days after birth, expressed in total pigs lost divided by total number of sows, the control group lost an average of 0.51 pigs per litter and the SmartGuard® group lost and average of 0.34 pigs per litter (Table 1). This difference represents a 31.37% improvement in crushing mortality with the implementation of the SmartGuard® technology.

While these results are encouraging, there is no significant difference in crushing mortality between the two groups. A challenge to this study is the number of litters represented in the study. This is an ongoing project and more data is being collected. With increasing sample size, we anticipate results may become statistically significant over time. Another limitation of this study is the layout of treatments within rooms. In an effort to remove environmental or location effects, we allotted both treatments within a room. The SmartGuard® technology is intended to be used in every stall throughout an entire room and is designed to work in a network. SmartGuard® uses acoustic volume measurements and communicates within the network of units to properly pinpoint the stall in which a layon event is occurring. By mixing non-guarded pens with guarded SmartGuard® pens within a single room, the network less reliably will find the correct pen where a layon is occurring. The full efficacy of SmartGuard® can be achieved when an entire room of stalls is protected by the SmartGuard® technology, thus granting the highest confidence at reacting to the appropriate event location.

Overall, these results are encouraging and elucidate the possibilities of technology in improving the effectiveness of farrowing house staff and overall farm productivity by saving more pigs. This technology is feasible to use within a commercial setting and in the near future will contain even more animal health functionality in that it will be able to monitor the sow for impending signs of farrowing as well as temperature and standing/lying patterns to notify caregivers of when to intervene.

Figure 1: The SmartGuard® system in use showing the control unit and wearable attached to the sow.

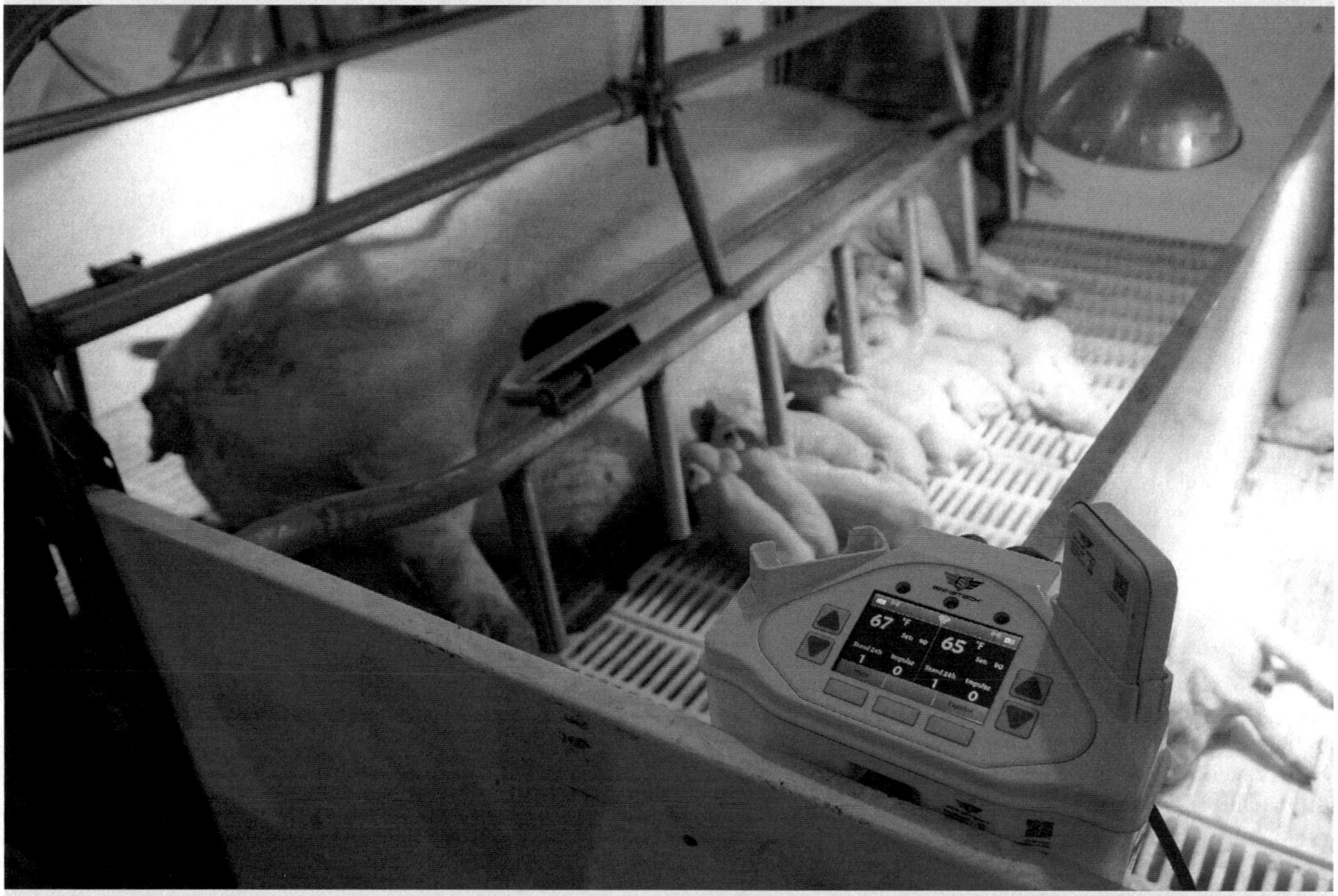

Table 1: Laid on piglet mortality difference between litters utilizing the SmartGuard® system and unguarded litters

Treatment	Number of litters	Total pigs	Day 1 laid on mortality	Day 2 laid on mortality	Day 3 laid on mortality	Day 4 laid on mortality	Total laid on mortality	Laid on mortality per litter
SmartGuard®	120	1749	1	26	9	6	42	0.35
Unguarded (Control)	111	1674	4	37	11	5	57	0.51
							Percentage Difference	31.37%

Study of testicles from boars eliminated from artificial insemination centers

Raquel Ausejo[1]; Noelia Mendoza[1]; Joaquín Miguel[1]; Pedro Soler[1]; M Victoria Falceto[2]; Olga Mitjana[2]
[1]Magapor S. L., Ejea de los Caballeros (Zaragoza);
[2]Animal Pathology Department, Veterinarian Faculty of Zaragoza , Zaragoza

Introduction

Post-mortem collection of boar genitalia in the slaughterhouse and its later study in the laboratory is a useful diagnosis tool for the veterinarian because help us in the decision taking to solve a reproductive problem. To check the reproductive tract of the boars culled from boar studs contribute to the understanding about the alteration mechanisms of the seminal quality because there is a lack of studies that give results about the frequent elimination of young boars because of these problems.

In the past 10 years, an increase of slaughtered young boars due to bad semen quality has been recorded. This fact makes the animal amortization worse and decreases the productive capacity of the boar stud.

This routinely diagnosis strategy could help us in the clinical exercise in the swine business to take future decisions in other subfertile or infertile animals. To quickly diagnose the testicular pathology could increase the retention rate, as we already said.[1]

The spermatocytogeneses-spermiogenesis and epididymal maturation could last from 45 to 47 days. Any alteration during this process due to genetic problems, hormones, circulatory problems or due to any pathology will make the seminal quality to be altered.

The aim of this study was to see if there is any link between anatomical, histological study (in vivo biopsies and slaughter sampling) and reproductive problems or semen quality able to make an early diagnosis of treatable diseases, thus increasing the boar retention rate.

Materials and methods

100 testes from boars were collected in the slaughterhouse. Boars were culled due to the bad quality of their ejaculates, disease or genetic progress. The following parameters were assessed: macroscopic examination, histological evaluation and testes biopsy.[2] Moreover, the last semen doses are sent for analysis to the laboratory: routine analyses, besides flow cytometry test are performed in all samples. With this technique, viability, acrosome integrity, mitochondrial activity and early apoptosis are evaluated.[3]

In the macroscopic evaluation, a thorough exam of all the external characteristics is recorded and pictures taken.

If a not normal content is seen, a microbiological study is performed directed to the identification and pathogen quantification, besides the antibiogram analysis.

For the histological evaluation, samples of 6 × 6 were taken from the testicle, head, body and tail from the epididymis, in addition to all the macroscopical lesions previously seen. They are introduced in formaline 10% and sent to the pathologist for examination.

Results

From 100 boars, 85% were slaughtered due to bad semen quality (abnormal forms, or low volume/concentration) or libido, 5% due to lameness and 10% due to genetic progress. In 90% of cases, macroscopic lesions were identified.

In the majority of the culled boars, eliminated without a specific cause, different grades of fibrosis in the testicles were recorded. It is not uncommon to find inflammatory focus, hematomas, and testicle hemorrhages. Most common lesions were oedema, inflammation, fibrosis and varicocele (in more than 40% of the cases[4]) (Figure 1).

The epididymis was the most frequently injured area (Figure 2). Sometimes a tumor that partially or totally affects the parenchyma is found (Figure 3).

The microscopic study of injuries found is essential to confirm the macroscopic diagnosis. Microbiological study of the contents allows the identification and the sensitivity analysis to diverse antibiotic of the isolated strain.

Conclusions

The testicle's pathology is produced by a multifactorial process.[5] The study results in the slaughterhouse allow us to identify the anatomical origin of the pathology and its possible main cause, differentiating pathologies, in addition to guide for a treatment or acting measures for other animals in the boar stud. When the technician considers it for the final diagnosis, water, feed, urine, blood, semen or others would have to be taken. The study of seminal quality will allow us to find a correlation between testicular pathology and the affected sperm parameter, in order to predict testicular damage through seminal quality in situ in the same insemination centers.

References

1. Falceto (2004) La Universidad al Servicio de la sociedad: Servicios de Asesoría y Diagnóstico porcino en la Facultad de Veterinaria de Zaragoza. Avances en Tecnología Porcina Volumen 1, Marzo.pp: 81-88.

2. Falceto, M.V.; Bascuas, J.A.; Lara,C.; Ubeda, J.L (2007) Servicio de Asesoria y Diagnostico Reproductivo Porcino (SARPOR): una ayuda en la solución del fallo reproductivo de la explotación. Albéitar nº 110. Noviembre.

3. Kęstutis Mažeika, Neringa Sutkevičienė, Henrikas Žilinskas, Vita Riškevičienė1, Albina Aniulienė, Nomeda Juodžiukynienė (2012) Relationship between sperm quality and testicular lesions in culled ai boars. VETERINARIJA IR ZOOTECHNIKA

4. Ubeda, J.L.; Bascuas, J.A.; Falceto, M.V. (2006) El varicocele en el verraco. Suis nº 26. Abril.

5. Ubeda, J.L.; Dahmani, Y.; Ausejo, R. (2011) Patología del testículo y evaluación del eyaculado. Suis nº 77. Mayo.

Figure 1: Varicocele.

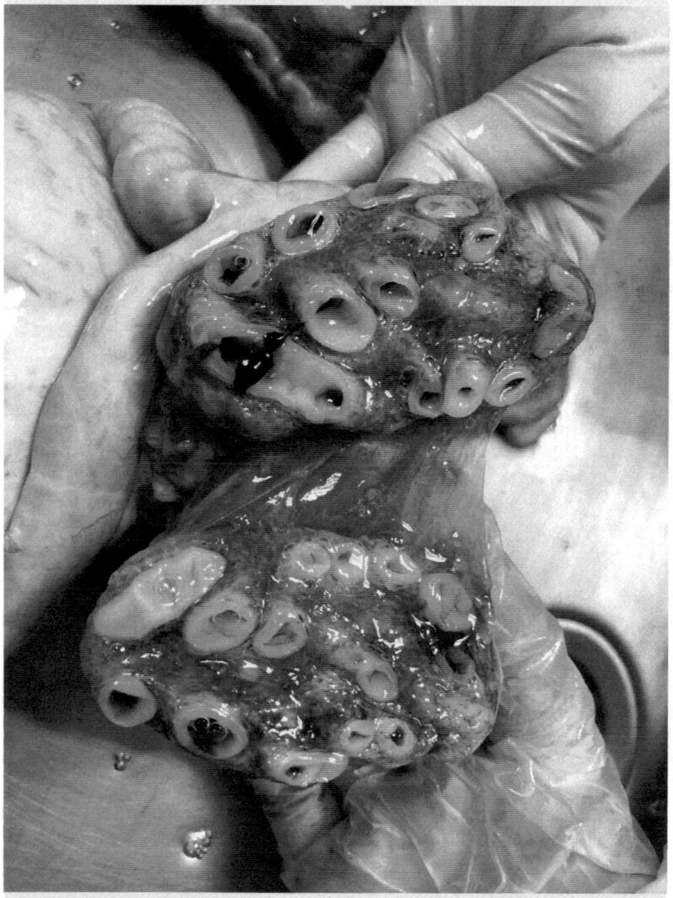

Figure 2: Epididymis.

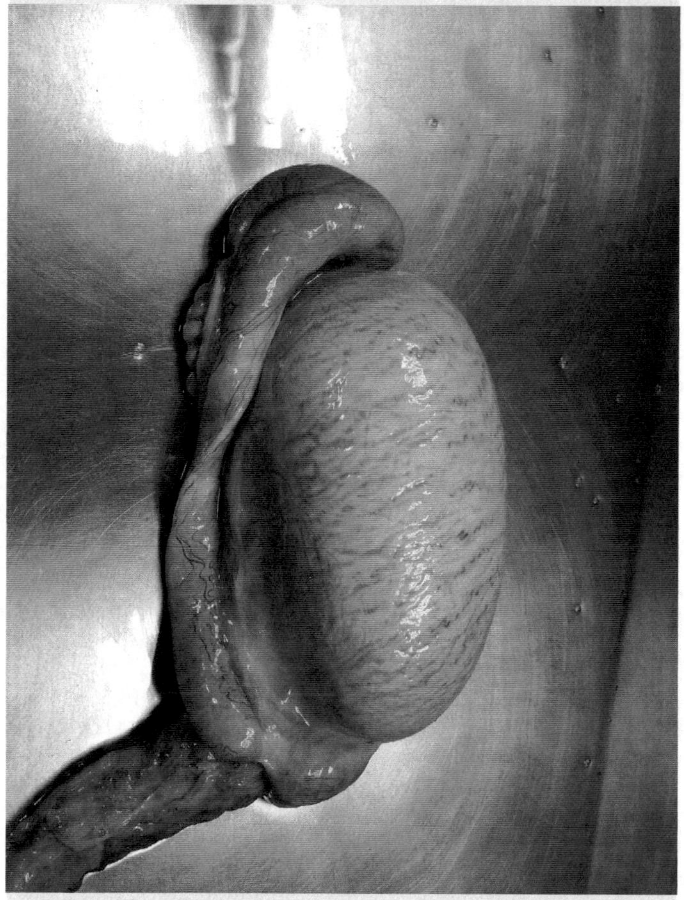

Figure 3: Tumor.

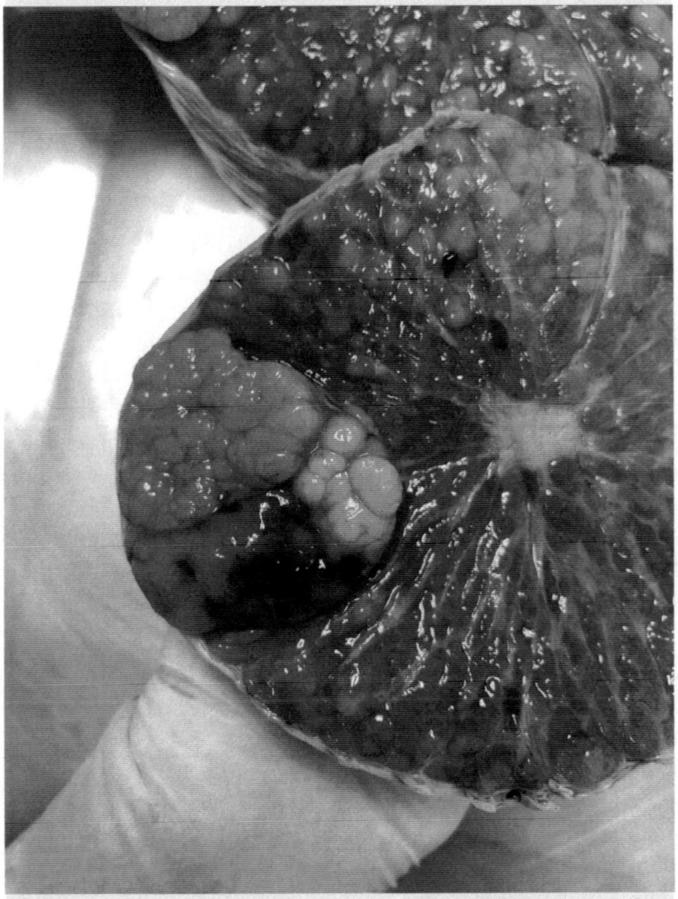

Trends of mycotoxin contamination in US corn harvests

E. G. Hendel[1], VMD, PhD; S. Ramirez[1], PhD; P. N. Gott[1], PhD; U. Hofstetter[2]; G. R. Murugesan[1], DVM, PhD
[1]BIOMIN America, Inc., Kansas City, Kansas; [2]BIOMIN Holding GmbH, Getzersdorf, Austria

Introduction

Mycotoxins are harmful secondary fungal metabolites and are of key concern to food and feed safety globally. These toxins are detrimental to animal health and can compromise animal performance. Classic signs such as reduced feed intake and vomiting used as indicators for exposure overlook other costs of mycotoxicosis. Even at low levels mycotoxins act as predisposing factors for enteric[1] and respiratory diseases by causing inflammation,[2] modulating the gastrointestinal environment,[2] and immune dysfunction[3] which can contribute to vaccine failure.[4] Additionally, mycotoxins cause reductions in reproductive performance and may contribute to sow pelvic organ prolapse. BIOMIN's annual surveys have indicated increasing trends of prevalence and co-occurrence of toxins produced by *Fusarium* spp. fungi in US corn in prior years.[5] The data herein presents the most current information regarding mycotoxin contamination of samples from the 2019 corn harvest in comparison to the previous five years of the BIOMIN survey.

Materials and methods

Corn samples (corn, corn silage, corn by-product) from the United States used in multiple livestock industries (cattle, swine, and poultry) were submitted for mycotoxin analysis utilizing the LC-MS/MS method. Samples were marked with harvest year, with the 2019 harvest sample submission starting in October of 2019. Six major mycotoxin groups were analyzed: aflatoxins (Afla), type A trichothecenes (A-Trich), type B trichothecenes (B-Trich, i.e. DON; vomitoxin), fumonisins (FUM), zearalenone (ZEN), and ochratoxin A (OTA). The 2019 sample pool data were compared to previous years to examine mycotoxin contamination trends (Figure 1, 2, and 3).

Results and discussion

The delay in harvest across the Midwest has resulted in fewer samples submitted for analysis. Higher numbers of samples were submitted by this time in the previous crop year (Nov 2018: 176 samples vs. Nov 2019: 65).[5] A high proportion of samples of the 2019 corn harvest from the US contain at least one detected mycotoxin type (96%), which is similar to the 2018 sample pool (97% positive, 764 samples). The prevalence of co-occurrence in samples thus far is lower than 2018 (2019: 51%, 2018: 75%, % of samples with ≥ 2 mycotoxins). Currently in this year's sample pool, B-Trich is the most prevalent mycotoxin group, and this year's crop appears to have a lower prevalence, but similar average contamination levels compared to last year (2019: 75%, 1,376 ppb, 2018: 83%, 1,686 ppb). FUM has lower prevalence and similar average contamination levels compared to last year's crop (2019: 54%, 2968 ppb 2018: 68%, 2,905 ppb), while ZEN contamination levels and prevalence have decreased (2019: 42%, 386 ppb, 2018: 54%, 604 ppb). As of yet, no Alfa has been detected; however, corn stored with higher moisture content has increased risk for storage toxins. Producers should be vigilant for storage toxins such as Afla and OTA. throughout this production year. The percent of positive samples above threshold levels for the three *Fusarium* toxins are similar for B-Trich (2019: 47%, 2018: 43%, % positive samples over 1000 ppb B-Trich), decreased for FUM (2019: 49%, 2018: 57%, % positive over 1000 ppb FUM), and equivalent for ZEN (2018 and 2019: 78% of ZEN, % positive over 100 ppb).

The preliminary results from the 2019 corn harvest suggest a continued risk due to mycotoxins produced by *Fusarium* fungal species. Mycotoxin risk of this harvest season is still coming into focus as harvest delays have influenced sample submission and is likely to shift the risk profile of the harvest as additional samples are analyzed. Due to continued risk of multi-mycotoxin contamination, multiple mitigation strategies are required beyond adsorption, including biotransformation, support of the immune system, and liver function.

References

1. Oswald, I.P., C. Desautels, J. Laffitte, S. Fournout, S. Y. Peres, M. Odin, P. Le Bars, J. Le Bars, and J. M. Fairbrother. 2003. Mycotoxin Fumonisin B1 increases intestinal colonization by pathogenic Escherichia coli in pigs. Applied and Environmental Microbiology. (2003). 69:10. Pp. 5870-5874.

2. Pinton et al. Toxicity of Deoxynivalinol and its acetylated derivatives on the intestine: differential effects on morphology, barrier function, tight junction proteins, and mitogen-activated protein kinases. *Toxilogical Sciences*. 2012; 130(1): 180-190.

3. Pierron et al. Impact of mycotoxin on immune response and consequences for pig health. *Animal Nutrition*. 2016; 2: 63-68.

4. Savard, C. et al. Deoxynivalenol (DON) naturally contaminated feed impairs the immune response induced by porcine reproductive and respiratory syndrome virus (PRRSV) live attenuated vaccine. *Vaccine*. 2015; V. 33, Issue 32, pp. 3881-3886.

5. E.G. Hendel, S. Curry, P.N. Gott, U. Hofstetter, G.R. Murugesan. Trends in mycotoxin contamination in US Corn. AASV 2019, pp. 177.

Figure 1: Prevalence of mycotoxins.

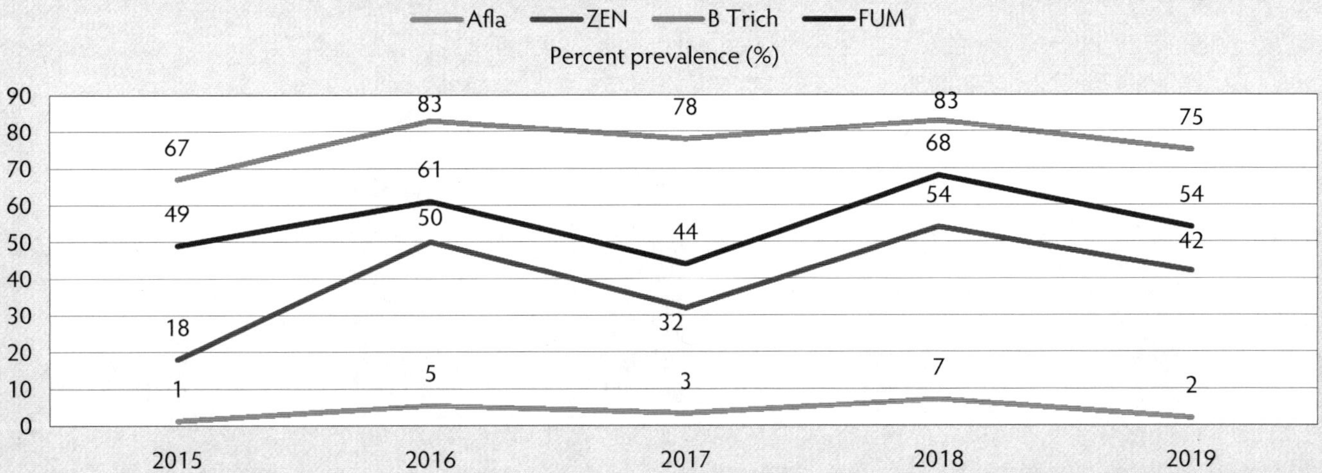

Percent positive samples for Afla, ZEN, DON and FUM are presented by year (due to small numbers of positive samples, T2 and OTA are not represented graphically).

Figure 2: Co-occurrence of mycotoxins.

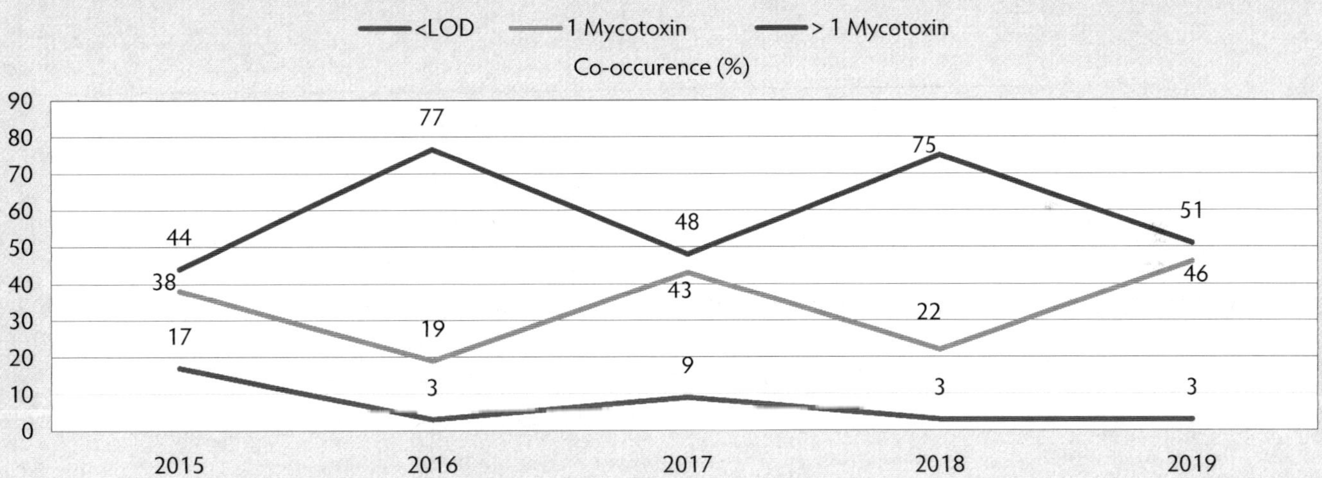

Percent of samples below the limit of detection (LOD), with one, or more than one mycotoxin for each year.

Figure 3: Sample number per year.

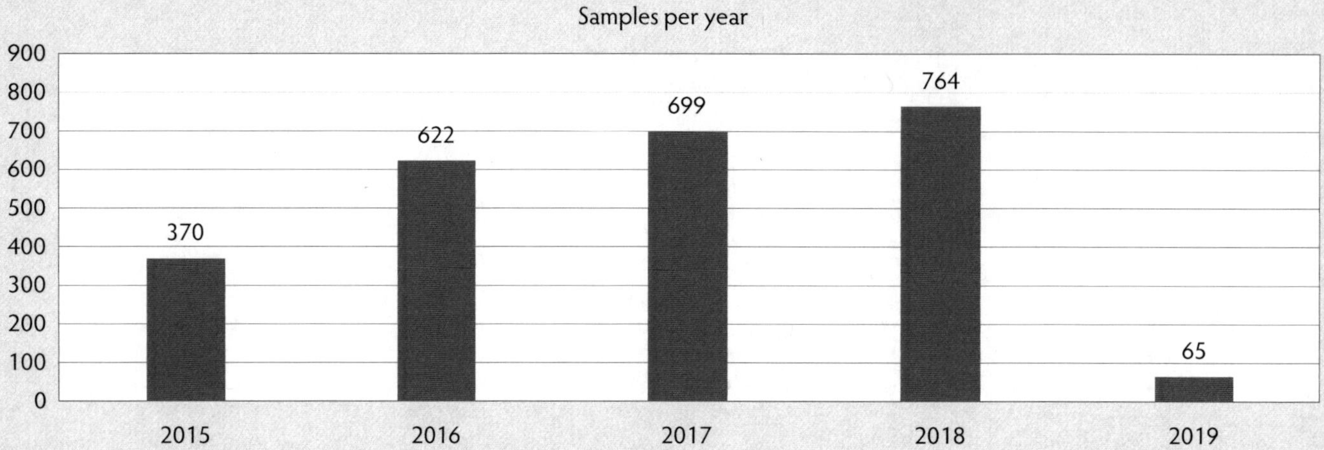

Number of US corn samples for each year from 2014 through the 2018 harvest.

Ensuring that a company's vision and mission and core values are living on farm via pre-training, targeted training, and post-training animal care competency measurements

Sarah Probst Miller, DVM

Introduction

Many companies have a Vision, Mission, and Core Values. Writing these statements and values is one thing. Ensuring that they are living on farm is another. The first goal of Borgic Farms was to create custom Vision and Mission statements that reflected their desire to have a culture of care for both pigs and people while being mindful of Core Values. The second goal was to measure where they were regarding living their Vision, Mission, and Core Values on farm. They wanted to create a strategic training map that acknowledged competency opportunities, alongside remeasurement of desired competencies to ensure progress towards a verified living culture of care. Although this author has employed this methodology in the face of identified abuse or accused abuse in undercover video situations, she was eager to apply the same approach proactively. This study reports animal care competency levels pre-training and post-training in a system wanting to make their newly created Vision and Mission statements alive on farm to quantitatively assess their living culture of care. This study shares expressed competency results of employees post-training compared with the baseline expressed competency results.

Materials and methods

Establishing guiding principles and identifying key competencies

First, a session was held to define the company's Vision and Mission statements. Competencies critical to supporting the farm's Vision and Mission were identified. The farm owner, farm management, and farm veterinarian worked together to determine the company's Core Values, articulate the farm's Vision, and write a concise action plan to achieve the farm Mission.

From the initial brainstorming session, a Vision and Mission were written, and Core Values were identified. The final draft of these is as follows:

Our Vision (where we want to be, the end destination): We respect people and pigs. We get better every day, responsibly.

Our Mission (how we are going to get there, the map): We professionally communicate. We offer good food, air, water, and care to pigs. We measure where we are, set goals on where we want to be, and take action to get there.

Our Core Values: Family, Teamwork, Respect, Integrity, Continuous Improvement, Accountability

Key competencies that needed to be living on farm to ensure the Core Values were alive alongside the Vision and Mission were then identified. It was determined that competencies related to low pressure handling, welfare, and euthanasia should be the focus of the Phase 1 competency assessment (Table 1). Baseline competencies were observed and measured per employee via 3rd party on-site evaluation prior to any training or education of staff regarding the newly created Vision, Mission, and Core Values.

The first external competency assessment was on August 5, 2019. The external verifier worked alongside employees as a new employee asking to be taught the task to determine which of the competencies on the list were living and which were deficient or rather opportunities. Any opportunities were determined to be a training need and the training plan was created to address all identified opportunities (Table 2).

Training plan

This targeted plan included a specific retraining plan on low pressure animal handling, welfare, euthanasia, and other competencies that impact a culture of care for both people and pigs (Table 2). The farm's training was implemented from August through November utilizing Pork Avenue Training Portal (PATP)[1] to administer and track training. PATP is a subscription-based product which includes a library of learning experiences as well as a training record keeping system on online learning management software. The program components per learning experience include online training, graphical SOPs, and In-barn Verification checklists which employ the *See it. Do it. Teach it.* methodology.[2-8] Simultaneously, an internal and external verification plan was detailed and executed post retraining to measure expressed competency results using the lessons' In-Barn Verification checklists as guiding documents. All training accomplishments, including In-Barn Verifications, were recorded on PATP's online learning management software and a transcript of learner accomplishment was created per learner.

See it.

The farm chose to administer most videos and quizzes as a group and after each lesson have a discussion on skills they needed to express differently on farm. Groups were formed based on preferred

Table 1: Action list and baseline assessment scores versus post training assessment scores. Opportunities for immediate actions and employee training were identified from the on-farm assessment by an external veterinarian.

External verification- all farms Observe the following actions	Baseline assessment score	Post training assessment score
Training and handling boars		
Handler properly positioned sort-board between self and boar	1	1
Two people present if boar in pen with people	1	1
Farm has plan for boar culling	1	1
Farm safely loads boar to boar movement equipment	1	1
Moving pregnant sows and gilts		
Hallway is prepped and distractions removed	0	1
No dead pigs in hallway	1	1
Sows are moved calmly and slowly	1	1
Sows are moved in groups of 3 to 5	0.5	1
Pressure is appropriate	0	1
Noise is intermittent, calm and not constant	1	1
Employees use appropriate language when moving animals	0	1
Weaning sows		
Hallway is prepped and distractions removed	1	1
No dead pigs in hallway	1	1
Sows are moved calmly and slowly	0.5	1
Sows are moved in groups of 3 to 5	1	1
Pressure is appropriate	1	1
Noise is intermittent calm and not constant	1	1
Employees use appropriate language when moving animals	1	1
Weaning pigs		
Hallway prepped and not slippery, distractions removed	1	1
No dead pigs in hallway	1	1
Pigs are moved calmly and slowly	0	1
Pigs are moved in properly sized groups per farm protocol	0	1
Pressure is appropriate	0	1
Noise is intermittent, calm, and not constant	0	1
Employees use appropriate language when moving animals	0	1
Euthanasia		
No pigs with BCS of 1 observed in random sample of farrowing	1	1
No pigs with BCS of 1 observed in random sample of gestation	1	1
Farm has approved method of euthanasia in place	1	1
There is a log for cleaning of captive bolt gun	0	1
Captive bolt gun is cleaned, disinfected, and stored properly when not in use	1	1
Designated euthanasia employee can demonstrate method to confirm death	0.5	1
Total score	20.5	31
Total possible score	31	31
Competency percentage	66%	100%
Percentage point increase from baseline		34%
Percentage increase from baseline		52%

language of English or Spanish. Learners were informed prior to watching the videos that they needed to learn the topics well enough that they could teach the content to a verifier. The *See it.* portion of the experience was marked as complete on individual employee transcripts.

Do it.

Then the learners continued to complete skills which were manager or mentor observed prior to verification. If desired, pictorial SOPs were available for review in English and Spanish.

Teach it.

An internal and external verification plan was detailed and executed post-training to measure expressed competency results. The internal verification was completed by a member of the farm crew with leadership potential. He was trained how to assess by submitting himself to being taught. After coordinating the video-based training session with the team, he scheduled one-on-one sessions with employees. Using an In-Barn Verification Checklist for guidance, he would say, "Pretend I am a new employee and teach me how to do this task." He then listened for desired competencies on the In-Barn Verification Checklist and marked them as present or absent. He could ask questions to prompt teaching on competencies if needed. He spent most of the verification experience listening. When individuals successfully passed their In-barn Verifications, Verifiers marked this part of the learning experience complete on the learner's online transcript.

On November 18, 2019, the key competencies were remeasured by the external verifier going to the farm and approaching random employees with the following request: "Pretend I am a new employee and teach me how to do this task." The verifier when applicable, then proceeded to do the task on farm with the employee to observe competencies. Competencies were then marked present or absent.

Results

The baseline expressed competency as determined by an external verifier was at 66%. Post external verification of training, staff averaged a 100% competency expression rate. This is a 34 -percentage point difference and a 52% increase in handling competency expression on farm (Table 1). While subjective, the verifier discerned "a palpable change in attitude" of workers on farm from the first verification to the second.

Conclusion and discussion

Using the "See it. Do it. Teach it." methodology[2-8] has effectively improved competency expression in other studies from a 25.8% level of competency expression to a 95% competency expression level.[9] This case study also achieved improved levels of expressed competency. Specifically utilizing this educational methodology alongside external verification, it increased expressed competency 34 percentage points, which was a 52% increase. Of note, in this study the baseline competencies were specifically chosen to represent the competencies the farm owner, managers, and veterinarians wanted to ensure were living on farm.

During the first baseline external verification, a high number of predatorial movements and expressions and vocalizations were observed in handlers by the external verifier. Some handlers were even "growling" as they moved pigs. Very little inter-staff communication occurred. All vocalizations were directed at pigs. Occasionally curse words were used regarding pigs and people. Post training, handlers were visibly more relaxed albeit working hard. Staff smiled. A couple sang or hummed with a positive tone. Verbal communication was practiced. Although occasional foul language was used, it was not directed in a slanderous way towards pigs or people. Figuring out how to measure presence and absence of human behavior and expressions that represent a risk to the culture of care could be an interesting way to more objectively assess this perceived change in culture.

Of note: During the first baseline external verification, behavior of a department lead was noted as a deep concern by the verifier based upon aggressive behavior towards a stalled sow during movement. This same employee had received warnings regarding animal care in the past. The farm manager and owners decided it was time to dismiss this employee if the person was not going to show their commitment towards the Vision, Mission, and Core Values. The farm manager and the external verifier reported that making the decision to dismiss the person of concern was critical to the success of the program.

Besides reassessing the competencies focused on during the August 2019 assessment, the verifier also took baseline competencies of additional skills that Borgic Farms wanted to see living to create a Phase 2 Strategic Training Plan. The employees scored a 61% on second phase competencies present as a baseline. Reassessment will occur in February prior to delivery of this presentation and will be reported at that time.

References

1. Pork Avenue Training Portal. 2018. Available at `www.porkavenuetraining.com`.

2. Bear, Anne, Technology, Learning, and Individual Differences, Journal of Adult Education, 2012.

3. Cercone, K. (2008). Characteristics of adult learners with implications for online learning design, AACE Journal, 16(2), 137-159.

4. Dale, Edgar, Audio visual methods in teaching. Holt, Rinehart and Inston, 1965.

5. Dale, Edgar, Cone of Experience, Education Media, 1969.

6. Geaetano, Kristian, 5 ways adults learn differently than children, Learn Kit, 2016.

7. Probst Miller, Sarah, Applying the science and physiology of how adults learn to our on-farm approach as practitioners, AASV Proceedings, 2016.

8. Ryland, Bruce, Fostering Critical Reflection in Adulthood, by Jack Mezirow & Associates, San Francisco: Jossey-Bass, 1990.

9. Probst Miller, Sarah, Krantz, Seth, Inspiring a verified culture of care post-identification of abuse on an undercover video using a baseline assessment of expressed competencies compared to post training measurement of expressed competencies, AASV Proceedings, 2019.

Table 2: A strategic training plan post baseline assessment was implemented and internally and externally verified via "teach it back" experiences

Action	Responsible party	Completion
4.01eng - Safe Pig Handling - Pig Behavior and You-- lesson	Employees and internal verifier	10/21/19
4.01eng - Safe Pig Handling - Pig Behavior and You-- internal verification	Internal verifier	10/22/19
4.01eng - Safe Pig Handling - Pig Behavior and You-- external verification	External verifier	11/18/19
4.05eng - Safe Pig Handling - Weaning Pigs-- lesson	Employees and internal verifier	10/23/19
4.05eng - Safe Pig Handling - Weaning Pigs-- internal verification	Internal verifier	10/24/19
4.05eng - Safe Pig Handling - Weaning Pigs-- external verification	External verifier	11/18/19
4.03eng - Safe Pig Handling - Moving Pregnant Sows and Gilts-- lesson	Employees and internal verifier	10/25/19
4.03eng - Safe Pig Handling - Moving Pregnant Sows and Gilts-- internal verification	Internal verifier	10/30/19
4.03eng - Safe Pig Handling - Moving Pregnant Sows and Gilts-- external verification	External verifier	11/18/19
4.02eng - Safe Pig Handling - Training and Handling Boars-- lesson	Employees and internal verifier	N/A
4.02eng - Safe Pig Handling - Training and Handling Boars-- internal verification	Internal verifier	N/A
4.02eng - Safe Pig Handling - Training and Handling Boars-- external verification	External verifier	11/18/19
5.04eng - Take 5 - Captive Bolt Maintenance-- lesson	Employees and internal verifier	N/A
5.04eng - Take 5 - Captive Bolt Maintenance-- internal verification	Internal verifier	N/A
5.04eng - Take 5 - Captive Bolt Maintenance-- external verification	External verifier	11/18/19
14.01eng - Take 5 - Vaccination Guidelines-- lesson	Employees and internal verifier	N/A
14.01eng - Take 5 - Vaccination Guidelines-- internal verification	Internal verifier	N/A
14.01eng - Take 5 - Vaccination Guidelines-- external verification	External verifier	11/18/19
2.22eng - Take 5 - Body Condition Scoring with Calipers	Optional for managers	N/A

N/A = Not available

Where are all the low viability pigs coming from?

C. Bruns, PhD; B. McNeil; J. Sonderman, PhD; T. Rathje, PhD
DNA Genetics, Columbus, Nebraska

Introduction

Low birth weights are associated with low viability pigs. Lower birth weight pigs have been shown to have greater pre-weaning mortality and poorer lifetime performance compared to heavier pigs at birth.[1,2] The negative relationship between total born in a litter and birth weight has been well-documented for many years.[3,4] Since the average litter size has increased over time, there is concern that the number of low viability pigs is increasing at a faster rate than the number of viable pigs; however, there are still more pigs with an acceptable birth weight produced in larger litters.[3] The objective of this study is to examine birth weight distribution differences between litters with varying numbers of pigs born alive. Understanding the impact of increased litter size on the birth weight distribution is important for optimizing herd management and production efficiency. Additionally, the trend in birth weight over time within a given litter size will be examined. A positive trend in birth weight within a given litter size could suggest that genetic selection for heavier birth weights can be effective at reducing the number of low viability pigs within a litter.

Materials and methods

Data for this study were obtained from 2 sow farms with similar genetics and management protocols. All sows included in the study were from a purebred Yorkshire population. Some of the sows were mated to a purebred Yorkshire boar producing a purebred litter, and some sows were mated to a purebred Landrace boar producing a crossbred litter. The birth weight was recorded for every pig born alive in 2,511 purebred and 715 cross breed litters from January to November 2019. A total of 34,875 purebred and 10,160 crossbred pigs were included in the dataset. The average born alive was 14.5 ± 2.8 for the purebred litters and 14.8 ± 2.7 for the crossbred litters. The relationship between born alive and average birth weight (AVGBW), standard deviation of birth weight (SDBW), the number of pigs heavier than 800 g at birth (N800), and the percent of liveborn pigs heavier than 800 g at birth (PER800) was estimated using a model with litter breed and sow parity as fixed effects, gestation length as a covariate, and month within herd as a random effect. The threshold of 800 g was chosen based on a common industry practice for defining a minimum acceptable birth weight.

A second dataset comprised 109,868 purebred pigs from 5,334 litters born between January 2017 to November 2019 at 3 sow farms with similar genetics and management protocols was used to evaluate the trend over time in AVGBW and PER800. This population has been under genetic selection for maternal birth weight, the dam's contribution to birth weight, since March 2017. In this dataset, the average number born alive was 14.3 ± 2.9 pigs. The trends were estimated using a model with an interaction between born alive and month, a covariate of gestation length, and a fixed effect of parity. This model allowed for a different regression estimate across time for each number born alive.

Results and discussion

The predicted means and standard deviations across born alive values for AVGBW, SDBW, N800, and PER800 are displayed in Tables 1 and 2 for purebred and crossbred litters, respectively. Similar results are found in both purebred and crossbred litters; however, the crossbred pigs tended to be heavier at birth compared to the purebred pigs. While the average birth weight of pigs in a litter was estimated to decrease as the litter size increases, little change in birth weight variation within a litter was estimated as litter size increases. The number of pigs classified as low birth weight should not significantly increase with litter size. The estimated number of pigs within a litter born weighing more than 800 g increased with born alive; therefore, the evidence within this population suggests that increasing number born alive results in more quality pigs at birth, not fewer.

The linear trend from January 2017 to November 2019 for AVGBW and PER800 in purebred Yorkshire gilts litters are shown in Figures 1 and 2, respectively. For ease of understanding, the trends in the figures were grouped for born alive sizes 14 to 15, 16 to 17, and 18 to 20 based on the similar slopes within the categories. The genetic selection for increased birth weight has resulted in a positive trend for ADGBW and PER800.The average birth weight has increased ~50 g and the percent above 800 g has increased ~1.5% for litters where the born alive was 16 to 17 in under 3 years. Similar trends were found for P2 litters (results now shown); however, the average birth weight was 16 g heavier ($P < 0.0001$) with 4.0% more pigs ($P < 0.001$) weighing more than 800 g compared to gilt litters. It is interesting to note that while the AVGBW trend in the largest litters (18 to 20 born alive) is lower compared to the smaller litter sizes, the PER800 trend in the largest litters is better than the PER800 trend in the lower litter sizes. This suggests that a significant increase in AVGBW is not required to increase the PER800 within a litter.

The results of this study indicate that increasing born alive does increase the number of quality pigs at birth, even though the average birth weight decreases with increasing litter size. Additionally, since beginning selection for birth weight, a positive trend in AVGBW and PER800 was observed. The positive trend in PER800 implies that the positive trend in AVGBW did not result in more variation in birth weight. Thus, fewer liveborn pigs

Table 1: Predicted[a] means (SD) across litter size for birth weight traits for purebred litters farrowed by Yorkshire sows

Born alive	Birth weight (kg)[b]	Std. dev. (kg)[c]	N > 800 g[d]	Percent > 800 g[e]
5	1.65 (0.04)	0.23 (0.02)	4.73 (0.36)	95.9% (2.4%)
6	1.51 (0.04)	0.23 (0.02)	5.77 (0.33)	95.2% (2.2%)
7	1.55 (0.04)	0.21 (0.01)	6.88 (0.29)	97.5% (2.0%)
8	1.54 (0.03)	0.22 (0.01)	7.82 (0.28)	97.6% (1.9%)
9	1.51 (0.03)	0.24 (0.01)	8.73 (0.25)	96.6% (1.7%)
10	1.46 (0.03)	0.25 (0.01)	9.46 (0.24)	94.5% (1.6%)
11	1.47 (0.03)	0.25 (0.01)	10.65 (0.23)	96.6% (1.6%)
12	1.42 (0.03)	0.26 (0.01)	11.44 (0.22)	95.2% (1.5%)
13	1.38 (0.03)	0.27 (0.01)	12.33 (0.22)	94.7% (1.5%)
14	1.36 (0.03)	0.26 (0.01)	13.15 (0.22)	93.9% (1.5%)
15	1.33 (0.03)	0.27 (0.01)	13.92 (0.22)	92.8% (1.5%)
16	1.30 (0.03)	0.28 (0.01)	14.43 (0.22)	90.2% (1.5%)
17	1.27 (0.03)	0.28 (0.01)	15.22 (0.22)	89.4% (1.5%)
18	1.25 (0.03)	0.28 (0.01)	16.08 (0.23)	89.2% (1.6%)
19	1.20 (0.03)	0.29 (0.01)	16.41 (0.24)	86.2% (1.7%)
20	1.19 (0.04)	0.29 (0.01)	17.55 (0.30)	87.4% (2.0%)

[a] Estimates from a model with additional fixed effects of sow parity and litter breed, a covariate of gestation length, and a random effect of month within herd
[b] Average birth weight of pigs born alive in a litter
[c] Standard deviation of pig birth weight within a litter
[d] Number of pigs in litter weighing greater than 800 g
[e] Percent of pigs in litter weighing greater than 800 g

born weighing less than 800 g were observed over time. Being able to increase quality pigs while increasing litter size will enable producers to improve production efficiency and throughput through their operation.

References

1. Bergstrom et al. Effects of piglet birth weight and litter size on the preweaning growth performance of pigs on a commercial farm. Kansas State University Swine Day 2009. Report of Progress No. 1020.

2. Jourquin et al. Pigs at risk: birth weight impact on survivability and days to market. International Pig Veterinary Society Belgium Regional Meeting. 2015;Poster 5.

3. Lush et al. Factors affecting birthweights of swine. Genet 1934;19:329.

4. Fredeen and Plank. Litter size and pre- and post-weaning performance in swine. Can J Anim Sci 1963;43:135.

Table 2: Predicted[a] means (SD) across litter size for birth weight traits for crossbred litters farrowed by Yorkshire sows

Born alive	Birth weight (kg)[b]	Std. dev. (kg)[c]	N > 800 g[d]	Percent > 800 g[e]
5	1.70 (0.04)	0.24 (0.02)	4.86 (0.36)	97.0% (2.5%)
6	1.55 (0.04)	0.24 (0.02)	5.90 (0.33)	96.3% (2.3%)
7	1.60 (0.04)	0.22 (0.01)	7.01 (0.30)	98.6% (2.0%)
8	1.59 (0.04)	0.23 (0.01)	7.95 (0.28)	98.7% (1.9%)
9	1.55 (0.03)	0.26 (0.01)	8.86 (0.26)	97.7% (1.8%)
10	1.51 (0.03)	0.26 (0.01)	9.59 (0.25)	95.5% (1.7%)
11	1.51 (0.03)	0.26 (0.01)	10.78 (0.23)	97.6% (1.6%)
12	1.46 (0.03)	0.28 (0.01)	11.57 (0.22)	96.2% (1.5%)
13	1.42 (0.03)	0.28 (0.01)	12.46 (0.23)	95.7% (1.5%)
14	1.40 (0.03)	0.28 (0.01)	13.28 (0.22)	94.9% (1.5%)
15	1.38 (0.03)	0.28 (0.01)	14.05 (0.22)	93.9% (1.5%)
16	1.34 (0.03)	0.30 (0.01)	14.57 (0.23)	91.3% (1.5%)
17	1.31 (0.03)	0.29 (0.01)	15.35 (0.23)	90.5% (1.6%)
18	1.29 (0.03)	0.29 (0.01)	16.21 (0.24)	90.3% (1.6%)
19	1.25 (0.03)	0.30 (0.01)	16.55 (0.25)	87.3% (1.7%)
20	1.24 (0.04)	0.31 (0.01)	17.68 (0.30)	88.5% (2.1%)

[a] Estimates from a model with additional fixed effects of sow parity and litter breed, a covariate of gestation length, and a random effect of month within herd
[b] Average birth weight of pigs born alive in a litter
[c] Standard deviation of pig birth weight within a litter
[d] Number of pigs in litter weighing greater than 800 g
[e] Percent of pigs in litter weighing greater than 800 g

Figure 1: Average birth weight of purebred Yorkshire pigs in gilt litters.

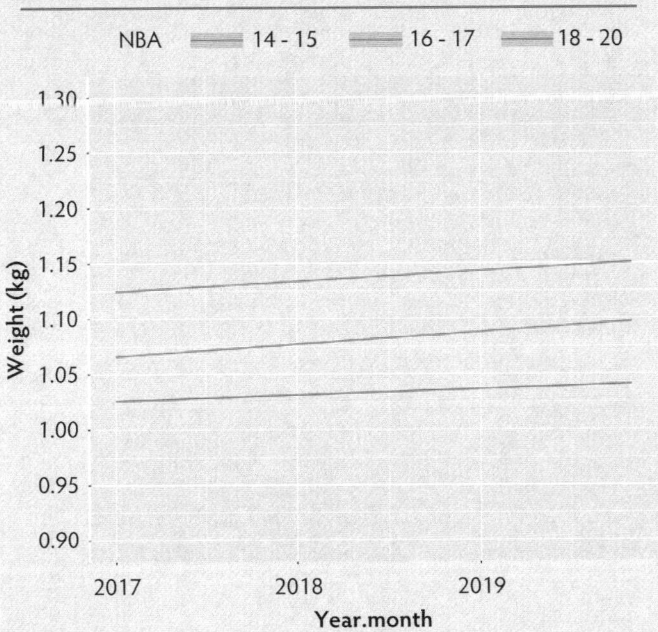

Figure 2: Percent of purebred Yorkshire pigs in gilt litters weighing > 800 g.

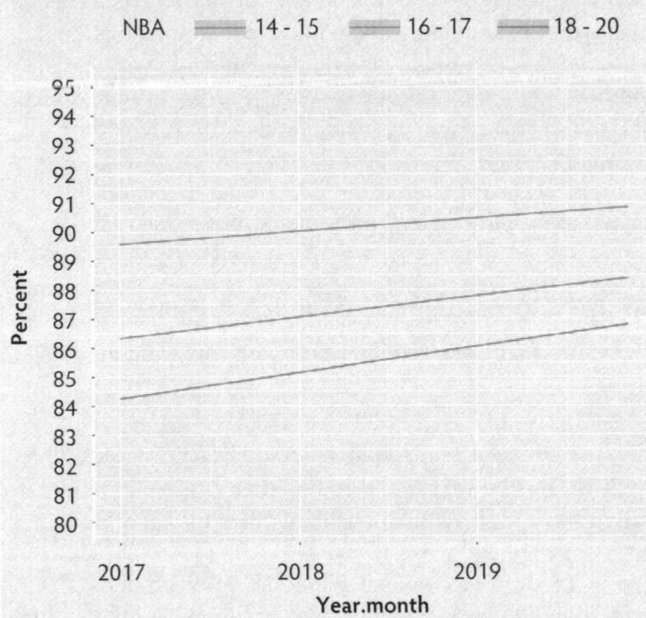

Tonsil scrapings for porcine reproductive and respiratory syndrome virus detection

H. L. Walker[1]; A. S. Bowman[1], MS, DVM, PhD, DACVPM; J. B. Ferreira[2], DVM, MSc, DVSc; A. G. Arruda[1], DVM, MS, PhD

[1]Department of Veterinary Preventive Medicine, College of Veterinary Medicine, The Ohio State University, Columbus, Ohio; [2]Department of Population Health and Pathobiology, College of Veterinary Medicine, North Carolina State University, Raleigh, North Carolina

Introduction

Porcine reproductive and respiratory syndrome (PRRS) is the most costly disease currently affecting the North American swine industry, costing over 600 million dollars annually in financial damages.[1] It has been shown under experimental conditions that PRRS virus (PRRSV) can persist in lymphoid tissues for long periods of time;[2] but for practicality reasons, determination of herd-level PRRS status does not commonly consider lymphoid tissue sampling in the recommended strategies.[3] Even though tonsil scrapings (TS) is the recommended sample of choice for foreign diseases such as classical swine fever,[4] its use for PRRS virus detection has not been investigated under field conditions. The objectives of this study was to compare the use of TS, oral fluids (OF), nasal swabs (NS), and environmental swabs (ES) to detect PRRSV in growing pig farms under field conditions; and investigate whether farm vaccination status had an effect on PRRSV detection in TS.

Materials and methods

Two PRRS positive growing pig farms using different PRRS control strategies were enrolled in this study. The first was a 3,500-head wean-to-finish facility in which unvaccinated pigs were weaned from a newly PRRS positive sow farm. The second was a 2,800-head finisher in which pigs were vaccinated with a modified live vaccine at processing. Sampling began approximately 52 and 21 days post-exposure for farms 1 and 2, respectively. OF and ES samples were collected at the pen level and TS and NS samples were collected at the individual level from each farm using fixed spatial sampling[5] for 5 months. Samples were tested using RT-PCR, and a Ct value < 37 was used to declare PRRSV positivity.[6] The association between farm vaccination status and TS PRRSV detection was tested using Pearson's chi-squared test in STATA 14.2 using $P < 0.05$ to declare statistical significance.

Results

A total of 256 samples were collected: 64 TS, 64 OF, 64 NS, and 64 ES. The majority of tonsil scraping samples (n = 37, 57.8%) tested positive for PRRSV throughout the study period. During the study period, 2 (3.1%) OF samples, 6 (9.4%) NS samples, and 1 (1.6%) ES sample tested PRRSV positive, as seen in Table 1. During the sampling period, all fixed sampled areas for both farms tested positive at least once on TS, with three areas testing positive consecutively for 3 months. TS samples tested positive up to 147 and 203 days post-exposure for vaccinated and unvaccinated farms, respectively. The overall prevalence of PRRSV, considering all sample types, in the unvaccinated and vaccinated herds was 50.0% and 75.0% respectively, and there was no association between farm vaccination status and a positive TS PCR ($P = 0.102$).

Conclusion

There are currently no established protocols for PRRSV sampling and/or proof of elimination for growing pig populations; even though this represents a large and important portion of swine herds. Our results showed that commonly used sampling methods such as oral fluid sampling may yield false negatives; and therefore, an inaccurate PRRS herd status. While tonsil scrapings may be a more invasive and time consuming sampling method, it may be a suitable method for PRRSV detection in growing pig populations, especially when attempting to prove freedom of infection or PRRSV negative status.

References

1. Holtkamp et al., *J Swine Health Prod.* 2013, 21:72-84
2. Wills et al., *J Clin. Microbiol.* 2003, 41:58-62
3. Holtkamp et al., *J Swine Health Prod.* 2011, 19:44-56
4. USDA-APHIS. Procedure Manual for Classical Swine Fever (CSF) Surveillance. 2007. Available at: https://www.aphis.usda.gov/vs/nahss/swine/csf/CSF_procedure_manual_2007.pdf . Accessed 29 January 2019.
5. Rotolo et al., *Vet. Microbiol.* 2017, 209:20-29
6. Gerber et al., *J Clin. Microbiol.* 2013, 51(2):547-556.

Table 1: Proportions (total number) of PRRSV positive samples for the different sample types assessed in this study; calculated based on the number of PRRS positive samples divided by the total number of samples tested by each method for each month, by farm

		Tonsil scrapings	Oral fluids	Nasal swabs	Environmental swabs
Farm 1 (Unvaccinated)	April	87.5% (8)	12.5% (8)	0% (8)	0% (8)
	May	62.5% (8)	0% (8)	0% (8)	0% (8)
	June	50% (8)	0% (8)	0% (8)	0% (8)
	July	25% (8)	0% (8)	12.5% (8)	0% (8)
	August	25% (8)	0% (8)	12.5% (8)	0% (8)
	Overall	50% (40)	2.5% (40)	5% (40)	0% (40)
Farm 2 (Vaccinated)	April	75% (8)	12.5% (8)	0% (8)	0% (8)
	May	83.3% (6)	0% (6)	0% (6)	0% (6)
	June	0% (4)	0% (4)	25% (4)	0% (4)
	July*				
	August**	100% (6)	0% (6)	50% (6)	16.7% (6)
	Overall	70.8% (24)	4.2% (24)	16.7% (24)	4.2% (24)

* No pigs available to sample for this month
** New group of pigs

Evaluating the detection of porcine circovirus type 3 in individual and pooled processing fluids before, during and after sow herd vaccination

R. Stika[1], BS; H. Johnson[2], DVM; A. Betlach[2], DVM; P. Yeske[2], DVM, MS; D. Holtkamp[1], DVM, MS
[1]Iowa State University, Ames, Iowa; [2]Swine Vet Center, St. Peter, Minnesota

Introduction

Since 2015, porcine circovirus type 3 (PCV3) has been detected with increased frequency in the United States.[1] Porcine circovirus type 3 has been detected in multiple sample types, including sera, fetal tissue and processing fluid (PF).[2,3] Processing fluid is obtained at processing from testicles and tails and has been considered the sample of choice for detecting PCV3 in sow herds due to presumed sensitivity and feasibility. However, the swine industry still lacks specific information about the detection of PCV3 in PF, the impact of pooling PF from multiple litters, and the variability of PCV3 detection that exists within herds. The objective of this study was to evaluate the detection of PCV3 and influence of parity using pooled and individual PF samples in a sow herd infected with PCV3 before, during and after two, whole herd PCV3 vaccination. Furthermore, the herd's time to negative pathogen status following vaccination was also evaluated.

Materials and methods

This study was conducted in a 2500 head Midwestern farrow-to-wean farm that vaccinated for PCV3, with the goal of decreasing mummies and improving downstream pig production. This farm had increased gilt matings to correct a deficit in breed numbers and to alter the parity structure of the herd.

Sixty individual litter PFs (testicles only to reduce probability of environmental contamination) were collected from the sow farm prior to and on the day of the first whole herd administration of a PCV3 vaccine (Sequivity®, Merck Animal Health, Kenilworth, New Jersey, USA). These individual litter and 5:1 pooled samples were tested for PCV3 by PCR to assess PCV3 status prior to vaccination. After the first injection of the PCV3 vaccine, 30 individual litter PFs (testicles only) were collected weekly, pooled by 5:1 and tested for PCV3 by PCR. If a pool was PCR positive, individual litter PFs from that pool were tested. Sow ID and parity were recorded for all litters. Parity results were assigned groups: parity-0 (gilts), parity-1, parity-2, parity-3-5 and parity 6+. Data was collected until PCV3 results ensued negative for three consecutive weeks. All diagnostic testing was completed at the University of Minnesota's Veterinary Diagnostic Laboratory, St. Paul, MN, USA.

The first PCV3 vaccination was administered intramuscular (1 mL dose) to all sows, boars and incoming gilts on the farm. The second PCV3 vaccination was administered via the same procedure 4 weeks after the first vaccination.

Pearson's chi square was used to analyze the percentage of positive PCR results by parity group. Tukey t-test was used to analyze the pairwise comparisons of cycle time (Ct) values of litters in each parity group.

Results

Figure 1 presents the percentage of positive results of pooled and individual litter samples by collection time point until negative herd status. The Pearson's chi square test was significant (P-value = 0.0240), indicating that the percentage of positive results was affected by parity. Ct values were significantly lower (higher virus levels) for gilts compared to all other parities except parity-2 females (P-values < 0.05).

Discussion

Gilts were shown to be a good subpopulation to assess PCV3 herd prevalence. The PCV3 pre-vaccination and post-vaccination prevalence of detection levels varied each week. This may influence the ability to assess PCV3 detection due to the wide variety of exposure demonstrated. Inconsistent detection of PCV3 may be attributed to the increase in gilt matings or represent herd health instability, as potential slow viral transmission could create negative subpopulations within the population. The ability to detect PCV3 was similar between individual and pooled samples. Based on this finding, the collection of pooled PF samples for PCV3 detection is a cost-effective and sensitive approach that can be applied in the field. However, as the herd begins to stabilize, individual PF samples may result in a more precise detection. In weeks 14, 15, 16 and 17-post vaccination, PCV3 was consistently undetected, suggesting negative status. The findings reported here are consistent with results from two additional sow farms (Stika, unpublished data, 2019), which all demonstrated that implementation of the Sequivity® PCV3 vaccine into a sow herd resulted in decreased pathogen prevalence within 4 months of vaccination.

References

1. Palinski R, Pineyro P, Shang P, et al. 2016. A novel porcine circovirus distantly related to known circoviruses is associated with porcine dermatitis and nephropathy syndrome and reproductive failure. *J Virol.* 2016; 91(1):01879-16.

2. Byers E. 2018. Porcine circovirus type 3: Practitioner perspective. Proc. 49th AASV Annual Meeting, 445-447.

3. Riley A, Betlach A, Johnson H. 2019. Prevalence of porcine circovirus type 3 based on PCR in processing fluids or piglet serum collected from Midwest sow farms. Proc 50th AASV Annual Meeting, 289-290.

Figure 1: Percent positive PCV3 PCR results for individual & pooled processing fluid samples by week post-vaccination.

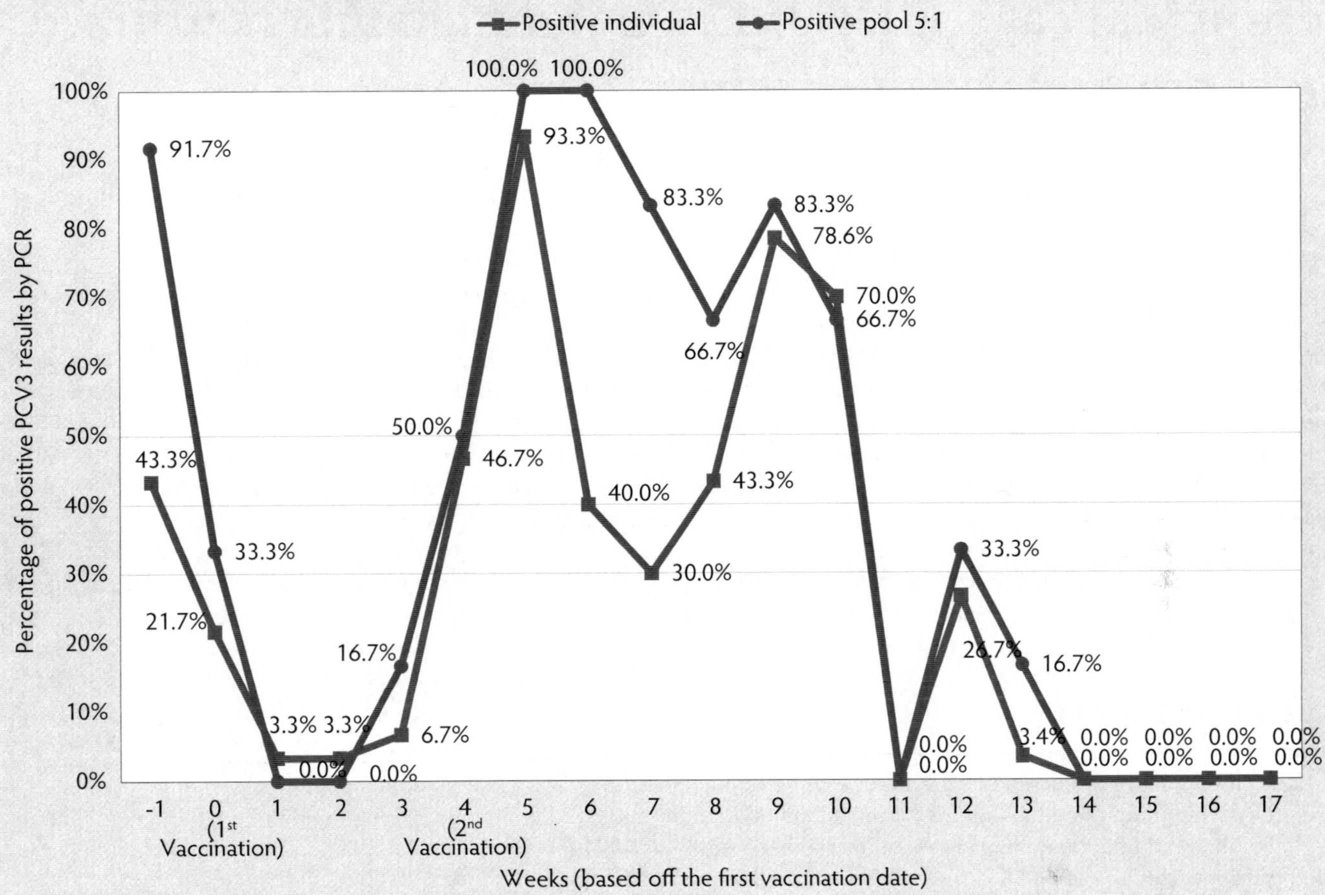

Assessing the impact of organic matrices on the germicidal capability of UVC light chambers utilized on commercial swine farms

T. Pieper[1]; B. Blair[1]; J. Lowe[1,2]
[1]College of Veterinary Medicine, University of Illinois, Urbana, Illinois; [2]Lowe Consulting Ltd., Mahomet, Illinois

Introduction

Many techniques exist to limit the risk of disease introduction through the entry of tools and supplies to swine farms. One such technique is the utilization of ultraviolet (UV) disinfection chambers, a common biosecurity practice for the disinfection of small items. While germicidal capabilities of UV-C light are well established within the scientific literature,[1] mixed reports exist as to its efficacious application within a farm.[2] The presence of shadowing or organic materials have been suggested to impact the germicidal activity of UV-C on a surface, yet little research exist testing the application of UV-C disinfection under these complex conditions. This study attempts to evaluate the impact of organic material on the efficacy of UV-C decontamination of a surface.

Materials and methods

Compost, a model for manure, and brain heart infusion broth (BHIB), a model for proteinaceous bodily fluids were selected to test the decontamination of a complex organic material by UV-C. Sterilized compost was mixed with sterile water at 5 different dilutions. 1mL of each dilution was then inoculated with one dose of Enterisol® Salmonella T/C. BHIB was mixed at a 1:1 and 2:1 dilution and inoculated in the same manner. The inoculated sample was placed on a smooth plastic surface and directly subjected to UV-C in a model UV chamber designed from a 30L plastic bin lined with reflective material and two UV-C lights directly above the samples. UV-C treatment was applied for 16, 32, or 64 minutes for 9 replicates per dilution. Treated samples were swabbed and plated on salmonella selective XLD agar. Due to the results of salmonella decontamination testing of only the lowest dilutions of BHIB and compost (1:32) were tested for decontamination of PRRSV. 1 mL of the organic material dilution was inoculated with one dose of Ingelvac PRRS MLV before placed in the UV-C chamber for 8, 16, 32, and 64-minutes (N = 9). Plates were swabbed and samples used to infect MARC-145 cells cultured in 24-well plates. Cytopathic effect of each well was assessed at 3- and 6-days post infection.

Results and discussion

Complete decontamination of salmonella within compost or BHIB was not achieved within any dilution regardless of the length of UV-C exposure. Although complete decontamination was not achieved, UV-C was seen to impact the growth of salmonella. As the dilution of organic material and the length of UV-C exposure increased, the amount of salmonella growth observed decreased (Table 1). Complete decontamination was achieved for all PRRSV samples irrespective to dilution or length of exposure. These results prove that the presence of organic material has a negative impact on the decontamination of a surface by direct UV-C exposure. While complete decontamination of PRRSV was achieved, the lack of decontamination of salmonella generates concerns about the efficacy of UV-C on hardier pathogens. The findings of this study suggest that items entering the farm through a UV chamber must be free of all organic material to maximize the decontamination potential of direct UV-C exposure.

References

1. Gostine, A., Gostine, D., Donohue, C., Carlstrom, L. & Words, K. American Journal of Infection Control Evaluating the effectiveness of ultraviolet-C lamps for reducing keyboard contamination in the intensive care unit : A longitudinal analysis. AJIC Am. J. Infect. Control 44, 1089–1094 (2016).

2. Rieland, K., Yang, M., & Torremorell, M. (2019). *Evaluation of Efficacy of Ultraviolet Germicidal Chambers in Swine Farms. Evaluation of Efficacy of Ultraviolet Germicidal Chambers in Swine Farms*. Swine Disease Eradication Center. Retrieved from https://vetmed.umn.edu/sites/vetmed.umn.edu/files/shmp_2018119.48_evaluation_of_ultraviolet_germicidal_chambers-sciencepage.pdf.

Table 1: The percentage of surfaces remaining positive for pathogen after exposure to UVC for varying time

Matrix	Pathogen	Dilution	Time(min)	% Positive
Compost	Salmonella	1:32	64	88.9%
		1:32	32	100%
		1:16	64	100%
		1:16	32	66.7%
		1:8	64	100%
		1:8	32	100%
BHIB	Salmonella	High	64	100%
		High	32	100%
		Low	64	66.7%
		Low	32	77.8%
Compost	PRRSV	1:32	64	0%
		-	32	0%
		-	16	0%
BHIB	PRRSV	Low	32	0%
		-	16	0%
		-	8	0%

A comparison of sow fecal PRRS antibody levels to PRRS antibody levels measured in serum, colostrum, or milk samples

S. Baker[1]; T. Schwartz[2], DVM; D. Baumert[3], DVM; D. Amodie[3], MS
[1]Iowa State University College of Veterinary Medicine, Ames, Iowa; [2]Suidae Health and Production, Algona, Iowa; [3]Zoetis, Parsippany, New Jersey

Introduction

Porcine reproductive and respiratory syndrome virus (PRRSV) is a significant swine pathogen estimated to cost the US swine industry $664M annually.[1] The current industry "gold standard" for PRRSV antibody detection is serum sampling. While a sound method of sampling, it requires technical skill that involves time, effort, and training. Colostrum and milk samples have also been used to a lesser degree to determine sow PRRSV antibody levels. Additionally, facilitating ease of sample collection while maintaining confidence in the test results is of interest to veterinarians. The objective of this study was to compare the detection of PRRSV antibody in traditional samples to PRRSV antibody detected in fecal samples collected at similar time points.

Materials and methods

This was a cross-sectional study which began with an enrollment phase of 60 due-to-farrow sows at a 6,000 head, PRRSV positive, stable farm (Category II). The first phase consisted of serum and fecal collection one-week pre-farrowing. Based on the serum PRRS ELISA test results, a subset of 15 sows with low S/P ratios and 15 sows with high S/P ratios were identified. As the subset of thirty sows farrowed, colostrum and feces were collected from each sow within 24 hours of farrowing; additionally, milk and feces were collected within 72 hours post-farrowing, which concluded the second phase. Statistical analyses looking for an association between milk, serum or colostrum antibody levels and fecal antibody level were performed for this study. Using the SAS Proc Corr procedure (SAS 9.4, Cary, NC), correlation analyses were performed using the Pearson's correlation coefficient to test the statistical relationship between the level of PRRSV antibody detected in several biological samples

Results

The relationship of PRRSV antibody levels in feces, serum, milk, and colostrum is reported in Figure 1. PRRSV antibody levels in feces were poorly correlated to PRRSV antibody levels in pre-farrow sow serum ($R^2 = 0.238$) and early lactation milk samples ($R^2 = 0.138$) as measured by ELISA S/P ratios. PRRSV antibody levels in feces were moderate to highly correlated with PRRSV antibody levels in post-farrowing colostral samples ($R^2 = 0.597$) as measured by ELISA S/P ratios. When only PRRSV ELISA-positive pre-farrow serum samples were considered, the level of PRRSV antibody in the fecal samples was moderately correlated ($R^2 = 0.54$) to the level of PRRSV antibody in the serum samples.

Discussion

Based on the correlation between the level of PRRSV antibody identified in colostrum and matched fecal samples, and the correlation between the level of PRRSV antibody identified in ELISA-positive serum and matched fecal samples, fecal samples may be an appropriate, easier-to-obtain sample than serum or colostrum, when a determination of the level of PRRSV antibody in a sow herd is desired. Additionally, fecal ELISA testing may provide a more consistent PRRSV antibody S/P ratio than serum, colostrum, or milk sample testing as these may be affected by the individual target animal's hydration status, phase of colostrum/milk production, or suckling activity before colostrum/milk sampling. Due to the relatively low number of samples tested in this trial, additional larger-scale testing may be desired. Finally, investigation into fecal antibody levels for other significant swine pathogens may be of interest.

Reference

1. Holtkamp DJ, Kliebenstein JB, Neumann EJ, et al. Assessment of the economic impact of porcine reproductive and respiratory syndrome virus on United States pork producers. *J Swine Health Prod.* 2013;21(2):72-84.

Figure 1: Correlation of fecal antibody S/P ratios to traditional sample antibody S/P ratios.

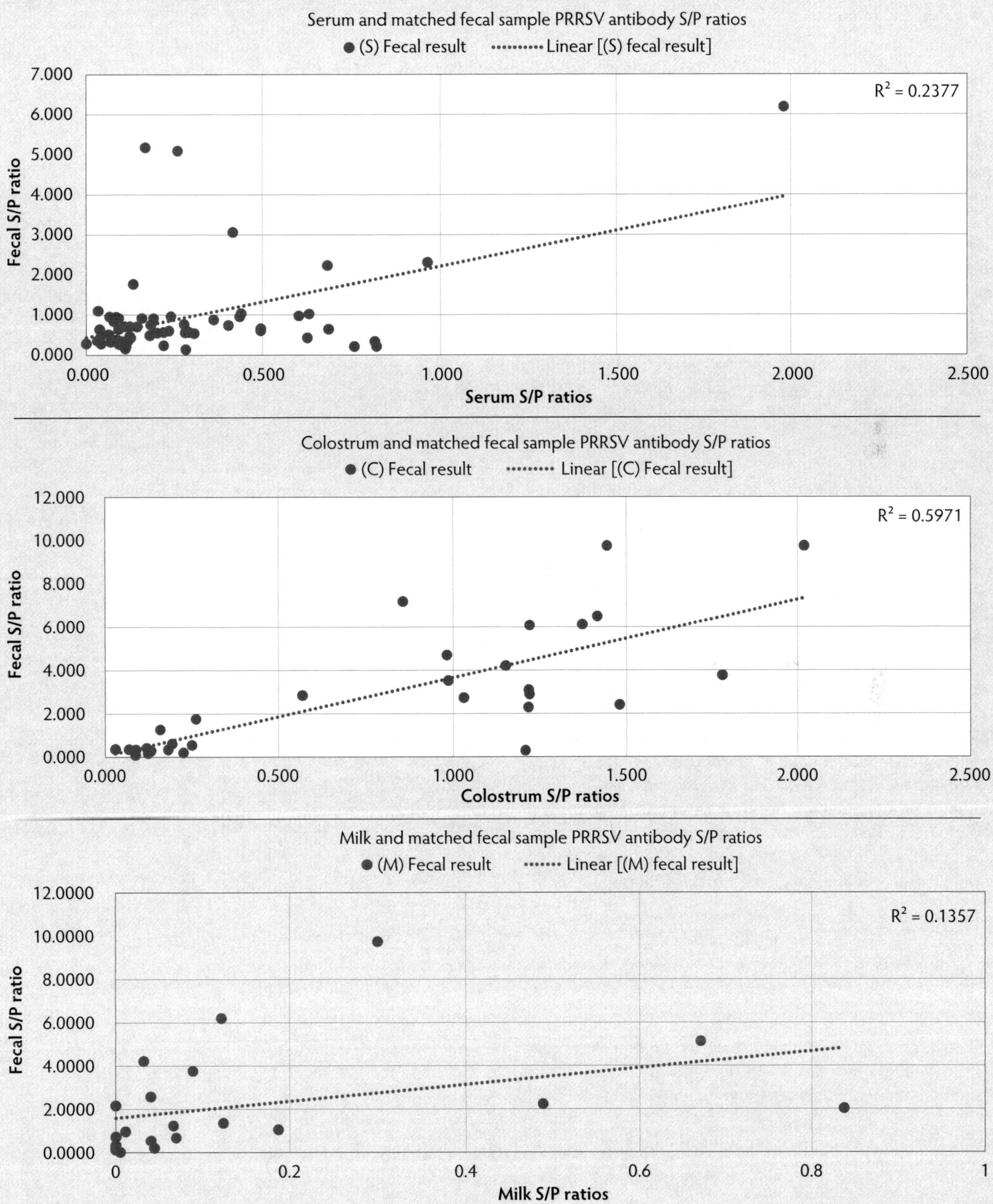

Application of time of flight cameras for sow lameness detection

Matthew Boulanger; Thomas Parsons, VMD, PhD
University of Pennsylvania School of Veterinary Medicine

Introduction

Growing public concern about the welfare of gestating sows is driving changes in husbandry practices, as producers are being asked to move away from individual housing of sows to group housing. Increased incidence of sow lameness is often reported with group housed sows. While group housing can make lameness more noticeable, inter-animal aggression at group formation can also lead to increased leg injuries and lameness. Lameness is both a production and welfare problem, but early detection remains challenging as existing metrics are notoriously unreliable. Thus, reliable early detection of lameness promises improved therapeutic outcomes, increased productivity, and improved welfare.

Objective

To test and further an approach to automated lameness detection on sows that was developed for dairy cattle, where the topography of the vertebral spines are detected with infrared cameras and computerized image analysis.

Materials and methods

Four gilts (DNA Line 241) were recruited from an ongoing behavioral study where the animals were observed daily. Two sound sows, and two sows diagnosed with lameness were removed from the original study. A Time of Flight (TOF) infrared camera (O3D313, ifm Efector) was attached to the ceiling of a barn hallway with a solid concrete floor. It was mounted 77 inches from the floor and pointed vertically down to generate a topography of the sow's back (Figure 1a). The camera was controlled through Matlab (Mathworks) to alternate collecting images at 5 Hz and 50 Hz. Study sows walked under the camera 4 consecutive times. The resulting images were processed two different ways:

1 – The mathematical concept of "curvedness" was used to identify the topology of a sow's spine and then fit with a polynomial (longitudinal perspective Figure 1b). Based on work in cattle, lame animals are proposed to have flatter backs and lower coefficients in a fit polynomial.

2 – The average height of the sow's shoulder and hips were calculated as the animal walked (top-down perspective Figure 1b), and asymmetry was tracked between left and right sides.

Results

Spine topography was successfully fit to a polynomial with R^2 values above 0.85 by averaging across each transverse section. However, the coefficients generated in the first method exhibited no statistically significant difference between sound and lame sows. The second approach yielded data which fit amplitude coefficient differences of 148 and 62 (cm) between the shoulders of sound gilts. Lame gilts fit coefficient differences of 0.8 and 16 (cm), which is consistent with more constrained movement.

Discussion

Adoption of TOF cameras into mass marketed facial recognition systems has driven down its price. For example, the Microsoft Kinect v2 images up to 4 meters away and cost around $400. While this study used an 8 meter camera for flexibility, the 4 meter camera could have also worked. Automated detection of lameness promises to be a boon for the management of group housed sows, especially in farms utilizing electronic sow feeding where the sows could be easily monitored by the camera as they exit the feeder each day. The failure of the first algorithm to detect lameness may relate to anatomical differences between cattle and swine. However, preliminary findings from the second approach suggested that a threshold could be determined to differentiate between lame and sound sows. This warrants further development towards barn implementation- from recording more gilts to increasing parity representation. Furthermore, this study has used guilts which were monitored daily and confirmed to be chronically lame in a front limb- not recovering from their lameness. Moving forward, a broader range of lameness (both in terms of location in severity) will be enrolled into the study.

Figure 1: Depictions of the collected data. To the left, Figure 1a, is a sample frame of footage of a pig as she walks under the 3D camera. To the right, Figure 1b, are three different perspectives of the data generated. Figure1b displays different ways the data can be considered, and the successful recording of 3D topography.

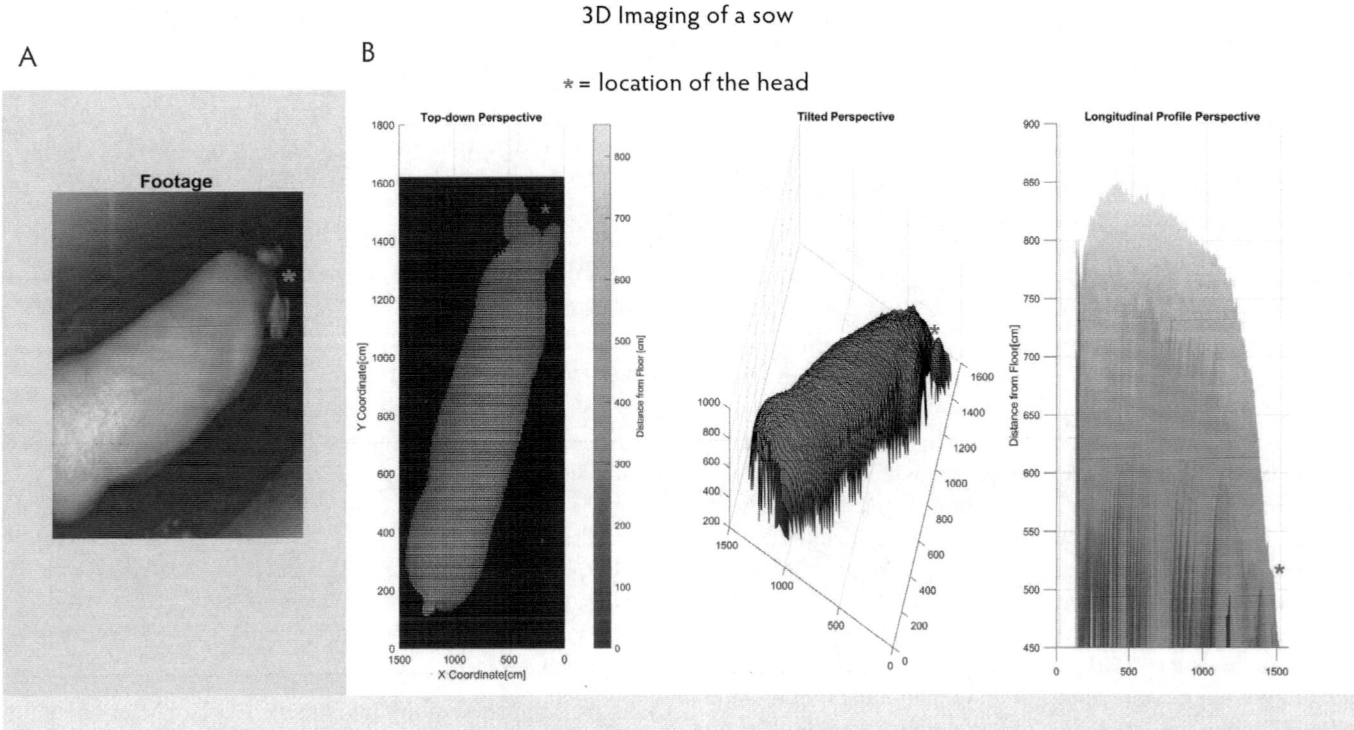

Investigation of area spread of *Mycoplasma hyopneumoniae* during aerosol acclimation of gilts

N. Benge[1], BS; A. Betlach[2], DVM; J. Morgan[3], DVM; H. Johnson[2], P. Yeske[2], DVM, DVM; E. Fano[3], DVM, PhD
[1]Iowa State University College of Veterinary Medicine, Ames, Iowa;
[2]Swine Vet Center, St. Paul, Minnesota; [3]Boehringer-Ingelheim Vetmedica, Duluth, Georgia

Introduction

Mycoplasma hyopneumoniae (Mhp) continues to be a prevalent and economically significant respiratory pathogen in the swine industry. This pathogen has a low transmission rate (R_n = 0.56-1.16) and can shed up to 240 days.[1,2] Primarily transmitted by nose-to-nose contact, Mhp may shed from dam to offspring with gilts being the primary shedders within a sow herd.[3] To promote cessation of Mhp shedding at first farrowing, gilt acclimatization strategies have been performed with the intent to minimize dam-to-piglet transmission.[2] Furthermore, these strategies have been utilized to help determine day "0" for Mhp elimination programs. Techniques involving infected seeder animals, or intra-tracheal inoculation with herd specific, Mhp-positive lung homogenate have been used to acclimate gilts.[4] Recently, controlled exposure of gilts to a herd-specific, Mhp lung homogenate (via aerosol) has been attempted due to presumed high, exposure success and ease of administration. Since airborne transmission of Mhp has been described (9.2 km from naturally positive site),[5] the potential risk for airborne spread of the pathogen during this acclimation technique remains unclear. Such information is important to understand, especially when considering neighboring biosecurity and risk for exposure to neighboring herds. Therefore, the objective of the study was to assess area spread of Mhp during gilt acclimation strategies utilizing aerosol exposure.

Materials and methods

Five gilt development units (GDUs) undergoing gilt acclimation for Mhp using aerosol exposure were selected for the study. Four GDUs were managed via continuous flow, while one GDU was all-in/all-out. To confirm Mhp airborne detection, air collectors were placed inside one GDU during and three-weeks post-aerosol exposure to serve as a positive control. Next, air samples were collected immediately outside the exhaust fans, upwind (~100m), and downwind (~100, 200, and 400m) from the daily prevailing wind. Samples were collected at four time points including: prior to and during aerosol exposure, during maximum ventilation post-exposure, and at three weeks post-exposure to determine airborne spread at peak of shedding. Dry filter and cyclonic air collectors (INNOVAPREP, Drexel, MO; Midwest MicroTek, SD, USA) with a 30 min run-time were used to collect air samples. For three GDUs, maximum ventilation occurred during the exposure period. To evaluate potential risk factors for Mhp airborne detection, weather conditions (i.e. humidity, temp, wind speed), distance(s) to neighboring site(s), and ventilation parameters were recorded. Three weeks post-exposure, 30 tracheal samples were collected from each site to confirm Mhp exposure success. All samples were tested for the presence of Mhp genetic material using real-time PCR.

Results and discussion

A total of 101 air samples were collected throughout the study. During aerosol exposure, Mhp was detected in 100% (8/8) of the air samples collected inside the GDU (avg Ct Value =31.3(29.8-33.3)). At three weeks post-exposure, 100% (0/16) of the air samples collected inside the GDU were negative. At each sampling event, 100% of the air samples collected outside each GDU were negative for Mhp. (Table 1) Overall, 78% of the deep tracheal samples were Mhp positive, in which 50-100% were positive at each GDU.

Under the conditions of this investigation, lack of Mhp detection outside GDUs during exposure suggests aerosolized Mhp lung homogenate, though infective and present in aerosol within the barn, poses minimal risk of neighboring, airborne spread. However, additional research is needed to further understand the potential impact of variables such as strain variation, co-infection with other respiratory pathogens, weather conditions, ventilation parameters, geography, air sampling protocol, and acclimation protocol on aerosol spread of Mhp.

References

1. Meyns T, Maes D, Dewulf J, Vicca J, Haesbrouck F, de Kruif A. Quantification of the spread of Mycoplasma hyopneumoniae in nursery pigs using transmission experiments. *Preventative Veterinary Medicine*. 2004; 66(1-4) 265-275.

2. Pieters M and Fano E. Mycoplasma hyopneumoniae management in gilts. *Veterinary Record*. 2016; Jan: 122-123

3. Garza-Moreno L, Segales J, Pieters M, Romagosa A, Sibila M. Survey of Mycoplasma hyopneumoniae gilt acclimation practices in Europe. *Porcine Health Management*. 2017; 3(21): 1-8.

4. Yeske P. Mycoplasma hyopneumoniae: Lateral transmission and gilt exposure methods. *Proc AASV*. San Diego, CA. 2018:482-484.

5. Otake S, Dee S, Corzo C, Oliveira S, Deen J. Long-distance airborne transport of infectious PRRSV and Mycoplasma hyopneumoniae from a swine population infected with multiple viral variants. *Veterinary Microbiology*. 2010; 145: 198-208.

Table 1: *Mycoplasma hyopneumoniae* positive PCR results from air samples during gilt acclimatization at GDUs

	Air samples prior to acclimatization	Air samples upwind during acclimatization and max ventilation	Air samples outside exhaust during acclimatization and max ventilation	Air samples downwind during acclimatization and max ventilation	Air samples inside of barn during acclimatization	Air samples 3-weeks post aerosol exposure
GDU A	0/6	0/2	0/2	0/8	N/A	0/4
GDU B	0/5	0/2	0/2	0/4	N/A	0/6
GDU C	0/5	0/2	0/2	0/4	N/A	0/5
GDU D	0/5	0/2	0/2	0/4	N/A	0/5
GDU E	N/A	N/A	N/A	N/A	8/8	0/16
Total	**0/21 (0%)**	**0/8 (0%)**	**0/8 (0%)**	**0/20 (0%)**	**8/8 (100%)**	**0/36 (0%)**

* N/A = not available.

Analysis of porcine parvovirus maternal antibody decay in replacement gilts by HI and ELISA assays

A. Anderson[1], BS; M. Herring[1]; S. Mehling[1], BS; S. VanderPoel[2], DVM; J. Nerem[2], DVM
[1]Iowa State University, Ames, Iowa; [2]Pipestone Veterinary Services, Pipestone, Minnesota

Introduction

Porcine parvovirus (PPV) is a significant cause of reproductive disease in gilts. Fortunately, commercial vaccines have led to PPV stability in many of today's commercial swine farms. However, it is important to administer the vaccine in the absence of maternal antibody (Mab) to ensure an adequate immune response. Previous research has has suggested that Mab persists until an average of 21.2 weeks of age in replacement gilts from clinically infected dams when tested by hemagglutinin inhibition (HI).[1] Recent data have indicated a much shorter Mab duration based off enzyme-linked immunosorbent assay (ELISA).[2] In this study, only 1.5% of pigs were positive for PPV antibody by 12 weeks of age. Due to these differences, the objective of this study was to determine the PPV maternal antibody decay curve for HI and ELISA and elucidate the youngest age at which replacement gilts could be vaccinated.

Materials and methods

This study was conducted on two 5000 sow commercial sow farms, which were sourced from the same multiplier. Sows in the multiplier were vaccinated with Zoetis FarrowSure® GOLD by label as gilts. A cross-sectional study was designed to address the research objective. 182 replacement gilts ranging from 4 to 29 weeks of age were enrolled in the trial. Gilts received vaccinations of Zoetis FarrowSure® GOLD at 21 and 23 weeks of age, along with exposure to mummified fetuses. Blood was collected from each gilt and serum was assayed by HI and ELISA to detect PPV antibody. For this study, the INgezim PPV ELISA kit from Ingenasa Biotech-Spain was used. HI titers were adjusted to allow for comparison to ELISA s/p ratios.

Results

The attached graph depicts the mean adjusted HI titer and the mean ELISA titer by age group (Figure 1). Both tests demonstrated titer decay over time. By 13 weeks of age, 86% of gilts were negative for PPV antibody by both HI and ELISA, with 90% negative by 16 weeks of age. Titers remained low until the time of vaccination/mummy exposure, followed by a sudden rise and gradual reduction over time. Seventy of 182 (38.5%) HI tests were positive while 65 of 182 (35.7%) ELISA tests were positive. The percent agreement of the two tests was 73.1% and determined by categorizing samples according to their HI/ELISA status (23.6% positive/positive, 14.8% positive/negative, 12.1% negative/positive, 49.5% negative/negative).

Discussion and conclusions

Overall, there is moderate agreement between HI and ELISA tests. Since the HI test is currently available and affordable, it is still the best method for PPV antibody detection. Regardless of test, this study revealed that it is possible to vaccinate replacement gilts for PPV earlier than previously thought; however, the maternal antibody decay curve may vary from farm to farm. Practitioners can immediately implement this information to make decisions regarding farm-specific testing methods and the best time to vaccinate replacement gilts in their herds.

References

1. Paul PS et al. Duration and biological half-life of passively acquired colostral antibodies to porcine parvovirus. *American Journal of Veterinary Research.* 1982;43(8):1376-1379.

2. Gava D et al. Dynamics of vanishing of maternally derived antibodies of Ungulate protoparvovirus 1 suggests an optimal age for gilts vaccination. *Tropical Animal Health and Production.* 2017:49(5):1085-1088.

Figure 1: Line graph depicting the average adjusted HI titer and average ELISA titer for replacement gilts at various ages.

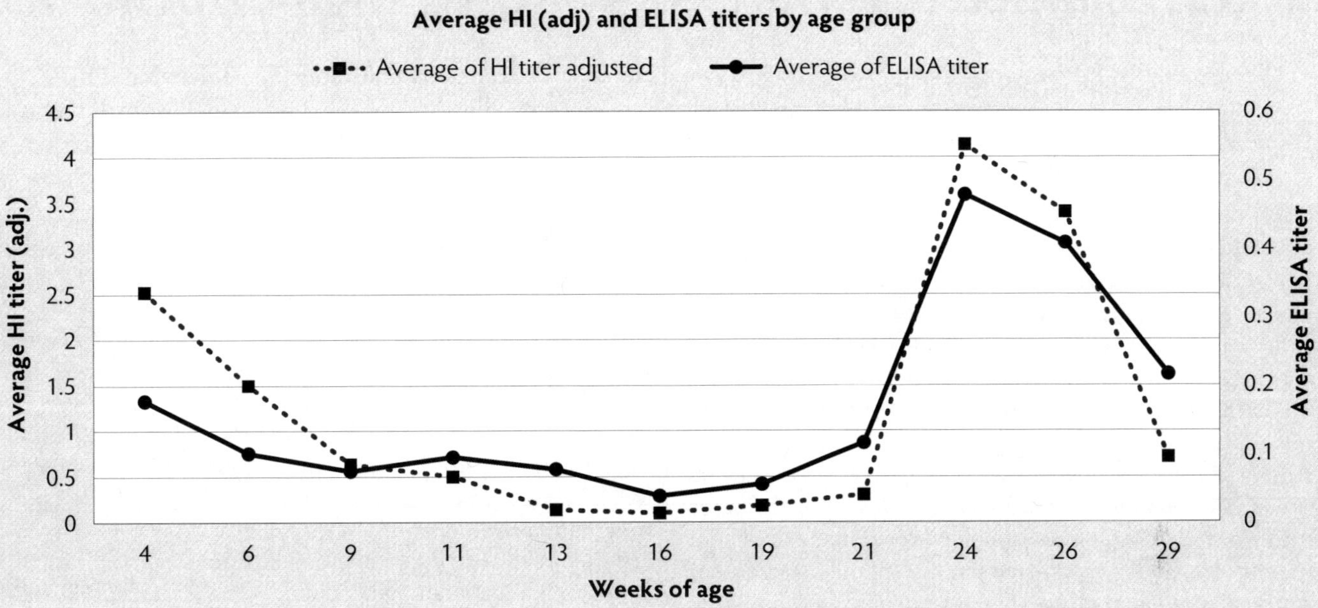

Comparison of prefarrow rotavirus vaccine and natural planned exposure on suckling pig performance

W. Boyd[1], BS; J. Pittman[2], DVM, MS, DABVP; Z. Wolfe[3], BS; D. Kumar[4], DVM, MS; D. Marthaler[4], PhD
[1]Virginia-Maryland College of Veterinary Medicine, Blacksburg, Virginia; [2]Smithfield Hog Production – North Region, Waverly, Virginia; [3]North Carolina State University College of Veterinary Medicine, Raleigh, North Carolina; [4]Kansas State University Veterinary Diagnostic Lab, Manhattan, Kansas

Introduction

Rotavirus (RV) is known to cause diarrhea in young pigs resulting in reduced weight gain and increased mortality. There are three main RV types in pigs: RVA, RVB and RVC. Natural planned exposure (NPE) such as feedback of feces collected from farrowing has been a longstanding practice designed to boost maternal immunity and protect suckling piglets. But this method does present the risk of not knowing what pathogens are in the material with minimal quality control. To mitigate this risk, preparation of NPE material through inoculation of colostrum deprived piglets with known RV types, commonly called the "ice cube" or Master Seed Method, presents a safer, more consistent alternative.[1] However, this process is not without risk as well as being labor-intensive and time-consuming. A vaccine platform would be much more effective in these measures but currently a commercial vaccine is only available for RVA, as RVB and RVC have been difficult to adapt to cell culture. Recently, RNA particle vaccines (RPV) can be made for any of the three swine rotavirus types using Sequivity™ (Merck Animal Health) where sequenced genes are inserted into RNA particles. In a previous study, we found a 4% decrease in preweaning mortality and a 1.5 lb increase in weaning weights in litters from gilts that were exposed to rotavirus NPE prior to farrowing compared to non-exposed controls.[2] The study reported here was conducted to compare the efficacy of NPE to an RPV for controlling RVA and RVC in the suckling period.

Materials and methods

The project was conducted at an 1,800-head commercial gilt-only farm with group housed gestation. All gilts in a breed group (n = 80) were exposed using the NPE method during acclimation and prior to breeding as the priming exposure to live rotavirus. Pens of 6 gilts were randomly assigned to either NPE exposure only or a combined RPV and control pen (with gilts then randomized to receive one of these two treatments). This was done so that non-NPE treatment animals did not receive exposure to the NPE. The RVA and RVC for the NPE, RPV and piglet challenge were all developed from the same RV strains. The RPV types were sequenced and then constructed using the Sequivity™ technology. The NPE group received NPE at 5, 4, and 3 weeks prior to farrowing (wpf). The RPV group was injected with 1 mL of the RPV IM, at 4 and 1 wpf. The CNT group served as controls and were neither exposed nor vaccinated in the prefarrow period. Gilts were fecal sampled in gestation prior to vaccination and at loading into farrowing crates. Gilts that died, aborted, or farrowed less than 8 pigs were excluded from the study. Standard animal care and treatments were allowed but cross-fostering was not allowed in study litters. All pigs were challenged with a 1 mL oral dose of a 1:8 dilution of NPE at 2 days of age. Litters were weighed at 2 days of age and 3 days prior to weaning. Litter weights were transformed into average piglet weights and wean weights were adjusted to 21 days of age. At challenge, three pigs in each litter were identified by ear notch and a rectal swab was collected at challenge, two days post challenge and at 3 days prior to weaning. Fecal swabs were pooled by litter and tested by PCR for RVA, RVB and RVC. Statistical analysis was performed by Kruskal-Wallis ANOV, One-way ANOV, Chi Square or Fishers Exact Test depending on data type and distribution.

Results and discussion

The results are summarized in Table 1. Total mortality rates were numerically, but not significantly different among the three treatment groups. However, the mortality rate due to scours was significant ($P < 0.05$) and was lowest in the RPV group, intermediate in the NPE group and highest in the control group. Per pig adjusted weaning weights and weight gains showed a similar significant difference ($P < 0.05$) pattern. Gilt and piglet fecal samples prior to challenge were nearly all PCR negative for all RV types (data not shown). The rates of PCR positive piglets after challenge and prior to weaning were similar among the groups for RVA and RVC. After challenge, the NPE group had the highest levels (lowest CTs) of RVA and RVC, controls were intermediate and RPV were the lowest (significantly different for RVC). At weaning, there was no significant differences between the three groups although the control had the highest level of virus (both RVA and RVC). These results indicate that the RPV was beneficial in controlling rotavirus while the benefit of NPE was not apparent as compared to controls.

References

1. Pittman JS. Field Experiences with Interventions for Rotavirus Control. 2016 ISU James D. McKean Swine Disease Conference, November 3–4, 2016; 30-36.

2. Anderson et al. Evaluating natural planned exposure protocols on rotavirus shedding patterns in gilts and the impact on their suckling pigs. AASV 2018; 64-66.

Table 1: Piglet mortality, weight gain, and fecal shedding data

Parameter		Value	Control	NPE	Vaccine
Mortality and weight data					
			Treatment group		
Parameter		**Value**	**Control**	**NPE**	**Vaccine**
No. litters		Count	22	25	23
Mortality	Total	Pct.	16.95	14.42	12.68
	Scours	Pct.	6.44[a]	4.81[ab]	2.17[b]
Pig start weight		Lb., Avg.	3.59	3.69	3.78
Pig wean weight		Lb., Avg.	12.42[a]	12.93[ab]	13.76[b]
Gain per pig		Lb., Avg.	8.83[a]	9.24[ab]	9.98[b]
Piglet fecal swab testing by PCR					
Parameter			**Treatment group**		
Time	**Type**	**Value**	**Control**	**NPE**	**Vaccine**
After challenge	A	Rate	18/22	25/25	21/23
		CT	27.2	26.1	30.3
	B	Rate	1/22	0/25	0/23
		CT	39.6	40.1	40.1
	C	Rate	20/22	24/25	19/23
		CT	26.1[ab]	23.8[b]	29.6[a]
At weaning	A	Rate	21/22	24/25	19/23
		CT	23.5	25.0	26.4
	B	Rate	0/22	0/25	0/23
		CT	40.1	40.1	40.1
	C	Rate	14/22	12/25	8/23
		CT	33.2	36.1	36.9

[ab] Superscripts within row differ statistically ($P < 0.05$)

Comparing processing fluid to serum for antibody detection of porcine reproductive and respiratory syndrome virus, porcine circovirus type 2, influenza A virus of swine and *Mycoplasma hyopneumoniae* on commercial sow farms

Brian Johnson, BS; Channing Sebo, DVM; Patrick Graham, DVM, MS; Brad Thacker, DVM, PhD, MBA; Kelly Greiner, DVM
University of Illinois College of Veterinary Medicine, Urbana, Illinois;
Pike Pig Systems/Ghrist Veterinary Clinic, Pittsfield, Illinois; Merck Animal Health, De Soto, Kansas

Introduction

The US swine industry is increasingly dependent on diagnostic testing of weaned piglets for disease monitoring of sow farms to improve the effectiveness of health-related interventions. Veterinarians are always investigating new techniques to develop more practical and more robust methods for herd health level monitoring. This study used enzyme-linked immunosorbent assay (ELISA) and immunofluorescent antibody (IFA) assay to compare the serological status (positive vs. negative) and antibody levels of processing fluid versus serum for porcine reproductive and respiratory syndrome virus (PRRSV), porcine circovirus type 2 (PCV2), influenza A virus of swine (IAV-S) and *Mycoplasma hyopneumoniae* (M hyo) in piglets on sow farms at the time of processing. Serum is most commonly used to measure antibody status for diseases. Processing fluid has become a common sample type for molecular diagnostics, but application for antibody testing has been limited. Processing fluids can be collected by farm staff and does not require the expertise of a veterinarian or trained technician for collection. Validation of processing fluids for antibody testing could enable accurate assessment of the immunological status of the whole sow herd, reduce the time for sample collection, improve the welfare of sampled piglets, and ultimately improve disease monitoring options. The objective of this study was to compare processing fluid to serum for monitoring the serological status and antibody levels of piglets prior to weaning for PRRSV, PCV2, IAV-S and M hyo.

Materials and methods

Four commercial sow farms were selected based on their disease statuses. All farms were stable for IAV-S and PCV2. Farm A was stable for PRRSV and positive for M hyo. Farms B and C were negative for PRRSV and M hyo. Farm D was positive for PRRSV and M hyo. At each farm, processing fluid (testicles only) and serum were collected from four male piglets per litter from 15 litters, 60 litters total. All samples were tested using commercially available ELISA for PRRSV, PCV2, IAV-S and M hyo. Also, for PCV2, IFA assay was used. All serum samples were pooled by litter. For each disease, the percent agreement for each test was calculated and correlations between antibody levels were performed by linear regression and Pearson correlation. The analysis was done across all farms regardless of their disease status.

Results and discussion

The results are presented in Table 1. For M hyo, the percent agreement for serostatus and antibody level correlations were relatively high, although the average antibody levels based on S/P ratios were lower for processing fluid compared to serum. For PRRSV, the percent agreement was relatively low, correlations were relatively high and antibody levels (S/P ratios) were lower for processing fluid. For IAV-S, the percent agreement was high, correlations were high, and the antibody levels were lower (higher S/N ratio) for processing fluids. For these three tests, the results suggest that the dilution of processing fluid may need to be lowered, which is done with testing oral fluids by ELISA. For PCV2 ELISA, the serostatus results are difficult to interpret because all litters were positive for both sample types. Correlations were low and antibody levels were lower in processing fluid. Refinement of this assay for testing processing fluid may be more complicated. PCV2 IFA showed a high percent agreement, but correlations were low with lower antibodies in processing fluid. In summary, these results demonstrate the potential utility for using processing fluid instead of serum for determining serostatus and antibody levels in nursing piglets. Refinements of the assays may be needed to provide similar acceptable results.

Table 1: Comparison of serum versus processing fluid for determining antibody levels of four common swine pathogens

Test	Value	Sample type: Serum	Sample type: Proc. fluids	Correlations: R square	Correlations: Pearson	Agreement
M hyo ELISA	Avg. SP	0.651	0.403	0.959	0.979	
	No. positive	26	23			91.67%
	No. negative	34	37			
PRRS ELISA	Avg. SP	0.344	0.153	0.979	0.989	
	No. positive	24	6			70.00%
	No. negative	36	54			
PCV2 ELISA	Avg. SP	2.129	1.707	0.692	0.832	
	No. positive	60	60			100.00%
	No. negative	0	0			
IAV-S Nucleo-protein ELISA	Avg. SN	0.28	0.37	0.884	0.940	
	No. positive	53	49			93.33%
	No. negative	7	11			
PCV2 IFA	Geom. Titer	513.9	272.2	0.727	0.853	
	No. positive	49	53			93.33%
	No. negative	11	7			

Environmental sampling for porcine epidemic diarrhea virus and porcine delta coronavirus to emphasize the importance of biosecurity practices

C. Peterson[1], MS; K. Dion[2], DVM; A. Sponheim[3], DVM
[1]Iowa State University CVM, Ames, Iowa; [2]The HANOR Company of Wisconsin, LLC., Enid, Oklahoma; [3]Boehringer Ingelheim Animal Health USA Inc., Duluth, Georgia

Introduction and case history

Swine enteric coronaviruses (SeCOVs) are stable in the environment and can be readily introduced into farms during animal and employee entry. Furthermore, SeCOVs can be difficult to eliminate once introduced into swine facilities. Biosecurity practices including secure entry at the line of separation (LOS) and cleaning and disinfection are commonly utilized to prevent and eliminate the viruses from the environment. During the winter of 2018-2019 a large production system experienced increased numbers of introductions of SeCOVs and difficulty in eliminating the viruses from facilities following cleaning and disinfection. The objective of this investigation was to evaluate high-risk areas at the LOS for the presence of SeCOVs on farms to understand the risk of virus reintroduction and emphasize sound biosecurity practices.

Case description

Environmental samples were collected from fourteen nurseries where the previous group of animals were diagnosed SeCOVs positive by PCR via fecal sampling. All farms were depopulated, cleaned and disinfected, and all but two were repopulated with coronavirus negative animals, confirmed by fecal PCR. Six sample points were identified at the LOS where SeCOVs could enter the farm and sampled when available. These locations included the floor on the dirty side of the shower/entry, the dirty boot tray at the entry, loading chute on the outside of the building, dirty boot tray at the mortality door inside the building, mortality door handles inside and outside, and the cement pad by the incinerator. Environmental sampling kits containing Dey and Engley (D/E) neutralizing broth and Swiffer cloths (Procter & Gamble, Cincinnati, OH) in plastic bags were used for sample collection. 10mL of D/E broth was poured into the plastic bag to wet cloths. Cloths were used to wipe designated areas and the liquid sample from the cloth was squeezed into the plastic bag. Three milliliters of sample was collected from the bag with a pipette and placed into microcentrifuge tubes on ice. Gloves were changed between each sample location to prevent cross contamination. Samples were tested using a multiplex PCR for porcine epidemic diarrhea virus (PEDV), porcine delta coronavirus (PDCoV), and transmissible gastroenteritis virus (TGEV).

Results

PEDV and PDCoV PCR results are presented in Figure 1. No samples tested positive for TGEV. All areas sampled had at least one positive sample. The sample location with the highest percentage of positives (50%) was on the floor of the dirty side of the shower/entry. Of the 22 positive samples, 10 tested positive for PEDV, 11 tested positive for PDCoV, and one sample tested positive for both PEDV and PDCoV.

Discussion

Environmental sampling was useful to detect SeCOVs in this investigation. A positive PCR on environmental samples does not necessarily indicate infectious virus, but indicates the presence of genetic material at detectable levels near the LOS, consistent with previous research.[1] Based on this investigation, the recommendations and actions taken by this production system were to install Danish entry benches to further define the LOS and increase employee training on the importance of secure entry at the LOS. Further emphasis was put on disinfecting LOS areas when cleaning and disinfecting the facility. Environmental sampling for SeCoVs can be a useful tool to identify areas of risk for potential routes of virus reintroduction and areas to emphasize solid biosecurity practices.

Reference

1. Radke et al. Evaluation of disinfectants to neutralize porcine epidemic diarrhea virus. AASV. 2015;117-118.

Figure 1: Diagnostic data for detection of PEDV and PDCoV at different environmental locations.

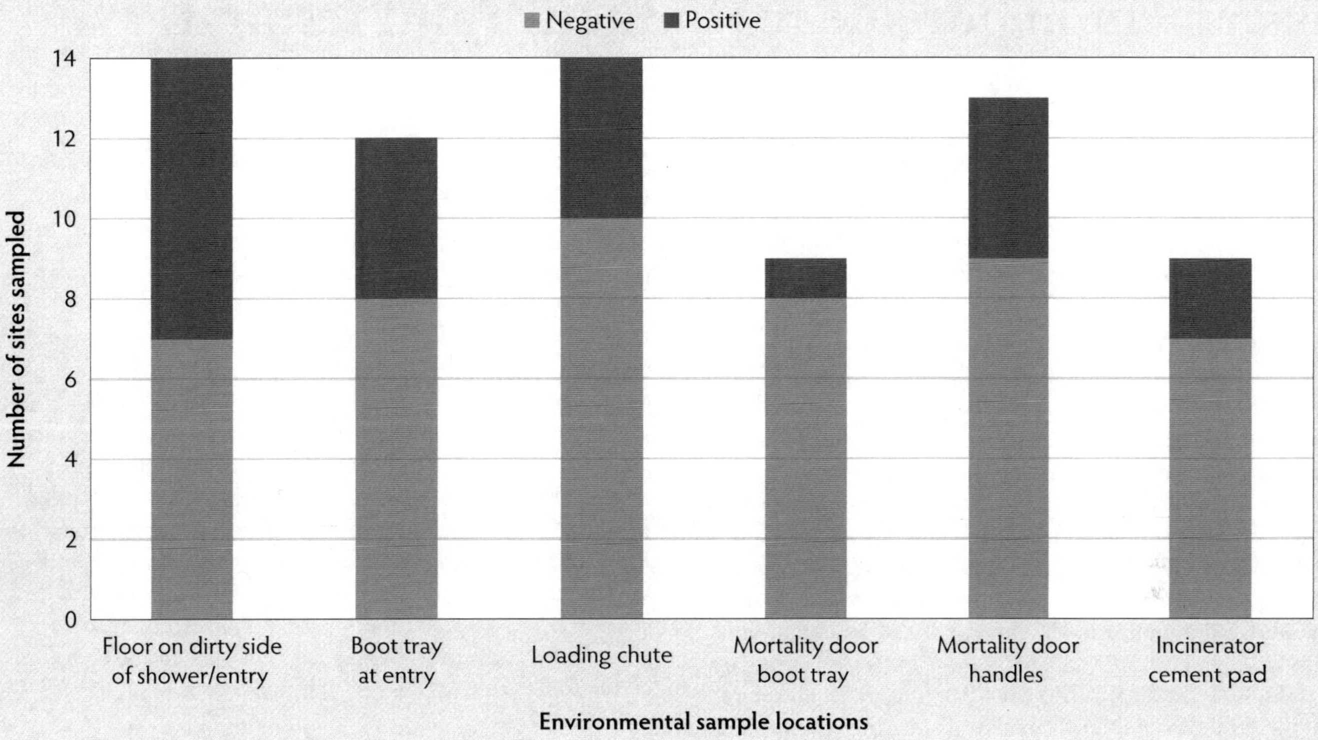

Comparison of tulathromycin vs. tilmicosin at placement in porcine reproductive and respiratory syndrome virus positive pigs

K. Carroll[1], BS; M. Hoogland[2], DVM, MS; R. Swalla[3], DVM; N. Lauterbach[2], DVM;
D. Amodie[3], MS; A. Szczotka[1], BS
[1]Iowa State University College of Veterinary Medicine;
[2]Smithfield Hog Production Division, Midwest Region, Algona, Iowa; [3]Zoetis, Parsippany, New Jersey

Introduction

Porcine reproductive and respiratory syndrome virus (PRRSV) is estimated to cost the swine industry $664 million a year.[1] The major cost of PRRSV is due to the secondary bacterial infections that accompany PRRSV infections. Tulathromycin and tilmicosin are two antibiotics that are labeled for the control or treatment of many secondary pathogens associated with PRRS. Both antibiotics show benefits in reducing the severity of PRRS related morbidity and mortality,[2,3] but there are no data in the literature comparing the effectiveness of tulathromycin and tilmicosin in weaning age pigs for the control of secondary infections. This study compared the mortality, morbidity, and average daily gain (ADG) of PRRSV positive pigs administered either tulathromycin or tilmicosin at placement.

Materials and methods

Wean age pigs (from a PRRSV positive sow farm/flow) were evenly distributed by sex (same number of barrows and gilts per pen) into pens, totaling 30 pens per treatment. Average weight of each pen was similar across all pens. Pens were randomly assigned to either a single 0.75cc (2.5 mg/kg) intramuscular tulathromycin (Draxxin 25,® Zoetis, Parsippany, NJ) injection at placement or tilmicosin (Pulmotil,® Elanco Animal Health, Greenfield, IN) administered through the water at placement. Tilmicosin was used in an extra label manner using the tilmicosin feed-grade dosing rate. Tilmicosin was given through the water system but was intended to follow the label for feed-grade tilmicosin. This was done to simulate pigs receiving tilmicosin in the feed. Pigs given tilmicosin received 17.6 mg/kg for 21 days. Doses of tilmicosin were adjusted at the start of every week to account for increases in weight. Ceftiofur crystalline free acid (Excede,® Zoetis, Parsippany, NJ) was used to treat sick pigs, and all additional treatments and mortalities were recorded on a pen basis. On day zero (d0), treatments were administered and pigs were weighed, in groups of 10-15 pigs, to nearest tenth of a pound. On d42 pigs were weighed and an average weight per pen was used to calculate average daily gain. Any pigs that died throughout the study were excluded for weights. Mortality (defined as a binary variable), ADG and number of treatments were all analyzed by a Generalized Linear Mixed Model (GLMM) approach using SAS (SAS 9.4, Cary, NC). Both SAS-Proc Glimmix (binary) and Proc Mixed (continuous) used a model that considered week as a covariate, treatment as a fixed effect, and barn along with pen (barn*trt) and the residual error as random effects. Treatment Least Squares Means (LSMeans) were calculated and compared by two-sided t-test at 5% level of significance.

Results and discussion

Averages of mortality, number of additional treatments, and average daily gain (ADG) for the 42 day period were compared between treatments. These values can be seen in Table 1. Based on the data presented, tulathromycin showed a significantly larger number of pigs alive ($P = 0.0096$), and amount of additional treatments tended to be lower ($P = 0.0751$) compared to tilmicosin. There did not appear to be any difference ($P = 0.2674$) in ADG between treatment groups. The economic impact of this data is not currently known, and additional data, such as feed intake, and feed and drug costs, is needed to be able to evaluate the benefit of giving one of these treatments over the other.

References

1. Holtkamp DJ et al. Assessment of the economic impact of porcine reproductive and respiratory syndrome virus on United States pork producers. Journal of Swine Health and Production. 21(2). 2012.

2. Benfield DA et al. An evaluation of the effects of tilmicosin in feed on nursery pigs inoculated with porcine reproductive and respiratory syndrome virus. AASV. 2002.

3. Desmonts DJM et al. Immuno-modulating and anti-viral properties of Tulathromycin in porcine reproductive and respiratory syndrome. The FASEB Journal. 31(1). 2017.

Table 1: Mortality, individual pig treatments, and average daily gain of pigs treated with tulathromycin and tilmicosin over a six-week post-weaning period.

	Tulathromycin 2.5 mg/kg	Tilmicosin 17.6 mg/kg (feed-grade dosage)	Standard error	*P*-value
Mortality, %	5.65[a]	8.44[b]	0.74	0.0096
Individual pig treatments, %	19.9	22.3	1.21	0.0751
ADG, lbs/d	0.85	0.83	0.03	0.2674

[ab] Superscripts within row differ statistically ($P < 0.05$).

Predef® (isoflupredone acetate) post-partum in sows may reduce inflammation

Y. Yan[1], BS; L. Greiner[2], PhD; D. A. Nelson[3], PhD; L. Baumgard[2], PhD; J. Ross[2], PhD; S. Becker[2], BS

[1]College of Veterinary Medicine, Iowa State University, Ames, Iowa; [2]Department of Animal Science, Iowa Pork Industry Center, Iowa State University, Ames, Iowa; [3]Zoetis, Parsippany, New Jersey

Introduction

Numerous lactating cow studies have reported a positive relationship between pre-calving body condition score (BCS) and excessive systemic inflammation resulting in poor lactation performance.[1,2] This relationship has not been well documented in sows. The objectives of this study were to evaluate the impacts of pre-farrow sow BCS and Predef® 2X (Zoetis, Parsippany, NJ) on metabolic and inflammatory markers and sow lactation performance.

Materials and methods

Eighty sows were sorted into 4 groups (20 sows/group) upon entry into farrowing: (1) BCS = 3 without Predef; (2) BCS = 3 with Predef; (3) BCS > 3 without Predef; and (4) BCS > 3 with Predef. Pedef was given at 5 mg/136.36 kg body weight, intramuscularly for 2 consecutive days starting on the day of farrowing. Sows were on test until d14 post-farrowing. Total sow feed intake, sow weight (estimated via heart girth circumference), and total litter weight on d0 and d14 were recorded. Ten sows per group (40 total) were randomly selected to sample blood on d0, d3, d7, and d14. Serum was analyzed for glucose, non-esterified fatty acids (NEFA), beta-hydroxybutyrate (BHBA), insulin, and lipopolysaccharide binding protein (LBP), as indicators of metabolic and immune function.[3-7] Data was analyzed as a 2 × 2 factorial using Proc Mixed in SAS.

Results

Estimated sow body weights on d0 tended ($P = 0.10$) to be greater for BCS > 3 sows. The BCS > 3 sows had higher ($P \leq 0.05$) insulin on d0 and d3 and tended to have higher NEFA ($P = 0.11$) and glucose ($P = 0.10$) on d14. The BCS > 3 sows also had higher ($P < 0.05$) BHBA on d7 and consumed less ($P < 0.01$) lactation feed. The BCS > 3 sows had lower ($P \leq 0.05$) LBP on d3 and d7. Piglets nursing BCS > 3 sows had lower ($P = 0.01$) average daily gains than piglets nursing BCS = 3 sows. Predef tended ($P < 0.07$) to lower NEFA in both sow BCS groups on d3. Predef increased ($P < 0.03$) BHBA in BCS = 3 sows but reduced BHBA in BCS > 3 sows on d14. Predef reduced ($P < 0.04$) lactation weight loss in BCS = 3 sows, but increased weight loss in BCS > 3 sows. Data of sow and litter performance as well as metabolic and inflammatory markers are shown in Table 1.

Discussion

These data are consistent with the literature showing that BCS > 3 sows at farrowing perform poorer than BCS = 3 sows. The tendency for greater NEFA on d14 of BCS > 3 sows may reflect their lower feed intake and negative energy balance status; however, there are no detectable differences between the BCS categories during the transition period. While BHBA, NEFA and glucose are indicators of negative energy balance, BHBA, NEFA and insulin may also play a role in systemic inflammation;[3-7] in addition, LBP can mediate host responses to inflammatory processes.[8-10] Although there were several interactions between Predef and BCS, the overall data demonstrates that overweight sows do not demonstrate ketosis and inflammation during transition similar to a dairy cow. Additional research is needed to define the appropriate timing and duration of Predef use in relation to parturition and interactions with sow BCS.

References

1. Roche JR et al. Invited Review: Body Condition Score and Its Association with Dairy Cow Productivity, Health, and Welfare. Journal of Dairy Science. 2009;92:5769.

2. Bradford BJ et al. Invited Review: Inflammation during the Transition to Lactation: New Adventures with an Old Flame. Journal of Dairy Science. 2015;98:6631.

3. Contreras GA et al. Nonesterified Fatty Acids Modify Inflammatory Response and Eicosanoid Biosynthesis in Bovine Endothelial Cells. Journal of Dairy Science. 2012;95:5011.

4. Ospina PA et al. Using Nonesterified Fatty Acids and β-hydroxybutyrate Concentrations during the Transition Period for Herd-level Monitoring of Increased Risk of Disease and Decreased Reproductive and Milking Performance. The Veterinary Clinics of North America. 2013;29:387.

5. Kasimanickam RK et al. Associations among Serum Pro- and Anti-inflammatory Cytokines, Metabolic Mediators, Body Condition, and Uterine Disease in Postpartum Dairy Cows. Reproductive Biology and Endocrinology. 2013;11:103.

6. Sun Q et al. New insights into insulin: The anti-inflammatory effect and its clinical relevance. World Journal of Diabetes. 2014;5:89.

7. Bicalho MLS et al. The association of plasma glucose, BHBA, and NEFA with postpartum uterine diseases, fertility, and milk production of Holstein dairy cows. Theriogenology. 2017;88-270.

8. Zweigner J et al. The Role of Lipopolysaccharide-binding Protein in Modulating the Innate Immune Response. Research in Virology. 2006;8:946.

9. Stasi A et al. Emerging role of Lipopolysaccharide binding protein in sepsis-induced acute kidney injury, Nephrology Dialysis Transplantation. 2017;32:24.

10. Dunzendorfer S et al. Toll-like receptor 4 functions intracellularly in human coronary artery endothelial cells: roles of LBP and sCD14 in mediating LPS responses. Federation of American Societies for Experimental BiologyJ. 2004;18:1117.

Table 1: Sow and litter performance and metabolic and inflammatory markers

	Treatment				*P* Values		
	BCS = 3 -Predef	BCS = 3 +Predef	BCS > 3 -Predef	BCS > 3 +Predef	Predef	BCS	TRT × BCS
Sow and litter performance							
Sow weight d 0, kgs	207	201	216	215	≥ 0.12	0.10	≥ 0.12
Sow weight loss to d 14, %	7.0	1.9	4.0	11.0	≥ 0.12	≥ 0.12	0.04
Total feed intake	168.0	171.0	123.0	127.0	≥ 0.12	0.01	≥ 0.12
Daily litter weight gain	2.7	2.6	1.8	2.1	≥ 0.12	0.02	≥ 0.12
Metabolic and Inflammatory markers							
Insulin, d0 (mU/L)	6*	8*	22*	10*	≥ 0.12	0.02	≥ 0.12
Insulin d3 (mU/L)	18	21	31	25	≥ 0.12	0.03	≥ 0.12
Glucose d3 (mg/dl)	71	85	90	84	≥ 0.12	≥ 0.12	0.05
Glucose d 7 (mg/dl)	85	95	97	83	≥ 0.12	≥ 0.12	0.06
NEFA d3 (ueq/L)	329	240	355	276	≥ 0.12	≥ 0.12	0.07
NEFA d14 (ueq/L)	210	243	320	352	≥ 0.12	0.11	≥ 0.12
BHBA d3 (mmol)	0.396	0.344	0.215	0.221	≥ 0.12	0.06	≥ 0.12
BHBA d7 (mmol)	0.151	0.140	0.193	0.241	≥ 0.12	0.05	≥ 0.12
BHBA d14 (mmol)	0.141	0.171	0.198	0.132	≥ 0.12	≥ 0.12	0.03
LBP d3 (ng/mL)	21	19	16	12	≥ 0.12	0.05	≥ 0.12
LBP d7 (ng/mL)	17	10	7	7	≥ 0.12	0.02	≥ 0.12

* Pred × BCS = *P* = 0.06

Evaluation of temperature variation within tissue boxes during shipment

Kris Kovach[1]; Katie Wedel[2], DVM; Nathan Schaefer[2], DVM
[1]Iowa State University, Ames, Iowa; [2]Iowa Select Farms, Iowa Falls, Iowa

Introduction

Tissues are a common sample type for disease diagnosis and are routinely shipped by mail to reach a Veterinary Diagnostic Lab (VDL) for diagnostic case work-up. There are concerns that during extreme temperatures, especially during the warm summer months, autolysis can occur during shipment, leading to poor tissue quality, impaired diagnosis, and compromised pathogen isolation. The objectives of this study were to monitor the variation in temperature detected during the shipping process when tissues were sent from a production system to a VDL and determine opportunities to improve tissue case submissions.

Materials and methods

This observational study was performed using the Sensitech TempTale® Direct temperature monitor. Prior to shipment, one temperature monitor was placed directly on top of the tissue samples within each shipping container. The temperature monitor was activated and the shipping container was sent overnight to a VDL. Throughout the duration of shipment, the temperature monitor recorded a temperature reading every minute. Average ambient temperature was observed and recorded from the US climate data website. Upon receipt, the temperature monitors were retrieved and inactivated. The data were recovered and analyzed to determine the temperature variation throughout transportation. Several observations were noted once the shipping containers were received, including: 1) the size and weight of the shipping container, 2) the number and size of ice packs, 3) the type of insulation, and 4) the shipment duration. Tissues were also examined by a pathologist to determine if autolysis had occurred.

Results and discussion

Table 1 depicts the mean temperature, high and low temperature extremes, duration of the shipping process and average ambient temperatures.

Autolysis is time and temperature dependent. Under ideal circumstances, diagnostic labs would receive tissues that are harvested and chilled to refrigerator temperature (1.6 degrees Celsius) immediately post-mortem and throughout the duration of shipment.[1] The results of this study demonstrated that none of the tissues reached refrigerator temperature before or during shipment, despite being refrigerated for 17 hours prior to shipment on average. Additionally, shipping temperatures ranged from 2.1-21.7 Celsius during shipment, with 67% of cases recording temperatures 11.7 Celsius or higher. However, despite several elevated temperature readings, no autolysis was noted on any of the pathology reports.

The findings suggested modifications to the tissue collection and shipping process could minimize temperature variation and elevated temperature recordings during shipment, including a renewed focus on tissue collection, cooling tissues prior to shipment and including additional ice packs within each shipping container. These results also encouraged the system to evaluate and modify refrigerator temperatures in the sample storage freezer. The ambient temperature throughout the entirety of this study averaged around 21.8 Celsius; therefore, this study will need to be repeated during the winter months to observe if there are temperature differences and whether colder ambient temperatures could negatively affect tissue samples.

Reference

1. Pomponio et al. (2012). The effect of temperature on the activity of μ- and m-calpain and calpastatin during post-mortem storage of porcine *longissimus* muscle. *Meat Science, 91*(1), 50-55.

Table 1: Temperature variations during the shipping process

Trip length	Temperature on arrival (Celsius)	Mean +/- Std deviation (Celsius)	Low extreme (Celsius)	High extreme (Celsius)	Average ambient temperature (Celsius)
16 hours 15 minutes	11.7	9.9+/-1	7.1	11.7	23.3
16 hours 16 minutes	2.8	4.3+/-0.5	2.1	5.8	17.8
16 hours 20 minutes	6.1	5.5+/-1.3	4.5	7.2	19.7
16 hours 40 minutes	13.8	16.3+/-1.7	13.2	18	23.6
16 hours 45 minutes	12.5	13.4+/-1.1	11.1	15.6	22.8
17 hours 20 minutes	14.1	15 +/-1.2	11.8	16.7	24.4
17 hours 40 minutes	20.7	20.3+/-0.4	19.3	21.7	20.8

Evaluation of the T-cell immune response of Prevacent PRRS® vaccinated pigs compared to non-vaccinated control following challenge with a wild type PRRSV strain

A. Jergens[1]; T. Schwartz[2], DVM; B. Evelsizer[3], DVM; B. Arruda[1], DVM; P. Gauger[1], DVM; C. L. Loving[4], PhD
[1]College of Veterinary Medicine, Iowa State University, Ames, Iowa; [2] Suidae Health and Production, Algona, Iowa;
[3] Elanco Animal Health, Greenfield, Indiana; [4] National Animal Disease Center, USDA, Ames, Iowa

Introduction

Porcine reproductive and respiratory syndrome virus (PRRSV) is a major swine disease, costing the industry more than $600 million annually.[1] Vaccines have been widely used with varying success and correlating PRRSV-specific immunity to protection is difficult. T-cells likely play a role in anti-PRRSV immunity, as T-helper cells inducing antibody production and inducing antiviral IFN-γ mediated responses. The evolution of PRRSV-specific T-cell responses over time following infection has been primarily assessed by the IFN-ELISPOT assay, which detects the number of T-cells producing IFN-γ following re-stimulation with PRRSV. Circulating PRRSV-specific T-cells are induced following vaccination, though numbers in circulation following a single application of vaccine may be low. Anamnestic T cell responses may provide insight on the priming of cell-mediated immunity following a single done of PRRSV vaccine. The purpose of this study was to evaluate the number PRRSV-specific IFN-g producing cells induced following vaccination with a MLV PRRSV vaccine (Prevacent PRRS®) and challenge with a characterized 1-7-4 RFLP field isolate. Non-vaccinated/challenged pigs were included as controls.

Materials and methods

Thirty twenty-one day old pigs were obtained from a gilt multiplier and confirmed PRRSV-negative by RT-PCR and X3 ELISA. Pigs were assigned to two groups, Group 1 and 2, and housed separately until comingled on day of challenge.

Group 1 received Prevacent PRRS® vaccine and Group 2 received sham vaccine on D0. On D35, the groups were merged into 1 pen and challenged intranasally with 1 mL of 1-7-4 wild type PRRSV (2.3 × 10^5 TCID50/mL and 12.69 ct value). Samples were collected every 7 days from individual pigs from D0-D70 and included serum and whole blood. Blood was collected into CPT separation tubes for isolation of porcine peripheral blood mononuclear cells (PBMC) to assess cell mediated immunity by enumerating T-cells that produced IFN-γ by ELISpot assay after 18 h restimulation with media-alone, Prevacent virus, or challenge virus (1-7-4).

Results and discussion

Prevacent vaccination induced PRRSV-specific IFN-g secreting cells. On day of challenge (D35 post vaccination), there was not a significant response to PRRSV from PBMC collected from non-vaccinated (NV) pigs (Fig 1A). However, PBMC from the majority of Prevacent vaccinated pigs (Vx) produced IFN-g following re-stimulation with Prevacent virus. PBMC from Prevacent vaccinated pigs also produced IFN-g following restimulation with the 1-7-4 challenge virus, indicating cross-reactive responses to a different virus. Specifically, there were approximately 64 ± 13 IFN-g secreting cells following restimulation with Prevacent virus and 41 ± 8 IFN-g secreting cells following restimulation with 1-7-4 virus.

Prevacent vaccination primed for enhanced PRRSV-specific IFN-g response following challenge. Anamnestic immune responses are a hallmark of primed immunity, and may be indicative of protection to infection in vaccinated populations. While PRRSV-specific IFN-g secreting cells were detected in circulation following Prevacent vaccination, the numbers were considered low, and changed minimally at 7 days post-challenge (Fig 1B). However, the number of PRRSV-specific IFN-g secreting cells in vaccinated pigs was increased significantly over non-vaccinated pigs on day 14 post-challenge (Fig 1C). In non-vaccinated pigs at day 14 post-challenge, 56 ± 11 IFN-g secreting cells were detected following 1-7-4 re-stimulation. However, 180 ± 20 IFN-g secreting cells were detected in vaccinated group. Thus, the number of PRRSV-specific IFN-g secreting cells was significantly increased in the pigs that had previously received Prevacent vaccine. Collectively, Prevacent primed PRRSV-specific IFN-γ T-cell based immunity, and responses were enhanced after challenge.

References

1. Holtkamp DJ et al. Assessment of the economic impact of porcine reproductive and respiratory syndrome virus on United States pork producers. J of Swine Health Prod. 2013. 21:72-84.

2. Loving CL et al. Innate and adaptive immunity against Porcine Reproductive and Respiratory Syndrome Virus. Vet Immunology and Immunopathology 167 (2015) 1-14.

Figure 1: Number of PRRSV-specific IFN-g secreting cells A) 35 days after Prevacent PRRS® vaccination (prior to PRRSV challenge); B) 7 days after challenge, or C) 14 days after challenge (Ch) in PBMC from non-vaccinated (NV) or vaccinated (Vx) pigs. Cells were exposed to media-alone, Prevacent virus, or 1-7-4 virus as indicated. Data are expressed as the mean for each group, with each dot representing a single animal in the group, statistically significant differences between NV and Vx groups are indicated with a bar ($P < 0.05$).

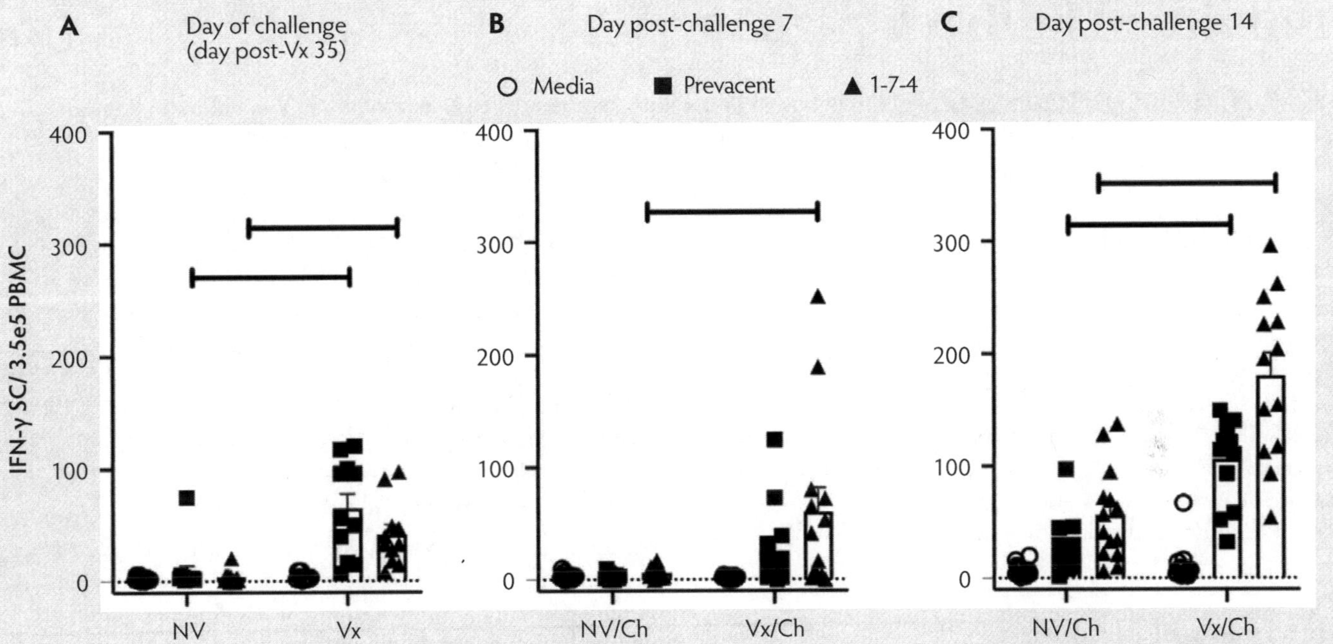

Pen-side comparison of biochemical, blood gas, and hematological analytes between two common clinical conditions and case-controls: nursed-off sows and individual fallback piglets

W. Miller[1], BS; B. Mason[2], DVM; N. Garbes[3], DVM; TH. Nguyen[3], PhD; B. Payne[3], DVM; D. Amodie[3], MS
[1]Michigan State University College of Veterinary Medicine, East Lansing, Michigan;
[2]Country View Family Farms, Middletown, Pennsylvania; [3]Zoetis, Parsippany, New Jersey

Introduction and objective

Individual nursed-off sows (NOS) and fall-back piglets (FBP) have a negative economic impact on sow farms. In high health herds, traditional swine antigen and antibody diagnostics have been unrewarding at determining a direct cause of these two clinical cases. The objective of this study was to ascertain whether a metabolic or infectious component was associated with these clinical cases by utilizing on-site point of care biochemical, blood gas, and hematological diagnostic equipment (Zoetis, Parsippany, NJ). In addition, feedback of the practicality and workability of on-site diagnostic equipment was evaluated.

Materials and methods

Blood samples were collected at two 5,600 head breed-to-wean sow farms, each with sows from two genetic lines. Case definitions for NOS and FBP were: 1) sows off feed for a minimum of two days or poor milk production 3-15 days post-farrowing and 2) piglets within 3-7 days of age that had poor body condition compared to their littermates. Identification of NOS and FBP was performed each morning through routine room observations. Rectal temperatures were collected before restraint for sample collection. NOS (n = 71) and FBP (n = 100) were bled from the jugular vein utilizing 16g 3" or 18g ¾" needles respectively. Blood was placed into 1.3 mL Lithium Heparin and EDTA blood tubes which were gently inverted 10-15 times. Control sows (CS, n = 71) were matched by parities (same), total born (± 3 pigs) and farrow dates (± 2 days). Two control piglets selected from the same farrowing crate were healthy, visibly larger in size and were the same (CP1, n = 100) or opposite gender (CP2, n = 100) of the FBP. The VETSCAN® VS2 large animal profile, i-STAT® Alinity® v CG8+ profile, and VETSCAN® HM5 (Zoetis, Parsippany, NJ) three-part differential were utilized to measure quantitative biochemical, blood gas, electrolyte and hematological analytes. Sow and piglet data were analyzed using the SAS Proc Mixed Procedure with a Generalized Linear Mixed Model approach (SAS 9.4, Cary, NC). Tests were performed at the 5% level of significance.

Results and discussion

Analyte differences and demographic-analyte interactions are reported in Table 1. Findings in NOS and FBP may be associated with an inflammatory/infectious process, a possible etiology for the clinical presentations. Interestingly, fevers were not associated with increased white blood cells parameters in this study. Elevated blood urea nitrogen and decreased glucose can be associated with metabolic processes and therefore, could be suggestive of more than one physiological component in FBP.

Identifying an inflammatory, infectious, or metabolic etiology is important because there are medications and treatment plans that can possibly improve the conditions of NOS and FBP. Indicators of dehydration due to the clinical presentations of NOS and FBP, as in this study, are also significant when practitioners develop and caretakers implement individual sow and piglet treatment plans.

This information will allow the practitioner to narrow the investigation to etiologies of such processes and, in-turn, put preventative measures in place at a faster rate. This project shows that on-farm diagnostic equipment can play an integral role in speeding up the determination of individual treatment and preventative plans for farms and should be investigated further for best-use applications in production facilities and swine veterinary clinics.

Table 1: Analyte identification key and directional differences of significance between treatment and control groups for LS means of all analytes.

Test groups	Significance	VS2	i-STAT	HM5	Blood smear	Rectal temp
NOS vs CS	NOS > CS Significant $P \leq 0.05$	GGT, GLOB, TP	-	MON, RBC, WBC	BAS, MYELO, Band, WBC	-
	NOS > CS Significant $P \leq 0.05$	ALB, Mg	K+	-	-	✓
	Not Significant	ALP, AST, BUN, Ca, CK, PHOS	BE, GLU, HGB, HCO_3, HCT, iCa, pCO_2*, pH, pO_2, sO_2, tCO_2	HCT, HGB, LYMP, MCH, MCHC, MCV, NEU, PLT	EOS, LYMP, META, MON, NEU	-
FBP vs CP	FBP > CP Significant $P \leq 0.05$	BUN, GGT, GLOB, TP	BE, HGB, HCT, Na^+, pH, pO_2*, sO_2*!	HCT, HGB, NEU, PLT, RBC, WBC	META, NEU, Band, WBC	-
	FBP > CP Significant $P \leq 0.05$	ALB, ALP, Ca, CK, PHOS	GLU, iCa, pCO_2	MON	EOS, LYMP#	✓
	Not Significant	AST, Mg	HCO_3*, K+, tCO_2*	LYMP, MCH, MCHC, MCV	BAS, MON, MYELO	-

Identification Key:

ALB= Albumin
ALP= Alkaline Phosphatase
AST= Aspartate Aminotransferase
Band= Band Neutrophil
BAS= Basophil
BE= Base Excess
BUN= Blood Urea Nitrogen
Ca= Calcium
CK= Creatine Kinase
EOS= Eosinophil
GGT= Gamma-Glutamyl Transferase
GLOB= Globulin
GLU= Glucose
HCO_3= Bicarbonate
HCT= Hematocrit
HGB= Hemoglobin
iCa= Free calcium
K^+= Potassium
LYMP= Lymphocyte
MCH= Mean Cell Hemoglobin
MCHC=Mean Cell Hemoglobin Concentration
MCV= Mean cell volume
META= Metamyelocyte
Mg= Magnesium
MON= Monocyte
MYELO= Myelocyte
Na^+= Sodium
NEU= Neutrophil
pCO_2= Partial pressure of carbon dioxide
PHOS= Phosphorus
PLT= Platelet
pO_2= Partial pressure of oxygen
RBC= Red Blood Cell
tCO_2= Total carbon dioxide
sO_2= Oxygen Saturation
WBC= White Blood Cell

* significant group-parity interaction
! significant group-genetic interaction
\# significant group-gender interaction

Evaluation of an RNA-particle platform vaccine for porcine circovirus type 3 (PCV3) in gilts prior to entry

J. Buchan[1], BSc; E. Byers[2], DVM
[1]Ontario Veterinary College, Guelph, ON, Canada; [2]Prestage Farms Inc., Clinton, North Carolina

Problem

Porcine circovirus type 3 (PCV3), first diagnosed in North Carolina in 2015, is a novel, genetically divergent strain of porcine circovirus.[1] It is associated with porcine circovirus-associated disease (PCVAD) and reproductive failure in sows, both of which have significant economic consequences on swine production.[1] Within a large production system, twelve farrow-to-wean herds were determined to be PCV3 positive based on PCR testing of aggregate weekly processing fluids, elevated mummy and stillborn rates, piglet quality concerns, and nursery PCVAD and post-weaning multi-systemic wasting syndrome (PMWS). Due to the reproductive impact of PCV3 in these herds, implementation of a vaccine is being considered to combat these losses using an RNA-particle platform (RP) vaccine. RP vaccines can be customized to the pathogen of interest, creating a highly specific, immunogenic, and safe vaccine.[2] Previous work in this production system demonstrated the benefit of a PCV3/2d RP vaccine (SEQUIVITY™, Merck Animal Health, Ames IA) to combat PCV3 in breeding females, mainly in younger parity sows.[3] Accordingly, administration of a monovalent PCV3 RP vaccine to gilts prior to entry into the sow herd was evaluated in the study reported here. This study aims to evaluate the impact of a monovalent PCV3 RP vaccine in PCV3 positive herds with regard to reproductive fertility and fecundity in vaccinated gilts compared to non-vaccinated control gilts.

Materials and methods

Twelve sow herds received approximately an equal number of vaccinated gilts (n = 612) and unvaccinated control gilts (n = 607) during one of their regular gilt shipments. Enrolled herds were mixed parity and continuous farrowing herds with a total inventory of 21,800 sows. Gilts receiving vaccination were given two 1mL doses of PCV3 RP vaccine administered intramuscularly four weeks apart. The first dose was given 4 weeks prior to shipment from a single gilt development unit (GDU), and the second dose was given 1-day post-shipment. Upon administration of their first dose, farm standard numbered tags were applied to all gilts. Vaccinated gilts also received a yellow button ear tag to identify them as vaccinates. Unvaccinated controls received no placebo and no additional ear tag. All gilts were bred as per the farms' standard breeding procedures. Fertility (conception data to-date) was extracted from the company's PigKnows database for the analysis reported here. Farrowing (litter size) data will be extracted as it becomes available. Statistical analysis using Statistix was/will be used to determine any differences in reproduction parameters between treatment groups.

Results

Final results are currently pending. The preliminary data reported here was limited to events collected from entry to the sow farm through pregnancy diagnosis at 4-5 weeks of gestation (Table 1). Control gilts had 1.7% more non-serviced females (failed to reach or exhibit puberty) than vaccinated gilts. When this value was adjusted by removing culled, dead, and destroyed gilts, control gilts had 1.3% more non-serviced females compared to the vaccinated gilts. Of those that were serviced, 0.6% more vaccinated gilts became pregnant, with a pregnancy rate of 91.8% compared to the rate of 91.2% in controls. The difference between the adjusted pregnancy rates was 0.5%, with the vaccinate group having the greater rate. However, these differences were not significant based on Chi Square tests. Further statistical analysis of these groups will be conducted to identify farm differences when all data becomes available by December 2019.

References

1. Palinski R et al. (2017). A novel porcine circovirus distantly related to known circoviruses is associated with porcine dermatitis and nephropathy syndrome and reproductive failure. *J Virol 91*(01), 1-13.

2. Chackerian B. (2007). Virus-like particles: flexible platforms for vaccine development. *Expert Rev Vaccines 6*(3), 381-390.

3. Vennekotter A, Byers E, Thacker B, Cagle J. (2019). Evaluation of PCV3/2d prevalence in placental umbilical cord blood and processing fluids and reproductive outcomes following mass vaccination with an RNA-particle platform PCV3/2d vaccine. *AASV 2019 Proceedings,* 89-93.

Table 1: Service results from vaccinated and control gilts.

Parameter		Group	
		Vaccinated	Control
% Served vs. Entered	All	89.9%	91.6%
	Adjusted*	91.2%	92.5%
% Pregnant** - 1st service	All	91.8%	91.2%
	Adjusted*	92.5%	92.0%

* Removals (cull, died, destroyed) not included.
** Based on ultrasound examination at 4-5 weeks of gestation

Telomere length as biomarker of welfare in breeding sows

Katherine Dorph; Sanne Roelofs; Thomas D. Parsons
Swine Teaching and Research Center, School of Veterinary Medicine,
University of Pennsylvania, Kennett Square, Pennsylvania

Introduction

Growing public concern about the welfare of gestating sows is driving changes in husbandry practices. However, advancing sow welfare to meet societal expectations is complicated by shortcomings in the tools available to assess animal welfare. Current measures tend to address physical rather than psychological health, struggle to identify positive states of welfare, and often only report the impact of recent/acute stressors rather than their cumulative effects.

Objective

To compare two novel measures of animal welfare. One is based on a judgement bias task, a behavioral task that is a putative measure of both positive and negative welfare states. The second is telomere length, a biomarker that has been proposed as an integrative and cumulative measure of welfare in humans.

Materials and methods

Nine bred gilts (DNA Line 241), around the same age and stage of gestation, were selected to be trained in a judgement bias task. First, animals were taught a go/no-go spatial discrimination task that required them to differentiate a rewarded positive cue from an unrewarded negative cue. The number of training sessions required to learn the task was noted. After discrimination training, gilts were exposed to intermediate, ambiguous cues. Go-responses to ambiguous cues were considered an expectation of reward and indicative of a positive bias. Gilts were tested on ambiguous cues for 5 daily sessions and their go-responses quantified. A generalized linear model (GLM) using a quasibinomial distribution was fit to the behavioral data with percentage of go-responses as the outcome variable and gilt and cue as explanatory variables. A post-hoc Tukey multiple comparison tested differences between individual cues and gilts. At the end of the behavioral trial, blood was collected and peripheral blood leukocytes isolated. Telomere length of interphase nuclei from individual leukocytes was measured using high-throughput Q-FISH (Quantitative Fluorescence In-Situ Hybridization). A second GLM was employed to examine the ability of telomere length to predict behavioral outcomes.

Results

Eight of the nine gilts completed judgement bias training and were tested with ambiguous cues. The type of cue presented (negative, positive or ambiguous) significantly impacted the percentage of go-responses ($\chi^2_{(4)} = 79.5$, $P < 0.001$). Gilts exhibited more go-responses to the positive cue compared to the ambiguous and negative cues. This validates the ability of the gilts to discriminate between the positive and negative cues. Gilt identity also impacted the percentage of go-responses ($\chi^2_{(7)} = 52.3$, $P < 0.001$), suggesting inter-individual variation in judgement bias. One gilt performed significantly less go-responses (averaged across cue-types) compared to the rest of the animals, whereas another animal performed significantly more go-responses. The remaining gilts did not differ from one another. No relationship was observed between telomere length and go-responses. However, telomere length did predict the number of sessions required for training ($\chi^2_{(1)} = 38.1$, $P < 0.001$; Figure 1), with gilts that required a higher number of training sessions having shorter median telomere length.

Discussion

The two proposed novel indicators of animal welfare demonstrated variability across this population of gilts. However, there appears to be a lack of correlation between these metrics. There are several possible explanations for this result: it could be that one or both metrics are not measuring what they are proposed to quantify, that the animals' short- and long-term welfare experiences differ, or perhaps a larger sample size is needed. Additional experiments promise to help elucidate these initial discrepancies as well as explain why telomere length was able to predict the number of training sessions required for gilts to learn the judgement bias task.

Figure 1: Average number of training sessions required for gilts to learn the spatial discrimination task plotted against median telomere length in base pairs.

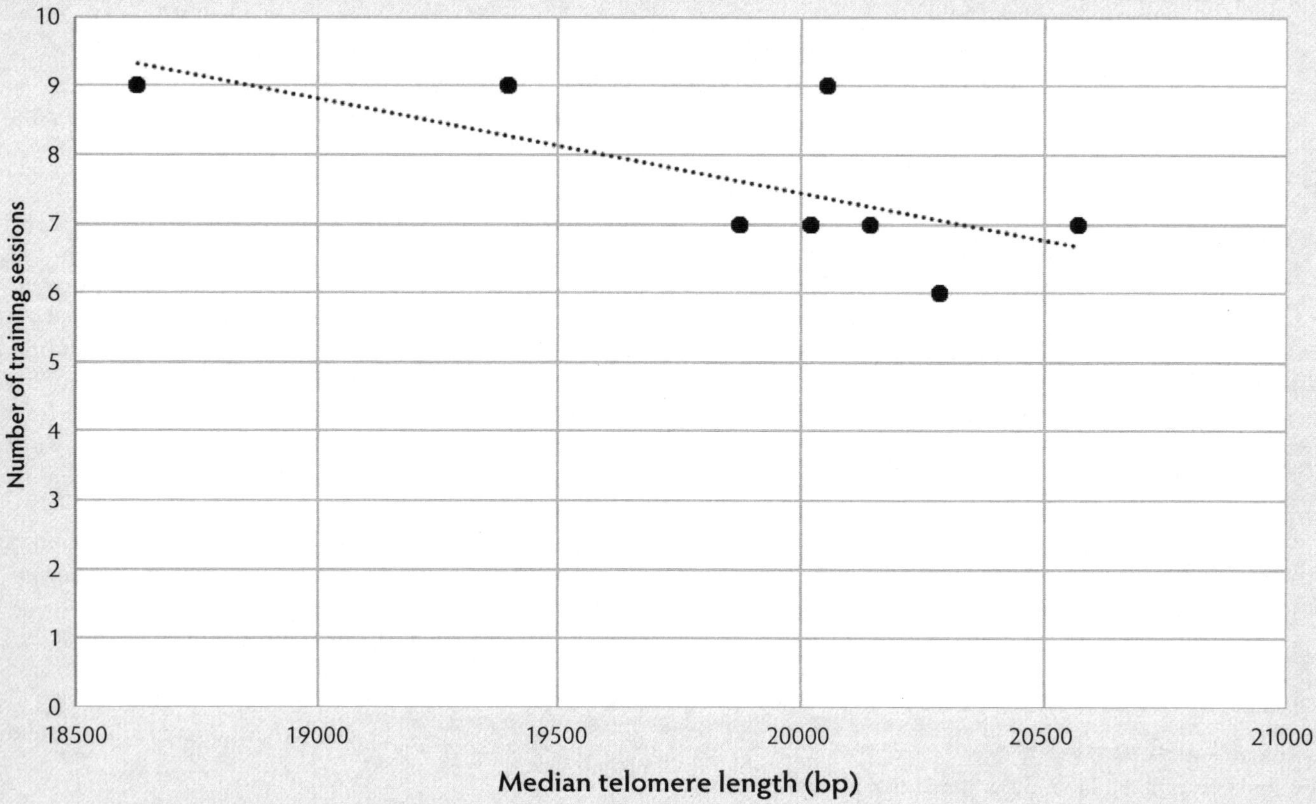

Comparison of standard and novel precision agriculture vaccination strategies and the effect on vaccine responses in weaned pigs

G. Doughan[1], BA; J. Brown[1], DVM; K. Skoland[1], BS; M. Nickel[1], DVM; H. Kittrell[1], DVM; C. Ruston[1], DVM; M. Breuer[1], BFA; J. Ellingson[2], DVM, MS; P.R. Thomas[2], DVM, MS; M. Petersen[3], BS, L. Karriker[1], DVM, MS, DACVPM

[1]Swine Medicine Education Center, Iowa State University Ames, Iowa; [2]AMVC Management Services, Audubon, Iowa; [3]Iowa State University College of Veterinary Medicine, Ames, Iowa

Introduction

Effective vaccination strategies remain important to continue to reduce industry reliance on antimicrobial use. The objective of this study is to utilize both standard and novel injectable devices to evaluate the post-vaccination antibody response to a commercially available, killed, Porcine Parvovirus, *Erysipelothrix rhusiopathiae* (ERY) and *Leptospira* spp. bacterin (PLE) (Parvo Shield L5E®, Elanco US Inc.). Weaned pigs are vaccinated using two main methodologies: 1) picking up and individually restraining each pig and 2) crowding the pigs for mass vaccination. Advantages and disadvantages to these methodologies have not been thoroughly examined. In addition, new precision agriculture devices can be used to administer vaccinations using these methodologies. Understanding this novel technology along with the potential effects of current injection strategies are pivotal components to continue effective antimicrobial reduction. Understanding antibody response between the different methodologies and devices could influence producer decisions on equipment and vaccination methods.

Materials and methods

Vaccination devices were calibrated to develop a performance baseline and verify accurate dosing prior to use on-farm. 495 cross-bred, mixed gender, seven-week-old pigs located in a wean-to-finish barn in central Iowa were enrolled in the study. Pigs were selected at random and sorted into three treatment groups of 165 from four randomly selected pens in the barn. Each pen included representatives from each treatment group. Selected pigs were ear tagged and individual blood samples were collected to establish baseline ERY antibody levels. Pigs were vaccinated utilizing three strategies: individually restraining pigs for standard syringe vaccination (Prima Tech® Vaccinator) (individual restraint), crowding pigs for mass standard syringe vaccination (Prima Tech® Vaccinator) (crowd) and crowding pigs for mass vaccination with a novel, precision agriculture inspired, injectable device (Automed® Delivery Device) (precision). Blood was collected at three time points for ERY Fluorescent Microbead Immunoassay[1] (FMIA) testing to determine individual antibody levels. Three entry level vaccinators were randomized and assigned to a sequence of three groups of 55 pigs per vaccination strategy; individual restraint, crowd and precision vaccination. Each pig was vaccinated with a dose of 2 mL. Three weeks later, blood samples were collected from enrolled pigs prior to booster vaccination with the same dose and PLE product. A final blood collection occurred three weeks after booster administration. All blood samples were centrifuged at 3100 rotations per minute (RPM) for eight minutes by a Thermo Scientific ST16R© centrifuge and stored at -80° C. Samples were individually submitted in a blinded manner to the Iowa State University Veterinary Diagnostic Laboratory for ERY FMIA testing. FMIA results were log transformed and analyzed by analysis of variance (ANOVA) using SAS software (9.4, Cary, NC, USA).

Results and discussion

FMIA responses from 489 pigs were analyzed. The remaining six pigs were lost to follow-up. There are significant differences between serum collection dates ($P < 0.0001$), indicating a response to vaccine. There were no significant differences between crowd, precision and individual restraint vaccination strategies ($P = 0.93$). Under study conditions, the tested vaccination strategies produced similar responses. There were limitations to the present study: a vaccination crew was not used and the vaccinators were not blinded to the treatments. Further analysis could include utilizing different vaccines and confirming protection with disease challenge. Two observations that warrant follow-up: a) individual animal variation in response within treatments which may be relevant for other vaccines such as PCV2, b) differences in the mechanical administration and process of vaccination between treatments and these ergonomic differences may impact vaccination of larger populations. Vaccinating via the crowd method by ensuring correct administration of the vaccine can result in a similar response as the individual restraint method. Precision vaccination does not require individual animal restraint and could be an effective tool in accurate vaccine and antimicrobial dosages.

Reference

1. Gimenez-Lirola, L.G, Xiao, C.T., Halbur, P.G, Opriessnig, T., 2012. Development of a novel fluorescent microbead-based immunoassay and comparison with three enzyme-linked immunoassays for detection of anti-*Erysipelothrix* spp. IgG antibodies in pigs with known and unknown exposure. J. Microbiological Methods, 91, 73-79.

Efficient establishment of multi-site Secure Pork Supply plans for swine operational management

C. Found[1], BA; J. Ellingson[2], DVM, MS
[1]Iowa State University College of Veterinary Medicine, Ames, Iowa; University of Iowa College of Public Health, Iowa City, Iowa; [2]Audubon-Manning Veterinary Clinic – Management Services, Audubon, Iowa

Introduction

With the significant increase in public concern for African swine fever, and enduring concerns for other foreign animal diseases such as classical swine fever and foot-and-mouth disease, the protection of pigs on American soil has become fundamentally critical. The Center for Food Security and Public Health, with assistance from the USDA and Pork Checkoff, has developed the Secure Pork Supply (SPS) Plan for swine facilities to implement in a foreign animal disease (FAD) control area.[3] Preliminary incorporation of these plans is in its preliminary stages as swine owners continue to understand and assimilate the plan components.[1]

Currently we find that one person or company is responsible for constructing numerous SPS plans; local veterinarians may formulate these plans for their client groups, or a swine management company may be responsible for doing their contracted operations. Each plan is extensive and unique, being at least thirteen pages immersed with action items, including a detailed map pertaining to the site at hand. With the time and effort of these plan authors in mind, there needs to be an efficient way to create these plans, with a focus on creating realistic bio-exclusion channels at each site. If planning for FAD appears too daunting, and plans are too confusing, we risk the potential to be at serious risk and underprepared for an outbreak.

Clinical case description

Out of the previously described issue came the development of a complete and efficient method for constructing a large number of SPS plans in a short amount of time. From June to August 2019, data were collected on ninety-six grow-finish operations in western Iowa. Data collection included biosecurity red flags at each site, site-specific aerial view components, contract grower concerns regarding FAD and SPS integration, and pooled resources from policy-makers involved in the SPS incorporation. Heavy research went into fully understanding the components required by the SPS plan website, with the consequent intention of creating plans for ninety-six grow-finish operations. Out of this, a well-constructed template was developed for future plan authors to use in accordance with their specific site(s). The template is designed to be edited in a step-wise manner over time as persons involved (i.e. veterinarians, biosecurity managers, farm hands) complete action items (i.e. washing station is established, new compost area is developed). The SPS plan also requires a detailed map of the site. Project author expertise in Geographic Information Systems aided in the establishment of a map template to be used by future plan authors. Plan authors can simply add in their aerial pictures of specific sites and easily maneuver map aspects. In an effort to create a more useful and comprehensive plan, items were added to the map template beyond what is required by the SPS plan, including a rendering contingency plan, load-out location(s), and directionality guidance. These additional components make better use of the map while ensuring biosecurity measures are clear and consistent.

Discussion

Time is of the essence in a foreign animal disease outbreak.[2] SPS plans must be allocated efficiently, and to do this, a streamlined method has been created to:

- Allow fast integration of SPS plans to specific hog sites
- Apportion easy updates and changes to the SPS plan
- Ensure maps are comprehensive, user friendly, and maintain a high standard of biosecurity

Future work involves making this template available and easily accessible to all SPS authors, alongside straightforward online instructional materials. In tandem, consistent, user-friendly communication from FAD outbreak authorities to swine owners will allow better preparation, efficient execution, and protection of time in the case of a foreign animal disease outbreak.

References

1. Iowa Pork Industry Center. African Swine Fever (ASF). 2019. Available at **www.ipic.iastate.edu/AfricanSwineFever.html**.

2. Normile, Dennis. African swine fever marches across much of Asia. *Science*. 2019;617-618.

3. Secure Pork Supply. Secure Pork Supply Plan. 2019. Available at **www.securepork.org/**.

Evaluating the use of processing fluids for sow herd monitoring of porcine circovirus type 2

Karissa Frealy[1], BS; Jon Ertl[2], DVM; Vitelio Utrera[3], DVM
[1]Oklahoma State University Center for Veterinary Health Sciences, Stillwater, Oklahoma;
[2]Sioux Nation Ag Center, Sioux Falls, South Dakota; [3]Zoetis, Parsippany, New Jersey

Introduction

Porcine circovirus type 2 (PCV2) is found throughout the world and is associated with a number of disease syndromes, collectively termed porcine circovirus-associated disease (PCVAD). The ubiquity of the virus in the environment and its resilience to commonly used disinfectants makes it an especially problematic pathogen for both producers and veterinarians. Monitoring of herd status is important for determining stability and may be accomplished by sampling placental umbilical cord serum, pre-suckling serum, and/or colostrum. These samples can be time-consuming and difficult to collect. A previous study demonstrated that Ct values from testing processing fluids (PF) for PCV2 by polymerase chain reaction (PCR) Ct may be useful for monitoring PCV2 sow herd stability.[1] The present study evaluates the use of PF, the serosanguinous fluid collected from piglet castration and tail docking tissues, as a tool for monitoring sow herd PCV2 stability and its association with PCVAD observed downstream at the nursery.

Materials and methods

Six sow farms (A, B, C, D, E, and F) located in eastern South Dakota and western Iowa were enrolled in this study. Farm A vaccinates pigs for PCV2 at weaning and boosters gilts pre-breeding. Farms B and F vaccinate at weaning, two weeks post-weaning, and booster gilts pre-breeding. Farm C vaccinates pigs at weaning and two weeks post-weaning. Farm D does not routinely vaccinate for PCV2. Farm E vaccinates pigs at processing and weaning. PF samples were collected from individual litters and tested for PCV2 using quantitative PCR (qPCR). At downstream nurseries, 30 serum samples were collected from pigs (21 to 34 days of age) and tested for PCV2 using qPCR and quantitative immunofluorescence assay (IFA). For qPCR, serum samples were pooled by five; IFAs were performed individually. Nursery pigs that were showing signs of potential PCVAD were euthanized – in accordance with the American Association of Swine Veterinarians guidelines – and necropsied. Any sudden death pigs were also necropsied. Fresh and formalin fixed tissue samples, including spleen, lung, mediastinal and inguinal lymph nodes, were collected from all necropsied pigs and evaluated for PCVAD using PCV2 qPCR and histopathology. Tissue samples were collected at all farms with the exception of D and E due to high health status of pigs.

Results

Results from PCV2 qPCR for PF, tissue, and serum samples are summarized in Table 1. Histopathology revealed no PCVAD lesions on any of the tissues collected. No pigs were dead or euthanized on farms D and E, so no tissue samples were collected. The IFA geometric mean titers for farms A-F were 1/130.0, 1/87.7, 1/63.5, 1/83.8, 1/570.2, and 1/236.98, respectively.

Discussion

All of the farms enrolled had found evidence of PCV2 clinical disease during general diagnostic submissions in the previous year. However, a low proportion of PF positive samples were detected by PCR in this study. Considering the ubiquity of PCV2, these results are remarkable, as PF is considered a highly sensitive test. These results are consistent with the previous study recommending interventions when PF PCR Ct values are found lower than 30 to continue monitoring for PCVAD downstream. This study showed that when PF Ct values were above 30, little PCVAD was observed in the nurseries. The low proportion of PCV2 detected by qPCR in PF and in serum samples may be associated with vaccination protocols applied in these farms. The variable proportion of IFA titers detected across farms warrants further evaluation in this area.

Reference

1. Castevers, K. Use of processing fluids to monitor sows for porcine circovirus type 2 and determine parity influence on viral detection. AASV. 2019.

Table 1: Summary of PCV2 qPCR for processing fluids, serum, and tissue samples

Farm	PF samples	Ct	Serum samples (30 pooled by 5)	Ct	Tissue samples	Ct
A	1/1 (-)	-	6/6 (-)	-	2/2 (-)	-
B	2/3 (s), 1/3 (-)	38.1, 38.0	1/6 (s), 5/6 (-)	38.2	3/4 (+), 1/4 (s)	35.0, 35.0, 35.5, 36.1
C	1/1 (s)	36.1	6/6 (-)	-	1/3 (s), 2/3 (-)	36.4
D	1/1 (+)	34.0	2/6 (+), 2/6 (s), 2/6 (-)	19.3, 29.2, 37.1, 38.0	None	-
E	1/3 (+), 1/3 (s), 1/3 (-)	35.7, 37.9	6/6 (-)	-	None	-
F	1/1 (-)	-	6/6 (-)	-	4/4 (-)	-

Note: (+) indicates positive, (-) indicates negative, (s) indicates suspect.

An attempt to determine within sample variation of the swine fecal microbiome

L. Glazik[1], BS; J. Lowe[1,2], DVM, MS; B. Blair[1], DVM; A. Megahed[1], PhD
[1]College of Veterinary Medicine, University of Illinois, Urbana, Illinois; [2]Lowe Consulting Ltd., Mahomet, Illinois

Introduction

The gastrointestinal microbiota has garnered much scientific and popular media attention because of the immense implications that it may have on animal and human health. Understanding the role of the gut microbiome and prevalence of antimicrobial resistance genes (ARGs) within these populations could lead to advances in the prevention and treatment of disease. However, the intrinsic complexity of the fecal microbiome and the multi-step work flow to evaluate it begets many opportunities for biases to alter the outcome. These confounding variables have caused a "scientific reproducibility crisis" that is impeding further progress. Recognizing the factors that lead to bias is the first step to controlling them. It is well documented that fecal sample preservation technique,[1] DNA extraction method,[2,3] and choice of RNA primer[4] results in different 16S rRNA outcomes. However, very little has been done to understand variations of the microbiome within a single fecal specimen. The purpose of this study was to observe the 'within sample variation' of swine fecal samples using two different collection methods.

Materials and methods

Three healthy growing pigs (A, B, C) were randomly selected, and a single fecal specimen was collected from each. Each fecal sample was swabbed in 10 different locations (S1-10). In addition, ten 1-gram portions of the same fecal sample were collected and homogenized with a 3:1 dilution of PBS (F1-10). The homogenized samples were split into two groups, one group that the DNA isolation was performed immediately (F1a-10a) and another that sat at room temperature for two hours before isolation (F1b-10b). Microbial DNA was extracted from each sample using the Power Fecal DNA Isolation kit. The hypervariable region V3-V4 of the 16S rRNA gene along with three antimicrobial resistance genes primers were amplified and sequenced on the Illumina MiSeq.™ After preprocessing, the 16S rRNA gene sequence data was analyzed using the open source Quantitative Insights into Microbial Ecology (QIIME) software. Relative proportions of the dominant phyla and ARGs were calculated and compared. Hierarchical clustering of all samples was performed to observe the relation of samples to each other.

Results and discussion

When observing the relative proportion of the three most abundant phyla, a pattern emerged showing that swabs sampled approximately 10% more Firmicutes in comparison to the respective homogenized samples. In addition, the swabs of each of the pigs had less Tet C resistance genes than the corresponding homogenized samples. Most shockingly, hierarchical clustering revealed that samples from each fecal specimen did not all cluster together by pig (Figure 1). The swabs of Pig C were more related to pig A's samples that homogenized samples from Pig C. In addition, one of Pig B's homogenized samples was more related to pig A than any other pig B sample.These results reveal the significant variation within a single fecal sample and between sampling methods. With the amount of variation observed, a singular sample of the microbiome does not give a complete or accurate representation of an individual's fecal microbiome. This is distressing because in microbiome research, it is customary to take a single fecal swab. Therefore, microbiome changes previously attributed to treatment effect may actually be normal variation within the sample.

References

1. Song, S.J. et al. Preservation methods differ in fecal microbiome stability, affecting suitability for field studies. mSystems https://dx.doi.org/10.1128/mSystems.00021-16 (2016).

2. McOrist, A.L., Jackson, M. & Bird, A.R. A comparison of five methods for extraction of bacterial DNA from human faecal samples. *J. Microbiol. Methods* 50, 131–139 (2002).

3. Kennedy, N.A. et al. The impact of different DNA extraction kits and laboratories upon the assessment of human gut microbiota composition by 16S rRNA gene sequencing. *PLoS One* 9, e88982 (2014).

4. Gohl, D.M. et al. Systematic improvement of amplicon marker gene methods for increased accuracy in microbiome studies. *Nat. Biotechnol.* 34, 942–949 (2016).

Figure 1: Hierarchical clustering shows the relation of samples to each other. Each color represents a unique cluster of samples.

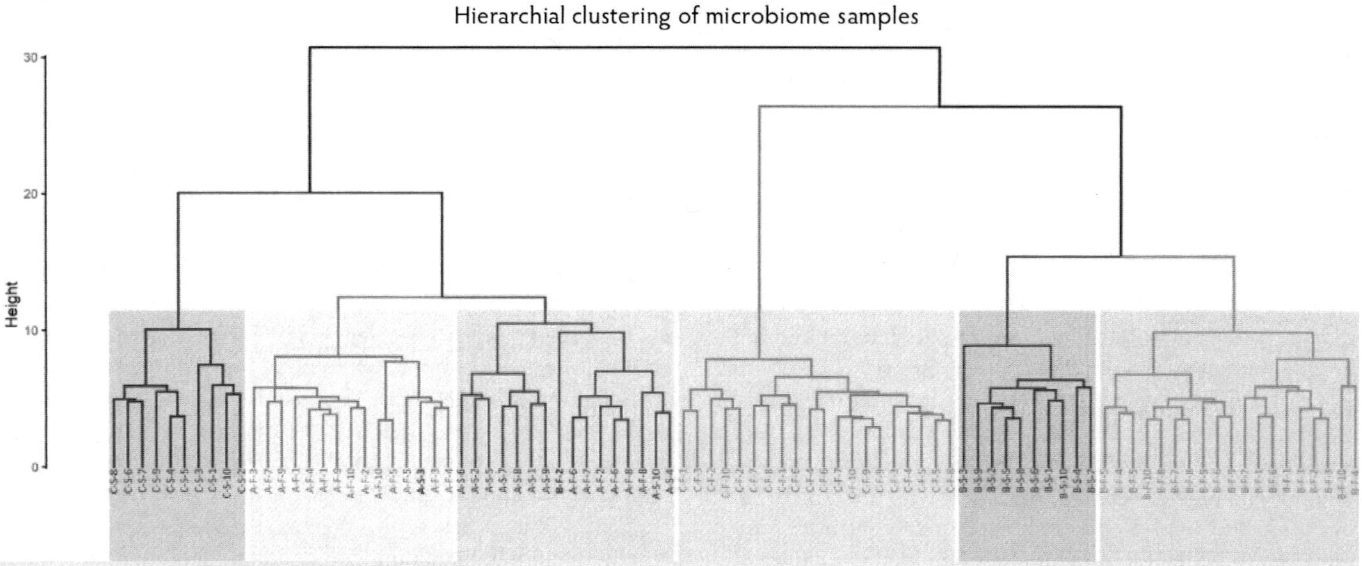

Effects on longevity and productivity for sows in pen gestation removed for lameness

A. Hallowell, MS; M. Pierdon, VMD, DACAW
New Bolton Center, University of Pennsylvania, Kennett Square, Pennsylvania

Introduction

Sows can be removed from a herd for many reasons. Lameness has been a problem in swine production for many years and becomes more important as the use of pen gestation increases. Lameness is an economic concern representing an involuntary removal and a welfare concern as it is a painful process for the animal.[1] It is important to evaluate the effects on longevity and productivity for sows removed for lameness compared to other reasons. These effects have not been evaluated for sows in pen gestation. Better comprehension of these relationships will aid swine producers in understanding the costs of lameness and the need for strategies to reduce removals due to lameness. The objective of this study is to determine the impact of lameness on sow productivity and lifespan in herds using pen gestation.

Materials and methods

Production records, including longevity and farrowing performance, were collected from 11 farms using pen gestation for the removal of 109,672 sows over 5 years (2014-2019). Farms were enrolled if recording reasons for over 80% of removals. Housing (static versus dynamic pens) and mixing type (pre versus post implantation) were recorded. Statistics were performed using Stata IC v.15. Cox proportional hazard analysis was performed to determine differences in median survival time. All data were non-normal and Mann Whitney U-test was used to assess differences in productivity. Pearson Chi Square test was used to determine the odds ratios. Results are reported as median and interquartile range. $P < 0.05$ was the level of significance.

Results and discussion

Top removal reasons included reproduction (54.5%), age (17.3%), lameness (12.5%), body condition (3.9%), and sudden death (2.8%) (Figure 1). Lame sows had significantly shorter days in the herd (219) compared to sows removed for other reasons (504) resulting in fewer litters for lame sows (1+/-2) compared to others (3+/-5) ($P < 0.001$). Lame sows in dynamic pens had fewer days in the herd (192) compared to static pens (259) ($P < 0.001$). Lame sows mixed pre-implantation lived fewer days (213) compared to post-implantation (263) ($P < 0.001$). Non-productive days per parity were higher for lame sows (44.0+/-44) compared to sows removed for other reasons (24.8+/-60.7) ($P < 0.001$). Born alive per litter was lower for lame sows (12.7+/-3.0) compared to other reasons (12.8+/-2.8) ($P < 0.001$). The odds of being removed as a gilt were higher for lame animals compared to other reasons (OR = 2.28; CI 2.20-2.37). The odds of a lame animal being euthanized or dying were higher than animals removed for other reasons (OR = 6.82; CI 6.56-7.09).

Discussion

Time in the herd for lame sows is significantly less generating two fewer litters in their lifetime. Sows removed for lameness are more likely to be removed before even having a litter. While the magnitude of the difference is small, lame sows are having fewer piglets born alive. Importantly, lame sows are accruing more nonproductive days, increasing the cost of lameness. Lame sows have higher odds of being removed by euthanasia or death, removing the salvage value of the animal. The percentage of sows removed for lameness in this study was lower than some other studies.[2] It may be underestimated here because it is based on the reports of farm staff and may be impacting other factors such as reproduction. Recognizing the cost of lameness is important because it helps swine producers and veterinarians understand how much they can invest in fixing the problem. Efforts at prevention should include housing and mixing strategy and other methods that were not investigated, like flooring type and early treatment.

References

1. Anil, S.S., Anil, L. and Deen, J., 2005. Evaluation of patterns of removal and associations among culling because of lameness and sow productivity traits in swine breeding herds. Journal of the American Veterinary Medical Association, 226(6), pp.956-961.

2. Zurbrigg, K. and Blackwell, T., 2006. Injuries, lameness, and cleanliness of sows in four group-housing gestation facilities in Ontario. Journal of Swine Health and Production, 14(4), pp.202-206.

Figure 1: Survival estimates for the top five removal reasons. Entry date, removal date and removal reason were collected for 109,672 sows. Each line indicates a removal reason. The y-axis shows the proportion of animals remaining in the herd and the x-axis shows the time in days. Lameness has the lowest median survival time at 219 days. Sudden death had a median survival time of 313 days and body condition score 314 days. Reproduction was the most common removal reason and median survival time was 385 days. Other reasons not in the top 5 had a median survival time of 398 days. The longest survival time was for animals removed for age, with 1047 days median survival.

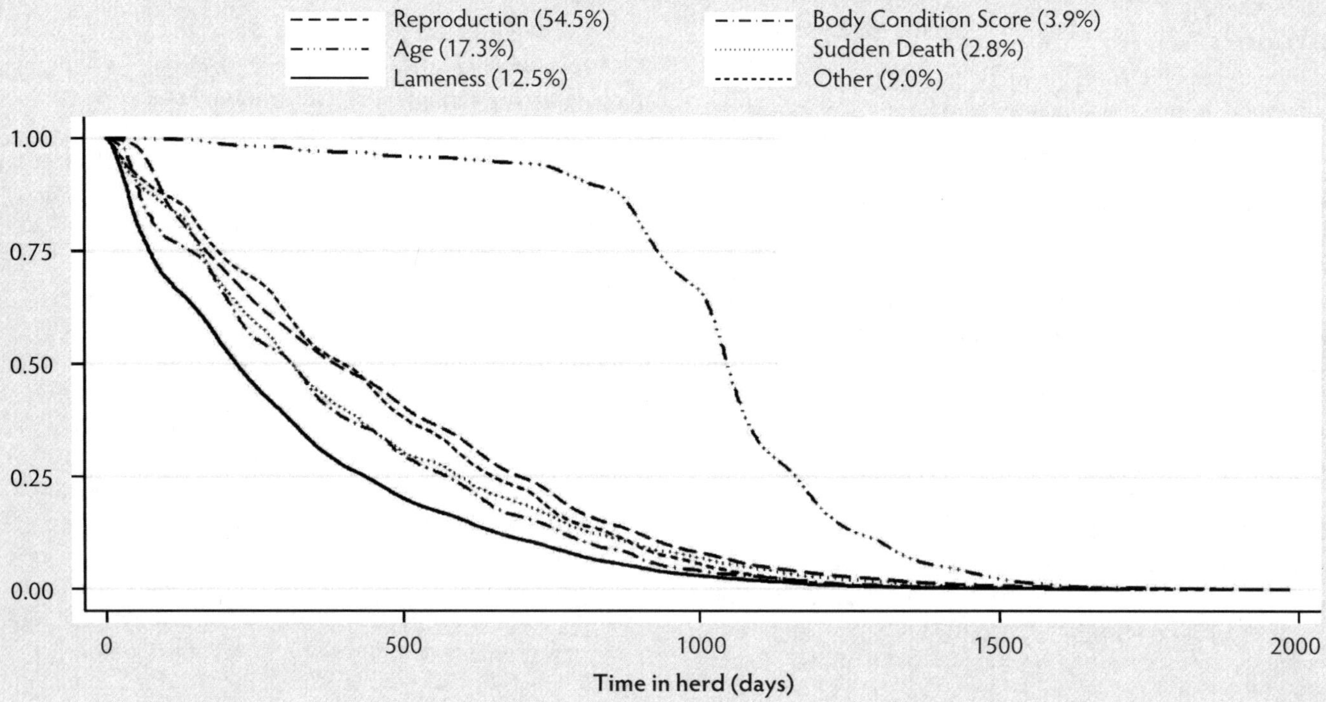

Quantifying the effect of population-based vaccination of the breeding herd on various productive parameters

M. Herring; J. Nerem, DVM; S. VanderPoel, DVM; R. Edler, M.S.; E. Little, MS
Pipestone Veterinary Services, Pipestone, Minnesota

Introduction

Population based vaccination is popular among many veterinarians as a method to decrease viral shedding within a swine breeding herd and therefore improve health stability.[1] However, many production supervisors and farm employees have a negative perception of this practice because they report decreased performance in the form of increased abortions, decreased total born pigs, and anorexic events in the sows. There is a lack of research into the true impact of vaccination on sow performance. The objective of this project was to quantify the differences in rectal temperature, anorexic behavior, abortions, mummified fetuses, stillborn piglets, total born piglets and piglets born alive between vaccinates and non-vaccinates when using a killed bacterin product for a whole-herd vaccination event.

Materials and methods

This study was conducted on one 5,000 sow commercial farm. The population was split into two groups. Even-numbered sows served as the non-vaccinated controls, while odd-numbered sows served as vaccinates and received one labeled dose of commercial parvovirus, *Leptospira*, erysipelas and swine influenza virus killed vaccine. Unbred gilts were excluded from the trial. Sows in the farrowing barn on the day of vaccination were also excluded to remove any bias due to the farrowing interval. Researchers collected baseline rectal temperatures from 5% of the population prior to administering the vaccine. Researchers returned to the farm the day after vaccination to collect post-vaccination rectal temperatures and record identification numbers of anorexic sows. To identify anorexic behavior with as little bias as possible, farm employees who regularly recorded anorexic events at the farm marked each anorexic sow and researchers went through the barn later and recorded the IDs. Farm employees were blinded in that each vaccinated sow had a paint mark, but employees were blinded to the treatment group. The farm's production data was used to track reproductive performance of the herd for 18 weeks following vaccination.

Results and discussions

Analysis of variance models were used to fit each parameter where treatment and trimester were held as fixed effects. Treatment sows on average had a 0.4 degree Celsius higher rectal temperature than the non-vaccinated controls ($P = < 0.001$, Figure 1). Regarding anorexic events, 12.4% of treatment sows were anorexic 24 hours post vaccination, compared to 1.6% of non-vaccinated control sows ($P = < 0.001$). Data for the remaining parameters were analyzed by treatment group as well as the sow's stage of gestation at time of vaccination: early, middle or late. There was no significant differences between the treatment groups for abortions ($P = 0.21$), mummified fetuses ($P = 0.83$), stillborn piglets ($P = 0.05$), total born piglets ($P = 0.58$) and piglets born alive ($P = 0.60$). In conclusion, one can expect anorexic sows with elevated rectal temperatures 24 hours after a population-based vaccination event with this killed commercial vaccine. Veterinarians can use this data to advise production managers that a population-based vaccination event may elicit a strong immune response 24 hours post vaccination, however in the long run it does not show detriment to litter performance.

Reference

1. Corzo CA et al. Observations regarding influenza A virus shedding in a swine breeding farm after mass vaccination. *J Swine Health and Prod.* 2016;20:6.

Figure 1: Mean rectal temperature (° C) of treatment sows and non-vaccinated control sows prior to vaccination and 24 hours post vaccination.

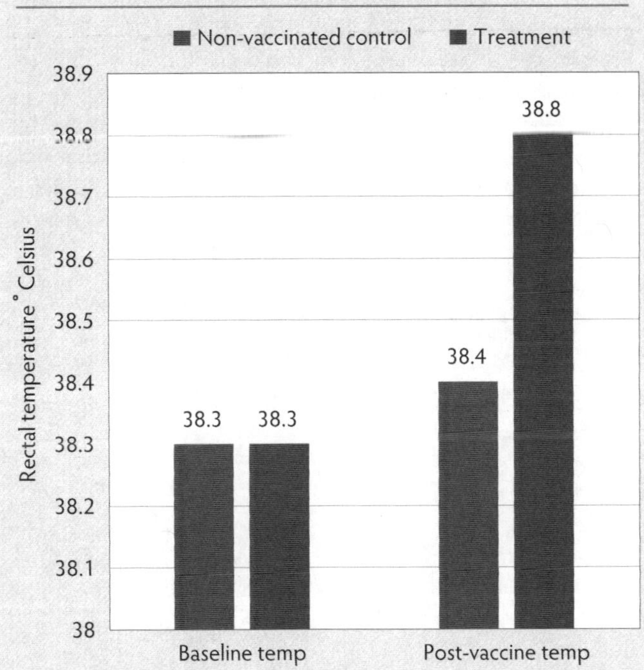

Impact of parity on aggression between gestating sows upon entry into large-pen group housing

Nicole Johnson Valladares; Sanne Roelofs; Thomas D. Parsons
Swine Teaching and Research Center, School of Veterinary Medicine,
University of Pennsylvania, Kennett Square, Pennsylvania

Statement of the problem

Growing public concern about the welfare of gestating sows is driving changes in husbandry practices through both legislative and market initiatives. Producers are being asked to move away from individual housing of sows to group housing of sows. However, establishment of social hierarchy during formation of these groups remains a concern for producers as little is known about what impacts inter-animal aggression during these times.

Objective

The objective of this study is to examine how parity (as a proxy for size and experience in the pen) impacted aggression when sows entered a large, dynamic pen of sows fed by electronic sow feeders (ESF).

Materials and methods

DNA Line 241 sows were inseminated in individual breeding crates and then moved at 8 days post-weaning to a large dynamic pen for gestation. The pen housed ~120 sows that ranged from 2nd to 8th parity and contained 2 ESFs. Individual sows were identified, and their aggressive and exploratory behavior recorded for 1 hour after their introduction into the pen. This experimental paradigm was replicated 5 times with ~10 sows per replicate over a period of 10 weeks, resulting in a total of 49 sows. Skin lesions were recorded immediately prior to entering the pen and 72 hours post-entry. Video recordings were coded off-line by a single observer to score the frequency or duration of specific behaviors. Behaviors examined included fighting, resting, chasing, retreating and different types of sow-to-sow contact. A linear mixed model was employed to assess how predictor variables impacted these different outcome variables. Parity was used as a fixed effect in the model and replicate was entered as a random effect.

Results

Mean scores for behavioral variables and skin lesions can be found in Table 1. Higher parity sows spent more total time fighting than lower parity sows ($X^2_{(2)} = 6.17$, $P < 0.05$), initiated more "head to body knocks" ($X^2_{(2)} = 8.88$, $P < 0.01$), but had less total skin lesions after mixing (11.5; $X^2_{(2)} = 7.95$, $P < 0.02$). No other effects of parity were found.

Discussion

Group housing of gestating sows leads to increased inter-animal aggression during mixing. The findings of the current study highlight how age and experience may shape behavior (and perhaps ultimately production) of sows in group gestation. Higher parity sows (that are older and more experienced in the pen) appeared to be more aggressive based on behaviors scored during mixing. However, these sows also had less skin lesions. This suggests that higher parity sows know how to engage in the establishment of a social hierarchy while minimizing injury. On the other hand, lower parity sows appear less likely to actively engage in fighting, but gain more skin lesions. Future studies are necessary to assess whether this increase in received aggression also affects performance (e.g. food intake, weight gain) of lower parity sows. Overall, these findings may support the consideration of greater parity segregation in large pen gestation facilities. Limiting the range of parities housed within the same pen may allow for younger sows to gain the necessary social experience to establish a hierarchy without eliciting unnecessary attacks from older sows.

Table 1: Mean (± SEM) scores for behavior during mixing and skin lesions of parity 1-2 sows (low parity sows) and parity 3-6 sows (high parity sows)

Variable	Type	Low parity sows (n = 26)	High parity sows (n = 23)
Fighting (s)	Duration	67.72 ± 23.13	225.14 ± 73.94
Resting (s)	Duration	476.07 ± 98.66	746.95 ± 135.80
Chase	Frequency	0.96 ± 0.33	1.26 ± 0.40
Retreat	Frequency	1.65 ± 0.33	1.39 ± 0.31
Head to body knock	Frequency	0.35 ± 0.14	0.96 ± 0.56
Skin lesions*	Count	11.46 ± 1.26	9.41 ± 1.44

* Skin lesion counts were performed 3 days post-mixing and corrected for presence of skin lesions prior to mixing.

Utilizing colostrum and processing fluids for monitoring porcine circovirus type 2 on sow farms

Madison Kapraun[1], BS; DVM; Katherine Wedel[2], DVM
[1]University of Illinois College of Veterinary Medicine; [2]Iowa Select Farms

Introduction

Porcine circoviruses (PCV) are evolving viruses that impact the commercial pork industry. As a non-enveloped, single-stranded DNA genome, PCV have substantial ability for clinical manifestation.[2] Porcine circovirus type 2 (PCV2) is a ubiquitous pathogen, with prevalence being the determiner of clinical signs. The PCV2 virus has been linked to post-weaning multi-systemic wasting syndrome, weight loss, slow growth, death, and evidence of Porcine Dermatitis and Nephropathy Syndrome.[3] Overall, the economic consequence leads to fallout pigs, commonly seen downstream after contraction at the sow farm. Vaccination against PCV2 has proven effective in reducing viremia and inducing immunity.[1,2] With an applicable monitoring method, sow farms can routinely assess the level of viremia to proactively address vertical transmission, and manage clinical signs of PCV2 associated diseases.

Objective

The objective of this study was to determine which sample, colostrum or processing fluids, is a better sample to detect PCV2 on a sow farm as a means to monitor PCV2 in order to optimize health of downstream pigs in a commercial flow.

Materials and methods

An initial PCV2 screening was done, through sampling of processing fluids, and five sow farms were chosen with PCR results lower than 37 cycle times. Paired colostrum and processing fluid samples were obtained from the same dam/litter. A fragrance-free wipe was used to clean each teat before a sample of 5-10mL colostrum was collected from actively farrowing gilts and sows using a conical tube. When piglets were 1-2 days old processing fluids consisting of serum from testicles and tails from each litter were collected. Gloves were changed between each sow and litter when collecting each sample. Colostrum and processing fluid samples were submitted to the Iowa State University Veterinary Diagnostic Laboratory and tested individually for PCV2 using PCR.

Results and discussion

See Table 1. Results indicate that when comparing colostrum and processing fluids, based on percent positive PCV2, processing fluid samples detected a higher prevalence of PCV2 via PCR. When parity of dam was included in analysis, no difference in PCV2 found in colostrum and processing fluids was noted across all parities. Overall, PCV2 is 2.5 times more likely to be found in processing fluids (31%) when compared to colostrum (12%) for all farms involved. However, processing fluids potentially have a higher means of contamination from the crate, feces, and handling, and may not reflect the true prevalence of PCV2 viremia levels of the piglets themselves. Processing fluids focus on the piglet, and not necessarily the sow; therefore, more diagnostic testing may be needed in order to justify intervention. If using colostrum to monitor the sow farm, interventions, such as mass vaccinations, could have an effect on altering prevalence. Both samples may have limitations due to ease of getting sample on farm and evaluating potential environmental contamination. Information provided by these samples from sow farms could be used to assess PCV2 status and vertical transmission to offspring.

Table 1: PCV2 real-time PCR positive results by farm and sample type

Farm	Sample size	Sample type: colostrum	Sample type: processing fluids
A	31	35%	97%
B	32	0%	3%
C	29	10%	7%
D	36	0%	42%
E	34	18%	9%
All farms	162	12%	31%

References

1. Fort et al. One dose of a porcine circovirus 2 (PCV2) sub-unit vaccine administered to 3-week-old conventional piglets elicits cell-mediated immunity and significantly reduces PCV2 viremia in an experimental model. Vaccine. 2009; 30:4031.

2. Murtaugh. Natural infection, immunity, and prevalence of PCV2. AASV BIVI Symposium. 2014; 189:86.

3. Segales J. Porcine circovirus type 2 (PCV2) infections: clinical signs, pathology and laboratory diagnosis. Virus Res. 2012;164:10.

Investigating the shedding and transmission of Enterisol® Ileitis

E. Kettelkamp[1], BS; C. LeFevre[2], DVM; F. Leite[3], DVM, MS, PhD; J. Waddell[3], DVM, MBA
[1]College of Veterinary Medicine, University of Illinois, Urbana, Illinois; [2]Carthage Veterinary Service, Carthage, Illinois; [3]Boehringer Ingelheim Animal Health USA Inc., Duluth, Georgia

Introduction

Ileitis, an endemic disease caused by *Lawsonia intracellularis,* costs the US swine industry over $100 million per year,[1] and is responsible for mortality and marginalized average daily gain in growing pigs.[2] Enterisol® Ileitis (Boehringer Ingelheim Animal Health USA Inc.) is a live attenuated vaccine against *Lawsonia*, administered via drinking water or drench. Previous research has found little to no shedding of the vaccine strain in swine.[3] However, the development of more sensitive diagnostic tests warrants investigation to update expectations and aid in interpreting diagnostic results from vaccinated animals. There is also lack of information as to the potential transmission of the vaccine strain.[3] Along with importance from a diagnostic perspective, understanding transmission of the vaccine is also important for research trial design to minimize vaccine exposure between treatment groups. The objective of this study was to investigate if Enterisol® Ileitis is shed in feces, and if there is transmission of the vaccine strain to non-vaccinated animals in contact with vaccinated animals.

Materials and methods

Forty-one pigs were blocked for litter, sex and randomly assigned to either vaccinated (n = 21) or non-vaccinated treatment groups (n = 20). All animals were housed in one pen to allow commingling between treatment groups and maximize chances of detecting transmission. At three weeks of age, vaccinated animals were administered a 2 mL oral dose of Enterisol® Ileitis, the other half of animals in the pen were left unvaccinated. Fecal samples were collected from all animals at 0, 1, 3, 7, 14, 21, 28, 35, and 49 days post-vaccination (dpv) for quantitative *Lawsonia* PCR. Serum sampling occurred at day 0, 35 and 49 dpv to investigate seroconversion. Environmental samples were collected using D/E Neutralizing Broth moistened Swiffer® Sweeper dry cloths, contacting multiple points within the pen at each fecal collection time point. The purpose was to investigate if *Lawsonia* levels could be detected in the pen environment by PCR and how they varied over time.

Results and discussion

Fecal and environmental shedding results are outlined in Figure 1. Investigating fecal shedding over time, we observed among vaccinated animals that most fecal shedding occurred at 1 dpv with 76% (16/21) of animals with positive fecal samples with values ranging from < 1,400 to 7.92×10^3 organisms per gram of feces. This coincided with the greatest level of *Lawsonia* detected in environmental samples (Figure 1). The 1 dpv time point was also when the greatest number of non-vaccinated animals was found to shed *Lawsonia*, with 15% (3/20) of fecal samples being positive with values at <1,400 organisms per gram of feces. Following 1 dpv, we observed a decline in fecal shedding in vaccinated animals with only 5 animals sporadically shedding *Lawsonia* between 28 and 49 dpv with levels ranging from < 1,400 to 1.04×10^5. The non-vaccinated group only had one animal with a positive fecal sample after 1 dpv, detected at 49 dpv with a value of 9.15×10^3. Although all animals were fecal PCR negative prior to vaccination, the possibility of exposure to a field strain of *Lawsonia,* especially after 1 dpv cannot be ruled out. No seroconversion was observed amongst treatment groups. This data suggests that while shedding of *Lawsonia* can occur in vaccinated animals, in the time points evaluated and conditions of this study, there was minimal to no transmission of Enterisol® Ileitis from vaccinated to non-vaccinated pigs.

References

1. McOrist S. Defining the full costs of endemic porcine proliferative enteropathy. Vet J 2005;170(1):8 9.

2. Zimmerman, J. J., & Karriker, L. A. (2019). Diseases of Swine (11th ed.). Hoboken, NJ: Wiley-Blackwell.

3. Fangman, T., Hartsook, G., Schnarre, K., Clemons, K., & Cline, G. (2014). Lack of fecal shedding of Enterisol® Ileitis following vaccination in 8 week old pigs. *Allen D. Leman Swine Conference*, 62.

Figure 1: *Lawsonia intracellularis* fecal and environmental pen detection.

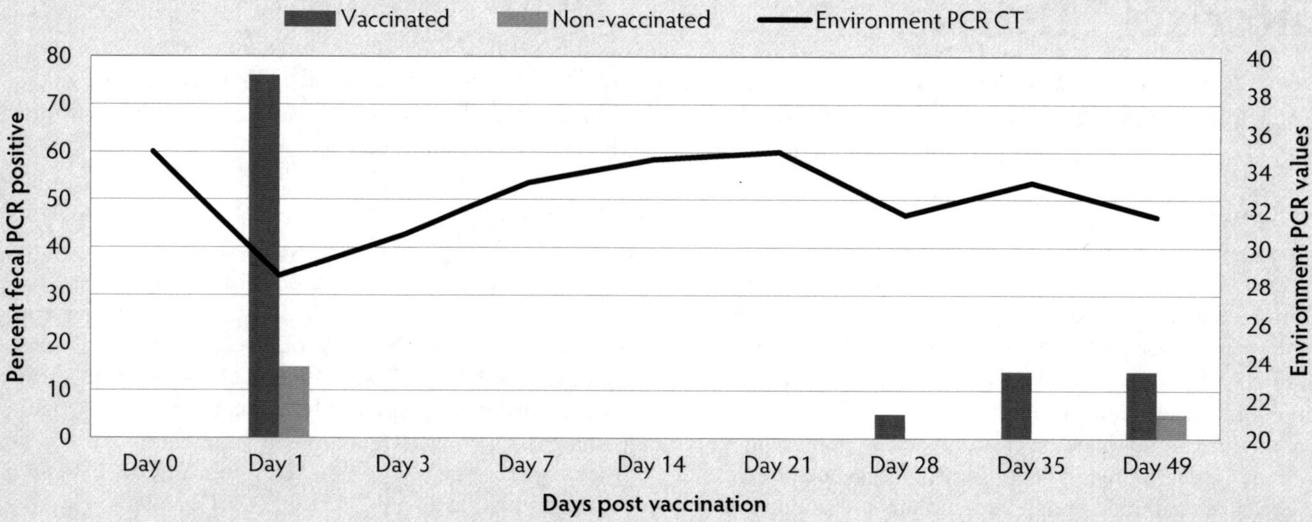

Porcine circovirus type 2 (PCV2) stabilization of replacement gilts by vaccination and the impact on offspring PCV2 status

Kate Mathes[1]; Ethan Spronk[2], DVM; Noel Garbes[3], DVM; Deborah Amodie[3], MS
[1]College of Veterinary Medicine, Iowa State University, Ames, Iowa; [2]Eichelberger Farms, Inc., Wayland, Iowa; [3]Zoetis, Parsippany, New Jersey

Introduction

Porcine circovirus type 2 is the principal etiological agent of porcine circovirus associated disease (PCVAD). PCVAD is an economically damaging disease that costs US producers an average of $3-4 per pig.[1] Stabilizing a sow herd for PCV2, in an effort to reduce vertical transmission, has become a positive management practice. The objectives of this study were to evaluate the association of a gilt's PCV2 status at pregnancy diagnosis with their offspring's PCV2 status, and to determine if an additional gilt PCV2 vaccination would impact their offspring's PCV2 status.

Materials and methods

A total of 100 gilts from two sources, source A (n = 50) and source B (n = 50), were enrolled in this study. Gilts 25 to 28 weeks of age were randomly selected from the first two gilt deliveries to a newly constructed sow farm. Gilts from both sources received two full doses of a PCV2 vaccine prior to their arrival to the facility and a third dose upon arrival. Blood samples were collected from each gilt upon arrival and again at pregnancy diagnosis. At pregnancy diagnosis, gilts were randomly allocated to one of two treatment groups: T01 and T02. T01 (n = 50) were re-vaccinated with a full dose of a PCV2 vaccine. T02 (n = 50) did not receive an additional vaccine. Post-farrowing, processing fluids (PF) were collected from each litter, and three piglets from each litter were subjectively selected by size (small, medium, large) and ear-tagged. At weaning, blood samples were collected from the tagged piglets. Gilt serum samples were tested individually at pregnancy diagnosis, and the piglet serum samples were pooled by litter for PCV2 detection via polymerase chain reaction (PCR). PF were analyzed by litter via PCR for PCV2. As the study progressed, 40 gilts were removed for various reasons unrelated to treatment (T01 = 28, T02 = 32). Gilt data, and their litter data, were not used in the statistical analysis if they were removed. Cross-fostering was not allowed in enrolled litters. A cycle time (Ct) value of greater than or equal to 37 was considered negative for PCV2 and was recorded as 37 for analysis. PF and Ct data was analyzed via SAS 9.4 (Cary, NC) utilizing a Generalized Linear Mixed Model at a significance level of $P \leq 0.05$.

Results and discussion

Upon arrival, 69% of source A gilts and 88% of source B gilts tested PCV2 PCR positive. There was no significant effect of vaccination protocol (T01 vs T02) on PCV2 status for PF ($P = 0.1219$) or wean pig serum ($P = 0.1540$). Moreover, there was no statistical association between the PCV2 status of the gilts at pregnancy diagnosis and the individual litter PCV2 status at processing based on the odds ratio 2.97; 95% Cl of 0.89 and 12.83 ($P = 0.1747$). This indicates no association between gilt PCV2 status at pregnancy diagnosis and offspring PCV2 status (Table 1). In this study, additional PCV2 vaccination administered to pregnant gilts at 42 days of gestation had no significant impact on their offspring's PCV2 status. Adding a vaccination post-breeding and seeing no significant impact on offspring PCV2 status correlates with previous work done using placental umbilical cord serum and colostrum sampling.[2] Future studies should consider gilt status prior to breeding and the role of environmental contamination in PF collection.

References

1. Gillespie, J., Opriessnig, T., Meng, X., Pelzer, K., Buechner-Maxwell, V., 2009. Porcine circovirus type 2 and porcine circovirus-associated disease. *Journal of Veterinary Internal Medicine* 1151-63.

2. Knox, A., Ochoa, L., Greiner, L., 2017. Correlating sampling methods to porcine circovirus type 2 (PCV2) viremia status of sows and piglets in a sow herd with or without PCV2 vaccination during gestation. AASV. 2017; 69-71.

Table 1: Gilt PCV2 serum status at pregnancy check compared to offspring PCV2 processing fluid status

	Litter positive	Litter negative	Total
Gilt positive	7	3	10
Gilt negative	22	28	50
Total	29	31	60

Investigating distance to slaughterhouses and weather parameters in the occurrence of PRRS outbreaks

J. Moeller[1]; J. Mount[2]; E. Geary[3]; C. A. Corzo[3]; A. G. Arruda[1]
[1]Department of Veterinary Preventive Medicine, The Ohio State University, Columbus, Ohio;
[2]Department of Hydroscience and Engineering, The University of Iowa;
[3]Department of Population Medicine, University of Minnesota

Introduction

Porcine reproductive and respiratory syndrome (PRRS) remains one of the most challenging and economically devastating diseases impacting pig producers and veterinarians in the US. From 2005 to 2010, the estimated total US annual cost of PRRS caused by productivity losses in breeding and growing herds was $664 million.[1] There continues to be a lot of unanswered questions from producers regarding how they can reduce their risk of PRRS virus (PRRSV) introduction; and insights on this issue will help producers maintain a healthy herd and an economically stable business. The objective of this study was to assess whether sow farm proximity to slaughterhouses and weather patterns were associated with PRRS occurrence.

Methods

This case-control study paired 104 farms that had reported a PRRS outbreak (cases) with farms from the same production system that had not reported an outbreak during that week (controls). All 208 farms were selected from the Morrison Swine Health Monitoring Project, a volunteer project that collects data from US breeding herds representing approximately half of the US sow population. Data on the distance to the closest swine slaughterhouse and weather conditions during the week prior to the PRRS outbreak were gathered and calculated in collaboration with the Center for Urban and Regional Analysis at The Ohio State University. Weather conditions, including temperature, wind, and precipitation were gathered from the closest weather stations through the Global Historical Climatology Network (GHCN) from NOAA's National Centers for Environmental Information, and averaged across the week. Distances between farms and their closest swine slaughterhouse was calculated using Open Street Map data and the Open Source Routing Machine, with slaughterhouse locations coming from USDA-APHIS's list of approved facilities. Other distance-related variables included distance to the main road and the presence of any farm within 1 and 3 km (s), which were gathered manually using Google Maps. In addition, variables including the presence of an air filtration system, swine density, and PRRS status the week before the outbreak were also included as potential confounders. A multivariable conditional logistic regression model was created using STATA 13 to investigate the association between variables of interest and the occurrence of a PRRS outbreak. Significance was declared at $P < 0.05$; and a trend was declared at $0.05 \leq P \leq 0.10$.

Results and discussion

The final model (Table 1) included swine density, presence of an air filtration system, presence of another farm within 1 km, and PRRS status the week before the outbreak. Farms in areas with a higher swine density had higher odds (OR = 2.48, $P = 0.034$) of reporting a PRRS outbreak, and farms located within 1km of another farm experienced a higher chance of being positive (OR = 3.72, $P = 0.002$), which is in accordance with other reports.[2] The presence of an air filtration system (OR = 0.0992, $P = 0.003$) and being either positive stable (OR = 0.107, $P = 0.102$) or negative (OR = 0.717, $P = 0.037$) decreased the odds of a farm having a PRRS outbreak compared to not having filters or to being PRRS positive unstable, respectively. The presence of an air filtration system in a farm would reduce the chances of PRRSV introduction through the airborne route but it also may be the case that these farms have higher biosecurity standards, also avoiding PRRSV introduction via fomite. In regards to disease status, our results were expected since as PRRS positive stable farms, there would be partial immunity to the virus, which potentially helps with avoiding outbreaks. Likewise, negative farms might have high biosecurity measures, and might be located in less swine dense regions, being less likely to report outbreaks. Surprisingly, none of the weather related variables nor the distance of the closest USDA-APHIS approved swine slaughterhouse were significantly associated with the occurrence of PRRS outbreaks. Limitations of this study include averaging out the weather data for the whole week prior to the outbreak, and using a level of judgement and manual assessment on Google Maps to determine neighboring farms and major roads. Further research is needed to better understand PRRS virus introduction in swine farms.

References

1. Holtkamp et al. *J Swine Health Prod.* 2013; 21(2):72-84
2. Velasova et al. *BMC Vet Res.* 2012; 8: 184

Table 1: Final multivariable conditional logistic regression model

Variable		OR	*P* - value
Pig density	> 4,800 pigs	2.48	0.034
Filtered		0.0992	0.003
Presence of farm within 1 km		3.72	0.002
PRRS status the week before the outbreak	Positive stable, undergoing elimination procedure	0.107	0.102
	Positive stable, ongoing field exposure	0.921	0.102
	Positive stable, ongoing exposure with live virus vaccine	0.288	0.329
	Provisionally negative	0.374	0.472
	ELISA negative	0.0717	0.037
	Depopulation		0.994

Efficacy of an additional *Mycoplasma hyopneumoniae* vaccination under field conditions

L. Nagel, BS; S. Hough, DVM; C. Smith, DVM

Introduction

An issue that often arises in production systems positive for *Mycoplasma hyopneumoniae* (Mhp) is that replacement gilts brought into a herd from an outside source are *M hyopneumoniae* naïve. Since *Mycoplasma* has such a variation and length in shedding time from gilt to piglets, it is difficult to determine the best timing and efficacy of medications and vaccination with an unstable flow downstream.[1,2] In these unstable flows it is common to use a 2-dose protocol for Mhp vaccination.[2,3] Now, 3-dose protocols are being explored to better protect growing pigs. The objective of this study was to determine the efficacy of an additional vaccination of a *Mycoplasma hyopneumoniae* monovalent vaccine and to also compare two known vaccines to a control.

Materials and methods

A flow of pigs that were historically positive for Mhp were chosen for this natural exposure trial. Using 3 barns at a single finishing site, 1872 pigs about 10 weeks of age were given an additional dose of Mhp monovalent vaccine. All pigs had previously been vaccinated with a PCV2/Mhp vaccine (PCVM- G2®, Merck) at weaning and 3 weeks later. The pigs were divided into three treatment groups. Two groups were vaccinated with different Mhp monovalent vaccines (MycoGard®, Pharmgate; Myco Silencer Once®, Merck), and a control group was injected with saline. The study pigs were enrolled by a systematic randomization where each pig received one of 3 ordered treatments represented in each pen. The tag color, number, sex, and starting weight were all recorded. All persons involved were blinded except for the student. About 10 weeks later (68 days post 3rd vaccination), all the pigs were weighed again and ADG and mortality were calculated. Oral fluids were collected at 1, 6 and 11 weeks post 3rd vaccination for Mhp, PRRSV and IAV detection (6 samples per barn per timepoint). Linear and logistic regression analysis were adjusted by barn, sex, days on feed and initial weight effects.

Results

Performance and safety results of each treatment are summarized in Table 1. There was no evidence of differences in mortality, vaccine reaction, or abscess rates between treatments. ADG was higher in the control group in comparison to Myco Silencer Once. ADG was similar when comparing MycoGard with the other groups. Finally, from 54 OF samples collected, 6%, 40% and 69% of them tested positive for Mhp, PRRSV and IAV respectively. Mhp was detected in one barn at 20 weeks of age (10 weeks post 3rd Mhp vaccination). IAV isolates were characterize as H3N2 human-like and H1N2alpha viruses.

Discussion

Even though low levels of Mhp infection were detected, there was evidence that ADG was higher in the control group. The lack of Mhp detection may be related to the diagnostic sample type, variation in Mhp infection shedding within the flow, or was possibly affected by the 3-dose protocol applied. The ADG differences found could be due to the immune system being stimulated and had a detrimental effect on ADG, or a random (statistical) finding that should be replicated to make sure it is a repeatable effect. More importantly, this study suggest that we need better strategies for Mhp vaccine field evaluations than just relying on natural disease occurrence and that we need to carefully interpret these findings considering statistical, biological and economical relevance altogether.

References

1. Holst S et al. Elimination of *Mycoplasma hyopneumoniae* from breed-to-wean farms: A review of current protocols with emphasis on herd closure and medication. *J Swine Health and Prod.* 2015, 23:6.

2. Maes D. *Mycoplasma hyopneumoniae* infections in pigs: update on epidemiology and control. *IPVS Congress Proceedings.* 2010.

3. Thacker, E. Mycoplasma vaccine: implications of age and timing changes. *AASV Proceedings.* 2008.

Table 1: Average daily gain and mortality of finishing pigs receiving vaccine or control

Measurement	Myco Silencer Once®	MycoGard®	Saline control	*P*-value
No. pigs	623	623	623	-
Initial weight, lb	57.4	57.4	57.4	-
Final weight, lb	198.8	200.3	202.0	-
ADG lb/day	2.08^{b}	2.10^{ab}	2.12^{a}	0.03
Mortality %	1.1	0.5	1.1	0.34
Reaction %	0.5	0.0	0.2	0.12
Abscess %	0.3	0.0	0.0	0.10

^{a,b} $P < 0.05$.

Analysis of farm management and intestinal parasite loads between antibiotic free sow farms and organic finisher farms

K. Newcamp, BS; M. Pierdon, VMD, DACAW
New Bolton Center, University of Pennsylvania, Kennett Square, Pennsylvania

Introduction

Organic and antibiotic free production represent a relatively small proportion of the number of pigs in the US swine herd but represent a larger number of farms. Often niche marketing schemes require alterations to parasite control programs while requiring outdoor access, certain floor types, and access to bedding. These requirements may predispose pigs to parasitic infection. The objective of this study was to determine the level of parasitism, type of parasites present, and effect of farm management in pigs on antibiotic free and organic farms. The number of organic farms in the United States was 130 in 2016[1] and niche market pork, including antibiotic free farms, are an active and growing segment of the US pork industry.[2]

Materials and methods

During the summer of 2019, surveys were conducted at 17 organic wean to finish farms ($\mu = 281$ hogs per farm) and 20 antibiotic free sow farms ($\mu = 131$ sows per farm) to investigate the impact of management techniques on overall parasite load. All farms were crate-free with solid concrete flooring and bedded straw. Fresh fecal samples were collected from 10% of sows on each sow farm, and 10 fecal samples were taken from each wean to finish farm. The samples (n = 508) were analyzed using the Modified Wisconsin Sugar Centrifugal-Flotation Method.[3] Saliva samples taken from 10 farms were also centrifuged and the pellet was processed to determine if rope tests could be used for parasite monitoring. Statistics were done with a Stata v.15 Kruskal Wallis test to assess the effect of treatment type and farm type on median eggs per gram.

Results

Multiple parasite types were found across the farms (Table 1). The pigs with outdoor access had significantly higher *Ascaris suum* counts than those that did not ($P < 0.05$). *Isospora suis* was more commonly found on sow farms than finishers ($P < 0.01$). Synthetic parasiticides significantly decreased the number of roundworms compared to the pigs that were treated with natural products or nothing ($P < 0.05$). Oral fluids, when compared to fecal samples, showed no false positives but a range of sensitivities (14% - 50%) depending on the parasite. Of the 37 farms only 3 reported problems with parasites.

Discussion

This study showed that though pigs may be asymptomatic, they maintain parasitic adults that continue to shed eggs. "Strongyle-type" nematodes were the most prevalent parasites detected. This may be due to their ability to undergo hypobiosis, evading the pig's immune system and anthelminthics. Further diagnostics are needed to determine if this is the case or if they are capable of anthelminthic resistance. This is of greater importance in farrowing sows that can transmit the parasites to piglets. This study supports that roundworms are effectively controlled by synthetic parasiticides such as Fenbendazole. The greater number of parasites in outdoor herds may be due to the difficulty of cleaning outdoor environments and the presence of paratenic hosts such as earthworms and beetles. When selecting anthelminthics, this study suggests that precedence should be given to parasiticides that target roundworm and bursate nematodes over whipworms, since whipworms were less prevalent. *Isospora suis* was more common on sow farms than on wean to finish farms, possibly due to the difficulty of disinfecting gestation pens, since wean to finish systems have a greater ability to disinfect barns in between herds. Further testing is needed to determine if the use of multiple oral fluid samples will increase the sensitivity of parasite detection.

References

1. United States Department of Agriculture. Certified Organic Survey. 2016. Available at: https://www.nass.usda.gov/Surveys/Guide_to_NASS_Surveys/Organic_Production/ Accessed 14 November 2019.

2. Honeyman, M.S., Pirog, R.S., Huber, G.H., Lammers, P.J. and Hermann, J.R., 2006. The United States pork niche market phenomenon. Journal of animal science, 84(8), pp.2269-2275.

3. Pittman, J., Shepherd, G., Thacker, B., and Myers, G., 2010. Modified technique for collecting and processing fecal material for diagnosing intestinal parasites in swine. Journal of swine health and production, 18(5):249-252.

Table 1: Fresh fecal samples were taken from 17 organic wean to finish farms and 20 antibiotic free sow farms. The samples were processed, and the types of parasites were determined as well as the eggs per gram in each sample. The table shows the parasite found in the first column, the prevalence of that parasite at the level of the farm in the second column, and the prevalence of the parasite in the individual fecal samples in the third column. Strongyle-type nematodes include *Hyostrongylus rubidus*, *Strongyloides ransomi*, and *Oesophagostoma dentatum*. Ectoparasites found in the fecal samples include *Sarcoptes scabiei*, *Haematopinus suis*, and *Ctenocephalides felis*.

Parasite	Farm prevalence	Sample prevalence
Trichuris suis	3/37	8/508
Ascaris suum	14/37	61/508
Ectoparasites	15/37	38/508
Isospora suis	19/37	83/508
Multiple parasites	29/37	80/508
Strongyle-type nematodes	36/37	170/508
Gastrointestinal parasite	37/37	198/508
Any parasite	37/37	260/508

Tracking pelvic organ prolapses in a breeding herd

Meredith B. Petersen[1], BS; Amanda Chipman[2], BS; Daniel Yonker[3], MS; Thomas J. Fangman[4], DVM, MS, DABVP
[1]College of Veterinary Medicine, Iowa State University, Ames, Iowa; [2]Iowa Pork Industry Center, Ames, Iowa;
[3]JBS Pork®, USA, [4]Boehringer Ingelheim Animal Health USA Inc., Duluth, Georgia

Introduction

Prolapses can contribute to over 20% of total sow mortality.[1] To better understand prolapse incidence and risk factors in a 30,000-sow batch farrowing system, an investigation was conducted in the summer of 2019. Previous work by the Iowa Pork Industry Center (IPIC) guided the creation of a data collection tool focused on pelvic organ prolapses (POP: vaginal (KV), rectal [KR], and dual [KD]).[1] The objective of this evaluation was to demonstrate data mining techniques to establish baseline prolapse rates for the system.

Materials and methods

Five sow sites in a batch farrowing system were tracked, beginning May 2019, and ending December 2021, with each site farrowing every five weeks. The sites utilized the same feed supply, genetic background, and labor. Weekly data entered into PigCHAMP® (PigCHAMP Inc., Ames, Iowa) was used to record prolapses for the system using KV, KR, and KD for vaginal, rectal, and dual prolapses respectively. A POP identification training video and training poster were developed to aid in the data recording. These materials were provided to the IPIC for use in the ongoing Improving Pig Survivability Project (`https://piglivability.org/`). A prolapse log was created in Microsoft Excel™ to establish system-wide baseline prolapse incidence. Following each batch, data from PigCHAMP® was populated into the prolapse log. The data was organized by site and batch and consisted of individual sow data for each prolapsed sow, including prolapse type, parity, gestation length, body condition score (BCS), live born, total born, farrowing date, and death date. From this prolapse log, both a system-wide and site-specific baseline prolapse incidence were estimated. Prolapse as a percentage of total mortality and as a percentage of total farrowing population were determined.

Results

The baseline prolapse rates for all five sites were less than 1.02% prolapses per sows farrowed in batch. The average prolapse rate across all five sites and all batches was 0.57% prolapses per sows farrowed in batch. (Table 1) Prolapses were identified as the cause of mortality in 8.09% of sow deaths. Variability between sites was observed for prolapse as a percentage of sows farrowed and prolapse as a percent of mortality.

Conclusions

The data collection framework demonstrated in this study can be used in large production systems to create baseline POP incidence rates and establish an intervention prolapse rate to stimulate investigations. Future efforts should focus on a response plan to be utilized for sudden spikes in the incidence of prolapses, which may be focused on feed, environment, seasonality, and health. As data collection continues in this system, the incidence of each type of prolapse will be analyzed to determine associations with feeding system (Gestal vs. Stanchion), BCS, gestation length, and parity. Further development of data collection techniques will allow this strategy to be used to track other production variables of importance and allow for more rapid response to problems.

Reference

1. Rademacher C. et al. Potential Risk Factors for pelvic organ prolapse: Survey of 104 US commercial breeding herds. AASV 2019.

Table 1: Prolapse incidence shown as a percentage of total farrowing population by prolapse type and batch ID

Site	Batch ID	Prolapse rate (prolapses/sows farrowed in batch) as a percentage				Site average (average prolapse rate of all batches in database)
		Vaginal	Rectal	Dual	Total	
1	3J	0.09%	0.34%	0.00%	0.43%	0.57%
	4J	0.42%	0.34%	0.42%	1.18%	
	1K	0.00%	0.34%	0.00%	0.34%	
	2K	0.08%	0.08%	0.17%	0.34%	
2	3J	0.26%	0.09%	0.00%	0.34%	0.43%
	4J	0.59%	0.08%	0.00%	0.68%	
	1K	0.25%	0.00%	0.00%	0.25%	
	2K	0.43%	0.00%	0.00%	0.43%	
3	3J	--	--	--	--	0.58%
	4J	0.00%	0.34%	0.00%	0.34%	
	1K	0.09%	0.61%	0.26%	0.96%	
	2K	0.09%	0.27%	0.09%	0.45%	
4	3J	0.26%	0.69%	0.00%	0.95%	1.02%
	4J	0.26%	0.34%	0.17%	0.77%	
	1K	0.70%	0.26%	0.79%	1.76%	
	2K	0.59%	0.00%	0.00%	0.59%	
5	3J	0.17%	0.09%	0.17%	0.44%	0.23%
	4J	0.00%	0.00%	0.07%	0.07%	
	1K	--	--	--	--	
	2K	0.17%	0.00%	0.00%	0.17%	
	Average rate Across all sites	0.25%	0.23%	0.13%	0.61%	0.57%

Evaluation of an injectable and water administered ileitis vaccines in finishing pigs

A. Szczotka[1], BS; M. Hoogland[2], DVM, MS; N. Lauterbach[2], DVM; K. Carroll[1], BS; W. Lincoln[2], DVM
[1]Iowa State University, College of Veterinary Medicine, Ames, Iowa;
[2]Smithfield Hog Production Division, Midwest Region, Algona, Iowa

Introduction

In the US, it is common for pigs to be vaccinated against *Lawsonia intracellularis* (*Lawsonia*), an organism found in swine around the world. Vaccination reduces the severity of this costly disease and may aid in decreasing treatment related antibiotic usage. Two vaccines, administered either orally or intramuscularly, are available in the US for controlling *Lawsonia* by reducing colonization and shedding, which in turn reduces the intestinal lesions that impact growth performance. The objective of this study was to compare the value of an avirulent live water administered vaccine versus an inactivated injectable vaccine versus non-vaccinated controls in growing pigs with naturally occurring *Lawsonia*.

Materials and methods

This study was conducted at a 4,000 head commercial wean-to-finish site sourced from a PRRS and PED virus negative flow. Pigs (n = 1800) were individually ear-tagged and allocated to three treatment groups (600 pigs per group): water vaccine (Ultra Frozen Enterisol Ileitis,® Boehringer Ingelheim, Inc., St. Joseph, MO), injectable vaccine (Porcilis Ileitis,® Merck Animal Health, Summit, NJ), and a non-vaccinated control. Water vaccine was administered per label instructions using a peristaltic water pump (M128 Pump, Stenner Pump Company, Jacksonville, FL) over a 6.25 hour period at 8 weeks of age. Injectable vaccine was administered intramuscularly with 2.0cc of vaccine per individual at 10 weeks of age. Control pigs did not receive any placebo. Additional biosecurity measures were implemented to prevent cross contamination of the avirulent vaccine to the other groups. Individual body weights were obtained at 11 weeks of age and pigs were randomly allocated to study pens in one of two barns (20 pens per barn). Each pen contained 15 pigs (45 total) from each treatment group of mixed gender and size. Throughout, pigs were treated according to the farm's standard procedures and remained in their assigned pen. Mortality was recorded by farm staff. Natural *Lawsonia* exposure was evaluated by periodic testing of fecal swabs and oral fluids from representative pigs and pens by PCR. The final weighing, fecal sampling, and collection of mortality occurred at 20 weeks of age.

Results and discussion

Results available to date are presented in Table 1. Both oral fluid testing and fecal swab results clearly demonstrate that natural exposure to *Lawsonia* occurred. Nearly all oral fluid samples were positive (35 of 36) with CT values decreasing over time; average by week was 27.9, 26.8, 22.3, 22.7 for study weeks 2, 4, 6, and 8, respectively. In summary, an important component of this project was natural exposure to *Lawsonia*, which can be unpredictable. For weight data and fecal results, statistical analysis is not available at this time.

Table 1: Weight, mortality, and fecal swab results available to date

Parameter		Treatment groups		
		Control	Water vaccine	Injectable vaccine
No. pigs start		600	600	600
No. pigs end		559	581	577
Average enrollment weight (lb)		63.7	63.4	64.7
Average end weight (lb)		202.6	206.0	209.0
Average daily gain (lb/day)		2.14	2.20	2.22
Mortality count		14	5	4
Mortality %		2.5%	0.9%	0.7%
Fecal swabs				
Weeks post enrollment	**Variable**			
2	No. samples	28	30	31
	% positive	17.9%	10.0%	12.9%
	Avg. CT of positives	28.4	32.2	31.7
4	No. samples	28	30	31
	% positive	14.3%	23.3%	3.2%
	Avg. CT of positives	27.53	31.5	29.7
6	No. samples	28	30	31
	% positive	71.4%	40.0%	45.2%
	Avg. CT of positives	28.8	30.6	30.9
8	No. samples	28	30	31
	% positive	67.9%	66.7%	51.6%
	Avg. CT of positives	26.2	28.8	29.7

A comparison of diagnostic sampling techniques used to assess transmission risk of *Mycoplasma hyopneumoniae* from positive sourced boar studs

Z. Talbert[1], BS; A. Maschhoff[2], DVM; M. Jansen[3], DVM
[1]College of Veterinary Medicine, University of Illinois, Urbana, Illinois;
[2]The Maschhoffs, Carlyle, Illinois; [3]Zoetis, Parsippany, New Jersey

Introduction

Mycoplasma hyopneumoniae (Mhp) plays an important role in porcine respiratory disease complex, causing high morbidity, decreased growth rate, and reduced feed efficiency. There are only a few studies in the literature testing different diagnostic methods for Mhp in boar studs or the transmission risk assessment between studs and breeding units. Typically, Mhp is transmitted by nose to nose contact, but systemic transmission has been speculated and documented. One of those studies was able to detect Mhp in 8 out of 15 semen samples.[1] Another study detected Mhp by polymerase chain reaction (PCR) in processing fluids.[2] The objective of this study was to evaluate four diagnostic sampling techniques and assess the transmission risk of Mhp from a positively sourced stud.

Materials and methods

In an effort to evaluate common Mhp detection methods used, a total of 90 boars from two studs known to previously receive Mhp-positive replacement animals were sampled for Mhp using four diagnostic sampling techniques: tracheobronchial swabs (TBS), frothy oral fluids (FO), semen (SEM), and serum (SM). TBS were taken using previously described techniques which consisted of using the inner catheter of a post-cervical artificial insemination (PCAI) catheter, with an 80 mm FloqSwab (Copan Diagnostics, Murietta, CA), attached at the end. After collection, the FloqSwab was removed from the PCAI catheter and placed into a 3 mL tube that contained 2 mL of phosphate buffered saline (PBS) solution. For FO collection, one 4" × 4" gauze square was placed on the soft rubber end of a PCAI catheter and secured via rubber band. Before sampling, each modified PCAI catheter was dipped into a small Ziploc bag containing PBS solution. The 4" × 4" gauze end was then used to swab the FO from the boar's lips and mouth. After swabbing, the gauze square was placed into a small Ziploc bag, squeezed to release the fluid, and transferred to a 3 mL tube. For SEM, 2 mL was placed in a 3 mL falcon tube prior to laboratory extension. Lastly, 3 mL of blood was collected, from the lateral saphenous vein. All samples were sent to the Iowa State Veterinary Diagnostic Laboratory. Blood samples were tested by ELISA (IDEXX, Westbrook, ME); PCR testing was ran on TBS, FO, and SEM samples.

Results and discussion

Both studs had seropositive boars on ELISA (S:P ratio > = 0.4) (Figure 1). Although positive ELISA tests were not surprising due to Mhp vaccination protocol in isolation, several boars had an S:P ratios > 1.0, indicating likely past exposure. All TBS, FO, and SEM samples were PCR negative for Mhp (Figure 1). Given the history of positive Mhp sourced boars into these studs, this study demonstrated that the transmission possibility of Mhp may be minimal in boar studs. Further research needs conducted to validate the isolation of Mhp in SEM. Given the lack of research in this area of Mhp, those owning and managing boar studs should continue to be diligent in understanding lateral transmission possibilities, elimination options and diagnostic testing methodology to assure classified health status. Many systems have undergone elimination within breeding herds and the negative health impact of transmission would be costly.

References

1. Milovanovic A et al. Boar Quality Semen Testing and Presence of Mycoplasma Organism. 2017. Available at: https://medwinpublishers.com/OAJVSR/OAJVSR16000137.pdf. Accessed 10 June 2019.

2. Sponheim A et al. Detection of Mycoplasma hyopneumoniae and Mycoplasma hyorhinis by PCR in processing fluids. 2018. Available at: https://www.aasv.org/library/swineinfo/item.php?22714. Accessed 10 June 2019.

Figure 1: Diagnostic results for TBS, FO, SEM (by PCR) and SM (by ELISA).

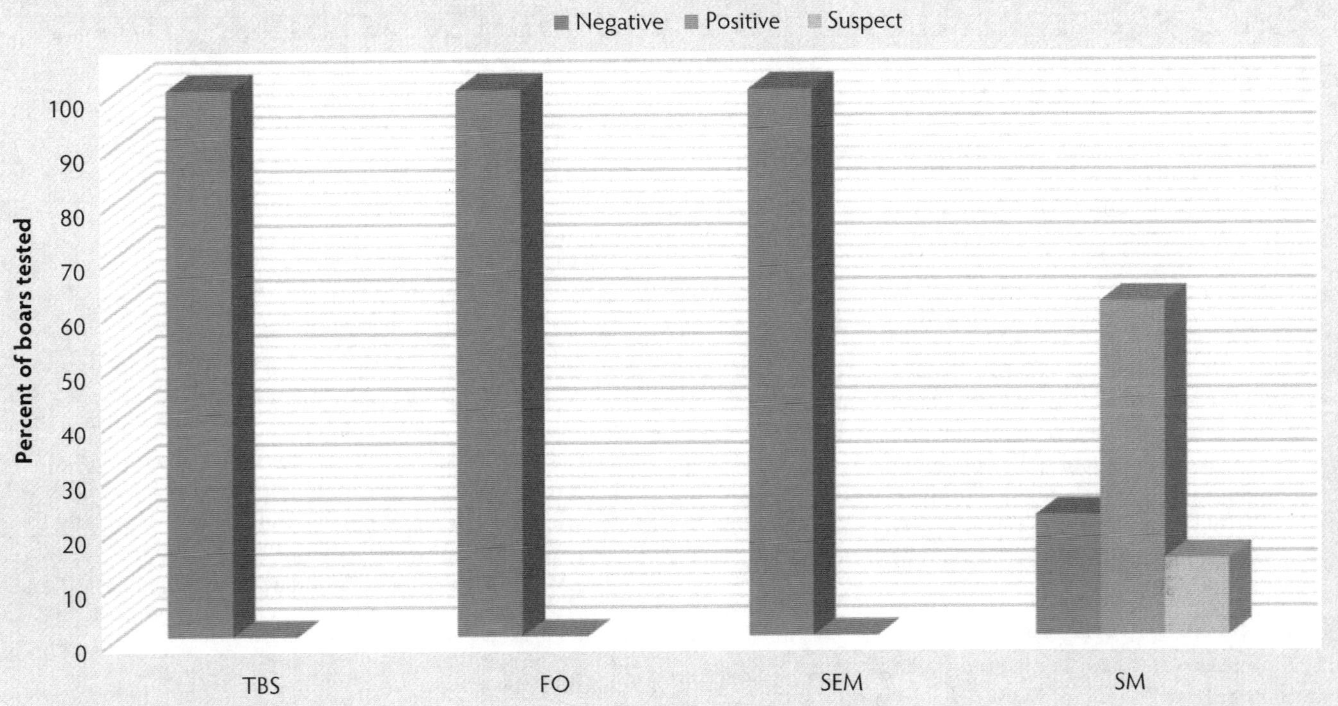

Longitudinal assessment of PCV2/PCV3 viremia in PCV2 vaccinated and non-vaccinated animals under PCV2d field exposure

A. Taylor[1]; M. Farber Billing[2], DVM; T. Specht[3], DVM; E. Fano[2], DVM, MSc, PhD
[1]The Ohio State University, Columbus, Ohio; [2]Boehringer Ingelheim Animal Health, Duluth, Georgia;
[3]Heimerl Farms Ltd, Johnstown, Ohio

Introduction

Porcine circovirus (PCV) types 2 and 3 are hardy DNA viruses associated with multiple clinical signs that are detrimental to overall pig health and viability and cause great economic losses to the swine industry. There are five known genotypes of PCV2 (a-e) and less is known about PCV3 disease associations and its genetic variants. The objective of this study was to analyze and compare PCV2 vaccine products and non-vaccinated pigs in relation to impacts on longitudinal PCV2 and PCV3 viremia.

Materials and methods

Ninety-three barrows were weaned at 18-22 days of age to a nursery facility prior to PCV2 vaccination. Each pig was randomly allocated to a treatment using a systematic blocked design and assigned to a pen so all treatments were comingled in each pen. At about 1 week post weaning, one of three treatments was administered to each barrow: Tx1: Ingelvac CircoFLEX® (BIVI) – one dose of 1mL per label (n = 31); T × 2: Fostera® Gold PCV (Zoetis) – one dose of 2mL per label (n = 30); or Non-vaccinated sentinels (n = 32). Serum was collected from all pigs in each group at weaning, 14 weeks of age, and 20 weeks of age, just prior to first marketing. Oral fluids were collected at the same time points. Tissues were collected from fall-behind animals at 14, 16 and 20 weeks of age. PCV2d circulation/exposure confirmed by sequencing prior the start of the study.

Results

At 14 weeks peak viremia for PCV2 was reached, with 70% of all animals positive whereas peak viremia for PCV3 was at 20 weeks with 14.64% of all animals positive (Figure 1). No correlation was observed between PCV2 and PCV3 viremia. As expected, neither PCV2 vaccine product provided viremia reduction of PCV3. Interestingly, four pigs were persistently infected with PCV3 (ct < 37) from weaning until market. PCR and histopathology on tissue samples from 2 different animals at 14 weeks tested positive for PCV2 and one positive for PCV3. Tissue samples and serum consistently tested positive for PCV2d. On day 0 of the trial 1 out of 4 oral fluid samples tested positive for influenza A and all were negative for PRRS. Oral fluids tested negative for influenza A, PRRS, and *M hyopneumoniae* at 14 & 20 weeks of age.

Discussion

This trial provides the industry with novel longitudinal data on PCV2 and PCV3 viremia combined with PCV2 vaccine comparison in PRRS and *Mycoplasma hyopneumoniae* negative pigs from wean to market. This trial provides evidence that current PCV2 vaccines containing genotype(s) a or combination of a and b provide equal reduction of viremia of PCV2 genotype d. Both vaccines significantly reduced PCV2d viremia compared to the non-vaccinates, similar to previous studies involving PCV2a and PCV2b challenge. An interesting finding from this study is that a small percentage of the population is persistently infected with PCV3 from wean to market. In regards to this emerging pathogen, the industry should continue to investigate how it manifests over the lifetime of the growing animal and its potential disease associations.

Figure 1: PCV2 and PCV3 viremia.

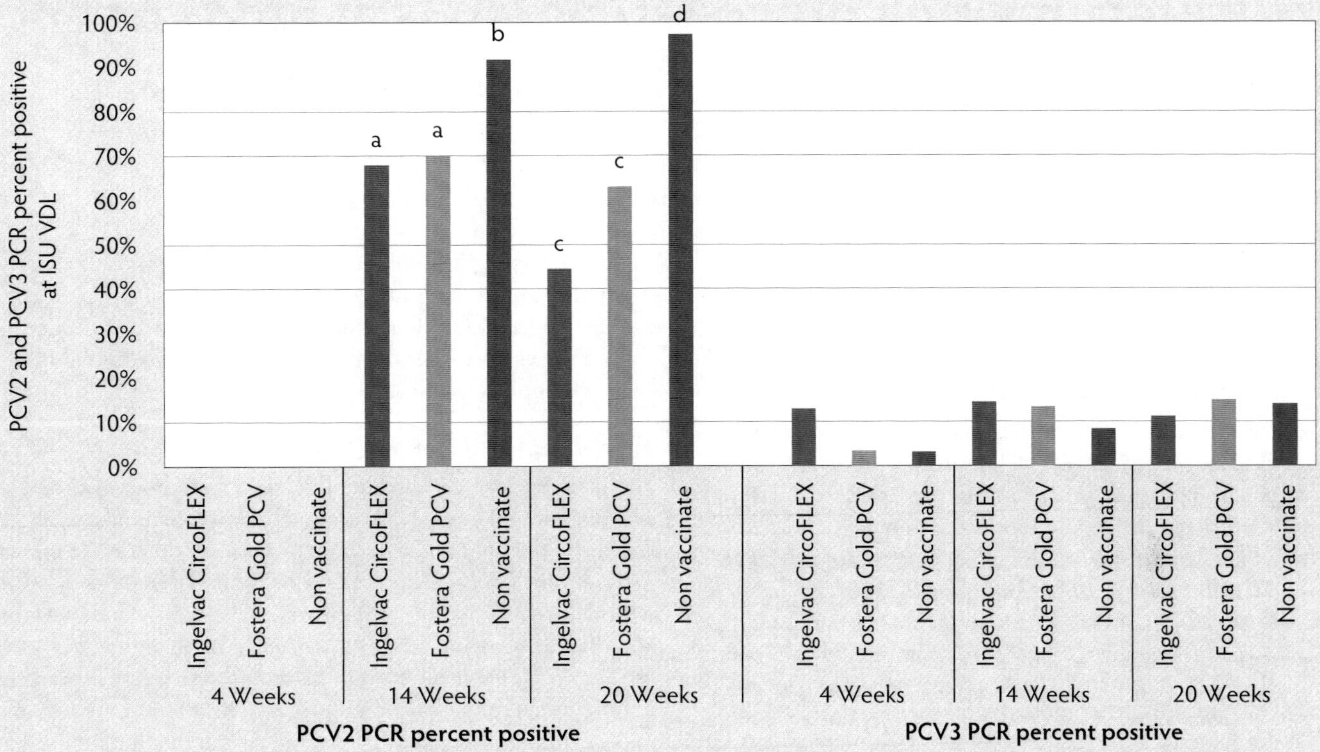

[a-b, c-d] Differing superscripts denote statistical differences within each time point sampled ($P \leq = 0.05$). Analysis was done using logistic regression model, using the Bonferroni correction for multiple comparisons.

Ultrasonographic observation of ovarian structures and evaluation of serum progesterone levels in breeding gilts prior to synchronization: a descriptive study

Jonathan Tubbs[1], MS; Julie Gard[1], DVM, PhD, DACT; Rick Tubbs[2], DVM, MS, MBA, DACT
[1]Auburn University CVM, Auburn, Alabama; [2]Tubbs Contract Research Organization, Jasper, Alabama

Introduction

Maintaining breeding targets and taking advantage of genetic advancement is a priority for sow farms. Issues with infertility, unobserved and silent heat, and disease challenges often negatively affect reproduction. Many tools of swine production help ensure breeding consistency such as feeding management in lactation and the wean-to-estrus period, synchronization protocols, and various breeding techniques. When implemented, these tools create reproductive predictability in the breeding population. However, replacement gilts can often present issues in hitting breeding targets and replacement rates, often being due to unpredictability in cycling, especially during summer, and the added labor needed for heat-no-service programs. One tool not commonly used in swine production that could potentially help bring consistency to breeding gilts is the use of real time ultrasound for reproductive evaluation. Therefore, the purpose of this study was to describe ovarian structures via real time ultrasound, and evaluate serum progesterone levels of replacement gilts to aid in insemination timing.

Materials and methods

An 80-sow unit in east central Alabama was utilized for the study beginning in early September so the late summer gilt breeding group was evaluated. Three replacement gilts and one open sow, used for comparison, were selected. Animals were snared in an open pen and blood samples taken for testing of serum estrogen and progesterone levels. Next, the rectum was digitally cleaned and 60 mLs of obstetrical lube was inserted via syringe. A SonoScape S8 EXP portable ultrasound machine with L741V 8.5 MHz transrectal probe was used to visualize the ovaries and ovarian structures. A 0.5 in. × 18 in. plastic pipe was wrapped and taped to the cord for rigidity and probe control. The probe was rectally inserted and advanced until ovaries were viewed and images captured. Two days following initial imaging and blood collection, study animals were started on a 14 day cycle of top dressed Matrix® in accordance with label directions.

Results and discussion

Per Table 1, gilts had follicular development on both ovaries. The sow showed luteal structures on the left ovary and a small follicle on the right and had the appearance of fluid in the uterus. Progesterone levels indicate estrus for animals 512, 514, and 480 while levels in 515 indicate anestrous. This study focused on gilts entering the breeding herd as this group is often unpredictable in estrous cycle. While ultrasound is not necessary in every breeding animal, use in those not showing signs of heat or not cycling normally would be beneficial. This is especially true in high value animals such as within genetic herds and multipliers. Economically, utilizing ultrasound can result in an approximate profit of $80/sow in commercial herds and a faster genetic turnover when the breeding potential of all replacement animals is met. Additionally, recent industry discussions concerning gilt infertility and silent estrus have emphasized the importance of the topic. Use of ultrasound is a valuable tool for reproductive assessment. While more research is warranted, use of ultrasound could be beneficial in finding solutions to this problem.

Table 1: Ultrasound and serum sample results by study animal

	Ovarian structures		Serum results
Animal ID	**Left**	**Right**	**Progesterone**
Gilt 512	3-5 F, 0.5 cm	5-7 F, 1.0 cm	0.2 ng/mL
Gilt 514	< 3 F, 0.5 cm	3-5 F, 0.7 cm	< 0.2 ng/mL
Gilt 515	7-10 F, 0.3 cm	3-5 F, 0.5 cm	11 ng/mL
Sow 480	LT	MSF	< 0.2 ng/mL

F – Follicle; LT – Luteal Tissue; MSF – Multiple small follicles

References

1. Knox R, Probst-Miller S. Using real-time ultrasound for identification of sources for reproductive failure in weaned sows. J. Swine Health Prod. 2004;12:71-74.

2. Flowers WL. Real-time ultrasonography and diagnosis of pseudopregnancy in swine. http://mark.asci.ncsu.edu/SwineReports/2001/02physflow.htm (Accessed June 20,2003). 2001.

3. Martinat-Botte F, Renaud G, Madec P, Costiou P, Terqui M. Ultrasonography And Reproduction In Swine. INRA, Paris. 2000.

4. Peet B. Tracking efficiency with NPD's. Western Hog Journal 2004; Winter.

5. Hollis WL. Practical perspectives on real-time ultrasound. Proc. Amer. Assoc. Swine Practitioners. 1998.

Resting heart rate of group-housed gestating gilts predicts judgement bias

Caitlyn R. Tukdarian; Sanne Roelofs; Thomas D. Parsons
Swine Teaching and Research Center, School of Veterinary Medicine,
University of Pennsylvania Kennett Square, Pennsylvania

Introduction

Growing public concern about the welfare of gestating sows is driving changes in husbandry practices. These concerns largely focus on the impact of confinement on the sow's psychological health. Unfortunately, our ability to measure affective state in sows is limited. Judgement bias has emerged as a possible measure of affective state in animals. In a judgement bias task, affective state (from positive to negative) is inferred from an animal's response to ambiguous stimuli that are intermediate to both rewarded and unrewarded conditioned stimuli. Expectation of reward in response to an ambiguous stimulus is considered indicative of a positive affective state, and thereby positive animal welfare. Heart rate and heart rate variability (HRV) have also been suggested as measures of animal welfare. These cardiac parameters reflect activity of the autonomic nervous system, with stress causing decreased vagal tone (increased heart rate at rest) and decreased HRV.

Objective

To examine if cardiac parameters predict responses in a judgement bias task.

Materials and methods

Nine group housed gestating gilts (DNA Line 241) were selected based on age and stage of gestation. Training and testing in the judgment bias task was conducted in an open, 2.4 × 8.3 m rectangular arena with five identical, equally spaced goal-boxes at one end. Each goal-box was fitted with a lid, with an open goal-box serving as a stimulus predicting either a positive or negative outcome. First, gilts were trained to associate the location of an open goal-box with either an accessible or inaccessible feed reward. For each gilt either the far right or far left goal-box was selected as the positive stimulus, with the opposite goal-box being the negative stimulus. Once a gilt could reliably approach the open positive goal-box and not approach the open negative goal-box with 80% accuracy, the three middle goal-boxes were introduced as ambiguous stimuli. Responses to the ambiguous goal-boxes were scored as go-responses (indicative of expectation of reward, i.e. positive bias) if the animal approached. Polar heart rate monitors and Elite HRV software were used to collect heart rate data. A five-minute reading was collected for each subject in their home pen to assess baseline stress. Several additional measures based on the recorded R-R intervals were derived using Kubios software. These values were compared with outcomes from the judgement bias task. Generalized linear models (GLM) using a Gaussian distribution were fit to the data with percentage of go-responses as the outcome variable and heart rate and time-domain based measures of HRV as explanatory variables.

Results

Eight sows completed judgement bias training and were tested with ambiguous stimuli. The type of stimulus presented significantly impacted the percentage of go-responses ($\chi^2_{(4)} = 79.5$, $P < 0.001$), and validates the ability of the gilts to discriminate between the positive and negative cues. Individual sows also exhibited significantly different numbers of go-responses ($\chi^2_{(7)} = 52.3$, $P < 0.001$). These differences in judgment bias were predicted by heart rate in the home pen ($\Phi_{(7)} = 8.5$, $P < 0.03$; Figure 1). Interestingly, other measures of HRV failed to significantly predict the judgement bias outcomes ($P > 0.05$).

Discussion

These finding suggest that gilts with greater vagal tone (lower resting heart rate) have a more positive judgement bias and by inference a more positive affective state compared to gilts with lower vagal tone. Parasympathetic activation, including heart rate at rest, might be a candidate for an easily accessible measure of an animal's affective state and thereby their welfare in group housing. Such a measure could be used to quickly identify individuals within the herd that are at risk of experiencing increased stress and lowered welfare in response to the many stressors present on commercial farms. Future studies are required to further validate resting heart rate as a measure of sow welfare.

Figure 1: Gilts' average percentage go-responses during judgement bias testing plotted against average heart rate in the home pen.

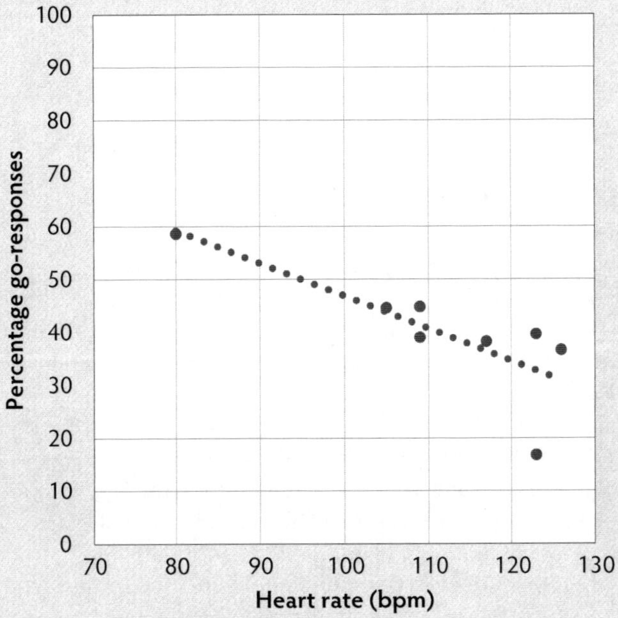

Experimental efficacy of a Merck RNA particle (RP) vaccine against porcine parainfluenza type 1 (PPIV-1) challenge in weaned piglets

M. Welch[1]; K. Harmon[1]; J. Zhang[1]; P. Pineyro[1]; L. Giminez-Lirloa[1]; E. Strait[2]; P. Gauger[1]
[1]Iowa State University, Ames, Iowa; [2]Merck Animal Health Inc, USA

Introduction

Parainfluenza viruses cause respiratory disease in many species including dogs, cattle, and humans.[1] Porcine parainfluenza virus type 1 (PPIV-1) has been detected in respiratory samples from swine of all ages and may be a cause of respiratory disease in production systems in the absence of common respiratory pathogens.[2] Vaccines may help reduce the impact of PPIV-1. The objective of this study was to evaluate the efficacy of an adjuvanted and non-adjuvanted replicon particle (RP) vaccine against challenge with a PPIV-1 isolate compared to live-exposure and non-vaccinated nursery pigs.

Materials and methods

Fifty, three-week-old piglets were screened prior to vaccination for PPIV-1 virus shedding and antibody by RT-qPCR and wv-ELISA respectively. The piglets were randomly allocated into five treatment groups designated non-vaccinated/non-challenged (NV/NC), non-vaccinated/challenged (NV/C), adjuvanted RP vaccinated/challenged (RPadj/C), non-adjuvanted RP/challenged (RP/C), and live exposure/challenged (LE/C). Piglets were administered two doses of their respective vaccine or placebo at -40 and -19 days post inoculation (dpi), designated "prime" and "boost", respectively. Piglets were challenged at 0 dpi with a 2 mL intratracheal and 1 mL intranasal dose of PPIV-1 isolate USA/IA/84915LG/2017 at a concentration of 1x10^4.8 TCID50/mL. Nasal swabs were collected on -47, -40, -19, -1, 0-5 dpi and oral fluids were collected on -40, -20, -15, -12, -8, -1, 3, and 5 dpi and tested by real-time reverse transcriptase polymerase chain reaction (RT-qPCR). Serum was collected on -43, -32, -25, -19, -12, -5, 0, and 5 dpi and evaluated with a whole virus enzyme linked immunosorbent assay (wv-ELISA) and neutralizing antibody assay. Pigs were evaluated at 0-5 dpi for clinical signs of respiratory disease. At 5 dpi necropsy, nasal swabs, proximal and distal tracheal swabs, cardiac lung lobe, nasal turbinate and bronchoalveolar lavage fluid (BALF) were collected and tested by RT-qPCR. Lungs were evaluated for percent consolidation and histopathology lesions. Data were analyzed using a generalized linear mixed-effect model in the program R with the "lme4" package comparing across treatment groups. The individual pig was considered the experimental unit.

Results and discussion

Clinical respiratory disease was not observed in any group post-challenge. At 2-5dpi, mean PPIV-1 log10 genomic copies were significantly lower ($P < 0.05$) in the RPAdj/C, LE/C, and NV/NC groups compared to the NV/C and RP/C pigs in nasal swabs (Figure 1). Gross and microscopic lesions were minimal with no significant differences, consistent with prior challenge studies.[3] However, mean PPIV-1 $\log_{10}$ genomic copies were significantly lower ($P < 0.0005$) in BALF, tracheal swab and turbinate in all vaccinated groups compared to the NV/C pigs. Significantly higher ($P < 0.0001$) neutralizing antibody titers were detected in the sera from RPAdj/C group at 5 dpi compared to RP/C and LE/C groups. Similar trends were observed in the wv-ELISA. Increased levels of IgA were detected in the BALF of the LE/C group, contrasting against higher levels of IgG collected from the RPAdj/C group. These data suggest swine production systems with clinical signs and diagnostic results supporting the presence of PPIV-1 respiratory disease may benefit from vaccination. Our data demonstrate that RP vaccines with adjuvant may reduce PPIV-1 viral shedding similar to challenged pigs with prior natural exposure. However, all vaccinated groups had reduced PPIV-1 in BALF and respiratory tissues at necropsy suggesting vaccination may reduce viral load throughout the respiratory tract. Pigs exposed intranasally to PPIV-1 develop a predominantly mucosal response in contrast to a systemic response from intramuscular vaccination. Both systemic and mucosal antibody responses appear to reduce PPIV-1 virus levels in tissues at necropsy and shedding in nasal swabs.

References

1. Henrickson KJ. Parainfluenza viruses. *Clinical microbiology reviews.* 2003;16(2):242-264.

2. Park JY, Welch MW, Harmon KM, et al. Detection, isolation, and in vitro characterization of porcine parainfluenza virus type 1 isolated from respiratory diagnostic specimens in swine. *Vet Microbiol.* 2019;228:219-225.

3. Welch M, Park J, Harmon K, et al. Pathogenesis of a porcine parainfluenza virus-1 isolate (USA/MN25890NS/2016) in conventional and CDCD piglets. AASV Annual Meeting; 2018.

Figure 1: Mean $\log_{10}$ genomic copies per milliliter of PPIV-1 RNA and standard error of the mean (SEM) detected by RT-qPCR in processed nasal swabs for NV/C, RP/C, RPAdj/C, LE/C, and NV/NC at 0 - 5 days post inoculation. Different letters within DPI indicate significant differences in shedding between treatment groups at $P \leq 0.05$.

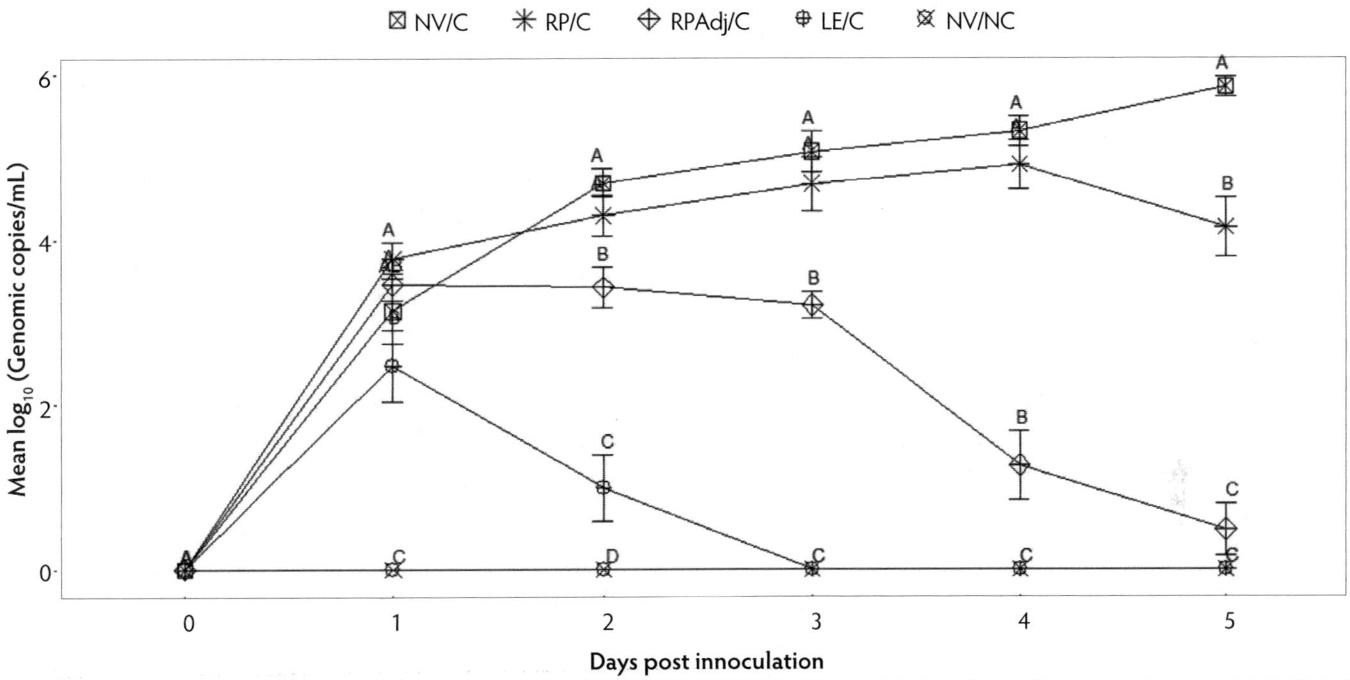

NV/NC: non-vaccinated/non-challenged
NV/C: non-vaccinated/challenged
RPAdj/C: adjuvanted replicon particle/challenged
RP/C: non-adjuvanted replicon particle/challenged
LE/C: live exposure/challenged

Streptococcus suis in swine nurseries of a commercial production system

K. White[1], BS; D. Boykin[2], DVM; A. Hintz[3], DVM
[1]Texas A&M University College of Veterinary Medicine and Biomedical Sciences, College Station, Texas;
[2]Cactus Family Farms, Osceola, Iowa; [3]Pharmgate Animal Health, Wilmington, North Carolina

Introduction

Streptococcus suis (S Suis) is the most economically important streptococcal infection of pigs for the commercial hog producer.[1] Measures used to control and prevent *S suis* infections include antibiotic treatments and vaccination of sows and piglets, as well as varied biosecurity protocols. Vaccination against *S suis* has resulted in variable success.[2] If vaccination of sows pre-farrow proves to be effective, it would be an economic alternative to vaccinating pigs.[3] The objective of this study was to evaluate the effects of the administration of an autogenous vaccine to sows and the implementation of biosecurity protocols on morbidity and mortality of *S* suis infections in pigs in a commercial swine operation.

Methods

Since 2018, a multi-site commercial pig flow experienced clinical signs of neurologic disease in post-weaned piglets. Four flows were followed for the presence of disease. Shortly after arrival to the nursery, piglets exhibited depression, lameness, incoordination, and sudden death. Mortality ranged from 4-20%. In April 2019, an autogenous *S suis* vaccine was used with isolates from the affected flows given six- and three-weeks pre-farrow. The vaccine was developed from *S suis* strains isolated from meninges, spleen, liver, and joints of acutely ill or rejected nursery piglets. Isolates were sent to Newport Labs to further identify the serotypes that caused disease and production of the vaccine. The vaccine utilized *S suis* isolates from affected flows including serovars 1/2, 7, and 10. Sows from herd one were given a half-dose, or 1mL, while sows from herds two, three, and four were given a full dose, or 2mL. Piglets from herds three and four were commingled due to health status prior to vaccination and after vaccination were separated by rooms. Percent mortality was calculated at the end of the nursery phase prior to and after administration of the pre-farrow vaccine in the four flows for six weeks of production in each flow and by group. One-way ANOVA was performed to test the differences in mortality before and after vaccination by flow. There was no statistical difference in herd one, while there was a statistical difference in the reduction in mortality from 4.4% to 2.2% ($P = 0.007$) in herd two and a reduction in mortality from 4.8% to 1.2% ($P < 0.0001$) in herd three. In this study, full dose vaccination is associated with a decreased mean mortality in herds two and three (Table 1).

Discussion

When administering an autogenous *S suis* vaccine, a full dose of vaccine to the sow pre-farrow may reduce the incidence of *S suis* in pigs. Further, strict biosecurity by separation of piglets by room significantly reduced *S suis* mortality. Further investigation is warranted to determine if the combination of vaccination and biosecurity measures would reduce or possibly eliminate *S suis* infections from herds. This study shows the importance of providing a high quantity of homologous bacterin in the vaccine for adequate protection for the herd. Although not conclusively demonstrated in this study, it appears that vaccine efficacy is improved by selecting herd-specific serotypes for vaccine development.

References

1. Zimmerman, J.J., Karriker, L.A., & Ramírez Alejandro. (2012). Streptococcosis. *Diseases of swine*. Hoboken, NJ: John Wiley & Sons, Inc.

2. Torremorell, M., Pijoan, C., & Trigo, E. (1997) Vaccination against *Streptococcus suis*: Effect on nursery mortality. *Swine Health and Production,* 5(4), 139-143.

3. Amass, S.F., & Stevenson, G.W., Vyverberg, B.D., Huxford, T.W., Knox, K.E., Grote, L.A. (2000). Administration of a homologous bacterin to sows pre-farrowing provided partial protection against streptococcosis in their weaned pigs. *Swine Health and Production, 8,* 217–219.

Table 1: Results of piglet mean percent mortality before and after vaccinating sows pre-farrow by herd

Herd	Vaccine dose	Piglet mean % mortality: before vaccination	Piglet mean % mortality: after vaccination	*P*-value (accepted *P*-value < 0.05)
1	1 mL	4.8	5	0.8751
2	2 mL	4.4	2.2	0.007
3 and 4	2 mL	4.8	1.2	< 0.0001

Dynamics of a rotavirus type A and C coinfection in colostrum deprived piglets

Z. Wolfe[1], BS; J. Pittman[2], DVM, MS DABVP
[1]College of Veterinary Medicine, North Carolina State University, Raleigh, North Carolina;
[2]Smithfield Hog Production, North Region, Waverly, Virginia

Introduction

Rotavirus (RV) causes diarrhea in young pigs and results in economic loss due to increased mortality and decreased weight gain. The common serogroups in swine are type A (RVA), B (RVB), and C (RVC). One method to control RV is immunization of sows through natural planned exposure (NPE) of live virus. RV antibodies are transferred to piglets through colostrum and milk. NPE can be made by harvesting intestines from colostrum deprived (CD) piglets inoculated with RV. Previous work by Madson et al[1]. demonstrated the dynamics of singular and coinfections with RV in CD piglets, but measurements were only qualitative, with sampling intervals of 12 hours over 3 days. The objective of this study was to quantify the viral shedding of a RVA-RVC coinfection in CD piglets to determine optimal harvesting time for NPE.

Materials and methods

On two occasions, ten CD pigs were collected via vaginal tract, placed into clean, disinfected totes, and transferred to the laboratory. Pigs were separated equally into two groups (2 replicates). Pigs were provided supplemental heat and administered 10 mLs of commercial replacement milk orally. RVA/C inoculum was from known tested frozen stock utilized in NPE protocols. Each pig was orally administered 1 mL of RVA/C. All pigs were orally administered 10 mL of replacement milk 4 hours post inoculation (hpi). Fecal swabs were taken at 0, 12, 16, 18, 26 and 30 hpi for replicates 1 and 2 and 0, 12, 18, and 24 hpi for replicates 3 and 4. All swabs were submitted for individual RV PCR.

Results

Data is presented in Table 1. In replicates 1, 2 and 4 all pigs tested positive for RVA and RVC at all pi time points. In replicate 3, 3 of 5 pigs tested positive for RVA at 12 and 18 hpi, and 5 of 5 at 24 hpi. All pigs in replicate 3 tested RVC positive at 12, 18 and 24 hpi. In replicates 1 and 2, the lowest average ct values for RVA (16.8 and 15.9) and RVC (19.9 and 18.2) were at 12 hpi, with values steadily increasing over time. For replicate 3, the lowest ct value for RVA (17.5) was not until 24 hpi, when all 5 pigs were positive, and for RVC 18 hpi had the lowest average ct value (15.7). Replicate 4, was similar to replicates 1 and 2 in pattern, with a low ct value for RVA (20.7) and RVC (17.9) at 12 hpi.

Discussion

The overall pattern of RVA and RVC shedding in all replicates was similar with pigs shedding by 12 hpi through 24 or 30 hpi. This work supports results by Madson et. al[1], where shedding was detected at 12, 24 and 36 hpi. This study provides additional information, reporting ct values for more frequent time points within the first 24-30 hours post inoculation. It must be noted that while PCR ct values for RV do not correlate to infectious virus, it is currently the only practical method to predict relative virus. Further work is needed to correlate fecal shedding values with total intestinal RV yield. This data can be used by practitioners to define the optimum time post inoculation to harvest CD piglet intestines for maximum yield RV NPE.

Reference

1. Madson D, Hoang H, Arruda P, Stevenson G, Yoon K-J. 2012. Rotavirus Coinfection: Diagnostic Overview. *Proc ISU Swine Dis Conf* 69–72.

Table 1: Number positive/total (average PCR ct value)

	Rotavirus type A			
	Replicate			
hpi	**1**	**2**	**3**	**4**
0	2/4 (35.3)	4/5 (33.7)	0/5 (36.0)	0/5 (36.0)
12	4/4 (17.1)	5/5 (15.9)	3/5 (30.7)	5/5 (20.7)
16	4/4 (16.8)	5/5 (17.0)	ND	ND
18	4/4 (18.2)	5/5 (19.0)	3/5 (27.1)	5/5 (21.6)
24	ND	ND	5/5 (17.5)	5/5 (18.1)
26	4/4 (18.6)	5/5 (20.0)	ND	ND
30	4/4 (19.0)	5/5 (18.8)	ND	ND
	Rotavirus type C			
	Replicate			
hpi	**1**	**2**	**3**	**4**
0	0/4 (≥ = 36.0)	0/4 (≥ = 36.0)	0/5 (≥ = 36.0)	0/5 (≥ = 36.0)
12	4/4 (19.9)	5/5 (18.2)	5/5 (17.1)	5/5 (17.9)
16	4/4 (21.8)	5/5 (21.1)	ND	ND
18	4/4 (23.3)	5/5 (23.6)	5/5 (15.7)	5/5 (19.0)
24	ND	ND	5/5 (19.6)	5/5 (21.5)
26	4/4 (24.8)	5/5 (25.4)	ND	ND
30	4/4 (25.4)	5/5 (25.0)	ND	ND

ND; testing not done

Frequency of wild-type PRRSV infections in growing pigs in Midwest US

Jose Angulo[1,2]; Samantha Jansen[1]; My Yang[1]; Albert Rovira[1]; Peter Davies[1]; Jeff Zimmerman[3]; Montserrat Torremorell[1]
[1]University of Minnesota; [2]Zoetis Inc.; [3]Iowa State University

Introduction

Porcine reproductive and respiratory syndrome virus (PRRSV) infection costs $4.67 per growing pig due to its effects on average daily gain, feed conversion, and mortality.[1] However, we know little about the distribution and frequency of PRRSV infections in growing pigs particularly in herds that use vaccination to control the disease. In this study, we report the frequency of wild-type PRRSV infections in wean-to-finish pigs originating from PRRSV negative or stable herds located in medium to high dense pig regions in the US Midwest.

Material and methods

Data were collected from wean-to-finish (W-F) sites (n = 63) across 10 production systems in the Midwest US. One cohort of pigs per site was enrolled and W-F site selection criteria included: 1) pigs sourced from PRRSV negative or stable breeding herds (based on the AASV PRRSV breeding herd classification[2]), 2) pig placement in all-in/all-out sites located in medium to high density areas, 3) willingness to collect monthly oral fluid samples for PRRSV testing, and 4) willingness to share site level information on production and management practices. Most cohorts studied (57 of 63) used vaccination to control PRRS, and 6 did not vaccinate. Oral fluid samples were collected at each W-F site at ~3, 8, 12, 16, 20 and 25 weeks post-placement using fixed spatial sampling (8 ropes per site). Oral fluid samples (n = 2,585) were tested individually by PRRSV RT-PCR. Sequencing (ORF5) was performed at each positive sampling event on the sample with the lowest RT-PCR cycle threshold (Ct) value. PRRSV-ORF5 sequences (n = 150) were aligned and classified as wild-type PRRSV using a cut-off of > 2.0% nucleotide difference from vaccine viruses references.

Results and discussion

PRRSV RNA was detected in 60 (95%) herds; and both vaccine and wild-type viruses were detected in 19 herds. Wild-type PRRSV RNA was detected in 44% of cohorts (28 out of 63) where a total of 45 sequences representing 18 distinct wild-type viruses were obtained. RFLP patterns from these sequences were 1-7-4, 1-8-4, 1-4-4, 1-3-2, 1-10-4, 1-26-2, 1-8-3. Wild-type PRRSV sequences were detected more frequently at 12 and 16 weeks post-placement than at other sampling times ($P = 0.02$). There were 3 cohorts that tested positive with wild-type virus at the first sampling event, and of the 57 cohorts using vaccination, 42% vaccinated in the breeding herd either at processing or at weaning, and the remainder (58%) vaccinated post-weaning, either at placement or within 1-3 weeks post placement. RNA sequences consistent with MLV vaccine viruses were detected in 78% of cohorts and detection was more frequent at 8 weeks post placement (105 sequences) and matched the 3 main MLV commercial products used in these W-F sites. There were no differences in wild-type detection rate or timing of detection between herds vaccinating in the breeding herd or at the W-F sites.

In summary, wild-type PRRSV infections were prevalent in growing pigs originating from PRRS stable or negative herds located in medium to high pig dense areas in the Midwest. Infections were distributed throughout the growing period, with higher detection rates at 12 and 16 weeks post placement. Although detection of wild-type PRRSV in vaccinated herds may be underestimated due to limitations in diagnostic methods to differentiate vaccine and wild type virus, co-circulation of vaccine and wild-type viruses was common.

References

1. Holtkamp DJ, Kliebenstein JB, Neumann EJ, et al. Assessment of the economic impact of PRRSV on United States pork producers. J. Swine Health Prod. 2013;3,21(2):72-84.

2. Holtkamp DJ, Polson DD, Torremorell M, et al. Terminology for classifying swine herds by PRRSV status. J. Swine Health Prod. 2011; 11;19(1):44-56.

Impact of within-farm movements of workers on production parameters for three US swine farms

N. J. Black[1], MPH; L. E. Moraes[2], MS, PhD; A. G. Arruda[1], DVM, MS, PhD
[1]Department of Veterinary Preventive Medicine, The Ohio State University, Columbus, Ohio;
[2]Department of Animal Sciences, The Ohio State University, Columbus, Ohio

Introduction

Many infectious diseases that impact the swine industry have high economic consequences due to decreased production and excess costs incurred from disease control and elimination efforts. Fomite transmission plays a major role in the spread and persistence of such diseases within a herd, thus a strong internal biosecurity protocol is integral.[1] While there are general internal biosecurity guidelines commonly implemented on farms, detailed quantification and characterization of within-farm employee movements is currently unavailable in the literature. The objective of this study was to utilize beacon-sensing technology and machine learning tools to identify specific types of movements associated with production performance parameters for three commercial US swine farms.

Materials and methods

Two farms located in Indiana and one located in Iowa, all with a PRRS history, were enrolled and an internal biosecurity system was installed. Farm 1 was a 1,500-sow farrow-to-finish operation with no shower and no filter system; farm 2 was a 4,500-sow farrow-to-wean operation with showers and not filtered; and farm 3 was a 4,400-sow filtered, shower-in farrow-to-wean operation. Internet services were optimized throughout the farms and sensors were placed in each room. Sensors were set up to detect Bluetooth-based beacon devices, which were individually distributed to farm employees. Movement data was collected for approximately one year, sent to a central database and collapsed weekly. A movement was recorded when an employee spent at least two minutes in one room after traveling from another room in the farm. Weekly production parameters were obtained for the three farms and included number of pigs weaned per sow (PWS) and pre-weaning mortality (PWM). To identify specific movements that were associated with these two parameters, LASSO regression models were fit in R using the glmnet package.[2] The optimal lambda (the variable shrinkage parameter) was selected using K-fold cross validation. Model performance was evaluated by the root-mean-squared error (RMSE).

Results and discussion

The analysis included 31, 13, and 27-weeks of movement data for farms 1, 2, and 3, respectively; after removing weeks for which area sensors were not working. Approximate number of workers for each farm was 11, 18, and 26 for farms 1, 2, and 3, respectively. Farm 1 had averages PWM of 14.12% (SD = 2.48%) and PWS of 11.66 (0.469). For this farm, for every 10 movement increase in one week from the finishing barns to the farrowing load out, there was an increase in PWM by 0.6% the following week (RMSE = 0.172), indicating a potential issue with pathogen transmission from the finishers to the farrowing rooms. Additionally, every 100 movement increase in one week between farrowing rooms was associated with a decrease in PWS by approximately 1 pig for every 10 sows (RMSE = 0.0393), suggesting that excessive movement between farrowing rooms may facilitate pathogen transmission; therefore, additional biosecurity measures between farrowing rooms may be beneficial. Farm 2 had averages PWM of 12.66% (5.05%) and PWS of 11.06 (0.498). For every 10 movement increase from the nursery to the farrowing area on this farm, there was a decrease in PWS by approximately 1 pig for every 2 sows two weeks later (RMSE = 0.0442). These results support the notion the nursery and load out areas could be particularly problematic with respect to pathogen transmission, as nurseries have been previously reported to be 'hot spots' for pathogens.[3] However, it is important to note the RMSE for this PWM model was 0.375, which is indicative of poor model performance. Finally, farm 3 had averages PWM of 11.11% (2.35%) and PWS of 12.35 (0.358). The results for this farm were not in agreement with expected findings; for example, an increase of 100 movements from the load out to the farrowing supply room was associated with a decrease in PWM by 2.1% two weeks later; and for every 100 movement increase in one week from the gestation supply room to the nursery there was an increase in PWS by approximately 1 pig/sow.

In conclusion, LASSO regression is a machine learning technique that was selected for this study due to its robust ability of variable selection on data containing a high number of predictors (movement types) and relatively few observations (weeks of production).[4,5] However, this method is currently unable to accommodate missing data, which was a major limitation in our study, due to periodic sensor inoperability. Therefore, only complete data was utilized, which greatly limited the sample size. Furthermore, farm-specific structure, sensor placement and proximity, and unavoidable employee movement based on typical production flow may have contributed to findings that were difficult to explain; which highlights the importance of careful analysis of machine learning model results before making inferences and recommendations. Nonetheless, this study showed that beacon-sensing technology can be implemented under commercial conditions in order to quantify and assess on-farm movements, which could be a short-term valuable tool to identify and target specific movements that could be leading to biosecurity breaches.

References

1. McCaw M.B., 2000; Swine Health Prod. 8(1):15-21.

2. Friedman, J. et al., 2010. J. Stat. Softw. 33(1).

3. White, L., Torremorell, M., & Craft, M. (2017). Prev. Vet. Med., 138, 55-69.

4. Fonti, V. et al. 2017. Feature selection using lasso. VU Amsterdam Research Paper in Business Analytics.

5. Hastie, T. and Qian, J., 2014. Glmnet Vignette. Available at: https://web.stanford.edu/~hastie/ glmnet /glmnet_alpha.html.

Improving *Mycoplasma hyopneumoniae* sampling techniques for detection, bacterial recovery, and diagnosis of disease

R. J. Derscheid, DVM, PhD, DACVP
Iowa State University Veterinary Diagnostic Laboratory, Ames IA

Introduction

Mycoplasma hyopneumoniae presents a diagnostic dilemma due to its slow growth, lag-time in immune response, and insidious and variable onset of clinical symptoms and lesions. Diagnostic samples for detection, culture, and diagnosis of *Mycoplasma hyopneumoniae* vary widely in collection technique and sample preservation en route to the diagnostic laboratory. Identifying best practices for sample collection and delivery to laboratory to increase consistency and reliability of diagnostic results for *Mycoplasma hyopneumoniae* would serve the swine industry to better target interventions and cost of disease management. To this end, our group has collected swabs for diagnostic testing as compared to lung tissue in both a research setting as well as in diagnostic cases.

Materials and methods

Samples were taken from studies for other purposes included: swabs and tissue from submitted diagnostic samples, per routine SOP for tissue and Eswab of airway (Study A); at necropsy from pigs infected with MHP as part of a vaccine study (Study B); at necropsy from pigs infected with MHP as part of a challenge study (Study C); three euthanized cull gilts as part of an effort to improve culture technique (Study D). All PCR samples were submitted through routine diagnostic service at ISU VDL, which uses a published assay.[1] Briefly, all samples were extracted via magnetic bead extraction with PCR run on an ABI 7500 Fast PCR System with a Ct cutoff of 37 cycles. PCR target was the Mhp183 gene, P97 cilium adhesin.

Results were compared using Wilcoxon matched-pairs signed rank and Bland-Altman method comparison.

Results and discussion

Comparing Ct values utilizing the Iowa State Veterinary Diagnostic Laboratory standardized diagnostic *Mycoplasma hyopneumoniae* PCR, swabs of airways yield significantly lower values than paired lung samples (Table 1). Values were compared using Wilcoxon matched-pairs signed rank test and Bland-Altman method comparison. This difference is slightly greater for research samples than diagnostic samples. This data provides evidence for an alternative sampling method, tracheobronchial swabs, which may be easier to replicate in the field for non-veterinary specialists. Preliminary data on recovery (culture) of *Mycoplasma* utilizing similar sampling methods has shown no difference between tissue and swab samples. Key limitations of the overall project that should be noted: sample collection was not performed by the same person for each substudy, nor was each person given the same training for all studies; Study A sampled largest available airways, which was not always primary bronchi. Entirety of lung was not always available for sampling; Different types of swabs were used for each sub-study. Additional sampling is required to determine accurate sensitivity comparison of the two methods as well as compare sample management en route to the laboratory.

Reference

1. Straight EL et al. Real-time PCR assays to address genetic diversity among strains of *Mycoplasma hyopneumoniae*. *J of Clin Microbiol.* 2008; 46:8.

Table 1: Paired PCR Cts for individual samples. The green value for each pair indicates the lower value. The purple boxes indicate samples that tested negative. For Studies B, C, and D, all swabs had lower Cts for swabs than lung. Average for each sample type for each study is in bold at the bottom of each representative column.

	Study A		Study B		Study C		Study D	
	Swab	Lung	Swab	Lung	Swab	Lung	Swab	Lung
	26.6	30	17	>37	18.1	24.7	16.7	21.5
	20.5	27.1	21.6	31.8	20.2	23.9	20	23.1
	32.8	>37	19	27.9	19.7	>37	19.6	24
	28.2	28	19.7	27.6	21	23.8		
	26	30.3	21.6	>37	20.2	25.5		
	25.5	26.1	18.4	26.6				
	31.8	28	26	28.4				
	32.7	30.5	17.9	28.7				
	28.5	30.9	19.6	28.6				
	22.8	26.1	16.1	28				
	26.7	>37	19.2	31				
	24.1	29.4	16.5	26.8				
	24.6	26.5	18.5	35.5				
	24.3	29.5	20.3	28.3				
	31.5	>37	21	28				
	26.5	24.7	26.6	32.7				
	27.6	26.6	20.7	31.7				
	28	31	19.7	31.9				
	24.4	26.5						
	27.8	27.8						
Average	**27.05**	**28.18**	**19.97**	**29.59**	**19.84**	**24.48**	**18.77**	**22.87**

Agreement between culture and PCR-based methods to detect *Brachyspira hyodysenteriae* and *B pilosicoli* in feces and oral fluids

Juliana B. Ferreira[1], DVM; Perle Zhitnitskiy[2], DVM, MSpVM; James Mark Hammer[3], DVM; Megan Jacob[1], MS, PhD
[1]Population Health & Pathobiology, North Carolina State University;
[2]Veterinary Population Medicine, University of Minnesota; [3]Swine Business Unit, Elanco Animal Health

Introduction

Swine dysentery and porcine intestinal spirochetosis are important diseases for the pork producer and are associated with the spirochetes *Brachyspira hyodysenteriae* and *B pilosicoli,* respectively. Disease causes a hemorrhagic and mucoid diarrhea in finisher pigs, resulting in reduced growth performance, higher feed conversion and economic loss.[1] Given their fastidious nature, culture-based detection of these organisms is specialized and time-consuming. Molecular-based detection of *Brachyspira* organisms, which may improve turnaround time, is highly appealing. In addition, although the use of oral fluids as a diagnostic specimen may offer several advantages, the sensitivity of using oral fluids has not been reported for *Brachyspira* detection. We determined the diagnostic sensitivity of both culture- and PCR-based assays for use in *B hyodysenteriae* and *B pilosicoli* detection from artificially inoculated and naturally occurring oral fluids and feces.

Materials and methods

Fresh feces (200 g) and oral fluids (200 mL) were collected from twelve-week old asymptomatic, *Brachyspira*-free pigs. Samples were divided into 10 g (feces) or 10 mL (oral fluids) aliquots and inoculated with 1 mL of a known concentrations of either *B hyodysenteriae* or *B pilosicoli* (10^1 -10^7 CFU/mL; determined by microscopic visualization and direct plating). Each spiked aliquot was evaluated in duplicate, twice by culture- and twice by PCR-based assays, to determine the detection limit and ensure repeatability. Additionally, we sampled four North Carolina commercial finishing farms, two of which were exhibiting clinical signs consistent with swine dysentery and two of which were not. Each of the four farms were sampled for four consecutive weeks at the end of finishing to detect *B hyodysenteriae* by culture and by PCR using oral fluids (n = 10 per week) or feces (n = 10 per week) collected from the ground. Culture for all samples included inoculation on both BJ and CVS media, anaerobic incubation at 37° C for five days, and observation of beta-hemolytic colonies. All colonies were confirmed by PCR. Molecular detection of *Brachyspira* species occurred after DNA extraction from feces or oral fluids using a commercially available kit. The PCR for both *B hyodysenteriae* and *B pilosicoli* was performed as previously described.[2] A culture was positive if a hemolytic colony from either CVS or BJ media was obtained and confirmed by PCR; a direct PCR was considered positive if a visible image of a gel product of appropriate size was observed.

Results and discussion

Culture-based methods had a significantly higher sensitivity than PCR in detecting *B hyodysenteriae* from both spiked oral fluids ($P < 0.0001$) and feces ($P = 0.0005$). The odds of detecting *B hyodysenteriae* in oral fluids were 3.9 times higher than in feces. There was no significant difference between testing methods or sample types for the detection of *B pilosicoli*. In samples from commercial farms, only 9 of 58 (15.5%) presumptive positive samples were simultaneously detected by PCR and culture-based methods; the kappa statistic indicated fair agreement and was 0.25. Three samples were considered positive by PCR only (2 oral fluids, 1 feces), whereas 46 samples were positive after culture. Regardless of detection method, prevalence was 14.4% and 22.8% in oral fluids and feces, respectively. Based on our results, a culture-based assay optimized the sensitivity of detecting *B hyodysenteriae,* and decisions regarding submitting oral fluids or feces should be made with a risk assessment.

References

1. Figi R et al. Eradication of swine dysentery as modified partial depopulation in a nucleus sow breeding farm. [Modifizierte Dysenterie-Teilsanierung in einem Kernzuchtschweinebetrieb.], Schweizer Archiv für Tierheilkunde. 2014; 156(8):373-80. German.

2. La T et al. Development of a duplex PCR assay for detection of *Brachyspira hyodysenteriae* and *Brachyspira pilosicoli* in pig feces, J. Clin. Microbiol. 2003; 41(7):3372.

Viremia and mortality in piglets born to sows during a PRRSV outbreak

M. Kikuti[1], DVM, MPH, PhD; C. Vilalta[1], DVM, PhD; J. Sanhueza[1], DVM, MS, PhD; R. Kienhe[2], DVM; K. VanderWaal[1], BS, PhD; D. Schroeder[1], BS, PhD; C. A. Corzo[1], DVM, MS, PhD
[1]Department of Veterinary Population Medicine, University of Minnesota, Saint Paul, Minnesota;
[2]Swine Vet Center, St. Peter, Minnesota

Introduction

Most of what is currently published regarding the porcine reproductive and respiratory syndrome virus (PRRSV) viremia in young piglets refers to post-natal infection and rarely to in-utero infections from experimental studies.[1,2] We aimed to describe viremia and mortality of piglets born during an ongoing PRRSV outbreak.

Methods

For this study, we recruited a PRRSV naïve 2500-sow farrow-to-wean farm located in the Midwestern United States (US), representative of U.S. current pig production that broke with PRRSV RFLP type 1-3-4 after maintaining a naïve status for over three years. As part of the PRRSV outbreak management protocol, live-virus inoculation (LVI) was performed in the entire herd nine days after identifying and characterizing the newly introduced PRRSV. Two weeks after the LVI, a cohort study was started by recruiting all litters of piglets 3-5 days old at the farm on the day of the visit. Piglets were individually ear tagged to ensure correct follow-up and a blood sample was collected. The same piglets were sampled again at 17-19 (pre-weaning), 40-41 (first sampling at nursery), 65-66 (second sampling at nursery), and at 109-110 (at finisher) days old. PRRSV viremia was individually assessed by RT-PCR and described in terms of RT-PCR Cycle threshold (Ct) values.

Results

A total of 127 piglets from 21 litters were included in the study and sampled at day 3-5 of age. All animals were RT-PCR PRRSV positive with a median Ct value of 14.7 (IQR: 13.1 – 16.1) with females having significantly higher median Ct value than males (15.5 vs 13.7, Kruskal-Wallis $P < 0.001$). At the pre-weaning sampling, 50.4% of the piglets had died. All the surviving piglets were still PRRSV positive with a median Ct of 18.3 (IQR: 17.0–19.7). Piglets that died between the baseline and the second sampling had significantly lower median Ct value at baseline than surviving piglets (14.0 vs 15.5, $P = 0.003$). At the first nursery sampling, mortality from the baseline was at 72%. All surviving piglets were still positive for PRRSV with a median Ct value of 24.5 ranging between 18.8 and 34.9. At the second nursery sampling, cumulative mortality was at 78%. Among survivors, 61.5% remained viremic with a median Ct value of 34.7 ranging from 28.4 to 39.8. Lastly, all remaining 22 surviving animals sampled at the finisher were RT-PCR negative for PRRS.

Discussion

Viremia seems to persist for months in pigs born to sows undergoing an outbreak of PRRSV as reported in previous experimental studies.[2] Ct values (as a proxy for viral load) were lower among males than among females at baseline, which has also been previously reported in an observational study at this age group.[3] Overall mortality among this cohort was 83%. This study not only highlights the huge economic impact that a PRRS outbreak can have on susceptible animals but also a prolonged population viremia which could have important implications in transmission.

References

1. Rowland R, Lawson S, Rossow K, Benfield D. Lymphoid tissue tropism of porcine reproductive and respiratory syndrome virus replication during persistent infection of pigs originally exposed to virus in utero. Vet Microbiol. 2003;96(3):219-235.

2. Pileri E, Mateu E. Review on the transmission porcine reproductive and respiratory syndrome virus between pigs and farms and impact on vaccination. Vet Res. 2016;47:108.

3. Vilalta C, Sanhueza J, Alvarez J, et al. Use of processing fluids and serum samples to characterize porcine reproductive and respiratory syndrome virus dynamics in 3 day-old pigs. Vet Microbiol. 2018;225:149-156.

Does topical flunixin meglumine have potential to treat pain in pre-wean piglets?

H. Kittrell[1], DVM; J. P. Mochel[2], DVM, MS, Ph.D., DECVPT; J. Brown[1], DVM; K. Hayman[1], DVM; A. Forseth[1], DVM; B. Schneider [2], MSc, ; S. Rajewski[2], PhD; B. Ratliff[1], BS; K. Skoland[1], BS; L. Karriker[1], DVM, MS, DACVPM
[1]Swine Medicine Education Center, Iowa State University College of Veterinary Medicine,
[2]Iowa State University College of Veterinary Medicine

Introduction

With the transdermal (TD) application of flunixin meglumine (FM) now approved for pain control in cattle, Animal Medicinal Drug Use Clarification Act (AMDUCA) algorithms suggest that the product should be considered when treating pain in other food animal species. The use of non-steroidal anti-inflammatory products may reduce the pain associated with castration and tail-docking during piglet processing. Confirming that FM reaches useful serum levels would allow development of dose regimens for pre-wean piglets. Oral (PO) or TD application of FM would reduce the number of needle injections. This study evaluated the pharmacokinetics and bioavailability of PO, TD, and intramuscular (IM) flunixin meglumine in pre-wean piglets. Serum concentrations over time were measured and pharmacokinetic parameters calculated using nonlinear mixed-effects compartmental modeling. This study determined the serum levels and bioavailability of TD and PO FM and informs veterinarians of the potential utility in piglets.

Materials and methods

41 pre-wean piglets, weighing 5 to 10 pounds, with no previous FM treatment, were randomly allocated into study groups according to route of administration. The study was divided into two phases: Phase I evaluated the pharmacokinetics of PO, TD and IM administration and phase II evaluated the bioavailability of TD FM using a crossover study design of intravenous (IV) and TD administration. The dose for IV, PO, and IM was 2.2 mg/kg while the TD dose was 3.3 mg/kg. Blood collection occurred at 27 consecutive time points from 0-9 days. All samples were collected via jugular venipuncture, placed in heparinized blood tubes and centrifuged at 1500g for 10 minutes. Plasma was pipetted into cryogenic vials, and stored in a -80 °C freezer until analysis. Plasma FM concentrations were determined using ultra-high performance liquid chromatography (Q Exactive Focus Orbitrap, Thermo Scientific, San Jose, CA, USA) coupled with mass spectrometry (Dionex Ultimate 3000, Thermo Scientific, San Jose, CA, USA) and nonlinear mixed-effects compartmental modeling was performed in Monolix 2018R2 (Lixoft, France).

Results

Mean FM peak plasma concentrations (C_{max}) of ~ 6.5 µg/mL (IM), 4.8 µg/mL (PO) and 0.03 µg/mL (TD) were measured. The time points of peak FM concentrations (T_{max}) were estimated at 30 min, 1 hr, and 24 hrs for IM, PO, and TD, respectively. The absolute bioavailability of oral FM was estimated at greater than 99%, while the bioavailability of transdermal FM was estimated to be low, 7.84%.

Conclusions

The reported C_{max} of FM after IM and PO administration is consistent with therapeutic concentration ranges that mitigate pain in other species and adult pigs. However, the low estimated concentration of FM after TD dosing, concurrent with poor absolute bioavailability, is not expected to mitigate pain in pre-wean piglets. Additionally, our findings on TD bioavailability suggest that expanding the surface area of application is unlikely to maintain therapeutically relevant systemic levels of FM for pain modulation.

Use of live attenuated and killed virus vaccination to reduce the risk of zoonotic influenza A virus transmission at the human-swine interface

Joshua N. Lorbach[1], DVM; Sarah W. Nelson[1], MS; Sarah E. Lauterbach[1], BS; Jacqueline M. Nolting[1], MS, PhD; Christa Goodell[2], DVM, MS, Dipl ACVPM, PhD; Andrew S. Bowman[1], DVM, MS, Dipl ACVPM, PhD
[1]Department of Veterinary Preventive Medicine, College of Veterinary Medicine, The Ohio State University, Columbus, Ohio; [2]Boehringer Ingelheim Vetmedica, Inc., Duluth, Georgia

Introduction

Interspecies influenza A virus (IAV) transmission is a fundamental mechanism fueling the emergence of novel IAVs with significant impact on human health in the form of influenza epidemics and pandemics. Cases of zoonotic influenza occur sporadically at the human-swine interface, with prolonged contact between humans and pigs at agricultural exhibitions permitting bidirectional transmission of IAV. There are no verified methods to reduce the risk of interspecies transmission at the swine-human interface. Our objective was to validate risk mitigation strategies aimed at reducing future spillover of IAV circulating in swine (IAV-S) to humans.

Materials and methods

We hypothesized IAV vaccination of swine could reduce the public health threat posed by IAV-S. We compared the effect of IAV vaccination in swine on post-challenge viral shedding and transmission. Influenza-naïve market age barrows were assigned to one of three IAV vaccine treatment groups: live attenuated influenza vaccine (LAIV; n = 5), killed virus vaccine (KV; n = 5), or sham vaccine (n = 5). Licensed, commercially available vaccines were used; single and multidose vaccine regimens were timed so all pig groups were fully vaccinated according to manufacturer instructions before IAV challenge. At 25 days post-vaccination, swine from each treatment group were challenged with a single intranasal dose (1×10^6 $TCID_{50}$) of a human-like H3 IAV strain (clade 3.2010.1). IAV-naïve ferrets were placed in simulated direct or indirect contact with each group of swine following IAV challenge. Nasal samples were collected every 48 hours from swine and ferrets following challenge and exposure, respectively. Specimens were screened for presence of IAV by RT-PCR, and infectious virus was quantified in PCR-positive samples.

Results and discussion

Pigs in all three vaccine treatment groups had similar viral shedding at the first post-challenge nasal sampling. LAIV and KV pigs had decreased peak viral titer in nasal samples and had a shortened duration of nasal shedding compared to sham-vaccinated pigs. Post-challenge IAV shedding was similar between LAIV and KV pigs. While use of either vaccine decreased viral shedding in pigs, the use of LAIV vaccine in swine had an impact on nasal virus shedding in exposed ferrets acting as human surrogates. Ferrets exposed to LAIV pigs had decreased viral shedding compared to ferrets exposed to KV or sham-vaccinated pigs. Overall, our findings indicate pre-exhibition vaccination of swine with LAIV or KV could reduce the public health risk posed by IAVs at the human-swine interface.

Identifying parameters associated with wean-to-finish mortality in a swine production system

E. S. Magalhães[1], DVM ; P. Thomas[2], DVM, MS; C. A. A. Moura[1], DVM; G. Trevisan[1], DVM, MBA; C. Rademacher[1], DVM ; D. C. L. Linhares[1], DVM, MBA, PhD
[1]Veterinary Diagnostic and Production Animal Medicine Department, Iowa State University, Ames, Iowa;
[2]Iowa Select Farms, Iowa Falls, Iowa

Introduction

Understanding parameters associated with cumulative wean-to-market mortality, which is a commonly measured key performance indicator in production systems, is a reliable approach to assist interventions and decisions in the swine industry [2]. The association between growing-phase mortality and multiple environmental and management risk factors was reported by Oliveira, et al. (2009)[3]. However, there is still the need to better understand how to take advantage of multiple risk factors to forecast grow-finish mortality. This study consisted of developing an automated framework to capture and merge multiple data streams utilized in a Midwestern USA production system, including breeding-to-wean (BTW) productivity parameters, closeout data of growing pig lots, BTW and growing pig health status, facilities structure information, and selected management factors. The main objectives were to develop an automated model to gather the aforementioned datasets in a consolidated database; and to report the extent that those variables can explain the variability of cumulative wean-to-market (here on referred to as WTM mortality) mortality of growing groups raised under field conditions.

Materials and methods

The experimental units for this retrospective study were growing pig groups originated from single and multiple sow farms, raised in wean-to-finish barns, or in nurseries followed by finishers, and marketed from January to June 2019 in a large production system in the Midwestern USA. SAS software (SAS 9.4) scripts were developed in order to aggregate and merge data related to each group of pigs weaned in the sow farm (structure, management, health status, environment and productivity), to their downstream groups in the growing phase (closeout performance report), generating a final cohort database. For closeouts originated from multiple sources (i.e. multiple sow farms, and/or multiple nurseries), a weighted average was applied for the continuous variables. For disease status the worse classification from all sources were used. Log-transformation on the response variable (WTM mortality) was applied to normalize the distribution. In order to measure the association of specific parameters with the dependent variable, univariate analysis (PROC GLM) was conducted for each desired parameter on WTM mortality. Thereafter a multivariate analysis was executed based on a hierarchical generalized linear mixed model (PROC GLIMMIX), and a determination coefficient (R^2) was calculated between the predicted and actual WTM mortality, considering as eligibility criteria variables having $P < 0.200$ at the univariate procedure. Likewise, multicollinearity, as proposed by Cohen, et al. (1988)[1], was investigated and selected variables with correlation above 0.50 (PROC CORR) were excluded to avoid overfitting. Random statement was applied in order to control for pig source.

Results

Data was obtained from 651 closeout groups originated from 41 sow farms. The geometric mean of the WTM mortality of this study population was 9.58% (95% confidence interval 9.19% - 9.96%). At the univariate analysis 39 variables were analyzed, from which 14 were chosen for the multivariate step. The p-value and R^2 of the variables included in the multivariate model were: abortions ($P = \leq 0.0001$ $R^2 = 0.0519$); non-productive days ($P < 0.0001$ $R^2 = 0.0452$); pigs/weaned/sow ($P < 0.0001$ $R^2 = 0.1236$); filtration status ($P = 0.0091$ $R^2 = 0.0365$); *Mycoplasma hyopneumoniae* status ($P < 0.0001$ $R^2 = 0.0309$); pre-weaning mortality ($P = 0.0005$ $R^2 = 0.0334$); projected annual death rate of sows ($P < 0.0001$ $R^2 = 0.06$); average total born per litter ($P < 0.0001$ $R^2 = 0.0913$); average birth loss per litter ($P = 0.0002$ $R^2 = 0.0373$); percentage of repeat service ($P < 0.0001$ $R^2 = 0.0831$); average farrowing rate ($P < 0.0001$ $R^2 = 0.0758$); average pigs weaned/female/year ($P < 0.0001$ $R^2 = 0.0603$); porcine reproductive and respiratory syndrome (PRRS) status ($P < 0.0001$ $R^2 = 0.05142$); PRRSV genotype based on restriction fragment length polymorphism (RFLP) ($P < 0.0001$ $R^2 = 0.1936$). Selected examples of variables that were excluded due to multicollinearity are: pigs weaned/female/year (corr = 0.62 with litters/female/year); parity of sows removed (corr = 0.69 with herd parity); average born alive per litter (corr = 0.77 with total born). The multivariate final model accounted for 82.28% of the variability in WTM mortality (Figure 1).

Discussion and conclusions

We were able to build SAS scripts to automate the process of importing multiple data streams, constructing pig flow from preweaning-to-finish and consolidating into a final dataset. The initial step of the analysis consisted of assessing the association of each factor with WTM mortality. The univariate analysis allowed to understand the magnitude of the association of each variable with WTM mortality. For example, PRRS status had a greater

R^2 compared to mycoplasma status, meaning a stronger association of the first trait with subsequent WTM mortality. The final model can be used by the production system to forecast subsequent WTM mortality, as well as to measure the association of specific variables on the outcome, after adjusting to other co-variates. In summary, this study demonstrated the ability and the value of aggregating information regarding BTW and growing health, production and management to forecast subsequent WTM mortality in one specific production system. Under the conditions of this study, the variables that had the strongest associations with WTM mortality were: pigs weaned per sow, PRRS RFLP, average total born per litter, percentage of repeated services and PRRS status.

References

1. Cohen, J. (1988). *Statistical power analysis for the behavioral sciences* (2nd ed.). Hillsdale, NJ: Lawrence Earlbaum Associates.

2. Larriestra, A.J.; Maes, D.G.; Deen, J.; Morrison, R.B. Mixed models applied to the study of variation of grower-finisher mortality and culling rates of a large swine production system. Can. J. Vet. Res. 2005, 69, 26–31.

3. Oliveira J, Yus E, Guitián FJ (2009) Effects of management, environmental and temporal factors on mortality and feed consumption in integrated swine fattening farms. Livestock Science 123, 221–229.

Figure 1: Correlation of 82.28% at the multivariate analysis.

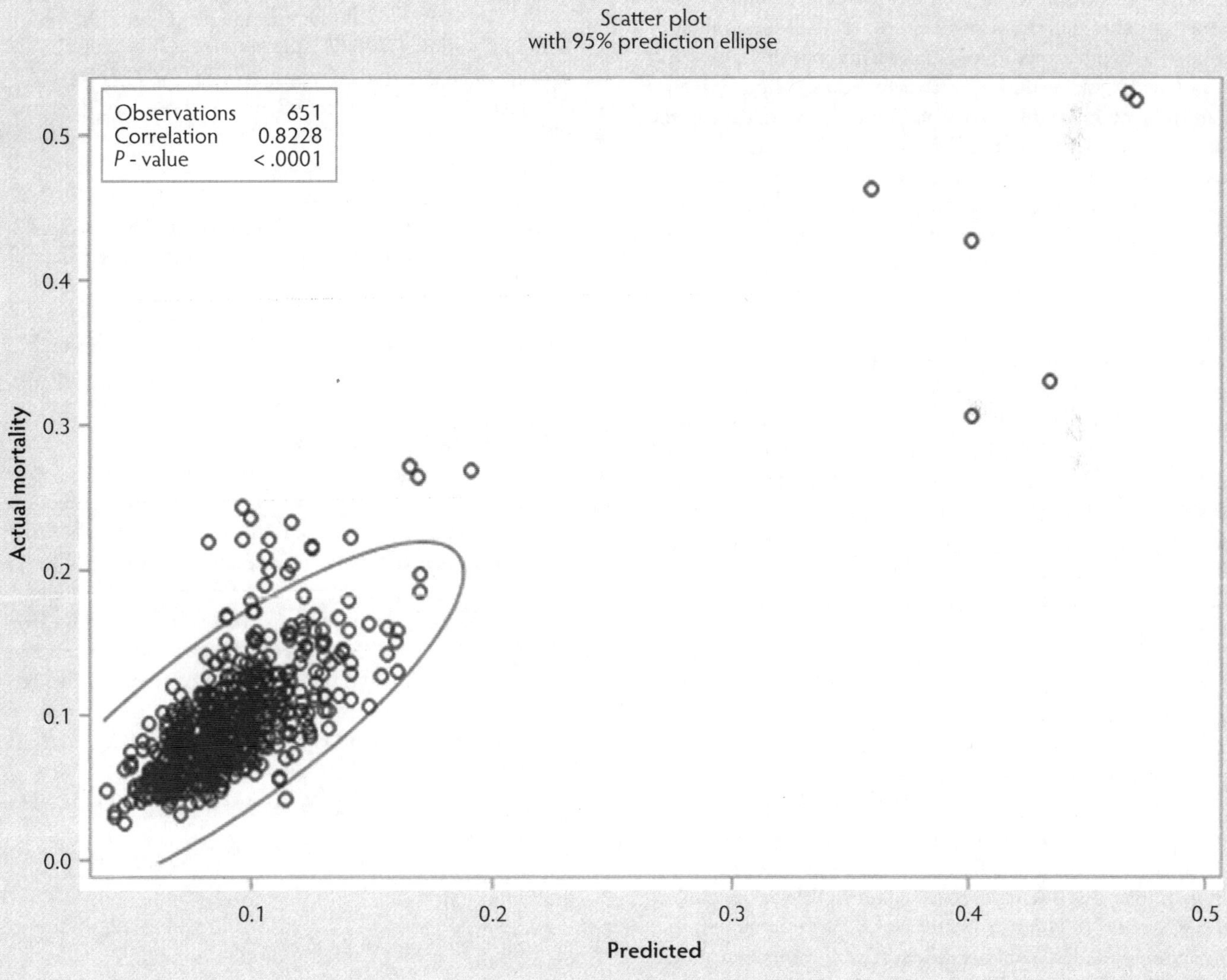

Effect of a lipid-based low pKa anti-pathogenic product on mitigating PRRSV in feed evaluated by quantitative RT-PCR and pig bioassay

L. Liu[1], PhD; A. Mahfuz[2], PhD; W. Yim-im[1], MS; L. Schumacher[1], DVM, MS, DACVP; P. Gauger[1], DVM, PhD; Q. Chen[1], DVM, PhD; A. Wang[1], PhD; M. Dasari[2], PhD; E. Weaver[2], PhD; J. Zhang[1], PhD
[1]Veterinary Diagnostic and Production Animal Medicine, Iowa State University, Ames, Iowa, USA;
[2]Feed Energy Company, Pleasant Hill, Iowa

Introduction

Contaminated feed has been confirmed as a vehicle for transmission of porcine epidemic diarrhea virus (PEDV).[1,2] Other viruses, such as African swine fever virus, Seneca Valley virus, porcine reproductive and respiratory syndrome virus (PRRSV), and so on, could maintain infectivity in feed ingredients even after 30-37-day transport, emphasizing the importance to reduce the potential transmission of these viruses via contaminated feed. In the current study, an *in vitro* experiment and an *in vivo* pig bioassay were conducted to evaluate the ability of a fat-based anti-pathogenic product, R2 product from Feed Energy Company that provides a source of essential fatty acids along with proven feed biosecurity benefits, to mitigate or inactivate PRRSV present in feed.

Methods

The *in vitro* experiment included 3 treatment groups with 3 replicates per group: T1 (NC) – untreated complete feed spiked with phosphate buffered saline (PBS, pH 7.4); T2 (PC) – untreated complete feed spiked with PRRSV MN184 (10^4 TCID50/g); T3 (Product) – complete feed treated with the R2 product one day before spiking with PRRSV MN184 (10^4 TCID50/g). At 0h, 1d, 4d, 7d, 14d and 21d post spiking, an aliquot of feed was taken from each replicate in each treatment group and processed into PBS to generate 20% feed supernatants. All of the feed homogenates were tested by a quantitative PRRSV RT-qPCR. Seven groups of feed supernatants from T1 (NC)-1d, T2 (PC)-1d, T3 (Product)-1d, T2 (PC)-4d, T3 (Product)-4d, T2 (PC)-7d, and T3 (Product)-7d were selected for pig bioassay. Twenty-one (10-day-old) PRRSV naïve pigs were divided into 7 groups with 3 pigs per group, one group per room. After 2-day acclimation, piglets were intramuscularly injected with 2mL and intranasally inoculated with 4mL (2mL per nostril) of feed homogenates. All pigs were necropsied on 7 days post inoculation (DPI). Serum samples collected on 0, 4, and 7 DPI and lung samples collected on 7 DPI were tested by PRRSV RT-qPCR. Formalin-fixed lung samples were subject to histopathology and immunohistochemistry (IHC) examinations.

Results

The results of the feed *in vitro* results are summarized in Figure 1. Feed supernatants from T1 (NC) remained PRRSV PCR negative throughout the 0-21d *in vitro* experiment The untreated control T2 (PC) was positive for the presence of PRRSV RNA by PCR with a mean Ct of 22.8 (range = 22.1-23.9). In contrast, the mean Ct value of the T3 (R2 Product) was 25.34 (range 23.5-26.4). The difference in the mean Ct levels between the T3 (Product) and the T2 (PC) was significant ($P = 0.00125$). For pigs inoculated with feed supernatants from T1 (NC)-1d, T3 (Product)-1d, T3 (Product)-4d, or T3 (Product)-7d, serum samples collected on 4 and 7 DPI and lung samples collected on 7 DPI were all PRRSV PCR negative, indicating that feed supernatants in these groups did not contain infectious PRRSV. In contrast, for pigs inoculated with feed supernatants from T2 (PC)-1d, T2 (PC)-4d, or T2 (PC)-7d, serum samples collected on 4 and 7 DPI and lung samples collected on 7 DPI were all PRRSV PCR positive, indicating that feed supernatants in these groups contained infectious PRRSV. Histopathology and IHC examinations on the lung samples supported the above results.

Conclusions and discussion

The results from the positive control group in this study provide evidence that PRRSV can survive in feed and result in infection of pigs, emphasizing that feed contaminated with PRRSV can be a risk factor for PRRSV transmission. Pre-treatment of feed with the R2 product can effectively render PRRSV, subsequently introduced into feed, non-infectious. Thus, the R2 product is capable of inactivating PRRSV present in feed and reducing the risk of PRRSV transmission via contaminated feed. This study also suggests that PCR testing in the *in vitro* experiment may not be as reliable as the pig bioassay to assess the infectivity of the PRRSV-contaminated feed samples.

References

1. Pasick et al. Investigation into the role of potentially contaminated feed as a source of the first-detected outbreaks of porcine epidemic diarrhea in Canada. *Transbound Emerg Dis*. 2014;61:397-410.

2. Dee et al. Survival of viral pathogens in animal feed ingredients under transboundary shipping models. *PLoS One*. 2018;13:e0194509.

Figure 1: Results of *in vitro* experiments testing the effects of treatment on PRRSV in feed at room temperature as examined by PRRSV RT-qPCR. The higher the Ct value, the less quantity of PRRSV RNA genetic material is detected.

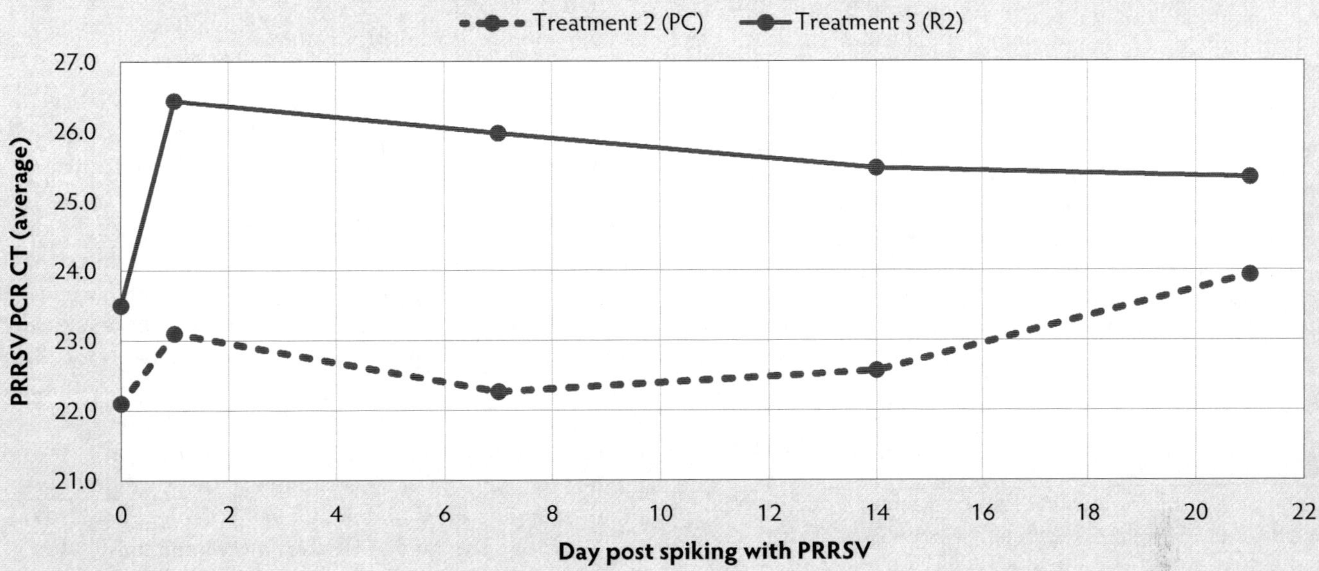

National Senecavirus A seroprevalence and risk factors

G. Preis; J. Sanhueza; C. Vilalta; C. Corzo
Veterinary Population Medicine, University of Minnesota, St. Paul, Minnesota

Introduction

Senecavirus A (SVA) is an emerging viral agent that belongs to the *Picornaviridae* family and has been associated with vesicular disease and neonatal mortality in swine herds in different countries. In the United States (US), SVA has been previously reported but currently there are no seroprevalence estimates and the epidemiology of this disease is not well understood. Therefore, the objectives of this study were to estimate the seroprevalence of SVA at the sow and growing pig level together with identifying risk factors for seropositivity.

Material and methods

Data on SVA farm prevalence is scarce, therefore, it was decided to conduct a national cross-sectional observational study to estimate a 50% prevalence with a 95% confidence level and a 10% allowable error. Based on this, 100 sow and 100 growing pig herds were required for testing. Within each herd, the number of samples collected per farm was 30, which allowed us to be 95% confident of detecting at least one positive sample when the within-herd prevalence was at least 10%. In sow farms, samples were collected across parities and in finishing sites samples were collected from pigs 20 weeks old or older. Samples collected were tested through an indirect immunofluorescence assay (IFA) for anti-SVA antibodies. Herd sensitivity and specificity were calculated to define the number of positive samples required to classify a farm as positive. Participating farms were also asked to answer a farm characteristics survey to determine herd-level risk factors associated with the presence of anti-SVA antibodies. Based on the test sensitivity of 90.3% and specificity of 100%, and after calculation of herd sensitivity (94%) and specificity (100%), it was decided to classify a farm as positive when having at least one positive sample.

Results and discussion

Our preliminary results show that 17 out of 95 (17.9%) sow herds and 7 out of 95 (7.4%) growing pig herds had at least one IFA positive sample. The apparent within-herd prevalence in positive farms ranged from 3.3% to 100%, with a mean within-herd prevalence of 37.2%. Two farms had one IFA positive sample, three farms had two positive samples and nineteen had three or more positive samples. The mean number of positive samples was 11.2 for both sow and growing pig farms, with a standard deviation of 10.1. Out of the sixteen represented states, seven had at least one farm with a positive sample. Univariate logistic regression preliminary results show that sow or finishing farms using a rendering service have 4.99 (95% CI 1.69, 14.74) higher odds of being seropositive ($P < 0.01$). Having farm-specific boots yielded a protective effect against seropositivity, with an odds ratio of 0.14 (95% CI 0.04 – 0.48) ($P < 0.01$), while having farm specific clothing yielded an odds ratio of 0.38 (95% CI 0.13 – 1.12) ($P = 0.08$). Having a downtime period for entering the farm also appeared to have a protective association against seropositivity as the odds ratio were 0.19 (95% CI 0.05 – 0.71) ($P = 0.014$) in sow farms only. The number of sows on the farm, the number of employees in the farm, having different load-outs for the cull sows and weaned piglets, having employees working on another farm and having replacement gilts born and raised on-site were not associated with being seropositive ($P > 0.05$). At the time of writing, samples and surveys were still being collected. Preliminary results show that the seroprevalence appears to be relatively low among US farms. Standard biosecurity procedures appear to have a protective effect against SVA whereas the use of a rendering service represents a risk. Final results and multivariate analysis will be shared during the conference.

References

1. Segalés J, Barcellos D, Alfieri A, Burrough E, Marthaler D; Senecavirus A: An Emerging Pathogen Causing Vesicular Disease and Mortality in Pigs?; Vet Pathol. (2016) Jul 1.

2. Sun D, Vannucci F, Knutson TP, Corzo C, Marthaler DG. Emergence and whole-genome sequence of Senecavirus A in Colombia. Transbound Emerg Dis. 2017; 64(5):1346-1349. doi:10.1111/tbed.12669.

3. Leme RA, Alfieri AF, Alfieri AA. Update on senecavirus infection in pigs. Viruses. 2017; 9(7):1-15. doi:10.3390/v9070170.

Comparison of sample types to diagnose Senecavirus A during later stages of infection

G. Preis[1]; E. Vasquez[1]; J. Garrido-Mantilla[1]; M. Sturos[1,2]; F. Vannucci[1,2]
[1]Veterinary Population Medicine, University of Minnesota, St. Paul, Minnesota;
[2]Veterinary Diagnostic Laboratory, University of Minnesota, St. Paul, Minnesota

Introduction

Senecavirus A (SVA) is a non-enveloped RNA virus, belonging to the family *Picornaviridae* and has been associated with cases of vesicular diseases in pigs. Common lesions found associated with SVA have high similarity with World Organization for Animal Health (OIE) reportable diseases such as foot-and-mouth disease (FMD), swine vesicular disease, vesicular stomatitis and vesicular exanthema. These characteristics have raised concerns in the pork industry worldwide due to potential confounding in the event of an FMD outbreak in an SVA endemic region. Because of its similarity to high consequence vesicular diseases, pork processing plants across the US are continuously facing false-alarms of OIE reportable disease, which takes time and a substantial amount of money to clarify. Most of the economic impact of the disease comes from the allocation of a state or credited veterinarian for sample collection, costs for sample testing, epidemiological surveillance on the farm site, segregation of affected animals in packing plants, and occasional neonatal mortality. The presence of viable virus was observed in the tonsils of mature boars naturally-exposed to SVA after 91 days post-clinical signs, as well as the persistence of SVA in the tonsils of experimentally-infected boars after 42 days post-inoculation. These findings led to the possibility of SVA inducing asymptomatic carrier state. The goal of this study was to evaluate and compare different strategies to detect SVA in the acute phase of disease infection as well as during later stages of SVA infection.

Material and methods

A total of 28 three-week-old piglets were divided into three groups: Group H, inoculated with the historical isolate from 1999 (n = 12); Group C, inoculated with contemporary isolate from 2017 (n = 12); and a negative control group (n = 4). Sampling methods included oral, fecal and tonsil swabs, as well as tonsil scrapings, sera collection, and oral fluids. Samples were tested by SVA RT-qPCR, and serological testing was done by indirect immunofluorescence (IFA). The time points for collections were on 1, 3, 7, 10, 14, 21, 28, 35, 42, and 48 days post-inoculation (dpi).

Results and discussion

Detection of SVA RNA was successful in all pigs from both inoculated groups by oral, fecal, and tonsil swabs at 3 and 7 dpi. Overall, oral swabs were less effective at detecting SVA RNA in inoculated pigs from both groups over time, followed by tonsil swabs and fecal swabs respectively. Tonsil scraping was the most efficient method to detect SVA over time. In the historical and contemporary groups, tonsil scrapings were able to detect SVA RNA in 58.3% (7/12) and 18.2% (2/11) of the animals on day 48, while none of the other individual sampling methods yielded positive results by PCR since day 28 and 35, respectively. Oral fluids from both historical and contemporary groups were able to yield positive PCR results from day 1 through 35 post-inoculation. Tonsil scraping has shown to be a promising method to detect SVA RNA in the tonsils of clinically recovered animals, during the late-phase infection. Virus isolation from positive samples and statistical analysis are being performed and will be presented at the poster session.

Reference

1. Segalés J, Barcellos D, Alfieri A, Burrough E, Marthaler D; Senecavirus A: An Emerging Pathogen Causing Vesicular Disease and Mortality in Pigs?; Vet Pathol. (2016) Jul 1.

Role of transportation stress on transmission of Senecavirus A in gilts

G. Preis[1]; E. Vasquez[1]; J. Garrido-Mantilla[1]; M. Sturos[1,2]; F. Vannucci[1,2]
[1]Veterinary Population Medicine, University of Minnesota, St. Paul, Minnesota;
[2]Veterinary Diagnostic Laboratory, University of Minnesota, St. Paul, Minnesota

Introduction

Senecavirus A (SVA) has been responsible for significant concern on the swine industry worldwide due to the similarities with other vesicular diseases, especially foot-and-mouth disease. The clinical identification of vesicular lesions in pigs triggers a foreign animal disease investigation, demanding time and resources for diagnostics and causing logistical problems for hog farmers and packing plants. Vesicular lesions are commonly detected in finishing pigs and sows after arrival at packing plants while often going undetected in the farms or buying stations. A hypothesis to explain these events and new outbreaks in sow farms is the potential persistence of the virus in the tonsil of infected animals after recovering from the clinical disease, and a possible effect of transportation stress on transmission and reoccurrence of lesions. The goal of the present study was to evaluate the ability of SVA infected animals to transmit the virus in three different stages of disease progression through direct contact with naïve animals after transportation stress. In addition, reoccurrence of vesicular disease and shedding of SVA after transportation stress in asymptomatic carrier animals was assessed.

Materials and methods

Eighteen gilts were allocated in four groups: SVA stressed (n = 4), naïve-contact day 7 (n = 4), naïve-contact day 21 (n = 4), naïve-contact day 35 (n = 4) and control (n = 2). SVA stressed animals were inoculated intranasally with 2.7×10^8 $TCID^{50}$. Naïve-contact animals were comingled with infected seeders on 7, 21 and 35 days post-inoculation (dpi) and kept in the same pen for 12 days until necropsy. Simulation of transportation stress was done on SVA stressed group using an experimental model based on National Pork Board's guidelines regarding temperature, space and feed and water restriction, with a duration of 8 hours. To mimic weather conditions in summer, the temperature of the room was raised to induce a low level of heat stress. To assess the effect of transportation stress in all timepoints, the transport model was performed in a manner that the end of the 8 hours coincided with the arrival of the naïve-contact animals on days 7, 21 and 35. Monitoring of viremia, serological IgG response and viral shedding by oral swabs, fecal swabs and tonsil swabs at 2, 5, 7, 9, 12 dpi or contact with infected seeder pigs was performed by RT-qPCR and IFA, and monitoring for clinical signs was performed daily. Infected seeder pigs were monitored on 14, 21, 28, 35, 36, 42 and 49 dpi, and tonsil scrapings were performed in this group of animals after 21dpi.

Results and discussion

Transmission occurred to naïve-contact animals from day 7, with 4/4 animals showing viremia on day 5 post-contact. All naïve-contact comingled at day 7 had positive IFA results 9 days post-contact. All SVA-stressed animals were shedding at 7 dpi, but only 1/4 seeders and 0/4 was shedding 21 and 35 days post-inoculation, respectively. Vesicles were detected in 2/4 of inoculated animals on days 4 and 5 post-inoculation, and reoccurrence of lesions did not happen after any mock transportation procedure. Interestingly, at 36dpi one seeder that had a constant detection of SVA RNA only through tonsil scraping since 21dpi, yielded positive PCR results in oral, fecal and tonsil swabs with Cts of 35, 33 and 33 respectively. This result highlights the potential transmission of the virus under field conditions from asymptomatic animals that were previously assumed to be clear of the virus. In conclusion, stress appeared not to be determinant on SVA transmission, and the decrease of fecal shedding in SVA-stressed animals over time coincides with the lack of transmission to naïve animals as they were put in contact on days 21 and 35. However, under field conditions, stress might play a role in shedding and transmission from asymptomatic animals.

Reference

1. Segalés J, Barcellos D, Alfieri A, Burrough E, Marthaler D; Senecavirus A: An Emerging Pathogen Causing Vesicular Disease and Mortality in Pigs?; Vet Pathol. (2016) Jul 1.

Forecasting outbreaks of PEDV for near real-time data-informed decision making

K. VanderWaal; I. A. D. Paploski; R. Bhojwani; C. A. Corzo
Department of Veterinary Population Medicine, College of Veterinary Medicine,
University of Minnesota, St. Paul, Minnesota

Introduction

Porcine epidemic diarrhea virus (PED) is among the most important endemic swine pathogens in the US, yet risk factors that drive viral circulation at regional levels are poorly understood. This is partly because risk is dynamic and changes rapidly both spatially and temporally. Here, we describe predictive machine-learning models that we have developed to estimate the weekly probability of a PED break, and second present the software platform that we have operationalized to perform real-time forecasting and reporting.

Materials and methods

Utilizing a subset of data from the Morrison Swine Health Monitoring Project (MSHMP) containing weekly PED breeding herd status, the forecasting model uses data on pig movements, geolocations of farms, environmental, and weather factors (Figure 1) to predict the probability that a sow farm will become infected two weeks in the future. A gradient boosting machine learning algorithm (XGBoost) with the SMOTE oversampling technique enabled the best performance of our predictive pipeline.

Results and discussion

The most important predictors are 1) the PED status at the farm of origin for pigs moved into the focal farm, 2) number of pigs moved into neighboring farms, and 3) environmental characteristics of the surrounding neighborhood.

This model currently performs with ~20% sensitivity, meaning that participants could potentially obtain advance warning for about 20% of PED breaks. More importantly, the model's positive predictive value (PPV) is ~70%, meaning that the model is quite accurate about the breaks it predicts. We have created a software platform for the analysis pipeline used for building the forecasting model and generating weekly predictions. This platform generates farm-level forecasts for two weeks in the future that are updated according to new data that emerge each week. These two-week projections will be delivered each week to participating production systems. In addition, the forecast platform has a built-in feature to periodically re-train the predictive model in order to remain as relevant as possible to current epidemiological situations. The platform can also be readily applied to other diseases, such as PRRS or a foreign animal disease. Because of its large spatial and temporal scope, incorporation of animal movements, and the use of innovative data science methods, we have the potential to generate new knowledge about the interacting processes that contribute to viral spread. Producers will have access to their own weekly farm-level forecasts and will be able to take action as needed. Ultimately, the near real-time estimation of forecasts will provide tools for data-informed actions by producers and practitioners to control outbreaks.

The authors would like to acknowledge the industry partners (production systems and veterinarians) who contributed to data for this analysis and to SHIC, as without their contributions this project would have not been possible.

Figure 1: Pipeline used in the forecasting model to generate weekly predictions of PEDV. Briefly, different types of data are entered (movement, farm list, spatial feature, outbreak status of farms) and merged to be run on a machine learning step. PEDV occurrence predictions are then forecasted 2 weeks into the future.

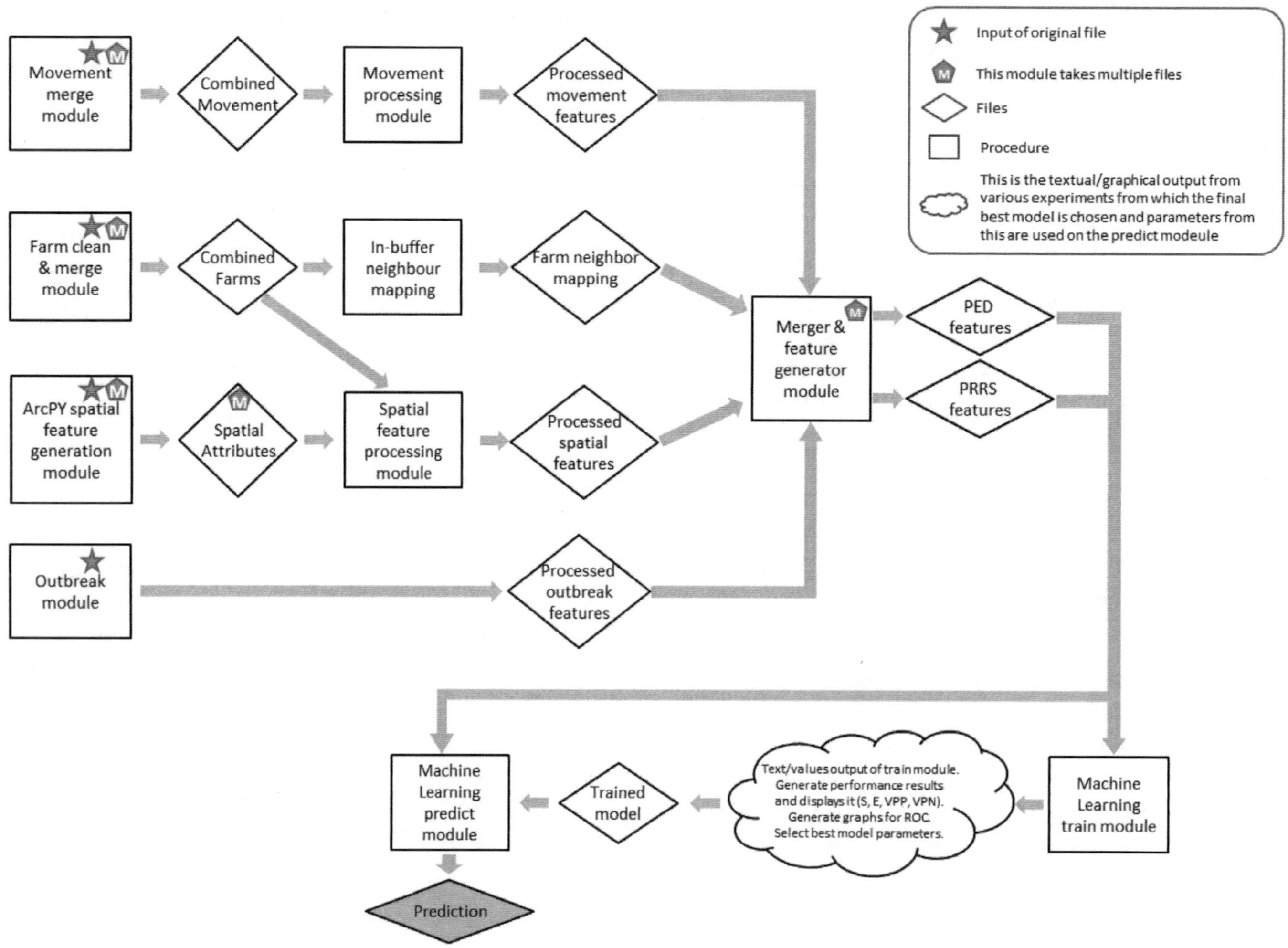

Evolution of PRRSV: lessons from the last 10 years and implications for the future

I. A. D. Paploski[1]; C. Corzo[1]; A. Rovira[1]; M. Murtaugh[2†]; J. Sanhueza[1]; C. Vilalta[1]; D. C. Schroeder[1,3]; K. VanderWaal[1]
[1]Department of Veterinary Population Medicine, University of Minnesota, St. Paul, Minnestoa;
[2†]*In memoriam*, Department of Veterinary and Biomedical Sciences, University of Minnesota, St. Paul, Minnsota;
[3]School of Biological Sciences, University of Reading, United Kingdom

Introduction

Despite control efforts involving improved biosecurity and different vaccination protocols, the control of porcine reproductive and respiratory syndrome virus (PRRSV) remains challenging and the virus continues to circulate and evolve. One of the foremost challenges in its control is high levels of genetic and antigenic diversity. Here, we quantify the co-circulation, emergence and sequential turnover of multiple PRRSV lineages in a single swine-producing region in the United States over a span of nine years (2009-2017).

Materials and methods

We classified over 4,000 PRRSV sequences (open-reading frame 5) into phylogenetic lineages and sub-lineages and documented the ongoing diversification and temporal dynamics of the PRRSV population, including the rapid emergence of a novel sub-lineage that appeared to be absent globally pre-2008.

Results and discussion

In addition, we report that the most prevalent lineage (L9) from 2009 to 2010 occurrence fell to 0.5% of all sequences identified per year after 2014, coinciding with the emergence or re-emergence of lineage 1 as the dominant lineage (Figure 1). The sequential dominance of different lineages, as well as three different sub-lineages within lineage 1, is consistent with an immune-mediated selection hypothesis for the sequential turnover in the dominant lineage. As host populations build immunity through natural infection or vaccination towards the most common variant, this dominant (sub-)lineage is replaced by an emerging variant to which the population is susceptible. An analysis of patterns of non- synonymous and synonymous mutations revealed evidence of positive selection on immunologically important regions of the genome, further supporting the potential that immune-mediated selection shapes the evolutionary and epidemiological dynamics for this virus. This has important implications for possible future patterns of emergence and re-emergence of genetic variants of PRRSV that have negative impacts on the swine industry. Constant surveillance on PRRSV occurrence is crucial to a better understanding of the epidemiological and evolutionary dynamics of co-circulating viral lineages. Further studies utilizing whole genome sequencing and exploring the extent of cross-immunity between heterologous PRRS viruses could shed further light on PRRSV immunological response and aid in developing strategies that might be able to diminish disease impact.

This paper has been published in full:

1. Paploski IAD, Corzo C, Rovira A, Murtaugh MP, Sanhueza JM, Vilalta C, et al. Temporal Dynamics of Co-circulating Lineages of Porcine Reproductive and Respiratory Syndrome Virus. Front Microbiol [Internet]. 2019 Nov 1;10. Available from: https://www.frontiersin.org/articles/10.3389/fmicb.2019.02486/full.

Figure 1: Stacked bar chart of the relative frequency of ORF5 sequences according to lineages over years and quarters.

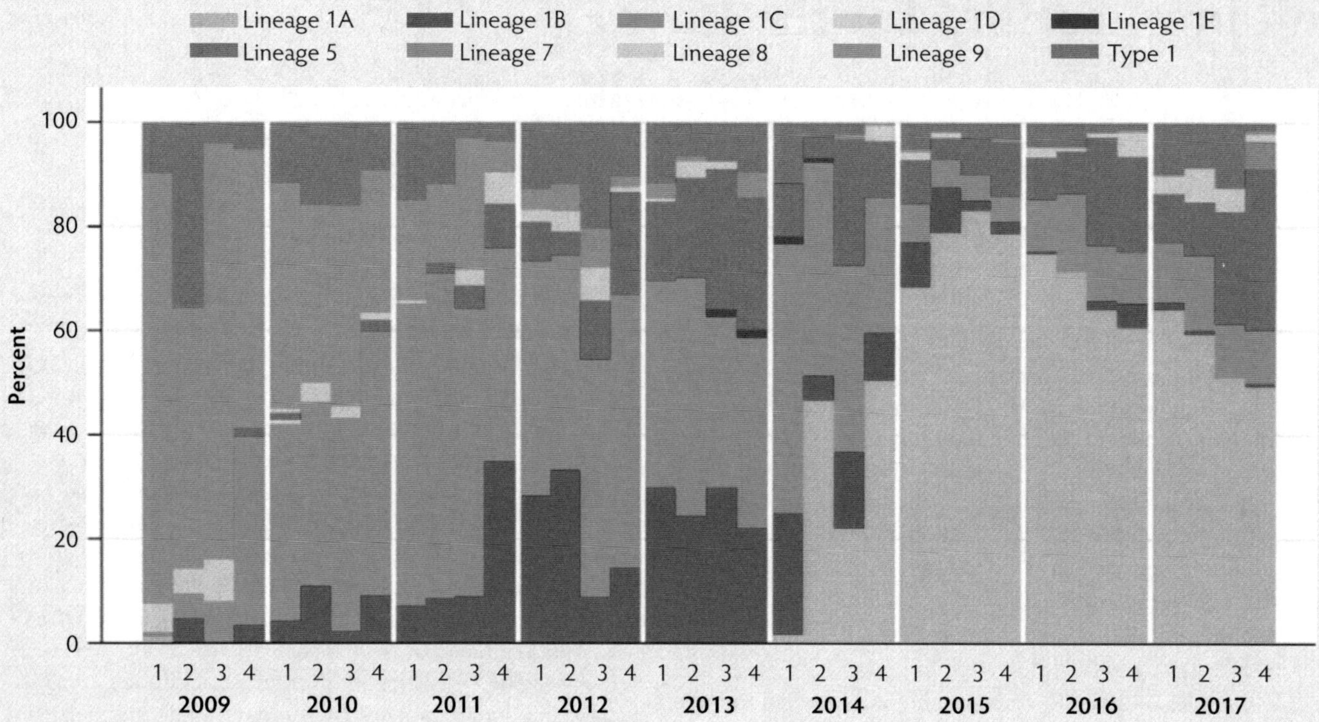

Validating pain behavior methodologies in castrated piglets

R. M. Park[1]; B. Wagner[1]; C. Cramer[2]; A. V. Viscardi[3]; L. E. Moraes[4]; H. Coetzee[3]; M. Pairis-Garcia[1]
[1]Department of Population Health and Pathobiology, College of Veterinary Medicine, North Carolina State University, Raleigh, North Carolina; [2]Department of Animal Sciences, College of Agricultural Sciences, Colorado State University, Fort Collins, Colorado; [3]Department of Anatomy and Physiology, College of Veterinary Medicine, Kansas State University, Manhattan, Kansas; [4]Department of Animal Sciences, College of Food, Agriculture and Environmental Science, The Ohio State University, Columbus, Ohio

Introduction

Surgical castration is a painful procedure routinely performed on commercial swine farms without pain relief.[1,2] However, in the US, there are no FDA approved drugs labeled for pain relief in swine. To gain FDA approval, a product must be shown to reduce pain in a target species utilizing validated pain assessment methods. To date, behavior sampling methodologies used to quantify castration pain have not been validated. Therefore, the objective of this study was to validate scan sampling methodologies to quantify pain behaviors expressed in castrated piglets.

Materials and methods

A total of 25 Yorkshire-Landrace × Duroc male piglets (5 days-old) were surgically castrated using a scalpel blade. Pain behavior frequency and durations (scratching, spasms, stiffness, tail wagging and trembling) of each piglet was continuously collected for the first 15 min of every hour (baseline pre-procedure; day 1: hour 1-8, hour 24) using the Noldus Observer® XT program. Sampling intervals (2-min, 3-min, 5-min) were compared against continuous sampling using PROC REG in SAS. Three criteria were used to determine if the sampling interval accurately reflected true duration and frequency for each behavior. Criteria included: 1) coefficient of determination (R^2) ≥ 0.9, 2) slope did not differ from 1 ($P > 0.05$) and 3) intercept did not differ from 0 ($P > 0.05$).[3]

Results and discussion

In the present study, no pain behaviors met all three criteria for scan sampling methodologies when compared to continuous sampling (Table 1). Pain behaviors occur in short durations[4] and at high frequencies.[5] This may inhibit identification of these behaviors with scan sampling methodologies when compared to continuous sampling. Overall, to fully capture pain behaviors for castration, continuous sampling is recommended. Utilizing validated behavioral methodologies in future research can assist in developing on farm pain protocols developed from objective, science-based recommendations.

References

1. Lonardi C et al. Can novel methods be useful for pain assessment of castrated piglets? Animal. 2015;9:871-877.

2. Gottardo F et al. Pain alleviation during castration of piglets: a comparative study of different farm options. J. Anim. Sci. 2016;94:5077-5088.

3. Chen JM et al. Technical note: Comparison of instantaneous sampling and continuous observation of dairy cattle behavior in freestall housing. J. Dairy Sci. 2016;99:8341-8346.

4. Viscardi, AV et al. Efficacy of buprenorphine for management of surgical castration pain in piglets. BMC Vet Research. 2018;14:318.

5. Van Beirendonck S et al. Behavior of piglets after castration with or without carbon dioxide anesthesia. J. Anim. Sci. 2011;89:3310-3317.

Table 1: R^2 (slope *P*-value, intercept *P*-value) for castration pain behavior duration and bout frequency at three sampling intervals (2-min, 3-min, 5-min) compared to continuous sampling

Behavior	2-min	3-min	5-min
Scratching			
Duration	0.82 (< 0.01, 0.17)	0.55 (< 0.01, 0.01)	0.45 (< 0.01, < 0.01)
Bouts	0.89 (< 0.01, 0.01)	0.74 (< 0.01, < 0.01)	0.41 (< 0.01, 0.02)
Spasms			
Duration	0.91 (< 0.01, 0.02)	0.07 (< 0.01, 0.01)	0.25 (< 0.01, 0.02)
Bouts	0.77 (< 0.01, 0.01)	0.16 (< 0.01, < 0.01)	0.48 (< 0.01, < 0.01)
Stiffness			
Duration	0.72 (< 0.01, 0.02)	0.66 (< 0.01, 0.01)	0.52 (< 0.01, < 0.01)
Bouts	0.84 (< 0.01, 0.08)	0.69 (< 0.01, 0.02)	0.69 (< 0.01, 0.05)
Tail wagging			
Duration	0.93 (< 0.01, < 0.01)	0.82 (< 0.01, < 0.01)	0.69 (< 0.01, < 0.01)
Bouts	0.88 (< 0.01, < 0.01)	0.79 (< 0.01, < 0.01)	0.64 (< 0.01, < 0.01)
Trembling			
Duration	0.92 (< 0.01, 0.08)	0.50 (< 0.01, 0.01)	0.67 (< 0.01, < 0.01)
Bouts	0.94 (< 0.01, 0.04)	0.80 (< 0.01, 0.01)	0.70 (< 0.01, 0.06)

Distribution of PCV2 genotypes in VDL clinical samples and its association with PCVAD

A. Rovira[1], DVM, MS, PhD; V. Utrera[2], DVM, PhD; L. Galina[2], DVM, PhD
[1]Veterinary Diagnostic Laboratory, University of Minnesota; [2]Zoetis

Introduction

Porcine circovirus associated disease (PCVAD) caused major losses to the US swine industry in 2003-2007. With the introduction of very effective commercial vaccines in 2006, PCVAD has been under control in most US farms for the last ten years. However, porcine circovirus type 2 (PCV2), the causative agent of PCVAD, is still circulating in swine populations and sporadic cases of PCVAD are experienced in some farms. Multiple reasons can explain the occurrence of PCVAD sporadic cases, from lack of vaccination, to maternal antibody interference, to vaccination of pigs during an outbreak of different disease (PRRSV or SIV). The potential role of the PCV2 strain in these cases of PCVAD has been a topic of debate. PCV2 is a rapidly evolving virus with wide genetic variability between different strains. A classification system based on the DNA sequence of the ORF2 gene of the virus has identified 6 different genotypes. Small-scale experimental studies have shown repeatedly that vaccination with one genotype confers protection against challenge with a different genotype. However, virological and immunological data suggest that cross protection may not be complete. Imperfect cross protection could therefore be responsible for the PCV2 genotype shifts observed over time. While PCV2a was the dominating genotype in the late 1990's, PCV2b became the main genotype in the early 2000's and a third shift was observed to PCV2d a few years later. PCV2c, PCV2e and PCV2f appear to be very infrequent. Unfortunately, there is little information on the distribution of the PCV2 genotypes currently circulating in the US and the available data is biased towards cases of PCVAD. Therefore, we designed a study to investigate the distribution of PCV2 genotypes in the US.

Materials and methods

One hundred PCV2-positive samples with Ct lower than 26 from diagnostic cases submitted to the Minnesota Veterinary Diagnostic Laboratory during 2017 and 100 submitted during 2018 were sequenced.

Results and discussion

The distribution of PCV2 genotypes from 2017 was as follows: 35% of the samples were PCV2a, 12% were PCV2b, 50% were PCV2d and 3% were PCV2e. The distribution of PCV2 genotypes from 2018 was as follows: 35% of the samples were PCV2a, 6% were PCV2b, 57% were PCV2d and 2% were PCV2e. Samples from 2017 were further analyzed to show that 25% of the sequences were from cases of PCVAD (based on characteristic lesions in lymphoid tissues), 62% of the sequences were from cases without PCVAD (no characteristic lesions), 4% of the sequences were from abortion cases and 10% of the sequences had unknown PCVAD status (lymphoid tissues unavailable). Interestingly, 43% of PCV2d cases were cases of PCVAD, while only 14% of PCV2a and 18% of PCV2b cases were cases of PCVAD. In PCVAD cases, the range of Ct values was between 5 and 22, whereas in cases without PCVAD the range was between 10 and 22. Overall, this study indicates that PCV2d is responsible for most of the recent cases of PCVAD in the US. However, PCV2a and PCV2b are still circulating in swine populations and can cause PCVAD. The role of PCV2 in non-PCVAD cases with very low Ct values remains unclear and warrants further investigation.

Investigation of gilts as a factor for delayed time-to-stability

J. M. Sanhueza; C. A. Corzo; M. Kikuti; C. Vilalta
Veterinary Population Medicine Department, University of Minnesota, St. Paul, Minnesota

Introduction

Recently, the fluid of processing tissues (i.e. tails and testicles) was observed to provide accurate information on the litter porcine reproductive and respiratory syndrome virus (PRRSV) status.[1] Additionally, a tendency for lower parity litters to have a higher percentage of PRRSV positive results has been reported.[1] Although other factors, as the method of virus exposure (i.e. live-virus inoculation or modified-live vaccine) used in the herd, the virus type associated with the outbreak, and the season when the outbreak occurred can affect the time-to-stability,[2,3] the role of gilt litters in maintaining PRRSV in the breeding herd has not been assessed to date. This study aimed to evaluate the role of parity one litters in maintaining PRRSV infection in breeding herds.

Methods

Fifteen litters of parity one (P1), parity two (P2) and parity three and plus (P3+) sows were sampled in nine farrow-to-wean herds at processing on a weekly basis during 20 to 30 consecutive weeks. All processing tissues (i.e. tails and testicles) of a litter were individually collected in Ziploc® bags and sent to the laboratory. Processing tissues were aggregated by parity group before extracting the processing fluid (PF) for rRT-PCR testing. The proportion of rRT-PCR positive results was compared among parity groups using the χ^2 test. A logistic regression model was built to assess the relationship between rRT-PCR result (positive/negative) and parity group (i.e. P1, P2 and P3+), after accounting for the effects of time after the PRRS outbreak (weeks) and farm.

Results

A total of 11,803 PF from individual litters from 9 farms were submitted for testing since March, 2018 (10 and 20 weeks after the beginning of the outbreak). Overall, a total of 246 out of 858 (28.7%, 95% CI 25.7% - 31.8%) aggregated PF were rRT-PCR positive. The percentage of positive rRT-PCR in P1 litters (31.2%, 95% CI 26.1% - 36.8%) was not different than the percentage of positive rRT-PCR in P2 litters (24.5%, 95% CI 19.95% - 29.8%) ($P = 0.09$) and the percentage of positive rRT-PCR in P3+ litters (30.3%, 95% CI 25.3% - 35.9%) ($P = 0.9$). Multivariable model results showed that the odds of rRT-PCR positivity decreased slightly more in P2 and P3+ litters compared to P1 litters. However, the adjusted proportion of rRT-PCR positive P1 was not significantly different to the proportion of rRT-PCR positive P2 or P3+ litters in any of the weeks assessed (Figure 1). Nevertheless, in one out of the nine farms, P1 litters were rRT-PCR positive up to week 38 after the outbreak. In contrast, the last rRT-PCR positive for P2 and P3+ litters was on week 27 and week 21, respectively in the same farm.

Conclusions

Processing fluids continue to show their value as a tool for monitoring PRRSV in breeding herds. rRT-PCR positive percentage did not vary significantly among parity groups over time. However, P1 litters did remain positive for 11 weeks longer than P2 litters and 17 weeks longer than P3+ litters in one out of nine farms. The proportion of herds having this type of pattern in the population is unknown. Despite the lack of association observed in this study, first parity sows may still play a role in maintaining PRRSV infection in some breeding herds in the population.

References

1. Vilalta, C., Sanhueza, J., Alvarez, J., Murray, D., Torremorell, M., Corzo, C., Morrison, R., 2018. Use of processing fluids and serum samples to characterize porcine reproductive and respiratory syndrome virus dynamics in 3 day-old pigs. Veterinary Microbiology 225, 149-156.

2. Linhares, D.C., Cano, J.P., Torremorell, M., Morrison, R.B., 2014. Comparison of time to PRRSv-stability and production losses between two exposure programs to control PRRSv in sow herds. Preventive Veterinary Medicine 116, 111-119.

3. Sanhueza, J.M., Vilalta, C., Corzo, C., Arruda, A.G., 2019. Factors affecting Porcine Reproductive and Respiratory Syndrome virus time-to-stability in breeding herds in the Midwestern United States. Transboundary and Emerging Diseases. DOI : 10.1111/tbed.13091 66, 823-830.

Figure 1: Proportion of rRT-PCR positive processing fluids by parity group between 20 and 65 weeks after a PRRS outbreak in nine breeding herds (four contributed data from week 20 to 45, and five from week 20 to 65).

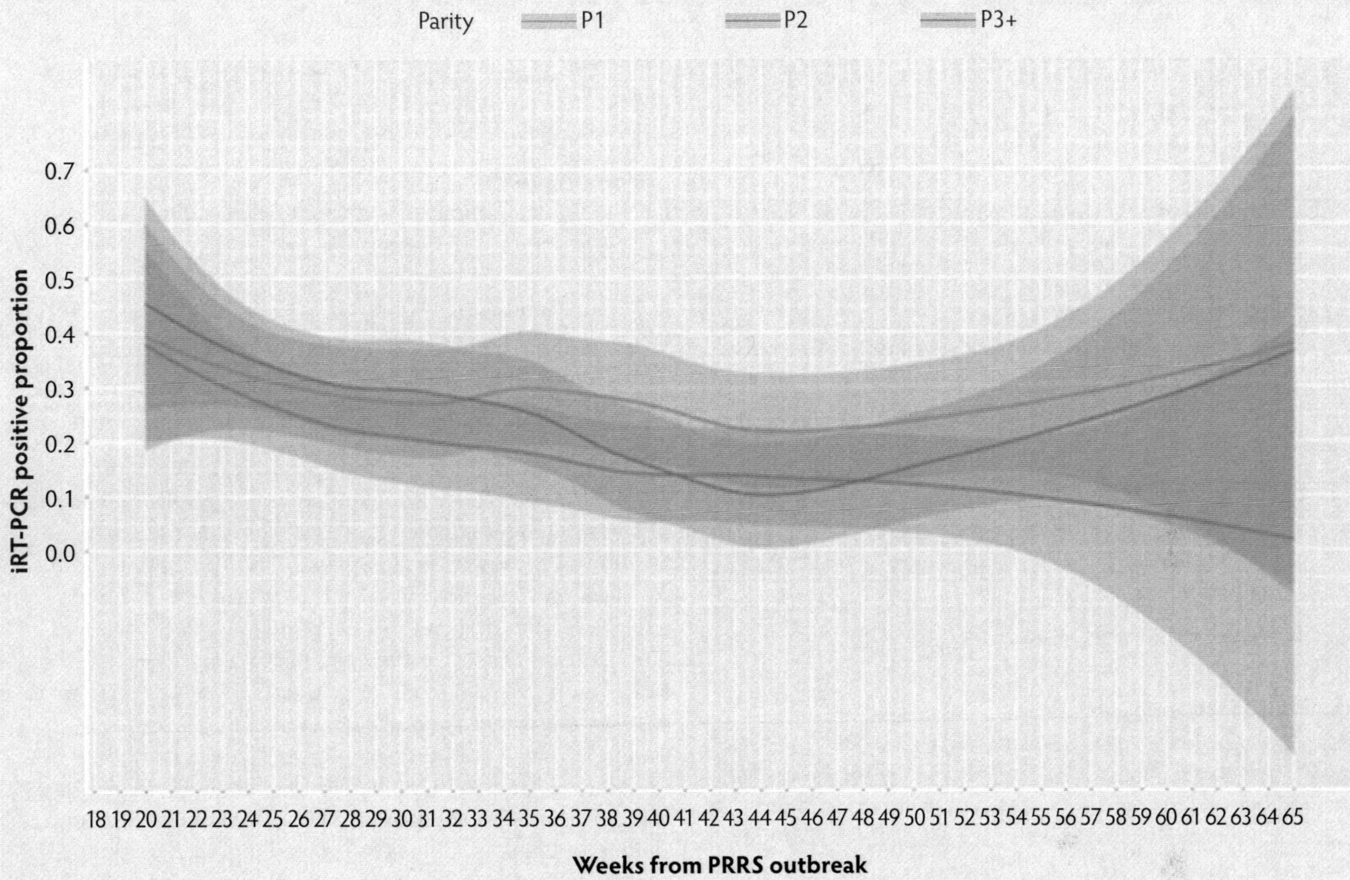

Virulence and antibody responses of different US PEDV strains in pigs of different ages

Loni Schumacher[1], DVM, MS, DACVP; Ashley Buerkley[1], MS; Qi Chen[1], DVM, PhD, Hai Hoang[1], DVM, PhD; Ronaldo Magtoto[1], Luis Gimenez-Lirola[1], MS, PhD; Wannarat Yim-Im[1], MS; Li Liu[1], PhD; Maria Merodio[1]; Rachel Derscheid[1], DVM, PhD, DACVP; Drew Magstadt[1], DVM, MS; Phillip Gauger[1], DVM, MS, PhD; Kent Schwartz[1], DVM; Min Zhang[2]; Jianqiang Zhang[1], MD, MS, PhD

[1]Department of Veterinary Diagnostic and Production Animal Medicine, College of Veterinary Medicine, Iowa State University, Ames, Iowa; [2]Department of Statistics, Iowa State University, Ames, Iowa

Introduction

Porcine epidemic diarrhea virus (PEDV) is an alphacoronavirus and is the causative agent of porcine epidemic diarrhea (PED), an enteric disease affecting global swine industries.[1] Currently, two main strains of PEDV circulate in US swine (US non S-INDEL and S-INDEL PEDVs).[2] While studies demonstrated U.S. S-INDEL PEDV had lower virulence than US non S-INDEL PEDV in approximately one-week-old conventional pigs;[3] little is known about their virulence in older pigs of different ages. Therefore, we aimed to compare virulence and antibody responses of these two PEDVs in three ages of pigs.

Materials and methods

Thirty 3-week-old ("weaned"), thirty 8-week-old ("grower"), and thirty 23-week-old ("finisher") PEDV naïve pigs were included with each age divided into 3 groups (10 pigs/group) and orogastrically inoculated with PEDV isolate USA/IN19338/2013 (Non S-INDEL), USA/IL20697/2014 (S-INDEL), or virus-negative medium. Half number of the pigs in each group were randomly selected for necropsy at 4 DPI and remaining pigs were necropsied at 28 DPI. Virus load from fecal swabs, oral fluids, and tissues were determined by quantitative PEDV N gene-based real-time RT-PCR. Serum neutralizing antibody was measured by a fluorescent focus neutralization assay using non S-INDEL PEDV as the indicator virus. Serum IgG and oral fluid IgA antibodies were measured by PEDV fluorescent microsphere immunoassay (FMIA) based on the N-terminal portion (S1) of the non S-INDEL PEDV spike protein.

Results and discussion

As shown in Figure 1, in "weaned" pigs, non S-INDEL PEDV had longer duration of fecal shedding and significantly higher fecal and oral fluid virus loads than S-INDEL PEDV. In "grower" pigs, S-INDEL PEDV had longer fecal shedding than non S-INDEL PEDV; S-INDEL fecal virus load was significantly higher than non S-INDEL PEDV at 7 and 14 DPI but was opposite at 10 DPI. In "finisher" pigs, the onset of non S-INDEL PEDV fecal shedding was earlier and fecal virus load was higher than S-INDEL PEDV. For non S-INDEL PEDV, the onset of fecal virus shedding was earliest and shedding level was highest in "weaned", followed by "grower" and "finisher" pigs. S-INDEL PEDV trended similarly when compared across age groups. The data suggest that virulence of PEDV is pig age-dependent (more severe in younger pigs) and virus strain-dependent. Non S-INDEL PEDV appeared to be more virulent than S-INDEL PEDV in "weaned" and "finisher" pigs, but virulence difference of two viruses was less distinct in "grower" pigs. Neutralizing antibody, serum IgG, and oral fluid IgA responses induced by non S-INDEL PEDV were greater than S-INDEL PEDV in both "weaned" and "finisher" pigs, while the difference of antibody responses induced by two PEDV strains in "grower" pigs was not a clear cut and it depends on antibody assay. Non S-INDEL PEDV induced similar neutralizing antibody responses in three ages of pigs, stronger serum IgG responses in "weaned" pigs than in "grower" and "finisher" pigs, and stronger oral fluid IgA responses in "finisher" pigs than in "grower" and "weaned" pigs. Interestingly, S-INDEL PEDV appeared consistently to induce stronger antibody responses in "grower" pigs than in "weaned" and "finisher" pigs. In summary, this study suggests that pathogenicity and antibody response of PEDV is both virus strain-dependent and pig age dependent. The data provide some guidance on selecting appropriate PEDV strain to induce antibody response in different age of pigs and on selecting PEDV strain for vaccine development.

References

1. Jung K, et al. Porcine epidemic diarrhea virus infection: Etiology, epidemiology, pathogenesis and immunoprophylaxis. *Vet J*. 2015;204(2):134-43.

2. Lin C-M et al. Evolution, antigenicity and pathogenicity of global porcine epidemic diarrhea virus strains. *Virus Res*. 2016;226:20-39.

3. Chen Q, et al. Pathogenesis comparison between the United States porcine epidemic diarrhoea virus prototype and S-INDEL-variant strains in conventional neonatal piglets. *J Gen Virol*. 2016;97(5):1107-21.

Figure 1: Virus shedding in rectal swabs of pigs at different ages inoculated with different US PEDV strains. Virus load was determined by quantitative PEDV N gene-based real-time RT-PCR. Virus titers (log_{10}[genomic copies/mL]) at each time point were the mean values of all available pigs in each group. Data labels (fractions) represent the number of PEDV PCR positive pigs out of all pigs examined at that time point (both PCR-positive and negative pigs). Standard error bars are shown in each figure. Significantly different values for each DPI are indicated by different superscripts ($^{a, b, c}$). The significance level was set to $P < 0.05$.

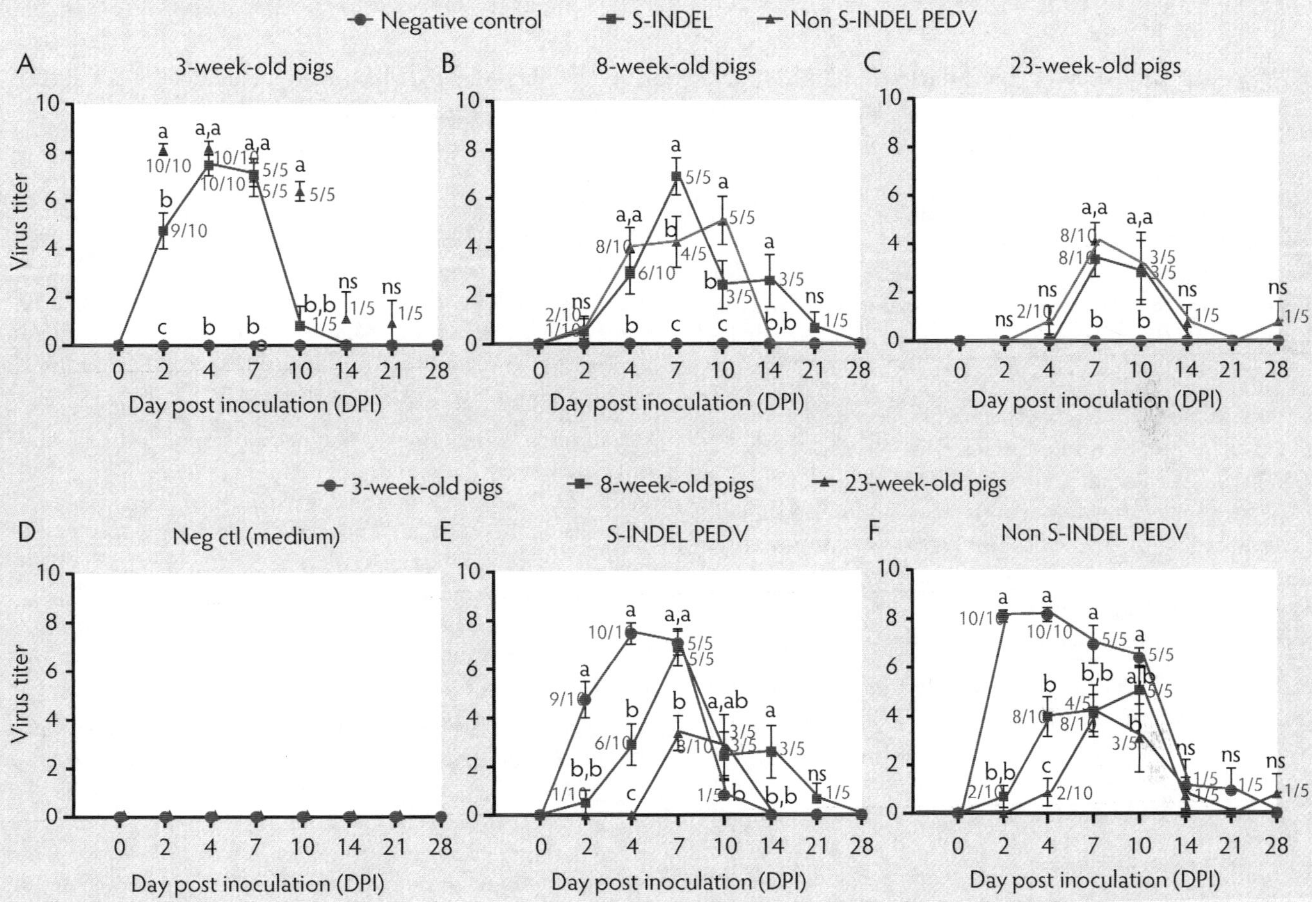

Genetic diversity of prevalent PRRSV RFLP patterns within and between US states during 2007-2019: implications for epidemiological investigations and control programs

G. Trevisan, DVM, MBA; A. Sharma, MS, PhD; P. C. Gauger, BS, DVM, MS, PhD; K. M. Harmon, 1 BS, PhD; J. Zhang, MD, MS, PhD; R. Main, BS, DVM, PhD; D. C. L. Linhares, DVM, MBA, PhD
Veterinary Diagnostic and Production Animal Medicine, Iowa State University, Ames, Iowa

Porcine reproductive and respiratory syndrome virus (PRRSV) genetic diversity is commonly reported based on restriction fragment length polymorphism (RFLP) patterns.[1] The purpose of this study was to characterize the genetic diversity of the most prevalent PRRSV RFLP patterns detected at the Iowa State University Veterinary Diagnostic Laboratory (ISU-VDL).

Open read frame 5 (ORF5) sequencing and metadata were obtained from diagnostic submissions from 01/01/2007 to 31/05/2019. ORF5 sequences were grouped by RFLP type and a comparative genetic analysis over time and within US states for the 4 most frequently detected RFLP types was performed using MEGA-X, R programing, and Dendroscope. Research cases, submissions having multiple sequences, and cases without an animal ID were removed from the study.

A total of 37,967 ORF5 sequences were retrieved representing 217 distinct RFLP patterns. The most frequent RFLP types and mean pairwise distances were: RFLP 2-5-2 = 0.74% (n = 8,041); RFLP 1-7-4 = 3.01% (n = 5,842); RFLP 1-4-4 = 8.58% (n = 4,396), and RFLP 1-8-4 = 9.72% (n = 3,603). A state-wise cluster analysis of RFLP 1-7-4 demonstrated 4 main clusters: Ohio (OH); North Carolina (NC); Minnesota (MN); and a group including Missouri (MO), Iowa (IA), Illinois (IL) and Nebraska (NE). There were also 3 intermediary clusters: Indiana (IN); Michigan (MI); and Pennsylvania (PA). For RFLP 1-8-4, there were 4 distinct clusters: OH and IN; MO; Oklahoma (OK); and IA, IL, and MN. There were 6 distinct clusters for RFLP 1-4-4: OH; IN; OK; NC; NE; and MN, IA, MO, and IL. RFLP 2-5-2 sequences formed a single cluster. Over time, RFLP 1-7-4 formed distinct clusters before and after 2014. RFLP 1-8-4 formed very distinct clusters before and after 2012. Within RFLP 1-4-4 a similar pattern of evolution over time was detected. Principal component analysis (PCA) shows divergent and distant genetic groups within recent years for RFLP 1-7-4 and 1-8-4. Considering the VR-2332 sequence (Genbank ID = U87392) as the common ancestor, RFLP-1-7-4 and RFLP 1-8-4 belongs to the lineage 1;[2] RFLP 1-4-4 to lineage 1 and 3; and RFLP 2-5-2 to the original sequence VR-2332 (Lineage 5.1).

For RFLP 1-7-4, NE and MO were the closest clusters with 3.10% mean genetic variability, while OH and MN were the most distant clusters with a genetic distance of 4.58%. For RFLP 1-8-4, OH and IN were the clusters most genetically similar at 4.68%, and the MO-OK clusters were separated by a mean distance of 13.21%. Within RFLP 1-4-4, MO and OH were the closest clusters with a genetic difference of 6.66%, while NC and OK were more distinct with a genetic difference of 14.39%. Within RFLP 2-5-2, KS and MN were the closest at 0.42% genetic difference, and CO and UT were less similar with a genetic distance of 2.59%.

In this study, we describe the magnitude of genetic divergence of PRRSV over time and from different US states sharing the same RFLP pattern. Additionally, a lineage comparison was performed. The RFLP method was originally developed to differentiate the 2-5-2 pattern (typically vaccine-like) from other PRRSVs. The overall mean genetic variability among 2-5-2's (0.74%), here presented, was relatively low compared to other RFLP types. Additionally, sequences from this RFLP type were descendants of the original VR-2332 sequence.

In summary, data from this study suggests PRRSV genetic evolution is constantly occurring over time and there is a distinct geographical genetic difference between PRRSV sharing the same RFLPs, especially between states. The presence of geographical differences in PRRSV genetic diversity (up to 14.39% for 1-4-4's) has important implications that may help develop PRRSV regional control programs and herd immunization strategies. The geographic genetic differences may aid producers and veterinarians in understanding events with success vs. failure to control and or eliminate from herds a PRRSV strain that was identified as having the same RFLP type. Additionally, understanding regional differences in PRRSV genetic diversity may help design appropriate animal movements across state borders and reginal placement strategies to minimize the risk of PRRSV transmission and minimize mixing of genetically different PRRSV strains that share the same RFLP type.

References

1. Wesley RD, Mengeling WL, Lager KM, Clouser DF, Landgraf JG, Frey ML. Differentiation of a porcine reproductive and respiratory syndrome virus vaccine strain from North American field strains by restriction fragment length polymorphism analysis of ORF 5. J Vet Diagn Invest. 1998;10(2):140-144.

2. Shi M, Lam TT, Hon CC, Murtaugh MP, Davies PR, Hui RK, Li J, Wong LT, Yip CW, Jiang JW, Leung FC. Phylogeny-based evolutionary, demographical, and geographical dissection of North American type 2 porcine reproductive and respiratory syndrome viruses. J Virol. 2010;84(17):8700-8711.

Brachyspira hyodysenteriae eradication with a non-antibiotic Zn-chelate: the approach of the future?

F. Vangroenweghe[1], DVM, MSc, PhD, PhD, Dipl. ECPHM; G. Lammers[2], PhD
[1]Elanco Animal Health, BU Food Animals, Plantijn en Moretuslei 1, 2018 Antwerpen, Belgium; [2]Intracare BV, Voltaweg 4, 5466 AZ Veghel, The Netherlands

Introduction

Swine dysentery (SD) caused by *Brachyspira hyodysenteriae* (*B hyo*) is an important intestinal disease with clinical signs typically consisting of muco-haemorrhagic diarrhea. Economic losses are due to mortality, diminished growth rates, deterioration of feed conversion and cost of medical treatment. Diagnosis is performed using pooled fecal samples (microbial culture, PCR test). The objective of the study was to eradicate *B hyo* from a closed swine herd using a novel Zn-chelate (Intra Dysovinol®; Elanco) in combination with a cleaning and disinfection (C&D) protocol, including rodent and fly control, manure management and improved external and internal biosecurity measures.

Materials and methods

A closed swine herd with 240 sows – having their nursery piglets and fattening pigs in a one-site production system – had been diagnosed positive for *B hyo* in December 2018. A first eradication was carried out in Spring 2019 that failed due to Ab-resistance and inconsistent internal biosecurity measures. The isolated strain demonstrated resistance to all currently used antibiotics (tiamulin, valnemulin) for eradication purposes. Subsequently, an audit – to identify critical control points to be improved – was performed. Several steps in C&D, rodent and fly control and biosecurity issues were optimized before a new eradication was started. A second eradication was initiated using a novel Zn-chelate (Intra Dysovinol®; Elanco) for oral drinking water application,[1] which has been proven efficacious in eliminating *B hyo* from the pigs following a 6-day continuous (24/24h) drinking water administration. The objective of the present study was to report the eradication of *B hyo* from a closed swine herd using a novel, non-antibiotic Zn-chelate through treatment of the sow population in combination with all necessary additional management measures related to *B hyo* eradication.

Results

The entire sampling and treatment schedule with subsequent results is given in Table 1. Overall, 4 sampling points were identified to check the *B hyo* infection status of the animals in scope for the Zn-chelate treatment. Following treatment, in combination with C&D (soaking with Intra Foamcleaner (Intracare BV) and disinfection with Intra Multi-Des GA (Intracare BV)), all animals tested PCR-negative and remained negative until the end of the protocol. Further follow-up is planned over the next months to confirm the long-term success of the *B hyo* eradication protocol.

Discussion and conclusions

Eradication of *B hyo*, the etiology of SD, on a swine unit can only be successful when all possible efforts are made to start-up as clean as possible through a combination of thorough C&D protocols, rodent and fly control, improved manure management and biosecurity measures.[2] The current case report describes *B hyo* eradication with an innovative non-antibiotic approach using a novel Zn-chelate (Intra Dysovinol®; Elanco) dosed through the drinking water. Earlier studies have demonstrated efficacy of the Zn-chelate to clear *B hyo* from the pig after a 6-day treatment.[1] In the current eradication protocol, treatment was continued for 15 days in

Table 1: Sampling and treatment protocol for *B hyo* eradication using a novel Zn-chelate (Intra Dysovinol®; Elanco) product. D0 – day of initiation of treatment

Timepoint	Activity description	Results
d-3	Fecal sampling (n = 21) prior to treatment	3 positive samples (3/21; 14.2%)
d0	Start of treatment with novel Zn-chelate through drinking water	
d6	Fecal sampling after 6 days treatment	0 positive samples (0/21; 0%)
d7	Start of C&D protocol, combined with sow washing and movement	
d10	End of C&D protocol	
d15	End of treatment with novel Zn-chelate through drinking water	
d16	Fecal sampling after end of treatment	0 positive samples (0/21; 0%)
d65	Fecal sampling at 7 weeks post-treatment	0 positive samples (0/21; 0%)

order to protect the sows from re-infection from the environment during the critical period of C&D, including sow washing and movement to cleaned zones within the sow compartment. Previous experiences using tiamulin have shown that this continued treatment increases the overall success of the entire *B hyo* eradication protocol.[2] Further follow-up during the next months, combined with a continued high level of internal biosecurity to prevent re-infection from the nursery or fattening compartments, will be necessary to confirm success of the eradication in the long term. In conclusion, a novel Zn-chelate has been successfully applied for *B hyo* eradication in a closed swine herd infected with an Ab- resistant *B hyo* strain. This approach might be interesting for future application under specific circumstances of Ab-resistant *B hyo* strains and in the light of further measures concerning continued reduction of Ab-use worldwide.

References

1. Lammers G., van Berkel R., Roijackers D., Vulders C. Brouwer-Middelesch H., van Hout J. 2019. Treatment of clinical *Brachyspira hyodysenteriae* with zinc chelate in pigs – a blinded, randomised controlled trial. Veterinary Record, in press.
2. Vyt P., Heylen P., Neven M., Castryck F. 2007. A practical approach to the elimination of swine dysentery (*Brachyspira hyodysenteriae*) from single-site, farrow-to-finish herds. Flemish Veterinary Journal, 76:124-129.

Biochemical evaluation of current erysipelas vaccines for emerging strains

Alan J. Young[1,2]; Martee Larson[1]; Ellys Johnson[1]; Dakota Weathers[2]; Thomas Halbur[2]; Ashley Peterson[2]
[1]Department of Veterinary and Biomedical Sciences, South Dakota State University, Brookings, South Dakota;
[2]Medgene Labs, Brookings, South Dakota

Justification

Erysipelothrix rhusiopathiae is a gram positive bacteria capable of causing disease in swine, as well as other mammals including humans. Standard presentation of the disease in pigs includes acute septicemia that can lead to mortality, and chronic proliferative lesions the lead to arthritis, endocarditis, or a combination of these. Numerous live attenuated and killed vaccines are available to address the problem, however swine producers are reporting a resurgence of disease that may not be covered by these existing products. We investigated the potential for target antigen variability as a potential cause for vaccine failure, based on current regulatory frameworks for licensing of these products. In the 1990s, it was discovered that a 65Kd surface protein, later defined as the Surface Protective Antigen A (SpaA), was a primary target for protection. As a result of this work, a monoclonal antibody directed against this protein (ERHU-B6091) was found to provide protection when passively administered to swine prior to challenge with virulent *Erysipelothrix rhusiopathiae*. Although the precise epitope associated with this antibody remains unknown, it has been demonstrated that it resides on the N-terminus of the SpaA protein. As a result of the known challenges associated with mouse potency tests for erysipelas vaccines, the USDA-CVB adopted an ELISA-based potency assay for all licensed erysipelas products based on quantifying the SpaA protein with the ERHU-B6091 antibody (SAM613). This guarantees that all licensed products will contain consistent levels of the SpaA protein to immunize against swine erysipelas. Unfortunately, since that decision, it has become clear that erysipelas strains are not uniform, existing in multiple serotypes. One characteristic of this diversity is that there are multiple forms of the Surface Protective Antigen found on swine erysipelas isolates, termed SpaA, SpaB, SpaC, and others.

Methods

To investigate the potential cross-reactivity of the existing SpaA specific monoclonal antibody for other SpaA serotypes, we produced recombinant forms of the SpaA protein in *E Coli*, as well as the SpaB and SpaC proteins. All sequences were produced with a terminal His-Tag sequence for use in later analysis. The sequences for these proteins were cloned into a previously described expression vector, and transfected into *E Coli* for production. Protein lysate was harvested, and analyzed by Western Blot for reactivity with with anti-HisTag antibody or ERHU-B6091.

Results

All 4 recombinant proteins (Full-length SpaA, Truncated SpaA, SpaB, and SpaC) reacted well with the anti-HisTag antibodies (Figure 1A). Both full-length and shortened forms of the recombinant SpaA protein reacted well with the ERHU-B6091 antibody in vitro, indicating that the reactive epitope for the antibody lies in the N-terminal domain. In contrast, the monoclonal antibody failed to recognize the SpaB and SpaC proteins (Figure 1B). This would seem to indicate that the current established vaccines will only provide coverage against SpaA-expressing strains of erysipelas and may be selecting for virulent strains expressing alternative forms of the Spa (p65) antigen. Future work will focus on evaluating SpaA/B/C expression on field isolates from affected swine, and its potential applicability to designing targeted vaccines to protect against erysipelas.

Conclusions and significance

Additional work will be presented to further indicate the specific reactivity of the USDA-recognized monoclonal antibody is directed ONLY against the SpaA protein, and fails to recognize other serotypes of *Erysipelothrix rhusiopathiae*. Although strains expressing the SpaA isoform remain the major strain infecting the US Swine Herd, the appearance of non-SpaA producing strains may result in expanded risk even in erysipelas vaccinated animals. Future vaccine efforts should directly monitor the appearance of Spa-antigen expression in diagnostic samples, as well as target vaccination strategies to incorporate target antigens represented by SpaB and SpaC positive strains.

Figure 1: (A) Reactivity of Full-length SpaA, Truncated SpaA, Spa B, and SpaC constructs with anti-His antibody; (B) Reactivity of Full-length SpaA, Truncated SpaA, Spa B, and SpaC constructs with ERHU-B6091 antibody.

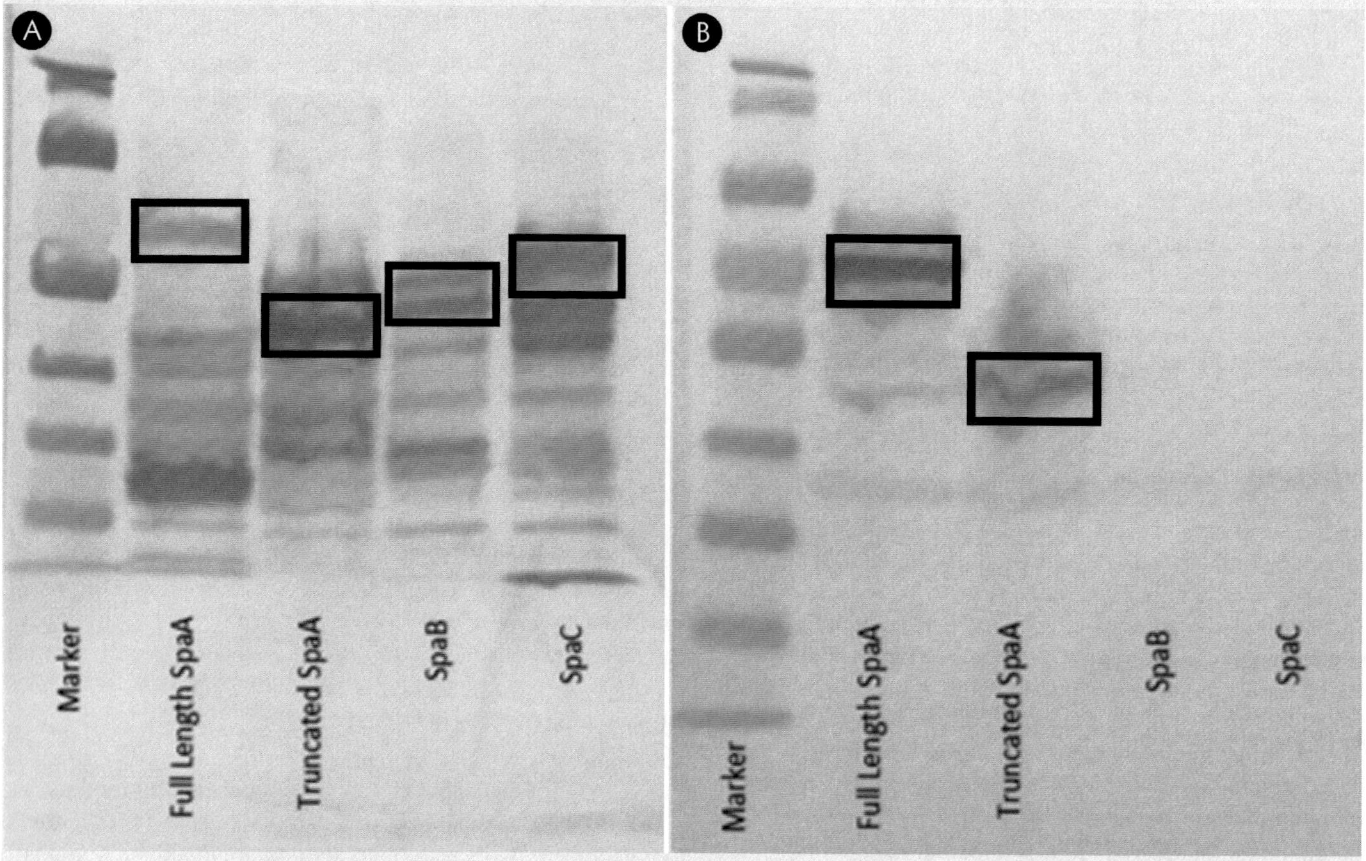

AllBite™, a novel solution to reduce vice behaviors

Bents, DVM; L. Swalla, MS
Hubbard Feeds, Mankato, Minnesota

Background

Despite the many improvements the swine industry continues to make in areas of health, nutrition, productivity, efficiency, and welfare, the challenge of tail biting and other vice behaviors in pigs continues to persist. As the industry continues to change and grow, new solutions for these old challenges must continue to be evaluated.

The AllBite block was developed in response to continued frustration with the current protocols to combat tail biting and other vice behaviors. Traditionally, the approach to manage tail biting once it has started in a pen has taken many forms. Techniques include sorting off affected animals if there is space and the inciting animal can be identified, re-evaluations of the diet, managing tail length, and introduction of some form of environmental enrichment for the pigs. Enrichment materials have included rubber hoses, paper bags, sod, logs, tires, bowling balls, chains, and plastic toys, all of which have their own pros and cons. For best results, enrichment materials must be "edible, chewable, manipulatable, and destructible." [1] These materials must also have the ability to be introduced into a facility in a biosecure manner, should be easy for farm staff to implement, and should come in at a reasonable cost.

The AllBite block was designed to fulfill all of these requirements. Production and packaging offers a product that has been developed in a biosecure manner, and it comes ready to use out of the package to make implementation easy for farm staff. It's core design as a molasses block fulfills the four requirements of adequate enrichment, and it has been produced in a manner that makes the cost palatable to producers.

We continue to evaluate potential uses for the AllBite block on farms to combat aggression and vice behaviors among pigs. The block was developed as a response to tail biting, but further uses in gilt developers, in sow group housing, and even in research settings has opened many doors for investigation. The first of these studies was designed to demonstrate if an objective reduction in tail biting could be appreciated through use of the AllBite block in response to a tail biting outbreak in a commercial finishing barn.

Material and methods

This project was conducted at a commercial grow-finish barn south of Mankato, MN. The farm manager identified nine pens from which he had removed pigs who were victims of tail-biting. Tail damage was assessed on every pig in these pens using a modified 0 to 4 scoring system based on previously defined methods[2]: 0 = no damage; 1 = healed lesions without visible blood; 2 = open wounds without signs of infection; 3 = open wounds with signs of infection, or partial loss of the tail without signs of infection; and 4 = partial or total loss of the tail with signs of infection. The tail damage scores of every pig were added together by pen to obtain a total pen score. After tallying the baseline tail damage scores, two AllBite blocks were added to every pen, each of which contained an average of 31 pigs. The tail damage scores were assessed again at both three days and six days following the introduction of the AllBite blocks into the pens.

Results and discussion

Results can be seen in Table 1. The tail scores were reduced by an average of 0.5 points six days after introduction of the AllBite blocks into the affected groups. This initial project was developed to demonstrate a beneficial effect on tail biting scores in affected pigs following introduction of the block into the pens. Further work is underway to evaluate these effects against a control group to further understand the value of utilizing the AllBite blocks in finisher, gilt, and sow production.

References

1. Frobose, H. "Modern sow feeding opportunities: A utopian view in an industry hesitant to change." Lecture, Hilton Orlando Buena Vista Palace, Orlando, FL, March 10, 2019.

2. Kritas, S.K., and R.B. Morrison. 2004. An observational study on tail-biting in commercial grower-finisher barns. J. Swine Health Prod. 12:17-22.

3. Gonyou HW. The influences of group size on behavioral vices and cannibalism. Proc AD Leman Swine Conf. 1998;237–239.

4. Geiger, J. "Dealing with tail biting." Lecture, Hilton Orlando Buena Vista Palace, Orlando, FL, March 9, 2019.

Table 1: Incidence of tail damage in growing-finishing pigs after introduction of AllBite blocks

Average pen score	Baseline score 1.22	After 3 days 1.0	After 6 days 0.71
Individual pig scores			
Score 0	274 (97.5%) *	274 (97.5%)	277 (98.6%)
Score 1	3 (1.07%)	5 (1.78%)	3 (1.07%)
Score 2	4 (1.42%)	2 (0.71%)	1 (0.36%)
Score 3	0 (0 %)	0 (0%)	0 (0%)
Score 4	0 (0 %)	0 (0%)	0 (0%)

* Percentage of the total number of pigs given that respective tail score.

Effect of an oral live vaccine against *Lawsonia intracellularis* on the performance of pigs at a production system in Mexico

J. Ochoa[1], DVM; F. Pinal[1], DVM; F. Leite[1], DVM, MS, PhD; F. Contreras[2]; A. Herrera[3], DVM, MVA
[1]Boehringer Ingelheim Animal Health; [2]Productores Pecuarios San Carlos SC de RL de CV; [3]PORKCONSULTING

Introduction

Porcine reproductive enteropathy (PPE) is caused by *Lawsonia intracellularis* and has a considerable economic impact on swine production worldwide, affecting pigs during the grow-finish period.[1] The objective of this study was to evaluate the impact of applying the only live oral vaccine against *Lawsonia intracellularis* (Enterisol® Ileitis) on pig performance in a production system in Mexico.

Materials and methods

This study was performed in an 1800 multi-site sow farm located in Jalisco, Mexico. The sow farm works as a 3-week batch production system. At 24 days of age, piglets are weaned and sent to a wean to finish site. The production system decided to implement a vaccination program with an oral live vaccine (Enterisol® Ileitis), after constantly observing lesions related to *Lawsonia intracellularis* at the slaughterhouse, and exposure to *Lawsonia* was confirmed by serology. In this study, a comparison was made before and after implementation of vaccination. A total of, 39975 pigs were included and divided into two groups: Group A with 25921 animals without ileitis vaccine and Group B with 13794 vaccinated animals, given a 2 mL oral dose of Enterisol® Ileitis prior to wean. For this study, it was decided to evaluate the following parameters: average daily weight gain (ADWG), feed conversion rate (FCR), market weight (MW) and market age (MA). These parameters were analyzed by the Two-sample T-test method (Minitab 17) and an economic analysis was done using BECAL (Boehringer Ingelheim Economic Calculator).

Results

After the implementation of Enterisol® Ileitis, we observed a numerical difference in all the parameters evaluated and a statistical difference in ADWG ($P < 0.05$), when comparing the vaccinated with that non-vaccinated groups (Table 1). Group B (vaccinated) demonstrated a better production performance compared to Group A (non-vaccinated). The difference in ADWG and FCR was 0.031 kg/day more and 0.08 less, respectively, which represent a gain of $2.32 and $3.61/pig, respectively. Using BECAL we obtained a ROI 5.61:1.[1,2]

Conclusion

In this study, we observed that using a live oral ileitis vaccine (Enterisol® Ileitis) it was possible to improve pig performance and obtain a more efficient production system.

References

1. Jacobson et al., 2010. Vet J, 184, 264-268.
2. A. Ibanez. et al; 2016. 24th International Pig Veterinary Society Congress.

Table 1: Parameters evaluated

Parameters	Group A	Group B
ADWG	0.699[a]	0.730[b]
FCR	2.19	2.11
MA	168.8	163.5
MW	108.1	107.9

[a,b] Different letters indicate differences at $P < 0.05$.

PEDV detection in manure pits confirmed 841 days to 1,949 days after disease outbreaks

Grant Allison[1], DVM; Phil Gauger,[2] DVM, MS, PhD; Jianqiang Zhang[2], MD, MS, PhD, DVM; Gene Spellman[3], BS
[1]Walcott Veterinary Clinic, Walcott Iowa; [2]Vet Diagnostic and Production Animal Medicine, Veterinary Diagnostic Laboratory, Iowa State University, Ames, Iowa; [3]Central Life Sciences, Schaumburg, Illinois

Introduction

Despite extensive investments in biosecurity measures and protocols, farms still have unexplainable cases of viral diseases caused by porcine epidemic diarrhea virus (PEDV). Many potential sources and vectors of viruses for this disease has been studied in the past. Investigating and finding the source of a PEDV disease re-break is difficult and rarely accomplished. Manure pits have been identified as a risk for harboring PEDV as a source for a disease re-break. Studies have found manure pits to harbor infective PEDV from four months[1] to nine months[2] in duration. The purpose of this study is to determine if PEDV can be detected in manure pits for extended periods of time post disease and to determine if an end-point can be reached where manure pits are not considered a risk for a PEDV disease re-break.

Materials and methods

Manure pit samples were collected from swine producers in Eastern Iowa. The exact dates of the most recent PEDV disease outbreak was known and noted at each location before pit sampling. Locations sampled were typical gestation barns, gilt development units, and finishers, each with separate deep manure pits. Samples were collected using Sterile Amies Agar Gel Transport Swabs (JorVet J0867). A piece of twine sufficient in length to reach below the surface of the pit was tied to the handle of the swab. A sinker-style weight was taped to the swab. The swab was lowered between the slats and care was taken to only contact and sample below the top surface of the manure pit. Samples were analyzed by real-time reverse transcription polymerase chain reaction (rRT-PCR) per routine protocols at the Iowa State University Veterinary Diagnostic Lab (ISU VDL).

Results and discussion

Results are reported in Table 1. Positive PEDV PCR results were detected from 46 days to 1,949 days post disease outbreak in this study. PCR test levels ranged from Cycle Theshold (Ct) values of 19.9 Ct (55 days post-outbreak), to 29.3 Ct (803 days post-outbreak) to an extremely long time after disease outbreak of 1,949 days (Ct 34.3). The PEDV S1 (first 2.2 kb portion of the spike gene) sequences were determined from the Farm 5 Finisher, Southeast location. The S1 sequence from this case has about 99% nucleotide identity to the U.S. PEDV prototype strains and 94% nucleotide identity to the U.S. PEDV S-INDEL-variant strains (Table 1). Sources of PEDV re-breaks continue to frustrate producers and veterinarians. This study indicates that manure pits can continue to be a risk that producers need to evaluate as a source for a PEDV disease outbreak for a much longer time, either through direct exposure to the manure pit's contents, or vectored to susceptible animals by insects such as flies.[3]

References

1. Steve Tousignant, DVM, Swine Vet Center P.A., St. Peter, MN | Nov 12, 2014, American Association of Swine Veterinarians (AASV)

2. Hein M. Tun, Monitoring Survivability and Infectivity of Porcine Epidemic Diarrhea Virus (PEDV) in the Infected On-Farm Earthen Manure Storages (EMS), *Front Microbiol.* 2016; 7: 265.

3. Allison, G.A. et al, PEDV positive bioassay reveals houseflies (Musca domestica) can transmit infectious PEDV to pigs, *Proceedings of the American Association of Swine Veterinarians* (AASV) 2019: 206-208

Table 1: PCR results and days post infection outbreak.

Farm (outbreak)		Gestation	Date	Ct	Days post-infection
1	(12/28/16)	North	4/26/18	33.5	478
		South	4/26/18	32.5	478
		North	4/29/19	32.8	841
		South	4/29/19	31.1	841
		South	6/03/19	31.1	875
		South	7/23/19	31.6	925
1	Pre – pumping	North	3/21/19	30.7	803
		South	3/21/19	29.3	803
	Post – pumping	North	4/30/19	32.8	842
		South	4/30/19	31.1	842
2	(2/23/18)		6/10/19	34.1	467
			6/24/19	33.2	481
3	(3/05/18)	GDU	5/01/18	30.0	46
4	(3/31/18)	GDU	5/23/18	30.2	53
		GDU	7/09/18	32.3	99
		South	7/09/18	31.4	99
		North	7/09/18	28.7	99
		GDU	7/08/19	31.0	458
		GDU	7/15/19	35.8	465
		GDU	10/28/19	32.0	568
Finisher					
5	(5/16/19)	North	9/03/19	26.4	107
		Southeast	9/03/19	26.6	48
5	(7/15/19)	Southwest	9/03/19	32.7	48
		Southeast	9/10/19	20.3, 19.9	55*
6	(2/27/14)	Gestation	8/26/19	34.3	1,979

* The PEDV S1 (first 2.2 kb portion of the spike gene) sequences were determined. The S1 sequence from this case has about 99% nucleotide identity to the US PEDV prototype strains and 94% nucleotide identity to the US PEDV S-INDEL-variant strains.

Elimination of *Mycoplasma hyopneumoniae* in a multiplier herd to initiate elimination on all commercial sows

F. Vangroenweghe[1], DVM, MSc, PhD, PhD, Dipl. ECPHM; R. Evelsizer[2], DVM; J. M. Hammer[2], DVM
[1]Elanco Animal Health, Antwerpen, Belgium; [2]Elanco Animal Health, Greenfield, Indiana

Background and objectives

Older *M hyopneumoniae* (M hyo) elimination programs close the breeding herd until the breeding population is 10 months of age or older.[1-4] Newer programs use gilt exposure with a 240 day herd closure period.[5] The objective of this report was to assess if a 140 day herd closure period would allow the production of M hyo-negative gilts.

Materials and methods

A 1,700 sow nucleus/multiplier herd with an on-site nursery, on-site gilt replacement barns, and an off-site gilt farm was selected. Gilt replacements were stopped on week 0. All breeding stock was at least 10 months of age on week 16. Tilmicosin-medicated feed with concentrations of 363 g/ton and 181 g/ton in breeding and gestation or farrowing, respectively, was provided from week 18 to 21. Phased weekly piglet lincomycin injections at 100 mg/piglet were administered from week 16 to 21. The on-site and off-site gilt replacement barns were depopulated by week 16 and 25, respectively (Table 1). Clinical examinations, serological tests (Idexx and Dako ELISA), and necropsies to collect broncho-alveolar lavage fluids (BALF) for M hyo PCR were performed.

Results

No M hyo clinical signs (i.e. dry non-productive cough) were observed in the off-site gilt farm through week 88. At week 41, the detection of serologically positives pigs for both ELISA tests (double positives) necessitated further investigation. Because of M hyo vaccination at weaning, collection of BALF for M hyo PCR on all double positive pigs was initiated until non-vaccinated sentinels were available. Six collections of BALF consisting of sixteen pigs in total were PCR negative over six months. Six monthly non-vaccinated sentinel pig tests of various aged pigs detected no double positives.

Discussion and conclusion

The M hyo elimination was successful to date. These M hyo negative gilts are currently supplying M hyo negative breeding herds without clinical signs.

References

1. Baekbo, P. et. al. (1994). Eradication of *Mycoplasma hyopneumoniae* from infected herds without restocking. 13th International Pig Veterinary Symposium.

2. Baekbo, P. et. al. (1995). Eradication of *Mycoplasma hyopneumoniae* from a herd without restocking. American Association of Swine Practitioners

3. Luim, B. et. al. (1992). An Attempt to eradicate *Mycoplasma hyopneumoniae* form selected Norwegian farrowing to finishing herds. 12th International Pig Veterinary Symposium.

4. Schneider, P. (2004). Experiences with *Mycoplasma hyopneumoniae* and Transmissible Gastroenteritis Eradication from Sow Herds. American Association of Swine Veterinarians.

5. Holst S, Y. P., Pieters M. (2015). Elimination of *Mycoplasma hyopneumoniae* from breed-to-wean farms: A review of current protocols with emphasis on herd closure and medication. J Swine Health Prod. **23**(6): 321–330.

Table 1: Timeline of events for the M hyo elimination

Timeline (weeks)	Action
Week 0	Last gilt entry into herd
Week 15	Depopulation of pigs < 10 months of age completed
Week 16	Farrow last gilts breed – begin injection* on newborn piglets
Week 17	Continue inject* all piglets at birth adding week 2 Wean pigs off-site START medication herd with Pulmotil® as follows Lactation – 200 ppm Gestation – 400 ppm
Week 21	STOP medicated sow feed Resume pig flow at on-site nursery M hyo negative gilts placed in breeding herd (not M hyo vaccinated)@
Week 28	Upgraded health piglets exit on-site nursery to gilt finishing farm
Week 48	M hyo replacement gilts available for commercial herds#

* lincomycin injection weekly 100 mg/pig with phased initiation
@ vaccinated 3 months after entry
\# M hyo vaccinated at placement and weaning

Compliance and efficiency in swine health documentation

Christopher James Rademacher, DVM
Associate Director of Iowa Pork Industry Center, Iowa State University

Introduction

It is estimated that over one million pigs are being transported between herds each day in the US, many of those traveling across state borders. GlobalVetLink (GVL) alone estimates that a total of about 55,000 swine CVIs will be created in their software platform in 2019 (Figure 1). The risk of disease transmission with these movements is in part controlled by the oversight of veterinarians.

Veterinarians are charged with inspecting the herds of origin as mandated by federal and state regulations and providing a Certificate of Veterinary Inspection (CVI), also known as a health certificate. The requirements vary for transportation within and between states and can change based upon current disease issues, both endemic and the risk of a foreign animal disease. It can be difficult for a veterinarian to stay current with the specific diagnostic testing and/or vaccination requirements established for each state. This paper will discuss the requirements and an easy-to-use, cloud-based solution for digital CVIs.

CVI requirements for swine[1]

Most states require a valid CVI that includes an inspection date within the last 30 days prior to transport of any animals from that herd. A certificate must show individual animal identification numbers including: the official ear-tag number(s), breed association registration tattoo or brand numbers, breed association registration number and any other official individual identification on each animal to be moved. Exceptions to individual animal identification may be allowed.

Required information on the CVI includes:

1. Number of animals covered by the certificate.
2. Purpose for which the animals are to be moved.
3. Points of origin and destination.
4. The consignor and the consignee.
5. A statement by the issuing accredited or State or Federal veterinarian that the animals were not exhibiting clinical signs associated with disease of concern for that species at the time of examination.
6. Swine entering certain states must have remarks to show the animals have been examined and found free of disease and lack of exposure to diseased premises within the last 30 days.

Import permits

Some states also require a permit number issued by the animal health office prior to shipment. Check your state requirements (see "State movement requirements" section). This permit number may be obtained by the veterinarian by calling or emailing the state of destination.

Official identifications

For interstate movements, all swine must be individually identified with official identification unless the swine are kept as a group. The CFR part 71.19 explains the requirements for group shipment of swine. Swine moving within a production system do not need to be individually identified if there is a swine production health plan agreement, as defined in 9 CFR Part 71.1. Official Identifications should be listed on the CVI (NUES Tag, AIN (840) Tag, RFID, Location ID (PIN)).

Swine production health plan

Producers moving swine via interstate commerce within their production system may qualify to use Swine Production Health Plans, or commuter herd agreements. To qualify the producer would need to develop a swine production health plan and have it approved by both the sending and receiving swine state animal health officials. There are also additional requirements on moving swine within a production system explained in 9 CFR 71.19(g). Commuter Herd Agreements may become null and void in the case of a foreign animal disease outbreak.

State movement requirements

AnimalRegs.com, built by GVL, provides regulations and requirements for moving any animal species within, to or from a state. GVL works with state animal health officials to ensure this information is accurate. GVL's software platform for veterinarians also incorporates these movement requirements into the creation process of CVIs and auto-checks for errors.

GVL digital health certificates

GVL provides veterinarians with an easy-to-use, cloud-based software solution for digital CVIs for all species of animals. GVL CVIs are approved for all US states and territories, and include automatic submission to state animal health officials. For more information on GVL and the online CVI system, visit **globalvetlink.com**.

Figure 1: GVL digital swine CVIs.

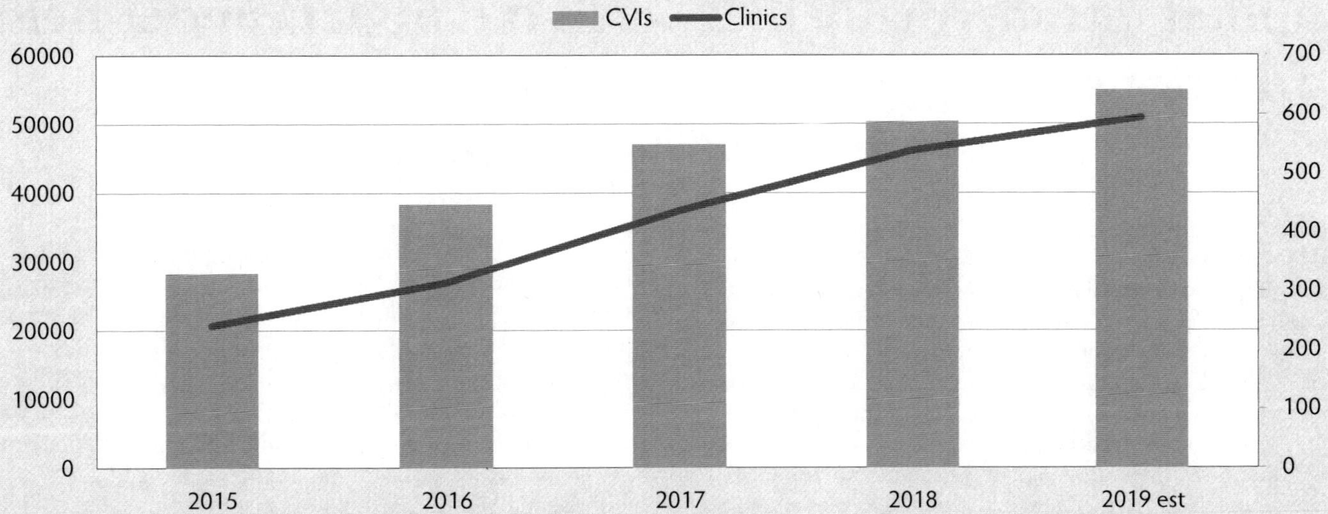

Reference

1. USDA Interstate Movement of Cattle, Horses, Swine, Sheep and Goats. 2017. Available at: https://www.aphis.usda.gov/aphis/ourfocus/animalhealth/nvap/NVAP-Reference-Guide/Animal-Movement/interstate-animal-movement. Accessed 11 November 2019.

Efficacy of a type 1 MLV PRRSV vaccine when applied intradermally in 2-week-old piglets under field conditions

J. Miranda, DVM; L. de Lucas, DVM; A. Camprodon, DVM
HIPRA, Amer, Spain

Introduction

The economic losses of PRRSV in growing animals are caused by the decrease of average daily weight gain, mortality rate, feed conversion rate and antibiotic consumption. PRRSV piglet vaccination during lactation provides benefits to control the disease from a virological and clinical point of view. Intradermal (ID) administration using needle-free devices has proved to improve the efficacy of vaccines, showing a stronger antibody and CMI responses compared to the IM vaccination.[1] Moreover it reduces the spreading of viraemic diseases such as PRRSv and cause less stress and pain for the animals. The aim of this study was to demonstrate that a PRRSV1 vaccine (UNISTRAIN® PRRS, HIPRA) administered at 2 weeks of age (woa) by ID route can aid in the control of the PRRSV infections in the early nursery phase in European pig farms.

Materials and methods

663 piglets from a farrow-to-finish farm suffering losses associated with PRRSV were selected. The animals were distributed in two different groups; Group A (N = 322) was vaccinated at 2 woa with UNISTRAIN® PRRS intradermally with Hipradermic®, whereas Group B (N = 341) remained non-vaccinated. Vaccinated and non-vaccinated pigs were housed separately in different rooms. Clinical signs and mortality were recorded weekly until the entrance in the fattening unit (10 woa). Blood samples were collected from 12 randomly selected pigs per batch at 2, 6, 8 and 10 woa for measuring total antibodies against PRRSv and the presence of the virus. Statistic comparison was performed using Mann-Whitney U, Kruskal-Wallis and the Chi-square tests.

Results

Statistical differences were found between groups regarding the presence of clinical signs, showing Group A less total cases of clinically ill pigs compared to group B ($P < 0.01$). Mortality observed after the supposed onset of immunity (30 days post vaccination) and during the entire study was higher in group B (7.3%) compared to group A (3.1%) (Figure 1). Antibody levels and viraemia until the exit of the nursery phase didn´t show statistically significant differences between groups.

Figure 1: Dynamics of the cumulative mortality during the study. The slashed line represents the movement from lactation to the nursery phase. Asterisk indicates statistically significant differences ($P < 0.05$).

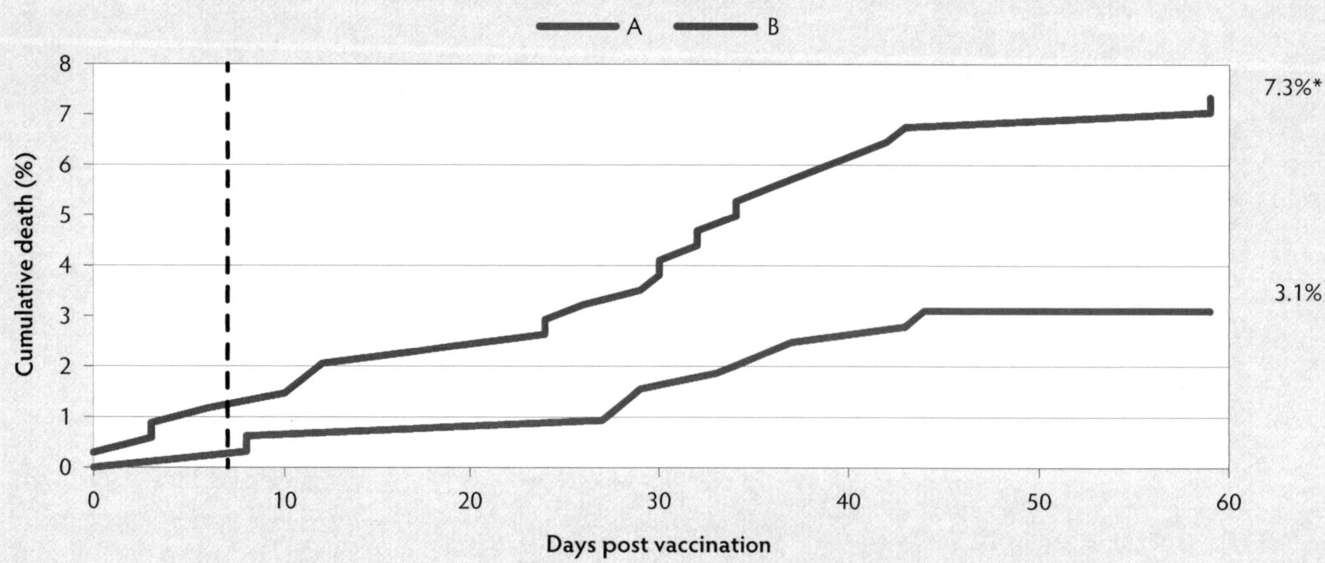

Conclusions

UNISTRAIN® PRRS administered in piglets at 2 weeks of age by intradermal route with Hipradermic® device in presence of maternally-derived antibodies against PRRSV has demonstrated to reduce the number of clinically ill pigs and mortality until the end of the nursery phase.

Reference

1. Martelli P, et al. Protection and immune response in pigs intradermally vaccinated against porcine reproductive and respiratory syndrome (PRRS) and subsequently exposed to a heterologous European (Italian cluster) field strain. Vaccine. 2007;25:3400-8.

Evaluation of three commercial ELISAs for detection of antibodies against *Mycoplasma hyopneumoniae* in serum samples

A. Hidalgo, DVM, PhD; S. Koller, BSc
IDEXX Laboratories, Inc., Westbrook, Maine

Introduction

Serological tests are commonly used to monitor the health status of pig herds. ELISA is a rapid, inexpensive and easily automated diagnostic method for *Mycoplasma hyopneumoniae* (M hyo) that provides useful information on the presence of antibodies and seroconversion. However, differences in sensitivity and specificity of M hyo ELISA kits have been reported.[1] This study compares the performance of three commercial M hyo antibody ELISAs using serum samples from pigs of known status.

Materials and methods

A total 1,216 serum samples were used to evaluate the sensitivity and specificity of three commercial M hyo antibody ELISAs. For the specificity study, 500 serum samples from M hyo naïve pigs were included. For the sensitivity study, 671 serum samples from M hyo positive pigs were used. The presence of M hyo was confirmed on tracheal swabs at the pen level before sampling. In addition, serum samples ($n = 45$) from ten pigs experimentally infected with M hyo were collected at 14-, 23-, 35-, 41- and 56-days post infection (dpi) and used for the onset of detection study. The experimental design of the infection and transmission study has been described previously.[2] Serum samples were tested with (i) IDEXX M hyo Ab Test [IDEXX Laboratories, Inc., Westbrook ME, USA] (IDEXX M hyo ELISA), (ii) another commercial M hyo antibody indirect ELISA (I-ELISA) and (iii) a blocking M hyo ELISA (B-ELISA) according to the manufacturer's recommendations. For the sensitivity calculations, positive and suspect results were grouped together.

Results and discussion

Both indirect ELISAs, IDEXX M hyo ELISA and I-ELISA, correctly identified known negative serum samples (Specificity = 100%). The B-ELISA failed to correctly identify four negative serum samples (Specificity = 99.2%), with three positive and one suspect results.

Following the experimental infection, the IDEXX M hyo ELISA detected 70% of positive pigs at 14 dpi and 100% at 23 dpi. However, the I-ELISA was only able to detect 20% and 66.7% of positive animals and the B-ELISA 40% and 66.7% at 14 and 23 dpi, respectively (Figure 1). The better onset of detection showed by the IDEXX M hyo ELISA, with a higher percentage of positive pigs detected as early as 14 dpi, support its use for early diagnosis of M hyo infection and detection of index cases compared to the other ELISA kits evaluated.

The sensitivity of the IDEXX M hyo ELISA was higher than the other indirect ELISA (I-ELISA) included in this study, detecting 58 positive samples more (8.1%).

In summary, the IDEXX M hyo Ab Test showed 100% specificity, better than a blocking M hyo ELISA, higher sensitivity than another M hyo indirect ELISA and better onset of detection than the indirect and blocking ELISAs evaluated in this study.

References

1. Sibila M et al. Current perspectives on the diagnosis and epidemiology of *Mycoplasma hyopneumoniae* infection. Vet J. 2009; 181:221-31.

2. Clavijo MJ et al. Spatial spread of *M. hyopneumoniae* in pig populations. Proceedings of ISU James D. McKean Swine Disease Conference. 2019, pp 64-65.

Figure 1: Onset of detection of three M hyo antibody ELISAs.

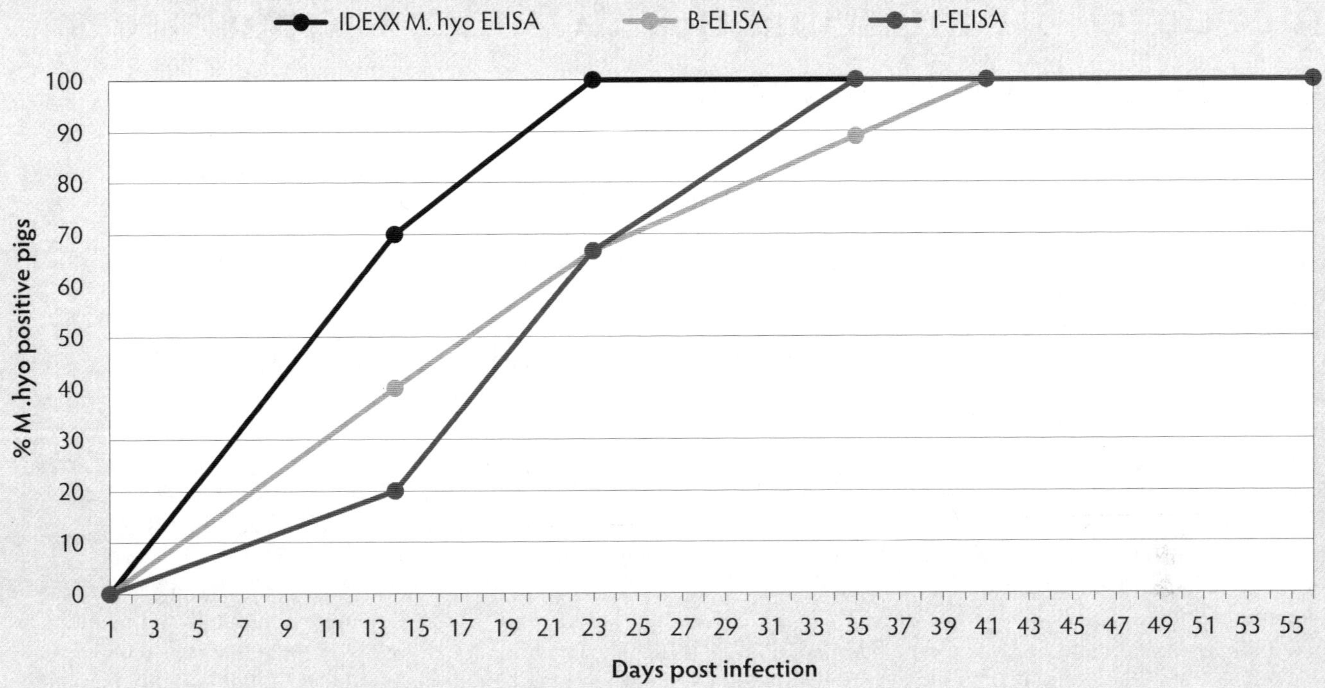

ButiPEARL™ Z: a novel encapsulated butyric acid and zinc source for intestinal health under various stress conditions for pigs – a review

Tom Marsteller, DVM; Venkatesh Mani, PhD; Paul Matzat, PhD; Brian Kremer, PhD
Kemin Industries, Inc., Des Moines, Iowa

Introduction

The gastrointestinal tract (GIT) is an important organ system that directly influences production animal performance. Since the GIT is composed of only a single layer of epithelial cells, it is easily compromised by stress events and disease challenges.[1] Both butyric acid and zinc help to maintain intestinal integrity through various mechanisms, in addition to butyric acid acting as an antimicrobial at higher concentrations.[2] ButiPEARL™ Z (BPZ) is an encapsulated source of zinc and butyric acid, and the patented encapsulation technology delivers these ingredients, in a controlled manner, throughout the intestinal tract thereby not only improving gut health but also performance. This abstract is a review of three studies conducted to test the efficacy of BPZ under three different growth scenarios in pigs: (1) no challenge; and exposure to (2) non-infectious and (3) infectious stressors.

Materials and methods

Trial 1: This trial was conducted without any applied challenge for 42 days and contained two treatments (Control and BPZ 0.7 kg/MT, corn-soy pellet diet). There were 9 pens/trt and 11 pigs per pen (start wt. 6.0 ± 0.2 kg).

Trial 2: To test the efficacy of BPZ against a non-infectious stressor, a heat stress (HS) swine trial was conducted with 24 pigs in individual pens (n = 12 pigs/trt, start wt. 35 ± 1 kg). Pigs were supplemented with either BPZ (7.5 kg/MT) or fed a control diet for 24 d with a corn-soy mash diet before exposure to biphasic HS, 32°C for 8h and 28°C for 16h, for 7d. Performance and intestinal health were measured at the end of the HS period.

Trial 3: To test the efficacy of BPZ on reducing porcine epidemic diarrhea virus (PEDV) associated intestinal damage, a nursery trial was conducted with 24 weanling piglets housed individually (n = 16 pigs/trt, 2 pigs/pen, start wt. 7.5 ± 1.2 kg). There were two treatments, PEDV and PEDV + BPZ. Piglets were given a control diet or diet supplemented with BPZ (5 kg/MT) mash feed and gastric gavaged with 5 ml of 10^3 $TCID_{50}$ of the PEDV plaque-cloned isolate from Iowa (18984/2013). At 5 and 20d post-infection (dpi), intestinal health and performance were measured.

Results and discussion

All the performance results from the three trials are presented in Table 1. In Trial 1, under no challenge conditions in nursery pigs, BPZ significantly improved the average daily gain (ADG), average daily feed intake (ADFI) and Feed:Gain (F:G) compared to control ($P < 0.05$). In Trial 2, HS trial, ADG and F:G were significantly better in BPZ group (0.53 vs 0.79 and 3.45 vs 3.19 respectively, $P < 0.05$). Macromolecule permeability and transepithelial resistance in the ileum were numerically lower, and villus height and villus height to crypt depth ratio was numerically greater in the BPZ group. In Trial 3, PEDV trial, d0-20 ADG was not different, but BPZ had a greater ADG on d10 (0.47 vs 0.53 kg, $P < 0.05$) and d20 (0.53 vs 0.66 kg, $P < 0.05$). BPZ had a greater ADFI on d20 (1.04 vs 1.21 kg, $P < 0.05$) and overall (0.77 vs 0.82 kg, $P < 0.05$). F:G was not different on d20, but BPZ F:G was significantly better between d6-10 and d15-20. Lesion score in the jejunum was significantly lower on d5 (1.4 vs 2.1, $P < 0.05$). Villus height to crypt depth ratio was greater in d5 jejunum in the BPZ group (0.9 vs 1.1, $P < 0.05$). The data from the three trials provide evidence that BPZ may have the potential to minimize the negative effects associated with non-infectious and infectious stressors in swine and help the intestine with faster recovery.

References

1. Blander, J.M., et. al.2017. Regulation of inflammation by microbiota interactions with the host. Nature Immunology. 18:8:851-860.

Table 1: Effect of ButiPEARL™ Z on swine performance under no challenge, heat stress and porcine epidemic diarrhea viral (PEDV) disease conditions*

Trial 1: No Challenge§ - 42 d nursery trial

Days	CON-ADG (kg)	BPZ-ADG (kg)	CON-ADFI (kg)	BPZ- ADFI (kg)	CON-F:G	BPZ-F:G
D 0-14	0.12	0.13	0.19	0.19	1.67^{x}	1.47^{y}
D 0-21	0.20^{a}	0.22^{b}	0.28	0.29	1.41	1.34
D 0-28	0.26	0.27	0.36^{c}	0.38^{d}	1.39	1.38
D 0-35	0.33^{a}	0.35^{b}	0.47	0.49	1.43	1.42
D 0-42	0.40^{a}	0.43^{b}	0.59^{c}	0.61^{d}	1.48^{x}	1.46^{y}
SEM	0.02	0.02	0.04	0.04	0.01	0.01

Trial 2: Heat Stress Challenge† - 7 d grower trial

	Control	BPZ	SEM	*P*-Value
ADG (kg)	0.53	0.79	0.06	< 0.05
ADFI (kg)	1.61	1.82	0.07	0.2
F:G	3.45	3.19	0.06	< 0.05
Ileum Villus Height, μm	362	377	± 26.5	0.707
Ileum Crypt Depth, μm	245	204	± 17.5	0.141
Ileum Villus: Crypt Ratio	1.59	1.87	± 0.140	0.192
Ileum FITC-Dextran Permeability	4.80	2.11	± 1.50	0.15
TER Ω.cm^2	113.61	124.40	± 20.00	0.18

Trial 3: PEDV Challenge‡ – 20 d nursery trial

DPI	CON-ADG (kg)	BPZ-ADG (kg)	CON-ADFI (kg)	BPZ-ADFI (kg)	CON-F:G	BPZ-F:G
D 0-5	0.18	0.09	0.35	0.32	2.17	2.15
D 6-10	0.47^{a}	0.53^{b}	0.58	0.56	1.24^{x}	1.05^{y}
D 11-15	0.58	0.57	0.83	0.93	1.42	1.58
D 16-20	0.53^{a}	0.66^{b}	1.01^{c}	1.21^{d}	2.03^{x}	1.69^{y}
D 0-20	0.42	0.47	0.77^{c}	0.82^{d}	1.59	1.53
SEM	0.03	0.03	0.04	0.04	0.08	0.08

* The superscripts a and b, c and d, and x and y represent statistically significant differences at $P < 0.05$ for the parameter measured (ADG or ADFI or F:G) within the same trial. BPZ-ButiPEARL™ Z, TER-Transepithelial resistance.

§ Nursery piglets were raised under no challenge conditions and supplemented with 0.7 kg/MT of BPZ.

† Pigs were supplemented with 7.5 kg BPZ/MT and were exposed to biphasic HS, 32°C for 8h and 28°C for 16h.

‡ Piglets were supplemented with BPZ (5 kg/MT of feed) for 20 days and gastric gavaged with 5 mL of 103 $TCID_{50}$ of the porcine epidemic diarrhea virus (PEDV) plaque-cloned isolate from Iowa.

The use of a digital biosecurity system on three swine farms - preliminary results

László Búza[1]; László Ózsvári[2]
[1]MSD Animal Health, Hungary; [2]University of Veterinary Medicine Budapest, Hungary

Objectives
Reliable biosecurity protocols are essential to intensive pig production systems. The aim of this research was to evaluate if use of the B-eSecure™ (PigCHAMP Pro Europa) digital system, which detects the movements of individual workers within farms, could improve compliance with biosecurity protocols.

Material and methods
The B-eSecure system was installed on three farrow-to-finish pig farms in Hungary. Because of the national legislation these farms run PRRS elimination programs. Before the installation there was a detailed farm biosecurity audit in August 2017 by Merck Integrated PRRS Solution System. After the installation there was a three months long learning process wherein the system learned the workers' shifts, movement patterns and buildings involved. The baseline data collection period ran from 01 January 2018 until 30 June 2018. For analysis of the data, personnel movements were categorized into three biosecurity risk categories: safe, risky and unsafe. For better protection of farms health status Lares Agri digital (PigCHAMP Pro Europa) external biosecurity system was used too.

Results
The maximum number of movements detected for each farm and the percent in the unsafe category, also the maximum and minimum percent unsafe movements by farm and by month are captured in Table 1. On farms A and B most of the unsafe movements were related to the transportation of dead animals. The analysis also revealed that specific employees on Farms B and C were responsible for the majority of the unsafe movements. The results of the biosecurity analyses were presented to the farm managers and biosecurity protocols were modified. Data is now being collected to measure the impact of the protocol modifications.

Conclusions
This digital biosecurity system was able to detect risky and unsafe biosecurity movements within the farms and was used by farm managers to evaluate their workers' compliance to the protocols. More time and data is required to determine if improvements in worker movements will occur on these farms.

References
1. Larrochelle R. et al. 2003. Virus res. 96: 3-14
2. Vangroenweghe F. et al. 2009. Varkens-Gezond-Hounden, University of Gent

Table 1: Maximum number of personnel movements and percent of maximum and minimum unsafe movements by farm and month

Movements of workers	Month	Farm A	Farm B	Farm C
Maximum # personal movements (% unsafe)	January			18,089 (19.0%)
	February		9,380 (5.0%)	
	May	21,478 (3.0%)		
Maximum unsafe personal movements (%)	January	5.0		
	March			22.0
	May		15.0	
Minimum unsafe personal movements (%)	April	1.0		
	June		9.0	16.0

Evaluation of the porcine gut microbiome response to *Lawsonia intracellularis* infection

F. Leite[1], DVM, MS, PhD; B. Weber[2], PhD; T. Johnson[2], PhD; T. Gould[2], MS; E. Vaskez[2], DVM; D. Beckler[3], DVM,MS; F. Vannucci[2], DVM, Msc, PhD; R. Isaacson[2], MS, PhD
[1]Boehringer Ingelheim Animal Health, Duluth, Georgia; [2]Univeristy of Minnesota, St. Paul, Minnesota; [3]Gut Bugs Inc., Fergus Falls, Minnesota

Introduction

Lawsonia intracellularis is a Gram-negative intracellular bacterial pathogen of swine that primarily infects the small intestine but may also colonize the large intestine.[1] Infection has been estimated to occur in more than 90% of US swine farms, and can lead to significant production loss.[1] The gut microbiome is known to influence health as well as production performance of swine.[2] It is also known that the composition of the microbiome varies between different portions of the intestinal tract.[2,3] To better understand how *L intracellularis* impacts the gut microbiome during infection, we evaluated the microbiome response of the small and large intestine of pigs experimentally challenged with *L intracellularis.*

Materials and methods

Two treatment groups were evaluated: one treatment group received *L intracellularis* challenge and the other treatment group served as non-infected control. Animals that received challenge were inoculated with a gut homogenate containing approximately 1×10^6 *L intracellularis* organisms at six weeks of age. Fecal samples were collected at time of challenge and sequential necropsy was performed on 6 pigs per treatment group at 7, 21 and 28 days post infection to evaluate microbiome response at different stages of infection. At each necropsy, samples collected were: digesta of the terminal ileum, mucosal scraping of the terminal ileum, and feces. After DNA extraction, microbiome analysis was performed by analyzing the V4 region of the 16S rRNA gene using DADA2 and phyloseq packages in R.[4]

Results and discussion

Preliminary results indicate *L intracellularis* challenge led to a change in the microbial composition of ileal digesta, mucosa and feces. Beta diversity (community composition) differences were evident between challenged and non-challenged animals especially at the time points of 21 and 28 days post challenge.

Conclusion

Evaluating differences in microbial composition further will lead to a better understanding of the interactions of *L intracellularis* with other bacteria present during infection and their potential interaction in promoting disease and its negative consequences on health and production performance.

References

1. Vannucci et al. Proliferative Enteropathy, Diseases of Swine, 11th edition, 2019.
2. Isaacson R, Kim HB. The intestinal microbiome of the pig. Anim Health Res Rev. 2012 Jun;13(1):100-9.
3. Holman DB et al., Meta-analysis To Define a Core Microbiota in the Swine Gut. mSystems. 2017 May 23;2(3).
4. McMurdie P; Holms S., phyloseq: An R Package for Reproducible Interactive Analysis and Graphics of Microbiome Census Data. PLoS One. 2013 Apr 22;8(4):e61217.

The PRRS Calculator – an economic tool assessing the impact of PRRS and mitigation strategies

Rob Musser[1], PhD; J. Tyler Holck [2], DVM, MS, MBA
[1]NutriQuest, Mason City, Iowa; [2]Feed His People, LLC, Gilbert, Iowa

Introduction

Porcine reproductive and respiratory syndrome (PRRS) is a viral disease which by the name of the virus has two clinical presentations, complications with reproduction and respiratory disease in pigs of any age. The PRRS virus (PRRSv) can affect all age groups of pigs and all types of herds including high or ordinary health status, irrespective of size. PRRS is the most economically significant disease to affect US swine production and costs the United States pig industry around $650 million annually.[1]

The struggle often comes on how to blend a rapid response to the health challenge and a rapid financial review of the expected impact and options on how to react. This need for a rapid response will typically result in a quick plan to stabilize health in the long term and many times will stop at this point. The ability to quickly input information and blend health and financial models together can allow for a deeper understanding of potential expectations. While different PRRS viruses can have a wide range of impacts, if with working with the health care advisor and potential experiences in the region with similar strain you can make the most educated guess possible.

PRRS Calculator

Inputs: Farm performance parameters; farm financial parameters; depopulation and herd closure assumptions; health outbreak assumptions. Calculations: Impact of PRRS outbreak on sow herd; impact of PRRS outbreak on WF performance; total financial loss by farm, by sow and by pig. PRRS mitigation options: do nothing; focus on producing PRRS neg pigs in the future; technologies to have immediately mitigate financial loss; combinations of immediate and long-term mitigation strategies.
(Figure 1: NutriQuest PRRS Calculator™)

Discussion

The initial response to a PRRS outbreak is typically focused on stabilizing the sow herd to expedite the time frame for the farm to return to baseline reproductive performance and producing PRRS negative weaned pigs. However, there are management techniques and technologies that can help reduce the devastating financial loss in the immediate timeframe by augmenting herd immunity and improving weaned pig performance in the face of PRRS viremia. The financial costs and values of these technologies often have limited financial review due to the desire to get a plan in place and lack of tools to rapidly compare the costs, input views on efficacy and compare return on investment levels.

The tool uses many input numbers that are built in from survey of leading production systems and veterinarians. The inputs allow for a clear understanding of the size of operation, current level of production, financial markets of pigs and feed, experiences with the pathogen (both at sow farm and downstream) and inputs on options for other interventions. The tool then looks at how these individual impacts are affecting financial output of the operation both in the sow farm level and if ownership of the pigs is maintained what can be expected downstream.

In the second tab of the tool it allows the user to input the costs of some of the interventions or tools that can allow for estimates of impact the technology will have on the immediate challenge. The categories are outlines as: Do nothing and let it become endemic, list of choices available will make for long term health of herd and goal to get negative (stay negative), list of choices available to deal with the impact the immediate challenge is having on the sow farm productivity and finally a list of choices available to deal with the impact on the pigs being produced from the period of the challenge.

Then the last tabs on the tool is established to be a list of executive summaries to share with team members to outline the combination of plans and what can be expected from a performance and financial impact.

Conclusions

The NutriQuest PRRS Calculator can be a beneficial tool when assessing the economic impact of various PRRS mitigation strategies. This tool can be rapidly completed and allow for immediate analysis of the plans; as well as, assist in understanding the levels of financial loss that will still be incurred. This tool is built to be an effective way to allow owner/veterinarian/financial institution to all get on the same plan rapidly to minimize and understand the impact.

Reference

1. Holtkamp DJ, Kliebenstein JB, Neumann EJ, et al. Assessment of the economic impact of porcine reproductive and respiratory syndrome virus on United States pork producers. *J Swine Health Prod.* 2013;21(2):72-84.

Figure 1: NutriQuest PRRS Calculator™.

NutriQuest

Reset to Default Values | Share as PDF in email

PRRS Calculator: Farm Name: Test Farm

Inputs		Value	Units
Farm Performance Parameters			
Health Challenge		PRRS	
Sows on Farm		6,000	# sows
Pigs / Sow / Year Weaned		28	# pigs
Wean Weight		13.0	lbs
Market Weight (live)		280.0	lbs
Percent Yield		75.0%	when pigs are sold
ADG W-F		2.00	lb/day
FCR for W-F		2.6	lb/lb
Wean-to-Market Death/Cull Loss		7.0%	total cull & dead %
Average Weight of W-F Mortality		45.0	lbs
Feed Consumed per Sow per Year		2,200	lbs
Sow Cull Weight		350.0	lbs
Farm Financial Parameters			
Transfer Value of Weaned Pig	$	40.00	$/pig
Carcass Value	$	0.80	Market in 12 month
Cost of W-F Feed	$	0.10	$/lb
Cost per Day of Finishing Space	$	0.12	$/pig/d
Sow Cull Value (live)	$	0.30	$/lb
Cost of Sow Feed	$	0.11	$/lb
Yardage Cost of Sow Farm (based on $4.50/pig = $.35/s/d)	$	0.35	$/sow/day
Value of Replacement Breeding Sow	$	180.00	$/sow
Depopulation and Herd Closure assumptions			
Down Time for Herd Closure, days closed		240.0	days
Able to Stock with Gilts (1 = Yes, 2 = No)		1.0	Answer
Cost of Clean Up, Disenfection Replacement Supplies	$	-	$/sow
Down Time for Depopulation (offsite breeding projects with regulated feed intake)		45.0	days*
Lost Margin and Expenses (during downtime)	$	25.00	/pig (seasonal)
Number of Years between Depopulation Events		3.0	years
Health Outbreak Impact (based on "mild strain")			
Sow Performance Impact During Break			
Impact of Challenge on Pigs Born Live per Litter		2.5	pigs/litter decrease
Impact of Challenge on Preweaning Mortality		7%	% increase
Duration of Challenge		120	days
Impact on Sow Mortality (actual %)		5%	% increase
Impact on Abortion Rates (actual %)		10%	% increase
Pig Performance Impact Wean to Finish on Pigs Placed During Break			
Mortality rate difference (W-F)		6%	% increase
ADG (lb / day difference)		0.1	lb/d
FCR impact		0.15	lb/lb
Difference in Pigs to Primary Markets (% marketed)		5%	% more culls
Revenue Difference/Pig between Primary and Secondary Markets	$	50.00	$/pig

Total Return	Space Short		
PRRS Challenge Impact Calculation			
Impact of Challenge on Sow Herd			
Impact on Born alive	$	32.88	$/sow/year
Impact on Preweaning Mortality	$	24.88	$/sow/year
Impact on Sow Mortality (assume no litter)	$	2.96	$/sow/year
Impact on Sow Abortion Rates	$	36.16	$/sow/year
Cost of Clean Up Supplies and Replacement Supplies	$	-	$/sow/year
Impact of Challenge on Financial Loss at Sow Farm	**$**	**96.88**	**$/sow/year**
Impact of Challenge on Financial Loss of W-F Performance (assuming no comingling with clean pigs)			
Impact on ADG			
Lost Weight lbs/pig		13.35	lbs/pig
Lost lbs/sow/year		112.25	lbs/sow/year
Net Loss Accounting for Less Feed	$	38.16	$/sow/year
Impact on FCR	$	1.32	$/pig
Poorer FCR	$	33.67	$/sow/year
Impact on Mortality			
Fewer Incremental pigs/sow mktd		0.54	
Loss of pig and opportunity value	$	41.61	$/sow/year
Impact on Primary Market Opportunity			
Fewer full value pigs		0.82	/pig
Value of the impact	$	21.02	$/sow/year
Impact of Challenge post weaning, financial loss due to challenge	**$**	**134.47**	**$/sow/year**
Total Financial loss from Challenge - Sow farm to finish	**$**	**231.35**	**$/sow/year**
Total Financial loss from Challenge - Sow farm to finish	**$**	**8.41**	**$/pig**
Total Financial loss from Challenge - Sow farm to finish	**$**	**1,388,092**	**$/per event**

DISCLAIMER: All numbers, projections, and the like contained herein regarding costs, expenses, production results, etc., are for ILLUSTRATIVE PURPOSES ONLY. The assumptions used and the resulting projections may not be appropriate for a given situation. ACTUAL RESULTS COULD VARY SIGNIFICANTLY. NUTRIQUEST MAKES NO WARRANTIES, EXPRESSED OR IMPLIED, REGARDING ANY SUCH PROJECTIONS, AND PARTICIPANTS AGREE THAT NUTRIQUEST LLC SHALL HAVE NO LIABILITY IN CONNECTION THEREWITH.

The use of medium chain fatty acids to reduce porcine reproductive respiratory syndrome virus replication in MARC-145 cells

S. A. Crowder[1], PhD; R. M. Pogranichniy[2], DVM; B. De Rodas[3], PhD; T. P. Karnezos[4], PhD; J. S. Radcliffe[1], PhD
[1]Dept. of Animal Science, Purdue University, West Lafayette, Indiana; [2]Department of Diagnostic Medicine and Pathobiology, Veterinary Diagnostic Laboratory, Kansas State University College of Veterinary Medicine, Kansas; [3]Purina Animal Nutrition LLC., Arden Hills, Minnesota; [4]Land O'Lakes Inc., Arden Hills, Minnesota

Introduction

The swine industry has been affected by the porcine reproductive respiratory syndrome virus (PRRSV) for over 30 years. Genetic diversity in the PRRSV has been an issue with vaccine efficacy and cross protection between strains. Recent studies have examined the antiviral properties of medium chain fatty acids (MCFAs) with the most recent studies focusing on feed biosecurity and viral diseases. PRRSV is of highest economic importance to the Pork Industry today with $660 million in annual production losses.[1] Activated or free MCFAs have been used in different combinations and concentrations to optimize pig performance. Therefore, the objective of this study was to evaluate the efficacy of MCFAs to reduce viral replication of the PRRSV.

Materials and methods

MARC-145 cells, which are a subclone of the African green monkey kidney MA104 cells that are highly permissive to PRRSV, were maintained in Dulbecco modified Eagle medium (DMEM) containing 10% fetal bovine serum and 50µg of gentamicin. Type II North American PRRSV strain P129 and Type I European PRRSV Strain Lelystad were propagated in MARC-145 cells to prepare the virus stocks for the present study. MCFAs were used in varying concentrations ranging from 1000µg/mL to 1µg/mL and compared to control cells and $TCID_{50}$/mL was calculated for all treatment concentrations and non-treated controls. PPRSV replication was detected using the FITC labeled IgG anti-PRRSV monoclonal antibody (SDOW- 17F, RTI Labs, Brookings, South Dakota).

Results

The results of this study demonstrated that MCFAs were able to reduce replication of both strains (type I Lelystad and type II P129) of PRRSV in the MARC-145 cells. Data from this study revealed a statistical difference in the efficacy of individual MCFAs to reduce viral replication of PRRSV strains. MARC-145 cells incubated with caproic acid prior to and after inoculation with type I Lelystad strain or type II P129 strain of the PRRSV did not affect PRRSV replication ($P > 0.05$). Incubation of MARC-145 cells with caprylic acid prior to and after inoculation with type I Lelystad strain or type II P129 strain of PRRSV resulted in a 2.5 to 3.1 Log reduction in PRRSV replication ($P < 0.05$). MARC-145 cells incubated with capric acid prior to and after inoculation with type I Lelystad strain or type II P129 strain of the PRRSV resulted in a 3.6 to 4.2 Log reduction in PRRSV replication ($P < 0.05$). Capric acid reduced replication of PRRSV type II P129 virus strain at a lower concentration than other MCFAs. Additional studies are needed to confirm these results in an in-vivo model. Further research is also needed to determine the mode of action of individual as well as combinations of MCFAs on PRRSV replication.

Reference

1. Holtkamp DJ, et al. Assessment of the economic impact of porcine reproductive and respiratory syndrome virus on United States pork producers, *Journal of Swine Health and Production*, 2013;21(2):72-84.

The value of LeeO, an individual animal identification system adding value to the pork supply chain in the Netherlands

P. ter Linde[1]; T. Holck[2], DVM, MS, MBA; J. Ley[3]; J. Stave[3]
[1]LeeO, Deventer, NL; [2]Feed His People, LLC, Gilbert, Iowa; [3]PrairiE Systems, Spencer, Iowa

Introduction

LeeO is a precision livestock farming system utilizing Radio Frequency Identification (RFID) technology to track animal inventories, movements, medicinal treatments, as well as other on farm activities. LeeO, an application for mobile devices, utilizes Bluetooth technology to wirelessly connect to accessories such as scales, and RFID scanners to digitally record and track on farm activities. LeeO is currently deployed on 140 farms averaging 350 sows per farm throughout The Netherlands. Integration with Westfort Meat Products, NL and the return of carcass data on an individual pig basis to producers allows those producers to adjust farming practices to increase revenue and improve pig care.

Summary of commercial data collection

At the direction of Westfort Meat Products (Westfort) all 140 participating farms applied Ultra High Frequency (UHF) RFID tags to the ears of newly farrowed piglets. The RFID tags were then scanned with the use of a UHF reader, recording the identity of the piglet and associating the piglet with daily on farm activities in the LeeO mobile application. After birth piglets were scanned and registered with the sow to keep genealogy history linked. Users then scanned the ear and recorded data such as; mortality, medicinal injections, weights, cross fostering, illness and pig movements.

As pigs were transported from one location to another, data ownership also had to be transferred to each producer in the supply chain culminating when the pigs reached the harvest location, Westfort Meat Products. A secure data sharing process was created that allowed data to be transferred to only the desired recipients, by scanning the RFID tag upon arrival at the new location. This scan of the ear tags requested historical information from the cloud hosted LeeO database to be available for the pig at the new location which is authorized by a pre-determined digital hierarchy.

Upon reaching Westfort RFID tags were scanned on the harvest line and paired with an internal pig ID generated by the harvest information system. Using data recorded in LeeO carcasses were positioned for the retail market, by antibiotic usage or absence of (No Antibiotics Ever; NAE) as the sorting metric. Carcass weight, muscle and fat thickness, and meat ratings metrics were then recorded throughout the harvesting process. This data was then made available to producers for application in their production systems.

Conclusion

This matrix of connected data collected with LeeO has allowed for the discovery of deviations within the quality of meat products that is correlated with genetic, parity, feed consumption, environment, and health status metrics that occurred during the production cycle of the pigs. These correlations combined with adjusted production practices has resulted in production savings

Table 1: Financial impact of data recorded with LeeO

		Financial impact	
Criteria	**Avg. sow heard size**	**Euros (€1.00)**	**USD ($1.10)**
Time to market[1]	300	€ 35,000	$38,500
Sow ultilization[2]	300	€ 9,000	$9,900
Feed administration[3]	300	€ 4	$4.40
Health status[2]	300	€ 12,000	$13,200
Pig utilization[2]	300	€ 6,000	$6,600

[1] Study conducted by Wageningen University
[2] Estimated annual savings per herd
[3] Premium per pig marketed

Efficacy of Vigilex™ feed ingredient in reducing PRRSV contamination risk in complete feed

Sara Ebarb, MS; Kari Saddoris-Clemons, PhD
Provimi, Cargill Animal Nutrition & Health, Brookville, Ohio

Introduction

Porcine reproductive and respiratory syndrome virus (PRRSV) is an enveloped virus that has unfavorable impacts on herd health and causes detrimental losses in production at multiple levels of swine production.[1] Recently, highly pathogenic viruses have been shown to survive on feed and feedstuffs following a transboundary transportation simulation.[2] A mitigation strategy that reduces the risk of viral transmission through feed could be an important addition to a production system's biosecurity program. Cargill's Vigilex™ feed ingredient, a proprietary oil blend, has been optimized to help reduce the risk of porcine epidemic diarrhea virus contaminated feed. Due to PRRSV also being an enveloped-RNA virus, the efficacy of Vigilex on reducing the viral contamination in complete sow feed intentionally infected with PRRSV was evaluated.

Materials and methods

A commercial sow diet was formulated to meet or exceed Swine NRC 2012 requirements and manufactured by Provimi in Lewisburg, OH. Using a KitchenAid 5-quart stand mixer, approximately 1,000 g of mash sow feed was mixed with each respective Vigilex™ feed ingredient treatment for approximately 5 minutes. The feed treatments used in the current trial were as follows: 1) Complete sow feed with no product added and no inoculation (negative control; NC); 2) Complete sow feed with no product added but inoculation (positive control; PC); 3) PC with 0.2% Vigilex; 4) PC with 0.4% Vigilex; and 5) PC with 0.6% Vigilex. Treatment diets were sent to Iowa State University, a BSL-2 laboratory, for all PRRSV viral inoculation procedures. Upon arrival, a 20 g sub-sample of each treatment was collected to verify feed was initially PRRSV negative, then all other material was stored at 4°C overnight. Following the overnight rest period, 90 g sub-samples (n = 3 replications/treatment) of each feed treatment were placed in a 650 mL sterile bottle. Samples were spiked with 10 mL PBS (Treatment 1) or 10 mL of PRRS isolate MN184 P13 to provide a final titer of 10^5 $TCID_{50}$/mL (Treatments 2-5) using a 20 mL syringe connected to a nebulizer with intermittent shaking. Following addition of the virus/PBS to the feed samples, bottles were vigorously shaken to ensure an even mixture. At time point 0 hr (immediately following mixing process) and 24 hr post-inoculation (stored at 4°C), 20 g sub-sample of spiked feed was collected and added to 80 mL of PBS and allowed to rest overnight at 4°C. Feed supernatants were collected from each treatment replications and transferred into 5mL snap-cap tubes, then stored at -80°C until processed for quantitative real-time RT-PCR analysis.[3]

Results and discussion

Following inoculation with PRRSV, all feed treatment replications were detected as PCR positive except the non-inoculated negative control (Treatment 1). Reductions in PRRSV genomic copies are shown in Table 1. Over the 24 hr *in vitro* time course, PRRSV genomic copies were reduced ($P < 0.05$) by 0.6 logs in the PC (Treatment 2). Increasing levels of Vigilex™ feed ingredient linearly reduced ($P < 0.05$) PRRSV genomic copies by 1.03, 1.68, and 1.88 logs for each incremental increase in concentration of Vigilex (Treatment 3, 4, and 5, respectively). It is important to note the qPCR assay, cannot distinguish between viable, infectious virus and non-viable, or inactivated virus. In this study, the inclusion of Vigilex in a commercial sow diet reduced the concentration of PRRSV in feed following a 24 hr *in vitro* time course.

Acknowledgments

A special thanks to Dr. Jianqiang Zhang and the Iowa State University Veterinary Diagnostics Laboratory for their work on this study.

References

1. Nathues, H. et al. Cost of porcine reproductive and respiratory syndrome virus at individual farm level – An economic disease model. *Preventive Veterinary Medicine*. 2017;142:16-29.

2. Dee SA et al. Survival of viral pathogens in animal feed ingredients under transboundary shipping models. *PLOS ONE*. 2018;14(3):e0214529.

3. Schumacher et al. Evaluation of the minimum infectious dose of porcine epidemic diarrhea virus in virus-inoculated feed. Am J Vet Res. 2016;77(10):1108-13.

Table 1: Effects of Vigilex™ feed ingredient on reducing feed viral contamination of PRRSV following an *in vitro* time course.

Variable	NC[1]	PC[2]	Vigilex™ feed ingredient[3], %		
			0.2	0.4	0.6
0 hr genomic copies/mL	0	1.21×10^7	3.09×10^6	9.41×10^5	1.19×10^6
24 hr genomic copies/mL[4,5]	0	3.05×10^6	1.11×10^6	2.55×10^5	1.60×10^5

1 NC = non-inoculated commercial sow feed with no product added.
2 PC = PRRSV (10^5 $TCID_{50}$/mL) inoculated commercial sow feed with no product added.
3 Vigilex™ feed ingredient was added to commercial sow feed at each respective dose and feed was inoculated with PRRSV (10^5 $TCID_{50}$/mL).
4 Linear slope reduction, $P < 0.001$.
5 Quadratic slope reduction, $P = 0.183$.

Speculating trends in domestic pork consumption during a global swine health crisis

D. D. Boler[1], PhD; J. R. Dunkelberger[1], PhD; J. M Eggert[1], PhD; E.F. Knol[2], PhD
[1]Topigs Norsvin USA, Burnsville, Minnesota; [2]Topigs Norsvin Research Center B.V., Beuningen, The Netherlands

Introduction

African swine fever (ASF) is a viral disease that is impacting the health of pigs but does not affect the health of people. There is no public health threat or food safety concern due to ASF. Still, it is a concern that will impact the US pork industry in a multitude of ways. The World Organization for Animal Health considers ASF to be a trade limiting foreign animal disease. This means that even if a country does not have pigs infected with ASF, like the US, they are still impacted due to trade implications. The US are working diligently to prevent ASF from entering the country. In 2007, ASF, spread from Africa to the Republic of Georgia. Recently, the virus spread to Southeast Asia and parts of Europe. Now, the virus is currently active in 25 countries, and has implications for approximately 75% of the world's pig population.

The concern

In the last year, China reported nearly a 40% decrease in the number of live pigs. Some estimates predict the reduction may exceed 50% by the end of 2019. To put into perspective the importance of the Chinese pork industry on the global market, it is important to know the scale of the Chinese pork industry. China produces 98% of its pork internally. That means they only import 2% of the pork consumed there. However, that 2% represents approximately 20% of global pork trade. The projected 50% swine herd reduction in China, coupled with a greater than 30% reduction in Vietnam, equates to 14% reduction in global pork supply. These numbers do not include pigs and pork implicated in North and South Korea, Cambodia, or the Philippines.

Pork (40.4%) represents the greatest percentage of global meat consumption, followed by chicken (32.4%), beef (21.8%), and mutton and goat (5.3%). The law of demand states that when the price of a good increases, the quantity demanded of that good will decrease. Because of the reduction in global pork supply, pork prices are expected to increase. In the short run, the pork industry will rely on frozen storage reserves to satisfy the global demand for pork. Inevitably, however, frozen storage reserves will diminish, and pork prices should increase. In fact, pork prices in China have doubled in 2019.[1] Even though demand for meat, regardless of species, is generally less sensitive to price changes, there is still a potential for consumers to redirect their purchasing decisions to chicken rather than pork. This is phenomenon is referred to as cross price elasticity of demand and measures the demand responsiveness of a particular good following a price change of a related good. Pork and chicken are considered "substitute products" meaning that an increase in the price of one good can lead to an increase in demand for a rival product. This raises the question: as the price of pork increases, will the demand for pork decrease?

Already, projections in the production increase for other meat animal proteins has ranged from 1% to 10%. Another predictor of demand is income elasticity of demand. Meat is generally considered a "normal" good meaning that as consumer income increases, more money is spent on that good. In cases of a normal goods, quality (meeting consumer expectations) becomes especially important. Consumer expectations for meat are largely based on eating experience traits like tenderness and juiciness. Because of this, some experts have expressed a necessity to focus on eating experience as a means of sustaining pork demand during the period of depressed global pork supply. This has created a need for swine genetic companies, such as Topigs Norsvin, to emphasize meat quality, in addition to historical production indicators such as growth rate and feed efficiency.

Conclusion

During this period of health crisis, it becomes especially important to maintain meat quality and deliver pork to consumers that exceeds expectations for tenderness, juiciness, and flavor rather than focusing exclusively on production traits in order to satiate global supply. Eventually, the health crisis will subside and pork supply will normalize. It is imperative for long term benefit of the swine industry that meat quality not be sacrificed in the short term to maintain confidence and acceptability of US pork during the global health crisis.

Reference

1. Strak,, J. 2019. Quarterly update: The global pig price cycle moves up. Available at: pigprogress.net Accessed 12 November 2019.

DFM Pak®, a novel combination of *Bacillus* strains selected for corn fiber utilization and biofilm activity, shows beneficial impacts on environmental and storage characteristics of manure

J. Lee[1], MS; E. Galbraith[2], MS; J. Spencer[1], PhD; S. Son[2], MBA.; M. King[2], PhD
[1]United Animal Health, Sheridan, Indiana; [2]Microbial Discovery Group, Franklin, Wisconsin

Introduction

Including a *Bacillus*-based, direct-fed microbial (DFM) in swine diets has historically been effective in changing the composition of manure, influencing cleaning characteristics,[1] pumpability and ammonia emissions.[2] However, previously conducted studies were performed primarily with traditional corn/soybean meal diets and fail to reflect the high amounts of corn fiber found in modern diets containing dried distillers grains with solubles (DDGS). DDGS have changed the properties of manure in the pen and in storage, resulting in increased ammonia production, increased long chain fatty acid (LCFA) production, impacting manure pH, and creating challenges with solids degradation, manure foaming and crusting. It is well documented that *Bacillus* strains are genetically diverse and show wide variation in intrinsic physiological capabilities. Therefore, selecting *Bacillus* strains with digestive activities on modern diet components is a positive strategy for developing contemporary DFM solutions. In this study, proprietary *Bacillus* strains sourced from high performing animals were evaluated for their ability to thrive with modern diet substrates and their effect on environmental, safety and sanitation challenges caused by manure from modern, high corn fiber diets.

Materials and methods

A combination of *in vitro* lab testing and field case studies from animal or municipal waste containment facilities were utilized to support the *Bacillus* strain selection process.

Corn fiber and long chain fatty acid utilization: *Bacillus* isolates were screened for their digestive capabilities of DDGS or fatty acids which are abundant in DDGS. Samples of minimal medium containing DDGS or each fatty acid were inoculated with *Bacillus* and incubated for 72 hours at 25°C or 48 hours at 37°C for DDGS or LCFA experiments and monitored for growth.

pH and ammonia reduction: Manure collected from deep pit swine facilities were divided and inoculated with various combinations and dose rates of *Bacillus* strains. Inoculated samples were incubated and monitored for pH and total ammonia nitrogen (TAN) at 24h, 48h and 72h.

H_2S reduction: Poultry litter was collected into separate containers and replicate containers were inoculated with a *Bacillus* strain combination or remained uninoculated (control). 3L of water was added to each container, litter was thoroughly mixed, and containers were sealed with a modified lid to allow for Drager-Tube® attachments. H_2S was measured at 48h,120h and 140h.

Crust, solids and biofilm reduction: Field evaluations were conducted at multiple agricultural lagoons and municipal wastewater containment facilities to evaluate the effect of specific *Bacillus* strains on the physical properties of manure in storage.

Results and discussion

Of all strains evaluated, DFM Pak® *Bacillus* strains exhibited superior growth and digestive activity in the presence of DDGS, as well as utilization of foam producing LCFA. Furthermore, a reduction in manure pH was observed and NH_3 emissions were decreased by 24% and 19% when compared to the control and a competitor product, respectively. H_2S emissions were reduced by 58% compared to the control. A reduction in manure solids, crust formation, liquification of existing crust, reduced biofilm build up and odor were observed in agricultural lagoons and/or municipal wastewater field evaluations. Feeding *Bacillus* DFM strains capable of utilizing substrates found in contemporary swine manure can provide a mechanism for influencing manure composition and show promise for improving downstream environmental and storage challenges associated with modern manure management.

References

1. Davis, M.E., T. Parrot, D.C. Brown, B.Z. de Rodas, J.B. Johnson, C.V. Maxwell, and T. Rehberger. (2008). Effect of a Bacillus-based direct fed microbial feed supplement on growth performance and pen cleaning characteristics of growing-finishing pigs. J. Anim. Sci. 86:1459.

2. Petersen, S.T., (2010). The potential ability of swine nutrition to influence environmental factors positively. J. Anim. Sci. 88: E95-E101.

Meta-analysis of porcine circovirus type 2 (PCV2) vaccines used in growing pigs

M. A. Mellencamp[1], PhD; B. Poulsen Nautrup[2], DVM, PhD; I. Van Vlaenderen[3], BS
[1]Zoetis, Parsippany, New Jersey; [2]EAH-Consulting, Aachen, Germany; [3]CHESS, Bonheiden, Belgium

Introduction

Porcine circovirus type 2 (PCV2) vaccination is one of the most commonly used interventions in growing pigs. The first commercial PCV2a vaccines, which were licensed in 2006, showed decreased morbidity and improved performance in vaccinated herds.[1] However, PCV2 disease pressures are changing and recent identification of PCV2a/PCV2b and PCV2a/PCV2d in clinically ill pigs emphasizes the need for new vaccines with broader antigenic coverage. Fostera Gold® PCV MH (Zoetis) is a new vaccine with two genotypes: PCV2a and PCV2b. The aim of this study was to use meta-analytic techniques to compare the effects of Fostera Gold PCV MH with commercial PCV2a vaccines on average daily gain (ADG), mortality, and full value pigs.

Materials and methods

Fostera Gold PCV MH, Circumvent® PCV-M G2 (Merck Animal Health) and Ingelvac CircoFLEX® MycoFLEX® (Boehringer Ingelheim Vetmedica) were investigated in pigs that were either naturally exposed to environmental PCV2 challenge (n = 4 trials) or experimentally exposed to PCV2d with *Mycoplasma hyopneumoniae* or porcine reproductive and respiratory syndrome (PRRS) virus as cofactors (n = 2 trials). Natural challenge studies were performed in large production systems in the US Southeast and Midwest. Experimental trials were conducted in the Midwest under field conditions with a dual challenge to mimic disease patterns seen on many farms. Hedges'g standardized mean differences were used to compare effects of vaccination on ADG (total n = 9,379 pigs). Risk ratio differences were used to compare vaccines for mortality (total n = 9,715 pigs), full value (total n = 9,715) and cull (total n = 8,989) pigs. Meta-analyses were conducted using Comprehensive Meta-Analysis v. 2.2 (Biostat, Englewood, NJ). Random-effects model was used for pooled analyses, and mixed-effects model was used for subgroup analyses. $P < 0.05$ was considered statistically significant.

Results and discussion

Although all 6 studies were unpublished and provided by Zoetis, heterogeneity was comparably low, as estimated by Q- and I-squared statistics, even when pooling all comparisons. This is likely because the meta-analyses were based on differences and

Figure 1: Forest plot showing the effects of vaccination with Fostera Gold PCV MH, Circumvent PCV-M G2 or CircoFLEX-MycoFLEX on average daily gain from placement to end of finisher for experimental (EXP) and environmental (ENV) PCV2 challenge studies.

Group by ENV / EXP	Study name	Hedges's g	*P*-value
ENV	17PPTB10-01-01	0.023	0.700
ENV	17PPTBIO-01-02	-0.056	0.163
ENV	18PPTBIO-01-04	0.019	0.660
ENV	18PPTBIO-01-08	0.037	0.441
ENV		-0.001	0.959
EXP	16PRGB10-01-01 a	0.442	0.079
EXP	16PRGBIO-01-01 b	0.346	0.167
EXP	16PRGBIO-01-01 c	-0.396	0.110
EXP	16PRGBIO-01-01 d	-0.491	0.048
EXP	16PRGBIO-01-01 e	0.285	0.247
EXP	16PRGBIO-01-01 f	0.191	0.436
EXP	18PRGBIO-01-03	-0.132	0.014
EXP		0.014	0.908
Overall		-0.001	0.977

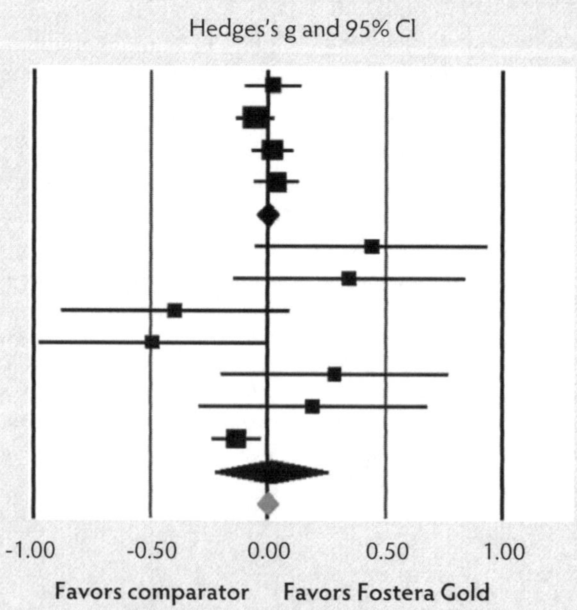

risk ratios that than absolute values, which adjusts for different baselines.[2] Despite the small number of studies, and lack of independently-run studies, analysis showed that the results were robust and there was no evidence of selection bias. Results showed that Fostera Gold PCV MH was equally effective as comparator vaccines for overall ADG ($P = 0.997$) as well as for nursery ($P = 0.323$) and finisher ($P = 0.788$) phases (Figure 1). Similarly, Fostera Gold PCV MH was equally effective as comparator vaccines for overall mortality ($P = 0.825$), as well as nursery ($P = 0.531$) and finisher ($P = 0.467$) mortality. Fostera Gold PCV MH-vaccinated pigs had the same probability of reaching full market value as comparator vaccines ($P = 0.885$), and there was no difference in risk of becoming a cull pig ($P = 0.190$). These results demonstrate that Fostera Gold PCV MH with PCV2a and PCV2b improved ADG, reduced mortality and increased the number of full market value pigs as well as competitor vaccines under field and experimental challenge conditions.

® Registered trademark of Zoetis
® Registered trademark of Merck Animal Health
® Registered trademark of Boehringer Ingelheim Vetmedica

References

1. Karuppannan AK et al. Porcine circovirus type 2 (PCV2) vaccines in the context of current molecular epidemiology. Viruses 2017; 9:99.

2. Kristensen CS et al. A meta-analysis comparing the effect of PCV2 vaccines on average daily weight gain and mortality rate in pigs from weaning to slaughter. Prev Vet Med 2011; 98:250.

Challenges and opportunities in modern swine veterinary education

G. Patterson[1], VMD, MPH, DACVPM; L. Karriker[2], DVM, MS, DACVPM; G. Almond[3], DVM, PhD; G. Althouse[4], BS, MS, DVM, PhD; A. Bowman[5], MS, DVM, PhD, ACVPM; M. Pieters[6], DVM, PhD; A. Ramirez[7], DVM, PhD, ACVPM; P. Zhitnitskiy[6], DVM, MSpVM

[1]Lincoln Memorial University College of Veterinary Medicine, Harrogate, Tennessee; [2]Swine Medicine Education Center, Iowa State University College of Veterinary Medicine, Ames, Iowa; [3]North Carolina State University Veterinary Medicine, Raleigh, North Carolina; [4]University of Pennsylvania, School of Veterinary Medicine, Philadelphia, Pennsylvania; [5]The Ohio State University College of Veterinary Medicine, Columbus, Ohio; [6]University of Minnesota College of Veterinary Medicine, St. Paul, Minnesota; [7]Iowa State University College of Veterinary Medicine, Ames, Iowa

Introduction

The importance of swine medicine to a modern veterinary education cannot be overstated. With pork being the most widely consumed animal protein in the world, swine production is intimately tied to billions of human lives as a source of food, income, and well-being. The US swine industry itself produces an estimated 2.2 million metric tons of pork and pork products on an annual basis, worth approximately $23.4 billion in output. Nearly 26% of all pork and pork products produced in the US is exported to other countries, and the industry directly supports an estimated 550,000 jobs within the US. Millions of acres of high quality farmland throughout the US are dedicated to growing corn and soybeans to be fed to pigs to convert into pork products (NPPC). Additionally, pigs play critical roles in public health, through their involvement in the transmission of zoonotic diseases such as influenza and serving as ideal comparative species for the research and treatment of human diseases.

Additionally, food supply systems in the US have undergone rapid and extensive change in the last twenty years. The complexity of animal protein production systems and the variety of considerations that must influence veterinary decisions has increased exponentially. Societal and consumer influences and expectations of the system have evolved to include emphasis on sustainability, environmental stewardship, animal welfare and reducing public health risk. This is occurring in an environment of shrinking financial returns at the farm level, increased feed and energy costs, rising global demand for safe and consistent meat products, and rapid information exchange among stakeholders, activists and consumers. As a consequence, swine veterinarians find themselves in a diverse, rapidly shifting, and highly demanding environment. Disease diagnosis and treatment is no longer an insular activity, but rather an integrated component of the management complex in dynamic food supply systems. A single pig, harvested at 265 pounds live weight, with a 70% carcass yield generates 371 meals consisting of 8 ounce portions. On a farm site with 5,000 animals destined for market, there are 1.8 million meals. These can be impacted by a single treatment decision.

Meanwhile, the process of curriculum refinement represents a continuous challenge for colleges and schools of veterinary medicine in North America. On top of maintaining the high standards needed for their accreditation, schools must also continuously modify their curricula in order to ensure that their graduates are well prepared to integrate into modern veterinary practice. This requires frequent scrutiny of topics, techniques, and clinical skills that are taught to future veterinarians during their education and to evaluate which of these are most applicable to the realities of current veterinary practice. As new scientific discoveries and technologies become available and animal industry standards evolve, it is inevitable that topics that once were considered mainstays of veterinary education are not allocated the same priorities and resources that they once were and are slowly pushed toward the curricular fringes before being phased out entirely.

For many years now, such has been the fate of topics related to swine production medicine when evaluated by veterinary curriculum committees. As curriculums increasingly represent the needs of a companion animal practitioner, food animal topics, in particular swine, receive less attention. This has resulted in not only a reduced number of dedicated swine hours in modern teaching curricula but also a dwindling number of veterinary colleges where students can gain substantial hands-on swine experience. The results of this process are predictable, as one only needs to observe an AASV meeting to see that apart from the occasional alumnus from a non-traditional swine school, the majority of the next generation of swine veterinarians hail from a select handful of veterinary colleges.

Given the increased consolidation and integration of the modern swine industry, it is unsurprising that the veterinary colleges that retain substantial swine-teaching resources are located in states that have significant swine populations. While such concentration of resources has resulted in a few noteworthy veterinary colleges where a world-class swine-focused veterinary curriculum can be offered, the contrasting reality is that the vast majority of newly minted veterinarians in North America have little to no swine exposure during their time in school. This reality has wide implications for how modern swine production is perceived and understood by the greater veterinary community, prompting the need to seriously consider the role of swine education as a crucial part of the twenty-first century veterinary curricula.

Statement of problem

The polarization and consolidation of swine veterinary education resources should be concerning to the AASV membership for multiple reasons. Firstly, with limited swine opportunities available at a majority of North American veterinary colleges, it is predictable that many potential swine veterinarians are far less likely to gain the exposure, experiences and professional connections necessary to create an interest in and successfully join the swine industry. While not impossible to overcome, as there are many examples of veterinarians within the AASV membership that have done so, this lack of exposure and opportunity creates a significant barrier to the swine-interested or undecided veterinary student. The extent to which the swine industry is dispossessed of potential intellectual talent as a result of this void in exposure is unknowable. It is intuitive that consolidation means that programmatic or budget threats at remaining programs would present a larger risk for the profession.

Secondly, in the face of increasing complexity and accountability for the efficiency of academic budgets, veterinary training programs have recognized the need for new models of training food supply veterinarians. Veterinary colleges face a unique challenge when they consider local legislative funding and regional demographics in light of the national need for advancing modern food production training programs. At most veterinary colleges, student interest in food supply training represents a minority of the students that are served by the college. Relative to the number of students that are served, the cost of maintaining teaching herds and broad food animal faculty expertise can be significant. As a consequence, food animal training is appealing as an area to de-emphasize. The loss of swine faculty expertise outside of the major swine producing states has already been observed. The sum of these local decisions is a smaller national pool of swine medicine training opportunities, even in the face of growing food animal populations, increasing challenges to food security and public health. In fact, the proportion of the veterinary curriculum committed to food supply veterinary medicine is in the process of being downsized substantially in several CVMs. Specifically, in reference to swine production medicine, the argument that CVMs should substantially down size the time and resources allocated to swine medicine in the curriculum is increasingly common and perhaps justifiable since only 4% of the questions on the North American Veterinary Licensing Exam are about pigs (vs 17% bovine, 3% small ruminant). When there is less swine production medicine in the curriculum, there is a perception of less need for swine focused faculty and eventually those positions are eliminated and that informed voice of influence is lost on the faculty at that institution. Despite a clear national need for robust food animal medicine training programs, most states do not have the necessary faculty expertise or case load to provide a comprehensive swine medicine program. Without a significant portion of the swine industry in their state, most colleges will not be motivated to expand or even maintain swine training opportunities. Among the accredited colleges of veterinary medicine (30) and colleges with a "letter of reasonable assurance" (2), only 5 are located in states with more than 5% of the US pig crop: Iowa, North Carolina, Minnesota, Illinois and Indiana. It is reasonable that the remaining schools and states would have insufficient industry influence to convince state legislatures to invest in expensive modernizing of infrastructure technology and maintenance of larger herds to teach swine medicine in the necessary context.

Fifty of Iowa's ninety-nine counties are part of a Critical or High Priority Veterinary Shortage Situation (IA 191 –IA 197) that identify swine as a species that must be served. These issues are exacerbated when transboundary or foreign animal disease concerns are heightened. However, Veterinary Shortage Situations listing swine as a "must serve" species span many states including the top hog producing states of North Carolina, Minnesota, Illinois, Indiana and Ohio. Shortage Situations in need of swine veterinary expertise also exist in a diversity of states including Louisiana, Maine, Tennessee, Alaska, Montana, Nebraska, Texas, Pennsylvania, New Jersey and others.

Third, it is critically important that all veterinarians, no matter their specialty, have at least some fundamental knowledge of modern swine production practices and management. Veterinarians, being widely regarded as animal health experts among the public, media and lawmakers, are often called upon to set standards for food safety and animal well-being. In order to fairly offer insight or perspective on these topics as they relate to pork production, it is important that veterinarians have at least some baseline knowledge of how the modern industry functions. With the understanding that for the majority of veterinarians in practice, the extent of their swine knowledge is limited to a few core-curriculum lecture hours, it is important that the AASV consider some of the key messages that are important for all veterinarians to comprehend. It is also important for swine veterinarians to be recognized as experts in practical application of population medicine including great expertise in biosecurity, disease elimination and eradication programs, as well as the applications of diagnostic programs for herd health. All of these skills have a prominent value for all veterinarians; especially those interested in population medicine, including shelter medicine. It is these messages that will serve as the basis of mutual trust and respect between swine veterinarians and their colleagues throughout the profession.

Finally, if swine medicine continues to consolidate as it is trending, students will either be excluded from access to swine medicine and clinical training or a new paradigm about funding will need to be adopted. Successful models in several species, including swine, where students travel to locations that have clinical training available exist at their home schools. These programs have solved concerns about safety, professional liability and the coordination of academic schedules. What remains to be solved is the lack of sustainable of funding. It is clear from these models collectively that college and university budget structures limit or prohibit the transfer of tuition funds when students seek additional clinical training at other schools in the form of preceptorships/internships or participation in a clinical course. As budgets tighten, parochial questions arise at host schools about how local resources are being used to cover the actual costs of students that are hosted.

Moving forward

In the interest of ensuring that swine education remains a fundamental piece of modern veterinary education, the AASV Collegiate Activities Committee has established a working group to investigate the state of swine veterinary education in the United States. By pooling our resources, the group has identified five broad themes that will, with the help of the organization's membership, further develop into action: 1) development/refinement of day one core competencies in swine for veterinarians, 2) preservation of swine clinical training opportunities for students, 3) recruitment of students into swine medicine, 4) preservation/expansion of swine faculty FTEs and 5) increase awareness of current educational situation among non-academic members of the profession. Preliminary action items include:

1. Work with AASV membership to build relationships with veterinary colleges so that swine-interested students can more easily be exposed to and find swine related opportunities.
2. Develop a "Roadmap to become a Swine Veterinarian" and present it to 1st year vet students. Provide contacts and opportunities to get involved.
3. Support Operation Mainstreet talks at all schools and consider other methods of "distance mentorship" to schools with a limited swine program.
4. Develop a core message for all vets to understand and identify how to get core messaging out to all schools.
5. Call attention to and prioritize the creation of opportunities for students with no previous swine exposure. Easy to keep swine focused students seeking opportunities (they know they don't know). Hard for students who don't know about swine to realize what they don't know (don't know they don't know).
6. Enlist the broader membership in developing a sustainable mechanism to keep this moving forward, which must include consideration of dedicated funding.
7. Identify challenges to recruiting swine faculty and preserving their ability to focus on clinical teaching.
8. Develop a core swine curriculum that could be "best practice" for veterinary curricula and could then be shared with the AAVMC and other stakeholders.
9. Investigate and describe the processes in place for swine interested students to complete externships at other universities with larger swine programs?
10. Summarize and serve as a contact point for funding sources to support travel and tuition for students coming from a different program?

How can AASV membership help?

1. Be open to providing practice experiences for swine interested students and interested students that may not plan to focus exclusively on swine but are needed as advocates. Is biosecurity becoming an easy excuse to prevent students from gaining exposure?
2. Advocate at all levels for swine medicine courses, faculty, and training programs.
3. Exert influence in your organization to keep consideration of how we develop human capital part of future planning.
4. Consider if it is worth a financial investment to ensure the future development of human capital in swine medicine.

In addition to the important technical considerations above, what is the solution to the financial challenge? It would only take $0.01/pig and sow in the US herd annually to build and maintain a complete swine medicine curriculum (introductory, advanced, clinical rotations, on-line courses) available to every interested veterinary student and veterinarian in the United States free of charge including room and board at the host schools of clinical courses (67.948 m market hogs + 6.349 m breeding herd = 74.297 m swine = $742,970 annual budget). Every school could access (and contribute to according to desire and expertise) a current, multi-media curriculum that incorporates new information ***as it is developed*** (processing fluids, tracheal swabs, etc.). This could be accomplished without adding to student debt loads (which would perhaps give swine medicine some competitive advantage when students consider specialization). Considering the cost to produce a pig and the economic engine that pork production is for much of the nation, is there a more effective, multi-etiology health investment for $0.01/head?

Summary and call to action

The AASV membership has long valued education and explicitly includes "increase the knowledge of swine veterinarians," "promote the development and availability of resources that enhance the effectiveness of professional activities" and "mentor students, encouraging life-long careers as swine veterinarians" in our mission statement. Moving forward in the modern realities of veterinary education, we need to recognize the elimination and consolidation of swine medicine training opportunities and be deliberate in our efforts to preserve access to swine medicine training for all veterinary students.

A vision for the future of global markets

Steve Meyer

No paper submitted

Reset to positive

Elizabeth "Betsy" Charles, DVM, MA
Veterinary Leadership Institute

Introduction

One does not have to look far to find all kinds of evidence to support the fact that veterinary medicine is facing a mental health crisis. Headlines about the rate of suicide among veterinarians, clients making the practice of veterinary medicine challenging, burnout, moral stress and compassion fatigue are not difficult to find. In fact, the mental health crisis within veterinary medicine is mainstream national news with coverage in Time magazine,[1] The Washington Post,[2] NPR,[3] and many other news outlets. The recent focus on the multiple issues veterinarians must deal with in practice has brought much needed awareness to the problem. Though awareness is key and an excellent first step, little has been done to implement solutions that can help eliminate the struggle so many veterinarians face each day. The purpose of this paper is to offer a framework for thinking about how to move the veterinary profession toward well-being by resetting to positive.

Discussion

Much research in the social sciences has been done with regard to how to create exceptional work environments that minimize stress, encourage collaboration, and build skill sets that allow for effective communication that minimizes conflict among team members as well as with clients. Using the work of Dr. Brene Brown,[4] Dr. Shawn Achor[5] and Simon Sinek,[6] we can build a framework veterinarians can use to implement strategies within practice in order to reformat the profession so veterinarians can thrive. This framework includes, but is not limited to, the following components: embrace vulnerability as strength, implement practical tools of positive psychology (connection, gratitude, reflection), and understand the importance of viewing practice as an infinite game.

Embrace vulnerability as strength

The culture of medicine is one where failure is frowned upon, asking for help is considered a sign of weakness, and showing emotion indicates a lack of personal control. Essentially, the medical professions discourage any kind of vulnerability. All of these things can lead medical professionals, whether human or veterinary, to adopt unhealthy coping strategies that ultimately result in mental health issues, burnout, moral stress and compassion fatigue. Dr. Brown's work in the areas of vulnerability and shame has shown that the idea that vulnerability as weakness is actually a myth.[4] Her research has shown that when we are willing to talk about our failures, understand and share our emotions and ask for what we need to be successful, our teams are stronger and more effective. Her research debunks the myth and illustrates exactly the opposite, that vulnerability is actually strength.

Gratitude, reflection and connection

The field of positive psychology has given us many tools for creating successful work environments. Dr. Achor's work within organizations has shown us that gratitude is a powerful driver of success.[5] When employees focus on what they are thankful for, it changes the lens through which they look at other situations. Instead of being influenced by the rampant negativity and stress in the world and the workplace, gratitude helps team members reframe to positive. The skill of reflection on the positive things that happen during the day is an important tool for resetting one's outlook to positive which is correlated with better performance. The importance of connection is critical in creating work environments that are positive and people can thrive. Incorporating these skill sets is a great first step toward creating positive work environments where veterinarians can thrive.

Understand the importance of veterinary medicine as an infinite game

Significant changes in company structures over the last century have shifted the emphasis away from long-term viability toward short-term gains. Metrics are now focused almost solely on shareholder value instead of employee and customer satisfaction and whether the organization is benefitting society in a positive way.[6] Unfortunately, this short-term mindset has had a devastating impact on employee health and satisfaction. Veterinary medicine, especially in an age of corporatization, is not immune to this shift. Simon Sinek describes this change as a movement away from viewing business as an infinite game, a game that has no end and its goal is long tern organizational success that outlives the players toward viewing business as a finite game where the game is over when someone wins and someone loses. In this new approach to business, those who work for the company suffer. We must begin to implement leadership strategies that empower veterinarians to play the infinite game.

Conclusion

If veterinary medicine is to survive as a profession and veterinarians are to thrive, we must reset our mindset to positive using the strategies outlined in this paper. Veterinarians can redefine the culture of veterinary medicine so it says failure is an opportunity to learn, understanding our emotions is a critical skill for thriving, asking for help makes the profession stronger and vulnerability is strength. We focus on gratitude, practice the skill of reflection and connect with each other. When we begin to do these things, it is much easier to prioritize a long-term infinite game.

References

1. Veterinarians Face Unique Issues That Make Suicide One of the Profession's Big Worries, Time Magazine, 2019. Available at: https://time.com/5670965/veterinarian-suicide-help/. Accessed 29 November 2019.

2. Suicides Among Veterinarians Become a Growing Problem, The Washington Post, 2019. Available at: https://www.washingtonpost.com/national/health-science/suicides-among-veterinarians-has-become-a-growing-problem/2019/01/18/0f58df7a-f35b-11e8-80d0-f7e1948d55f4_story.html. Accessed 29 November 2019.

3. Veterinarians Are Killing Themselves. An Online Group is There to Listen and Help. NPR, 2019. Available at: https://www.npr.org/2019/09/07/757822004/veterinarians-are-killing-themselves-an-online-group-is-there-to-listen-and-help. Accessed 29 November 2019.

4. Brown, B. (2018) Dare to Lead. New York, NY: Random House.

5. Achor, S. (2010) The Happiness Advantage. New York, NY: Random House.

6. Sinek, S. (2019) The Infinite Game. Portfolio/Penguin.

Field experiences managing PRRS through control, elimination, and prevention

Kylie Glisson, DVM
The HANOR Company, Enid, Oklahoma

Introduction

The pork industry has been impacted by porcine reproductive and respiratory syndrome virus (PRRSV) for over three decades, but there are still many unanswered questions about the disease. Arguably, the most contested is, should we control or eliminate PRRS virus? This question is highly case dependent, however there are many tools that are useful in deciding to control or eliminate: outbreak investigations, biosecurity surveys and risk evaluations, and creating biological and economic performance models. Correct methodology should target your client's goal, whether that is to produce a PRRS negative pig, to introduce PRRS negative gilts, or to reduce the prevalence of the disease within their operation. In the course of this pursuit, focusing on biosecurity and employee training is critical to success.

Background

The HANOR Company has approximately 100,000 sows in 7 states and markets 1.7 million pigs/year. For HANOR, 2015 was a challenging year and it completely altered how the company handled new virus introductions. The Oklahoma system broke with a 1-8-4 PRRSV strain while the NC system broke with a 1-7-4 PRRSV strain. HANOR continued to see clinical signs in the sow farm and in grow-finish, in conjunction with a PED break in the Oklahoma system. The result of these outbreaks caused HANOR to shift strategy to elimination of both PED and wild-type PRRSV from the sow system. The wild-type PRRSV elimination program for the Oklahoma system, consisting of 48,000 sows, 36,000 GDU spaces, and 90,000 nursery spaces, started week 37 of 2018.

Evaluating risk events

The first step was to evaluate and decide whether to control and stabilize the herd(s) for PRRS virus or eliminate it. The 2018 Swine Health Information Center project comparing low and high incidence PRRS farms highlighted four risk event areas that separated the high and low incidence farms:

1. Monthly event frequency
2. Down time requirements
3. Swine density
4. Operational connection with other sites

Ultimately, the 2018 Swine Health Information Center project suggested that the goal in managing PRRSV should be to reduce the number of risk events by reducing the number of pig movements and people entering into the farm.[1]

A follow up 2018 project developed a scoring system using multi-criteria decision analyses to assess the vulnerability in biosecurity practices that could result in a PRRS virus introduction. Swine movements, transmission of PRRS by air or water, and people movements were identified as significant risk events for PRRS virus introduction.[2]

HANOR, decided to focus biosecurity upgrades on people entry, supply entry, and mortality removal. In addition, based on these studies, it was apparent that the company needed to make a change to the replacement animal program.

Biosecurity upgrades

People entry

While it is understood that people entry is a risk when considering PRRSV introduction and control, employees have to enter the site daily. Therefore, to mitigate this risk, HANOR invested in building bench entries in sow farms in Oklahoma. After building the benches, it was important to train all staff and visitors on how to properly use the facility. The HANOR veterinarian team created informative posters on how to enter using the bench and had them posted prominently in all facilities.

Supply entry

As with people, bringing supplies to any unit is unavoidable. HANOR elected to build fog rooms and to purchase UV chambers for sow farms. Both of these enhancements also required further employee training and development of training materials.

Mortality removal

Monthly event frequency has been identified as a major difference between low and high PRRS incidence farms. Within the frequent events, rendering was the most significant difference where 64.3% of high PRRS incidence farms employed rendering services, while only 31% of low incidence farms used rendering.[1] To eliminate this risk, HANOR converted the entire Oklahoma system to composting. The Oklahoma operations also built designated mortality areas on the farm with a clear line of separation to eliminate cross traffic between the clean side, where on farm employees deposit the mortality, and the dirty side, where internal mortality management employees enter to pick up the mortality. For the upgrades to mortality program, employee education focused on the risks that mortality pose and defining line of separation.

Gilt source

When deciding on control or elimination a practitioner must consider the goal for their operation. This could include the ability to introduce naïve non-vaccinated gilts, vaccinated gilts, or to expose gilts to wild-type PRRS virus for a stabilization program. Prior to the decision to eliminate PRRSV, HANOR used on-site internal multiplication as a gilt source. Although this approach was successful at stabilizing PRRSV, it was not ideal for eliminating PRRS virus. HANOR was concerned with the risk of creating persistently infected animals and identified this as an issue with eradicating PRRS. Consequently, the HANOR company elected to build two new sow farms to act as a negative gilt source.

Testing schedule and MCREBEL procedures

The introduction of processing fluids dramatically changed the sensitivity for detecting PRRS virus in neonatal pigs. Testing for PRRSV in the sow farm should be designed with one of two goals, either to control or eliminate. Furthermore, for control, a practitioner must determine whether the goal is to produce a PRRS negative pig, or to return to baseline performance? These questions are important when deciding how many litters are included in each processing fluid tube and when determining how to pool samples. As the goal for HANOR was to eliminate PRRS virus, over time the process became more strict on how many litters were allowed per tube to further increase sensitivity. Ultimately, HANOR collected processing fluid from all litters, 15 litters per tube, with no pooling until the farm completed the elimination program.

Testing methods in wean pigs can include several options, such as blood, blood swabs, family oral fluids, and oral swabs. During the elimination program employed by HANOR, all sampling protocols were evaluated. HANOR personnel were unable to identify one sampling protocol to be more successful. However, HANOR gravitated towards blood swabs or oral swabs for ease of collection.

MCREBEL procedures are critical in any control or elimination program for PRRS virus. MCREBEL is necessary to stop the transmission of PRRS from the environment, sows, and wean pigs to the neonatal pig. Unfortunately, MCREBEL procedures often result in increased pre-wean mortality. To help reduce pre-wean mortality, HANOR elected to allow farms to operate under normal farrowing procedures until processing fluids were consistently negative for PRRS virus. Once processing fluids returned negative results, it was assumed that pigs were being born negative and therefore, if they were positive at weaning, they acquired the infection from the farrowing environment. After four consecutive weeks of negative results in processing fluids, sow farms enforced strict MCREBEL and started to collect wean pig samples. Employee training was again critical as MCREBEL procedures are inconvenient and often require extra labor. Employees must understand why they are being asked to perform these protocols. HANOR noted that giving employees a basic understanding of how PRRS virus was being transmitted was necessary for effective management of herd health protocols.

Keys to success

While the HANOR system has made substantial investments in upgrading biosecurity, those upgrades would not be valuable without employee training. Regular communication and follow-up training with all farm staff is the single most critical portion of effective biosecurity protocols for PRRS, or any virus. Many biosecurity practices increase the workload of the employee and explaining why the steps are necessary to prevent infection tends to increase compliance.

It is easy to get let down and discouraged when there is a positive processing fluid after weeks of negative results. Communication with staff on the sensitivity of processing fluids is critical to develop appropriate expectations from employees and senior level stakeholders.

Given that time is an essential component for all control and elimination projects, it is important to remain fluid and dynamic in the approach to achieving a client's goals. During the control or elimination program, a practitioner will learn new information and find new challenges. Be prepared to alter a course of action and understand that each farm comes with its own set of unique challenges.

References

1. Linhares et al. Description of biosecurity aspects of herds with low or high PRRS incidence and comparison within and between production systems. Swine Health Information Center. 2018.

2. Silva et al. Development and validation of a scoring system to assess the relative vulnerability of swine breeding herds to the introduction of PRRS virus. Preventative Veterinary Medicine. 2018; 160, 116-122.

Which route of exposure is best for gilt acclimatization to *Mycoplasma hyopneumoniae*?

AP Poeta Silva[1], DVM, MS; C. Alonso[2], DVM, PhD; B. Arruda[1], DVM, PhD; D. Linhares[1], DVM MBA PhD; Paul Yeske[3], DVM; L. Gimenez-Lirola[1], BS, PhD; R. Derscheid[1], DVM, PhD; L. Karricker[1], DVM, MS; J. Zimmerman[1], DVM, PhD; M. Clavijo[1], DVM, PhD

[1]Veterinary Diagnostic and Population Animal Medicine, Iowa State University, Ames, Iowa; [2]Barcelona, Spain; [3]Swine Vet Center, Minneapolis, Minnesota

Introduction

Mycoplasma hyopneumoniae (MHP) is a chronic respiratory infection that impacts pig production by easing co-infections, increasing use of antimicrobials, and decreaing growth performance. Since most replacement gilts are MHP-naïve, once introduced into endemically infected farms will represent the most probable source for maintance of the bacteria in the breeding herd and spread to offspring. Minimizing vertical transmission from gilts and young sows to piglets would significantly reduce disease downstream and this is achieved by successful and early gilt acclimation. Due to the slow transmission and lenghtly shedding period of MHP, exposure than hinges on pig-to-pig transmission can be inneficient and not timely enough to secure reduced shedding of gilts at farrowing. Therefore, MHP intentional exposure strategies during acclimation provides the opportunity for early, effective and repeteaded MHP exposure in time for gilts to develop a protective immune response and stop shedding MHP before their first farrowing. The purpose of this study was to compare three different MHP exposure strategies, intratracheal, intransal and aerosol, in terms of production of infection and the establishment of immunity.

Materials and methods

Six-week-old gilts (n = 78) were randomized by weight to four groups: 1) aerosol exposure (cold fogger; n = 24; 2 pens); 2) intranasal exposure (mucosal device, n = 24; 2 pens); 3) intratracheal (feeding tube, n = 24; 2 pens); 4) negative control (Friss media intratracheally, n = 6; 1 pen). Lung homogenate with *MHP* strain 232 was used in exposed groups. Body weight (3 measures per gilt), serum (8 measures per gilt), tracheal (7 measures per gilt), and pen-based oral fluid samples (26 measures per pen) were collected. At 49th day post-exposure (DPE), lung tissue was evaluated for gross lesions and histopathology (H&E).

Results and discussion

Tracheal, serum and oral fluid samples remained negative until 49 DPE in control group. MHP infection was achieved by all three routes in all gilts (22/22 at 7 DPE intratracheal, 24/24 at 14 DPE intranasal, and 24/24 at 21 DPE aeorosol). Seroconversion were firstly observed in the intratracheal (14 DPE), followed by intranasal and aerosol groups (21 DPE). MHP DNA detection in pen-based oral fluid samples were incositent over DPE (8/26 intransal, 10/26 intratracheal and aerosol). No differences in the overall amount of MHP PCR Cts were observed in exposed groups, but earlier and higher S/P responses were observed in intratrachel ($P < 0.05$). All MHP exposures significantly decreased body weight gain compared to negative control group ($P < 0.05$). Gross and/or histopathologic lung lesions were observed in 17/22, 19/24 and 19/24 gilts from the intratracheal, intranasal and aerosol groups.

All three routes of exposure succeeded in MHP infection. However more cost for producers, earliest detection to confirm exposure was reached by tracheal swab (< 7 DPE). Intratracheal exposure produced earliest DNA and antibody detection but this approach requires skilled personnel and causes more stress for pigs and people. This study pointed out that intranasal and aerosol exposures, which are more animal welfare friendly, resulted in similar performance as intratracheal, and they can be considered alternatives to achieve a successful gilt acclimation.

Are we there yet? The future of bacterial pathogen surveillance

Maria J. Clavijo[1,2], DVM, PhD; Ganwu Li[1], MS, PhD; Bailey Arruda[1], DVM, PhD; A.W Tucker[2,3], DVM, PhD; Marcelo Gottschalk[4], DVM, PhD; Orhan Sahin[1], DVM, MS, PhD; Beatriz Martínez-López[5], DVM, MPVM, PhD; Nubia Macedo[1], DVM, MS, PhD; Jean Paul Cano[2], DVM, PhD; Rodger Main[1], DVM, PhD
[1]Veterinary Diagnostic and Production Animal Medicine, Iowa State University, Ames, Iowa; [2]PIC®, Hendersonville, Tennesee; [3]University of Cambridge, Department of Veterinary Medicine, Cambridge, United Kingdom; [4]University of Montreal, Department of Pathology and Microbiology, College of Veterinary Medicine, St-Hyacinthe, Quebec, Canada; [5]University of California Davis, School of Veterinary Medicine, Davis, California

Introduction

Molecular epidemiological analysis allows us to study the distribution and determinants of diseases with the aid of molecular microbiology methods.[1] DNA-based typing methods have been widely employed by the swine industry for well over a decade to increase our understanding on sources and routes of infection, detecting virulent strains, evaluating the effectiveness of control measures and overall improving our surveillance methods for these agents. For example, sequencing of the hemagglutinin gene (HA) of Influenza A virus or the ORF5 gene in PRRSV is carried out on a daily basis in diagnostic settings to characterize these pathogens and provide useful information for their prevention and control.[2]

Swine endemic systemic bacterial agents, such as *Streptococcus suis, Glaesserella parasuis, Mycoplasma. hyorhinis, Mycoplasma hyosynoviae, Actinobacillus. suis* and *Streptococcus equisimilis* remain a global challenge for swine production by reducing productivity, impacting welfare and leading to an over-dependence on antimicrobials. For these agents, while a myriad of DNA-based typing tools have been developed, they are based on information from a handful of genes, thus missing a more complete assessment of their genomes. This is in contrast with viruses, where single gene sequencing is informative given their small genome size. While whole genome sequencing (WGS) of bacterial isolates is commonplace in research settings, these technologies have yet to make a leap into routine surveillance programs for swine systems. This is mainly due to the high cost of WGS and the complexity of the data outputs. However, the continued decrease in sequencing cost and improved efficiencies in bioinformatic pipelines have allowed these technologies to be more accessible for bacterial typing in veterinary diagnostic laboratories. *The objective of this initiative is to develop a routine surveillance platform for systemic endemic bacterial agents of pigs.*

Materials and methods

Flows, isolates and genomic analysis

The current WGS database is comprised of a set of contemporary clinical isolates (2015-2019) from the following species: 283 *S suis,* 50 *G parasuis,* 31 *S equisimilis,* 20 *A suis,* 6 *M hyorhinis* and 6 *M hyosynoviae.* Isolates originate from various tissue/sample types depending on the bacterial species. The isolates have been obtained from a total of 45 different swine systems (20 from the top 30 US pork producers) retrospectively or prospectively by the attending veterinarian, who follows a specific case definition for each bacterial species. For each isolate, the following epidemiological data are recorded: date, state, GPS coordinates, flow, site/farm, age, organ, predominant clinical sign and clinical outcome. Samples and isolates are submitted to the Iowa State University Veterinary Diagnostic Laboratory (ISU-VDL), and bacteriological culture is attempted. The broth microdilution method was used to determine the minimum inhibitory concentrations (MIC) for all isolates, expect for *M hyorhinis* and *M hyosynoviae.* Genomic DNA is extracted using the ChargeSwitch gDNA mini bacteria kit (Life Technologies, Carlsbad, CA). Multiplex whole genome libraries are prepared using the Nextera XT DNA library preparation kit (Illumina, San Diego, CA) with dual indexing. Bacterial genomes are sequenced using Illumina MiSeq platform (Illumina) with 250x2 read length in the Next Generation Sequencing (NGS) Unit of the ISU-VDL. Pre-processed reads are assembled using SPAdes Genome Assembler version 3.11.1-Linux. From the WGS data, the following information is extracted from each isolate (when available/applicable): phylogeny (pan or core genome), serotype, multi-locus sequence typing (MLST) profile, putative virulence genes and antimicrobial resistance genes (AMR) profiles. Phylogenetic trees based on the core or pan genomes are generated for each bacterial species.

Applications

There are three main areas that can be refined by using information obtained from WGS pipelines:

1. **Pig flow management**
   a. **Determining ideal gilt replacement sources:** with the objective of minimizing negative health consequences when introducing breeding stock, it is essential to select a candidate source that harbors highly similar endemic strains as the receiving herd. These pipelines have allowed for the in-depth characterization of endemic agents from multiplier farms, and such information has allowed swine practitioners to determine the most appropriate source for their replacement gilts.

   b. **Strategic commingling:** within swine systems multiple-source commingling is common practice. Information gained from WGS pipelines can help determine which flows are more compatible for mixing depending on their profile of endemic agents.

2. **Prevention**
   a. **Selection of vaccine candidates:** Vaccine candidate selection for agents such as, *S suis* and *G parasuis* has mostly relied on data obtained from serotyping and MLST. WGS pipelines can expand the information by adding relevant information on phylogeny, putative virulence factors and AMR genes. This information is coupled with essential epidemiological and clinical information from the strain, allowing for the identification of the most appropriate vaccine candidate. In addition, information generated can also help to assess the effectiveness of these products. In multiple cases completed so far, information derived from these platforms have revealed that strains included in the current vaccine did not cluster with the clinical isolates obtained from the farms, evidencing that improvements in the vaccine selection process is warranted.
   b. **Antimicrobial resistance profile:** The identification of AMR genes present in a strain and correlation with clinical and susceptibility data can potentially help to identify the most suitable antimicrobial to use on farms and can help track changes in prevalence of resistance genes and clones.

3. **Surveillance:**
   The use of WGS can allow for vast improvements in on-going surveillance programs for these agents. For example, a baseline of current strains in a flow can be obtained and routine incorporation of clinical strains can help identify the emergence of new pathotypes or changes in prevalence. It can also aid epidemiological investigations by identifying potential sources of introduction, risk factors for disease occurrence and novel transmission routes.

Whole genome sequencing for swine systemic bacterial pathogens is now offered as a routine service at the ISU VDL. The genomic databases are maintained with routine submission of cases and the most frequent and relevant contemporary pathotypes are included within each dendrogram to support epidemiological analysis. Future efforts are targeted towards the development of an online swine bacterial visualization, analysis and sharing platform that can be utilized by swine practitioners and diagnostic laboratories to improve control and prevention programs for these endemic pathogens, similar to that is available for Influenza A virus at ISU VDL website or for PRRS and other swine viruses using the UC Davis Disease BioPortal.[2,3] Furthermore, continued inclusion of diverse strains from clinical and non-clinical cases will aid in the identification of global genetic markers, thus further improving rapid and accurate diagnosis for the disease-associated endemic bacterial strains in the field.

References

1. Riley LW, Blanton R. Advances in Molecular Epidemiology of Infectious Diseases: definitions, approaches, and scope of the field. Microbiology spectrum. 2018 Nov;6(6).

2. Zeller, M. A., Anderson, T. K., Walia, R. R., Vincent, A. L., & Gauger, P. C. (2018). ISU FLUture: a veterinary diagnostic laboratory web-based platform to monitor the temporal genetic patterns of Influenza A virus in swine. *BMC bioinformatics*, *19*(1), 397. (data retrieved 03 Dec, 2019).

3. Martinez-Lopez Beatriz, Polson Dale, Lowe Erin, Whedbee Zack, Main Rodger (2019). Accelerating data sharing, visualization and analysis to support efficient disease monitoring and more real-time decision making in the swine industry using the Disease BioPortal platform.Frontiers in Veterinary Science 00102 DOI=10.3389/conf.fvets.2019.05.00102.

A practitioner's perspective on managing bacterial pathogens

B. Leuwerke, DVM, MS
Swine Vet Center, St. Peter, Minnesota

Introduction

There are a variety of bacterial pathogens that cause health complications and mortality in both adult and growing pig herds. Three of these pathogens that we commonly focus on as part of a herd health management plan, are *Streptococcus suis, Haemophilus parasuis* and *Actinobacillus suis*. Transmission of these organisms is dynamic with the most frequent being vertical movement from sow to piglet during lactation. It's likely that the more shedding (colonization) that takes place prior to weaning the more significant disease challenges will be for groups of growing pigs post-weaning. Although, elimination of these bacteria from herds may one day be reality, much of the focus on control involves reducing colonization and in turn limiting the level of challenge post-wean. Within herds, areas of focus for bacterial pathogen control include: Gilt sourcing and acclimation, stabilizing or eliminating viral and Mycoplasma hyopneumoniae co-infections, following good sanitation procedures, using autogenous vaccines and strategic placement of antibiotics.

Gilt sourcing and acclimation

Herds that can maintain the same source of gilts over time tend to have the best stability to bacterial pathogens. This isn't to say that herds maintaining a consistent source of replacement animals don't have challenges. When we avoid the introduction of new bacterial serotypes or strains with live animal introductions, we reduce the bacterial disease challenges those herds have. What is the best gilt acclimation for bacterial pathogens? Ideally, replacement gilts should be brought in at an age that allows multiple months of exposure and acclimation to herd pathogens before those replacements reach breeding age. Acclimation involves co-mingling the replacement gilts with live animals from the herd so that transmission, colonization and an active immune response can take place. Just because gilts are brought into a herd at a young age doesn't mean automatic acclimation occurs. Shedding of these bacteria is sporadic, so allowing multiple animals (younger gilts from the destination herd or off-quality weaned pigs as examples) to be introduced to replacements, increases the chance successful transmission of these pathogens occurs.

Stabilizing or eliminating viral and *Mycoplasma hyopneumoniae*

There is no doubt, herds experiencing viral (PRRSV, influenza, porcine circovirus) or *Mycoplasma hyopneumoniae* instability will have increased risk of disease caused by bacterial pathogens. Stabilization or elimination of these infectious cofactors reduces bacterial challenges both within the sow herd and nursery-grow/finish populations.

Good sanitation practices

Even though these bacterial pathogens aren't often found surviving well outside the pig, poor sanitation will lead to transmission of bacterial pathogens. Processing (castration and tail-docking) equipment has direct contact with incisions and should be routinely cleaned during use and completely washed and sanitized after use. Good barn clean-up, disinfection and drying are all important parts of any disease control including bacterial organisms.

Vaccination

Autogenous vaccines have been beneficial in reducing the impact of bacterial disease. Autogenous vaccines are used in a couple different ways: first, to reduce shedding by gilts and sows in lactation and second, to build active immunity in growing pigs post wean. For pathogens that have multiple serotypes, like *Streptococcus suis* and *Haemophilus parasuis* it's important that ongoing monitoring be carried out. Cross-protection may be limited so when new serotypes are found, often due to the introduction of a new gilt source, updates or additions to current autogenous should be made. A common vaccination program for a herd would include one or two doses to gilts during development and acclimation, a pre-farrow vaccination as an aid to reduce shedding during lactation and vaccination of pigs pre or post-wean depending on when the downstream challenges are occurring. The vaccine adjuvant does matter so working with autogenous vaccine companies on the correct adjuvant is also important.

Antibiotic strategies

Antibiotics still are an important tool for treating bacterial diseases but aren't as effective if the above areas aren't covered also as part of a bacterial control program. Overused, especially in young pigs, we risk that pigs never become colonized and are set-up for worse disease. Often bacterial challenges occur with a similar timing so many antibiotic treatments can be timed closely with the onset of disease.

Conclusion

Clearly, there isn't a simple, straightforward answer to controlling bacterial pathogens. Diseases caused by *Streptococcus suis, Haemophilus parasuis* and *Actinobacillus suis* can cause tremendous losses in both adult populations and growing pigs. Our goal has been to control the colonization of pigs with these pathogens through gilt acclimation and autogenous vaccination as well as manage exposure by practicing good sanitation and appropriate use of antibiotics. Alone these practices aren't highly effective but when put together can do a lot to control disease caused by bacterial pathgens.

Effect of different natural planned exposure (NPE) strategies on the shedding of rotavirus A, B, C, and pre-wean morbidity and mortality in an endemic sow farm

A. Farkas[1], DVM; L. Ochoa[1,2], PhD; L. Greiner[1,3], MS, PhD
[1]Carthage Veterinary Service, Ltd.; [2] Furst McNess; [3] Iowa State University

Introduction

One of the most common viral infectious causes of neonatal piglet diarrhea is rotavirus (RV) and every pig experiences a rotavirus infection, if not multiple infections within its lifetime. RV infections in piglets cause high mortality and significant intestinal damage, resulting in reduced weight gain and absorption of nutrients, translating to increased days to market and economic losses to the pig producer. Due to the limited efficacy of current commercial vaccines, natural planned exposure (NPE) is used to expose the dams prior to farrowing with the objective of boosting passive immunity via colostrum to piglets. Bottlenecks that can impact the success of NPE consist of understanding what type of NPE material to use (fresh scours or amplified RV ice cubes); concentration of NPE; frequency and timing of NPE; route of administration; and the absence of serological essays to measure the immune response and antibody levels in response to NPE protocols. This study evaluated different timing of second NPE of pregnant dams (first NPE occurring 8 weeks pre-farrow) to reduce the impact of rotavirus scours in piglets during the suckling period.

Materials and methods

The study was performed on a commercial sow farm in West Central Illinois. A total of forty pregnant sows and forty pregnant gilts from the same breed week were assigned to the study. NPE was administered via oronasal route using frozen aliquoted isolates (farm specific RVA and RVC) that were collected from previously infected piglets on farm 8 weeks pre-farrowing to all dams on test. Dams were randomly allocated within parity to 1 of 4 second NPE treatments: Treatment 1) NPE 4 weeks pre-farrow; Treatment 2) NPE 2 weeks pre-farrow; Treatment 3) NPE 3 weeks pre-farrow; and Treatment 4) NPE 1week pre-farrow). Upon entry to the farrowing rooms, dams were placed blocked by farrowing dates within treatments. Piglets were individually ear-tagged within 24 hours of farrowing. Cross-fostering was allowed after colostrum intake within treatment group. The number of pigs scouring in each litter was recorded daily throughout the study. Rectal fecal samples were collected from dams 3 weeks pre-farrow, upon entry to farrowing crates, day of farrowing (day 1), and on days 2, 3, and 17 of study. On days 1, 2, 3, and 17, rectal swabs were collected from each pig, and were analyzed in pools of 4 swabs. Fecal samples and swabs were sent to the Iowa State Veterinary Diagnostic Lab to test for RV by PCR. Pre-weaning morbidity and mortality, scours incidence and shedding rate was measured. Statistical analysis was done using a $\chi 2$ rank-based transformation using the Proc Rank procedure of SAS; transformed data was then be analyzed using the MIXED procedure of SAS.

Results

Results are shown in Table 1. Statistically, there were no treatment differences for pre-wean morbidity and mortality. Treatment 4 showed the lowest percentage of pigs scouring ($P = 0.001$) throughout the study. There were no statistical differences between treatments for percentage of litters with scours ($P = 0.21$), however, Treatment 4 had a lower number of scouring litters throughout the study than the other treatments. With regards to the shedding rate, there was no statistical difference between treatments ($P = 0.69$), however, Treatment 2 had the highest incidence of litters shedding rotavirus within the first 3 days of age. The majority of litters (≥ 93%) were shedding rotavirus at weaning regardless of treatment. There was no indication that timing of NPE changed the rotavirus profile in pigs.

Discussion and conclusion

Despite lacking statistically significant differences in morbidity and mortality, these results suggest that lactogenic immunity of sows can be manipulated by the use of different NPE strategies. In this study, exposing sows to NPE material 1 week before farrowing reduced the negative impact of rotavirus as seen in the lower number of pigs with scours, the number of litters scouring, and the number of litters shedding RV at 3 d of age. In this study, giving a second NPE 1 week before farrowing improved piglet and litter conditions during the suckling period. Disease severity and mortality is greatly influenced by levels of maternal antibodies (nursing pigs), pig age, weaning age and the level of virus challenge exposure, therefore a multifactorial program approach is needed. Further research is needed to understand how to stimulate lactogenic immunity of the sow.

Table 1: The effect of a second NPE to pregnant sows on the pre-wean morbidity and mortality, presence of scours, and shedding of RV in piglets.

Treatment:	1	2	3	4	
Time of second NPE (weeks pre-farrow)[1]:	**4**	**2**	**3**	**1**	***P*-value**
Pre-wean morbidity and mortality, % of pigs.	19.9	18.8	14.4	17.3	0.48
3 days of age measurements:					
Pigs scouring, % of pigs.	13.5	23.1	18.5	9.3	0.001
Litters scouring, % of litters.	71.4	66.7	60.0	42.1	0.21
Litters shedding RV, % of litters.	46.2	71.4	62.5	50.0	0.69

[1] First NPE was administered 8 weeks pre-farrow.

Ileitis prevention and elimination: we have the tools!

Nathan L. Winkelman, DVM
Swine Services Unlimited, Inc.; Rice, Minnesota

The US swine industry has the information and tools today to try to either prevent ileitis in sow herds or pig-flows and to also eliminate it from currently positive systems. As swine herds eliminate economically important disease pathogens such as PRRS, PEDV, and M hyo, ileitis should be the next economically important endemic bacteria to tackle. Veterinarians must first understand and be convinced of the significant cost of both clinical and subclinical ileitis and be willing to monitor pig-flows to be sure neither form of the disease exists.

Our current control programs with vaccine or antibiotic pulses, although very cost effective, do not prevent subclinical infection and are never as good as no infection at all. In randomized controlled *Lawsonia intracellularis* challenge trials with vaccines and/or antibiotics, this is proven time and again. The strict negative control groups (non-challenged, non-treated) almost always have significantly better performance than the vaccinated or antibiotic groups (treated, challenged).

Four key elements. There are four key elements ("tools") swine vets must understand and apply to successfully prevent and/or eliminate ileitis from a production site or system:

1. **Antibiotics** – Carbadox (Mecadox®) is a unique antimicrobial compound that can be used to both prevent *Lawsonia intracellularis* colonization and also eliminate *Lawsonia* carriers when fed for at least 14 days at 50 g/ton.[1,2]
2. ***Lawsonia* immunity** – *Lawsonia intracellularis* is a unique intracellular bacterium that lends itself to elimination because once exposed, a pig obtains potential life-long immunity. Pigs are resistant to re-infection after initial exposure to low or high doses of the bacteria.[3] This knowledge can be useful to eliminate both subclinical and clinical infection in sow herds, replacement gilts, and grow-finish pig-flows with controlled live exposure followed by an antibiotic pulse.[4]
3. **Diagnostic tools for analysis** – Ante-mortem fecal PCRs and IPMA serology both are useful to determine the ileitis status of a herd and differentiate ileitis vaccination vs infection titers.
4. **Biosecurity** – Needless to say, only herds with very good biosecurity will remain free of ileitis. Ileitis is not known to be spread through the air or the feed, so it should actually be easier to prevent than some other diseases.

Preventing Ileitis with Carbadox (Mecadox®). "Prevention" here means not having subclinical or clinical *L intracellularis* exposure at all, i.e. negative fecal or oral fluid PCRs.

Although it does not have FDA approval for ileitis treatment or control, it has been thoroughly studied in randomized controlled ileitis challenge studies. It is the only known antimicrobial compound that when fed at 50 g/ton for 14 days eliminates *Lawsonia* from clinical or subclinical carriers.[1] It is also the only antimicrobial that prevents colonization in pigs challenged with low or high doses of *Lawsonia intracellularis*.[2] There are five FDA approved antibiotics in the feed that effectively control and treat clinical ileitis. However, they typically will not prevent *Lawsonia* colonization or eliminate the carrier pig.

Wean-to-finish example – In a large pig-flow with 8 separate sites of 4,000 to 10,000 pigs, potential subclinical weaned carriers are treated with Mecadox 50 (and OTC 400) in the Nursery 1 and Nursery 2 diets for about 17 days post-arrival. This gives us confidence *L intracellularis* is not "leaking" from the sow farm in subclinical carriers. In three of those sites, we do not vaccinate for ileitis because they have excellent biosecurity and historically have not had any bacteria present. In the other sites we do vaccinate for ileitis to get enhanced immunity in case of exposure to *Lawsonia*. All sites are monitored via fecals and/or oral fluid PCRs at 120-280 lbs B.W. for the presence of *Lawsonia* to be sure the programs are working.[5]

Sow herd prevention – Similarly, preventing ileitis from establishing itself in a new or repopulated sow site on subclinical carrier gilts could be achieved by having Mecadox at 50 g/ton for at least 14 days in the gilt diets prior to moving them into the new or cleaned facilities. Once a negative sow herd is established, replacement gilts will need to be monitored via fecal PCRs to be sure subclinical carrier gilts are not present.

Establishing ileitis free sow herds and/or monitoring replacement gilts to ensure a negative *Lawsonia* status could and should be done in breeding stock multipliers. Once a sow herd is established negative it will be providing a weaned pig free of *Lawsonia*. Subsequent downstream pig flows can also be established free of ileitis. This could be very beneficial in antibiotic-free or NAE (No antibiotics Ever) niche markets.

Elimination in grow-finish pig-flows with controlled *Lawsonia* exposure and antibiotics.

In all finish pig-flows, the first goal for ileitis control programs is to not ever have any clinical signs of ileitis. Clinical ileitis will cost $10-$22 per pig in lost performance, mortalities, and treatment costs depending on the clinical severity.[5] The second important goal is to not have any subclinical ileitis – i.e. no positive *Lawsonia* fecals in the grow-finish stage. This can be accomplished with the DLI® Immunity Program.

DLI® (Diluted *Lawsonia intracellularis*) Immunity is a patented method of administering a specific controlled dose of oral live *Lawsonia* for enhanced immune stimulation compared to commercial vaccine. The program requires an autogenous source, VCPR oversight, and is followed up with an antibiotic pulse two weeks post-DLI.® When done properly, the pigs are completely resistance to colonization and subsequent bacteria fecal shedding found in both subclinical and clinical ileitis.[3,4]

DLI® Immunity live exposure is used to eliminate both subclinical and clinical ileitis and not have any fecal shedding at all or any clinical signs of diarrhea, or positive fecal color scores in grow-finish pigs. It is a good tool to use in problematic ileitis barns or sites.

The Enterisol® ileitis vaccine has been shown to reduce clinical disease and to increase weight gain. However, while the natural infection with *L intracellularis* can provided complete protection against re-infection, this has not been achieved by this vaccine. Cell-mediated immune responses are likely mediators of protective immunity against *L intracellularis*, with CD8+ effector cells, and CD4+, CD8+ positive memory T cells as main contributors to the antigen-specific IFN-y production.[6]

Elimination in replacement gilts with natural exposure. A likely source of *Lawsonia* exposure into sow herds is from asymptomatic subclinical replacement gilts. These gilts should be monitored via fecal PCR, oral fluids, and/or serology to be sure they are not shedding *Lawsonia* prior to entry into a sow unit. Occasional PHE (porcine hemorrhagic enteropathy) outbreaks occur in ileitis vaccinated gilts and also rarely in multiple parity sows. This could be prevented with the DLI® Immunity Program in replacement gilts followed by antibiotics 14 days later. This system works well in gilt isolation barns with AI/AO pig-flow. Eventually the entire sow herd would have complete *Lawsonia* protection as normal sow cull rates are replaced with ileitis immuno-protected gilts. This example has been used in a "high-health" commercial client for about 8 years which supplies ileitis negative pigs for research trial work.[5]

Antemortem diagnostic monitoring

***Lawsonia* PCRs** are useful in a number of applications:

- Ensure negative fecal shedding in weaned pigs from negative sow herds
- Monitor AIAO gilt pools to be sure they are no longer shedding after post-DLI® immunity and an antibiotic pulse (e.g. Carbadox, tiamulin, etc.)
- Monitor finish pigs post-DLI® to confirm no subclinical (or clinical ileitis) is present
- Determine the PCR presence and quantity of *Lawsonia* in fresh fecal samples or oral fluids

In general, fecal PCRs are a very useful diagnostic tool to be sure any mild diarrhea or fecal blood is *Lawsonia* positive or negative. If the ct value at the U of MN is ≤ 31 ct 40, the sample is associated with PPE intestinal lesions and economic performance loss.[7]

Oral fluid PCRs - Oral fluids are also useful to determine the presence and quantity of *Lawsonia* in the environment. The ct values are interpreted the same way as with fecal PCRs and have similar or slightly lower number values than with pooled fresh fecal PCRs. One advantage is that more pigs are usually chewing on the ropes so sensitivity may be increased.

Serology – IgG *Lawsonia* antibodies measure prior exposure to either field infection or *Lawsonia* vaccine. Know what to expect for *Lawsonia* titers relative to the antibody test. For example:

- Unvaccinated, negative herds will not have any antibody titers
- Porcilis® and DLI Immunity® will have ≤ 480 titers on IPMA
- Enterisol[R] oral vaccine does not illicit an ELISA titer response at all and an inconsistent IPMA titer response[8]
- Field exposure or ileitis outbreak titers are usually ≥ 480 and much higher (up to 15,000) depending upon the disease insult.

More on biosecurity. Knowledge regarding the epidemiology and transmission of *Lawsonia intracellularis* is imperative to prevent it from contaminating a clean site. Although there are still plenty of knowledge gaps, we know transmission is primarily through asymptomatic carrier pigs, and fomites (people, trailers, equipment, etc.) in contact with contaminated feces. *Lawsonia* can live in manure and pits for at least two weeks, probably longer. Cleaning and disinfection procedures should include a detergent step, hot water wash, and disinfection. It has been proven that insects (flies, cockroaches) and rodents are mechanical and biological vectors for *Lawsonia* transmission, respectively.[9]

Summary. It is always more economical to prevent and eliminate diseases than to treat and/or control them. The swine industry now has both the knowledge and tools available to prevent and/or eliminate *Lawsonia intracellularis* from pig-flows. Swine veterinarians need to begin to establish *Lawsonia* free pig-flows and systems and monitor their success to motivate others to do the same.

References

1. Kinsley K, Winkelman et al; Elimination of *Lawsonia intracellularis* from Pigs Fed Carbadox (Mecadox®); AASV 2006 pp 157-161.

2. Winkelman N., Gebhart C., Deen J., Gramin B.; Correlation of Clinical Signs and Diagnostic Indicators of Ileitis in Pigs Fed Carbadox; AASV 2003 pp 135-139.

3. Collins A.M., Love R.J.; Re-challenge of Pigs Following Recovery From Proliferative Enteropathy; Vet Micro 120 (2007) pp 381-386.

4. Winkelman N., Mueller A., Domanque R.; Effectiveness of DLI® (diluted *Lawsonia intracellularis*) "Vaccine" Followed by Denagard LC, Aivlosin®, or Mecadox® in Finishing Pigs Challenged with a High Dose *Lawsonia intracellularis* Mucosal Homogenate AASV 2014 pp 273-278.

5. Winkelman N., personal communication 2019.

6. Ulla Riber, Peter M.H. Heegaard, Henriette Cordes, Marie Stahl, Tim Kare Jensen, Gregers Jungersen; Vaccine of pigs with attenuated *Lawsonia intracellularis* induced acute phase protein responses and primed cell-mediated immunity without reduction in bacterial shedding after challenge; Vaccine 33 (2015) pp 156-162.

7. Winkelman N.; Subclinical Ileitis: Diagnostic Monitoring, R^2 and Economics; 2018 AASV pp 488-493.

8. Vasquez E.; Comparison of IPMA and Commercial Blocking ELISA for *Lawsonia intracellularis*; AASV 2018 pp.

9. Friedman M., Bednar V., Klimes J., et al 2008; Lett Appl Microbiol 47: pp 117-121.

Batch farrowing for disease control

Clayton Johnson, DVM
Carthage Veterinary Service, LTD, Carthage, Illinois

Population disease management for endemic pathogens requires the successful transition of individuals within the population from "Susceptible" to "Infected" and eventually "Resistant" status. In populations with well controlled endemic disease, the population moves as a collective group of individuals through the "Infected" status with minimal clinical signs, minimal performance impacts and minimal duration of infection. In populations with poorly controlled endemic disease, subpopulations of individuals exist in "Susceptible", "Infected" and "Resistant" categories simultaneously. The subpopulation of "Susceptible" individuals serve as a reservoir of naive hosts for the pathogen to replicate within. The population of "Infected" individuals serve as a reservoir of active pathogen replication, sustaining infectious pathogen in the population. The population of "Resistant" individuals may end up being overwhelmed by the high levels of pathogen in the environment, reentering the "Infected" category as their immunity wanes.

Principles of All-In-All-Out (AIAO) production were built with population disease management in mind. We place a group of contemporary animals all with the same individual health status and transition them from "Infected" to "Resistant", minimizing disease impacts through appropriate vaccination and medication strategies. This approach works well in growing pig populations, but its application in piglet populations at Breed-to-Wean (BTW) farms is often overlooked. BTW farms commonly operate breeding barns on a "Continuous Breeding" cycle which results in a "Continuous Farrowing" cycle. When farms farrow every day, by definition we have "Susceptible", "Infected" and "Resistant" individuals within the BTW piglet population, a recipe for poorly controlled endemic disease.

Continuous Breeding and Farrowing were facilitated by the industry transition to multi-site production several decades ago. On Farrow to Finish (FTF) farms, nursery space was always a bottleneck that prevented continuous weaning which prevented continuous breeding which prevented continuous farrowing. As the US industry moved to multi-site production, we were no longer limited by the available nursery space to regularly wean. Producers quickly recognized the value of their new system – breeding and farrowing every day decreases the percentage of time a farrowing crate isn't being used to lactate a litter of piglets. As such, both throughput (number of farrowing's per unit time) and wean age improved in Continuous Breeding and Farrowing BTW farms.

Endemic pathogen challenges have changed dramatically since the transition to multi-site production and continuous farrowing and breeding. Influenza and Rotavirus are tremendous examples of pathogens that have evolved from their previously less diverse and pathogenic ancestors. Novel and emerging pathogens such as PCV2, PRRSV and PEDV have further challenged endemic disease control. Simply put, the consequences of poorly controlled endemic disease challenges are much greater in 2018 than they were several decades ago. Batch Farrowing remains as a tool for producers currently challenged with poor endemic disease management. Batch Farrowing has its negatives – Less efficient use of lactation crates, lower wean age, lower throughput (lower PSY) and the surges in labor needs during breeding, farrowing and processing all must be considered. The negatives are generally stagnant, their impact was the same several decades ago as it is today. The positives of batch farrowing in terms of improved endemic disease control have grown more valuable with each novel and emerging pathogen challenge. Across the world, producers are asking themselves – even with less pigs produced from the farrowing house can I produce more pounds of pork out of my production system utilizing Batch Farrowing?

There are four common models of Batch Farrowing, 2 models that are AIAO in the farrowing house and 2 models that split the farrowing house into 2 units. Each model lets the producer choose whether they prefer to take more impact to throughput (less farrowing events per unit time) or wean age (more farrowings but younger wean pigs). There is no single model that is best for all producers, the best model for each producer must be determined independently.

5/4 and 4/5 Batch models are AIAO models that have 5 groups of sows farrowing every 4 weeks or 4 groups of sows farrowing every 5 weeks, respectively. The 5/4 model maximizes throughput with a greater sacrifice in wean age. The 4/5 model maximizes wean age with a greater sacrifice to throughput. As both models are AIAO they are generally thought of as creating the greatest health advantage.

7/3 and 10/2 Batch models split the farrowing house into 2 units and have 7 groups of sows farrowing every 3 weeks or 10 groups of sows farrowing every 2 weeks, respectively. The 10/2 model maximizes throughput with a greater sacrifice in wean age. The 7/3 model maximizes wean age with a greater sacrifice to throughput. While this model has 2 subpopulations in the farrowing house at any given time, its impact on health may or may not be different than the AIAO Batch models. For producers with two independent farrowing houses, this model may give the same health advantages as AIAO models with the increased flexibility of more frequent breeding weeks.

Batch Farrowing provides unique challenges for Producers and Veterinarians to overcome. Natural variation in a gilts 1st estrus age requires either the utilization of hormones to place gilts into an appropriate breeding week or a common gilt pool for multiple farms. Recycles similarly have a natural variation in their recycle dates and must be transferred to another breeding group using hormones or simply culled at the time of recycle identification, increasing the farms annual gilt replacement rate. Nurse sows are challenging to create in a Batch system and consideration should be given to the use of supplemental milk decks as a method of creating nurse sows. Farrowing and Breeding Dates create a surge in labor needs relative to continuous production. Batch farms must shift labor from other areas of the farm is they are unable to rotate their best labor across multiple facilities. There is an important theme to recognize when managing Batch-specific challenges, farms who can share common sources of gilts and/or employees have a unique advantage due to their ability to apply these resources to the farm currently needing animals ready to breed or employees with a specific skill set.

The importance of farrowing house sanitation in a batch production system cannot be underestimated. While there is a tremendous amount of work to do, labor to wash the farrowing house is generally not a rate limiting resource. In the days immediately following weaning, there are no piglets to care for, no sows to farrow and minimal if any breeding. As such, the time required to clean, disinfect and set up the farrowing house for the next batch is entirely regulated by the quality and quantity of washing equipment at the farm. Producers should aim to run multiple power washers simultaneously and stagger employee start/stop times to allow as close to 24 hour washing as possible. Routine maintenance checks of the washing equipment and supplies should occur 5-7 days prior to weaning, allowing any needed maintenance or supplies to be in place and all systems functional the day of weaning. The faster we wash and reload the farrowing house, the more we maximize wean age and throughput, minimizing the inherent negative sow farm impacts.

In summary, Batch Farrowing improves endemic disease control by taking groups of piglets through the infection and immune development process as a group, minimizing the duration and performance consequences associated with infection. In addition to improved endemic disease control, Batch Farrowing also more easily facilitates epidemic disease elimination, particularly when combined with MEW/SEW programs. Farms that have the opportunity to share gilt sources and/or rotate specialized labor have a significant advantage toward overcoming the unique challenges associated with Batch Farrowing. To minimize the Batch impact on wean age and throughput, producers should invest in additional power washing equipment and ensure this equipment is functional prior to weaning every batch. Batch Farrowing has potential costs to consider including the initial conversion from continuous farrowing, increased non-productive days, decreased wean age, decreased litters farrowed and increased hormone usage. Producers should evaluate their projected costs and revenues under both Batch and Continuous production models. Batch Farrowing may not produce the most piglets from your farrowing house, but it may produce the lowest cost of pork from your overall production system.

African swine fever response scenarios in Europe: effective strategies for control and eradication

Timothy Paul Snider
PIC Europe, Sant Cugat del Vallès, Spain

Introduction

A paper focussed upon successful responses to African swine fever virus (ASFV) outbreaks would be incomplete without a review of the preparedness and response measures that have been developed for the region of concern. Long before ASFV entered the country of Georgia in 2007, and during its subsequent spread across Russia, eastern Europe and much of Asia, the European Union (EU) passed a series legislation that guides the legal response of member states (MS) in the case of an ASFV incursion within their territory. In addition to other supporting legislation and OIE guidelines, the core EU legislation, Council Directive 2002/60/EC, provides the main framework which guides the MS efforts to successfully control and ultimately eradicate the disease from their territory.

This paper will outline the key components of the EU's legally defined ASFV response strategy and then review how the practical implementation of this framework has unfolded in ASFV outbreaks of the Czech Republic (June 2017) and Belgium (September 2018)

Chronological history of the European Union response framework[1]

2000 - Identification and registration of swine premises and pigs moving from their place of origin

The underlying principle of traceability of all pig movements is a key component for the implementation of an effective outbreak response when considering movement controls and the investigation of historic pig movements. The Council decision 2000/678/EC and the subsequent Council directive 2008/71/EC of the European Commission (EC) legislate these requirements across the EU.

2002 - Framework of specific provisions for the control of ASF

Amongst several additional aspects of the response framework, the Council Directive 2002/60/EC includes provisions regarding the disease notification process and measures to be taken in cases when either domestic or feral swine populations are suspected or confirmed. As well, this legislation also outlines the determination of protection and surveillance zones at a minimum of 3 and 10 km respectively. This regulation forms the main framework of the EU response to ASF.

2003 - ASF diagnostic manual

To ensure that the confirmation of ASF is consistently performed across the EU, a uniform set of diagnostic procedures, sampling methods and criteria are outlined in Commission Decision 2003/422/EC.

2009 - Swill feeding

As a result of issues related to foot and mouth disease virus (FMDV) and bovine spongiform encephalitis (BSE), this legislation regarding the use of animal by-products and animal derived products not intended for human consumption (i.e. Regulation No 1069/2009) was instituted and impacts rules regarding the feeding of 'swill' to swine.

2013 – Guidelines on surveillance and control of ASFV in feral swine and preventative measures in pig holdings

A guideline document entitled SANCO/7138/2013 was developed to give guidance to MS that are suspected or confirmed to have ASFV in feral swine. The guidelines aim to reduce the risk of transmission to domestic pigs and to prevent the disease from becoming endemic in the feral swine population.

2014 - Regionalization and the movement controls

To ensure consistently implemented risk management within all affected member states, regulation 2014/178/EU was enacted. This legislation outlines the classification of affected territories as: Part I – at risk with no cases; Part II – feral swine cases only; Part III – feral and domestic swine cases; Part IV – endemic situation. As well, the legislation defines swine related commodities (i.e. 1) Live pigs, semen, ova/embryos, & porcine animal byproducts & 2) Pig meat, meat preparations, and meat products) and if they move into, out of, or within each territory type. EU regionalization maps are routinely updated and available at the following websites as a pdf (`https://ec.europa.eu/food/sites/food/files/animals/docs/ad_control-measures_asf_pl-lt-regionalisation.pdf`) or in the form of an interactive online map (`https://eu-commission.maps.arcgis.com/apps/webappviewer/index.html?id=3db65168d4ad4d829a38560d7f868ace`).

2015 – Principles and criteria for geographically defining ASF regionalization territories

To facilitate a consistent approach across all affected MS in defining the territories defined in regulation 2014/178/EU, a set of principles and criteria was outlined. SANCO/7112/2015 lists several factors that are recommended for consideration when defining the size and shape of each territory include: a) administrative divisions, natural and artificial barriers, domestic and feral swine distributions, etc.

Review of effective control and eradication strategies in the Czech Republic and Belgium

Article 15 of European Council's Directive 2002/60/EC outlines the measures to be taken by a MS when an ASFV infection is suspected or confirmed in feral swine. Article 16 outlines the requirement to supply a plan for eradication of ASFV from the MS's feral swine population. This plan must be submitted within 90 days of the start of the outbreak.

Following the feral swine outbreaks of ASFV in the Czech Republic (June 2017) and Belgium (September 2018), both countries were required by EU law to follow a specific response plan. The pre-defined response plan for an ASFV outbreak in an EU MS includes the following steps: 1) Immediate reporting to the European Commission and other member states & 2) Naming an expert group that is tasked with: i) defining the infected area (IA), ii) the measures to be carried out in the IA, iii) movement controls and iv) the eradication plan.

The outbreaks in the Czech Republic and Belgium were remarkably similar in many ways. Both affected MS were located deep within the western European region and appeared to be a result of human mediated transmission since other populations of infected feral swine were not in the immediate geographic area. The infected areas of both outbreaks and their surrounding regions had very low domestic swine populations which made their depopulation financially feasible and easily accomplished. Both affected MS have been able to prevent the spread of the infection significantly beyond the initially defined borders of the first IA and this has prevented the spread to any other domestic herds or neighboring MS.

In addition to several active measures within and surrounding the IA, both Czech Republic[2] and Belgium[3] included the following key actions: 1) restrictions on hunting feral swine within the infected zone; 2) intensive hunting around the perimeter of the control areas; 3) efforts to rapidly remove feral swine carcasses; 4) physical barriers (e.g. fences) to minimize movements out of the IA; and 5) restrictions on personnel movements within the control zone.

As of April 2019, the Czech Republic was declared negative for ASFV with no cases detected for a full 12 months. At the time of writing (i.e. November 2019), Belgium has not detected a new case since April 2019 and is well on the way towards being the second country in Western Europe to successfully eradicate ASFV from its territory.

References

1. European Union Law. All European Union legislation. Available at: https://eur-lex.europa.eu/homepage.html. Accessed 15 November 2019.

2. Czech State Veterinary Administration. https://www.svscr.cz/africky-mor-prasat-aktualni-informace/. Accessed 15 November 2019.

3. Houdart, P. ESPHM May 2019 Keynote session: 'ASF updates and review of applied measures; practical experiences.'

African swine fever "top 5" biosecurity strategies and considerations

Clayton Johnson, DVM
Carthage Veterinary Service, LTD, Carthage, Illinois

Since the initial report of African swine fever (ASF) in China in August of 2018, the global swine industry has been monitoring the predictable spread of ASF throughout Asia with great fear and anxiety. While markets outside of Asia have taken advantage of new export opportunities, swine veterinarians in the US are keenly aware of the economic disaster for our farmers that could occur if ASF is found in our swine herd. For pig producers in the US, it is advantageous to watch the spread of this devastating disease in the Chinese and Asian swine herds, and to better understand the steps necessary to stop or prevent the spread of disease.

While ASF poses a few unique risks, it's important to remember that ASF will likely transmit throughout our industry in much the same way that PRRS, PED, IAV-S, SVA and other diseases have for years. The usual culprits of transportation, mortality management and personnel entry but be managed effectively to have any hope of preventing ASF introduction into our farms. We have to assume that similar to other recent disease introduction, ASF will be present at multiple sites, geographies and production systems by the time it is diagnosed. As such, we have to treat every risk with the assumption ASF is at other farms already.

This paper is not an exhaustive list of strategies to prevent the introduction of ASF into a US pig farm, rather it is meant to highlight some of the most important risks and suggestions for management strategies. All farms and situations are unique, the strategies listed below may not be practical for each farm and as veterinarians we should always strive to work with our producers to develop creative biosecurity approaches for risks not able to be managed through common industry practices. Remember the 3 basic strategies you can apply to disease introduction risks, Exclusion, Segregation and Hygiene and Sanitation. Exclude risk factors from farm entry where possible. Segregate risk factors to decrease the number of independent farms they contact. Apply hygiene and sanitation steps including detergents, cleaning and disinfection (chemical, thermal, pH) to all items that enter a farm.

Top 5 biosecurity risks and associated strategies:

1. **Contaminated pork** – The most unique risk associated with ASF is its ability to survive in the carcass post death or harvest. Fresh, frozen and processed pork all represent a risk. We have mostly focused on imported meat being a risk for contamination but we must remember that it is highly possible that actively marketing sites may ship infected pigs to slaughter before an official ASF diagnosis is attained. Domestic pork could be a risk before we know we are infected. As such, a "No Pork Allowed" policy at all US farms should be adopted.
2. **Contaminated visitors** – The ASF naive regions of the world are becoming smaller every day, making international visitor management more important than ever. Similar to pork products, a policy of complete exclusion in which no international visitors are allowed will provide the most assurance of risk management. In cases where visitor exclusion isn't possible or practical, a 5 day downtime is recommended, although science to support this is limited. In addition to downtime, it is recommended to provide visitors with US sourced clothing and footwear to wear to the farm visit and not allow them to bring anything with them to the farm.
3. **Contaminated transport vehicles** – Transportation biosecurity is a foundational component to any biosecurity program. Fortunately, the US has some of the best transportation biosecurity capabilities in the world. The virus is susceptible to cleaning and disinfection. Multiple disinfection processes can be effective including chemical disinfection, thermal disinfection and pH manipulation. Specifically, trailers should have at least one disinfectant step applied following cleaning and prior to moving pigs. Ideally, two disinfection steps should be applied to provide redundancy in the biosecurity program of this critical risk.
4. **Contaminated mortality management equipment** – One thing we know is true, the pigs representing the index cases of ASF will likely die before a diagnosis is made. Further, these mortalities will likely be managed using normal processes. Should mortality management risk be shared through rendering, this puts a large number of sites at risk for ASF introduction through mortality management equipment. If you use rendering it is critical that a Clean/Dirty side of the rendering transfer location be developed and implemented. Areas of cross traffic at the mortality pick up area are highly likely to be contaminated and represent a large risk for introduction. Ideally, composting should be implemented if at all feasible.
5. **Contaminated feed ingredients** – It has been well documented that ASF and other foreign animal disease (FAD) pathogens pose a significant risk for introduction into the US. Management strategies include limiting the importation of feed ingredients to only products which cannot be sourced from the US as well as the use of available chemical disinfectant agents including formaldehyde based products, medium chain fatty acids (MCFA) and acidifiers. It should be noted that "garbage feeding" still occurs within the US. While there are heat treatment processes that are capable of inactivating ASF, this practice still presents significant risk to our industry and should be discouraged.

Survival and transmission of foreign animal diseases in feed

M. C. Niederwerder, DVM, PhD
Department of Diagnostic Medicine/Pathobiology, College of Veterinary Medicine,
Kansas State University, Manhattan, Kansas

Introduction

Since the incursion of porcine epidemic diarrhea virus (PEDV) into the United States in 2013, feed and feed ingredients have been recognized as a risk for transboundary spread of swine diseases. The global distribution and export market for ingredients manufactured in countries with circulating foreign animal diseases (FAD) make feed a risk for disease introduction and transmission into naïve countries. African swine fever virus (ASFV), classical swine fever virus (CSFV) and variant pseudorabies virus (PRV) are three of the most significant threats to pork production in the US and around the world. The objective of this research was to investigate feed and feed ingredients as a risk for ASFV, CSFV, and PRV introduction and transmission. This objective was targeted with a three part approach, including 1) determining the survival of FAD in feed and feed ingredients subjected to varying environmental conditions simulating transoceanic shipment, 2) investigating the probability and dose necessary for FAD transmission through the natural consumption of contaminated feed, and 3) assessing tools for mitigating the risk of FAD transmission in feed and feed ingredients.

Materials and methods

All work with pigs and viruses was performed at the Biosecurity Research Institute, a biosafety level 3 laboratory housed on the Kansas State University campus. To investigate stability of FAD through imported feed ingredients, an environmental chamber was used to simulate fluctuating temperature and humidity conditions in transoceanic shipment models. Stability of ASFV Georgia 2007, CSFV Brescia strain, and variant PRV HeN1 were investigated in 12 feeds or feed ingredients in 30-37 day transoceanic models. Viral stability was determined through quantitative PCR, virus isolation and titration, and swine bioassays. Transmission of ASFV through feed was investigated in nursery pigs through the natural consumption of small volumes of liquid and feed contaminated with specific doses of ASFV. Doses administered ranged between 10^0 $TCID_{50}$ and 10^8 $TCID_{50}$. Pigs naturally consumed liquid in gravity-fed drinkers and naturally consumed feed in stainless steel creep feeders. ASFV infection was detected after 5 days using multiple diagnostic assays. Viral decay of ASFV was characterized in 9 feed ingredients over the course of a 30-day transoceanic shipment model to estimate viral half-life with SE and CI at moderate temperature and humidity conditions. Viral titrations were performed at 1, 8, 17 and 30 days post-contamination.

Results and discussion

ASFV, CSFV and PRV were all capable of surviving in feed ingredients subjected to transoceanic environmental conditions, with CSFV surviving in fewer ingredients compared to ASFV and PRV. All 3 foreign animal diseases survived in conventional soybean meal and pork sausage casings. ASFV was transmissible through the oral route after the natural consumption of contaminated liquid and feed, with higher doses required for infection through plant-based feed. Stability of ASFV was increased in all feed ingredients when compared to viral decay in media, with the average half-life of ASFV in all feed ingredients estimated at 12.2 days. Taken together, these studies provide evidence for the potential role of feed in the transboundary spread of foreign animal diseases of swine.

References

1. Dee, S. A., F. V. Bauermann, M. C. Niederwerder, A. Singrey, T. Clement, M. de Lima, C. Long, G. Patterson, M. A. Sheahan, A. M. M. Stoian, V. Petrovan, C. K. Jones, J. De Jong, J. Ji, G. D. Spronk, L. Minion, J. Christopher-Hennings, J. J. Zimmerman, R. R. R. Rowland, E. Nelson, P. Sundberg and D. G. Diel. Survival of viral pathogens in animal feed ingredients under transboundary shipping models. *PLoS One* 2018;13(3): e0194509.

2. Niederwerder, M. C., A. M. M. Stoian, R. R. R. Rowland, S. S. Dritz, V. Petrovan, L. A. Constance, J. T. Gebhardt, M. Olcha, C. K. Jones, J. C. Woodworth, Y. Fang, J. Liang and T. J. Hefley. Infectious Dose of African Swine Fever Virus When Consumed Naturally in Liquid or Feed. *Emerg Infect Dis* 2019;25(5): 891-897.

3. Stoian, A. M. M., J. Zimmerman, J. Ji, T. J. Hefley, S. Dee, D. G. Diel, R. R. R. Rowland and M. C. Niederwerder. Half-Life of African Swine Fever Virus in Shipped Feed. *Emerg Infect Dis* 2019;25(12): 2261-2263.

Update on feed mitigation research

Scott Dee[1]; Megan Niederwerder[2]; Gil Patterson[3]; Diego Diel[4], Eric Nelson[5]
[1]Pipestone Applied Research; [2]Kansas State University; [3]Lincoln Memorial University;
[4]Cornell University, [5]South Dakota State University

Summary

Contaminated feed has been documented as a risk factor for the transboundary and domestic movement of viral pathogens, including PRRSV, ASFV and Seneca Virus A (SVA, a surrogate for FMDV).[1-3] In an effort to reduce this risk, multiple feed additives, including medium chain fatty acid blends, organic acid mixtures, essential oils, acid/aldehydes and others, were tested to determine their ability to negatively affect viral survival. The test model was designed to represent field conditions, involving large groups of feed and tonnage of feed, and a point-in-time challenge model including PEDV, PRRSV 174 and SVA that was manually dropped into each feed bin to represent contamination of prepared feed on the farm. Under the conditions of the study, feed in all of the bins was successfully contaminated, resulting in delivery of all 3 viruses to all rooms. Transmission of PRRSV, PEDV and SVA via the feed was documented in the positive control group. In contrast, while no product sterilized the feed, no differences were detected across feed additives, all products proved to be beneficial to pig health and performance as compared to the positive control group. These data suggest that the use of validated additives may reduce the risk of viral infection via contaminated feed.

Introduction

In 2014, feed was first identified as a risk factor for PEDV infection of pigs. This observation was expanded across multiple viruses at both the domestic and transboundary levels, including PRRSV, ASFV and Seneca Virus A (SVA) as a surrogate for FMDV. In an effort to reduce this risk, multiple feed additives were tested to determine their ability to negatively affect viral survival. The identity of the additives will be disclosed at the AASV meeting. The test model was designed to represent field conditions and involved 100 pigs and 3 tons of feed/group, a point-in-time challenge model and multiple metrics to detect the presence of all 3 viruses in the feed, exposure and infection of pigs to the 3 viruses. The challenge model involved a 1 lb. "ice block" containing a mixture of PEDV, PRRSV 174 and SVA at equal concentrations that was manually dropped into each feed bin to represent contamination of prepared feed on the farm. Under the conditions of the study, feed in all of the bins was successfully contaminated using the ice block method, resulting in delivery of all 3 viruses to all rooms, transmission of PRRSV, PEDV and SVA via the feed was documented in the positive control group. In contrast, while no product sterilized the feed, no differences were detected across feed additives, all products proved to be beneficial to pig health and performance as compared to the positive control group. These data suggest that the use of validated additives may reduce the risk of viral infection via contaminated feed.

Objective

The purpose of this study was to evaluate whether feed additives could reduce the risk of transmission of viruses to pigs through contaminated feed.

Materials and methods

Site

The Pipestone Applied Research BSL-2 facility was the site of the study and the experimental unit was the room. This facility consisted of 6 self-contained rooms, each with its own feed bin. Five rooms were designated as treatments (mitigated feed was fed) while each replicate always contained a positive control room wherer non-mitigated feed was fed.

Mitigants

The feed additives tested in this project consisted of 18 products, several tested at different inclusion rates. The study is ongoing and the final identities of mitigants will be released at the AASV.

The study was based on the hypothesis that differences in the infection events will differ across groups, and utilized a 15-day period with room as the experimental unit with 100 pigs/room (6 pens/room) and a designated feed bin/room. Pigs (15 kg) originated from a source herd documented to be free of all 3 viral pathogens by monthly testing and clinical history. Viral challenge involved a 454g "Ice Block Model", consisting of 100 mL SVA (5 logs TCID50/mL, Ct = 20.72), 100 mL PRRSV 174 (5 logs TCID50/mL, Ct = 21.38), 100 mL PEDV (5 logs TCID50/mL, Ct = 24.25) and balanced with 154 mL MEM. Blocks were frozen at -80C and dropped into each feed bin on days 0 and 6 of the study. The blocks proceeded to melt, with liquid permeating the feed, which was then augured into room for pigs to consume via natural feeding behavior.

Metrics

Ante-mortem samples, including oral fluids and Swiffer samples of feeders, were collected across the 6 pens from each of the 3 rooms at 0, 6, and 15 days post-inoculation (DPI) of feed. For collection of feed samples, individual Swiffer cloths were drawn across the feed trough, contacting feed particles present. Cloths were immersed in sterile saline and a 3 mL aliquot was decanted for testing. Post-mortem samples were collected from 30 pigs from each room at 15 DPI. Samples collected included tonsil for SVA, serum for PRRSV and rectal swabs for PEDV. Samples were evaluated for the presence of viral nucleic acid by PCR and nucleic acid sequencing of the ORF 5 was performed on select samples, as needed. In addition, start and end weights were collected from all pigs (0 and 15 DPI). Differences in growth performance between groups were analyzed for significance ($P < 0.05$) using ANOVA. In addition, pigs were scored daily for the presence of the following clinical signs:

1. Lameness/vesicles (SVA)
2. Dyspnea/weight loss/rough hair coat (PRRSV)
3. Diarrhea (PEDV)

Results

To be shared at AASV.

Discussion

The strengths of the study included its novel approach, the use of controlled field conditions involving population size, the testing of multiple products, and the use of natural feeding behavior. In addition, a robust 3-way viral challenge model and multiple metrics, such as diagnostics and performance measures, were employed. Limitations included cost of the study that influenced study duration, the potential for a level of challenge that may exceed natural levels in the field, and the fact that only a single dose of each virus was utilized. Under the conditions of the study, the following preliminary conclusions were drawn as of this writing:

The challenge model was successful

- Feed in the bins was successfully contaminated using the ice block method, resulting in delivery of all 3 viruses to all rooms.
- Transmission of PRRSV, PEDV and SVA via the feed was documented in the positive control group.
- These data support the risk of contaminated feed as a means of viral entry to farms.

All products proved to be beneficial to pig health and performance

- Mitigants do not appear to "sterilize" the feed. Specifically, traces of viral RNA have been detected in most rooms (feeder samples, oral fluids, etc).
- Mitigation of the feed appeared to have an ameliorative effect on animal health, despite the presence of viral pathogens in the animals.
- If the results are consistent, these products should be strongly considered as routine additives to pig production feeding systems to reduce the risk and impact of viral disease transmission through feed.
- The fact that multiple products performed well provides options for producers and mills.
- Results are being shared with Dr. Niederwerder for possible evaluation using ASFV.

References

1. Dee S, Clement T, Schelkopf A, Nerem J, Knudsen D, Hennings J and Nelson E. An evaluation of contaminated complete feed as a vehicle for porcine epidemic diarrhea virus infection of naïve pigs following consumption via natural feeding behavior: Proof of concept. *BMC Vet Res* 2014,10:176.

2. Dee SA, Neill C, Singrey A, Clement T, Cochrane R, Jones C, Patterson G, Spronk G, Christopher-Hennings, J and Nelson E. Modeling the transboundary risk of feed ingredients contaminated with porcine epidemic diarrhea virus. *BMC Vet Res* 2016, 12:51.

3. Dee SA, Bauermann FV, Niederwerder MC, Singrey A, Clement T, et al. (2018) Survival of viral pathogens in animal feed ingredients under transboundary shipping models. PLOS ONE 13(3): e0194509. https://doi.org/10.1371/journal.pone.0194509.

Applying biosecurity to the feed supply chain

C. Jones[1], PhD; S. Stewart[1], PhD; J. Woodworth[1], PhD; S. Dritz[2], DVM, PhD;
J. Gebhardt[2], DVM, PhD; C. Paulk[3], PhD
[1]Department of Animal Sciences and Industry, Kansas State University, Manhattan, Kansas;
[2]Department of Diagnostic Medicine/Pathobiology, Kansas State University, Manhattan, Kansas;
[3]Department of Grain Science and Industry, Kansas State University, Manhattan, Kansas

Introduction

Environmental monitoring has been commonly used in food and other facilities manufacturing end-consumer products for years,[1,2] and has gained traction as a method to determine the presence of pathogens that typically indicate fecal presence (fecal indicators).[3] In addition to facilities producing direct-to-consumer goods, some healthcare systems have used environmental monitoring of both virus and bacteria to determine hygiene and biosecurity risk, including bacteria strains known to be resistant to antibiotics.[4] A correlation between the presence of *Salmonella* spp. and Enterobacteriaceae within feed mills has been demonstrated,[5] but little information is available on how the presence of Enterobacteriaceae correlates with viral pathogen presence, especially on farms or in feed mills. The potential use of environmental monitoring of viral pathogens within a farm environment has seen an increase in popularity with the growing pressure placed on production systems from diseases like porcine epidemic diarrhea virus (PEDV), Senecavirus A (SVA), and rotavirus. Environmental swabs have been shown to be effective when detecting viruses within feed manufacturing environments[6] and with on-farm use for swine operations. The ability for PEDV to be transmitted via contaminated feed ingredients and for contaminated feed to produce animal illness within research settings,[7,8] as well as the epidemiological evidence to support a historical animal feed transmission within North America[9,10] has brought increased levels of scrutiny on mills supplying feed to swine operations. The purpose of this study was to identify Enterobacteriaceae presence in the feed manufacturing facilities of a multi-farm system experiencing a viral outbreak as a method of identifying biosecurity gaps.

Methods

Swabbing method and location

Three separate feed manufacturing facilities (Sites 1-3) were evaluated and sampled for this study, with a biosecurity evaluation and audit (posted at `ksuswine.org`) performed during each visit. Each mill offered its own biosecurity challenges, either with the normal operating procedures or required tasks to be performed within facility limitations.

A total of 573 samples were taken over the course of four days, with 381 of those samples consisting of feed ingredient or finished feed, and the remaining 192 samples environmental swabs, collected across the 4 sites.

Feed ingredient and finished feed samples were collected using single-use plastic tubs. For each separate item, 10 individual samples were collected initially. For bulk-storage products, samples were either drop-collected or grabbed at multiple times while being conveyed. For bagged products, samples were obtained from each of 10 different bags onsite. Each sample was kept separate for individual analysis, with an additional blended composite sample created from the 10 samples analyzed.

For the environmental samples, two different collection methods were used. Cotton gauze swabs were utilized for areas within the mill that had easy access for swab collection. The gauze swabs were collected by swabbing a surface area of approximately 20cm × 20cm with a 10cm × 10cm cotton gauze square soaked in 5 ml of phosphate buffered saline (PBS) with a pH of 7.2. For areas without easy access, such as the interior of storage bins or truck trailers, a paint roller was utilized, as described by Dee et al., 2014.[10] Locations did vary based on each individual site, but within the 3 feed manufacturing facilities (Sites 1-3), similar locations were chosen. Each swab was assigned one of four zones, including direct feed or ingredient contact surfaces (Zone 1), close proximity (within 1m) non-contact surfaces (Zone 2), non-contact surfaces without close proximity (> 1 m of separation) (Zone 3), and transient surfaces, such as moveable tools, employees, and non-feed or ingredient delivery vehicles (Zone 4). Swabs taken from the fourth facility, the multiplier farm, were assigned zones based on proximity to pigs. This included direct feed-contact surfaces (Zone 5), direct pig-contact surfaces (Zone 6) including pen flooring, pen walls, feeders, and waterers (pig contact), and non-pig contact surfaces (Zone 7) including employee walkways, work areas, feed storage, and fans (non-pig contact).

Sample preparation and analysis

Feed and ingredient samples were collected individual. For each product, a composite sample was created by dividing and blending approximately 25 g from each individual sample. All product samples were stored at 4°C until shipped. Cotton gauze environmental swabs submitted for testing were initially prepared by adding 5 mL of PBS to a cotton gauze square in a conical tube prior to collection. After samples were collected, 20 ml of additional PBS were added. Swabs were kept at 4° C until shipped. The paint rollers used for sample collection were placed into large zipper-seal plastic bags immediately after use. To prepare them for shipment, 200 ml of 7.2 pH PBS was added to each roller. The sample was then agitated and allowed to set for 1 hour.

10 mL of the PBS was removed from each sample and stored at 4°C until shipped. After collection, samples were store and shipped on dry ice to the Iowa State Veterinary Diagnostic Laboratory. Samples were cultured on MacConkey agar, and the three types of bacteria with largest growth for each sample were identified and reported by assigning a growth index value.

Statistical analysis

Bacterial growth results were assigned an index value of either 0, 1, 2, 3, or 4 based on reported growth, with 0 representing a negative result and values 1, 2, 3, or 4 a few, low, moderate, or high positive result, respectively. Growth values were reported as individual bacteria, with each sample receiving an overall index sum. The data were analyzed using the GLIMMIX procedure of SAS (Version 9.4, SAS Institute Inc., Cary, NC) with the Tukey-Kramer adjustment using the assigned location zones as the levels with the response variables of total growth (sum of index values) and presence of bacteria typical used to indicate fecal matter is present (fecal indicators). Treatment means were separated using pairwise comparisons of means performed using the DIFFS option from the LSMEANS statement of SAS. Results were considered significant at $P \leq 0.05$. Data were also analyzed with the CORR procedure of SAS, with the variables including site, zone, and presence of fecal indicator bacteria.

Results and discussion

The US feed industry is not designed for these types of cleaning and disinfection processes, so the primary focus must be on keeping pathogenic viruses out of feed mills, particularly because feed mills are a central point of cross-traffic among multiple farms or sites. Figure 1 demonstrates just a snapshot of the normal traffic flow of feed mills. Depending on mill size, it can receive dozens of ingredient delivery vehicles daily. The mill mixes ingredients together and distributes them to dozens of farms. The swine and poultry industries have developed highly effective biosecurity procedures that prevent people or transport vehicles from serving as fomites for viral. These include protocols for changing shoes, clothing, and/or showering and dynamic biosecurity pyramids for transport vehicles. Many modern swine production systems implement biocontainment practices similar to BSL2 or 3 laboratories but on a much larger scale. However, these same procedures have not been implemented at feed mills and with personnel or feed trucks.

Maps of EBAC levels per facility are shown in Figures 2, 3, and 4. Audit scores for each facility were 83%, 67%, and 42% for Site 1, 2, and 3, respectively. Site 1 utilized locked exterior doors, required employees to change clothes and shoes prior to entry, and had handwashing stations located inside the doorway. The scale was located within a fenced perimeter, and was used to weigh

Figure 1: While biosecurity has evolved to prevent farm-to-farm contact, there is still frequent cross-traffic between farms due to feed delivery.

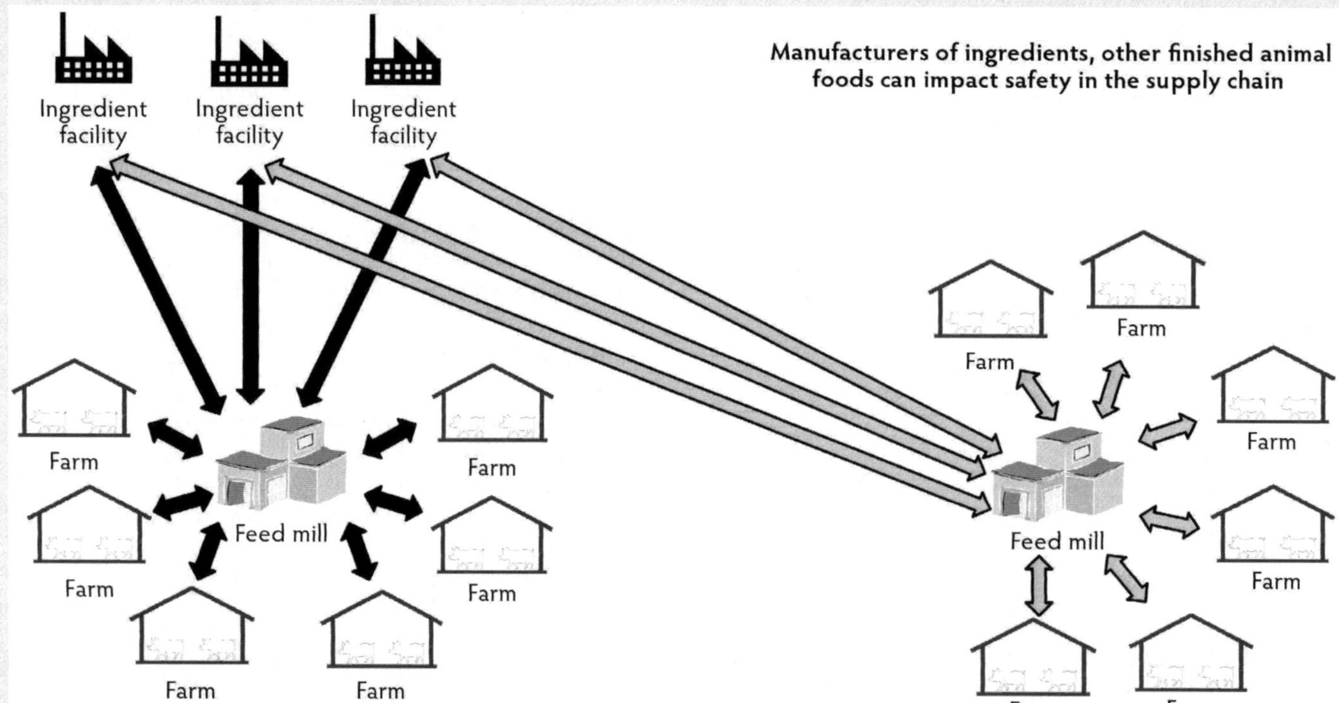

Figure 2: Hygiene map of Site 1, demonstrating relatively strong biosecurity.

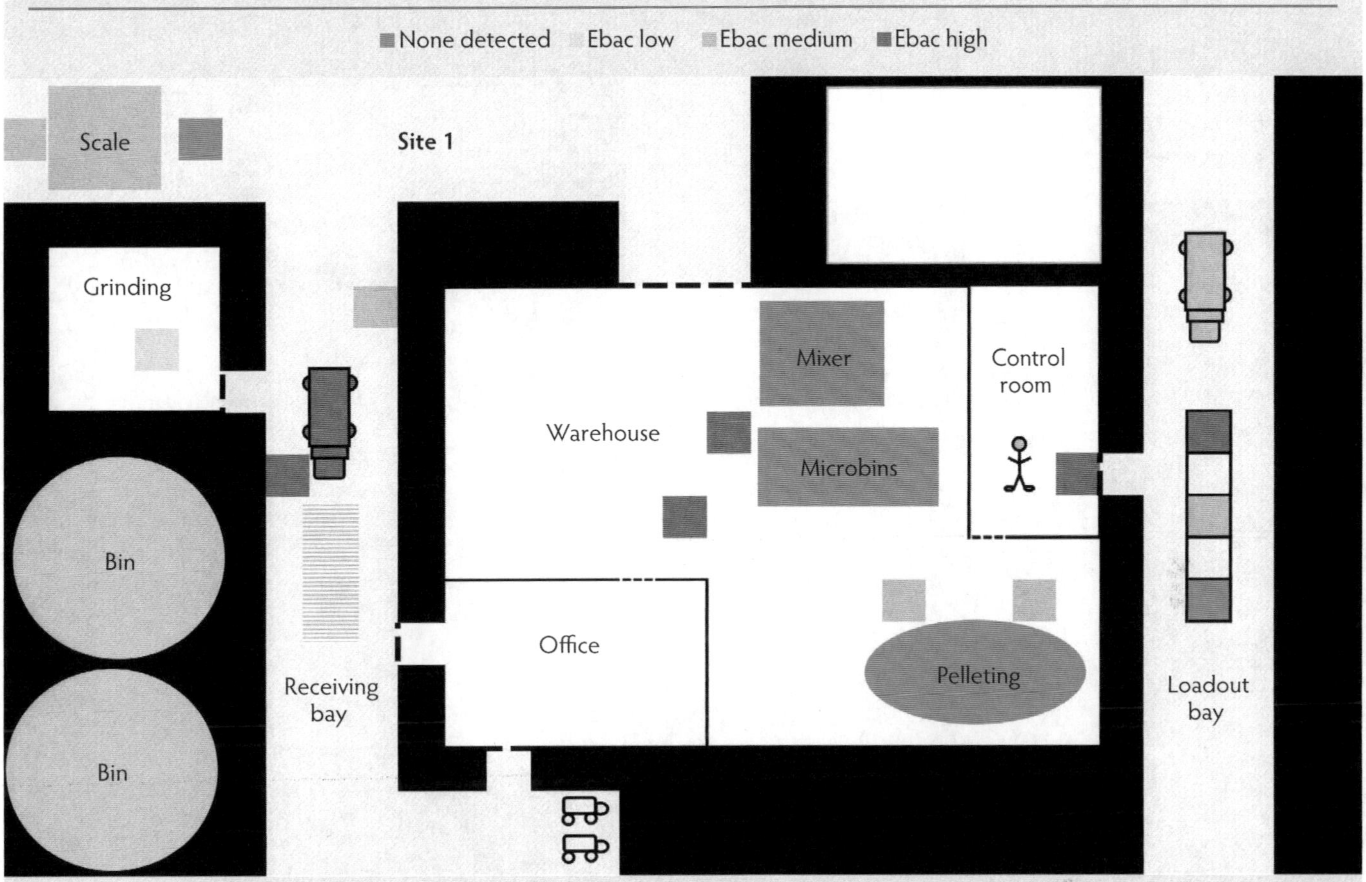

company-owned pigs occasionally. In Site 1, the scale, receiving pit, finished feed bin, and finished feed truck were the only feed-contact surfaces with detected EBAC. At Site 2, exterior doors were not locked and handwashing stations not used except for restroom purposes, but employees changed clothes and shoes prior to entry. There was no perimeter fence and the scale was routinely used to weigh animals. There was a moderate quantity of EBAC detected in all feed contact surfaces tested, with high levels on the floor of the manufacturing area. At Site 3, exterior doors were not locked and handwashing stations not used except for restroom purposes, but employees changed clothes and shoes prior to entry. There was no perimeter fence and the scale was routinely used to weigh company-owned animals, as well as those from other within the region. While it was difficult to obtain samples from feed contact surfaces in Site 3, those collected all had high levels of EBAC.

Across all facilities, zones 2, 3, 4, 6, and 7 had the similar reported mean growth values ($P \leq 0.05$) assigned, ranging from a mean index score of 18.1 for zones 2 and 3 to a mean index value of 21.4 for zone 7. Zones 1, 2, 3, 4, and 6 had similar mean growth values ($P \leq 0.05$). Zone 5 had the second-to-lowest demonstrated growth of all sample groups ($P > 0.05$), with the group of samples with the lowest growth index value being the raw ingredient and finished feed samples. There was evidence of moderate correlation noted ($r = 0.463$, $P \leq 0.0001$) between the zone and presence of EBAC, but no evidence of correlation ($r = 0.028$, $P > 0.05$) between zone a presence of fecal indicator bacteria. There was significant evidence of a weak correlation ($r = 0.201$, $P \leq 0.0001$) between EBAC presence and site.

Clearly, compliance with biosecurity protocols had a substantial impact of Enterobacteriaceae prevalence and distribution throughout the feed mill. As facilities begin to transition biosecurity from the farm to the feed mill, using environmental monitoring to evaluate risk for biosecurity gaps, as well as success in their mitigation, will be useful and necessary.

References

1. Rigarlsford J. Microbiological Analysis of Red Meat, Poultry and Eggs. *Woodhead Publishing Limited* 2007. ISBN; 978-1-84569-059-5.

2. Archibald F. The Presence of Coliform Bacteria in Canadian Pulp and Paper Mill Water Systems -a Cause for Concern?. *Water Quality Research Journal of Canada* 2000; 35. 1-22. doi: 10.2166/wqrj.2000.001.

3. Baylis C, Uyttendaele M, Joosten H, and Davies A. The Enterobacteriaceae and their significance to the food industry. *ILSI Europe Emerging Microbiological Issues Task Force* 2011. doi: D/2011/10.996/30. ISBN 9789078637332.

Figure 3 : Hygiene map of Site 2, demonstrating moderate to weak biosecurity.

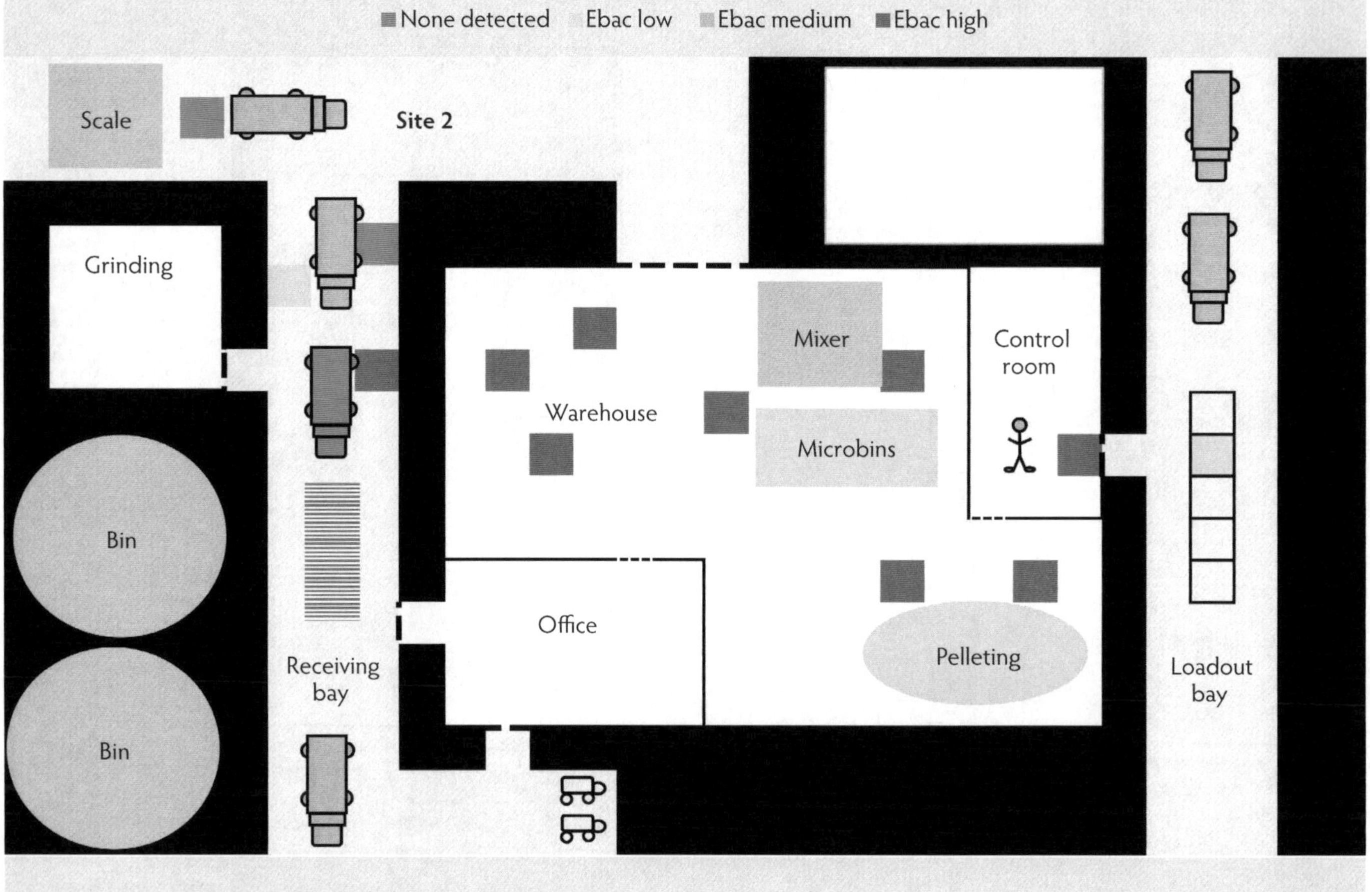

4. Palmore TN, Henderson DK. Managing transmission of carbapenem-resistant enterobacteriaceae in healthcare settings: a view from the trenches. *Clin Infect Dis*. 2013;57(11):1593–1599. doi:10.1093/cid/cit531.

5. Magossi, G, Cernicchiaro N, Dritz S, Houser T, Woodworth J, Jones C, and Trinetta V. Evaluation of Salmonella presence in selected United States feed mills. *MicrobiologyOpen*. 2019;8(5): e00711. doi: 10.1002/mbo3.711.

6. Muckey MB. Evaluation of the surface sanitation to prevent biological hazards in animal food manufacturing. *MS Thesis* 2016. Kansas State University, Manhattan.

7. Dee S, Neill C, Singrey A, Clement T, Cochrane R, Jones C, Patterson G, Spronk G, Christopher-Hennings J, and Nelson E. Modeling the transboundary risk of feed ingredients contaminated with porcine epidemic diarrhea virus. *BMC veterinary research* 2016:12(1), 51. doi: 10.1186/s12917-016-0674.

8. Schumacher LL, Woodworth JC, Jones CK, Chen Q, Zhang J, Gauger, PC, Stark CR, Main RG, Hessee RA, Tokach MD, and Dritz, SS. Evaluation of the minimum infectious dose of porcine epidemic diarrhea virus in virus-inoculated feed. *American journal of veterinary research* 2017;77(10), 1108-1113. doi: 10.2460/ajvr.77.10.1108.

9. Pasick J, Berhane Y, Ojkic D, Maxie G, Embury-Hyatt C, Swekla K, Handel K, Fairles J, and Alexandersen S. Investigation into the role of potentially contaminated feed as a source of the first-detected outbreaks of porcine epidemic diarrhea in Canada. *Transboundary and emerging diseases* 2014:61(5), 397-410. doi: 10.1111/tbed.12269.

10. Dee S, Clement T, Schelkopf A, Nerem J, Knudsen D, Christopher-Hennings, J, and Nelson E. An evaluation of contaminated complete feed as a vehicle for porcine epidemic diarrhea virus infection of naive pigs following consumption via natural feeding behavior: proof of concept. *BMC veterinary research* 2014;10(1), 176. doi: 10.1186/s12917-014-0176-9.

Figure 4 : Hygiene map of Site 3, demonstrating weak biosecurity.

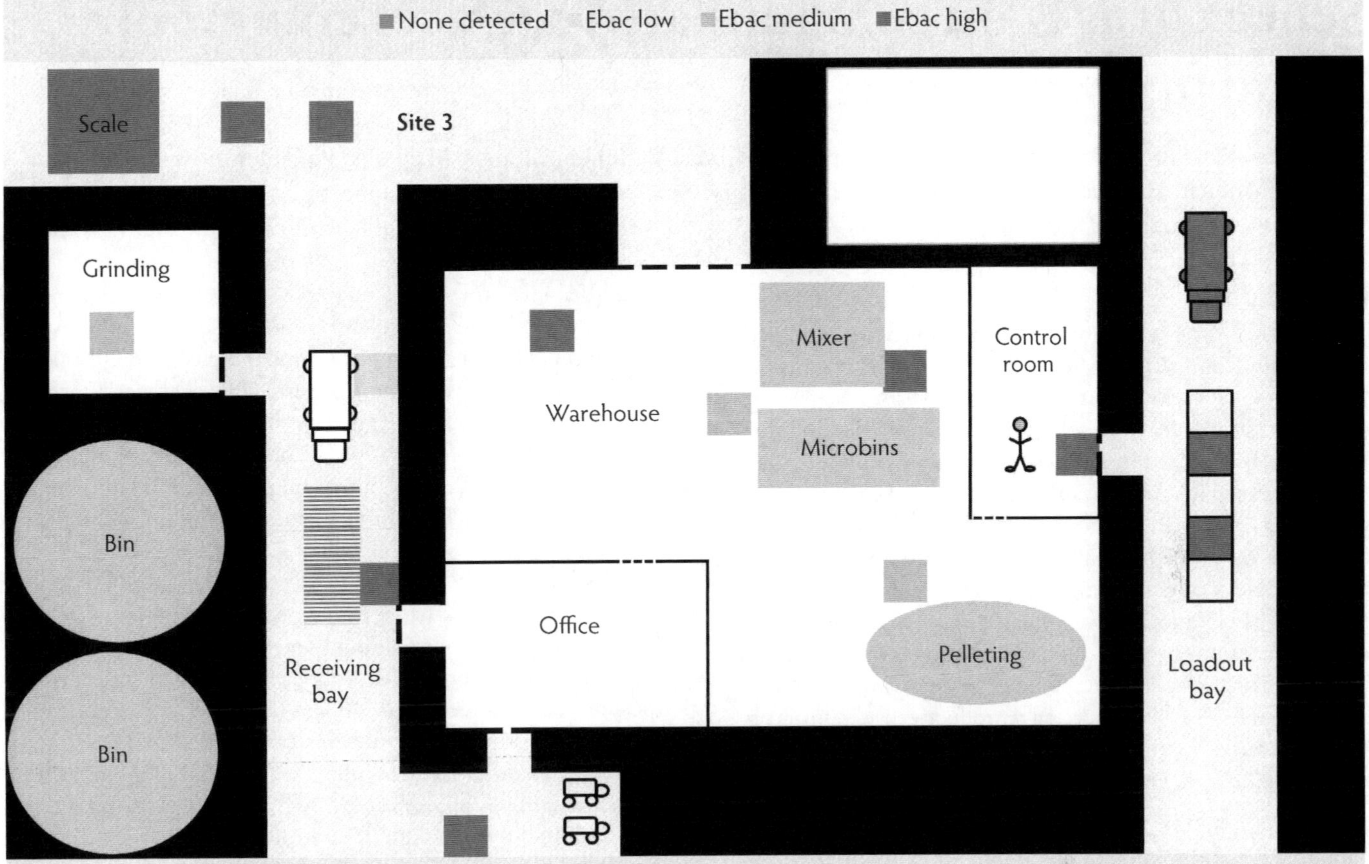

Efficacy of ultraviolet C disinfection for inactivating Senecavirus A on contaminated surfaces commonly found on swine farms

Chelsea Ruston[1,2], DVM; Derald Holtkamp[2], DVM, MS; Jianqiang Zhang[2], MD, MS, PhD; Jenna Scott[3], DVM
[1]Swine Medicine Education Center, Iowa State University, Ames, Iowa; [2]Veterinary Diagnostic and Production Animal Medicine Department, Iowa State University, Ames, Iowa; [3]Smithfield Hog Production Missouri, Princeton, Missouri

Introduction

The disinfecting properties of short-wave ultraviolet light, with wavelengths from 200 to 280nm (UVC), are well known and ultraviolet disinfection is particularly effective for inactivating single-stranded RNA viruses on surfaces.[1] The process by which UVC works is by the formation of thymine, cytosine, or uracil dimers in the DNA or RNA of viruses causing mutations and essentially inactivating the virus.[2] UVC light has been widely used for disinfection for a long time in many sectors, including human medicine and food processing. The practical application of the technology in livestock production is a more recent development and is increasingly being used on swine farms as producers look for ways to improve biosecurity for endemic diseases such as Senecavirus A (SVA) and porcine reproductive and respiratory syndrome virus (PRRSV) and in response to the recent threat of African swine fever virus (ASFV). The doses of UVC light required to inactivate human, and some swine pathogens, has been well studied. In general, several magnitudes of the doses required to inactivate the vast majority of bacterial and viral pathogens are practically achieved under the conditions that UVC is currently used on swine farms. However, the efficacy of UVC light when used under the challenging conditions found on swine farms, such as permeable surfaces, presence of organic matter and shadowing, is not well understood. Despite the lack of research on the application of UVC under conditions found on swine farms, the technology is increasingly being used to disinfect incoming supplies, employee lunches, medications, phones and other carrying agents entered into swine farms. Hence, research is needed to clarify whether UVC disinfection is efficacious under such challenging conditions.

The objective of this study was to evaluate the efficacy of UVC light for inactivating Senecavirus A (SVA) under the challenging conditions found on swine farms, such as permeable and porous surfaces, the presence of organic material and partial shadowing when objects to be disinfected are placed on wire shelving.

Methods

The effect of UVC on a SVA isolate inoculated onto two permeable and porous surface types (cardboard and cloth coupons) and one non-permeable and nonporous surface type (plastic coupons) was evaluated. Each coupon was inoculated with 1 mL of SVA plus 1 mL of PBS or 1 gram of feces on the top or bottom surface of the coupon and allowed to dry. Coupons were exposed to UVC in a commercially available pass through chamber (PTC) (Bioshift® Pass through germicidal chamber, Once Inc., Plymouth, MN) for 5 minutes or in a simulated supply entry room (SER) for 120 minutes. The UVC units utilized in the SER were ESP-DLuxTM units (Elevated Health Systems, LLC, Witchita, KS). The 3 surface types (plastic, cardboard and cloth), 2 inoculants (virus and PBS or virus and feces), 2 surface locations (top or bottom) and 3 treatments (UVC in PTC, UVC in SER and positive control) yielded 36 treatment groups. In addition, 3 negative control groups, one for each surface type (plastic, cardboard and cloth), without feces were included. After exposure, virus was recovered from each coupon and virus titration (median tissue culure infectious dose per ml, TCID50/mL) was performed on the sample collected. Viral reduction was measured as the difference between the $\log_{10}$ titers of each treated group and their relative positive controls.

Results and conclusions

The plastic coupons inoculated with SVA and PBS on either the top or the bottom surface of the coupon had more than a 7 log virus titer reduction in both the PTC and SER when compared to their relative positive controls. The cloth and cardboard coupons inoculated with SVA and PBS (on either the top or the bottom surfaces) had only a 2 log virus titer reduction or less in the PTC and SER. When the top or bottom surface of the coupons were inoculated with SVA and feces, all of the surface types, including plastic, had less than a 2 log virus reduction. There was no statistically significant difference in viral reduction between the top and bottom inoculated surfaces for any of the surface types or either inoculant (virus with PBS or virus with feces), suggesting that the shadowing caused by the wire shelving did not interfere with the UVC light.

The findings of this study suggest that UVC may be used to effectively disinfect surfaces of nonpermeable, non-porous materials, such as plastics that are free of organic material, such as feces. However, UVC will not effectively disinfect permable or pourous surfaces, such as cloth and cardboard or any surfaces contaminated with organic matter.

Acknowledgments

This study was funded by the Iowa Pork Producers Association (#18-114 IPPA). The authors would like to thank Once, Inc. and Elevated Health Systems for their in-kind contributions to the study. The authors also wish to thank Katyann Graham, Chris Mowrer and the staff of the Swine Medical Education Center at Iowa State University, Locke Karriker, Justin Brown, Heather Kittrell, Kristin Skoland, Mary Breuer and Kristen Hayman for assisting on the project.

References

1. Tseng, C. C. and Li, C. S. Inactivation of Viruses on Surface by Ultraviolet Germicidal Irradiation. *J. Occup. Environ. Hyg.* 2007;40:400–405.

2. Kowalski, JW. Ultraviolet germicidal irradiation disinfection theory. In: Kowalski, JW. *Ultraviolet Germicidal Irradiation Handbook: UVGI for Air and Surface Disinfection*. New York, NY: Springer Inc; 2009: 20-24.

Market transport: today and tomorrow

Michael C Eisenmenger, DVM
Swine Vet Center; Windom, Minnesota

Situation today

Every winter we are reminded of how efficiently disease is transferred to finishing sites via market trailers returning from packing plants. PED is a good biomarker to understand disease transfer due to the short incubation time from exposure to clinical diarrhea. A not infrequent scenario is 3-5 days after a marketing event PED occurs. While many explanations could be put forward why this occurs, industry surveys show contamination/previous exposure of both finishing sites and market concentration sites. When following the lines of transport in a typical week from packing plants to finishing sites in multiple states it is easy to understand the risk in the pork production system to transmitting disease quickly around the country. Each day in the US roughly 3000 market trucks back up to a packing plant, unload pigs and head back to states or origin.

In general, biosecurity at packing plants is a very difficult process. The lairage area functions as a comingled continuous flow finish site. Even though the areas are in general clean and rinsed down regularly it is an area for finishing disease agents to be delivered from multiple areas of the country in one day and be present in this area at any given time. Good news is little to no amplification should occur due to the short time pigs spend in the lairage area. However, much cross traffic occurs from truckers entering trailers and common plant areas. Unloading tools, boots, coveralls all have the opportunity to become contaminated during the unloading process.

Biosecurity of finishing facilities still needs improvement. Compared to typical biosecurity of sow farms the finish sites lack processes that stop infectious disease entry such as air filtration, effective mortality removal, and supervision/auditing of processes. A complicating factor is many facilities build 20 years ago, when biosecurity was less of a concern, are still in operation today due to investment in a new gate/feeder package makes the site very usable.

Even though new finish/wean to finish facilities are built in a manner that allows shower in, no outside exit during visit, clean supply entry and clean dirty line at chute the culture of biosecurity in finish is lacking compared to sow facilities. In most recent surveys of pigs placed PRRS negative at weaning greater than 50% end up PRRS positive at time of marketing. In addition, even though a finish turn is only 4 months, airborne transmission of PRRS occurs. Clear examples show the virus moving short distances without company to company interactions.

In addition, lack of biosecurity around mortality management makes it difficult to limit disease transfer from site to site of finishing sites. Many sites still have mortality storage area design and removal of mortality methods as it occurred decades ago.

Enhancing the disease transfer from market trucks to finishing sites occur due to:

- Truck design
- Skill level of market driver
- Lack of adequate supervision
- Barn design

These factors allow crossing of clean/dirty lines that should exist in the loading process. Loading personnel often cross over and enter trailers to speed up the process and often the truck driver is allowed in chute or barn.

Market truck sanitation is highly variable when polling the industry. It ranges from no wash events occurring on a regular basis to wash dry disinfect and heat on every truck that leaves a packing plant. Often, farms or companies have rules on how often the truck must wash. For example, a typical schedule is a wash and disinfect on all trucks except if the site closing out in 4-7 days.

It is easy to understand in our current environment, disease can easily move from finish sites to other finishing sites as well as finishing sites to packing plants and packing plants to finishing sites.

What should market transport look like?

What should the process look like if we wanted to stop infectious disease transfer form packing plants back to finishing sites? Or a good question to ask is, "what should the process look like if an FAD event occurred in US?" Clearly, the system we deploy should be robust enough not to require change if an FAD event occurred in this country.

All trucks that arrive to a finish site should be clean and free of infectious disease. If the trucks were as clean or cleaner than the site, then a clean/dirty line could be avoided. This would allow the driver to remain in the vehicle and site personnel could load the trailer. This would be an advantage as no "pass off" event would be needed at the back of the truck as pigs enter trailer and start to go up the internal ramp. This can be a baulking point during the loading process. Also, this would potentially negate the need for the driver of truck to be experienced in pig loading skills.

The plant itself must be considered infectious disease positive. We do not have the means or tools today to guarantee all loads to be specific pathogen free. We could limit contamination of transport vehicles by leaving the driver in the tractor and having plant personnel unload the pigs. The cab would stay clean and if unloading tools were part of the plant process no tools would be allowed to return to the trailer that would allow further clean up. This would also allow the truck driver to be a driver vs a pig handler.

How to clean the trucks and equipment: After the trucks have been at a plant, a flush, wash, and disinfect process needs to occur. An additional heating process would help ensure truck sanitation. Currently, we do not have enough public or private facilities to accomplish this on all market trucks today.

Location of a clean up facility has been discussed. Ideally it would be located near the plants to potentially limit transfer of disease to other parts of the country. But if internally owned and operated it would most likely be best if located near home base of the company or farm to allow for management and operation. In addition, the time to wash, clean and disinfect trailers is an issue especially if a long distance from home base.

Contract hauling of market pigs is a common situation in the industry. Some producers have now opted to own all tractors, trailers, wash/disinfect/heat facilities and use employee drivers. This creates a situation of biosecurity control at a very high level.

An advantageous situation could be a base of trailers owned by a common entity. Market loads would be unloaded, the dirty trailer dropped off at clean up station, tractor washed and disinfected, and a new clean and prepped trailer picked up. Much industry cooperation would be needed for this to happen.

Design of the facility should be in a flow through design to allow for dirty entry and clean exit typical for truck wash operations.

Trailer design is still an ongoing discussion. Easy to load and easy to wash designs are still needed.

Transportation biosecurity: dos and don'ts from a breeding stock company perspective

Jean Paul Cano, DVM, PhD; A. Pitkin, DVM, MS; D. Hemker, DVM, MS; J. Geiger, DVM, MS; J. Waddell, DVM, MS; M. Clavijo, DVM, PhD; M. Rotolo, DVM, PhD; T. Riek, DVM
Pig Improvement Company (PIC), Hendersonville, Tennessee

Introduction

Transportation is an essential activity in modern pig production. More than five million pigs get transferred every week in the USA. Tractors, trailers and drivers moving those pigs can become contaminated with infectious agents such as PRRSV, PEDV, *Brachispira* sp, FMDV, ASFV, SVA, etc. Transportation of breeding stock is especially critical because two main reasons:

1. Replacement gilts and boars are introduced to breeding herds and boar studs, the top of the production chain.
2. Genetic nucleuses and multipliers are usually closed for pig introductions and located in lower pig density regions. Therefore, the risk posed by contaminated transport vehicles represents a large component of the total risk of infection for these farms.

Transmission

Transmission probability is the likelihood that the contact between an infective source and a susceptible host successfully transfers the microbe causing an infection. An effective contact between the infectious and the susceptible individuals is required, and it is represented as:

$$\beta = \gamma \times \rho$$

The effective contact rate (β) is a function of the total number of contacts per time unit (γ) and the probability of infection in every contact (ρ). Therefore, the relationship between the frequency of an event and the likelihood that during that event a pathogen can infect a pig determines the actual risk for a production practice (transport, people/supply entry, feeding, mortality/manure removal, etc) to cause introduction of disease into a herd. For example, in a sow farm, loading weaned pigs happens several times per week but the likelihood of transmission is relatively low if the decontamination and loading process are consistently performed appropriately. In the other hand, culled sows are removed less often but if the tractor-trailer-driver contact a collection point or a slaughter facility exposure to pathogens is higher increasing the likelihood of transmission even if decontamination and loading procedures are similar.

Disease transmission can occur through direct contact between the infectious and the susceptible hosts, e.g. nose-to-nose, or indirect contact when biological or mechanical vectors, air, water, food, facilities, equipment or people become contaminated by an infectious host and get exposed to a susceptible host while the microorganism is still infective. Agents with the ability to survive longer in the environment such as PCV2 or *Ascaris suum* are potentially capable of infecting more susceptible pigs than more fragile agents like IAV or mycoplasmas because potential longer contact opportunity. Washing-disinfecting-drying protocols reduce the access of susceptible pigs to objects contaminated by infectious pigs.

Risk management

A biosecurity program aims to identify and mitigate the risk of introduction and dissemination of infectious agents to a population. The effective implementation of the program requires the integration of four main components:

1. **Risk assessment:** the continuous and systematic search for any potential opening for disease introduction is an opportunity to semi-quantify risks, audit routine processes, prioritize investments and educate teams.
2. **Policy and guidelines:** it is critical that the information and tools used to develop and communicate the biosecurity program are science-based, consistent, practical and easy to understand but complete.
3. **Education:** to drive engagement and accountability from all members of the organization, opportune and suitable training and retraining is required. Several systems have promoted a local biosecurity champion in each farm or flow to help dissemination of the information and keeping team engagement.
4. **Infrastructure:** all equipment and facilities that need to be used in pork production (trailers, loading chutes) or that can help to mitigate certain risk factors (air filters, TADD systems) represent an investment but play a significant role in disease prevention. Without the appropriate implementation of the three previous components, infrastructure becomes a heavy cost with no or little return on the investment.

Components of transportation biosecurity

PIC North American transport coordinates approximately 2,100 breeding stock shipments per year. In the last 27 years, only two disease breaks were attributed to transportation. This establishes a consistent record of biosecure transport to customer herds. This program consists of four components:

1. **Risk assessments:**
   a. Transportation decontamination facility assessment is used to audit all PIC associated truck wash and drying locations. This audit evaluates facility conditions, number of non-PIC washes, personnel biosecurity training, and wash procedures (water temperature, pressure, disinfectant class, detergent use, and drying).
   b. Transportation assessment maps transportation events and associated biosecurity by site. It evaluates the frequency and details of transport events per month, by category of movement, for the trailer, tractor and driver.

2. **Fleet segregation and downtime:** it has been documented for PEDV and more recently suggested for SVA that sale barns and slaughter plants are highly contaminated locations because the nature of their function. Due to the increased likelihood for contamination at collection points and slaughter facilities, tractors, trailers, and drivers are segregated between market and breeding stock movements. In the infrequent event where segregation is not feasible, an extended downtime of 72 hours is required after decontamination. Pig transfer and "drop and hook" procedures are used to facilitate segregation.
3. **Decontamination process (Wash-Disinfect-Dry-Inspect):** every trailer coming in contact with a PIC multiplier, for both internal and external movements, or gene transfer center (GTC) must be washed, disinfected and dried at a PIC approved facility that incorporates PIC biosecurity protocols. Third party inspections are also conducted on trailers. For each inspection, a mobile-based audit is completed and uploaded into a health database system. If a trailer fails inspection, the inspector has the ability to provide a photo and written comments on the failed area(s). This information can be used to train employees on locations commonly missed during washing. In addition to third party inspections, drivers, farm employees, and other PIC personnel have the ability to reject a trailer not meeting our decontamination standards.
4. **Loading and unloading process:** drivers wear plastic shoe covers before stepping down from the cab and put on coveralls and clean rubber overshoes or plastic booties as they step into the cargo area. The "sequencing" loading process occurs in phases between three clearly demarcated areas. Only unidirectional pig transit is allowed. One operator will be working in the barn designated as the clean area which ends on a fence leading to the loading chute. A second farm employee will move the pigs through the chute which is designated as the transition area ending on the trailer. The trailer is designated as dirty no farm personnel can step onto it. The operator working on the transition area (chute) wears transition boots-clothing and after loading needs to wash-disinfect the chute, remove clothing and shower-in before returning to the farm.

The risk assessments, the segregation of the fleet, the additional decontamination steps, the third party inspections and training-retraining of drivers, truck wash crews and farm personnel add significant cost to the transportation process; however, the potential impact of disease outbreak in multiplication or customer herds justify this investment.

Biosecurity lessons learned and action steps to reduce the risks associated with live animal transport

A. Maschhoff, DVM; M. Hensch, DVM; S. Stehlik, DVM; J. Miller DVM
The Maschhoffs, LLC, Carlyle, Illinois

Introduction

As the industry continues to understand and investigate disease outbreaks, continual improvement of biosecurity protocols take place system to system. Biosecurity protocols are designed to reduce the risk of disease transmission, but when those protocols are in practice within a large production system, there is not a one-size fits all approach to mitigating disease entry in a cost-efficient manner. Over the past several years, disease outbreak investigations within The Maschhoff system (TML) revealed gaps within transport biosecurity. Those gaps included trailer sanitation, driver protocols and connections between sites. Therefore, TML has implemented a variety of protocols to reduce risks associated with live animal transport.

Those actions steps center around the concepts of hygiene, segregation, dedication and audited execution. The following action steps have been variously applied between the commercial and genetics operating regions.

Hygiene and sanitation

As has been described by many systems, The Maschhoffs continue to implement both wash and disinfection protocols to clean trailers. TML has standard operating procedures (SOPs) that define the expectation and utilize continual internal and 3rd party auditing to ensure those SOPs are met within the system expectations. TML has actively utilized Thermo-Assisted Drying and Decontamination (TADD) at most company owned truck washes. Within the past year, the system added additional locations with dedicated TADD systems for trailers that travel a significant distance prior to reaching their first loading site. Within those TADD systems, the system utilizes temperature tracking stickers to monitor that trailers are reaching desired temperatures. Outside of the trailer itself, the system has integrated hygiene and sanitation protocols for the supplies utilized when moving pigs from site to site, such as expectations for sort boards, driver boots and clothing. TML also created SOPs for truck hygiene as well, making industrial vacuums and cleaning supplies readily available at truck wash locations. When the drivers drop off a used trailer, they are required to clean the interior of their cabs by removing trash, vacuuming out the cab, wiping down interior surfaces with disinfectant wipes prior to ending their day. The truck wash staff then washes and disinfects the exterior of the truck and interior floor mats prior to moving the truck to the "clean" side of the truck wash.

Segregation

Segregated truck washes

TML has approached segregation through multiple facets. First and foremost, the system committed to building a segregated truck wash that was dedicated to a portion of the nucleus and multiplier herds. This allowed for mitigating risk within those herds and separating them from commercial farms except in the case of delivering replacement breeding stock to commercial farms. This segregated truck wash includes segregated trailers, trucks, drivers and equipment. In situations where segregation was not feasible based on geography of sites, additional epidemic disease testing has been implemented at commercial sites that share a common truck wash. Additionally, the system has developed exclusive relationships with third party live haul contract systems to provide segregation of TML trucks and trailers from other producers. In both of these segregation steps, TML has been able to more rapidly control and adjust for risk between sites because when epidemic disease strikes adjustments can made within the internal supply chain. Within TML owned truck washes, clean dirty lines are also established to control driver footpaths at the beginning and end of day, keeping drivers from contacting clean trailers and trucks when they return to drop off used trucks and trailers.

Trailer classification system

To further segregate trailers and trucks, TML has implemented segregation of trailers through a classification system. That classification system clearly outlines where the trailer and truck can start, what site type the trailer and track can deliver animals too and outlines specific health designations within that class. For example, if a site is positive for a system defined epidemic disease, such as PRRSV, that site would be removed from high level classifications and moved to a lower tier. Specific requirements are then defined for wash, disinfection, TADD procedures and downtime between those trailer classes, so that the system can still balance biosecurity and utilization of system available trailer capacity. Those SOPs and downtimes apply to the driver, truck and trailer. In the future, the system plans to implement a program where the trailer and truck class are easily identified through exterior markings, which will allow for final inspection and confirmation of class by the team at the shipping or receiving herd.

Cull transfer chutes

The second step in segregation was through the implementation of cull transfer chutes. These transfer chutes are utilized to segregate trailers contacting sow farms from cull buying stations. Cull sows are loaded on specific sow farm dedicated trailers or within a specific classification of trailers and then moved from the breeding farm to the transfer chute. At the transfer chute, a contract hauler or internal trailer of a lower classification moves the sows to a different trailer and hauls them to the cull buying station. This step mitigates trailers from contacting sow farms and cull buying stations without an increased series of biosecurity steps such as additional washes or downtime. This process has been implemented in both genetic and commercial assets.

Driver segregation

Due to some internal investigations, TML additionally has taken steps to segregate live haul drivers from contacting the sites during pick up or delivery of animals. This has been implemented through a variety of site-specific steps, such as staged loading rooms or unloading the trailers by the farm staff – both leaving the driver in the cab and preventing additional pig contact with the live haul driver. Both those mitigation steps require a series of additional changes to the traditional SOPs and investment in capital and labor on the sow farm to allow for the farm staff or load or unload pigs onto a trailer. Those steps include, but are not limited to, asking the farm staff to take additional showers prior to re-entering the unit, one of those shower locations being designated as a "dirty" shower and one through the normal site entry shower. This mechanism also keeps the loading clothing separate on the "dirty" side of the farm and requires cleaning and disinfection prior to use again on the next load of pigs.

Dedication

Not only does a system need clean, dry trailers and segregation SOPs that support increased biosecurity, but dedication can also play a role in mitigating risks. For a few sites TML has invested in dedicated trailers. Those dedicated trailers only haul for a specific site or set of sites. Though this comes at an added cost, in a few situations the "insurance" cost was determined to be a good investment towards protecting those specific sites.

Conclusion

While these example action steps listed above are not foolproof in mitigating disease entry, they are applications of lessons learned from disease breaks with The Maschhoffs and are currently being applied today within the region(s) where those actions are most cost efficient and effective. As with most biosecurity protocols, human error is likely to occur, but as more research about effective mitigation steps becomes available, those steps should be continued to be evaluated and utilized to reduce costly epidemic disease breaks.

Evaluation of a staged loading procedure for the load-out of market pigs to prevent the transfer of swine pathogen-contaminated particles from livestock trailers to the barn

C. Ruston[1,2], DVM; D. Holtkamp[2], DVM, MS; D. Linhares[2], DVM, MS
[1]Swine Medicine Education Center, Iowa State University, Ames, Iowa;
[2]Veterinary Diagnostic and Production Animal Medicine Department, Iowa State University, Ames, Iowa

Introduction

In the swine industry, grow-finish biosecurity has received very little attention in comparison to swine breeding herd biosecurity. However, recent literature has shown that outbreaks in grow-finish herds can have significant impacts economically as well as serving as a source of virus for breeding herds. One study recently reported that 92% of 81 growing pig lots placed in high pig dense areas became infected with wild-type porcine reproductive and respiratory syndrome virus (PRRSV) during the growing phase.[1] Furthermore, data from the Swine Disease Reporting System demonstrates that significant increased detection of PRRSV in breeding herds are typically preceded by increased detection in growing-finishing pigs, supporting the hypothesis that the growing pig population is a major source of virus in the swine industry.[2] PRRSV infection is not the only disease that can have an impact during the grow-finish phase, one production company reported that a porcine epidemic diarrhea virus (PEDV) late in the grow-finish phase, caused a 21.4% reduction in average daily gain.[3]

One event that can pose a risk on introduction of virus into a grow-finish herd is marketing. It is well known that livestock trailers that load-out at harvesting plants are contaminated with PRRSV and PEDV.[4] There is a lot of variation in the swine industry with how livestock trailers hauling market pigs are handled after loading out at the harvesting plant. Some systems will scrape organic material, some will wash, some will wash, disinfect and bake, and others do not have the infrastructure to accomplish any sanitation and decontamination of trailers between market load-outs. There can be a high likelihood that a livestock trailer pulling up to a barn to load-out market pigs is contaminated. In the United States, pigs are typically marketed over several weeks, creating the opportunity for those pigs still on feed to become infected with virus that can be transferred from the livestock trailer to barn, during load-out of the first or second cut of market pigs.

Is there a transfer of contamination from livestock trailer to barn?

Little research has been done to assess how frequently contamination from the livestock trailer, driver or other carrying agents are transferred to the barn. A pilot study was conducted to determine whether Glo Germ could be used to study the transfer of contamination from livestock trailers to the center alleyway in grow-finish barns during a marketing event. In the pilot study 216 grams of Glo germ was mixed with obestrics (OB) gel and approximately dry wood chips in a plastic bag and spread evenly on the floor of a livestock trailer right inside the roll-up door that opens to the chute. After pigs were loaded onto the trailer, Glo germ was detected inside the load-out chute, load-out alleyway, center alleyway of the barn and the first 3 pens adjacent to the load-out alleyway, on both sides of the center alleyway.

What is staged loading?

If viral contamination can be transferred from contaminated livestock trailers to the pigs in the barn during loading, one way to possibly reduce the frequency of contamination particles from livestock trailer to the barn is a staged loading procedure during load-out. Staged loading is a procedure involving two lines of separation during the load-out process. Similar to the standard load out procedure shown in Figure 1, there is still the first line of separation between the livestock trailer and load-out chute, with the addition of a second line of separation between the load-out alleyway and the remainder of the barn (Figure 2). Essentially separating the livestock trailer, load out chute, load-out alleyway and barn into three zones. The driver stays in the livestock trailer or Zone 1. One member of the load-out crew (1) is stationed to Zone 2 (Figure 2), which includes the load-out chute, load-out alleyway and buffer zone where the load-out crew member stationed to Zone 2 can step to let pigs pass in the load-out alleyway. This member of the load-out crew (1) cannot step into the barn (Zone 3 in Figure 2). The remainder of the load-out crew in the barn (Zone 3 in Figure 2) cannot step across the second line of separation into the load-out alleyway or chute (Zone 2 in Figure 2).

The objective of this study was to evaluate if the staged loading procedure could reduce the amount of contamination transferred from livestock trailers to the barn where the pigs remaining on feed were housed.

Methods

Twenty commercial wean-to-finish barns located in Iowa were utilized for the study, with ten replicates per treatment group,

Figure 1: Diagram of standard load-out procedure utilizing a single line of separation between the livestock trailer and load-out chute.

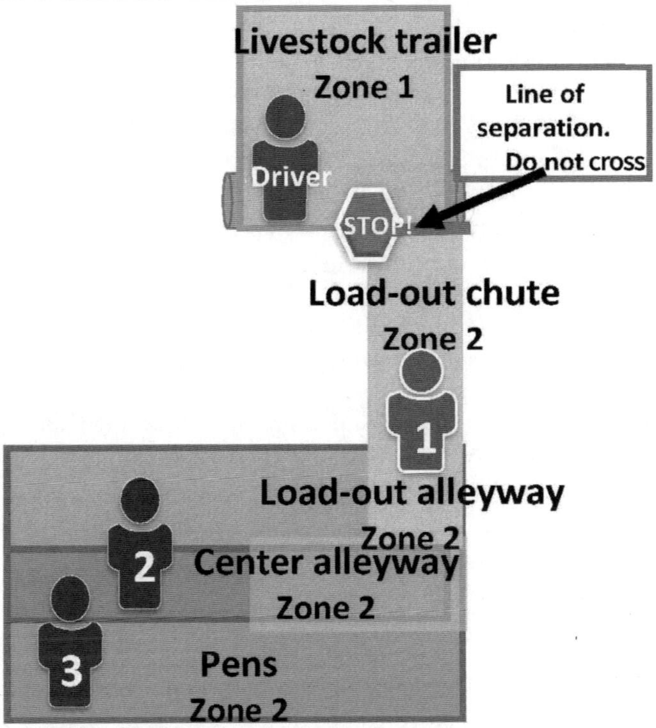

Figure 2: Diagram of the staged loading procedure utilizing two lines of separation, the first between the livestock trailer and load-out chute, and the second between the load-out alleyway and center alleyway of the barn.

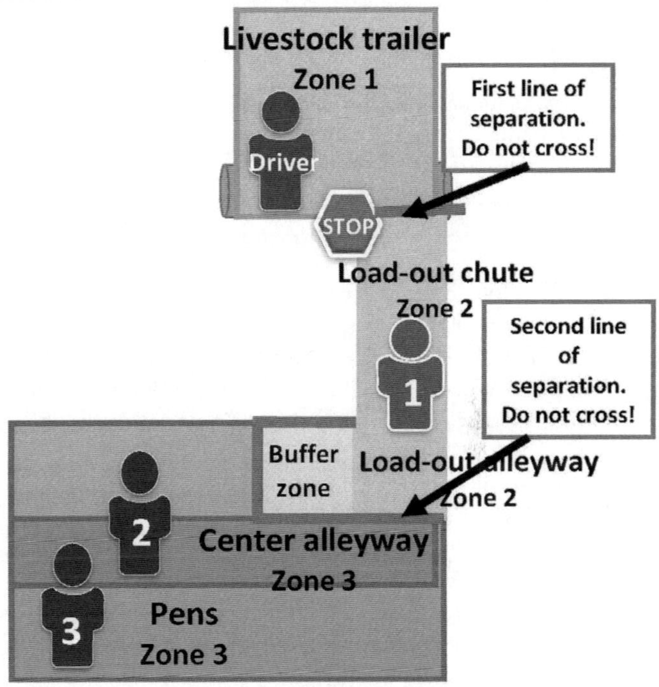

Staged vs Standard (Figure 1 and 2). On the last load-out of the night, 216 grams of Glo germ, OB gel and woodchips were mixed and spread evenly on the floor of a livestock trailer just inside the roll-up door. The loading procedure that was assigned to that night was observed. After the last load out was completed, the level of contamination was evaluated at several different measurement points within the load-out chute, load-out alleyway before the second line of separation and center alleyway, after the second line of separation.

Results

Compared to the standard loading procedure, the number of squares with Glo germ were lower for the staged loading at all of the measurement points taken in the center alley of the barn, after the second line of separation.

Under the conditions of this study, staged loading reduced contamination from the livestock trailer to the barn, and can serve as an additional layer of biosecurity so when contamination crosses the first line of separation, the second line of separation can serve as a back-up to reduce contamination transfer from load-out chute to the center alleyway in the barn.

Acknowledgments

This study was funded by the Swine Health Information Center (#19-219 SHIC). The authors also wish to thank Iowa Select Farms for their in-kind contributions and collaboration on this study.

References

1. Moura, C., Silva, G., Linhares, D. PRRSV detection dynamics in US. National Hog Farmer. October 30, 2019.

2. Trevisan, G., Linhares, D., 2019. Swine Disease Reporting System (Reports 18 and 19). https://www.swinehealth.org/domestic-disease-surveillance-reports/.

3. Pavlovic, L. Experiences managing wean-to-finish PEDV in a production system. 2018. Proc. From 49th Annual Meeting of American Association of Swine Veterinarians. San Diego, CA.

4. Lowe, J., Gauger, P., Harmon K., Zhang, J., Conor, J., Yeske, P., Loula, T., Levis, I., Dufrense L., Main, R. Role of transportation in spread of porcine epidemic diarrhea virus infection, United States. Emerg Infec Dis. 2014; 20:872-874.

Groundwater & livestock production and husbandry, part 1, biosecurity

Philip C. Olsen[1]; Steven J. Stone[2], DVM
[1]Founder, Water Think Tank, LLC; [2]Fairmont Veterinary Clinic

Forward

Misconceptions about what causes a disease and how it spreads have been very numerous. At times, so engrained in the scientific community, that to suggest anything quant air was branded as ludicrous. For instance, throughout history a prevailing idea behind disease transmission had been the miasma theory and "bad air" was thought to cause disease in areas that contained a bad stench or aura of sickness. John Snow was skeptical of this and when a cholera outbreak occurred in London in 1854, he had a chance to investigate. In his study, he was able to link the outbreak to contaminated well water. Yet, public health protection agencies continued to reject the possibility of groundwater containing pathogens until the introduction of the Ground Water Rule (GWR)[1] 152 years later in 2006. Today the agricultural industry may possess a similar level of hubris regarding groundwater and biosecurity.

A note of interest is in 1908 public water systems (PWS) began disinfection and treatment of drinking water. Over the next decade, thousands of cities across the United States followed suit. As they did, the number of deaths from infectious diseases declined markedly as shown in Figure 1.[2] This decline contributed to a sharp drop in infant and child mortality and a 29.2-year increase in life expectancy. Disinfection and treatment of public water supplies represents one of the greatest achievements in public health of the 20th century.[2]

Figure 1: Death rate (x 1,000/yr) for infectious disease - United States 1900 - 1996.

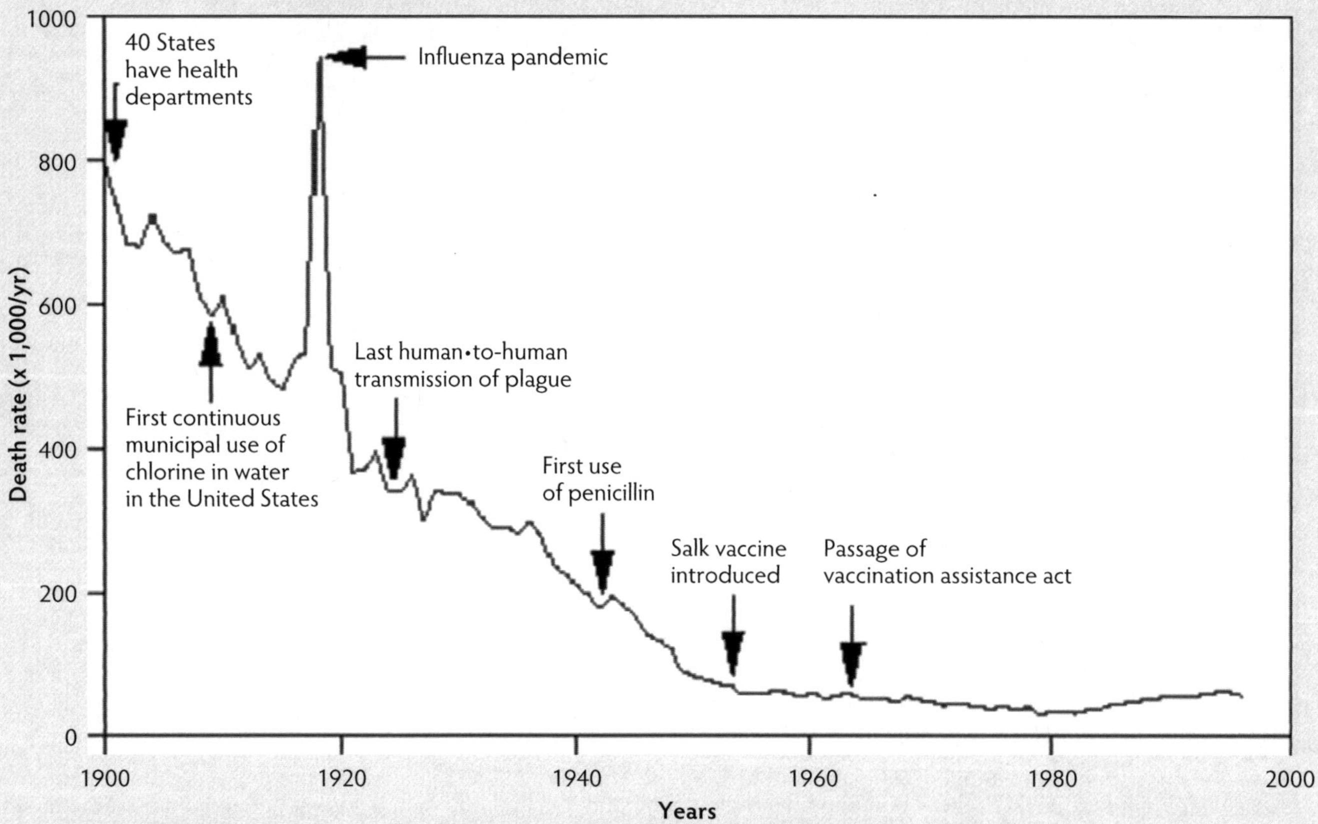

Center for Disease Control and Prevention (CDC): History of Drinking Water Treatment, A Century of US. Water Chlorination and Treatment: One of the Ten Greatest Public Health Achievements of the 20th Century. `https://www.cdc.gov/healthywater/drinking/history.html/`

Primary and secondary disinfection

During this period technologies, engineering design standards, and performance validation/verification protocols became well established to ensure water entering a drinking water distribution system is microbiologically safe, this practice is referred to as Primary Disinfection and addressed within this paper (Part 1). Secondary disinfection ensures water remains absent of pathogens as it travels through a water distribution system and will be addressed within a subsequent paper (Part 2).

Enteric viruses

Over 100 types of pathogenic viruses are excreted in human and animal wastes. These viruses can be transported in the environment through groundwater aquifers. Collectively known as enteric viruses, are transmitted via the fecal-oral route and primarily infect and replicate in the gastrointestinal tract of the host. Such are shed in extremely high numbers in the feces of infected individuals, typically between 10^5 and 10^{11} virus particles per gram of stool.[4]

Enteric viruses can be transmitted by food, water, fomites, and human contact. In addition to causing acute diseases, they are of public health concern due to their low infectious dose. For example, the probability of infection from exposure to one rotavirus is 31%, and no more than 1 plaque-forming unit (PFU) is required to cause infection in 1% of healthy adults. The risk of infection when consuming viruses in drinking water is 10- to 10,000-fold greater than that for pathogenic bacteria at similar exposures. Because of the potential for contamination from a variety of sources, enteric viruses in water are of particular concern. Since the 1980s, with significant advancements in the area of environmental virology, enteric viruses have been recognized as the causative agent in many nonbacterial gastroenteritis cases and outbreaks.[4]

Two of the most studied groups of enteric viruses as potential water quality indicators are the enteroviruses and adenoviruses. While the occurrence of human enteric viruses in the environment and their role in waterborne transmission have been studied extensively, little information is available on environmental transmission of enteric viruses in animals. Although, while 70% (62 serotypes) of nonpoliovirus enteroviruses have been associated with human infections, 30% have been associated with animal infections. Animal-specific enterovirus infections in hosts such as cattle and pigs are often asymptomatic but may cause diseases ranging from diarrhea to reproductive failure and neurological disorders. Two bovine enteroviruses (BEV), three porcine enteroviruses (PEV), 11 porcine teschoviruses (PTV) (10 were formerly classified as porcine enteroviruses), and 1 ovine enterovirus have been identified.[4]

While most studies are specific to pathogens infectious to human vs. swine populations. The question of address in this paper is related to the survivability and transport of virus in general. Though, PEV have a prevalence of 65% in pigs and wild hogs. PEV and PTV have been identified as the etiologic agents of the neurological disorder known as Teschen-Talfan disease, polioencephalomyelitis, vesicular diseases, myocarditis, pneumonia, diarrhea, fertility disorders, and dermal lesions in swine. Five porcine adenoviruses (PAdV), five bovine adenoviruses (BAdV), and six ovine adenoviruses have been classified under the genus *Mastadenovirus*. PAdV also has been isolated from pigs with encephalitis and pneumoenteritis.[4]

Presence of enteric viruses in groundwater

A plethora of peer-reviewed research conducted in advance of 2006 served to validate groundwater as a significant vehicle for disease transmission and subsequently served as the foundation from which the GWR was formulated and included within the US Safe Drinking Water Act (SDWA).[5] The GWR now represents an enforceable standard to protect the health of individuals connected to PWS's.

Although, the presence of pathogens in groundwater still accounts for more than 50% of disease outbreaks related to public drinking water supplies within the United States[6] and the CDC estimates up to 33 million cases of acute gastroenteritis illnesses occur each year due to the presence of pathogens within US PWS's. The following studies (by reference) provide further validation for the presence of infectious virus within groundwater aquifers.[7-22]

Virus entry, travel and spatial dispersion in aquifers

Due to their size (Table 1) relative to the interstitial (void) space(s) within soils and unconsolidated sedimentary layers they pass through (Table 2 & Figure 2) virus are easily transported through soils and throughout aquifers. Several researchers have reported viruses primarily travel through preferential pathways. To help emphasize how efficiently pathogens within surface water sources may be transported into aquifers, it is estimated that up to 96% of water is transported through only 0.32% of the soil volume via preferential pathways.[52] Pathways of preferential flow can develop due to structures present in surface and subsurface soils and geologic structure(s)[23] and serve serve to carry water and contaminants to depths very rapidly, resulting in greater impact on water quality than previously expected.[23]

Such hydraulic pathways serve to reduce potential interactions from occurring between particulates (virus) and soils related to

Table 1: Size of virus

	Size of virus
Virus	Average diameter (microns)
ASF	0.18
PRRS	0.06
PED	0.13

Britannica

Table 2: Size of hydraulic pathways within soils and aquafers

Size of hydraulic pathways within soils and aquafers				
Soil type	Diameter size range (microns)		Average (microns)	Interstitial space = diameter x 15.47% (microns)
Gravel	2,048,000	1,000	1,024,500	158,490.2
Sand	20	2,000	1,010	156.2
Silt	4	6	5	0.8
Clay	5	1	3	0.5

US Bureau of Soils

Figure 2: Comparative size of pathogens and hydraulic pathways.

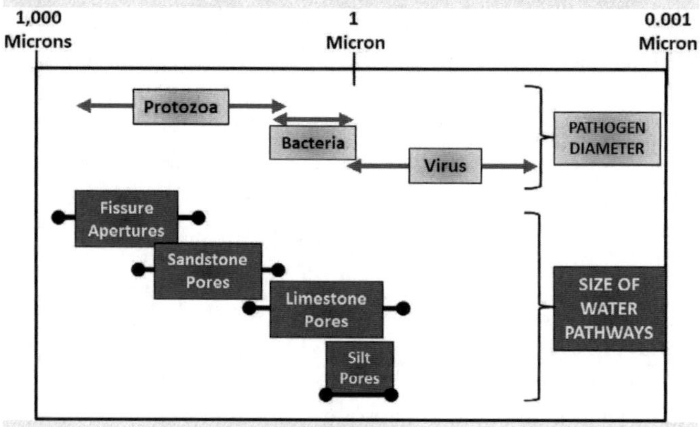

the following mechanisms: A.) Adsorption, B.) Brownian motion, C.) Ion exchange, D.) Sedimentation, E.) Size exclusion, F.) Surface charge (+/-), and E.) Van Der Waals (forces).

Due to the mechanics of fluid dynamics, preferential pathways also prevent particles (virus) from exiting channels of liquid flowage while other forces cause the extraction of water. These forces include: A.) Osmotic pressure (Low → High concentrate solutions); Effect –The amount of water extracted is dictated upon the difference in ionic concentration of groundwater verses the water being transported (typically surface water). As water is extracted, particulates and virus are not, causing their concentration within the remaining volume of water within such pathways to increase. Because ground water conductivity (ionic strength) is typically associated with higher concentrations of calcium, magnesium, sulfates and chlorides, susceptibility to this phenomenon is typically associated with higher concentrations of these ions. B.) Capillary Force; Effect – the smaller the interstitial spaces between (and within) granules of subsurface soils, the greater the force becomes to attract and retain fluids within them. This will affect water to migrate into areas where particulates and virus cannot. In turn, this also serves to dewater preferential pathways and concentrate virus. In addition, particulates such as microorganisms (virus) are transported as colloids.[52] Accordingly, they will not settle and cannot be removed from water by the mechanisms described above and facilitates their transport to (and throughout) drinking water aquafers.

These phenomena are further described within the following references. [10,24-29] Studies report virus may be transported from a contaminant source to municipal wells that were 220 to 300 meters deep within a matter of weeks[21] and on the order of hours in fractured bedrock.[30]

Penetration of pathogenic viruses through soils and geological strata into aquifers seems much more likely than for pathogenic bacteria and protozoa[31] and survive for extended periods of time[9,32] and transported over long distances.[33] In addition to what is described above, Table 3 provides a selection of factors that influence virus entry and travel within groundwater aquifers.

An additional note of interest is due to their unique genetic signatures, coupled with our current ability of detecting them in trace quantities, epidemiologists are now considering the use of clinical data as a means of tracing plumes of viral pathogens in aquifers over time (≤ 3 years). Such data sets produce virus "snaphots" of infected populations. Wherein, when correlated with the presence of viral pathogens, as measured within localized groundwater sources, serve as a marker for tracking virus and groundwater movement.[34]

The presence of viral genomes in groundwater demonstrate travel times in aquifers of two to three years.[35] A conservative estimate for virus survival in groundwater is three years, whereas[32] a reasonable estimate is one to two years.[9] Figure 3 provides an overview of factors influencing virus survival.

Hydraulics and sensitive geographies

One factor of note for livestock facilities is related to the volume of water they use in comparison to neighboring homes and farms. As larger volumes of water are extracted from the aquifer at such a facility (vs. other wells) the static water level is reduced at the farm's location. Accordingly, groundwater supplying their well system is pulled from greater distances and transported through and/or around numerous soils and geologic structure(s). Flow pathways are not uniform on either horizontal or vertical planes.

Table 3: Factors that influence virus entry and travel throughout an aquifer

Factor (in no order)	Comment
Location of fecal source	For a well to become contaminated with enteric viruses, there must be a fecal source. Sources of fecal contamination include, but are not limited to: leaking sanitary sewers, septic system effluent, landfills, field-applied sludge or septage, effluent holding ponds, wastewater irrigation sites, injection wells, reclaimed water recharge sites, surface water infiltration. The closer the source, the higher the risk potential.
Water table depth	Viruses released by a fecal contamination source directly into the saturated zone or at a depth where the water table seasonally raises will be the least attenuated. Subsurface fecal contamination sources, such as leaking sanitary sewers or septic systems, often discharge very close to the water table.
Groundwater pH	Viruses are generally less attenuated in water of neutral or alkaline pH compared to acidic water.
Aquifer material	Viruses are generally less attenuated in coarser material (coarseness continuum = gravel > sand > silt > clay) although positively charged mineral phases, such as iron, aluminum and manganese oxides or clays, can electrostatically adsorb viruses. For confined aquifers, it is important that the integrity of the aquitard be evaluated (i.e., maximum depth of open fractures and thickness) and preferential pathways through the aquitard be identified and characterized (i.e., local, extensive with window, extensive with fractures or unfractured).Water supply wells in karst and fractured bedrock aquifers are considered highly vulnerable to contamination; management of groundwater resources in karst and fractured bedrock should not be conducted in the same way as sand and gravel aquifers.
Ionic strength and rainfall	Rainfall may enhance virus transport because of its low ionic strength.
Dissolved organic matter	Fecal contamination sources with high concentrations of dissolved organic matter (i.e., septic system effluent, leaking sanitary sewers) present a greater potential for virus transport than fecal contamination sources with lower dissolved organic matter concentrations.
Virus survival	Temperature and time are important determinants of virus survival. Viruses survive much longer at cool groundwater temperatures. If the groundwater travel time is greater than the virus survival time, viruses are unlikely to be infectious when they reach the well. It is reasonable to assume that water with a travel time of two to three years or less is likely to transport infectious viruses. However, it is difficult to accurately determine travel times, particularly in fractured bedrock or karst formations.
Pumping rate	High capacity wells can create large hydraulic gradients and local groundwater velocities that draw in contamination and/or prevent virus attachment to the aquifer material.
Thickness of overburden	Viruses are less likely to be attenuated where a thin or shallow overburden exists. An increase in the vertical distance from a fecal contamination source to a well (i.e., overburden thickness) reduces the risk potential.
Well design and construction	Some considerations include well depth, well age, ingress prevention (e.g., adequate clearance from ground elevation, proper cap/cover, no cracks in casing, grouted annular space, ground condition within 10 m radius of the wellhead), multi-aquifer well.

Health Canada (2017). Enteric Viruses in Drinking Water: Document for Public Consultation—Prepared by the Federal-Provincial-Territorial Committee on Drinking Water, Health Canada, Ottawa, Ontario. Available at `https://www.canada.ca/en/health-canada/programs/consultation-enteric-virus-drinking-water/document.html`. Accessed 15 October 2019

Greater volumes will be pulled through more porous areas and gravitational force will cause greater amounts of water to flow from higher elevations of an aquifer that are closer in proximity to areas susceptible to surface water (pathogen) intrusion. These factors serve to increase the potential of microbial contamination of well systems proving drinking water for such facilities.

Understanding the mechanisms influencing entry, survival, transport and concentration of virus in groundwater it is accepted their distribution can be very rapid, broad, irregular and multidirectional (vertical/diagonal/horizontal) with higher viral inputs in areas of watersheds and/or events, that can be described as one, or a combination of the following: A.) Changes in hydraulic pressure in response to calendar season (i.e.: Winter, Spring runoff), B.) Climatic events (i.e.: rainfall/drought), C.) Areas of crop irrigation, D.) Change in elevation of (inflows) of localized geography(s), E.) Capacity, number and geographic locations of aquifer

Figure 3: An overview of factors influencing virus survival.

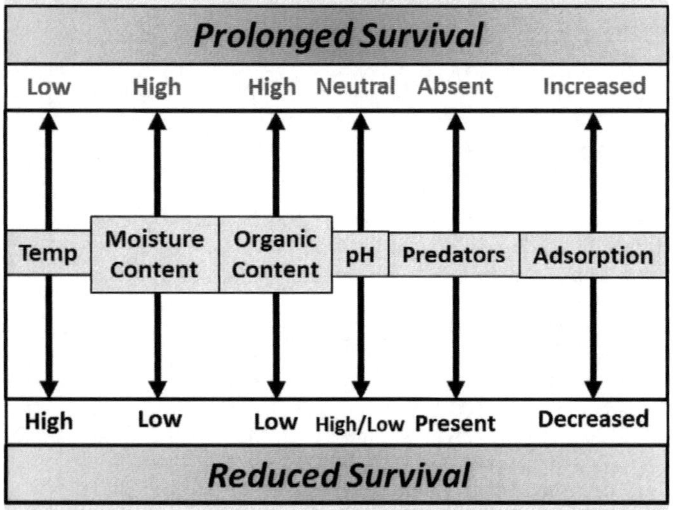

recharge/discharge in relation to changing (localized) hydraulic pressures, F.) Proximity to pathogen source(s) (i.e.: septic systems, wastewater treatment plants, feed lots, manure, landfills, lakes/ streams, etc., G.) Locations of unconsolidated, porous, fractured or cavernous geologies, H.) Uniformity, continuity, multiplicity and porosity (natural & manmade) of confining horizons, I.) Prevalent direction, speed and depths of groundwater flowage.

Survival, accumulation and release of virus within aquifers

As it is known temperature, protection from sunlight, presence of organic compounds, association with particulates and less microbial activity[37-39] influence survivability of virus in groundwater, speciation (virus) and soil composition do so as well. Clay particles are particularly effective at protecting virus from natural decay.[40,41] Some types of organic matter (i.e., proteins) are also reported to better protect viruses from inactivation.[42]

Of particular interest is the manner which virus accumulate within geologic strata by adsorption. Over time, virus in such areas become highly concentrated. Then subsequently desorbed unilaterally into the flowage of an aquifer, inherently creating highly concentrated plumbs of virus.[39,42] Since virus-soil interactions are very sensitive to surface charge, any water quality change that is enough to cause a charge reversal will result in the desorption of virus.[43,44] Water quality changes that can result in desorption include an increase in pH, a decrease in ionic strength, and the presence of organic matter.[40,45-47] For example, when alkaline septic effluent mixes with groundwater, the increased pH allows rapid desorption and transport of virus, especially under saturated flow conditions.[48] Rainfall recharge after a storm may decrease ionic strength and cause virus to desorb and transported. As such, viruses may continue to contaminate an aquifer at high concentrations on a periodic basis long after their initial entry.[40,49]

Due to the seasonality of porcine epidemic diarrhea (PEDV) and porcine reproductive and respiratory syndrome (PRRSV) outbreaks, it is apparent water temperature may serve as a critical factor regarding their survivability.

In a year-long survey of the occurrence of adenoviruses in drinking water in South Africa, adenovirus detection peaked in July (winter in South Africa), when up to 30% and 60% of treated and raw water samples were positive for adenoviruses, respectively. Enteroviruses were detected from an estuary in southwest Florida only when the water temperature was below 23° C. In an in vitro study, enhanced poliovirus survival and detection were observed at 22° Celsius (C) compared to 30° C in seawater. In artificial seawater, viruses were detected by reverse transcription polymerase chain reaction (RT-PCR) for at least 60 days at 22° C but for only 30 days at 30° C. Similarly, in seawater, it took 671 days to inactivate 90% of poliovirus and hepatitis A virus at 4° C and only 25 days at 25° C. In a study evaluating both human and bovine enteric viruses by polymerase chain reaction (PCR) in a mixed-use estuary, it was found that all virus types were correlated with cool water temperatures as shown in Figure 4.[23]

Concentration of virus in groundwater

Short-term peaks in pathogen concentration may increase risk of disease transmission considerably. Furthermore, results of water quality testing for microbes are normally unavailable in time for management to take action and prevent the supply of unsafe water from entering a livestock facility.[50] In general, virus occurrence and concentration in groundwater can be characterized as transient, intermittent or ephemeral, because wells are often not virus-positive for two sequential samples and the detection frequency is low on a per sample basis.[14,51] This may be attributable to A.) preferential flow patterns; B.) concentration of virus populations within diminishing volumes of water flowing through them; C.) adsorption-desorption processes. Accordingly, it is necessary to consider the extremely high numbers by which they are shed into the environment and how rapidly they may reach an aquifer. In fact, viral concentrations in deep municipal wells are generally as high as, or higher than virus concentrations in lake (surface) water.[21] The number of virus from infected individuals (and animals) range from 100,000 up to *100,000,000,000 per gram of stool[53] and can be representative of the potential concentration of virus within groundwater aquifers.

*Interpretation of aquifer input concentration potential: If a virus represented 1 inch, the lineal distance of the number of virus contained within 1 gram of stool would be equivalent to the distance of 63.4 times the circumference the Earth.

Current synopsis: disinfection of drinking water within the livestock industry

Viral illnesses such the Avian Influenza (Poultry), foot-and-mouth (Cattle) and PRRSV, PEDV & African swine fever (ASF) (Swine) continue to be a significant concern within the livestock

Figure 4: Human and bovine enteric viruses by PCR in a mixed-use estuary, correlated with cool water.

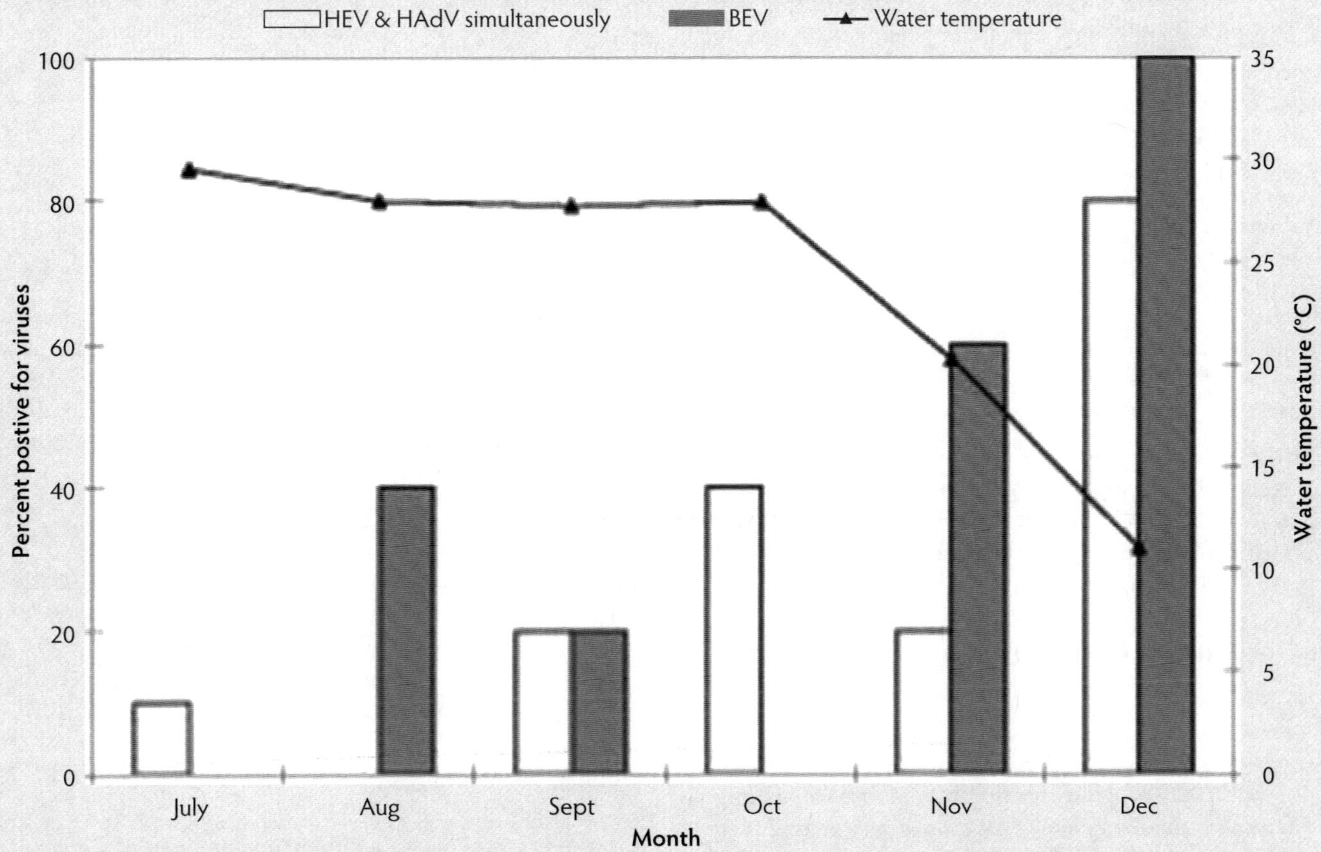

* Percentage of samples positive for human enteroviruses and human adenoviruses (detected simultaneously), as well as bovine enteroviruses, by month versus the mean monthly water temperature (°C) along the lower Altamaha River, Georgia, between July and December 2002 (n = 5).[4] Enteric Viruses of Humans and Animals in Aquatic Environments: Health Risks, Detection, and Potential Water Quality Assessment Tools. Theng-Theng Fong, Erin K. Lipp Microbiol Mol Biol Rev. 2005 Jun; 69(2): 357–371.

industry and have proven to be very resilient to an exhaustive list of mitigation strategies, including disinfection of drinking water. Although, a performance standard (level of disinfection required) for swine facilities remains elusive. Thusly, performance specifications required for the purchase of water disinfection systems remain absent, along with performance monitoring systems required to validate if an installation is, in fact working. A disinfection program, absent of a means to specify equipment and validate its performance on an ongoing basis lends itself to defeating its purpose. In this case, the worst-case scenario would include marketing such a program under the guise of providing a viable measure of biosecurity.

This has left one of the most recognized means for disease transmission; continue throughout the agricultural industry as being unaddressed, underestimated, and out-of-mind. Possibly, this is representative of, at least one of the underlying factors as to how it is possible for the avian influenza (poultry) and PRRSV & PEDV (swine), to persist over such a long period of time.

Is there a viable model employed for drinking water biosecurity elsewhere?

Within the public drinking water industry, disinfection of drinking water serving a municipal water distribution system must include a means to: a.) Inactivate or otherwise remove bacteria, viruses, and other potentially harmful organisms from entering; and is referred to as Primary Disinfection; b.) Maintain water quality by killing potentially harmful organisms and prevent the formation of biofilms as it flows through a drinking water distribution system and is referred to as Secondary Disinfection.

It is believed most water disinfection installations within the livestock production industry today simply use a chemical metering pump to inject a disinfectant directly into a water main serving a plumbing distribution system. Understanding the purpose of Primary and Secondary disinfection, this practice would lend itself to the latter and serves no relevance as a measure of biosecurity and preventing pathogens from entering a farms' drinking water supply. Presumably, the level of biosecurity, otherwise required to achieve this objective remains unknown and what has been predominately installed to date would be considered grossly insufficient as a measure to ensure drinking within a livestock facility will not serve as a causal agent for transmission of infectious disease.

Furthermore, it must be understood the term "disinfection" itself does not translate to a water supply as being free of pathogens. Rather, disinfection as it relates to drinking water, refers to the probability a water supply is absent of pathogens based upon efficacy of the technologies employed, the manner they are applied, installed, and a continuing performance verification to ensure that is. Whereas, in the event various drinking water disinfection systems are being considered for purchase, vendors whom suggest their technology disinfects water, and omits qualification (e.g. ? Log_{10} inactivation/removal virus at ? flow rate ? temperature and ? pH) and the means to verify this performance (via ASTM standards) on-site on a daily basis should be avoided.

This situation will continue until the agricultural industry, accepts groundwater as a vehicle for disease transmission. Thereafter, how critically important performance claims for equipment have been validated and ongoing performance verified will be understood and implemented.

Disinfection system design and performance validation

Today it is accepted virus are present within groundwater aquifers in varying concentrations, may remain infectious from days to years, travel at various speeds and direction in advance of being detected and then subsequently disappear (from a groundwater well system). In other words, to prevent waterborne pathogens from entering a livestock operation, water tests would need to be performed on a continuous basis. Plus, analytical methods capable of providing results in real-time become a necessity. In addition, due to the implications of a False-Positive result, analytical methods employed must be reliable and procedures (QA/QC) vigorously implemented.

Alternatively, the installation of a robust pathogen barrier may represent a more economical and practical approach. Public Water Systems (PWS) have been charged with the production and distribution of microbiologically safe drinking water for many years. Technology selection, engineering design and ongoing operation of such disinfection systems are well known and have proven to be effective and reliable. Utilizing such a tried and proven model may serve as a practical approach the livestock industry would consider.

Operative model for PWS's

In advance of the installation of a disinfection system a PWS is required submit an engineering plan for State (regulatory) review and acceptance. This plan must include equipment specifications and validation that it has been, or otherwise can be *validated to achieve and sustain a minimum of 4 Log_{10} (99.99%) inactivation/removal of virus over a defined range of water conditions, composition and flow rates. Understanding, higher removal and/or inactivation rates translate to a higher probability of maintaining a safe water supply. Within the PWS industry, 4 Log_{10} (99.99%) removal/inactivation represents the greatest balance in regard to performance (Biosecurity), cost and complexity with an acceptable level of protection against pathogens from entering and/or proliferating in public drinking water distribution systems.

*Via third-party testing under protocols, methodologies, and standards accepted specifically for this specific purpose.

This model is referred to as the CT Concept and is used for validation of disinfection efficacy in response to applied Disinfectant Concentration (mg/L) and Contact Time (minutes) and constituents of water composition, such as Temperature & pH and the means to confirm and record these parameters on a daily basis, via the use of accepted *Standard Methods* American Standard Test Method (ASTM).

Disinfectants

Currently this list includes chlorine, chloramine, chlorine dioxide and ozone. CT values for chloramine illustrate it is a much weaker disinfectant and recommended for secondary vs. primary disinfection or otherwise, if the formation of disinfection by-products present an issue. Alternative disinfectants can be used based upon proof of validity, safety, and ability to verify applied dose(s) for inactivation of select pathogens.

Ultraviolet Irradiation is accepted. Although performance (Ultraviolet (UV) dose) is dependent upon factors, including hydraulic profile within the reactor, flow rate, UV transmittance of the water, UV intensity, lamp output, lamp placement, lamp aging, fouling and microbe inactivation kinetics.[56,57] Extensive pretreatment technologies are typically required to ensure consistent performance and considered impractical for livestock applications.

Ozone must be generated on-site and adds to complexity, extensive site-specific engineering and sophisticated O&M practice. In addition, in atmospheric conditions ozone is gaseous, hazardous and only partially soluble in water. Its use is not recommended in the absence of a trained, licensed water treatment plant operator.

Hydrogen peroxide (H_2O_2) is currently employed within the livestock industry as a disinfectant. Unfortunately, the EPA, the governing body assigned enforce public drinking water quality standards, does not consider H_2O_2 a viable disinfectant and finds no reason to assign CT values for its use. Nor is the author aware if CT values have been developed by an independent 3rd party on behalf of a H_2O_2 manufacturer and/or distributor.

Chlorine dioxide (ClO_2) can be explosive. This presents significant issues in regard to shipping and storage. Similarly, the solid residue left from evaporated liquid ClO_2 presents a concern. ClO_2 must be generated on-site via mixing undiluted concentrations of Sodium Chlorite with an Oxidant (e.g. Chlorine) and/

or a Strong Acid (e.g. Hydrochloric Acid) in a controlled and safe manner. While there may have been an advancement in ClO_2 chemistry and it is now being shipped, stored, injected into a water line safely and measured thereafter at sufficient concentrations, this advancement remains unknown to this author. Accordingly, in the event you have, or currently use such a transportable ClO_2 product, the use of an ASTM standard or EPA approved method (www.hach.com) is recommended to verify measurable concentrations of ClO_2 are achieved and maintained throughout your plumbing distribution system.

Chlorination is an effective primary disinfectant for the inactivation of pathogens and commonly preferred for secondary disinfection as well. Further, A.) It is the easiest and least expensive disinfection method, regardless of distribution system size; B.) In a liquid state, such as sodium hypochlorite can easily be injected into water in advance of retention systems, measured and controlled; C.) The technology for chlorination is well developed as it is the most widely used and understood disinfection method throughout the world; D.) Although aqueous Cl solutions have a limited shelf life.

CT concept

The efficacy of chemical disinfectants can be predicted based on knowledge of the residual concentration of a disinfectant and factors that influence its performance, mainly temperature, pH, contact time and the level of disinfection required.[54] This relationship is commonly referred to as the CT concept, where CT is the product of "C" (the residual concentration of disinfectant, measured in mg/L) and "T" (the disinfectant contact time, measured in minutes) for a specific microorganism under defined conditions (e.g., temperature and pH). To account for disinfectant decay, the residual concentration is measured at the exit of the hydraulic contacting system.

Contact time T is calculated using a T_{10} value, defined as the detention time at which 90% of the water meets or exceeds the required contact time. The T_{10} value can be estimated by multiplying the theoretical hydraulic detention time by a baffling factor, dictated by the design of the retention system. Otherwise, a hydraulic tracer test may be performed to determine the actual contact time under expected maximum flow (rate) conditions. The T value is dependent on retention volume and the hydraulics related to the design of the retention system. Improving flow hydraulics to achieve CT requirements serves greater utility than increasing the disinfectant concentration and managed with physical modifications (such as to achieve laminar flow and/or increasing the distance of flow paths) within the contacting system. CT tables for 2 log, 3 log and 4 log inactivation of viruses can be found in an Environmental Protection Agency (EPA) Guidance Manual.[55] Selected CT values are presented for a range of 4 Log_{10} values using disinfectants that have been proven and acceptable for drinking water.

Determination of disinfection dose:

Log_{10} inactivation is based on the Delivered Dose, "CT"
"C" is the disinfectant residual (mg/L)
"T" is the exposure or contact time (minutes)
Multiply them:
$C \cdot T = mg/L \cdot min$ = (delivered dose)

CT Values can be found in US EPA tables to determine log inactivation based on specific monitored parameters (pH, disinfectant residual and/or temperature):

Example

Design a 20 gallon per minute (gpm) water disinfection system capable of inactivating up to 99.99% of all viruses on a continuous basis within the water supply for a Livestock operation. Chlorine is the preferred disinfectant water pH and Temperature is 7° and 10° C respectively.

From Table 4 a CT Value of 6 is identified to achieve 4 Log_{10} inactivation of virus with Free Chlorine within a source water with a pH between 6 and 9 and a temperature of 10° C.

A disinfection system of the design as described in Figure 5 can be validated as described below.

Hydraulic Factor (by Tracer Study @ 20 gpm) = 0.5
Net Hydraulic Retention Time (T) = (114 gallons/20 gpm) × .5 = 2.85 minutes

Calculation of Free Chlorine Concentration to achieve 4 Log_{10} Inactivation:

(CT Value/T) = mg/L
6/2.85 = 2.11 mg/L

With the performance parameters described above, when a chlorine solution enters the inlet of the first tank at a rate that will provide a concentration of Free Chlorine of ≥ 2.11 mg/L in the water exiting the last tank, performance is validated with an inactivation rate of 99.99% against viral pathogens.

Refer to Table 5 for concentrations (mg/L) providing less security (99% & 99.9%).

The same model applied for Chlorine Dioxide is shown in Table 6.

CT Value (mg/L*min) to achieve 4 Log_{10} inactivation of virus with Free Chlorine. With a water temperature of 10° C and a pH between 6.0 and 9.0 = 25.1 Net Hydraulic Retention Time (T) = (114 gallons/20 gpm) × .5 = 2.85minutes.

Calculation of Chlorine Dioxide Concentration to achieve 4 Log_{10} Inactivation:

(CT Value/T) = mg/L
25.1/2.85 = 8.81 mg/L

With the performance parameters described above, when a Chlorine Dioxide solution enters the inlet of the first tank at a rate that will provide a concentration of 8.81 mg/L in the

Table 4: Inactivation of viruses by free chlorine, pH 6.0-9.0

Concentration mg/L × Time (minutes) CT values for inactivation						
Disinfectant:	Chlorine (free)			pH: 6 - 9	VIRUS	
Safety level	Temperature (Celsius)					
	≤ 1	5	10	15	20	25
2-Log_{10} 99%	6	4	3	2	1	1
3-Log_{10} 99.90%	9	6	4	3	2	1
*4-Log_{10} 99.99%	12	8	6	4	3	2

* Required safety level (barrier) for potable drinking water systems

Figure 5: Example of retention system design.

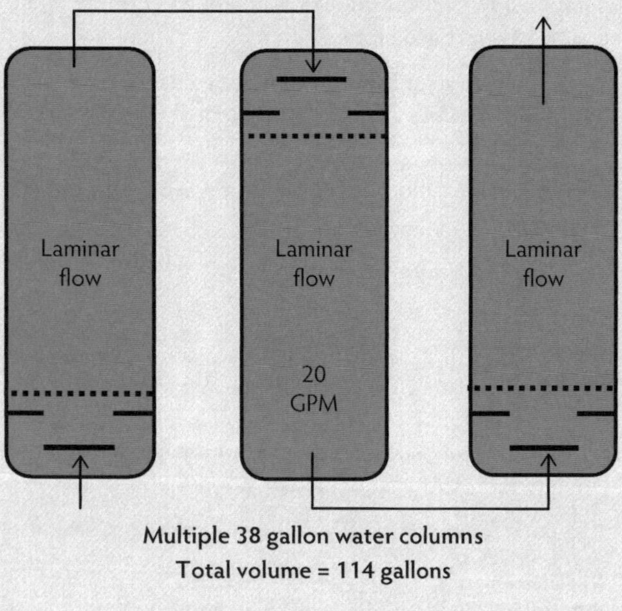

water exiting the last tank, the level of protection is validated with an inactivation rate of 99.99% against viral pathogens. Concentrations providing less security (99% & 99.9%) are shown in Table 7.

Water composition and pretreatment

The type of treatment prior to primary disinfection, and the manner it is prescribed, designed, and operated, will have a significant influence on the performance of each water disinfection system. Due to the vast array of groundwater chemistries and density of micro-particulates/microorganisms, to ensure expected performance outcomes are achieved and to avoid the promulgation biofilms within piping and fixtures, prefiltration is typically necessary (or required with the presence of constituents such as iron, arsenic, manganese and suspended solids). Reducing pH to < 7.5 becomes an important consideration with the use of sodium hypochlorite as a disinfectant.

The following parameters must be known and presented to the livestock facility manager(s) within the engineering design for each water treatment system:

1. Allowable concentration of disinfectant within the drinking water.
2. Lethality of the disinfectant in relation to targeted pathogens.
3. The amount of retention (contact) time (T_{10}) required to achieve 90%, 99%, 99.9%, 99.99% .... etc. inactivation/removal per measured concentration of disinfectant.
4. The means to validate T_{10}.
5. The means to validate an installations' expected performance in advance (of purchase), vs. what is claimed by the manufacturer, distributor, or vendor.
6. The means to accurately verify disinfection performance (ASTM test standards) in real-time and/or daily.
7. The means to accurately forecast an installations performance for the specific application(s) and site-specific conditions, including:
   a. A water analysis including all parameters that would serve to influence performance of the technologies to be installed.
   b. An understanding the degree the source water's composition may change beyond the first year of the well systems installation and thereafter.
8. The maximum flow rate available from the well system accounting for:
   a. Static water level.
   b. Flow capacity of the well (gpm)
   c. Flow capacity of the well's pump.

Table 5: Inactivation of viruses by free chlorine, providing less security

Concentration mg/L values for inactivation

For 20 GPM with 3 - 38 gallon columns @ laminar flow

Disinfectant:	Chlorine (free) (mg/L)**			pH: 6 - 9	VIRUS	
Safety level	Temperature (Celsius)					
	<1	5	10	15	20	25
2-Log_{10} 99%	2.11	1.40	1.05	0.70	0.35	0.35
3-Log_{10} 99.90%	3.16	2.11	1.40	1.05	0.70	0.35
*4-Log_{10} 99.99%	4.42**	2.81	2.11	1.40	1.05	0.70

* Required safety level (barrier) for potable drinking water systems
** Concentration limit for potable drinking water = 4.0 mg/L

Table 6: CT– Inactivation of viruses by chlorine dioxide, pH 6.0-9.0

Concentration mg/L × Time (minutes) cT values for inactivation

Disinfectant:	Chlorine Dioxide (mg/L)			pH: 6 - 9	VIRUS	
Safety level	Temperature (Celsius)					
	≤1	5	10	15	20	25
2-Log_{10} 99%	8.4	5.6	4.2	2.8	2.1	1.4
3-Log_{10} 99.90%	25.6	17.1	12.8	8.6	6.4	4.3
*4-Log_{10} 99.99%	50.1	33.4	25.1	16.7	12.5	8.4

* Required safety level (barrier) for potable drinking water systems

d. Total (actual) flow capacity available at the installation site based upon the following parameters:
   i. Calculated friction loss based upon static water level and piping (to installation site).
   ii. Specified performance curve (gpm vs. total friction loss + required operating pressure (water treatment plant and livestock operation).

9. With this information, a water treatment plant may be designed to ensure initial and ongoing performance of the disinfection system (primary) will be achieved with continued disinfection throughout the water distribution system (secondary).
10. Equally important is the availability of on-site personal that can be assigned responsibility to perform water analysis, record visual operational parameters of the water treatment plant, and monitor feed supply of chemical metering pump(s).

Economic considerations

The benefit of nearly all biosecurity systems are realized in response to events that occur on a periodic and unpredictable basis. Accordingly, the economic value they provide is based upon factors such as: a.) How well they perform during these events; b) The probability such an event will occur; c.) Potential of financial loss if it does; d.) Virulence and lethality of the pathogen it removes.

Therefore, the value of any biosecurity system is based on assessment of risk and cost. For water disinfection systems both capital and operating costs are directly dependent upon:

Table 7: CT– Inactivation of viruses by chlorine dioxide, providing less security

Concentration mg/L values for inactivation For 20 GPM with 3 - 38 gallon columns @ laminar flow						
Disinfectant:	Chlorine dioxide**			pH: 6 - 9	VIRUS	
Safety level	Temperature (Celsius)					
	≤1	5	10	15	20	25
2-Log_{10} 99%	2.94	1.96	1.47	0.98	0.74	0.49
3-Log_{10} 99.90%	8.98	6.00	4.49	3.02	2.25	1.51
*4-Log_{10} 99.99%	17.58	11.72	8.81	5.86	4.39	2.94

* Required safety level (barrier) for potable drinking water systems
** Concentration limit for potable drinking water = 4.0 mg/L

1. Process flow rate (gpm)
2. Water composition, temperature and pH
3. Reliability of equipment components
4. Selection of disinfectant
5. Desired level of
   a. Level of assurance (probability) water will be free of pathogens
      i. i.e.: 90% - > 99.99% inactivation
      ii. System efficacy and reliability.
   b. Automation and instrumentation
   c. Fail-safe options
6. Pretreatment requirements
7. If and to what degree, the installation may (+/-) affect:
   a. Livestock performance metrics
   b. Operating cost(s)

As described above, while most, if not all current installations within the livestock industry may not provide much in the order of primary disinfection, it is evident many have served to provide other favorable outcomes in regards to livestock performance metrics and operating costs (i.e. cleaning or replacement of plugged water fixtures, evaporative cool cells, etc.). Meaning, drinking water "disinfection" even with a minimal degree of efficacy has proven to be a component of a viable economic model beyond serving as a measure of biosecurity.

Cautionary notes

1. The Safe Drinking Water Act (SDWA) is enforceable for establishments either serving as a place of employment and/or accessible to the public.[7] Companies engaged within the livestock industry may remain unaware of this consideration, nor associated litigious issues they present. This becomes especially poignant if substances, intentionally added to a water supply may deleteriously affect its potability.
2. Administering solutions, (such as disinfectants, acids, sequestering agents, and/or medications) into a potable water distribution system (i.e. connected to a washroom, office area, bathrooms) at concentrations that may cause (directly or indirectly) water to become non-potable, is a concern. Similarly, is the practice of storing and handling concentrated (hazardous) substances within an area absent of appropriate ventilation, eye wash, physical separation of concentrates (such as acids and oxidants), and the manner they are transferred into a pressurized water line, (leaking metering pumps), etc. in the absence of appropriate safety protocols and supplies.
3. Note: If the purpose of a plumbing distribution system is solely dedicated for the conveyance of water to livestock and other non-potable uses, it is considered a non-potable water distribution system and must be well identified and remain isolated from any fixture that potentially could be used (i.e. inadvertently) as a potable drinking water source.
4. The use of pressurized gaseous disinfectants are generally not recommended within the livestock production industry.
5. Metering pumps
   a. Vapor or gas bubbles can form due to gasification (i.e. the degradation of sodium hypochlorite) produces a gas which is mostly oxygen, particularly if the solution is below atmospheric pressure. This can lead to gas locking of the suction line in a diaphragm pump. Therefore, pumps should be provided with a flooded suction (i.e. the pump inlet should always be located below the liquid level in a storage tank). Today this limitation has been addressed in the design of a number of newer diaphragm pumps.
   b. The use of peristaltic metering pumps have become prevalent within the livestock industry. Most have been

associated with a higher degree of maintenance and operator attention, resulting in chemical leakage and a continued state of disrepair. This is not representative of good practice in regard to biosecurity nor laws governing the use of hazardous chemicals. All metering pumps must be observed daily and maintained frequently. Further, the purchase of a quality pump is representative of a good investment and the acceptance of a pump used as a marketing ploy may not be.

6. Because Chlorine Dioxide serves to cause nervous system effects for infants and young children and anemia, the US EPA has a maximum contaminant level (MCL) of 1 mg/L for chlorite (from which ClO_2 is generated) and a maximum residual disinfection level (MRDL) of 0.8 mg/L for ClO_2.
7. Because Chlorine serves to cause eye/nose irritation and stomach discomfort the US EPA has assigned an MCL and MRDL of 4 mg/L for Cl.
8. The pH of sodium hypochlorite (a common chlorine solution) is high because sodium hydroxide is used in its manufacture to reduce decomposition and increase the stability of the product. Care is needed when dosing water with a high concentration of hardness ions or waters with carbon dioxide present as the highly alkaline product can lead to scale formation within plumbing systems leading to lower flow rates and plugged dosing quills.
9. While the following note may be painfully obvious, it is worth mentioning. Water containing either a high concentration of an oxidant and/or acid will serve toward the demise of any metallic component or elastomer within a plumbing distribution system and associated fixtures.
10. In the event Test Strips are used for water analyses, verify their accuracy with the use of an ASTM or EPA standard method frequently.
11. Installations as represented in Figures 6A, 6B, 6C, 6D, 6E provide examples of appropriate water treatment plant designs for biosecurity, process control, instrumentation and QA/QC for swine production facilities including; A.) Boar-Stud, B.) Sow/Gilt, C.) Nursery or Gilt, D.) Wean and/or Finish.
12. A water analysis providing verification of the absence of *E coli* or Coliforms cannot be used to suggest the absence of viral pathogens.

References

1. National Primary Drinking Water Regulations: Ground Water Rule. 2006. Available at **https://www.federalregister.gov/documents/2006/11/08/06-8763/national-primary-drinking-water-regulations-ground-water-rule. Accessed 15 October 2019.**

2. Center for Disease Control and Prevention (CDC): History of Drinking Water Treatment, A Century of U.S. Water Chlorination and Treatment: One of the Ten Greatest Public Health Achievements of the 20th Century. Available at **https://www.cdc.gov/healthywater/drinking/history.html/** Accessed 10 November 2019.

3. Center for Disease Control and Prevention (CDC), Control of infectious diseases, MMWR Morb Mortal Wkly Rep. 1999 Jul 30;48(29):621-9.

4. Fong, T. T., & Lipp, E. K. (2005). Enteric viruses of humans and animals in aquatic environments: health risks, detection, and potential water quality assessment tools. *Microbiology and molecular biology reviews: MMBR, 69*(2), 357–371. doi:10.1128/MMBR.69.2.357-371.2005.

5. United States Environmental Protection Agency: Safe Drinking Water Act (SDWA)(1974). Available at **https://www.epa.gov/sdwa**. Access 5 October 2019.

6. Craun, G.F., Brunkard, J.M., Yoder, J.S., Roberts, V.A., Carpenter, J., Wade, T., Calderon, R.L., Roberts, J.M., Beach, M.J., Roy, S.L. Causes of Outbreaks Associated with Drinking Water in the United States from 1971 to 2006. Clinical Microbiology Reviews Jul 2010, 23 (3) 507-528; DOI: 10.1128/CMR.00077-09.

7. Abbaszadegan, M., Stewart, P.W., LeChevallier, M.W. and Gerba, C.P. (1998). Application of PCR technologies for virus detection in groundwater. American Water Works Association, 90740, Denver, Colorado.

8. Abbaszadegan, M., Lechevallier, M. and Gerba, C. (2003). Occurrence of viruses in US groundwaters. J. Am. Water Works Assoc., 95(9): 107–120+12.

9. Banks, W.S.L., Klohe, C.A. and Battigelli, D.A. (2001). Occurrence and distribution of enteric viruses in shallow ground water and factors affecting well vulnerability to microbiological contamination in Worcester and Wicomico Counties, Maryland. U.S.G.S., Water resources investigations report 01-4147.

10. Banks, W.S.L. and Battigelli, D.A. (2002). Occurrence and distribution of microbiological contamination and enteric viruses in shallow ground water in Baltimore and Harford Counties, Maryland. U.S.G.S., Water resources investigations report 01-4216.

11. Lindsey, B.D., Rasberry, J.S. and Zimmerman, T.M. (2002). Microbiological quality of water from noncommunity supply wells in carbonate and crystalline aquifers of Pennsylvania. U.S.G.S, 01-4268.

12. Borchardt, M.A., Bertz, P.D., Spencer, S.K. and Battigelli, D.A. (2003). Incidence of enteric viruses in groundwater from household wells in Wisconsin. Appl. Environ. Microbiol., 69(2): 1172–1180.

13. Fout, G.S., Martinson, B.C., Moyer, M.W.N. and Dahling, D.R. (2003). A multiplex reverse transcription-PCR method for detection of human enteric viruses in groundwater. Appl. Environ. Microbiol., 69(6): 3158–3164.

14. Francy, D.S., Bushon, R.N., Stopar, J., Luzano, E.J. and Fout, G.S. (2004). Environmental factors and chemical and microbiological water-quality constituents related to the presence of enteric viruses in ground water from small public water supplies in Southeastern Michigan. U.S.G.S., Scientific investigations report 2004–5219.

15. Locas, A., Barthe, C., Barbeau, B., Carrière, A. and Payment, P. (2007). Virus occurrence in municipal groundwater sources in Quebec, Canada. Can. J. Microbiol., 53(6): 688–694.

16. Locas, A., Barthe, C., Margolin, A.B. and Payment, P. (2008). Groundwater microbiological quality in Canadian drinking water municipal wells. Can. J. Microbiol., 54(6): 472–478.

17. Hunt, R.J., Borchardt, M.A., Richards, K.D. and Spencer, S.K. (2010). Assessment of sewer source contamination of drinking water wells using tracers and human enteric viruses. Environ. Sci. Technol., 44(20): 7956–7963.

18. Gibson, K.E. and Schwab, K.J. (2011). Detection of bacterial indicators and human and bovine enteric viruses in surface water and groundwater sources potentially impacted by animal and human wastes in lower Yakima Valley, Washington. Appl. Environ. Microbiol., 77(1): 355–362.

19. Borchardt, M.A., Spencer, S.K., Kieke, B.A., Lambertini, E. and Loge, F.J. (2012). Viruses in nondisinfected drinking water from municipal wells and community incidence of acute gastrointestinal illness. Environ. Health Perspect., 120(9): 1272–1279.

20. Allen, A. S. (2013). Vulnerability of a fractured bedrock aquifer to emerging sewage-derived contaminants and their use as indicators of virus contamination. Master's Thesis, University of Guelph, Guelph, Ontario.

21. Bradbury, K.R., Borchardt, M.A., Gotkowitz, M., Spencer, S.K., Zhu, J. and Hunt, R.J. (2013). Source and transport of human enteric viruses in deep municipal water supply wells. Environ. Sci. Technol., 47(9): 4096–4103.

22. Pang, L., Close, M., Goltz, M., Noonan, M. and Sinton, L. (2005). Filtration and transport of bacillus subtilis spores and the F-RNA phage MS2 in a coarse alluvial gravel aquifer: Implications in the estimation of setback distances. J. Contam. Hydrol., 77(3): 165–194.

23. Guan T.T.Y., Holley R.A. (2003) Pathogen Survival in Swine Manure Environments and Transmission of Human Enteric Illness—A Review. In: Hog Manure Management, the Environment and Human Health. Springer, Boston, MA

24. Bales, R.C., Gerba, C.P., Grondin, G.H. and Jensen, S.L. (1989). Bacteriophage transport in sandy soil and fractured tuff. Appl. Environ. Microbiol., 55(8): 2061–2067.

25. Sinton, L.W., Finlay, R.K., Pang, L. and Scott, D.M. (1997). Transport of bacteria and bacteriophages in irrigated effluent into and through an alluvial gravel aquifer. Water Air Soil Pollut., 98(1–2): 17–42.

26. Berger, P. (1994). Regulation related to groundwater contamination: The draft groundwater disinfection rule. In: Groundwater contamination and control. Zoller, U. (ed.). M. Dekker, New York, pp. 645–659.

Figure 6 A: Boar stud facility – Water treatment plant design includes primary disinfection – filtration – secondary disinfection. Water used within laboratory and dilution semen also includes: residual disinfectant removal, 1 micron absolute pharmaceutical grade filtration, reverse osmosis and mixed bed DI with virgin, non-regenerated IX media, with continuous recycle through ultrapure water distribution system comprised of polyvinylidene fluoride (PVDF) materials and ultraviolet irradiation (254 nm @ 40+mJ/cm^2) disinfection (per each recycle) and online/continuous instrumentation for monitoring of step-processes, performance verification and QA/QC.

27. DeBorde, D.C., Woessner, W.W., Kiley, Q.T. and Ball, P. (1999). Rapid transport of viruses in a floodplain aquifer. Water Res., 33(10): 2229–2238.

28. Cherry, J.A., Parker, B.L., Bradbury, K.R., Eaton, T.T., Gotkowitz, M.B., Hart, D.J. and Borchardt, M.A. (2006). Contaminant transport through aquitards: A state-of-the-science review. AWWA Research Foundation, Denver, Colorado.

29. Hunt, R.J., Borchardt, M.A. and Bradbury, K.R. (2014). Viruses as groundwater tracers: Using ecohydrology to characterize short travel times in aquifers. Ground Water, 52(2): 187–193.

30. Levison, J.K. and Novakowski, K.S. (2012). Rapid transport from the surface to wells in fractured rock: a unique infiltration tracer experiment. J. Contam. Hydrol., 131(1–4): 29–38.

31. Schijven, J.F. and Hassanizadeh, S.M. (2000). Removal of viruses by soil passage: Overview of modeling, processes, and parameters. Crit. Rev. Environ. Sci. Technol., 30(1): 49–127.

32. Cherry, J.A., Parker, B.L., Bradbury, K.R., Eaton, T.T., Gotkowitz, M.B., Hart, D.J. and Borchardt, M.A. (2006). Contaminant transport through aquitards: A state-of-the-science review. Awwa Research Foundation, Denver, Colorado.

33. Keswick, B.H. and Gerba, C.P. (1980). Viruses in groundwater. Environ. Sci. Technol., 14(11): 1290–1297.

34. Borchardt, Hunt, Bradbury "Viruses as Groundwater Tracers: Using Ecohydrology to Characterize Short Travel Times in Aquifers" Groundwater Journal, January 16(2014)

35. Hunt, R. J., Borchardt, M. A. and Bradbury, K. R. (2014), Viruses as Groundwater Tracers: Using Ecohydrology to Characterize Short Travel Times in Aquifers. Groundwater, 52: 187-193. doi:**10.1111/gwat.12158**.

36. Keswick, B.H., Secor, S.L., Gerba, C.P. and Cech, I. (1982). Survival of enteric viruses and indicator bacteria in groundwater. J. Environ. Sci. Health Part A Environ. Sci. Eng., 17(6): 903–912.

37. John, D.E. and Rose, J.B. (2005). Review of factors affecting microbial survival in groundwater. Environ. Sci. Technol., 39(19): 7345–7356.

38. Gerba, C.P. (1984). Applied and theoretical aspects of virus adsorption to surfaces. Adv. Appl. Microbiol., 30: 133– 168.

39. Carlson, G.F.,Jr, Woodard, F.E., Wentworth, D.F. and Sproul, O.J. (1968). Virus inactivation on clay particles in natural waters. J. Water Pollut. Control Fed., 40(2): R89–106.

40. Sobsey, M.D., Shields, P.A., Hauchman, F.H., Hazard, R.L. and Caton III, L.W. (1986). Survival and transport of hepatitis A virus in soils, groundwater and wastewater. Water Sci. Technol., 18(10): 97–106.

Figure 6 B: Sow site A - water treatment plant design, primary disinfection – filtration – secondary disinfection. With a plumbing distribution manifold to accommodate each of multiple barns with multiple variations of water composition, medication(s) & pH, and provide independent water distribution lines for living quarters, utility, sprayer & compost applications.

Figure 6 C: Swine nursery or gilt site. (Comparative difference (before/after) in water treatment plant design).

41. Gordon, C. and Toze, S. (2003). Influence of groundwater characteristics on the survival of enteric viruses. J. Appl. Microbiol., 95(3): 536–544.

42. Bitton, G. (1975). Adsorption of viruses onto surfaces in soil and water. Water Res., 9(5–6): 473–484.

43. Song, L. and Elimelech, M. (1994). Transient deposition of colloidal particles in heterogeneous porous media. J. Colloid Interface Sci., 167(2): 301–313.

44. Pieper, A.P., Ryan, J.N., Harvey, R.W., Amy, G.L., Illangasekare, T.H. and Metge, D.W. (1997). Transport and recovery of bacteriophage PRD1 in a sand and gravel aquifer: Effect of sewage-derived organic matter. Environ. Sci. Technol., 31(4): 1163–1170.

45. Duboise, S.M., Moore, B.E. and Sagik, B.P. (1976). Poliovirus survival and movement in a sandy forest soil. Appl. Environ. Microbiol., 31(4): 536–543.

46. Bales, R.C., Li, S., Maguire, K.M., Yahya, M.T. and Gerba, C.P. (1993). MS-2 and poliovirus transport in porous media: Hydrophobic effects and chemical perturbations. Water Resour. Res., 29(4): 957–963.

47. Loveland, J.P., Ryan, J.N., Amy, G.L. and Harvey, R.W. (1996). The reversibility of virus attachment to mineral surfaces. Colloids Surf. a Physicochem. Eng. Asp., 107: 205–221.

48. Scandura, J.E. and Sobsey, M.D. (1997). Viral and bacterial contamination of groundwater from on-site sewage treatment systems. Water Sci. Technol., 35(11–12): 141–146.

49. DeBorde, D.C., Woessner, W.W., Kiley, Q.T. and Ball, P. (1999). Rapid transport of viruses in a floodplain aquifer. Water Res., 33(10): 2229–2238.

Figure 6 D: Wean-to-finish and Finish. (Comparative difference (before/after) in water treatment plant design).

Figure 6 E: Chemical metering station - example.

50. Guidelines for Drinking-Water Quality: Fourth Edition Incorporating the First Addendum. Geneva: World Health Organization; 2017. 7, Microbial aspects. Available from: https://www.ncbi.nlm.nih.gov/books/NBK442381/.

51. Allen, A. S. (2013). Vulnerability of a fractured bedrock aquifer to emerging sewage-derived contaminants and their use as indicators of virus contamination. Master's Thesis, University of Guelph, Guelph, Ontario.

52. Goss M.J., Unc A., Chen S. (2000) Transport of Nitrogen, Phosphorus and Microorganisms from Manure into Surface- and Groundwater. In: Balázs E. et al. (eds) Biological Resource Management Connecting Science and Policy. Springer, Berlin, Heidelberg

53. Farthing, M. J. G. 1989. Viruses and the gut. Smith Kline & French, Walwyn Garden City, Hertfordshire, United Kingdom.

54. Tun, H. M., Cai, Z., & Khafipour, E. (2016). Monitoring Survivability and Infectivity of Porcine Epidemic Diarrhea Virus (PEDv) in the Infected On-Farm Earthen Manure Storages (EMS). *Frontiers in microbiology*, *7*, 265. doi:10.3389/fmicb.2016.00265.

55. US EPA (1991). Guidance manual for compliance with the filtration and disinfection requirements for public water systems using surface water sources. US Environmental Protection Agency, Washington, DC.

56. US EPA (2001b). Method 1602: Male-specific (F+) and somatic coliphage in water by single agar layer (SAL) procedure. Office of Water, U.S. Environmental Protection Agency, EPA/821-R-01-029, Washington, DC.

57. Bolton, J.R. and Cotton, C.A. (2008). The ultraviolet disinfection handbook. American Water Works Association, Denver, Colorado.

Impact of regulation on future pharmaceutical use

Liz Wagstrom, DVM, MS, Diplomate ACVPM
National Pork Producers Council, Des Moines, Iowa

Introduction

The United States Food and Drug Administration (FDA) regulates veterinary pharmaceuticals while the United States Department of Agriculture (USDA) regulates veterinary biologics and diagnostics. FDA issues Guidance for Industry (GFI) documents that represent FDA's current thinking on a topic. They do not create of confer any rights for or on any person and do not operate to bind FDA or the public. Alternative approaches may be used if the approach satisfies the requirements of the applicable statutes and regulations. This approach creates a system of Guidance documents that in many cases are viewed by veterinarians and pork producers as "regulation". For the purpose of this paper we will focus on the regulatory environment for medically important antimicrobials.

Background

In September of 2018 FDA published the document Supporting Antimicrobial Stewardship in Veterinary Settings; Goals for Fiscal Years 2019-2023 (**https://www.fda.gov/media/115776/download**). This plan described three overarching goals for the agency: 1.) Align antimicrobial drug product use with principles of antimicrobial stewardship, 2.) Foster stewardship of antimicrobials in veterinary settings, and 3.) Enhance monitoring of antimicrobial resistance and antimicrobial drug use in animals. FDA published a report on progress toward meeting those goals in October, 2019 (**https://www.fda.gov/about-fda/fda-track-agency-wide-program-performance/fda-track-progress-fdas-support-antimicrobial-stewardship-veterinary-settings**).

The progress report reviewed steps taken under Guidance for Industry #213 and revisions to the Veterinary Feed Directive (VFD) rule that resulted in marketing status of medically important antimicrobials delivered in feed or water converting to either VFD or Rx. Of the 292 new animal drugs applications initially affected by GFI #213: 84 were completely withdrawn, 93 oral dosage form products intended for use in water were converted to prescription status, 115 products intended for use in feed were converted to VFD, and production (e.g. growth promotion) indications were withdrawn on all (31) products. This move resulted in 95%, by weight, of all medically important antimicrobials sold/distributed for use in food-producing animals requiring veterinary oversight.

The FDA progress report also stated that administration of medically important antimicrobial drugs for undefined periods may contribute to the development of antimicrobial resistance. Since 2003 the agency has required that all new unique indications for pioneer new animal drug approvals to include a defined duration of use. In 2016 FDA sought public input on establishing appropriately targeted durations of therapeutic use of medically important antimicrobial drugs in food producing animals. Additionally, FDA issued a Request for Applications funding opportunity for studies that can help target and define durations of use for certain medically important antimicrobial drugs. There are currently 85 approved drug applications that lack a defined duration of use.

FDA's progress report refers to work on new product development and GFI #152, Evaluating the Safety of Antimicrobial New Animal Drugs with Regard to Their Microbiologic Effects on Bacteria of Human Health Concern. GFI #152 was published in 2003 and lays out an antimicrobial resistance risk assessment process, including ranking of antimicrobials by their importance to human health. All new applications since 2003 have gone through the GFI #152 process.

Looking forward

The FDA has clearly signaled some expected regulatory actions to address concerns with the development of antimicrobial resistance. The agency has published draft GFI #263 which outlines a process that sponsors can take to move marketing status of the remaining dosage forms of medically important antimicrobials (e.g. injectable, intramammary) to by prescription. As was done with the conversion of feed and water dosage forms the FDA will work with sponsors to hemp them to voluntarily change the marketing status. Comments on the draft Guidance were due on December 24, 2019. There will be a period of time following the comment period for FDA to appraise the comments received and revise/respond to the comments in publication of a Final GFI #263. The timeframe given to make the marketing status changes is two years following the publication of the Final GFI #263. If the agency determines that adequate progress has not been made by the end of the 2-year timeframe they will consider further action under the Federal Food, Drug and Cosmetic Act.

FDA has stated that they will work with to establish targeted durations of use so that veterinarians have clear labeling instructions and instructions on how long to use a medically important drug. FDA has reported that they will publish a draft guidance that addresses establishing defined durations of use for affected new animal drug applications. This draft guidance will outline a proposed process and timeline for sponsors to revise approved conditions of use appropriately including updating product labeling. Implementation timeframes are uncertain as sufficient data to establish appropriately defined durations of use may not be available thus requiring new studies be conducted.

The agency has signaled that there is a need to update the GFI #152 including a potential revision to the Appendix A, the ranking of antimicrobial drugs according to their importance in human medicine to reflect current science and medical practice. This will help ensure that as part of the animal drug approval process there is an effective process for assessing antimicrobial resistance concerns. At this time, FDA is considering soliciting public input in early 2020, prior to initiating the formal public comment process on a draft guidance.

Summary

The United States has had an increasingly strict regulatory structure for animal antimicrobials since the 1980s. The FDA has signaled there will be additional regulatory actions taken to assure consideration of antimicrobial resistance during the approval process, that there is veterinary oversight of all uses of medically important antimicrobials and that labels clearly spell out defined durations of use for antimicrobials.

Customer pressure on future antibiotic use

Jarrod Sutton
National Pork Board, Des Moines, Iowa

Consumer and activist groups continue to pressure large retail and foodservice brands to eliminate the use of antibiotics in their supply chains. While the agendas of these groups may vary, their focus on residue and resistance is consistent on all fronts. Food brands are tasked with balancing science and the emotional needs of consumers. What will your customers demand next? Trend-spotting analytics combined with market intelligence gathered from interaction with food brands gives insight into where the major food companies are heading with their short- and long-term goals for antibiotic usage.

The future of antibiotic resistance pressures on pork production

Peter Davies, BVSc, PhD
Department of Veterinary Population Medicine, University of Minnesota, St Paul, Minnesota

Introduction

The specter of antimicrobial resistance has haunted human and veterinary medicine since the inception of antibiotic use in the 1940s, but clinically problematic resistance has become a critical concern in human medicine over the last twenty years. This has brought scrutiny of how antibiotics are used across all the prescribing professions, efforts to define what 'best practices' may be, and how they might be disseminated and implemented.

For over 60 years, antibiotics have been a key tool for maintaining the health and well-being of food animals. All decisions in livestock farming have trade-offs,[1] but the tradeoffs regarding antibiotic use in food animals are highly asymmetric. Farmers and veterinarians are asked to weigh tangible animal health benefits of antibiotic use against less tangible, and inconclusively debated, risks to human health that might ensue from that use. Although there are evolving efforts to advance antibiotic stewardship within all sectors of veterinary medicine, substantial changes in access to antibiotics may well be dictated from 'extra professional' sources. In food animal production, pressures to alter practices of antibiotic use are channeled via regulatory and commercial avenues. At this AASV meeting, the regulatory and commercial landscapes shaping antibiotic use in the US swine industry are being addressed by experts in those fields. The following discussion speculates about factors that may generate the pressures that motivate shifts in those regulatory and commercial landscapes regarding antibiotic use in production. It does not address the extent to which perceptions about antibiotic use and resistance may shape consumer behavior.

Antibiotic resistance is one of the core issues swept up into the epidemic of 'One Health' initiatives around the world. When confronting such complex problems, it is easy to sing 'Kumbaya' and laud the need for interdisciplinary interactions among experts in human, animal, and environmental health under a 'One Health' umbrella. But the devil always resides in the details, and translating well-intentioned hand wringing into constructive and evidence based action remains to be achieved. We are yet to see much enlightenment about how to identify and effectively implement stewardship approaches that objectively balance benefits to animal health and well-being, while minimizing potential risks to human health. Evidence based approaches require time which many in the medical field argue we simply do not have. Decision making cannot await perfect knowledge, and wherever science is slow to fill the void, philosophy, values and politics will occupy the driver's seat.

Scientific and data driven events

With the benefit of hindsight, I contend that few prescribing professionals would today argue that global over-the-counter marketing of antibiotics for purposes of production enhancement of food animals was a great idea. However, we have also long forgotten the history of food shortages and rationing around World War 2 that made food insecurity a major societal concern of that era. Efforts to 'put the genie back in the bottle' started in 1969 with the Swann Report in the United Kingdom.[2] The Swann report was significant in that it introduced the concept of regulation motivated by antibiotic resistance concerns; differentiated between antibiotics based on 'medical relevance'; and promoted the precautionary principle to recommend restriction of antibiotic use in food animals. Much later (1986), Sweden banned the use of all antibiotics for growth promotion, with a further decade passing before several antibiotics (avoparcin, bacitracin, spiramycin, tylosin, and virginiamycin) were banned across the European Union (EU). In 2006, the EU ban on growth promotion was extended to all antibiotics, again based more on the precautionary principle than convincing data indicating measurable impacts on public health. In the US, similar regulations (Guidance #213 and the new VFD regulations) were not implemented in 2017. While most of the momentum for these regulatory changes was arguably based on 'philosophy', the philosophy around the need to restrict use in animals was successively reinforced by epidemiologic or public health developments that raised alarms within the medical community and beyond. Some of the most influential events are listed below.

Multiple resistant salmonellae, including *Salmonella* Typhimurium DT104

A hallmark of *Salmonella* epidemiology in humans and animals is the occurrence of successive waves of particular variants that emerge to account for a relatively high proportion of cases before subsequently waning in importance.[3,4] This well recognized phenomenon remains poorly understood. Around 1964, an outbreak of antibiotic resistant *S* Typhimurium phage type 29 in cattle and humans was a key event that prompted the appointment of the Swann Committee in the UK in 1968.[5] In the mid-1990s, the DT104 variant of *S* Typhimurium emerged in many countries of the world including Denmark and the USA. DT104. The DT104 variant typically harbored resistance genes to five antibiotic classes (penicillins, tetracyclines, phenicols, aminoglycosides and sulfonamides) encoded together on the same cassette of DNA elements called integrons that were described in 1989. The emergence of

DT104 and other multiple resistant *Salmonella* variants was the driving force behind the broader banning of growth promotant antibiotics in pigs in Denmark in 2000, an action later adopted across the EU in 2006. DT104 became more prevalent among human and animal clinical isolates in multiple countries, and antibiotic use in food animals (particularly growth promotant use) was widely incriminated as the 'reason' for this emergence. In keeping with the long history of emergence and decline of *Salmonella* serotypes over time, DT104 is now relatively uncommon, and attention is now more focused on other emerging variants, notably the 1,4[5],12:i:- monophasic variant of *S* Typhimurium.[6] The extent to which antibiotic use in humans and animals influenced the emergence and diminution of DT104 and other resistant *Salmonella* variants globally remains uncertain.

Avoparcin and vancomycin resistant enterococci (VRE)

The experience with avoparcin use as a growth promotant in Europe is perhaps the best-established example demonstrating that antibiotic use in animals can lead to increased prevalence of resistant commensal organisms in animals, farm workers and the general community. This led to banning of avoparcin use in pigs and poultry in Denmark in 1995 and across the EU in 1997. Perhaps the most cited example of the 'human-animal link' in the literature, there remains little evidence that avoparcin use in animals led to any significant burden of human illness. Prior to the banning of avoparcin use in Europe, carriage of vancomycin resistant enterococci, particularly *E faecium*, was common among healthy humans and livestock, but human clinical disease remained rare. In contrast, in the USA, where no glycopeptides (including avoparcin) were ever approved for use in food animals, carriage of VRE by healthy people in the community was rare, yet clinical disease associated with VRE became prevalent from the early1990s. This was explained by much more intensive use of vancomycin in human medicine in the USA compared to Europe.[7] Moreover, clinical VRE infections did subsequently increase in prevalence in EU hospitals, but many years after the discontinuation of avoparcin in food animals. This was linked to the emergence of an invasive clone (CC 17) that appears to be derived from ampicillin resistant enterococci found in dogs.[8] Banning of avoparcin for growth promotion in food animals was certainly warranted based on the 'science', although there remains minimal evidence of a measurable impact on human illness. A recent review concluded that although avoparcin use in animals selects for VRE, other virulence factors are required before such selection would lead to clinical infections in humans.[9]

Livestock associated methicillin resistant *S aureus* (LA-MRSA)

The discovery of a high prevalence of ST398 MRSA in pigs in Holland in 2004 changed paradigms about *S aureus* epidemiology. Livestock were previously considered irrelevant to human *S aureus* infections, but suddenly moved to center stage particularly in Holland and Denmark, swine dense countries in which MRSA occur at very low prevalence in the human populations. It is now recognized that multiple species of livestock can harbor a range of sequence types of MRSA, including ST9 and ST5 in pigs in Asia in North America. A decade and a half since this became apparent, we know that clinical infections with MRSA of livestock origin do occur in immunocompetent people, but are generally mild and superficial infections. Although swine industry workers have high exposure to these organisms, events of clinical disease appear to be very uncommon. Although some European reports indicate LA-MRSA accounting for a quarter or more of human cases, these usually include screening samples (i.e., not clinical infections) from 'at-risk' individuals (e.g. farmers) that can represent 90% all the LA-MRSA reports.[10] Swine adapted variants of *S aureus* generally lack important virulence factors important for pathogenesis in human infections.[11,12] The LA-MRSA phenomenon was crucial in motivating changes in antibiotic regulations in some European countries, particularly Holland and Denmark. It has been widely assumed that antibiotic use (and by some specifically growth promotant use) in pigs has driven the emergence of LA-MRSA, but how and when is unexplained, and some anomalies are evident. Firstly, Denmark has been among the leaders in restricting antibiotic use in swine and banned all growth promotant use in 2000. However, LA-MRSA were first identified in pigs in Denmark in 2004, and remained at low prevalence through 2008, before rising to around 90% of herds currently. A detailed study indicated that pig movements among herds is the major factor in dissemination of LA-MRSA in Denmark.[13] Herd prevalence is similar in Holland, and apparently has not been influenced by the substantial reduction (> 50%) of antibiotic use in pigs in that country since 2009.[14] The fact that LA-MRSA are almost universally resistant to tetracyclines (tetracycline resistance is relatively uncommon in human medical isolates) has driven a specific effort to reduce tetracycline use in pigs in Denmark. This prevalent resistance to tetracycline is likely attributable to widespread use of these compounds over decades. However, methicillin susceptible *S aureus* (MSSA) in pigs in Denmark (and in the USA) are also mostly tetracycline resistant. Therefore it is hard to surmise that tetracycline use would specifically select MRSA variants over MSSA (*S aureus* being normal flora of swine) or, conversely, to expect that reduction in tetracycline use would have a rapid or marked impact on the prevalence of LA-MRSA (versus MSSA) in pigs. Although no systematic prevalence study has been conducted in the USA, available data indicate much lower prevalence of MRSA in both pigs and swine industry workers than seen in Denmark and Holland.[15,16] To date there is minimal evidence of livestock origin of clinically significant MRSA infections of humans in the USA. A multicenter study in Iowa reported S. aureus variants indicating a livestock origin comprised only 1% of S. aureus isolates, and 0.24% of MRSA isolates, from human clinical infections over 2 years.[17]

Colistin resistance and the *mcr* genes

Unlike in most of the pig-rearing world, colistin has never been used in swine in the USA. Until relatively recently, this 50-year old drug had minimal application in human medicine due to its renal toxicity, but has been used widely in swine and poultry in Europe and Asia since the 1960s. Colistin is now classed as 'critically important' by the WHO as a 'last resort' drug for treating multiple drug resistant (MDR) gram-negative infections. Until 2015, it was believed that colistin resistance was the result of chromosomal mutations and horizontal transmission had not been reported. A groundbreaking 2015 study from China reported colistin resistant *E coli* were prevalent in pigs and poultry at slaughter (and in retail pork and chicken), and that the resistance gene (*mcr*-1) was encoded on a plasmid, and was highly mobile.[18] A key feature of this report is its uncommon scenario where a medically important antibiotic used very sparsely used in human medicine was widely used in food animals. Therefore, perhaps unequivocally, animal use has been the predominant selective force driving the emergence of resistance at an ecological level. A subsequent flurry of research identified multiple mobile colistin resistance (*mcr*) genes (predominantly *mcr*-1, but also *mcr*-2 to -8) on diverse plasmids and in various bacterial species across most of the world.[19] These unfolding events have not directly affected the US industry, but led to very rapid changes (by July 2016) in recommendations of colistin use in animals in the EU. Spain reported a reduction of 97% in colistin use from 2015 to 2018. Although the clinical impact on human medicine is still unknown, this is a truly troubling event and increasingly there reports of the mcr genes in MDR organisms, including *Salmonella*. It is interesting that the mcr-1 gene is now known to have been present in China since the 1980s and Europe since 2005, but was not recognized to be a clinical problem. Also, its prevalence in pigs in Europe remained relatively low despite the widespread use of the drug for over 50 years.[20] In fact a 2015 review stated that 'despite extensive use in veterinary medicine, there is limited evidence for the development of resistance to colistin and no evidence has been found for the transmission of resistance in bacteria that have been spread from animals to humans'.[21] Another example of the rapidly shifting terrain of resistance ecology.

Other prominent issues relate to cephalosporin and fluoroquinolone resistance in major foodborne pathogens, and ESBL in uropathogenic *E coli*. Arguably, these have had greater impact on poultry (e.g., withdrawal of enrofloxacin due to resistance in *Campylobacter*) but add to the groundswell of concern about food animals in general. Any trends of increasing resistance to critically important classes of antibiotics, particularly in major foodborne pathogens, should be expected to provoke calls for withdrawal, or more restricted use, of entire classes of antibiotics, even for treatment of clinical disease in individual animals.

Advancements in the biology of antibiotic resistance

Apart from 'breakout' events such as those above, advancements in resistance biology and epidemiology concurrently disrupted the equilibrium of opinions on the benefits and risks of antibiotic use in animal agriculture. New discoveries of the role of horizontal transmission of resistance determinants on plasmids among zoonotic and commensal bacteria became widely known in the 1960s and contributed to the pressures leading to the Swann report. Concern about *S* Typhimurium DT104 was in no small part linked to the recent discovery of integrons providing mechanisms by which multiple resistance genes could be 'integrated' together in a section of DNA and transmitted among organisms. Reactions to colistin resistant *E coli* in food animals in China were magnified by the simultaneous demonstration of plasmid mediated transmission of the mcr-1 gene.

I think we are standing at the precipice of a whole new era of 'discoveries' that will apply considerable pressure on the food animal industries, although most of them will not involve epidemiologic data related to human health risk. As surveillance of bacteria in animals and humans becomes more extensive and powerful (due to technological advances in genome sequencing, microbiome, bioinformatics etc.), and in some cases 'agenda driven', we can anticipate a steady flow of reports of 'drug-bug' resistance pairs (or groupings) and novel mechanisms that will be uncovered in bacterial populations of food animals. Too many of these will be proclaimed to a hungry media as 'smoking guns' linking resistance in humans to antibiotic use in food animals, despite gaping voids in the relevant chains of causation. They will be reliably trumpeted as 'proof' that more restrictions on antibiotic use are needed. As in most of the examples mentioned above, that noise will not abate until the science shows that the 'damage' is remarkably small in the context of public health concerns. However, it should also be noted that more powerful tools also have the potential to help refute conclusions drawn from earlier and less comprehensive analyses. But the process of understanding these impacts will typically require a decade or more. Thus, we can expect the intervals between 'discoveries' to shrink and become miniscule in relation to the time required to objectively assess and understand and communicate the extent (often limited) of each potential concern. This will add up to a lot of time and resources playing defense.

'Not so scientific' factors and data free proclamations

Pressures arising from thorny scientific reports are probably not the greatest obstacle to progress towards an objective balancing of animal health benefits against human health risks. I have heard well-informed medical professionals offer the opinion that all antibiotic use in food animals should be abolished to help preserve antibiotic efficacy in human medicine. Less often (if ever) is this extended to companion animals. Those who afford any value to animal health and well being tend to differ. Some of the pressures we can anticipate fit in the following categories:

'Group think' and pressures to conform

For many aspects of swine production, suggestions that the US industry should emulate European (particularly northern European) initiatives have been a constant refrain for the last 30 years.

Examples include welfare standards and regulation (sow housing, weaning age, castration, tail docking,...), food safety initiatives (preharvest *Salmonella* control programs), and obviously regulation and surveillance of antibiotic use and resistance. Over time, there tends to be a general drift towards adoption of such measures, but the drivers towards such conformity are complex and often issue specific. Significant cultural differences between continents may mean that the pathways of change to similar ends can also differ. Regulatory driven initiatives tend to be the norm in Europe and market/industry led changes are more palatable in the USA (e.g., sow housing). Those who are less familiar with practical pig production often convey the belief that Europe possesses some superior wisdom and insight that provides a guiding light to the path of righteousness on all issues. For career industry critics, anything that looks like more regulation of industry holds appeal, irrespective of whether it achieves defined outcomes and/or has unwanted unintended consequences. A clear example of the fallibility of a well-intended but ultimately wasteful initiative is the implementation of preharvest control programs for *Salmonella* in swine, which were mandated in the EU following the example of Denmark. These programs are still largely in place, but have been 'de-emphasized' as they were not cost-effective and arguably led to misallocation of resources from more effective control options in the harvest and post-harvest sectors.[22]

Two much promoted European initiatives in relation to antibiotic use are mandatory reduction of antibiotic use as adopted in the Netherlands and Denmark among others, and elimination of growth promotant antibiotics. For the latter, to be clear I am in full support of the 2017 regulations which included increased veterinary oversight of all antibiotic use and I believe are positive for the industry and the veterinary profession. However, my support is philosophical, in that I do not believe over the counter availability of antibiotics to the public is philosophically acceptable in the modern era (and perhaps never should have been). My support is not based on demonstrated or expected improvements that such legislation will bring to public health outcomes. Banning growth promotants and reduction of antibiotic use in animal production are interventions to reduce antibiotic resistance, and not 'outcomes' per se. In the quest to reduce the human health impact of antibiotic use in food animals, at some point the definition of success must lie in some definable impact on human health. To my knowledge, this milestone is yet to be achieved. Dutch data estimated that a 54% relative reduction in antibiotic use in pigs was associated with a 22% relative reduction over 5 years in the proportion of *E coli* isolates of swine that were resistant to antibiotics (varying among antibiotics).[23] The upside of this is that impact on resistance was measurable, albeit limited to the bacterial flora of animals. The downside is that whatever relevance this measure has to clinical resistance in human medicine remains uncertain. In 2015, when seeing these data presented prior to publication, I asked the speaker whether they anticipated being able to document tangible public health benefits related to the changes. The response was 'maybe we will have something to show in 5 years'. I posed the same question to a Danish scientist with respect to the interventions they have made in restricting use in food animals since 1998 (growth promotion bans, various iterations of the 'yellow card' scheme), and the response was 'we believe we are doing the right thing'. My point is in no way to demean these initiatives which at some level are based on the need to 'do something' to address a major public health problem, or at least to convey responsiveness in an effort to appease public concerns. My point is to recognize that they remain 'faith based' initiatives and fall well short of being proven interventions that will benefit public health. In fact, my experience is that when asked to predict human health outcomes that might be expected from such interventions in animal production, responses seem to be vague (i.e., 'it must do some good sooner or later') rather than specific. Even in situations where interventions have been very targeted (e.g., withdrawal of enrofloxacin approval for poultry in the USA in 2005 due to emergence of resistant *Campylobacter*) outcomes can be underwhelming.[24] While MDR *Salmonella* were a large part of the impetus to ban growth promotant antibiotics in Denmark in 2000, subsequent surveillance data in DANMAP indicate stable or increased resistance in *Salmonella* to the antibiotics monitored. This does not mean there was no positive benefit (as there is no control group to indicate that resistance would not have increased more without the bans). But it serves as notice that likely modest 'signals' due to changes in antibiotic exposure of animals may not be readily detectable (or predictable) within the 'noise' of the ebbing and flowing of resistant bacteria and genes among complex environments. Nevertheless, pressures to conform with unproven (and usually expensive) initiatives that appeal to industry critics will continue to contribute to the groundswell of pressure that the industry and regulatory authorities will confront.

Competitive pressures

I will not dwell on the obvious here, as differentiation of protein products based on antibiotic exposure is already a familiar feature of the marketplace, and will become more so. The crux of differentiation will continue to be claims of exposure to antibiotics, rather than any indices that are translatable to human health risk. Interestingly, a recent study of US retail pork found no differences in bacterial contamination, resistant bacteria, or resistance gene abundance among sources of pigs having different claims of antibiotic exposure.[25] Broadly, competition will be positioned to differentiate plant from animal proteins, 'synthetic' (e.g., cell based) versus live-animal derived proteins, and between and within species for live-animal derived proteins. I have become something of a broken record in opining that it makes no biological sense to compare antibiotic use in different species and in different countries due to differing biology of species and health challenges that predominate in different geographic regions. The fact remains that comparing species and countries in the one thing that many people do like to do. Data from several countries indicate that the pig will not fare favorably in such comparisons in relation to other species (e.g., broilers). I regularly point out that antibiotic free pig production would be simple if the pigs hatched out of eggs in a machine, did not suckle their mother and be weaned, and only lived for 6 weeks. The extent to which competition between animal protein species plays out in the marketplace could have a substantial impact on customer and consumer behavior, and has the potential to be collectively detrimental to food animal industries.

Career critics – smoking guns and crying wolf

Last, but not least, the steady drumbeat of a familiar contingent of industry critics will not be going away. All industries need critics and criticism, and appropriate criticism can pave the way to doing a better job. Disappointingly, some of the louder voices can be seduced by the sensational at the expense of the accurate, or part of the story instead of the whole story. They rarely shy from apocalyptic language for each successive 'smoking gun', nor demonstrate compunction in crying wolf. Most 'senior' AASV members will be aware that if the industry had a major problem with antibiotic misuse, it was 30 to 40 years ago when about 1 in 7 market hogs had violative residues. Paradoxically, perceptions of the industry have tended to deteriorate over the intervening decades as industry performance has improved. Violative residues today are orders of magnitude below those of the 1970s. Much of this improvement is attributable to concerted efforts to educate producers (notably the PQA program), as well as industry structure, changing roles of veterinarians and other factors. The only answer is getting better – continuing to advance preventive medicine, evaluating and documenting processes of diagnosis, treatment and clinical outcomes, and being our own best constructive critics.

References

1. Lhermie G, Wernli D, Jørgensen PS, et al. Tradeoffs between resistance to antimicrobials in public health and their use in agriculture: Moving towards sustainability assessment. *Ecol Econ*. 2019, 166:106427. doi:10.1016/j.ecolecon.2019.106427.

2. Kirchhelle C. Swann Song: Antibiotic Regulation in British Livestock Production (1953-2006). *Bull Hist Med*. 2018;92:317-350.

3. Cherubin CE. Antibiotic Resistance of *Salmonella* in Europe and the United States. *Clin Infect Dis*. 2011. doi:10.1093/clinids/3.6.1105.

4. Branchu P, Bawn M, Kingsley RA. Genome variation and molecular epidemiology of *Salmonella* enterica serovar Typhimurium pathovariants. *Infect Immun*. 2018. doi:10.1128/IAI.00079-18.

5. Anderson ES. Drug Resistance in *Salmonella* typhimurium and its Implications. *Br Med J*. 1968. doi:10.1136/bmj.3.5614.333.

6. Mueller-Doblies D, Speed KCR, Kidd S, Davies RH. *Salmonella* Typhimurium in livestock in Great Britain - Trends observed over a 32-year period. *Epidemiol Infect*. 2018. doi:10.1017/S095026881800002X.

7. Bonten MJ, Willems R, Weinstein RA. Vancomycin-resistant enterococci: Why are they here, and where do they come from? *Lancet Infect Dis*. 2001. doi:10.1016/S1473-3099(01)00145-1.

8. Kahn LH. Antimicrobial resistance: A One Health perspective. *Trans R Soc Trop Med Hyg*. 2017. doi:10.1093/trstmh/trx050.

9. Ahmed MO, Baptiste KE. Vancomycin-Resistant Enterococci: A Review of Antimicrobial Resistance Mechanisms and Perspectives of Human and Animal Health. *Microb Drug Resist*. 2018. doi:10.1089/mdr.2017.0147.

10. van Alen S, Ballhausen B, Peters G, et al. In the centre of an epidemic: Fifteen years of LA-MRSA CC398 at the University Hospital Münster. *Vet Microbiol*. 2017. doi:10.1016/j.vetmic.2016.01.021.

11. Hau SJ, Sun J, Davies PR, Frana TS, Nicholson TL. Comparative prevalence of immune evasion complex genes associated with β-hemolysin converting bacteriophages in mrsa st5 isolates from swine, swine facilities, humans with swine contact, and humans with no swine contact. *PLoS One*. 2015;10(11). doi:10.1371/journal.pone.0142832

12. Mutters NT, Bieber CP, Hauck C, Reiner G, Malek V, Frank U. Comparison of livestock-associated and health care-associated MRSA-genes, virulence, and resistance. *Diagn Microbiol Infect Dis*. 2016;86(4):417-421. doi:S0732-8893(16)30255-3 [pii].

13. Sieber RN, Skov RL, Nielsen J, et al. Drivers and dynamics of methicillin-resistant livestock-associated staphylococcus aureus CC398 in pigs and humans in Denmark. *MBio*. 2018. doi:10.1128/mBio.02142-18

14. Speksnijder DC, Mevius DJ, Bruschke CJM, Wagenaar JA. Reduction of Veterinary Antimicrobial Use in the Netherlands. The Dutch Success Model. *Zoonoses Public Health*. 2015;62:79-87.

15. Sun J, Yang M, Sreevatsan S, Davies PR. Prevalence and Characterization of Staphylococcus aureus in Growing Pigs in the USA. *PLoS One*. 2015;10(11):e0143670.

16. Sun J, Yang M, Sreevatsan S, et al. Longitudinal study of *Staphylococcus aureus* colonization in swine veterinarians in the USA. *BMC Infect Dis*.

17. Nair R, Wu J, Carrel M, et al. Prospective multicenter surveillance identifies Staphylococcus aureus infections caused by livestock-associated strains in an agricultural state. *Diagn Microbiol Infect Dis*. 2016;85(3):360-366.

18. Liu YY, Wang Y, Walsh TR, et al. Emergence of plasmid-mediated colistin resistance mechanism MCR-1 in animals and human beings in China: A microbiological and molecular biological study. *Lancet Infect Dis*. 2016,16:161-168.

19. Wang X, Wang Y, Zhou Y, et al. Emergence of a novel mobile colistin resistance gene, mcr-8, in NDM-producing Klebsiella pneumoniae article. *Emerg Microbes Infect*. 2018, 7:122 doi:10.1038/s41426-018-0124-z.

20. Schwarz S, Johnson AP. Transferable resistance to colistin: A new but old threat. *J Antimicrob Chemother*. 2016, 7:122. doi:10.1093/jac/dkw274.

21. Catry B, Cavaleri M, Baptiste K, et al. Use of colistin-containing products within the European Union and European Economic Area (EU/EEA): development of resistance in animals and possible impact on human and animal health. *Int J Antimicrob Agents*. 2015, 46:297-306.

22. Dahl J. Controlling *Salmonella* in live pigs and at slaughter: the Danish experience. *Vet J* 2013, 197:529–530.

23.Dorado-García A, Mevius DJ, Jacobs JJH, et al. Quantitative assessment of antimicrobial resistance in livestock during the course of a nationwide antimicrobial use reduction in the Netherlands. *J Antimicrob Chemother*. 2016, 71:3607-3619.

24. Tang KL1, Caffrey NP2, Nóbrega DB et al. Comparison of different approaches to antibiotic restriction in food-producing animals: stratified results from a systematic review and meta-analysis. BMJ Glob Health. 2019, 31;4(4):e001710.

25.Vikram A, Miller E, Arthur TM et al. Food Service Pork Chops from Three US Regions Harbor Similar Levels of Antimicrobial Resistance Regardless of Antibiotic Use Claims. J Food Prot 2019, 82:1667-1676.

Veterinary Feed Directive: past, present and future

Chris Rademacher, DVM
Clinical Professor and Associate Director, Iowa Pork Industry Center
Iowa State University Ames, Iowa

On January 1, 2017, the U.S. Food and Drug Administration (FDA) began enforcement of Guidance 209, 213, and the Veterinary Feed Directive (VFD).[1-3] The intent of these regulations was to direct the use of medically important antibiotics (defined as antibiotics that are important for therapeutic use in human medicine) in livestock for therapeutic purposes only. Therapeutic purposes are defined as either treatment, control or prevention of disease.[1] The regulations are focused on use of medically important antibiotics given in mass medication formats as 95% of the medically important antibiotic use is delivered either in water or feed.

Use of medically important antibiotics in feed requires a Veterinary Feed Directive (VFD) order from the veterinarian to the producer and feed manufacturer under the guidance of a veterinarian with a valid veterinary-client-patient relationship (VCPR). This attributes the responsibility for making medical decisions for the farm to the veterinarian, while the producer bears the responsibility in following the medical directions of the veterinarian. Another aim of the regulations was to eliminate the use of medically important antibiotics for growth promotion use. Collectively, these new regulations have undoubtedly changed the ways that antibiotics are used in livestock production in the United States.

There were a lot of questions surrounding how swine producers, veterinarians and feed suppliers were going to implement these new regulations prior to the enactment date of January 1, 2017. FDA setup an email portal (`AskCVM@fda.hhs.gov`) for these stakeholders to ask the Center for Veterinary Medicine (CVM) any questions they may have regarding the VFD process. The answers would only go back to the submitter and then occasionally would be released in a guidance document, but that was not very timely in the time period leading up to the implementation date. The American Association of Swine Veterinarians (AASV) Pharmaceutical Issues committee were tasked with trying to collect VFD questions from practitioners, submitting them to Ask CVM, and once the answers were received, translating them into a frequently asked questions (FAQ) document. The document was placed on the Iowa Pork Industry Center (IPIC) website (`https://www.ipic.iastate.edu/info/VFDfaq.pdf`) and the link published in the AASV webpage (aasv.org) as well as the National Pork Board (pork.org). To date, that document has been downloaded over 5,000 times. The FAQ document continues to be updated as new questions or scenarios come in. Some examples of the more common questions included pulse dosing (using the same antibiotic more than once in the same group of animals), the ability to identify multiple groups of animals (flow based VFD) and how to handle the use of different feed distributors (one or multiple VFD's).

Schulz and Rademacher took a convenience sample of veterinarians who attended the Iowa State University James D. McKean Swine Disease Conference and administered a survey just prior to implementation of the new guidelines (November 2016) as well as approximately 1 year after implementation (November 2017). Several interesting trends were noted in comparing the two surveys. The pre-enactment survey seemed to indicate that majority of veterinarians (52.9%) were planning on moving from medically important to non-medically important antibiotics for growth promotion in the growing and finishing phases.[4] The post enactment survey demonstrated that the majority of veterinarians (58%) had actually advised their clients to eliminate all of their antibiotics used for growth promotion in the growing pigs.[5] The survey identified a higher percentage of veterinarians (42.9%) were choosing to write VFD's at the flow level, while 28.6% were planning to write VFD's to the site level, and another 21.4% were writing VFD's to the group or lot level.[5] It was also interesting to gauge veterinarians perception of the impact of the new regulations on the usage of medication in feed. Figure 1 demonstrates the difference between veterinarians' responses before and after the enactment of the regulations. Most expected a 10-30% reduction prior to the VFD rule change, but in the post-enactment survey, the reduction of in-feed antibiotics was more than 50%.

Much of the apprehension surrounding the VFD implementation came due to the fact that the responsibility of filling out the VFD correctly falls upon the veterinarian. Because of this, most veterinarians (82.9%) chose to utilize an electronic VFD service (e.g. GVL), which would use "smart engine" technology to ensure that all the legal requirements for each drug were properly displayed and formatted, once the drug was selected.[5] Of special concern was the cautionary statements and special instructions that could only be found on the labels. Electronic services enable the user to include all the appropriate statements and also allows for quick recall of VFD's in the case of an FDA audit.

The FDA requires annual reporting from every sponsor of an approved or conditionally approved animal drug application containing an antimicrobial active ingredient. These reports to the FDA must include, the amount of each ingredient in these drug products sold or distributed for use in food-producing animals. This summary report presents the sales and distribution data for actively marketed antimicrobial drugs approved for use in food-producing animals by drug class, medical importance, route of administration, indication, and dispensing status. For the first time, in 2017, they also included species-specific estimates of these sales and distribution for 2017. The 2017 data was published in December of 2018. Since the swine industry does not have a good methodology for reporting actual antibiotic usage, this report is often used as a proxy for antibiotic usage year over year. The impact of the regulations was very apparent in the data coming from this report. Figure 2 demonstrates the year over year use of medically important antibiotics by route of administration from 2009 to 2017.[6] It was apparent that there was some reduction (10%) in medically important antibiotic usage beginning in 2015. Most likely due to the implementation of the new regulations, that number decreased another 33% from 2016 to 2017 in all livestock. In swine alone, medically important antibiotic sales dropped 35% from 2016 to 2017. Tetracyclines, which represents the largest volumes of medically important antibiotic sales in livestock, decreased by 40% from 2016 to 2017.

FDA audits of the VFD process was also a concern by all stakeholders, specifically the potential for enforcement action being levied against them if any violations were discovered during the audit. The most common path for an audit of a veterinarian or producer begins with an FDA compliance officer or one of their designee's auditing a feed mill. Most feed mills in the US are FDA inspected routinely for other feed safety type programs (e.g. FSMA). During the inspection process, the inspector may randomly select 2-3 VFD's from the mill and inspect them. The compliance officer may choose to visit the veterinarian and producer and will inspect the VFD and associated feed delivery records, looking to verify that every item on a VFD has been filled out correctly. The officer will also want to ensure that the VFD (also feed delivery records for the VFD) has been retained for the 2-year time period, as well as if the VFD feed was fed to the authorized number of animals. Another item to be inspected will be if the withdrawal times were properly followed and make sure the VFD feed was not fed past the VFD expiration date.

Compliance to the labeled duration of feeding of a VFD can be challenging in swine production. Most feed is delivered in three ton increments to fit a compartment within a feed truck. Within the feed budgeting software, a diet may be formulated for a specified number of tons. Many times, this will be for a fraction of a compartment and most often are rounded up to the nearest 3 ton increment in order to maximize feed delivery efficiency. Depending on the drug, the number of pigs and their current feed consumption, this may put more VFD feed delivered than the amount written on the VFD order. Most compliance officers prefer to see the date the VFD feed was started and finished somewhere on the barn record sheets. From 2016-2018, the audits performed by the FDA or their designees were deemed to be of the education in nature. "Educational audits" are loosely defined as any discrepancies noted during an audit would be deemed educational and therefore not incurring fines or legal action, unless a severe, adulterating offense was noted. In a report shared by FDA-CVM (Table 1) of these audits, 91% of all livestock VFD inspections resulted in No Action Indicated (NAI) with the rest of them (except for one) were deemed to be Voluntary Action Indicated (VAI).

Looking forward, there will be continued pressure to evaluate and adjust antibiotic use policy. In September 2018, the FDA released a five-year blueprint detailing how the agency plans to build upon their current programs to advance antibiotic stewardship in veterinary settings. The perception of overuse of antibiotics in production animal species was specifically addressed within this document and described as contributing to the growing problem of antibiotic resistance.[7] One of the first steps in this blueprint is to ask veterinary antimicrobial sponsors to voluntarily withdraw their over-the-counter status, which would bring all veterinary antibiotics under the oversight of veterinarians (GFI #263).[8] One other method of antibiotic regulation is for individual states to put forth legislation that will impact veterinarians ability to use antibiotics. Legislative initiatives to monitor or restrict antimicrobial use in food animals are in various stages of development in several

Figure 1: Pre and post enactment perceived reduction in feed antibiotic usage.

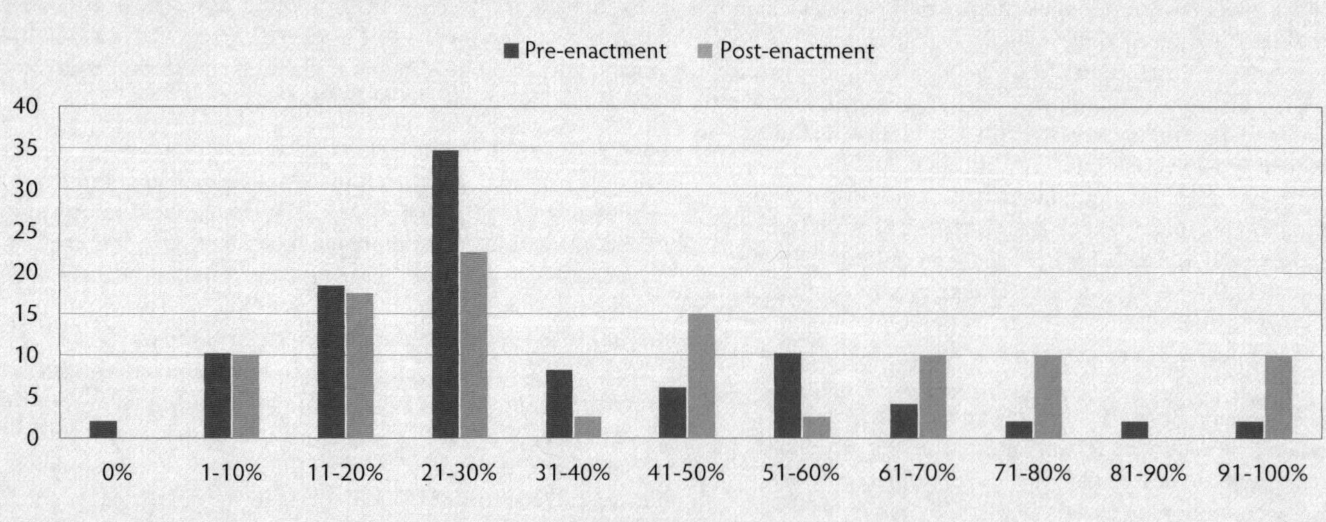

Figure 2: Medically important antimicrobial drugs approved for use in food-producing animals. Domestic sales and distribution data actively marketed 2009 - 2017. Reported by route of administration.

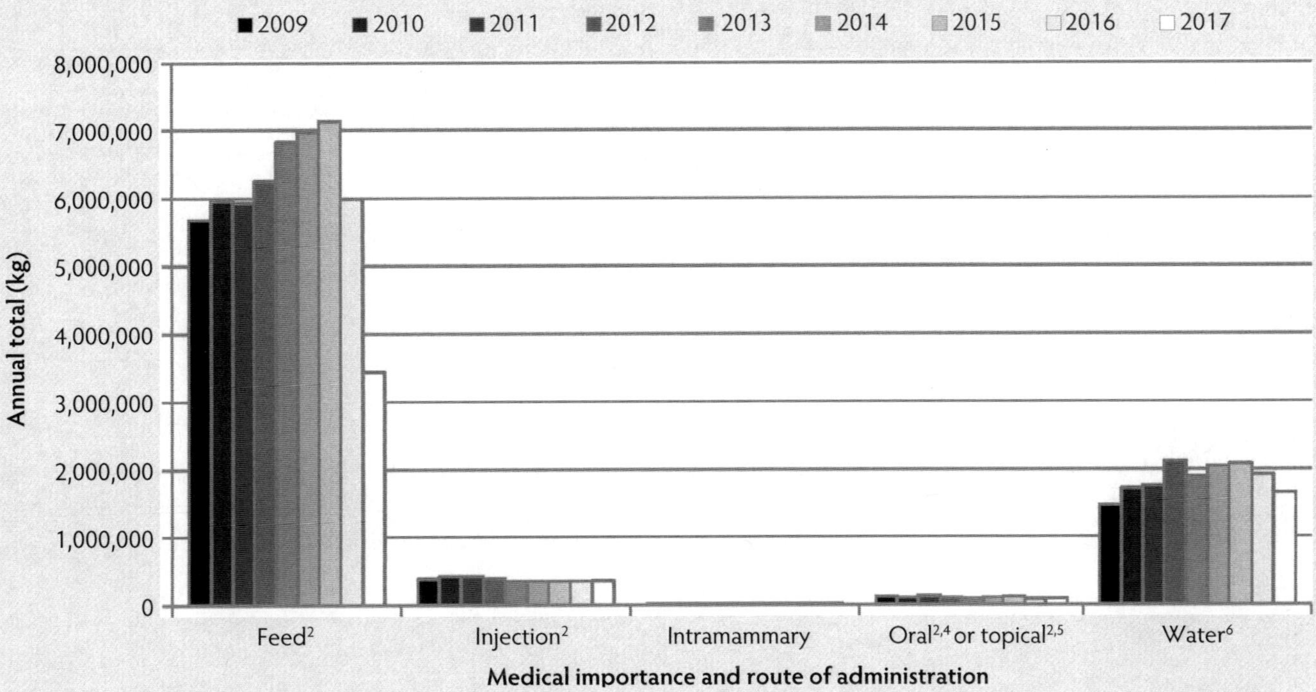

Table 1: Summary of FDA educational audits (2016-2018)

District decision	Fiscal year 2016	Fiscal year 2017	Fiscal year 2018	Total
No action indicated (NAI)	54	130	230	**414**
Voluntary action indicated (VAI)	3	0	38	**41**
Official action indicated (OAI)	0	0	1	**1**
Total	57	130	269	**456**

states. Senate Bill 27 in California was approved in 2015 and, among other provisions, "require(s) the Department of Food and Agriculture, in consultation with the Veterinary Medical Board, the State Department of Public Health, universities, and cooperative extensions, to develop antimicrobial stewardship guidelines and best management practices on the proper use of medically important antimicrobial drugs and would require the department to gather information on medically important antimicrobial drug sales and usage, antimicrobial resistant bacteria, and livestock management practice data." Legislation to prohibit antimicrobial use in food-producing animals, particularly prevention uses, and to require reporting by veterinarians was enacted in Maryland (HB 652/SB 471). A similar bill was introduced in New York (S. 5742) but failed to pass before the session ended. The Illinois State Senate also considered a bill with restrictions on use of antimicrobials in food-producing animals (SB 1186) but failed to pass it before the session ended. Bills in Pennsylvania and Oregon are also being drafted. It would be naïve to expect that these initiatives will stop or that states with agriculture-based economies are immune from such legislation at the state level. Veterinarians and producers will need to continue to evaluate their antibiotic choices and start to develop evidence of justification of routine use as continued scrutiny on mass medication of antibiotics will not cease anytime soon.

References

1. Guidance for Industry on the judicious use of medically important antimicrobial drugs in food-producing animals; Availability. Fed Regist. 2012;77(72):22328-22329.

2. US Department of Health and Human Services, Food and Drug Administration, Center for Veterinary Medicine. Guidance for Industry #209, The judicious use of medically important antimicrobial drugs in food producing animals. April 13, 2012. http://www.fda.gov/downloads/AnimalVeterinary/GuidanceComplianceEnforcement/GuidanceforIndustry/UCM216936.pdf. Accessed June 6, 2017.

3. Guidance for Industry on new animal drugs and new animal drug combination products administered in or on medicated feed or drinking water of food-producing animals: Recommendations for drug sponsors for voluntarily aligning product use conditions with Guidance for Industry #209; Availability. Fed Regist. 2013;78(239):75570-75571.

4. Schulz LL, Rademacher CJ. Food and Drug Administration Guidance 209 and 213 and Veterinary Feed Directive regulations regarding antibiotic use in livestock: A survey of preparation and anticipated impacts in the swine industry. J Swine Health Prod. 2017;25(5):247-255.

5. Rademacher CJ, Pudenz CC, Schulz LL. Impact assessment of new US Food and Drug Administration regulations on antibiotic use: A post-enactment survey of swine practitioners. J Swine Health Prod. 2019;27(4):210–220.

6. US Food and Drug Administration. 2017 summary report on antimicrobials sold or distributed for use in food-producing animals. **https://www.fda. gov/downloads/forindustry/userfees/animaldruguserfeeactadufa/ucm588085.pdf**. Published December 2018. Accessed November 2, 2019.

7. US Food and Drug Administration. Supporting antimicrobial stewardship in veterinary settings. Goals for Fiscal Years 2019-2023. FDA Center of Veterinary Medicine. **https://www.fda.gov/files/animal%20&%20veterinary/published/Supporting-Antimicrobial-Stewardship-in-Veterinary-Settings--Goals-for-Fiscal-Years-2019-2023.pdf**. Published September 2018. Accessed November 15, 2019.

8. US Food and Drug Administration. Recommendations for Sponsors of Medically Important Antimicrobial Drugs Approved for Use in Animals to Voluntarily Bring Under Veterinary Oversight All Products That Continue to be Available Over-the-Counter. **https://www.fda.gov/regulatory-information/search-fda-guidance-documents/cvm-gfi-263-recommendations-sponsors-medically-important-antimicrobial-drugs-approved-use-animals**. Published September 2019. Accessed November 2, 2019.

Future of the microbiome in the pig

B. Arruda, DVM, PhD
Department of Veterinary Diagnostic and Production Animal Medicine, Iowa State University, College of Veterinary Medicine, Ames, Iowa

Take home

- Disruption of the microbiota (dysbiosis) likely has a significant impact on production, and with the increased restrictions on the use of antibiotics, an applicable understanding of the swine microbiota is necessary for disease prevention and mitigation.
- Details are likely determinants of successful utilization of the swine microbiome, and microbiome analysis of feces only provides a very narrow perspective of the gut microbiome in pigs.
- We need to move beyond abundance and into a network-based approach utilizing numerous disciplines and –omics methodologies simultaneously to better understand the complex interactions and impact of the microbiota.

Introduction

Polymicrobial communities (microbiota) are key players in the health of organisms and ecosystems.[1] At a very basic level, the swine microbiota is every invisible organism in and on the pig at various anatomic sites including the respiratory, urogenital, and gastrointestinal tract. The pig gastrointestinal tract contains a diverse, malleable, and complex microbiota that is not a mere collection of independent microbes but an interconnected web of ecological communities that communicate, cross-feed, recombine, and co-evolve. The cumulative number of bacteria in the swine colon alone has been approximated to be 10^{10} to 10^{11} per gram of content.[2]

The gut microbiome is essential to the health of swine through the production of short-chain fatty acids and vitamins, inhibition and prevention of pathogen colonization, and development and maintenance of the immune system.[3] Due to the importance of these attributes and the advent of low-cost and high-throughput sequencing, the gut microbiome has received increasing attention over the last several years. However, analyzing and transforming microbiome data into meaningful biological discernments remains challenging.

Where are we now?

Mirroring the tremendous proliferation of human microbiome studies, the publication of swine microbiome studies has grown tremendously. Commonly these investigations evaluate prebiotics, probiotics, antibiotics, diet, age and/or disease state. However, variation is innate to biology; significance to health or performance is less clear. Recently, two studies found the biggest influencing factor on the bacterial communities was anatomic site.[4,5] One of these two studies also appears to support a core mucosal-associated microbiome in pigs that differs from the lumen-associated microbiome.[5] These kinds of body site-specific analyses are needed not only for the gastrointestinal tract but also for other body regions, such as the respiratory tract. Despite the increase in published studies, much of the work done in swine has been descriptive providing baseline data with primarily speculation in regards to the application aspects. Studies are greatly needed that utilize metagenomics, metabolomics, and metatranscriptomics approaches while concurrently testing specific hypotheses related to mechanism and function to best facilitate consistently statistically significant demonstrable health, welfare and economic benefit.[6] Although 16S rRNA sequencing reveals the complex taxonomic profile of a microbial community, it is also important to understand the functional capacities of the gut microbiota. This can now be done using whole-metagenome sequencing. Genome-focused methodologies have the ability to deliver organism-based metabolic information as well as investigate microbial interactions and microbe-phage interplay to elucidate ecosystem functions that later could be applied for "microbiome engineering."[7]

Quantity or activity of microbes?

Discord in the literature can facilitate the asking and answering of better questions. In human literature, there are numerous examples of discord in terms of association of microbes with disease status, which has given rise to questions concerning the relevance of phylum level analysis and relative abundance. A recent study in human medicine discovered that while most bacteria were well correlated when analyzing DNA and RNA abundances some bacteria that were abundant in terms of metagenomic data were 'inactive or dormant' in the gut while other bacteria were only detectable when analyzed at the transcriptional level.[8] This would suggest that it would be appropriate and/or more relevant to explore the metabolic capacity of the gut microbiota and the metabolites produced.[9]

Furthermore, some studies of the gastrointestinal tract microbiome have suggested that specific members interact directly with the host while others play more of a role in maintaining the microbial habitat.[10] New methodologies which can differentiate host tissue-associated microorganisms could offer new understandings into the presence of different ecological niches in specific regions and the roles of specific members in host-microbe interactions and should entail collaborations between a multidisciplinary team of nutritionists, microbiologists, immunologists and pathologists with expertise in mucosal tissue biology.

Where should we go?

Future studies should utilize network-based approaches, which have been used in human microbiome investigations but are only occasionally used for microbiome analyses in swine. Network-based approaches could provide novel insights into assembly, organization and functional role of the microbiome allowing for a greater understanding of the factors that influence the gut microbiome in pigs and possibly the selection of direct-fed microbials.[11]

There is also limited understanding of the nutrient requirements or other factors that may influence the microbial members in these communities. Novel methods that use radiotracers, stable isotopes, or metabolic inhibitors could lead to new knowledge regarding the specific nutrient requirements of these microbial communities and lead to the coherent implementation of microbiome-based therapeutics.[10]

It is abundantly clear that the microbiome plays a central role in health and can be highly influenced by dietary factors. Accordingly, research endeavors focusing on how dietary factors influence the microbiome could be highly insightful. Such avenues may provide meaningful information concerning the feeding of pigs under different physiological and metabolic conditions and could be translated into "precision nutrition" for swine.[11] Importance of multiomics approaches and how various genomic and metabolomics approaches are needed to further elucidate the interactions between microbes and host and network-based analytical approaches have the potential to help elucidate complex polymicrobial and microbe-host interactions to further the usefulness of microbiome research to production animal medicine. It is imperative that future investigations collect many types of phenotypic and "omic" information including transcriptomic, host genomic, and metabolomic data to develop models that surpass association.[12]

Conclusions

In many ways, the swine industry is better situated to make profound advances in the study and application of microbiota due the ability to regulate diet, perform controlled *in vivo* studies, follow pigs longitudinally over a relatively short period of time and collect samples from multiple locations along the gastrointestinal tract. As we undertake these ventures, we must go beyond simple associations and provide more complex analyses using multiple disciplines if we want to finally approach causality. As an industry we must learn from human missteps and misinterpretation and be more rigorous when drawing conclusions from papers that proclaim that the identification of differences in gut microbiota composition is strongly associated with a specific outcome. The utilization of new and proven methodologies concurrently promises to elucidate microbe-microbe and host-microbe interactions that will benefit swine production; however, a healthy dose of skepticism and critical thinking will likely be necessary to achieve meaningful applicability without paralyzing frustration and the squandering of resources.

References

1. Layeghifard M et al. Disentangling interactions in the microbiome: a network perspective. *Trends Microbiol.* 2017; 25:217-228.

2. Gaskins HR et al. Antibiotics as growth promotants: mode of action. *Anim Biotechnol.* 2002;13:29-42.

3. Brestoff JR, Artis D. Commensal bacteria at the interface of host metabolism and the immune system. *Nat Immunol.* 2013;14:676-684.

4. Holman DB et al. Meta-analysis to define a core microbiota in the swine gut. mSystems. 2017;2:e00004-17.

5. De Rodas et al. Microbiome profiling of commercial pigs from farrow to finish. *J Anim Sci.* 2018;96:1778-1794.

6. Guevarra RB et al. Piglet gut microbial shifts early in life: causes and effects. *J Anim Sci Biotechnol.* 2019;10:1.

7. Foo P, et al. Microbiome engineering: current applications and its future. *Biotechnolo J.* 2017;12:1600099.

8. Schirmer M et al. Dynamics of metatranscription in the inflammatory bowel disease gut microbiome. *Nat Microbiol.* 2018;3:337-346.

9. Cani PD et al. Human gut microbiome: hopes, threats and promises. *Gut.* 2018;67:1716-1725.

10. Proctor L et al. A review of 10 years of human microbiome research activities at the US National Institutes of Health, Fiscal Years 2007-2016. *Microbiome.* 2019; doi:10.1186/s40168-019-0620-y.

11. Aluthge ND et al. The pig microbiota and the potential for harnessing the power of the microbiome to improve growth and health. *J Anim Sci.* 2019;97:3741-3757.

12. Waldor MK et al., Where next for microbiome research? *PLoS Biol.* 13:e1002050.

Future issues of antibiotic-free production

Michael S. Pierdon VMD
Four Star Veterinary Service

Market pressures have spurred an increase in antibiotic-free (ABF) production systems for animal protein in recent years and demand for this type of product will continue to grow in the future. To meet this market demand the industry will face several challenges as it works to adapt production practices. Foremost is defining exactly what antibiotic-free production means for an individual producer. There are many different "antibiotic-free" marketing programs, however many have additional requirements within areas such as housing, husbandry, management, and nutrition that serve to differentiate meat marketed under competing programs. This will create two primary types of antibiotic-free protein production: "Commodity ABF" meat will be raised without antibiotics, but will offer few other differentiating characteristics and "Niche ABF" meat will be raised without antibiotics and will carry several additional requirements designed to attract consumers who desire more than just antibiotic-free meat. These two market categories will produce product in significantly different ways and therefore will find success with different strategies.

"Niche ABF" production requires significantly different practices from standard industry production. Traditionally, this product targets a smaller market of highly engaged consumers who want a truly differentiated product and are willing to pay a higher price for their meat protein. Requirements likely include, but are not limited to, older wean age, limitation on routine processing procedures, solid flooring, bedding access, and farrowing stall restrictions. These programs will become increasingly restrictive as they compete for consumers who value and demand the products. As a specialty market with a smaller consumer base, Niche ABF program requirements will be impractical for large-scale operations, meaning the ideal producer for this market will have a smaller herd size, be geographically diverse, and have a direct connection to a specific regional customer or outlet.

"Commodity ABF" production will focus on providing antibiotic-free meat for larger mass markets. This will be the product used for food service chains, grocery retailers, and box stores who desire to offer their customers antibiotic-free meat options at a modest premium and more affordable price level. This product will be produced by larger production systems using modern practices modified for compliance with program qualifications. Depending on the type of production an individual chooses, their path to success will be defined differently from the producer choosing the alternative option.

"Commodity ABF" producers must embrace new technologies that will reduce the need for antibiotics in a production setting with high risk of disease. These technologies include genetic modification; selective breeding; and nutrition/gut health. Genetic modification technology makes disease resistance a real possibility in the short- to medium-term. Selective breeding for animals with improved disease resistance and livability will be important for decreasing the need for antibiotics. Improved understanding of nutrition and gut health will help create stronger animals that are less likely to experience health challenges and more able to withstand them when they occur. Probiotics, prebiotics, acidifiers, medium chain fatty acids (MCFAs), botanicals and more will become our front-line approach to creating and maintaining healthy flows. Use of non-antibiotic interventions to prevent or treat disease outbreaks will be important. Vaccines, immune stimulants, anti-inflammatories, anthelmintics, antivirals and nutritional supplements can be used to manage some disease outbreaks without mass antibiotic use. Animal health companies must pursue development of these non-antibiotic medications. These and other technologies offer real promise in decreasing the need for antibiotics in commercial meat protein production. We must prevent consumers from getting the impression that we are substituting one "harmful" technology for another. The challenge for producers and veterinarians is to explain to customers and consumers the need for access to these technologies in pursuit of the goal for efficient, safe, affordable antibiotic-free meat.

"Niche ABF" production will likely not have access to all the same technologies available to "Commodity ABF." Consumers who are seeking label claims beyond simply "antibiotic-free" value a more traditional approach to meat production and are unlikely to embrace the use of new technology. By raising animals in isolated areas with smaller groups of lower density and increased oversight and management, these niche systems have the potential to decrease the need for antibiotics by reducing disease pressure. These changes will lessen the need for antibiotic use without embracing new technologies resulting in a substantial increase in the cost of production. Significant price premiums will allow these farms to succeed financially without the efficiencies offered by larger-scale production. Ultimately the success of these programs will depend on the market value of and demand for their products.

There is no question that production of antibiotic-free pork will continue to grow in the coming years. Veterinarians will continue to play an important role in helping producers implement health management protocols that protect animal health without routine antibiotic use, but the successful approach will vary depending on the market being served. Veterinarians will also be involved in education of production companies, marketing programs, and customers/consumers about emerging technologies and their potential to reduce antibiotic use. These parties will look to us for guidance on the value and necessity of these developments and the relative risk of technologies versus the risk of antibiotic use. We need to be prepared to engage in those discussions.

Future of pain medications for pigs

Johann (Hans) Coetzee
Department of Anatomy and Physiology, College of Veterinary Medicine, Kansas State University, Manhattan, Kansas

Introduction

Each year, more than 50 million piglets in the United States undergo processing procedures (e.g., castration, tail docking, and teeth clipping) that result in post-procedure pain. Pain-related behaviors have been shown to persist for up to 4 days after castration in 5-day-old piglets.[1] Consumer concern about the animal welfare implications of painful processing procedures has resulted in investigation of methods to provide analgesia and alleviate distress associated with these practices.[2] In the US, no drugs are labeled by the Food and Drug Administration (FDA) to mitigate pain in swine. Therefore, the development of pain mitigation strategies that address consumer concerns about animal welfare while still being safe, effective, practical and economical to administer would be beneficial to the swine industry.

The capacity to experience pain is considered to have a protective role by eliciting behavioral responses aimed at reducing further tissue damage and enhance wound healing.[3] However, persistent pain syndromes offer no biological advantage and are associated with suffering and distress.[3] Pathological pain states in swine occur as a result of tissue damage, nerve damage and inflammation and are frequently associated with pain hypersensitivity.[4] Pain hypersensitivity manifests as hyperalgesia (exaggerated responses to painful stimuli) and allodynia (pain resulting from normally innocuous stimuli).

Hyperalgesia has been reported to persist in dairy cattle and lame sheep for at least 28 days after the causal lesion has resolved.[5,6] Consequently, chronic pain associated with lameness is considered one of the most significant welfare concerns in livestock species, including sows. Inflammatory pain associated with lameness responds modestly to treatment with non-steroidal anti-inflammatory drugs (NSAIDs)[7,8] but neuropathic pain (due to nerve damage or neuronal dysfunction) is considered refractory to the effects of NSAIDs and many opioid analgesics. Therefore, there is also a need to identify novel drugs and drug targets for alleviating chronic pain of neuropathic origin in animals.

Current challenges associated with providing analgesia in swine

There are several challenges associated with providing effective analgesia in food animals in the United States. Firstly, there are currently no analgesic drugs specifically approved for the alleviation of pain in swine.[9] Therefore, use of any drug for pain relief constitutes extra-label drug use (ELDU).[10] Under the Animal Medicinal Drug Use Clarification Act of 1994 (AMDUCA),[11] ELDU is permitted for relief of suffering in swine provided specific conditions are met. These conditions include that (1) ELDU is allowed only by or under the supervision of a veterinarian, (2) ELDU is allowed only for FDA approved animal and human drugs; (3) ELDU is only permitted when the health of the animal is threatened and not for production purposes; (4) ELDU in feed is prohibited and (5) ELDU is not permitted if it results in a violative drug residue in food intended for human consumption. Therefore, use of an analgesic to alleviate pain associated with castration in calves in the United States would be required by law to comply with these regulations.

A second challenge to providing effective analgesia in swine is that there is often a delay between the time of drug administration and the onset of analgesic activity. For example, local anesthetics require 2 - 5 minutes before a maximal effect is achieved.[12,13] This may slow animal processing as producers must wait for local anesthesia to take effect. This delay may serve as a disincentive for them to provide routine preemptive analgesia. Furthermore, the requirement for large numbers of animals to be processed quickly may result in procedures being initiated before optimal analgesia is achieved. A third challenge is that the route or method of analgesic drug administration may require specialized training and expertise or may be hazardous to the operator. For example, the injection of large drug volumes to adult sows requires the animal to be adequately restrained and the operator to be proficient in IM administration. Similar issues are encountered with administration of local anesthesia into the scrotum which may be considered hazardous by some livestock handlers due to the risk of accidental self-injection and the injury. In addition, the majority of analgesic drugs that are available in the US have a short elimination half-life necessitating frequent administration in order to be effective.[13] This increases the stress on the individual animal and increases labor and drug costs.

In addition to the regulatory considerations discussed previously, certain drug classes such as the opioid and NMDA receptor antagonists are designated as Schedule 3 drugs and are subject to regulation by the US Drug Enforcement Administration (DEA).[14] Therefore administration of these compounds to provide pre-emptive analgesia is restricted to use by licensed veterinarians. Finally, the cost associated with providing preemptive analgesia contributes to the reluctance of producers to adopt these measures especially since there is no perceived economic benefit for doing so. It may also be difficult for producers and veterinarians to determine if analgesic compounds are effective because swine may not show overt signs of pain and distress. Thus determining the need for analgesia and the dose, route, duration and frequency of drug administration in swine can be especially challenging.

Pain assessment in swine

FDA Guidance Document 123 for the development of effectiveness data for non-steroidal anti-inflammatory drugs (NSAIDs) states that "validated methods of pain assessment must be used in order for a drug to be indicated for pain relief in the target species".[15] The identification and validation of robust and reproducible measurements of acute pain is therefore fundamental for the development and approval of effective analgesic drug regimens for use in livestock. This process is especially complex in a prey species, such as livestock that may inherently conceal pain. Pain is defined as "An aversive feeling or sensation associated with actual or potential tissue damage resulting in physiological, neuroendocrine, and behavioral changes that indicate a stress response". In previous research, markers for the evaluation of pain and distress associated with noxious animal husbandry procedures have focused on assessing behavioral, physiological and neuroendocrine changes. A change in animal behavior has been assessed using visual pen scoring, videography, vocalization, facial grimace scales, accelerometers,[9] mechanical nociceptive threshold (MNT) assessment and gait assessment using pressure mats. Physiological changes have been assessed using serum cortisol measurement, heart rate determination, feed intake and average daily gain. Neuroendocrine changes have been assessed through measurement of the neuropeptide substance P, infrared thermography (IRT), heart rate variability (HRV) and electroencephalography (EEG).

Of these outcomes, pressure mats and cranial thermography have shown the greatest promise for field applications.

A commercially available floor mat-based pressure/force measurement system (MatScan, Tekscan, Inc., South Boston, MA) can be used to record and analyze naturally occurring or experimentally-induced lameness. The pressure mat is calibrated daily and each time the computer software is engaged using a known mass to ensure accuracy of the measurements at each time point. Another benefit of this system is that video synchronization can be used to ensure consistent gait between and within calves for each time point and to correlate lameness scores with pressure mat data. Research grade software (HUGEMAT Research 5.83, Tekscan, Inc., South Boston, MA) is used to determine the contact pressure, contact area, and stance phase duration in the affected claws. Surface area is calculated by area only of the loaded or "contact" sensing elements inside a measurement box. Contact pressure is calculated as force on the loaded sensing elements inside a measurement box divided by the contact area.[16]

Recently, kinetic evaluation was used as the basis for regulatory approval for the first analgesic drug specifically labeled for alleviating pain associated with footrot in cattle (Banamine transdermal, Merck; NADA 141-450).[17] 30 Holstein steers (n = 15 calves/treatment), 8 months of age and weighing between 339 and 493 kg (site 1) and 326 – 470 kg (site 2). Animals were subjected to an induced lameness using *Fusobacterium necrophorum.* Lameness was scored prior to induction and once daily prior to enrollment (approximately 48 h after induction). Treatment success was designated as a statistically significant difference between treated and control calves at 6 (+/- 30 minutes) after administration of topical flunixin at 3.3 mg/kg. Improvement in average change in maximum total force and contact area for the right front foot between treatment and 6 h after treatment was also assessed using a real-time, gait analysis system (MatScan, Tekscan, Inc). Baseline data were included as a covariate in the model. At Site 1, mean change in maximum Force between baseline (pre-challenge) and enrollment was -31.29 kg ($P < 0.0001$) and mean change in contact area between baseline and enrollment was -10.89 cm^2 ($P < 0.0001$). At Site 2, mean change in maximum Force between baseline (pre-challenge) and enrollment was -32.95 kg ($P < 0.0002$) and mean change in contact area between baseline and enrollment was -15.51 cm^2 ($P < 0.0001$).[12]

Kinetic evaluation has also been successfully deployed to assess lameness and analgesic drug efficacy in swine. Specifically, Karriker et al (2013)[18] used a static force plate to demonstrate that injection of amphotericin B induced a predictable acute lameness that resolved spontaneously and is an effective method to model lameness in sows. Furthermore, Pairis-Garcia et al (2015),[19] reported that flunixin meglumine and meloxicam administration mitigated pain sensitivity in sows after lameness induction when pain sensitivity was evaluated with the embedded microcomputer-based force plate system and GAITFour pressure mat gait analysis walkway system. These findings support the use of kinematic evaluation as a validated method of pain assessment in swine after lameness induction.

Recently kinetic evaluation has been used to assess pain associated with castration in cattle and piglets and parturition in adult dairy cows. Specifically, Kleinhenz et al.[20] used a pressure mat to demonstrating that cows that received 1 mg/kg of meloxicam PO within 26 h of calving placed 48.9% (95% CI: 47.4% to 50.5%) of total force on the rear limbs compared to 46.3% (95% CI: 44.7% to 47.9%) in placebo-treated cows ($P = 0.02$). Total impulse on their rear limbs in the meloxicam-treated cows was 50.5% (95% CI: 48.6% to 52.4%) compared to 46.7% (95% CI: 44.8% to 48.7%) for cows in the placebo group ($P = 0.01$). No differences in contact pressure of the rear limbs were observed ($P = 0.27$). However, cows in the placebo-treated group had a longer stride length (101.3 cm with a 95% CI: 95.9% to 106.6 cm) vs. 90.8 cm (95% CI: 85.4% to 96.1 cm) ($P = 0.03$). Kleinhenz et al.[21] also demonstrated that calves undergoing surgical castration placed more force onto their fore limbs ($P = 0.02$) indicating a shift in their weight distribution to the front limbs. However, there were no measured differences in total step contact area and step contact pressure. Calves in the control group also had lower total impulses compared to surgically castrated calves ($P = 0.004$).

In summary, kinetic evaluation of animal gait represents a promising outcome to assess pain associated with lameness, castration and parturition in livestock. Further studies are needed to determine the extent to which these outcomes can be repeated across different age groups and analgesic drug interventions.

Analgesic options for use in swine

Pain perception involves the transduction of chemical signals at the site of injury into electrical energy. This is followed by transmission of the electrical signal via nerve fibers up the spinothalamic tracts where modulation may occur in the dorsal horn. Finally the impulse is projected to the brain where pain perception occurs. The initial response to a noxious stimulus is typically brief, well-localized and somewhat proportional to the intensity of the insult. The second phase of the response is prolonged, diffuse and often associated with hypersensitivity around the point where the initial stimulus was applied. This effect may lead to persistent post-injury changes in the central nervous system resulting in pain hypersensitivity or central sensitization ("wind-up"). These effects lead to hyperalgesia (increased pain from previously painful stimuli) and allodynia (a previously non-painful stimulus now produces pain).

Surgery-induced pain and central sensitization consist of two phases: an immediate incisional phase and a prolonged inflammatory phase that arises primarily due to tissue damage. The goal of administering analgesic compounds prior to castration is to mitigate both the incisional and inflammatory phase of the pain response. Effective analgesia therefore requires a multimodal approach using compounds that act on different receptor targets along the nociceptive pathway. This can be achieved through a combination of local anesthesia, nonsteroidal anti-inflammatory drugs (NSAIDs) and sedative-analgesic combinations of opioids, α2-agonists, and *N*-methyl *D*-aspartate receptor antagonists.

Local anesthesia

Local anesthetics are some of the most commonly prescribed pre-emptive analgesic drugs used in food animal practice. These compounds produce reversible loss of sensation in a localized area without causing loss of consciousness. Local anesthetics enter and block open sodium channels of nerve cells and prevent generation and propagation of nerve impulses. Repeatedly stimulated nerve cells are therefore more susceptible to the effects of local anesthetics. Furthermore, unmyelinated nerve fibers that transmit pain signals are preferentially blocked by local anesthetics compared with myelinated fibers that are responsible for pressure sensation and motor activity. The quality of local anesthesia in an acidic environment, such as infected tissues, is often poor because these compounds are weak bases that must dissociate in an alkaline environment to exert their effect. Lidocaine has a fairly rapid onset of activity (2 to 5 minutes) and an intermediate duration of action (90 minutes). Local anesthetic administration into the epidural space has also been shown to provide regional analgesia of the perineal region commencing 5 minutes after administration of 0.2 mg/kg lidocaine and lasting 10 – 115 minutes.

Non-steroidal anti-inflammatory drugs (NSAIDs)

NSAIDs produce analgesia and anti-inflammatory effects by reducing prostaglandin (PG) synthesis through inhibition of the enzyme cyclo-oxygenase (COX) in the peripheral tissues and central nervous system. COX exists in two isoforms. COX-1 is constitutively expressed in both peripheral and central nervous systems although expression is enhanced by pain and inflammatory mediators. COX-2 is ubiquitous in the central nervous system but only becomes the major enzyme for PG synthesis after induction by factors released during cell damage and death. It takes 2 – 8 hours for maximal COX-2 mRNA expression to occur in the peripheral tissues therefore initial release of PG is primarily due to COX-1. PG in the peripheral tissues lowers the activation threshold of sensory neurons and may initiate nociceptive activity. PG also works in concert with substance P, histamine, calcitonin gene-related peptide (CGRP) and bradykinin to lower the firing threshold of sensory nerves and produce inflammation. Therefore NSAIDs that inhibit COX-1 may have a more immediate impact on pain by inhibiting PG production in the periphery than COX-2 selective compounds. However, NSAIDs that inhibit COX-1 may be associated with increased risk for adverse gastrointestinal and renal effects.

Spinal PG, notably PGE_2, is responsible for increased excitability of the dorsal root ganglia leading to centrally mediated hyperalgesia. Given that COX-2 is constitutively expressed in the CNS, inhibition of spinal PGE_2 production by NSAIDs that inhibit COX-2 may be an important mechanism in preventing the establishment of hyperalgesia. The effect of NSAIDs on both central and peripheral PG synthesis suggests that these compounds have an important role in multimodal analgesic protocols.

Flunixin meglumine

Flunixin is a highly substituted derivative of nicotinic acid. Currently flunixin meglumine is the only NSAID approved by FDA for use in swine in the United States. A mean peak plasma concentration (C_{MAX}) after IM and PO administration of flunixin to mature swine (121–168 kg) at 2.2 mg/kg was 3748 ng/mL (range: 2749–6004 ng/mL) and 946 ng/mL (range: 554–1593 ng/mL), respectively.[22] T_{MAX} was recorded at 1.00 hour (range: 0.50-2.00 hours) and 0.61 hours (range: 0.17-2.00 hours) after PO and IM administration. Half-life ($T\,½\lambda_z$) for IV, IM and PO administration was 6.29 hours (range: 4.84-8.34 hours), 7.49 hours (range: 5.55-12.98 hours) and 7.08 hours (range: 5.29-9.15 hours) respectively. Therefore, once daily administration is likely required to maintain effective plasma drug concentrations. In comparison, bioavailability (F) for PO administration was 22% (range: 11-44%) compared to IM F at 76% (range: 54-92%). Although this drug class is recognized as having analgesic properties, flunixin is only indicated for control of fever associated with respiratory disease or mastitis, and fever and inflammation associated with endotoxemia, rather than for control of pain.

Salicylic acid derivatives

Salicylic acid derivatives, including aspirin (acetylsalicylic acid) and sodium salicylate, were the first NSAIDs to be used in modern medicine and are still widely used for their analgesic, antipyretic, and anti-inflammatory properties. Although the veterinary forms of aspirin are extensively marketed with label indications for the treatment of fever, inflammation, and pain relief, these have never been approved by the FDA Center for Veterinary Medicine for these indications. Therefore the legality of using salicylic acid derivatives in swine is questionable because these are technically compounded products.

Meloxicam

Meloxicam is an NSAID of the oxicam class that is approved in the European Union for adjunctive therapy of non-infective locomotor diseases and MMA when administered at 0.4 mg/kg IM. Mean peak plasma concentration (CMAX) after p.o. administration of meloxicam to mature sows was 1070 ng/mL (645–1749 ng/mL). TMAXwas recorded at 2.40 h (0.50–12.00 h) after p.o. administration. Half-life (T½kz) for i.v. and p.o. administration was 6.15 h (4.39–7.79 h) and 6.83 h (5.18–9.63 h), respectively. The bioavailability (F) for p.o. administration was 87% (39–351%).[23] These data suggest that meloxicam is well absorbed after oral administration in pigs.

Novel routes of analgesic drug administration

The transmammary delivery of analgesic drugs is an attractive option for providing long lasting pain relief to piglets and reducing labor costs because this reduces the need for individual injections and repeated handling events at the time of processing. A study evaluating a transmammary route of administration of the non-steroidal anti-inflammatory drug (NSAID), meloxicam, reported a significant reduction in expression of biomarkers associated with pain and stress (i.e., prostaglandin E_2 and cortisol) after processing in piglets nursing medicated sows. However, transmammary delivery of meloxicam required a dose of 30 mg/kg bodyweight to be administered orally to sows over several days. This is 75 times higher than the approved dose for swine in the European Union and is therefore not practical, safe or economically feasible for producers to implement in swine production systems.

Firocoxib is an NSAID with FDA-approval in the USA for control of pain and inflammation associated with osteoarthritis in horses and dogs. We conducted a study to compare the pharmacokinetics, efficacy, safety and tissue residue concentrations of four doses of firocoxib (0.5, 1.0, 1.5, or 2.0 mg/kg) administered to sows and delivered to nursing piglets prior to processing. Sixteen sows, 5 ± 2 d postpartum, were randomly assigned to one of four treatment groups.[24] On d 0 sows received a single intramuscular dose of firocoxib at 7 ± 1 h before piglet surgical castration, tail docking, and teeth clipping (males) or sham handling (females). Firocoxib and cortisol concentrations were determined from selected samples collected from sows and three piglets/litter (two barrows and one gilt) at 0, 2, 4, 6, 8, 12, 24, 48, 72, 96, and 120 h after drug administration. On d 21, piglets were weighed and all animals were euthanized and necropsied. Tissues were collected from 3 piglets/litter for histological examination and drug residue analysis. Mean (± SEM) peak plasma firocoxib concentrations (C_{max}) were 107.90 ± 15.18, 157.50 ± 24.91, 343.68 ± 78.89, and 452.83 ± 90.27 ng/mL in sows receiving 0.5, 1.0, 1.5, and 2.0 mg/kg firocoxib, respectively, and 9.53 ± 1.21, 31.04 ± 6.79, 53.30 ± 11.1, and 44.03 ± 7.47 ng/mL in their respective piglets. Mean plasma terminal half-life values ranged from 26 to 31 h in sows and 30 to 48 h in piglets. Barrows nursing sows that received 2.0 mg/kg firocoxib had a lower mean plasma cortisol concentration at 1 ± 1 h after processing compared to barrows nursing sows that received 1.0 mg/kg (P = 0.0416) and 0.5 mg/kg of firocoxib (P = 0.0397). From processing to weaning, litters of sows receiving 2.0 mg/kg firocoxib gained more weight than litters of sows that received 0.5 mg/kg (P = 0.008) or 1.0 mg/kg (P = 0.005). No signs of NSAID toxicity were observed on examination of the kidney, liver, stomach and small intestine and concentrations of firocoxib and the descyclopropylmethyl metabolite were below the limit of detection (0.01 µg/g) in all tissues examined from sows and piglets. These findings indicate that maternal delivery of firocoxib to suckling piglets before tail docking and castration may safely reduce processing-induced stress and enhance production by increasing weaning weights.

In July 2017, a formulation of flunixin meglumine for systemic absorption following topical administration was granted FDA approval for the control of pain associated with foot rot in beef cattle. This route of administration provides a convenient and cost effective method of analgesic drug delivery. However, when we examined this compound in adult sows, we reported that the percent mean bioavailability was only 1.55 ± 1.00%. These results suggest that topical administration is not an efficient route for delivering flunixin in mature sows.[25]

References

1. Hay, M., A. Vulin, S. Genin, P. Sales, and A. Prunier. 2003. Assessment of pain induced by castration in piglets: behavioral and physiological responses over the subsequent 5 days. Applied Animal Behaviour Science 82(3):201-208. doi: 10.1016/s0168-1591(03)00059-5.

2. Sutherland, M.A. 2015. Welfare implications of invasive piglet husbandry procedures, methods of alleviation and alternatives: a review. New Zealand Veterinary Journal 63(1):52-57. (Review) doi: 10.1080/00480169.2014.961990.

3. Woolf, C.J., Mannion, R.J., 1999. Neuropathic pain: aetiology, symptoms, mechanisms, and management. Lancet 353, 1959–1964.

4. Viñuela-Fernándeza, I., Jones, E., Welsh, E.M., Fleetwood-Walker, S.M., 2007. Pain mechanisms and their implication for the management of pain in farm and companion animals. The Veterinary Journal 147, 227-239.

5. Ley, S.J., Waterman, A.E., Livingston, A., 1996. Measurement of mechanical thresholds, plasma cortisol and catecholamines in control and lame cattle: a preliminary study, Research in Veterinary Science 61, 172–173.

6. Whay, H.R., Waterman, A.E., Webster, A.J.F., O'Brien, J.K., 1998. The influence of lesion type on the duration of hyperalgesia associated with hindlimb lameness in dairy cattle, The Veterinary Journal 156, 23–29.

7. Whay, H.R., Webster, A.J., Waterman-Pearson, A.E., 2005. Role of ketoprofen in the modulation of hyperalgesia associated with lameness in dairy cattle. The Veterinary Record 157, 729 – 733.

8. Flower, F.C., Sedlbauer, M., Carter, E., von Keyserlingk, M.A., Sanderson, D.J., Weary, D.M., 2008. Analgesics improve the gait of lame dairy cattle. Journal of Dairy Science 91, 3010-14.

9. Compendium of Veterinary Products, 13th Edition. Bayley, A.J. (Publisher). Port Huron, MI: North American Compendiums INC, 2010.

10. Smith, G.W., Davis, J.L., Tell, L.A., Webb, A.I., Riviere, J.E. 2008. Extralabel use of nonsteroidal anti-inflammatory drugs in cattle. J Am Vet Med Assoc 232(5): 697 – 701.

11.Animal Medicinal Drug Use Clarification Act of 1994 (AMDUCA). US Food and Drug Administration website. Available at http://www.fda.gov/RegulatoryInformation/Legislation/FederalFoodDrugandCosmeticActFDCAct/SignificantAmendmentstotheFDCAct/AnimalMedicinalDrugUseClarificationActAMDUCAof1994/default.htm. Accessed 18 March 2011.

12. Spoormakers, T.J., Donker, S.H., and Ensink, J.M. 2004. Diagnostic anaesthesia of the equine lower limb: a comparison of lidocaine and lidocaine with epinephrine. Tijdschr Diergeneeskd. 129(17):548-551.

13. Lemke, K.A., Dawson, S.D. 2000. Local and regional anesthesia. Vet. Clin. North Am. Small Anim. Pract. 30(4): 839-857.

14. DEA. Drug Scheduling. U.S. Drug Enforcement Administration. Available at http://www.justice.gov/dea/pubs/scheduling.html. Accessed 18 March 2011.

15. FDA-CVM. US Food and Drug Administration, Center for Veterinary Medicine. Guideline No. 123. Development of target animal safety and effectiveness data to support approval of non-steroidal anti-inflammatory drugs (NSAID's) for use in animals. Available at http://www.fda.gov/downloads/AnimalVeterinary/GuidanceComplianceEnforcement/GuidanceforIndustry/UCM052663.pdf. Accessed 29 August 2019.

16. Kotschwar JL, Coetzee JF, Anderson DE, et al. Analgesic efficacy of sodium salicylate in an amphotericin B-induced bovine synovitis-arthritis model. J Dairy Sci 2009; 92:3731-3743.

17. Freedom of Information Summary for Banamine Transdermal solution (NADA 141-450). Available at https://animaldrugsatfda.fda.gov/adafda/app/search/public/document/downloadFoi/1944, Accessed 30 August 2019.

18. Karriker LA, Abell C, Pairis MD, Holt WA, Sun G, Coetzee JF, Johnson AK, Hoff SJ, Stalder KJ. (2013). Validation of a lameness model in sows using physiological and mechanical measurements. Journal of Animal Science. 91(1):130-6. doi: 10.2527/jas.2011-4994

19. Pairis-Garcia, M. D., Johnson, A. K., Abell, C. A., Coetzee, J. F., Karriker, L. A., Millman, S. T. and K. J. Stalder. (2015) Measuring the efficacy of flunixin meglumine and meloxicam for lame sows using a GAITFour pressure mat and an embedded microcomputer-based force plate system. Journal of Animal Science. 93 (5): 2100-2110.

20. Kleinhenz MD, Gorden PJ, Burchard M, Ydstie JA, Coetzee JF. 2018. Rapid Communication: Use of pressure mat gait analysis in measuring pain following normal parturition in dairy cows. Journal of Animal Science 97 (2), 846-850. http://doi.org/10.1093/jas/sky450.

21. Kleinhenz , M.D. Van Engen N.K., Smith J.S., Gorden , P.J. Ji J., Wang C, Perkins S.C.B., Coetzee J.F.. (2018). The impact of transdermal flunixin meglumine on biomarkers of pain in calves when administered at the time of surgical castration without local anesthesia. *Livestock Science*. 212: 1 – 6. doi.org/10.1016/j.livsci.2018.03.016.

22. Pairis-Garcia., M.D., Karriker, L.A., Johnson, A.K., KuKanich, B., Wulf, L.W., Sander, S., Millman, S.T., Stalder, K.J., Coetzee J.F. (2013). Pharmacokinetics of flunixin meglumine in mature swine after intravenous, intramuscular and oral administration. *BMC Veterinary Research*, 13;9(1):165.

23. Pairis-Garcia, M.D., Johnson, A.K., KuKanich, B., Wulf, L.W., Millman, S.T., Stalder, K.J., Karriker, L.A., Coetzee J.F. (2015). Pharmacokinetics of meloxicam in mature swine after intravenous and oral administration. *Journal of Veterinary Pharmacology and Therapeutics*, 38: 265- 270. doi: 10.1111/jvp.12170.

24. J. F. Coetzee.,Sidhu, P.K., Seagen, K., Schieber, T., Kleinhenz, K., Kleinhenz, M.D., Wulf, L.W., Cooper, V.L., Mazloom, R., Jaberi-Douraki, M., and Lechtenberg, K. 2019. Transmammary delivery of firocoxib to piglets reduces stress and improves average daily gain after castration, tail docking, and teeth clipping, *Journal of Animal Science*, Volume 97, Issue 7, July 2019, Pg. 2750–2768, https://doi.org/10.1093/jas/skz143.

25. Cramer, M.C., Pairis-Garcia M.D., Bowman, A.S., Moeller, S.J., Zhang, Y., Sidhu, P.K., Magnin, G., **J. F. Coetzee***. 2019. Pharmacokinetics of Transdermal Flunixin in Sows. Submitted to *J Vet Pharmacol Ther*. Accepted April 2, 2019. In press.

What a pig wants: advances in animal welfare science

Meghann K. Pierdon, VMD, DACAW
New Bolton Center, University of Pennsylvania, Kennett Square, Pennsylvania

Introduction

Animal welfare science was launched with the formation of the Brambell Commission and their report in 1965. With the public concerned about animals in agriculture, and how they were being treated, came the five freedoms of animal welfare.[1] This framework is important because it provides a definition of animal welfare and foundation in science. Before the report though, there was still welfare. There was the farmer and the animals working together, husbandry and stockmanship the farmers contract with the animals. This is still a common sentiment among those of us who work with animals but now we have to ask: what does our obligation to animals mean for animal welfare? Animal welfare is "its state as regards its attempts to cope with its environment"[2] and can be defined in terms of "animal health and what animals want."[3] Animal welfare is fundamentally about the animal's perspective so how do we assess welfare in our modern facilities? After the Pig Park, where Stolba and Wood Gush released sows into a semi-natural environment, we had evidence that the behaviors of their wild ancestors were not gone in our commercial genetics,[4] but that the threshold for those behaviors may have changed and, their ability to perform them, curtailed by modern housing and production practices. Pigs are smart. Recently we have even seen the first documentation of tool use in pigs.[5] In another study, pigs learned what a mirror was and could go to a food source that was placed behind them instead of looking behind the mirror.[6] Given this and other studies of pig cognition, advances in animal welfare science are worthy of further investigation.

Assessment

Individual farms are assessed regularly. The commercial swine industry has implemented the Common Swine Industry Audit which mirrors the Pork Quality Assurance Plus Program assessment. The audit tool, and the assessment program that mirrors it, cover documentation of animal based and non-animal based measures.[7,8] About 50% of the audit focuses on animal-based measures including timely euthanasia, acts of abuse, body condition, lameness, hernias and lesions. It is worth asking however, how well these measures assess animal welfare. There is no assessment of behavioral freedom or behavioral diversity. The focus of the industry has been physiology and assessments of other realms of welfare. Emotions and natural behavior are left to niche market producers and smaller audit programs. Welfare programs proprietary to production systems allow companies to distinguish themselves from their competitors. Customers may be asking these questions of commercial production before long. A few areas that show potential for practical applications on farms are enrichment and precision livestock farming.

Enrichment

In the case of the Pig Park, an ethogram, or catalogue of behaviors, showed the amount of time in locomotion, and grazing at 59% of a pig's day.[4] When we look at the ethogram for pigs in a conventional finisher we see they spend 64% of their day lying inactive.[9] This points to a difference in the behavioral diversity and activity of pigs in a commercial environment compared to a natural environment. When used scientifically, environmental enrichment refers to environmental modifications which have enhanced the performance of strongly motivated species-specific behaviors or have led to the expression of a more complex behavioral repertoire.[10,11]

In pigs, enrichment is often aimed at attempting to provide substrate for foraging-like behavior. There are some practical examples in the literature that meet this definition. A recent study in sows showed that they interacted preferentially with cotton rope over other objects and continued to interact with it for at least 2 weeks.[12] In another study, boars preferred cotton rope and decreased stereotypic behaviors when it was provided.[13] Providing enrichment has not however been proven to decrease aggression when sows are mixed together.[12,14]

In piglets, provision of different types of enrichment has been shown to decrease manipulation of pen-mates. In a two-tiered system with different mirco-environments and a hiding area, pigs showed similar productivity and decreased tail biting compared to pigs in a conventional pen with the same square footage.[15] Piglets provided a small amount of peat are less likely to manipulate their pen-mates and if it was provided before and after weaning, the pigs were heavier than those not provided peat or who only had peat after weaning.[16] A 2007 study in early weaned pigs allowed weaned piglets to control the temperature in the room within certain parameters. The piglets decreased the temperature in the room at night.[17] Environmental control may be another way to provide enrichment by giving pigs control over their surroundings.

Precision livestock farming

Advances in technology are making their way into animal agriculture. Precision livestock farming (PLF) is the management of livestock using process engineering technology and principles.[18] Combining sensors for data collection and computer systems to aggregate the information for remote analysis is an application of PLF. A few developments are promising for the management of pig welfare. The first step is using human observers to teach the computer system what to look for. Researchers used this methodology to classify the phases of aggression by having a human observer map video recordings of aggressive interactions. They found that biting occurred predictably in the behavior sequence such that in the future a computer could predict the occurrence of such behavior and it could be prevented.[19]

Sound monitoring is particularly appealing because you can monitor many pigs with a few inexpensive sensors. Acoustic sensors have been used to detect estrous in sows[20] and the response to heat stress.[21] A useful tool for veterinarians is the research on coughing where different types of coughs have been distinguished, an algorithm created, and the functionality of the monitoring equipment tested.[22] Using technology to create objective measures of lameness, and eventually automate lameness detection, has long been under study. Multiple technologies are being tested including a Microsoft Kinect motion sensor[23] and force plate and image analysis.[24]

Conclusion

Conclusions about the benefits of providing positive welfare are often limited to the impact on the animal's behavior and physiology. It is logical that a technology or enrichment must be affordable, implementable, and beneficial to pig welfare. An added benefit though, could be employee satisfaction. In an industry searching for quality employees and aiming to keep them, welfare friendly practices may provide a means of improving job satisfaction. Altering the balance between positive and negative interactions between people and animals has the possibility of improving human animal interactions and possibly human animal relationships. More study should be done on whether the provision of enrichment or other welfare improvements increase job satisfaction for farm staff or alters the relationship between humans and the animals they are caring for.

References

1. Brambell, FWR. Report of the technical committee to enquire into the welfare of animals kept under intensive livestock husbandry systems. 1965.

2. Broom DM. Attempts to Cope with the Environment. *Acta Agric Scand Sec A Anim Sci* Suppl. 1996; 27:22-8.

3. Dawkins MS. Animal welfare with and without consciousness. *J Zool.* 2017; 301(1):1-0.

4. Stolba A, Wood-Gush DGM. The behaviour of pigs in a semi-natural environment. *Animal Science.* 1989; 48(2):419-425.

5. Root-Bernstein M, Narayan T, Cornier L, Bourgeois A. Context-specific tool use by *Sus cebifrons. Mammalian Biology.* 2019; 98:102-110.

6. Broom DM, Sena H, Moynihan KL. Pigs learn what a mirror image represents and use it to obtain information. *Anim Behav.* 2009; 78(5):1037-1041.

7. Pork Quality Assurance Plus® program. 2019. Available at: `https://lms.pork.org/Tools/View/pqa-plus/program-materials`. Accessed November 15, 2019

8. Common Swine Industry Audit. 2019. Available at: `https://www.pork.org/production/tools/common-swine-industry-audit/common-swine-industry-audit-materials/`. Accessed November 15, 2019

9. Guy JH, Rowlinson P, Chadwick JP, Ellis M. Behaviour of two genotypes of growing–finishing pig in three different housing systems. *Appl Anim Behav Sci* 2002; 75(3):193-206.

10. Newberry RC. Environmental enrichment: increasing the biological relevance of captive environments. *Appl Anim Behav Sci.* 1995; 44: 229–243.

11. van de Weerd HA, Day JE. A review of environmental enrichment for pigs housed in intensive housing systems. *Appl Anim Behav Sci.* 2009;116(1):1-20.

12. Horback KM, Pierdon MK, Parsons, TD. Behavioral preference for different enrichment objects in a commercial sow herd. *Appl Anim Behav Sci.* 2016; 184:7-15.

13. Sirovica LV, Creamer M, Horback KM. Preference for and behavioural response to environmental enrichment in a small population of sexually mature, commercial boars. *Anim Welf.* 2019; 28(3):271-278.

14. Greenwood EC, Van Wettere WH, Rayner J, Hughes PE, Plush KL. Provision point-source materials stimulates play in sows but does not affect aggression at regrouping. *Animals.* 2019; 9(1):8.

15. Fraser D, Phillips PA, Thompson BK. A test of a free-access two-level pen for fattening pigs. *Animal Science.* 1986; 42(2):269-274.

16. Vanheukelom V, Driessen B, Maenhout D, Geers R. Peat as environmental enrichment for piglets: The effect on behaviour, skin lesions and production results. *Appl Anim Behav Sci.* 2011; 134(1-2):42-47.

17. Bench CJ, Gonyou HW. Temperature preference in piglets weaned at 12–14 days of age. *Can J Anim Sci.* 2007; 87(3): 299-302.

18. Wathes CM, Kristensen HH, Aerts JM, Berckmans D. Is precision livestock farming an engineer's daydream or nightmare, an animal's friend or foe, and a farmer's panacea or pitfall?. *Comput Electron Agric.* 2008; 64(1):2-10.

19. Oczak, M., Ismayilova, G., Costa, A., Viazzi, S., Sonoda, L.T., Fels, M., Bahr, C., Hartung, J., Guarino, M., Berckmans, D. and Vranken, E., 2013. Analysis of aggressive behaviours of pigs by automatic video recordings. *Comput Electron Agric.* 2013; 99:209-217.

20. Rhim SJ, Kim MJ, Lee JY, Kim NR, Kang JH. Characteristics of Estrus related Vocalizations of Sows after Artificial Insemination. *J Anim Sci Technol.* 2008; 50:401–406.

21. Hillmann E, Mayer C, Schön PC, Puppe B, Schrader L. Vocalisation of domestic pigs (Sus scrofa domestica) as an indicator for their adaptation towards ambient temperatures. *Appl Anim Behav Sci.* 2004; 89: 195–206.

22. Vandermeulen J, Decré W, Berckmans D, Exadaktylos V, Bahr C, Berckmans D. The Pig Cough Monitor: from research topic to commercial product. *Proc of Precision Livestock Farming* Leuven, Belgium. 2013; 717–723.

23. Stavrakakis S, Li W, Guy JH, Morgan G, Ushaw G, Johnson GR, Edwards SA. Validity of the Microsoft Kinect sensor for assessment of normal walking patterns in pigs. *Comput Electron Agric.* 2015; 117:1-7.

24. Pluym LM, Maes D, Vangeyte J, Mertens K, Baert J, Van Weyenberg S, Millet S, Van Nuffel A. Development of a system for automatic measurements of force and visual stance variables for objective lameness detection in sows: SowSIS. *Biosyst Eng.* 2013; 116(1):64-74.

A glimpse inside the head of today's food shopper

David Fikes
Executive Director, FMI Foundation

Any sports fan can tell you that what happens at the beginning of the game is important, but it is what transpires in the final minutes or the last inning that creates the drama and attracts the attention. Because of this, coaches know they must have a great end game. While attending to thousands of details along the way, they cannot afford to lose sight of the end game. Closing the deal is key.

In the food supply chain, producers, manufacturers and retailers all have their roles to play, but we all have a common interest in having the best endgame we can and in this case the end game is the consumer. All our efforts are for naught if our product isn't selling – and it won't sell if we don't pay attention to the consumer

For the past 45 years, Food Marketing Institute has produced a report called *US Grocery Shopper Trends*. We do this because we deem it important to know as much as we can about changing consumer shopping patterns, so our end-game strategy is informed, smart and effective.

Before launching into the findings of the 2019 version of the *US Grocery Shopper Trends,* some context setting would be helpful. For the larger setting against which to examination this year's findings, I turn to the insights of futurist Alvin Toffler who at the end of the last millennium made what have turned out to be some solid predictions about life as we currently live it.

In his book, *The Third Wave,* Toffler identified the three stages of development in which the human species is engaged. The first of these he called the Agricultural Age and it started some 12,000 years ago, when our species transitioned from a hunter-gatherer structure and starting to form an agricultural society.

The second, stage, Toffler labeled as the industrial age, which had the modern, nuclear family in its center and was ignited by the Industrial revolution of the mid-1700s. This revolution was focused on the efficient mass production of products, which of course lead to mass everything: distribution, consumption, communication, media, entertainment, and education. Toffler's third wave is about the transition into the Information Age, which began in the late 1950's with the development of the transistor, the computer and the internet. This resulted in commodity of exchange – starting to move from produced material items to the development, gathering and exchange of information. This is where we find ourselves today, living in a world where the valued commodity of exchange is information and the focus has shifted from the masses and onto the individual. When we talk about the empowered consumer - it is the individual customer wielding a smartphone and able to source critical information about products, retailers, ingredients, everything, simply by pressing a few buttons or giving a few voice commands. This is why we at FMI constantly remind our members, you no longer just sell food, you also market information.

So, it is against this individual minded, information fueled back drop that we will look at some consumer shopping trends, beginning with some of the traditional elements that we've been tracking for 45 years.

Food is ubiquitous in the American retail landscape, and so are those who seek it. Where there are people there is food and where there is food, there are people buying it.

Changing patterns in consumers' lifestyles – the growth in non-traditional households, shifting family roles, along with more frequent eating habits and a generally deeper engagement with food and nutrition – have all combined to bring more consumers into food outlets.

Among all US adults, **86% claim at least half of their household's shopping responsibilities** and have shopped recently.

This has several huge implications, not the least of which is the day has long passed when we could refer to American grocery shoppers exclusively as "She." Today, everyone is a grocery shopper and the population of male shoppers continues to climb. But the fact that everyone shops also means a shift in the way households approach their grocery shopping.

The following statistics capture some of the changes in HOW households are shopping.

- About a ¼ of all shoppers come from single person households and handle all the household shopping for him or herself.
- This means that the remaining ¾ of shoppers are from multi-person households and are in some sort of co-shopping arrangement, meaning they share responsibility for the grocery shopping with others in their household.
- One-third of all shoppers are in an intentional 50/50 split with their household partner – and we have seen this number grow since we first started tracking it six years ago.

Now consider how well you communicate, share tasks and coordinate with those in your household. We can all use some help in this area to make life go smoother, so whatever the food retailer can do to help list building, list sharing and list execution run more efficiently is greatly appreciated.

Without a doubt, the increase in the number of food retail channels over the past 15 to 20 years has cut in the traditional supermarket **share as the primary store.** The first few years of the new millennium, the supermarket was at 70% share, then dropped into the 60%s some 15 years ago, fell further into the mid-50s ten years ago, and continued to slide to 52% in 2015. The traditional supermarket appears to have stabilized, with its share as the primary store hovering right at the 50% mark the past four years.

With the exception of 2010 when we were in recession recovery mode, supercenters have consistently registered in the 22/23% range. Other channels which includes pharmacies and online have grown the past five years and we'll be discussing that one more specifically, a bit later.

The other thing to note is that back in 2015 FMI began offering survey participants the option of "no primary store" for those who shop so many different outlets frequently that they could not designate any single one as their "primary store." We were surprised to learn that some 9% (or one shopper in ten) did not have a primary store. That number ebbed a bit last year but was back at 8% in this year's trends.

Two important points of traditional data FMI gathers include **weekly trips** to the grocery store and **average dollars spent on groceries per week**.

In 2019, trips are holding steady at the 1.6 trips per week level, however there are two observations to add.

1. That number only registers the Primary shopper's trips, when the other shared shopping partner gets added, the average household is visiting food retail outlet(s) some 2.5 times a week.
2. The second observation is that consumers who have online in their shopping repertoire still visit their brick and mortar outlet some 1.7 times a week. Online may be cutting into in-store basket size, but currently is not diminishing the number of trips to the store. We'll have more on that later when we get into online grocery shopping in detail.

The 2019 spend data reveals that the food retail industry experienced a slight uptick of $4.50 in this year's average household grocery spend, bringing the weekly average to $113.50.

Not only do most Americans grocery shop, they like shopping in multiple places. Shoppers visit an average of **4.4 retail banners** and **3.1 different channels** each month.

The number of different channels shopped each month is even higher among Gen Z (3.9) and Millennials (3.8). It is also higher in households with kids (3.8). Younger grocery shoppers tend to shop even more banners, with Gen Z'ers hitting some 6.2 a month, and Millennials (5.0). Retail banner average is also higher for Households with children (5.4) and households with $100K+ income. (5.5) Given that we view Millennials and Gen Zer's as the portents to the future, it appears channel fragmentation will be with us for a while.

Online shopping

Having noted all the channels that shoppers are using, we turn now to look a bit more explicitly at online, exploring how it is developing.

Back in 2015 when FMI first started tracking grocery shopper engagement with online, we found that about 16% of consumers were utilizing this channel. Online engagement continues its steady climb reaching the 33% level this year - officially doubling its audience in five years. While not a rocket trajectory, online adoption in the grocery world is showing steady growth.

Online use has always been an important area to examine by age group due to millennials leading the charge in embracing online grocery shopping. That appears to have cooled a bit with Millennial adoption of online leveling out a bit at around the 45% range. Even the level of online engagement - measuring how frequently a channel is used- reveals that millennial engagement is slipping just a bit, indicating that millennials continue using online as one of the channels they shop, but perhaps not as frequently as before.

However, there are three important developments in online engagement that should be noted:

- Gen X adoption in 2019 is beginning to rival millennial levels and even the *depth* of Gen X adoption picked up a bit.
- Boomers continue to pick it up a bit, but mature engagement with online shopping took a hit this year, falling back to near 2015 levels.
- This is the first year we started having Generation Z participants in our survey and discovered that they enter the online-grocery-shopping picture at a level right up there with their older millennial friends.

In short, all age categories that point to the future (Gen Z, millennials and Gen X) indicate a continued embracing of online shopping with adoption continuing its steady climb.

Here are some additional *Trends*-supported observations about the shopper who frequently use online channels:

First, in general, **they shop more than the average consumer**.

- They shop more banners, (7-8 and average is 4.4).
- They shop more channels (3).
- AND they make more weekly trips to the grocer. (1.7 VS average of 1.6).

Second, they **spend more in general**; ($160 vs. the average of $113), this includes online and in-store spend.

As previously pointed out, this means online has cut into basket size a bit, but **not** much given the higher spend amount and **not** at all in regard to number of trips. Frequent online shoppers are actually in stores more than those who do not shop online at all, but they are also spreading their higher grocery spend across almost eight grocery venues.

Analyzing all the demographic information, the emerging profile of the online grocery shopper could be personified as a 30-year old new Atlanta Dad who is college educated and earns six figures.

The quest for a food connection

So, let's turn our attention to the shopper quest for information and their strong desire for a deeper connection with his or her food.

If you are among the majority who think they get too many emails each day and wish you had someone to wade through your inbox helping you distill which emails you need to pay attention to and which you could delete, then you can sympathize with the dilemma facing the information age food shopper.

Today's shopper – like all of us – is both inundated with information and simultaneously wanting more information – but info that matters. As they map their grocery shopping excursions - choosing among the multiple options available to them - they are seeking those retailers that send clear messages helping them answer the growing number of questions they have, such as where do I want to shop, what to prepare for dinner, what do the others in my household need?

This question-plagued consumer appreciates those retailers that provide possible solutions to all their questions. They will engage with the retailer offering convenient options - from food delivery to meal preparation. And they will reward the retailer with their business if they can make it easier to coordinate shopping lists, weekly menus and snack options with the others in the household. And we should be clear; consumers don't expect everything done for them, but they do want retailers making it easier for them to execute their health, shopping and meal-prep strategy. They turn to their retailer for some help in sorting through and addressing these various values.

For the most part, consumers are satisfied with their primary store – indicated by a stable 8.7 satisfaction rating.

For those believing a perfect 10 score is possible, we now turn to explore what it takes to maintain a 8.7 score AND taking note of what it would take to gain those additional 1.3 points to rate a 10.

When asked to rate the most important attributes in a store, we find the fresh categories among the most customer appreciated:

- High quality fruits and vegetables – edging out high quality meat for top honors this year.
- Low prices tied for second.
- And the features of clean, neat store product selection and variety tied for fourth place.

The retail tendency is to pay attention to the top answers on the customer appreciation list. However, in this individual-over-the-masses information age world we operate in, there are some other considerations that should be highlighted.

Areas where there has been significant movement when compared to the 2018 scores, include:

- Accurate information displayed.
- Friendly courteous staff.
- Fast checkout.
- Convenient from work.

However, what really think warrants our attention are those areas where the Millennials and Gen Zers over-index or rated higher than normal.

The premise is if a store wants to maintain the status quo, pay attention to the items on the top of the list. If you want to thrive in the future, excel at the top and start concentrating on the areas the younger shoppers value.

These "lower ranking" attributes - honesty, sustainability and community involvement - are areas of significant importance to Millennials and Gen Z'ers. And even though they register rather low on our scale, it would be a costly mistake to ignore them, especially if a retailer is courting the business of the younger shoppers. This particular group of shoppers are looking for these concerns to be addressed AND they are looking now to see who is taking their concerns seriously.

What does it mean to eat well?

Over the past couple of years FMI has taken note that the consumer definition of health and wellness was expanding and starting to include concerns that went beyond traditional nutrition considerations. This prompted last year's exploration in *TRENDS* of what all shoppers included in their consideration of health and well-being. We asked consumers, "What does it mean to 'eat well'" and we were intentionally vague in how we asked the question, to not bias the direction they went.

The diversity of answers received was a bit surprising, but the answers fell into four categories: health, enjoyment, mindful connection and budget. The top vote getters -in order- were:

- I ate tasty food and beverages.
- I ate nutritious food and beverages.
- I ate within my budget.
- I ate high quality food and beverages.
- I ate in moderation.

Note that these answers spanned several categories.

Again, it should be noted that millennial and Gen-Z respondents over-indexed in the "mindful connection" category. While that answer did not place among the overall top vote-getters – it is a consideration in the minds of the younger generation when they are contemplating what it means to eat well. Of further interest, women over-indexed in the top option of "*I ate nutritious food and beverages,*" while men over-indexed in the category of "*I ate whatever I wanted.*" Clearly, there is a Mars and Venus lesson in there somewhere.

A bit of label-gazing

As a reminder we should not underestimate the power of product labels, it is worth noting that eight out of 10 shoppers indicate that they ACTIVELY look for at least one front of pack claim. Labels remain one of the strongest means of communicating with the consumer. When exploring what product claims shoppers seek when purchasing a food product, the 28-item list can be broken into six buckets of what consumers are looking for on package.

- 68% are looking to avoid negatives (the free-from movement).
- 62% are looking for claims that (to them) indicate minimal processing.
- 40% seek out good fiber products.
- 31% seek label claims conveying ethical practices.
- 23% seek positive nutrition claims.
- 22% look for heart health callouts.

When it comes to health and wellness concerns, we asked consumers who is working against you, and who is working for you? Again, this year, consumer's rated *family* highest among groups working for me, followed by *doctors, friends, farmers and fitness clubs. My primary store* fell a bit this year, scoring a 49% that was 6% points lower than last year's score. Retail in general follows. And *manufacturers, news media, fast food restaurants and the entertainment industry* are way down at the bottom of the list.

This is important data for food companies seeking to distinguish themselves in the health and wellbeing arena, providing the top influencers that should be engaged to help move the needle with shoppers.

Hit or myth- The art of personalization

In 2019, *Trend's* focused its research attention on the topic of personalization. What got uncovered were some common misconceptions - four myths in particular - about personalization that needed to be debunked.

FMI and our research partner The Hartman Group termed these misperceptions as myths, drawing on the common definition of a myth being a commonly accepted story that is not necessarily objectively true, but may contain elements of truth. Myths may have what Stephen Colbert coined as "a truthiness to them" but are dangerously overdrawn or misperceived.

We will now look at these four myths of personalization and seek to set the record straight about what is fact and what is fiction.

The first myth is that **shoppers want everything personalized for them**. This statement underappreciates the amount of work consumers are willing to do to personalize their shopping trip for themselves and overstates how much control they are willing to relinquish to the retailer in the name of personal service.

FACT: Differentiated treatment may definitely be a plus, and YES, most assuredly shoppers want help in making their shopping experience more pleasant and efficient, BUT shoppers aren't asking for everything to be done just for them on a silver platter. Shoppers have a strong degree of self-determination that must be respected in every attempt to personalize service. They want to control the experience and they expect do some of the work of personalization; in fact, that is what they ultimately think shopping is.

They just want grocery shopping to be easier, with flexibility to meet their families' changing needs. They don't want you making the choice for them, they want you helping them make their own choice.and there is a fine line that must be walked in this regard.

Myth number two is **Personal eating requirements mean that shoppers buy for individuals rather than a family.**

FACT: Families still want to eat together.

However, family meals and purchasing for snacks means paying attention to the individuals who constitute the family. It is a delicate balance of paying attention to both the needs of the one and the needs of the many. Especially when there is a collision of values at stake.

In the midst of the information age's stress on the individual and personalization, family meals continue to be an aspiration.

- 97% of households with kids say eating family meals at home is important.
- 84% of households with kids report wanting to eat more meals at home.
- 90% say that at-home meals are healthier.

The sum of these statistics adds up to a win-win scenario for both the customer and the food retail venue. In short, customers are saying they think family meals are healthier, have tremendous social value, and are something they expressly want to do more often. Grocery stores that inspire, support and help their customer overcome the obstacles to family meals will reap the reward of customer loyalty.

Myth number three is **Specialized eating habits are the domain of specialty retail. Mass-market food retail cannot compete on meeting personal eating aspirations.**

FACT: Retailers positioned around a specialty certainly find it easier to understand, make recommendations and provide guidance to shoppers coming into their stores. After all, by a customer's mere presence in a specialty store, they've tipped their hand about being interested in the area of the store's expertise. However, it is worth remembering that the stores garnering the highest rating from consumers get those marks through excelling at the benefits of personalization -- convenience, caring and enabling the shopper to meet their needs. Simply stated, specialty stores do not own these benefits. In fact, specialty stores may even be at a disadvantage in the key area of convenience.

So rather than specializing, mass marketers should focus on personalizing. This requires getting educated about what's driving consumers to the specialty store and better enable them to meet that need in the convenience of the more expansive store.

Myth number 4 is **Digital experiences are more easily customized, and therefore digital platforms are the best place for a retailer to start personalization initiatives.**

FACT: FMI has been saying for years that online versus bricks and mortar is not an either/or proposition, but one that is both/and. Nowhere does that truth hold firmer than in the area of personalization.

We know technology is embedded in the consumer shopping experience across all channels. We also know the consumer expectation is that tech-engagement must embrace more than sales and delivery. Shoppers crave interaction before a transaction and a great mobile experience provides an opportunity to stand out. In the case of providing support to shopping and if little- or low-stakes data is to be shared, a digital-first approach may be appropriate. However, anything beyond that kind of introductory information-sharing requires a **foundation of existing trust**. That kind of trust calls for face-to-face experiences that reinforces the positive impressions created virtually.

In this regard, FMI has four recommendations to better integrate online and off-line/in-store personalization efforts that will help build and retain customer trust;

1. **Empower user control** – Food retailers are encouraged to design new tools and platforms that allow maximum user control over data and information flow. Again, consumers want to be in the driver's seat regarding the shopping experience and that includes access to and use of their data.
2. **Verify privacy and security** – Be clear with shoppers and demonstrate accountability for data use, sharing and protection. Customers want to be assured that the banners they frequent have taken the necessary steps to safeguard the shopper's information before they will grant permission for it to be used – even if the company's use is exclusively for the customer's benefit.
3. **Establish value** – Clearly communicate the advantages of new personalization initiatives and demonstrate how they help shoppers meet household eating needs and preferences. Shoppers want to be able to customize the way they receive the information they need. Additionally, they want to understand how it addresses their families' needs without feeling their privacy is being needlessly invaded.
4. **Show optimal use** – Clarify how data is currently used and the benefits shoppers derive from it. Shoppers need to know what you're going to do with their data and how it will be gathered and used before granting you access to it.

To support personal eating and shopping in personal ways requires permission to gather data, interpret data and act upon that data. That level of permission is more readily granted when online messages are in sync with in-store experiences, with BOTH grounded in a sense of care, concern and trust-building respect.

In addition to the facts that debunk the myths surrounding personalization, what we learned this year, is that we may have been missing or at least underestimating the power of **personalization** as the operative motivation behind every major trend, consumer shopping behavior we've noted in recent years.

- The notion of shared shopping - while perhaps driven by a sense of sharing responsibility for a household chore, might also be motivated by "I need to make sure the groceries I prefer, and I items that I know our kids like are being purchased." In other words, shared shopping may be an act of shopping in self-defense to ensure MY personal interests are covered.
- The consumer demand for transparency likely has its base in the need for reliable information so the shopper can pressure test his or her values, preferences and judgments against those of the store and the company.
- As FMI explored in 2018, the shopper's notion of eating well is highly individualized and a unique mix of values and expectations that vary with each shopper. So, to accommodate those varying requirements, consumers shop multiple channels and a variety of stores seeking out those that will best enable, assist and empower the execution of each shopper's unique menu/diet strategy.

In short what FMI learned brought us back to a very important food retailing principle: **Serving the customer is more about what they want to buy, and less about what we want to sell.**

African swine fever: what's working and not working in China

Scott Dee, DVM, MS, PhD, DACVM; Gordon Spronk, DVM; Joseph Yaros, DVM

African swine fever virus (ASFV) was officially reported by the Ministry of Agriculture and Rural Affairs (MARA) in the domestic swine herd in China on August 3, 2018.[1] Pipestone staff have been visiting swine herds throughout China and Asia for over 20 years as requested by Chinese producers, vendors to the Chinese industry, governmental agencies, and non-governmental agencies (NGOs). These visits have given our staff extensive field observations of clinical disease in swine herds in Asia both prior to and after the official reporting of ASFV. It may be of interest that swine disease presentation in Asia in the field typically results in the same field observations of the same pathogens as in North American herds and the rest of the world. In the majority of cases, the pathogen presentation is as widely described and published. In a small subset of field observations however, the presentation may be different and vary widely due to co-infections, pathogen isolate differences, operational differences, and implemented interventions or medications. Additionally, several pathogens (described as Foreign Animal Diseases, FADs) including foot and mouth disease (FMDV), classical swine fever (CSFV), pseudorabies virus (PRV) that have been eliminated from swine herds in North America and Europe have been commonly observed in China swine herds. These field observations of isolated pathogens (FADs) or with co-infections may provide opportunity to educate future global veterinarians and health officials allowing improvement in the health of the global swine herd. This paper will focus on the field observations of ASFV in commercial operations in Asia.

ASFV in the world swine herd

Prior to official reports of ASFV in the largest swine herd in the world, ASFV was commonly known to have spread widely over the past nearly 100 years from Africa to the Iberian Peninsula, to eastern Europe, across Russia, and finally into Asia. Movement of the virus is commonly understood to be the result of "people or human" movement of the pathogen; the virus movement speculated to be by movement of either infected pigs or meat by producers or staff.

ASFV is known and reported to be one of the deadliest viruses (if not the deadliest as measured by mortality) to infect swine of all ages. In small stakeholder operations in Asia, the virus may result in death loss of 100% of the pigs in the herd (note that any surviving pigs are commonly sold prior to death resulting in under reporting of true mortality). In larger operations, a small percentage of pigs have been observed and reported to survive an ASFV episode. Clinical presentation of ASFV infection is widely published and commonly understood to include a hemorrhagic fever, slow spread in a population, and high mortality. There are variations from these published reports under large field conditions in Asia and will be reviewed during this session.

In China, from August 2018 until August 2019, there were official reports of 160 cases in 32 of the 33 provinces in China. The virus has spread throughout China (and now in 10 countries in Asia), affecting pigs of all ages and genetics. Additionally, all types of production systems have been infected and observed under field conditions including smaller traditional farms (less than 500 head) to modern western style barns (over 8,000 sows) of all designs and operational practices.

Currently, the pig population in China has been unofficially reported to be dramatically reduced with common concern that total losses documented by official reporting starting in August 2018 is understated. While official reports do report a decline in the size of the China herd, it is widely speculated that official reports are under-estimating the full extent of total loss due to ASFV. There is no accurate or commonly accepted method to determine the exact amount of loss, with various authors speculating a minimum of 30% reduction in herd size to many sources reporting losses at various levels approaching higher than 50% loss of the China sow herd. Regardless of the exact loss, field observations confirm the severe impact of this virus is as widely published.

Reports that vaccines for ASFV will be (or already are) available in China are widely described. The first tool that veterinarians and producers traditionally implement to control any pathogen is a vaccine which successfully results in both reduction of mortality in an affected population and stops the spread of the pathogen between naïve animals. One additional feature of a good vaccine is the ability to discern the difference between the vaccinated pig and a pig exposed and infected with field virus. Researchers have struggled for decades to develop an effective ASFV vaccine and as of this writing, there is no officially approved vaccine for ASFV to prevent infection of naïve pigs.[2]

ASFV epidemiology

The epidemiology of ASFV is widely published under controlled research conditions: the virus is shed from all porcine bodily fluids, with blood containing the highest virus concentrations. Pigs can be infected via nose to nose contact, semen, milk, fecal/oral, injections under the skin containing viral particles, or ingestion of tissue from infected pigs. The incubation time of the virus is 3-14 days, with clinical signs including lethargy, erythema, piling of pigs, febricity, bloody scours, and sudden death.

More epidemiological research and publication is needed as field reports from China in large commercial operations with potential co-infections dispute some of these reported results. Some of the variation in field reports may be explained by differences in virus isolate commonly reported as High, Moderate and Low virulence. These ASFV isolates also may have different field presentations and pathogenesis described as Peracute, Acute, Subacute and Chronic. Field observations support these reported differences in clinical presentation.

Differential diagnosis for ASFV include, but are not limited to, classical wwine fever, porcine reproductive and respiratory syndrome, pseudorabies virus, salmonella, and post-weaning multisystem wasting syndrome and other bacterial pathogens including *Clostridium* and *Streptococcus*.

ASFV is highly pathogenic, but not reported to be highly transmissible. Some authors describe this as "contagiosity" (there may be high mortality with low infection rate when compared to other pathogens like FMDV).[3] In clinical field cases in large populations of commercial pigs where the virus has been left uncontrolled and epidemiological data has been monitored, total mortality can more than double in the matter of days until the majority of the population is deceased.

Investigations into the source of the infections into a population in Asia have been conducted to understand the routes of entry which may allow for future prevention of movement of the virus to other area farms. In general, field observations in Asia support rapid area spread with speculated routes of movement of the virus attributed to movement of infected pigs, movement of infected meat, contaminated fomites, rats, flies, mosquitoes, staff, feed, and lack of implementation of proper biosecurity protocols. The role of contaminated feed cannot be overlooked in the transmission dynamics. Improper disposal of the carcasses and movement of infected pigs are high on the list of methods that may have moved this virus across all of China. Pigs that have died from ASFV require proper disposal and enhanced biosecurity practices to limit the spread of ASFV to naïve populations since carcasses are shedding high levels of virus.

Site depopulation and cleanup protocols have been implemented and observed under field conditions and vary widely. It is critical that following depopulation of a site, a cleaning and disinfection protocol that removes all organic material and viral particles is completed to prevent viral transmission to the next group of pigs and for the site to remain healthy and free from ASFV. Commonly accepted protocols for other viruses (including but not limited to PEDV or PRRS) have been successfully used in the field. Attempts to eliminate ASF virus from sow populations using test and removal techniques are being reported in China.[4] This elimination technique using herd closure, partial depopulation and whole herd sampling methods may hold promise and needs further research and documentation to understand if effective for this pathogen.

To ensure that ASFV does not enter other sites, a full biosecurity evaluation should be conducted to ensure that protocols regarding personnel entry, trucking, loading, feed, downtime, and other possible routes of entry are properly written, staff are educated on the topics, and standard operating procedures are properly implemented. A critical step is to be sure daily monitoring is also completed to maintain daily review and correction of any deficiencies in implementation and education by staff on the farm.

Additional capital expenditure projects to improve biosecurity are often necessary to prevent the infection and loss of more sites and farms. The only country that has successfully eliminated ASFV is Spain, in 1995, after years of unsuccessful attempts to eliminate ASFV from their national herd.[2] Asia now faces this decision as the virus has not only spread through China, but also into adjourning countries and may continue to spread without careful implementation of interventions to stop viral spread between farms and additional countries. All across the world it is the responsibility of the individuals with the greatest expertise in the swine industry to provide assistance and knowledge on preventing the spread and circulation of ASFV. In the United States, with leaders in the swine industry leading the charge to keep ASFV out of the country, the government will follow.

References

1. 2019. ASF Situation in Asia Update. *FAO ASF Situation Update - African Swine Fever (ASF) - FAO Emergency Prevention System for Animal Health (EMPRES-AH).* Food and Agriculture Organization of the United Nations.

2. Sanchez-Vizcaino, Jose M., et al. 2019. African swine fever virus. *Diseases of Swine. 11:443-452.*

3. Depner, K. 5 Aug. 2018. The World of ASF A World Apart. 7th ECPHM Resident Workshop.

4. Yan Jason 2019. Why New Generation Testing and Removal is Winning ASF Control in China. Proceedings, 2019 Leman China University of Minnesota. Zhengzhou, China. October, 2019.

US Customs and Border Protection: keeping foreign animal diseases out

Kevin C. Harriger
Executive Director, Agriculture Programs and Trade Liaison
Office of Field Operations, US Customs and Border Protection

Abstract

The US Customs and Border Protection (CBP) is America's frontline agency tasked with managing and enforcing trade, immigration, drug and human trafficking laws and regulations, while facilitating trade and travel. CBP operates in over 325 ports of entry spanning the air, sea, and land environments processing cargo, conveyance, passenger, and international mail pathways. Integrated within CBP's mission is safeguarding American agriculture managed by the Agriculture Programs and Trade Liaison (APTL) Directorate; with border inspections of agricultural and biological imports conducted by CBP Agriculture Specialists. With regulatory direction from US Department of Agriculture (USDA), Animal and Plant Health Inspection Service, APTL manages adjudication of USDA regulations, provides situational awareness on emerging agriculture and biological threats and deploys operational policy and field guidance on best practices in screening and examining imports of animal and plant products throughout all approaching pathways. This presentation will focus on CBP's comprehensive activities and programs aimed at excluding African swine fever, among other foreign animal diseases, from entering the country as part of the dynamic foreign animal disease enforcement program, thus posturing the United States to remain resilient against potential threats.

Prior to 2003, agricultural inspections traditionally focused on unintentional introduction of pests or diseases – those unnoticed in someone's luggage or hitchhiking on the walls of a container. Since 9/11, in addition to the unintentional, we also focus on the deliberate introduction of these threats along with nefarious biological introduction. With the added danger of biological threats, the role of the CBP agriculture specialists at our ports of entry is more crucial than ever. Trade is growing at an unprecedented rate in the land, air and sea environments as well as through the acceleration of small packages through e-commerce. International mail has more than doubled in the past few years due to the global rise of e-commerce. Enhancing agriculture security and compliance requires overcoming critical and complex challenges, from identifying bad actors across global supply chains to efficiently targeting and conducting risk-based triage of the large and growing volume of international trade. CBP continues to execute a steady-state mission through a fully integrated approach to national security, trade, and travel that making the world safer, facilitating international travel and trade, and promoting the continuous improvement of CBP's operations.

National Swine Disease Council

Patrick Webb, DVM
National Pork Board

Introduction

The US pork industry is vulnerable to introductions of diseases that can threaten the health and welfare of pigs, disrupt trade and commerce and impact economic health and well-being of pork producers and their employees. Rapid identification and response to diseases that threaten the US pork industry is considered a high priority requiring significant investment and focus. In an effort to address this issue, the pork industry has organized the National Swine Disease Council.

Mission and objectives

The mission of the council is to provide recommendations to animal health officials and industry stakeholders to mitigate threats and negative impacts to the US pork industry from diseases of concern. While the recommendations do not carry regulatory authority; they will be developed in coordination with regulators familiar with the industry in an effort to harmonize response activities. The primary objectives will be to coordinate industry response activities, protect trade and interstate commerce of pigs, pork and pork products, and build capacity to rapidly detect diseases of concern and limit the scope of a disease outbreak.

Membership and meetings

The council is made up of key influencers in the pork and allied industries and advisors from state and federal animal health officials and academia. The National Pork Board, National Pork Producers Council, the American Association of Swine Veterinarians and the Swine Health Information Center worked cooperatively to identify members and national staff provide administrative support to the council. The council meets in person at least once annually and calls are scheduled quarterly unless there is an issue that requires more frequent communications.

Areas of focus

The council's recommendations on detection and response focus primarily on outbreaks of foreign and emerging diseases in the US, outbreaks of non-regulatory diseases in the US and foreign or emerging diseases that are not present in the US, but pose a significant threat to US pork production if introduced.

Outbreaks of foreign animal and regulatory diseases

In the event of an outbreak of an OIE-listed foreign animal disease or regulatory disease of swine, state and federal animal health officials have statutory authority to lead the response. The council will play a supportive role, developing recommendations prior to or during an outbreak that will help authorities achieve their response goals. In general, these goals are to detect, control, and contain the disease in pigs as quickly as possible, eradicate the disease using strategies that seek to stabilize animal agriculture, the food supply, the economy, and protect public health and provide science- and risk-based approaches and systems to facilitate continuity of business for non-infected animals and non-contaminated animal products.[1]

Outbreaks of non-regulatory emerging diseases

In the event of an outbreak of a non-regulatory emerging disease of concern, the council will use a standardized process to coordinate state-federal-industry cooperative efforts to identify, characterize, prioritize and respond to the outbreak. The council will facilitate the development of recommended response actions and identify the responsible parties, and funding mechanisms necessary to implement the recommended actions needed to better protect the US swine herd.

Threats to US pork production from foreign or emerging diseases

In the event there is a disease of concern that threatens the US pork industry, but is not present in the country, the council will analyze available information to develop the context and situational awareness to determine appropriate response recommendations. The council will identify the responsible parties and funding mechanisms necessary to implement the recommended actions needed to better protect the US swine herd.

Reference

1. United States Department of Agriculture, Animal Plant Health Inspection Service, Veterinary Services. Foreign animal disease preparedness and response plan. 2014;42.

Regionalization, compartmentalization, and maintaining exports during an FAD outbreak

E. Jensen, DVM, MAM, DACPV
Aviagen North America, Huntsville, Alabama

Background

Aviagen Group is the global leader in broiler breeder genetics with a distribution network serving customers in over 100 countries. Sixty percent of global meat-type chicken breeders originate from the USA. The ability to supply breeding stock globally during foreign animal disease (FAD) outbreaks is critical to maintain the food supply chain. During 2014-2015 highly pathogenic avian influenza (HPAI) caused the largest animal health emergency in history. Two hundred thirty-two premises were infected including 211 commercial poultry farms in nine states. Over 7.5 million turkeys were lost (3% of annual production) and 42 million commercial pullets and layers died (11% of the population). The federal emergency response cost $700 million and over $4.2 billion of exports were lost.

Impact on interstate trade

The poultry industry is a complex supply system with significant movement of hatching eggs, pullets and products that cross state boundaries on a daily basis. Secure poultry supply (SPS) plans have been developed to provide guidance on maintaining movement during disease outbreaks. The USDA National Poultry Improvement Plan (NPIP) is a voluntary federal-state cooperative testing and certification program for important diseases impacting the poultry industry. The NPIP program for low pathogenic AI (LPAI) provides for indemnity and compensation for LPAI positive flocks. For indemnity to be considered each participating state must have an approved initial state response and containment plan (ISRCP). The ISRCP are based on SPS plans and participation in NPIP classifications for avian influenza. States produce these independently which may lead to inconsistencies between state plans. Some inconsistencies are the result of differing state regulations. Therefore, it is highly desirable to develop common ISRCPs between states where movement of animals and product is occurring. During a FAD the USDA APHIS will utilize the ISRCPs as a component of their response plan. International trading partners monitor interstate movement of poultry and products, and trade barriers between states will impact their decision on regionalization.

Impact on international trade

In 2015 Aviagen's largest export markets included Canada, Asia and Latin America with Asia representing over 60% of sales and China accounting for 20%. The initial response by international trading partners was restrained until commercial poultry flocks in multiple states were infected with HPAI. Many countries banned the entire US until the USDA provided additional epidemiological information. Most countries implemented regionalization relatively quickly. The European Union and Canada regionalized to county level while other countries typically regionalized by state. In many cases export health certificates (EHC) required modification to change the language regarding country freedom from HPAI. For other countries an additional certification was needed to verify the state of origin was free of HPAI and/or include negative test results for avian influenza. Two countries required shipment of hatching eggs instead of day-old chicks (seen as a more secure alternative). This presents two challenges; potential theft of genetic lines and the loss of byproduct. China and South Korea maintained a ban on the entire country; South Korea did not have regulations allowing for regionalization. With encouragement from commodity trade groups South Korean officials managed to pass new regulations in 2019 that permit regionalization. China has maintained the ban for what appears to be political reasons.

Live animals versus animal products

According to OIE day-old chicks, hatching eggs and fresh products must originate from a country, zone or compartment that is free from infection with HPAI viruses in poultry.[1] Commodities may be exported from an infected region if they have been processed to ensure the destruction of avian influenza virus. Specific guidelines for processing are listed in Article 10.4.26.[1] For African swine fever (ASF) the OIE requires that fresh meat come from animals which originated from herds located in a country or zone free of ASF or under surveillance for ASF demonstrating no case of ASF in the past three years. There are provisions that allow establishment of a compartment free of ASF.[1] For live animals or fresh animal products the only realistic options to maintain exports is through regionalization or compartmentalization.

Regionalization and compartmentalization

When trade is disrupted by disease the ability to meet demand for livestock and poultry becomes challenging. Keeping trade open through the use of internationally recognized strategies such as regionalization (zoning) and compartmentalization enables companies to provide FAD assurances and maintain exports. Regionalization is the recognition of geographical zones of a country that can be identified and characterized by its absence of a specific FAD. The zone must be clearly and effectively delineated by natural, artificial or legal boundaries. The zone must have a common control and surveillance policy for the specific disease. Compartmentalization is one or more establishments under a common biosecurity management system containing an animal subpopulation with a distinct health status to a specific disease or diseases for which required surveillance, control and biosecurity measurements have been applied for the purpose of international trade. Regionalization is a geographical separation whereas compartmentalization is separation by common management and biosecurity measures.

US avian influenza clean compartment

In February 2014 representatives of the US poultry primary breeders formed a working group with USDA officials and a facilitator from the US Poultry and Egg Association to establish a compartment program specifically for poultry primary breeders (animal subpopulation) and avian influenza (specific disease). The compartment program was designed to meet OIE guidelines.[1] This program was incorporated into the NPIP Plan as it serves as a framework for poultry health programs and an avian influenza clean classification already existed for primary breeders. The working group completed the compartment program in October 2015. The NPIP Plan was revised to incorporate provisions for compartmentalization in 9 CFR 145.45, 145.74 and 145.84 and these were published in Federal Register in July 2016.[2] And finally in September in 2016 the compartmentalization guidelines and audit checklist were approved to be added to the Program Standards at the NPIP biennial conference.[3] Aviagen required an additional eight months to implement structural upgrades and management protocols to meet the compartment requirements. The industry was also closely involved with USDA in the implementation of a program for training official auditors. Aviagen applied for registration as a compartment in May 2017 and the facility and records audit occurred in August 2017 and was certified as a US avian influenza clean compartment in October 2017. The compartment requires annual recertification where records, feed mills, hatcheries, egg depots and 25% percent of farms are audited.

What next?

After a compartment is established then the first step is to renegotiate EHCs to conform to the OIE standard paragraph "The flocks of origin and the hatchery are located in either a country, a region of the country or a compartment within the country which is officially free from HPAI according to the criteria of the OIE." Primary breeding companies strongly promote regionalization and do not agree to an EHC that only mentions compartmentalization. The option must therefore be for the importing country to choose between regionalization and/or compartmentalization during a HPAI outbreak in the US.

The second step is for USDA APHIS to develop standards that allow for recognition of compartments for animal disease status. Current regulations in 9 CFR part 92 only provide regulations for recognition of regions. A proposed rule for the evaluation and recognition of the animal health status of compartments was published in the Federal Register in June 2019. It is anticipated that the final rule will be published in January 2020 and will become effective 30 days after publication.

The third step is for the USDA APHIS to negotiate with the importing country to agree and accept how compartmentalization could be used in the event of an outbreak. Importing country officials must have a good understanding of the principles of compartmentalization and recognize how compartments prevent any import of an FAD. It is essential that acceptance of compartmentalization (and/or regionalization) takes place prior to a HPAI outbreak in the exporting country.

FAD and trade

Experiencing an FAD in the US can have a tremendous negative impact on exports. The best strategy is to plan for the long game and prepare in advance. It is important to develop close working relationships with federal and state animal health officials and define how to respond to an animal health emergency within the US. It is also critical to work with USDA APHIS and foreign animal health officials to develop trade agreements and revise EHC in advance of an FAD. Regionalization is the preferred method to facilitate trade during an FAD but compartmentalization provides an alternative when establishments are located within an affected zone or when regionalization is not accepted by the importing country.

References

1. World Organization for Animal Health (OIE) - Terrestrial Animal Health Code 2019.

2. Federal Register – Vol. 81, No. 133, Notices – July 12, 2016.

3. United States Department of Agriculture, National Poultry Improvement Plan – Compartmentalization for Protection against Avian Influenza Disease in Primary Poultry Breeding Companies in the United States of America, Specifications for: Management Guidelines and Protocols 2016.

The importance of transboundary animal diseases economically, socially, and politically

Peter J. Fernandez, Consultant, PJF AgroStrategies Consulting

The first descriptions of livestock diseases we still recognize to this day can be found in biblical and ancient Greek texts. Trade in animals and animal products also has an immensely storied past. And with the rise of trade in animals and their products also has come the risks of transferring illnesses across borders thus providing our first examples of transboundary animal diseases. As animal agriculture has contributed to national revenue, the occurrence of maladies in livestock has also had important negative economic effects on productivity. In developing countries where animals represent a direct means of daily sustenance, the impacts of disease are more pronounced, threatening food security. Access to food is a fundamental human need and its interruption can lead to social and political strife. In addition, well over half of all human pathogens have an animal-origin source as do three quarters of new and emerging diseases in humans.

Rising global populations with a growing middle class is driving increased demands for animal protein and expanded international trade. With this a myriad of factors can contribute to the emergence, maintenance and spread of transboundary animal diseases, including; deficient animal production systems, poor sanitation and inadequate biosecurity practices, ecological changes, greater human-livestock-wildlife interaction, and civil unrest with displacement of humans and animals. Convergence of these conditions can lead to animal health crisis affecting human health, global economies and societies at large. The very nature of transboundary animal diseases requires cooperation between countries in their control, management, and exclusion. Solutions to persistent animal health threats will require a more holistic and inclusive approach as is defined by the *One Health* approach.

The globalization of the international agricultural marketplace has provided consumers with easier access to a wider variety of consumables but has also provided easier movement of many zoosanitary concerns. To thwart these trade-related animal disease risks, countries have established import regulations and compliance protocols related to agricultural trade. To avoid the use of sanitary measures as technical barriers to trade, intergovernmental standards have been established. Through the World Trade Organization (WTO), the World Organization for Animal Health (OIE) has provided a transparent and science-based means to set standards allowing competent authorities to protect their national herds while facilitating trade. With a total of 182 member countries, the OIE annually generates and/or updates as many as 20 Terrestrial Animal Health Code (TAHC) texts. The US is an active contributor to the OIE standard-setting process and through USDA's Animal and Plant Health Inspection Service (APHIS), coordinates the drafting of submitted suggestions to the TAHC from stakeholders.

In addition to these multilateral global standards, on-going efforts by USDA to build and maintain solid regional and bilateral relations with official veterinary counterparts has maintained trade flows in the face of US disease occurrence and also opened and expanded access for animal agriculture. Historic cooperative programs have not only eliminated over 15 major animal illnesses from the US in the Twentieth Century but also assisted neighbours in the Americas in combating livestock diseases, thus, reducing risks regionally.